T0320944

HYDROPOWER '92

PROCEEDINGS OF THE 2ND INTERNATIONAL CONFERENCE ON HYDROPOWER
LILLEHAMMER / NORWAY / 16 - 18 JUNE 1992

Hydropower '92

Edited by
E. BROCH & D. K. LYSNE
The Norwegian Institute of Technology, University of Trondheim

A.A. BALKEMA / ROTTERDAM / BROOKFIELD / 1992

The texts of the various papers in this volume were set individually by typists under the supervision of each of the authors concerned.

Published by
A.A.Balkema, P.O.Box 1675, 3000 BR Rotterdam, Netherlands
A.A.Balkema Publishers, Old Post Road, Brookfield, VT 05036, USA

ISBN 90 5410 054 0
© 1992 A.A.Balkema, Rotterdam
Printed in the Netherlands

Hydropower'92, Broch & Lysne (eds) © 1992 Balkema, Rotterdam. ISBN 90 5410 054 0

Table of contents

2 Reservoirs and environmental aspects in water resources management

3 Dam safety

4 Condition monitoring of hydropower systems

5 Optimization of electricity production on regional, national and international levels – Considering analytic, operational and environmental issues

6 Electricity supply – Choosing the energy source for electric power production

7 Hydropower schemes

Late papers

Hydropower'92, Broch & Lysne (eds) © 1992 Balkema, Rotterdam. ISBN 90 5410 054 0

Preface

These Proceedings contain the papers presented at the International Conference on Hydropower Development in Lillehammer, Norway, June 16-18, 1992.

The first conference on Hydropower with emphasis on underground plants, was held in Oslo in 1987. The 1992 conference is the second conference in this series. This conference has a wider scope, covering technical, environmental and economic issues.

The initiators of this second conference have had in mind that the consumption of electric energy is increasing rapidly throughout the world. Hydropower is the only renewable energy source derived from solar radiation of any consequence for electricity supply at the present time. It is a mature technology where the positive and negative impacts are well explored. At present, hydropower is the most important commercially available renewable energy. Correctly developed, it fits the label 'sustainable resource' better than any of its competitors.

Nevertheless, the planning of hydropower developments is still a great challenge covering a wide range of technical, economic and environmental issues. The objective of this conference has been to address these issues and highlight the ways in which hydropower can be developed in a flexible manner to meet varying demands and changing conditions. These are important issues both for professionals and for the general public.

The papers were selected on the basis of a general invitation, except for a few specially invited lectures.

The editors thank all contributors, who have made it possible to collect documentations on many recent scientific and technical advances in hydropower engineering. We hope that the proceedings will form a valuable basis for further progress of hydropower development.

The proceedings have been produced by the offset printing method. All papers are typed by the authors in accordance with given instructions. The Editors are therefore not responsible for misprints or errors in the text. The opinions expressed are those of the authors and not necessarily those of the Editors.

Trondheim, 20th March 1992
Einar Broch and Dagfinn K. Lysne
Editors

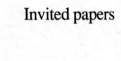
Invited papers

Hydropower'92, Broch & Lysne (eds) © 1992 Balkema, Rotterdam. ISBN 90 5410 054 0

Hydropower: Needs, challenges and opportunities

J. A. Veltrop
Harza Engineering Company, Chicago, Ill., USA

ABSTRACT: Many parts of the world are in desperate need of water and energy to increase food production, improve living conditions and support economic development. Hydropower inherently contributes in an important way towards increasing water supplies. Large resources remain undeveloped, in particular in the developing countries. Relevant data are presented and comparisons made between industrialized and developing countries. The growing importance of pumped storage and small hydro is illustrated. Impacts of hydropower projects are introduced without elaborating on threats to water supplies and environmental and social aspects. Future directions of hydropower are discussed with emphasis on multipurpose development and the need for water resource management. Constraints on the construction of new projects are identified. It is concluded that storage of large additional quantities of water is essential, because conservation and increased efficiencies are not sufficient in the long run. Man cannot do without dams.

1. INTRODUCTION

Only ten countries, of which Norway is one, depend on hydropower to supply over 95% of their electrical energy needs. As is well known, hydropower is a clean, renewable resource and its engineering technology is well developed. Why then is it that over 80% of its potential in the world is undeveloped? What are the constraints on utilizing the very large water resources which continuously circulate (and we trust forever) in the hydrological cycle? For comparison, the undeveloped potential is nearly 100 times the total hydroelectric energy currently produced in Norway.

Development of this large hydroelectric potential requires markets for energy, financing to cover high initial capital costs, strengthening institutional aspects, and rationalizing political processes. Such development is now subject also to a host of other, more recent considerations. The fast growth in the past 20-30 years in the rate of water use and contaminated discharges has exceeded the adaptative capacity of ecological systems, resulting in environmental degradation. Other major effects concern social and cultural impacts, such as resettlement of tribal people and inundation of productive agricultural lands.

Developing hydroelectric power entails manipulating the earth for the benefit of man. Subserving nature to man has been part of the Judeo-Christian tradition and has inevitably been strengthened by stunning technical achievements over the past several centuries: new sources of energy have been developed, new machines and industrial processes were introduced, modern agricultural techniques are literally feeding billions more people, and life expectancy has doubled. Man can now be liberated from drudgery and privation and enjoy a quality of life of a much higher order. Regrettably, only a minority of mankind has been able to take advantage of this enhanced quality of life.

With our greater awareness of the planet's fragility, we must ask: does hydro development lead to changes in land-use patterns, degradation and contamination of arable lands, destruction of wetlands, deforestation, expanding deserts, and soil erosion? In specific cases it has, in most others it has not. For example, it is difficult to evaluate the full consequences of the very large High Aswan Dam with its broad impact on physical, biological and human systems (White, 1988). (See Box 1). The crucial point now is how to balance the pros and cons with the need for damming rivers to provide water and power. However, even if dams sometimes have adverse impacts, mankind cannot do without them.

2. WATER RESOURCE DEVELOPMENT AND HYDRO POWER

Between 1940 and 1980 global water use doubled, and it is expected to double again by the year 2000. Water consumption is increasing several times faster than the spiralling population growth. The pressure on the finite water supplies in the world is growing due to rising expectations for economic development and improved living conditions, vast expansion of irrigated agriculture, increased use of natural resources, and huge requirements for discharging waste products. Theoretically there is enough water for an increase of the present world population to 10 - 11 billion. The problem, however, is its uneven distribution in time and place.

At the UNDP sponsored conference in New Delhi in September 1990, it was pointed out that there is neither a water supply system for 65% of the world's rural population, nor for 35% of the urban population. Despite the fact that during the past decade safe drinking water was brought to an additional 1.3 billion people, 1.2 billion out of the present 5.3

are still without safe water. Water withdrawal is approaching critical proportions in many arid and semi-arid areas. This paper will deal only with the multipurpose aspects of water use, with emphasis on hydropower. It does not cover the threats to water supply resulting from chemical pollution from industry and use of pesticides, herbicides and fertilizers for agriculture, nor will it consider the effects of global warming on sea level rises, loss of low-lying coastal cities and changing patterns of precipitation. It does not present the consequences of the rapid settlement of people in areas which are historically short of water, such as Southern California.

Water consumption can be reduced through conservation, improved efficiencies in industry and agriculture, and recycling of waste water. Supplies can be augmented by increased pumping from renewable groundwater sources, as well as by desalination of sea water. However, the enormous growth in demand necessitates capturing additional amounts and storing these for future use at appropriate times. Hydropower inherently contributes in an important way towards increasing water supplies.

Hydropower uses large amounts of water but returns it to the rivers. Hydro developments regulate seasonal and annual fluctuations in river flows. Increasingly large dams are the cornerstones of multipurpose projects providing desirable combinations of water supply for hydroelectric energy, as well as for domestic, agricultural and industrial uses, for protection of life and property from floods and droughts, for deepening navigation channels, for recreation in the form of fishing, boating and swimming, and for ensuring adequate water supplies to preserve wildlife and maintain ecological systems. The use of reservoirs in Europe (exclusive of the USSR) is shown

---

Box 1

Summary of Effects of the High Aswan Dam (up to about mid 1988)

Basic Aims

Three were met:

- Curbing river flows by controlling floods.
- Storing water for regulated releases to enlarge irrigated cropping.
- Generate hydroelectric energy.

One was not:

- Expansion of irrigated acreage.

Side Effects

- Most were anticipated, such as: relocation of Nubians from reservoir areas, creation of reservoir fishing, improvement of navigation, enlargement of downstream agricultural production, support of fertilizer manufacture, decreased sediment flow downstream with river degradation, and reduction in Mediterranean fisheries.

- Not anticipated were: change in water quality, need for an upstream emergency flood outlet, and reduction in traditional material for brick making, increases in waterrelated diseases.

Monitoring of environmental impacts has been spotty:

- With care: silt load, reservoir fish and evaporation
- Rather casually: river fish and extent of irrigated land
- Neglected until late 1970's: water quality and schistosomiasis

---

Box 2

Use of European Reservoirs in 1990

Use	Single Purpose	Multi Purpose	Percent of Total
Hydropower	1836	409	51%
Water Supply	838	427	32
Irrigation	547	327	22
Flood Control	84	313	10
Recreation	43	229	6.8
Navigation	23	66	2.2
Fish Breeding	-	13	.3
Other	23	-	.6
Sub Total	3194	1784	
Multipurpose	784		
Total	3978	(created by 4038 dams)	

in the Box 2.

Hydropower development must be integrated with environmental, social and biological aspects, sanitation and water quality control, management of drainage areas, reduction of reservoir sedimentation, protection of wetlands and creation of fish farms. It is undoubtedly a formidable task to supply fresh water to all of mankind, to provide electricity to alleviate poverty and stimulate economic growth, and at the same time advance waste management techniques to prevent man from drowning in his own waste products.

3. HYDROPOWER NEEDS AND POTENTIALS IN DEVELOPING COUNTRIES

Developing countries need affordable energy to (1) increase agricultural productivity and food distribution; (2) deliver basic educational and medical services; (3) establish adequate water supply and sanitation facilities, and (4) build and power new job-creating industries.

Worldwide, only 15.2% of the technically possible hydroelectric energy was developed by 1990, as shown in Table 1 (Handbook 1992) and in Figure 1.

The study "Energy for a Sustainable World" (Goldemberg, etal 1987) developed a scenario for the global energy mix for the year 2020. These data were divided into estimates for industrialized and developing countries (Besant-Jones 1989). The result indicates a need to increase hydroelectric energy production in developing countries from 489 TWh/year in 1980 to 2430 in 2020 (actual in 1989 was 661 TWh/year). Actual installed capacities were 110 GW in 1980, 154 in 1986, and is estimated 565 GW for 2020.

The natural resources for this seemingly large increase are available. For the 40-year period (1980 to 2020) the increase of 455,000 MW is equivalent to 36 Itaipu power stations, each with a capacity of 12,-600 MW. How does this goal compare with recent construction? In the 6 years between 1980 and 1986, 44,000 MW was added or 3.5 Itaipu's, which is well below the required rate of approximately one per year.

Table 1. Hydroelectric Generation in 1990 (in TWh/year)

Continent	Technical Potential (1)	Generated in 1990 (2)	(2) as % of (1)
Africa	1344	50	3.7
Asia	4212	387	9.2
Austr./Oc.	203	38	18.7
Europe	836	483	57.8
North Am.	969	573	59.1
Latin Am.	3486	380	10.9
USSR	2950	223	7.6
World	14000	2134	15.2%

The large differences in hydro development between industrialized and developed countries is further illustrated in Table 2.

Table 2. Hydroelectric Generation in 1990 (in TWh/year)

Countries	Techn. Poten.	In Oper.	Under Constr.	Plan.	Sum in %
Industr.[1]	1926	1135	35	126	61
Develop.[1]	9123	776	367	834	22
USSR	2951	223	50	13	10
World	14000	2134	453	974	25

[1] For definition see Appendices A and B

4. IMBALANCES IN HYDROELECTRIC GENERATION

The simplified calculation above shows that the recent rate of increase in installed capacity is slightly less than one half of what is required to reach ESW's goal in 2020. That is only one aspect of the development of hydro potential in developing countries. Another is the large variation which exists among the developing countries. In Table 3, 107 developing countries have been combined into 6 groups based on generation data for 1990 (Handbook 1992).

Variations are even larger when the three countries with the greatest activities are taken separately and compared with the remaining countries, as shown in Table 4. The three countries, Brazil, China and India have 41% of the hydroelectric potential among the developing countries, 52% of that in operation, 69% under construction, and 46% planned.

Hydroelectric energy in operation, under construction and planned in the most active countries in the world is listed in Table 5 in the order of largest amount under construction, with Norway and USA added for comparison. The results are not surprising because by far the largest hydroelectric resources occur in South America, Asia and the USSR as shown in Figure 2.

5. PUMPED STORAGE

Operation of many electric power systems has shown that pumped storage is the most economical way to store energy of secondary value for use in peak demand periods. It also brings dynamic benefits to system

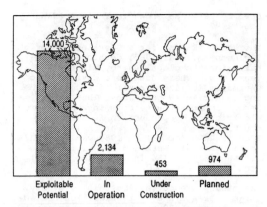

Fig. 1 Hydro Resources of the World (TWh/Year)

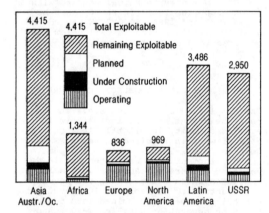

Fig. 2 Hydro Resources by Continent (TWh/Year)

Table 3. Hydroelectric Generation in 107 Developing Countries in 1990 (in TWh/year)

Region (Number)	Techn. Poten.	In Oper.	Under Constr.	Plan.	Sum in %
LAC 26	3486	380	194	347	26
East Asia 20	3026	187	98	208	16
EMENA 16	338	87	21	60	50
South Asia 7	938	79	49	161	31
East Africa 21	1015	29	2	17	5
West Africa 17	320	14	3	41	18
Subtotals 107	9123	776	367	834	22

operation: capabilities for voltage, frequency and power factor corrections, and the opportunity to minimize the cycling of thermal plants.

Table 4. Hydroelectric Generation with the three Largest Separated Out (in TWh/year)

Number of Countries		Country or Group	Techn. Poten.	In Oper.	Under Constr. Plan.
China	1	1923	126	96	164
Brazil	1	1195	213	120	70
India	1	600	58	39	157
Subtotal	3	3718	397	255	391
Remaining	104	5405	379	112	443
Subtotal	107	9123	776	367	834

Table 5. Hydroelectric Generation for the Most Active Countries in 1990 (in TWh/year)

Country	In Oper.	Under Constr.	Plan.
Brazil	213	120	70
China	126	96	164
USSR	223	50	13
India	58	39	157
Canada	293	13.2	41
USA[1]	280	3.0	-
Norway	122	3.5	11.5
Africa	50	6	61

[1] In the USA in 1990, 74 GW out of a total of 147 GW remained to be developed. However, 32 GW has been excluded by the Wild and Scenic Rivers Act and similar legislation.

The number of pumped storage plants in the world has increased rapidly since the 1960's. Of a total of 323 plants 65% are located in Europe, as shown in Table 6. (Handbook 1991).

The largest number of operating plants can be found in:

USA	39	Spain	22
Germany	38	Italy	21
Japan	38	Switzerland	18
France	29	Austria	17
Norway	22		

Worldwide installed capacity for all pure pumped storage plants is 80.4 GW, another 23.7 GW is under construction, while 47.8 GW is being planned. In the USA alone 49 new plants are in the planning stage for a total of 18.7 GW. (WP & DC 1991)

Table 6. Number of Pumped Storage Plants

Continent	In Oper.	Under Constr.	Plan.	% in Oper.
Africa	3	-	1	1
Asia	50	10	5	16
Austr./Oc.	6	-	-	2
Europe	211	8	20	65
North Am.	40	1	33	12
Latin Am.	11	-	7	3
USSR	2	2	3	<1
World	323*)	21	69	100

*) If pumping-only plants are excluded, this number reduces to 285

Table 7. Potential and Operation of Small Hydro Plants (in TWh/year)

Countries	Exploitable	In Operation
Industrialized	374	40.6
Developing	195	4.5
USSR	337	-
China	250	39.5
World	1,156	84.6

Table 8. Number of Small Hydro Plants and Installed Capacity

Countries	In Operation #	In Operation GW	Under Constr. #	Under Constr. GW	Plan. #
Industr.	16,980	20.5	92	.21	3,541
Developing	846	1.7	125	.31	674
USSR	-	-	100	.22	-
China	58,000	13.3	600	2.20	2,000
World	75,826	35.5	917	2.94	6,215

6. SMALL HYDRO (<15 MW)

When an isolated community needs power for household and community use, for driving small agricultural processing plants or other small industries, one of the now frequently used energy sources is a small hydroelectric plant. It can be part of an overall rural development program.

The Chinese experience (Zhu 1991) indicates significant positive social and economic benefits:

- Promotion of local industry
- Use of indigenous labor and materials
- Reduction of firewood consumption
- Raising income level of rural population
- Mitigation of population drift towards urban areas
- Development of potential for tourism

The general mode of operation in China has been to blend local and national involvement with private initiative and public cooperation. Local manufacturing and innovations have reduced capital costs of small hydro from the $3000-6000/kW range to $1000-2000/kW.

There are nearly 76,000 small hydro plants in the world with the bulk (58,000) in China and the trend shown in Tables 7 and 8.

Impediments for broader application of small hydro vary from country to country, but generally include cost, lack of inventory of potential sites and hydrological data, lack of local technical expertise and management skills, lack of long-term government support because of bias towards large projects, and lack of finances.

Opportunities for cost reductions include: standardize equipment, manufacture equipment locally, control cost of civil works, limit scope of pre-feasibility studies, implement a large number of projects together, and train local personnel in all phases of planning, design, construction, operation and maintenance, as well as management.

7. IMPACTS OF HYDROELECTRIC PROJECTS

The area of project influence extends from the upper limits of the reservoir through the downstream valley into the low-lying land areas, coastal and offshore zones. Damming a river has profound effects on the hydrology and the limnology of the river system, water temperature, and downstream degradation. These events impact soils, vegetation, agriculture, wildlife, fisheries, local climate, transportation, recreation, and human populations. In order to be socially and environmentally acceptable, the benefits of hydroelectric projects are subject to more limitations than was the case in the past 40 or so years. Minimizing adverse impacts will increase cost, adversely affect optimum operating conditions and reduce efficiency.

The greatest impacts are the result of flooding of reservoir lands. Most serious among the social effects is the removal of people from future reservoir areas. Usually they do not receive a proportional share of the benefits, which most often go to communities and industries some distance downstream, even in another watershed. This is true for irrigation water supply as well as electrical energy.

Timing of water releases may significantly affect downstream waterflow, groundwater levels, water quality and hence the ecology of the entire downstream regime. Water related diseases may increase if endemic to the area. Lack of deposit of nutrient-rich silt downstream leads to the need for artificial fertilizers to maintain agricultural productivity. Although riverine fisheries usually decline, in some cases more productive fisheries have been created in the new reservoir. Numerous other impacts occur during construction, some of which may take on a permanent character, such as new access roads and the construction of camps and an operators village.

Significant side effects on the lifespan of dam, reservoir and power project are caused by agriculture, settlements and deforestation in the upstream catchment areas. These also affect sediment deposits in the reservoir, as well as water quality and quantity. Decomposition of organic materials in the reservoir and deposition of fertilizers used upstream may stimulate growth of weeds, which can cause problems with water outlet structures and irrigation canals, and affect fisheries, recreation and navigation. Depletion of oxygen levels can lead to fish kills.

8. FUTURE DIRECTIONS FOR HYDROELECTRIC ENERGY SUPPLIES

This paper will not elaborate on increasing the efficiencies of energy generation, nor on reducing consumption through conservation, demand management and greater efficiency in energy use. Nor will it cover appropriate pricing of water and energy, reduction of energy generated from polluting carbon sources in favor of photo-voltaics, wind or wave energies, nor the future role of nuclear fusion.

With the rapid increase in population in the developing countries, and because of the obvious desire to raise living standards and eliminate abject poverty, mankind cannot wait for all the above measures to take effect. Besides, there is a definite limit to what can be accomplished with conservation * and by increasing efficiencies. Furthermore, the scale and the time in which new technological innovations will make significant contributions is not known. Therefore it is wise to develop sustainable hydroelectric power resources at the moderate rates envisaged in the report "Energy for a Sustainable World".

The following specific opportunities must be pursued wherever applicable:

1. Adding capacity to existing dams and hydroplants, in some cases by raising dams.**
2. Rehabilitating and upgrading older plants.
3. Developing hydro at new sites with priorities for already dammed rivers.
4. Increasing peaking capabilities by means of pumped storage projects.
5. Increasing construction of small-hydro projects.
6. Establishing regional interconnections.
7. Promoting multipurpose projects.
8. Continuing technological advances for conventional hydro, especially for methods to reduce sediment deposits in reservoirs.
9. Developing hydraulic machinery to operate efficiently at very low heads with large flows.

Hydroelectric developers must identify, recognize and deal with a variety of constraints imposed on the construction of dams and hydroelectric power projects. These include uncertainties about:

1. Future oil prices, available financing in light of existing debts, and market demands.

2. Adverse and beneficial environmental and social impacts which must be quantified in order to be incorporated in the economic evaluation of projects. A balanced approach with multi disciplinary participation in project planning promises to be the most effective way to resolve these difficult and contentious issues, which include resettlement of indigenous populations, salination and water logging of irrigated fields, and health issues resulting from water-related diseases.

3. A range of uncertainties exists in planning new projects: demand uncertainties include population growth, future economic growth, improvements in energy intensity, and the increasing needs of the Third World; uncertainties in supply include environmental aspects (such as ecological concerns related to CO_2, nuclear risk and waste disposal), the use of resources, the need for demand-management policies, and the transition to renewable energy sources.

4. Superimposed on these are geopolitical uncertainties which directly effect technology, economics and finance.

9. IN CONCLUSION

Water and energy needs are increasing rapidly. Very large hydropower resources in the world remain to be developed. The challenge to hydroelectric and dam engineers is to achieve stability and harmony with the environment. There is an opportunity to resolve today's environmental,

---

* No increase in energy consumption occurred in the USA between 1973 and 1986, while the economy grew by 30%.

** In the United States less than 5 percent of 67,00 dams over 25 ft high have power generating facilities, as these dams were originally built for flood control, recreation and water supply.

social and cultural problems with the cooperation of other disciplines beyond the
traditional group which designed, constructed, operated and financed these projects. It is possible to pursue further
advances in technology and at the same
time assure improvement of living standards for those who are now deprived.

Engineers and scientists will continue to
make significant contributions to sustainable development by assuring increasing
supplies of freshwater and hydroelectric
energy through the construction of dams.
Of prime importance are dam safety and
durability, reservoir sedimentation, refurbishing and upgrading hydro installations, promoting small hydro and pumped
storage, with emphasis on multipurpose
exploitation of reservoirs and management
of water resources.

Sustainable development implies that man
can prosper and survive with wise steward-
ship of technology, natural resources and
the earth's ecosystem.

REFERENCES

Besant-Jones, J. 1989, The future Role of
    Hydropower in Developing Countries.
    The World Bank Industry and Energy De-
    partment. Energy Series Paper No 15.
    Washington D.C. USA.

Goldemberg, J., Johansson, T.B., Reddy,
    A.K.N. and Williams, R.H. 1987. Energy
    for a Sustainable World. Washington:
    World Resources Institute.

Handbook 1992. The world's pumped storage
    plants, page 46-53, Water Power & Dam
    Construction.

Handbook 1992. The World's hydro resourc-
    es pp 34-37. Water Power and Dam Cons-
    truction.

ICOLD 1990. Dams in Europe and USSR - A
    Geographical Approach. International
    Water Power and Dam Construction, UK.

Sadik, N. 1991. The State of World Popu-
    lation. UNFPA/United Nations Popula-
    tion Fund. New York; Nuffield Press,
    Oxford, UK.

UNDP, 1990. Global Consultation on Safe
    Water and Sanitation for the 1990's.
    Background paper. New Delhi, India,
    September 10-14.

WP & DC 1991. Installed capacity of
    pumped storage plants by country. Jou-
    rnal of February 1991, page 8. UK.

White, G.F. 1988. The Environmental Ef-
    fects of the High Dam at Aswan. Envi-
    ronment, Vol 30, No7, September 1988.

Zhu Xiaozhang 1991. Small Hydro 1991,
    Experience in China. Water Power & Dam
    Construction, February 1990, page 38.
    UK.

APPENDIX A

List of 28 Industrialized Countries

North America (2)
Canada
USA

West Europe (17)
Austria
Belgium
Denmark
Finland
France
Germany
Greenland
Iceland
Ireland
Italy
Luxembourg
Netherlands
Norway

Spain
Sweden
Switzerland
United Kingdom

East Europe (6)
Albania
Bulgaria
Czechoslovakia
Greece
Hungary
Poland

Austr./Oc. (2)
Australia
New Zealand

Japan (1)

APPENDIX B

List of 107 Developing Countries

West Africa (17)
Burkina Faso
Cameroon
Central African
Rep.
Congo
Equatorial Guinea
Gabon
Ghana
Guinea
Guinea-Bissau
Ivory Coast
Liberia
Mali
Nigeria
Sao Tome &
Principe
Senegal
Sierra Leone
Togo

East Africa (21)
Angola
Burundi
Comoros
Ethiopia
Kenya
Lesotho
Madagascar
Malawi
Mauritius
Mozambique
Reunion
Rwanda
South Africa
Somalia
Sudan
Swaziland
Tanzania
Uganda
Zaire
Zambia
Zimbabwe

Korea (Rep.)
Laos
Malaysia
Mongolia
New Caledonia
Papua New Guinea
Philippines
Polynesia
Solomon Islands
Tahiti
Thailand
Vanuatu
Vietnam
Western Samoa

EMENA (16)
Afghanistan
Algeria
Cyprus
Egypt
Iran
Iraq
Israel
Jordan
Lebanon
Morocco
Portugal
Romania
Syria
Tunisia
Turkey
Yugoslavia

Latin America (26)
Argentina
Bolivia
Brazil
Chile
Colombia
Costa Rica
Cuba
Dominica
Dominican Rep.
Ecuador

<u>South Asia (7)</u>
Bangladesh
Bhutan
Burma (Myanmar)
India
Nepal
Pakistan
Sri Lanka

<u>East Asia &
Pacific (20)</u>
Cambodia
China
China (Taiwan)
Fiji
Indonesia
Korea (DPR of)

El Salvador
Grenada
Guatemala
Guyana
Haiti
Honduras
Jamaica
Mexico
Nicaragua
Panama
Paraguay
Peru
Puerto Rico
Suriname
Uruguay
Venezuela

Hydropower'92, Broch & Lysne (eds) © 1992 Balkema, Rotterdam. ISBN 90 5410 054 0

How can developing countries contribute to clean hydroelectric energy production?

O. Hoftun
United Mission to Nepal, Kathmandu, Nepal

ABSTRACT: Much of the world's remaining unexploited hydro power potential is found in the developing countries. These resources of clean energy can meet part of the growing demand for electricity in the third world. The industrialized countries have realized that their own dependence on fossil fuel for energy generation is a threat to the global environment. But the developing countries are still moving in the same direction. The negative impact on the environment of increased energy demand in these countries would be reduced if they made full use of their big potential for hydroelectric energy production. This is possible only if industrialized countries provide capital on reasonable terms. An effort must also be made to establish an internationally recognized basis for joint development of common rivers by neighbouring countries.

1 THE POTENTIAL IS WHERE THE DEMAND IS

Much of the remaining unexploited potential for hydro power development is found in the developing countries. In fact, some of the poorest and least developed among them are most richly endowed with water resource not yet put to use. This is particularly true in the case of countries situated at the upstream end of the big rivers, like Laos or Bhutan or Tibet or Nepal and the countries in the upper - reaches of the Amazonas.

At the downstream end are the bigger and stronger countries, like India and Brazil, with a fairly well established infrastructure and a large, fast growing population of which a majority are very poor people.

Perhaps two thirds of the population in these downstream countries do not presently have satisfactory access to electricity supply. In the future an increasing stream of people from the rural areas are likely to be moving into the cities. One effect of this will be a drastic growth in the demand for electric energy.

It has been said that during the next 20 years or so some one and a half billion of new people will require electricity supply. The majority of these will be living in cities where dependence on electric energy is much greater than in the rural areas. The demand may increase by as much as 10% a year, which in twenty years could mean a billion kW of added generating capacity.

This may be on the high side. But there can be no doubt: The demand for energy in the developing countries will be growing fast. And the question is: Will it be met by new coal fired thermal plants or by nuclear power? Or can this increased demand to some extent be met by new clean hydro power?

2 CLEAN ENERGY AND THE ENVIRONMENT

The environmental aspects of the increasing energy demand in the developing countries are alarming. To preach conservation of energy makes sense only where there is already high consumption, i.e. in the industrialized countries. The masses in the developing countries have nothing to cut, they can only add to the demand. They will need more electric energy which will have to come from somewhere.

Developing countries naturally feel that it as much their right to use the cheapest energy sources available today as it has been for the developed countries in the past. Thermal power plants require less initial capital and will be preferred even though they will increase pollution and add to global warming. Warnings from the industrialized countries will have little effect since these countries have too many past and present sins to make up for.

Certainly, the international community needs to address the global problems related to energy and environment. There may be those who question whether the effects of global warming are so serious as said to be. But nobody can brush aside the real danger in a continuing dependence on non-renewable energy sources like oil and gas and coal.

These environmental concerns are, of course, shared by the informed people in the developing countries. No where is pollution so bad as in the large cities of the third world. To live there is on the point of becoming unbearable. But still, these cities will continue to grow.

The environmentalists have to accept this as a fact: The growing cities in the third world will demand more electricity. They cannot function without it. The energy they need will have to be provided in one way or another. If the utilities cannot supply it, then private diesel generators will be installed in every city block. People will cook their food on coal fires, using the cheapest and dirtiest coal which may be at hand.

If people in the industrialized world want to restrict the use of fossil fuel for electric energy production in the developing countries, then they must come up with alternatives. There is no use in talking about solar energy or wind power as long as these sources of energy are not economically competitive nor practical to use.

There is no longer time for talk, but for action. And it is clear where the action is needed: It is among the consumers and the tax payers in the rich countries who continue to waste energy. They must reduce their own consumption and they must pay to provide clean energy for the developing countries.

What cannot be accepted is when environmentalist pressure groups make hydro power in developing countries more expensive by forcing donor countries and international agencies to impose rigorous environmental standards as a precondition for financial assistance, with the recipient having to pay for it as part of loan capital.

It should also be noted that rural electrification will have no effect in controlling deforestation and erosion in hilly areas like the Himalayas unless the cost of electricity is subsidized enough to enable poor villagers to cook electric instead of burning fire wood.

3 ECONOMIC DEVELOPMENT AND HYDRO POWER

It is fortunate that some of the least developed countries do have rich water resources. That makes it possible to combine hydro power development with general economic development. In this there is great scope for integration and secondary benefits.

Firstly, the exploitation of hydro power represents a never ending source income which these countries need badly.

Secondly, hydro power development can serve as a valuable input for rural development in the areas where projects are located. Such inputs should be recognized as worthy objects for investment by themselves. Compared with the heavy cost of hydro projects, the extra cost of securing maximum overall effect for rural development in the area surrounding a project is small. But the important thing is: This type of integrated development requires a different approach and new attitudes both in planning and implementation.

Thirdly, at the national level, hydro power projects can be combined with industrial development, by setting up workshops which can manufacture and repair equipment for power generation and distribution. Equally important are programmes for training and building up local expertice, especially when closely interlinked with the projects.

The training aspect usually comes into project planning more or less as an afterthought, if included at all. It ought to be a basic part of any aid package.

It does seem almost like a crime to invest large sums of money on construction work in developing countries, looking at the work only in narrow financial terms of cost effectiveness and efficiency, when just a little thoughtfulness and flexibility could make wonders with regard to the local impact on development.

4 INTER-DISCIPLINARY SPECIALISTS

Hydro power development requires the input
of specialists in many fields. But often
missing is the input of inter-disciplinary
specialists who are able to put things
together and see the wholeness of a project.

When hydro power projects are seen as part
of overall economic development, then things
become even more complex. It may be hard
to say what will work and what will not
work. Still, to carry out projects blindly
without attempting to investigate their
impact for good or bad on the local society
and environment is irresponsible.

Often is there lack of imaginations, lack
of willingness to go new ways. Utility
people are said to be conservative, tied
up with rules and regulations. Projects are
often made unnecessary expensive and complicated. Things required and justified
in an industrialized society, may be in-
appropriate or wrong in a developing country. This is not only in order to reduce
cost, but even more because too sophisticated designs often give cause to future
maintenance problems.

Outside expert input in the planning of
hydro power projects in developing countries
is usually required. But local input in
the planning process is just as important.
The dominance of foreign consultants is
one of the curses of the developing countries.

Inter-disciplinary specialists are difficult to find. In the absence of such
people, the different professions must be
brought together to consider important
basic issues. But experts are expensive and
should not waste time on talking. Professional people are also usually reluctant
to move outside their speciality. When con-
fronted with critical questions, they
tend to say: This is not my field. Please
don't disturb the process by bringing up
such complicated and irrelevant issues.

I think there is much need to have representatives of the various specialities
come together to study development aspects
as well as invironmental problems in depth,
trying to learn each others languages and
establish a common understanding of basic
issues.

Take the question of large dams: Anybody
with a little insight will realize that
one cannot talk about such projects in
general terms as good thing or bad things.

Large dams are not all bad, nor are they
all good. Advantages must be weighed against
disadvantages, including such aspects which
cannot be quantified in simple technical
or economic terms.

Nothing is accomplished with people shout-
ing at each other through the media, arguing
out of their insulated professional com-
partments. One must try to build bridges
of understanding across disciplinary boundaries, and try to agree on basic facts
identifying different choices and their
consequences.

5 ACROSS THE BORDERS

A large part of the potential for hydroelectric energy production in developing
countries is found in those rivers which
cross international borders. Whether it
is question of energy or irrigation or
flood control or ecology in the watershed
areas, the countries along these rivers
have no choice but to cooperate if they
want to use and control their water resources for common good. They must reach
an understanding about sharing of costs
as well as benefits, both upstream and
downstream.

It is clearly an inter-disciplinary type
of effort which is needed. Professional
people in different fields must get together
and build that foundation of facts and practical methods for quantifying things in
economic terms which can serve as a basis
for rules and legal framework for international cooperation in this area and help
the politicians to make necessary decisions.

Another important aspect of cooperation
across borders is exchange of energy, where
there is much to gain from integration
between hydro power and thermal or nuclear
power systems. Also here the difficulties
in putting price tags on benefits and services is hindering agreement where advantages otherwise are obvious.

It seems so often impossible to talk sen-
sibly across national boundaries. The outlook tends to be so selfishly narrow, so
loaded with mutual suspicion. The result
is that nothing happens.

One reason may be the lack of generally
accepted norms and procedures for this
type of international cooperation. In addition there may be need of some kind of
supernational authority to help individual
countries in this process. This may especially be true in relationships between big

and small neighbouring countries. This sort of international institutions do exist in the field of monetary matters, trade, etc. Should there not be something similar for water and energy?

6 THE ACTION IS NEEDED WEHRE THE MONEY IS

How can developing countries contribute to clean hydroelectric energy production?

Cooperation is a key word: Cooperation across international borders. Cooperation across disciplinary boundaries. And more than anything else, cooperation between those who have capital and those who don't have.

Money is handled by bankers and economists. Theirs is a different world. A world which lay people do not understand, where papers and lengthy calculations cover up the simple facts which ordinary people can comprehend,

How can it be that the short time perspective of economists and the narrow outlook of financial experts is permitted to dominate international relationships as much as is the case?

Take the case of hydro power financing: It is often said by hydro power people that the banks' approach to hydro power financing is completely wrong. They apply short term financing on long term investment and kill good projects with heavy debt servicing during the early years in their lifetime. It does not make sense. But nobody seem to be able to change this.

Another problem with bankers in connection with projects in developing countries, is that projects are considered apart from each other, on narrow financial terms. The value of integration and secondary benefits are not taken into account.

Bankers and other decision makers love feasibility studies. In Nepal local newspapers just wrote: The experts have now proved that such and such hydro power project is feasible. The fact is that the feasibility was based on a series of assumptions which have no real documented basis.

What is it that makes a project feasible? And what makes a feasibility report bankable? Is it the thickness of the volumes or the nice, glossy covers? Or the mass of details?

It really is ridiculous how experts can discuss the fine points up and down, and go to great lengths in examining details of this or that kind, but are satisfied to leave the larger questions unanswered, and hardly mention the enormous uncertainties that do exist.

7 WHAT CAN THE DEVELOPING COUNTRIES DO TO PRODUCE CLEAN HYDRO POWER?

The developing countries have this one major constraint which makes them feel helpless: They lack the capital. That puts them at the mercy of others in a way which is very humilating. It tends to bring out the attitude of the beggar: There nothing one can do except to make the most out of the art of begging.

The socalled donor countries have a good deal of responsibility for this. The reality is that they are not donors but debtors. They have no reason to behave like donors, but should consider themselves partners in cooperation, as it is being said so nicely, but seldom practiced.

When it comes to financing hydro power projects in developing countries, the industrialized countries have a debt to pay as the polluters of this earth. In their own interest they have to do something about this. And one of the most sensible things they can do is to help finance hydro power projects in developing countries.

The industrialized countries should get rid of this attitude of superiority towards the developing countries which poisons the relationship between them. Likewise the international financial institutions should lay off their paternalistic manners since their arrogance works against the very purpose of their activities.

But the developing countries can do and must do things themselves to a much larger degree than often is the case. They must establish national institutions able to handle the development of their water resources. The must work out suitable legislation, and integrate their planning within different sectors of society. They must put their own house in order, and learn to cooperate with their neighbours so that clean hydroelectric energy can be put to maximum use.

14

Hydropower'92, Broch & Lysne (eds) © 1992 Balkema, Rotterdam. ISBN 90 5410 054 0

Development of hydropower in China

Gu Zhao-Qi

Tsinghua University, Beijing, People's Republic of China

ABSTRACT: Situations about Chinese hydropower are introduced in detail in this paper. Some features of large dams and power stations which are finished, or being built and to be built, are given in this paper. At the same time, some questions in Chinese hydropower constructions are discussed as well. In this paper only the hydropower projects in Chinese main land are dealt with.

1 HYDROPOWER RESERVES IN CHINA

It is well-known that, China's hydropower resources rank first in the world, with a total reserves of 678 GW, among which some 380 GW can be exploited technically.

Most of these resources are distributed in south-west areas of China. On the upper reaches of Yangtze River -- Jingshajiang River, there will be a series of 18 key hydropower projects, with a total generating capacity of 60,000 MW, and 5 projects which capacity surpass 5,000 MW among them.

On the branch river of Yangtze River--Yalong-jiang River, there will be a series of 11 key hydropower projects near 20,000 MW in total capacity, among them there are 3 projects which generating capacity surpass 3,000 MW.

On other large rivers, such as Daduhe, Nujiang, Lanchangjiang, HongShuihe, Wujiang, Minjiang, Jialingjiang, etc., there are also a large number of hydropower projects to be built, with a lower capacity of 5,000 to 6,000 MW or a higher one of 20,000 to 30,000 MW on each river.

The second hydropower resources concentrated region is North-west China. A series of more than 40 power stations can be built on the upper and middle reaches of Yellow River, with a total capacity of about 41,000 MW.

The Three-Gorge Power Station is a especially large one on the middle reaches of Yangtze River, with a generating capacity of 17,680 MW and has been designed in detail.

The largest power station in the world may be built on Yaluzhangbu River in Tibet, with a capacity of at least 45,000 MW.

Hydropower resources in North China, North-east China and East China are not rich, only take a small proportion in China as a whole, see Table 1. These three regions are more developed

Table 1. Distribution of hydropower reserves in China

Region	Technically exploitable hydropower reserves		
	generating capacity (MW)	annual electricity production (Twh)	proportion(%)
North China	7000	20	1.2
North-east	12000	40	2.0
East China	18000	70	3.6
Central & South	67000	300	15.5
South-west	233000	1300	67.8
North-west	42000	190	9.9
Total	379000	1920	100

in China, with higher ratio of exploited hydropower resources.

From the situations briefly mentioned above,
we can see that, China has extremely abundant
hydropower resources which can not be fully-
exploited during the next several generatios.
Therefor, any hydropower constructor around the
world, if only he will like to invest in China's
hydropower industry, he will be warmly welcome.
On the other hand, almost all of these large
power stations in China are surrounded by large
mountains, the construction and operation
conditions are of extreme arduousness, Chinese
builders will face serve challenge.

2 SITUATIONS ABOUT EXPLOITATION OF CHINESE HYDRO-
POWER

Up to the end of 1991, in our country, hydropower
capacity more than 36,000 MW had been exploited
and utilized to generate 123.5 Twh of electricity
a year. The exploited generating capacity accounts
for about 9 percent of that exploitable, and the
annual electricity generation only accounts for
6 percent of that exploitable, so the ratio of
exploitation is very small.

There were only several presentable hydropower
stations in China in 1949. Over the past 42 years,
hundreds of thousands of Chinese hydropower
builders have made great achievements from
nothing. During the first Five-Year Plan, only
several small power stations being dozens of MW
in capacity had been built. With the developing
of our country, we have gradually been able to
build stations with the generating capccity from
less than 200 MW to several hundreds MW and then
to more than 1000 MW. By now, more than 80,000
various sizes of reservoirs have been completed,
and 11 hydropower stations which capacity are
more than 500 MW have been finished as well, see
Table 2.

Table 2.

Name	Capcity (MW)
Gezhouba	2715
Longyangxia	1280
Liujiaxia	1160
Baishan	900
Danjiangkou	900
Fengman	700
Gongzhui	700
Xinganjiang	657.5
Wujiangdu	630
Lubuge	600

In addition, several other stations such as
Tongjiezi, Yantan, Tianshenqiao (2nd-stage),
Shuikou, Geheyan, etc., will be put into operation
soon, the generating capacity of these stations
vary from 600 to 1200 MW. China has completed a

large number of embankment dams, of which the
highest one reaches 114 M, and many concrete
gravity dams as well, of which Longyangxia is
178 M in height. At the same time, A number of
gravity-arch dams and thin-arch dams have been
built. Dams more than 100 M in height are listed
in Table 3.

Table 3. Chinese dams more than 100 m in height

Name	Height (M)	Type
Longyangxia	178	gravity-arch dam
Wujiangdu	165	gravity-arch dam
Dongjiang	157	gravity-arch dam
Baishan	149.5	gravity-arch dam
Liujiaxia	147	gravity dam
Hunanzheng	128	buttress dam
Angang	115	gravity dam
Yunfeng	113.8	hollow gravity dam
Shitouhe	114	embankment dam
Fengtan	112.5	hollow arch dam
Panjiakou	107.5	hollow gravity dam
Huanglongtan	107	gravity dam
Shuifeng	106	gravity dam
Sanmengxia	106	gravity dam
Xinganjiang	105	hollow gravity dam
Xingfenjiang	105	buttress dam
Zhaxi	104	buttress dam
Lubuge	103	embankment dam
Jinshuitan	102	hyperbolic arch dam
Bikou	101	embankment dam
Qunying	100.5	masonry gravity-arch

The main reason of only a few high embankment
dams in this table is that, many river reaches
are canyons with big flow, therefore, building
concrete gravity dams has more advantages on
diversion than building embankment dams.
Dozens of high dams are being built and will
be finished in several years.

The largest underground power house built in
China is that of Baishan power station, with an
excavating dimension of $121.5 \times 25 \times 55$ m, while
the longest hydrotunnel in China is Pandaoling
Tunnel in Gansu province, with ordinary cross-
section, 15.7 KM in length, bad geologic
conditions and hard construction conditions.

From what mentioned above, we see that, China
has had the ability to build some high dams and
large power stations. Of course, China is a
developing country, still has some disadvantages
compared with advanced countries, especially in
some fields there exist large gaps, such as in
this field of underground engineering, we are
fairly weak and have many shortcomings as
compared with Nordic countries. Although we have
got a lot of achievements, we are not rest
content with that, we would like to learn from
foreign countries, import their advanced
technology and experiences and thus do more good
for chinese hydropower construction.

Table 4. Large sized stations being built in China

Name	Capacity (MW)	Annual electricity production (Twh)	Type	Year
Ertan	3300	17	hyperbolic arch dam	1988
Lijiaxia	2000	5.9	arch dam	1988
Shuikou	1400	4.95	gravity dam	1987
Xiaolangdi	1800	5.1	embankment dam	1991
Manwan	1250	7.8	gravity dam	1985
Yantan	1210	5.66	gravity dam	1985
Wuqiangxi	1200	5.37	gravity dam	1986
GuangZhou P.S.	1200	2.38	concrete face-slab dam	1988
Geheyan	1200	3.04	gravity arch dam	1986
Tanshenqiao	1320			1982

3 HYDROPOWER STATIONS BEING BUILT OR TO BE BUILT IN CHINA

According to the state government's plan, hydropower generating capacity will have reached 80,000 MW by the end of twentieth century, except the local midium and small sized power stations, in the next 8 years, a group of large sized hydropower stations with total capacity of more than 34,000 MW will be completed, that is, we will build large, midium and small sized stations with total capacity 4,000 to 5,000 MW per year, that is an extremely arduous task.

By the end of 1990, hydropower stations being built in China had a total capacity of 18,500 MW, during the period between 1990 and 1995, hydropower stations which capacity are more than 100 MW will be started and form a total capacity of 24,400 MW.

The large sized power stations, being built and more than 1,000 MW, are listed in Table 4.

In the near future, a large group of large sized hydropower stations will be begun construction, such as Laxiwa, Wanjiazhai, Heishanxia or Daliushu on Yellow River, Longtan on Hongshuihe River, Dachaoshan and Xiaowan on Lanchangjiang River, Hongjiadu, SHilin, Pengshui or Goupitan on Wujiang River, Shuibuya and Kaopazhou on Tsingjiang River.

At the same time, some pumped storage power stations may be begun to be constructed as well. Such as 2nd-stage Guangzhou pumped storage power station with a capacity of 1,200 MW, Tianhuangpin pumped storage power station with a capacity of 1,800 MW and some other stations with very nice sites.

Three-Gorge hydropower station has suitable location, favourable indexes and great benefits. Its earlier stage work has been done in every detail and particualar, at present, this project is waiting for the decision of Central government and State Council, the authorization by National People's Congress. If it is decided to be built, the construction may be begun some year between 1995 and 2000.

These above-mentioned hydropower stations are especially large, have many difficulties in technology and construction, therefore, it is necessary for Chinese builders to step into a new stage.

4 SEVERAL PROBLEMS OFTEN ENCOUNTED IN CHINESE HYDROPOWER STATION CONSTRUCTION

Lots of technical problems have been encountered in China's hydropower construction, many of them have been overcome through hard-working. But, some non-pure technical problems often puzzle Chines builders.

1. Funds for construction

China is a developing country, has a lot of work to do, thus is often short of funds, that results in that many projects with extremely favourable technical conditions can not be started as soon as possible. In the past, the price for electricity was set so low that it was difficult to raise funds to develop hydropower. After reforming and opening, we implement some flexible policies resulting in some improvements for hydropower developing, especially for midium and small sized power stations. But, for large sized stations, there are not yet radical changes, the price for electricity can not be set too high for general customers to bear. Of course, asking for foreign loans is one of the methods to solve this qestion, but there is lots of work to do on paying debts.

2. Submergence and moving people

Our country has a large population. Even in south-west regions of China, building a big reservoir often results in thousands of hectares of land submerged and thousands of people moved. Re-settling these migrators needs large amount of money and land, that is a big problem.

For Three-Gorge hydropower project, the migratory population will reach 1 million, that is one of the reasons why the decisions on the construction of Three-Gorge project can not be made soon.

With economic developing, this question will be slightly relaxed. In addition, the pattern of moving people can be improved.

3. Construction equipment and management

For a long time, in China the standard of construction mechanization has been low and there has been short of complete set of advanced equipments and managing experiences. Since the reform and open, importing a large number of advanced equipments greatly improve the construction situation. But, because the expensiveness of some imported equipments and high fee for each crew, some contractors prefer to use manual labours rather than use the advanced machines. Because of the popular using of manual labours, there are often thousands of employees on the construction site, addition to the families the construction site is just like a small town, that give rise to many difficulties for managing and supplying. On the other hand, there exists some wrong things in management system, the enthusiasm of all employees are not aroused to maximum extent, that is also a question needing to be resolved by reform.

Since the reform and open, many foreign civil engineering companies win the bid to take part in the construction of some Chinese hydropower stations, co-operate with Chinese construction bureaus, bring many advanced equipments and experiences to China. Both Chinese and foreign sides get benefits from the co-operation, the potentiality of co-operation between China and foreign friends will be enormous in the future.

5 SEVERAL QUESTIONS TO BE RESOLVED IN CHINESE HYDROPOWER CONSTRUCTION

In the construction of Chinese hydropower, some new questions will be encountered in the future. The key qestions are:
1. Design and construction of high concrete dams more than 200 M in height.

Such as Ertan and Longtan, both them are more than 200 m in height, there are new questions in many facts, such as, kinetic and static stress analysis, temperature control, high-speed construction, treatment of dam foundation and energy dissipation.
2. Design and construction of concrete face-slab rock-filled dams and roller-compacted concrete dams (RCCD) more than 100 m in height

We have completed a number of RCC dams and concrete face-slab rock-filled dams less than 100 m in height, have got a lot of nice experiences. But, for higher dams, we still lack experiences, many questions need deep studying, such as settlement and seepage control of concrete face-slab dams, crack prevention for concrete face-slab, the joining between the abutment and bank slopes, seepage control in the joints of RCC dams, high-speed construction techniques and crack prevention for RCC dams.

3. Design and construction of long and deep embeded tunnels, of big underground caverns

Each of the 3 hydro-tunnels in 2nd-stage Jinping power station which is to be built, is 20 KM long, with a maximum embeded deep of 2000 m. In Longtan and Ertan, big underground power house exist with unit capacity of 500 to 700 MW and large excavation span. Therefore, there still exist many techcinal questions.

4. In the near future, some pumped storage power stations with unit capacity of about 300 MW will be built in China, the water heads are very high, even to 1,000 m. Some reversible turbines which cannot be made by our country will be used. Some questions exist in turbines and other control equipments.

5. If the Three-Gorge power station is to be built, there of course exist a lot of new questions. Such as, high water head multistage shipping lock, large-diameter Francis turbine with unit capacity of 700 MW, giant pressure steel pipe more than 12.5 m in inner diameter, etc..

6 CONCLUSIONS

After all, China is extremely rich in hydropower resources. We have made great progress, however, we will face more serve challenges. We desire to absorb any advanced experiences from foreign countries, warmly welcome foreign experts and engineers to co-operate with us.

Hydropower'92, Broch & Lysne (eds) © 1992 Balkema, Rotterdam. ISBN 90 5410 054 0

New approach to hydropower schemes

P.Chr.Gomnæs & H.Nakling
Berdal Strømme a.s., Partner of Norpower A.S., Norway

ABSTRACT: Significant flexibility in Hydro Power Scheme Design has been obtained over the last decades based on
 - Developing the hydro power schemes underground
 - A highly developed tunneling technique which also allows for a substantial cost reduction
 - Modern engineering geology.
The flexibility affords a new approach to plant layout, - the search for alternative solutions, the search for alternative intakes, reservoirs and power house. The whole scheme may be studied more freely, looking for the environmentally and economically most favourable solutions.
Even very complex tunnel systems will have only minor impact on the overall cost.

1. INTRODUCTION

From nature Norway is blessed with a topography well suited for hydropower development. High mountains, natural lakes and deep valleys combined with an abundant precipitation has formed the natural basis for an impressive development of hydro electric power.

No wonder, that Norwegian hydro power engineering has been in the forefront, opening up new and advantageous design principles, based on new applied science in rock mechanics and a remarkable development of tunnelling techniques.

A common approach to hydro power development has been - and too often still is - the search for the concentrated lay out. This principle tends to locate the plant at the most apparent damsite, ie: the dam, spillway diversion and power house closely linked together and a great, artificial lake upstream. And with a remarkable lack of flexibility when environmental issues and conflict of interests arise.

An alternative approach to lay-out design requires the freedom offered by advanced engineering geology and modern tunnelling techniques, to search for:
 -Intakes and reservoirs located where the conditions are most favourable from a cost and environmental point of view
 -The power house located by a separate optimizing process, considering the relevant rock parameters, not necessarily linked to the damsite.
 -The most favourable alignment for the connecting tunnels.
 -The possible pick-up of run-offs from actual additional catchment areas. (Even under glacier branches, covered by hundreds of meters of ice, the tunnellers have succeeded in tapping the run-off from melting ice).

2. CASE STORIES

A. Borgund - Øljusjø, a 240 MW Hydro Power Scheme

The project is shown in plan and in a schematical vertical section on Fig 1 and Fig 2. Located in Western Norway, the catchment area is approx. 400 km².

One main characteristic is the lack of possible sites for establishing a major reservoir. As can be seen from the plan, the few existing lakes are all located in the outskirts of the catchment area, affording only moderate regulating capabilities.

The western part of the catchment area was evaluated separately and found not economically exploitable. From that point of view that area had no production value. Likewise, if regarding the eastern part isolated, that too would be of inferior value.

A high dam on Eldrevatn could of course improve the production scenario, but only at a prohibitive price. Furthermore, heavy environmental impacts in the midst of one of the most popular mountain areas for tourism would have stopped such an alternative immediately. The conclusion was that, evaluated isolated and separately, neither the eastern nor the western sides would yield viable hydro power projects.

But the planners achieved a completely different and economically favourable picture by connecting the two sides with a 20 km long tunnel from the Eldrevatn in the east to the Dyrkoll in the west. The tunnel was equipped with 9 creek intakes, picking up the run-off from both sides and leading the water to the central intake at Vassetvatn. In this way an increased and concentrated water flow was secured. But the need for regulation was not yet satisfied.

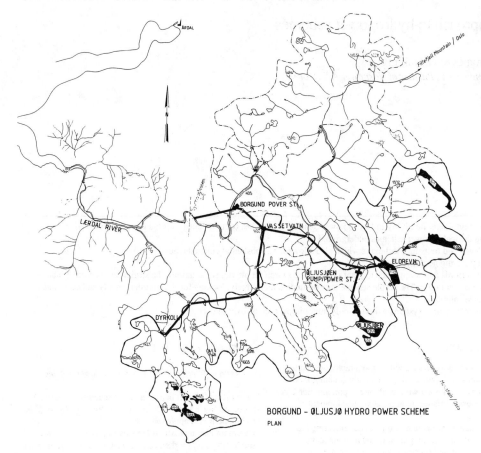

BORGUND – ØLJUSJØ HYDRO POWER SCHEME
PLAN

Fig 1

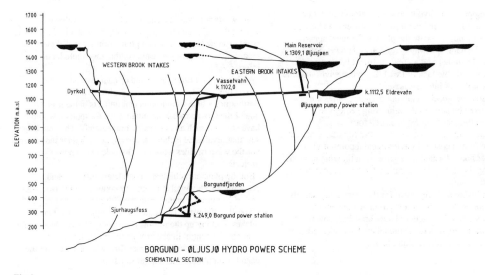

BORGUND – ØLJUSJØ HYDRO POWER SCHEME
SCHEMATICAL SECTION

Fig 2

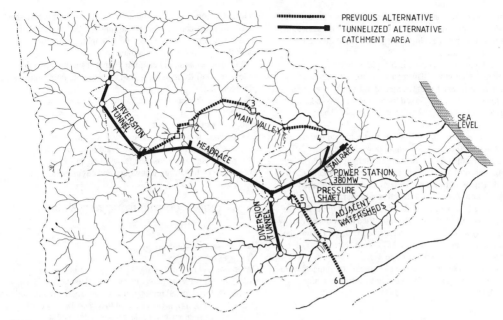

Fig 3

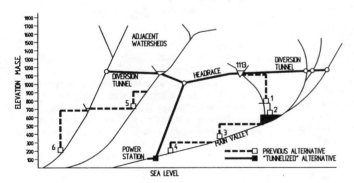

Fig 4

In a remote valley however, at a level some 200 m higher than the tunnel, favourable conditions for a major reservoir was found in the lake Øljusjø, and a reservoir of about 3 times the natural inflow was established.

A reversible pump turbine unit was introduced as part of the scheme, directly connected to the eastern tunnel branch. This lifted water by pumping surplus flow from the long collector-tunnel, as well as from the Eldrevatn area, to the Øljusjø reservoir, the minor reservoir in Eldrevatn serving as a buffer.

The remote location of the Øljusjø reservoir gave only minor environmental impacts, and the reservoir could be established at very moderate costs.

B. Overseas, - A 380 MW Hydro Power Scheme

One could easily state, that this combination of collecting unregulated run-offs with pumping to a high level reservoir (that is a two-level lay-out), turned the whole project from being not viable to an exceedingly economical one.

Another special feature of this project, although not very important, is the location of-, and access to- the powerhouse. The topography would normally indicate a powerhouse location at the Borgundfjord level. However, to utilize more of the total head available, the powerhouse was located some 150 m directly below the valley bottom near Borgundfjord and linked to the surface by a 1,5 km long serpentine shaped access tunnel. (Fig 2).

The project offered a rugged landscape consisting of steep valleys and creeks, giving only limited possibilities of establishing reservoirs with more than daily pondage.

According to previous plans the water-course was meant to be exploited with 4 separate power plants along the main river, including intake dams with heights up to 100 m. In

21

addition, two more plants were planned, utilizing the run-off from an adjacent water shed. The main characteristics would be a run-of-the-river development, with a feasibility in the medium to lower range.

See Fig 3 and Fig 4. These Fig also demonstrate a "tunnelized" alternative. Through an approx. 35 km long tunnel at high level (the same level as utilized by the upper plant) the run-off was collected by 6 creek intakes, conducting two branches to the central point at the pressure shaft and underground powerhouse.

Thus a total head of approx 1000 m was achieved and the whole run-off utilized in one single powerhouse.

It should be noticed that this tunnel alternative has to delete a certain part of the catchment area of the lower plants shown in Fig 3. This will, however, be compensated by the fact that more of the run-off can be utilized at a higher head.

The project is still a run-of-the-river type, but the feasibility has improved substantially. The unit price of produced energy is reduced by approx. 20%. In addition, the project offers the advantage of reduced maintenance costs as the number of power stations is reduced from 6 to 1.

Moreover, only minor negative environmental impacts were introduced, as the main valleys could be left practically undisturbed.

C. Ulla Førre - A 2000 MW Hydro Power Scheme

The Ulla Førre project, recently put into operation, is schematically shown in Fig 5.

As will be seen from the sketch, the scheme is a rather complex and sophisticated one, and in the context of this paper only a few, major features will be brifly presented.

The first thing to be specially noticed, is the two-level layout. The lower level at approx El. 600 m is suitable as intake level for the Kvilldal power station. At this level an intercepting tunnel system is provided, collecting the run-off from the major catchment area and conducting it to the central point at Saurdal. But only modest regulation would be possible at this level, because of inadequate reservoir capacity.

At the upper level of el 1000 m, however, a number of natural lakes were located at a relatively flat plateau. A communication between these lakes was made possible by an interconnecting tunnel system which also includes several challenging lake piercings.

Once this communicating tunnel system was established, the entire area was converted into one single 3100 hm³ reservoir by a system of 13 dams, up to 145 m height, closing the natural river run-offs from the plateau.

This huge reservoir is being fed mainly from the lower 600 m level run-off through the 640 MW reversible pump-turbine plant Saurdal.

The fact that this plant in turn is partly fed by two other smaller pumping stations, see Fig 5, just adds to the complexity of the concept.

As a whole, this hydroelectric power scheme is believed to be an excellent example of creative and innovative design. This proven design philosophy is based on thorough knowledge of tunnelling techniques and engineering geology.

3. COMPONENTS CONTRIBUTING TO FLEXIBILITY

Underground location

One major factor in obtaining flexibility is going underground. With the power house as well as the rest of the complex safely located underground obviously the maximum of freedom in design is provided. In that case the layout of the high pressure sections can be developed independent of surface topography, and economical and environmental advantages achieved.

Fig 6 will illustrate these different approaches.

Brook intakes

As described in the case stories, a major feature of a tunnelized design approach is the interception of river run-offs over wide areas, collecting and conducting them to reservoirs located favourably in the scheme. Flexibility is again a keyword, and the run-offs can be collected from remote catchment areas, even from neighbouring watersheds.

Reservoirs

Design flexibility is also required in the search for convenient storage possibilities. Reservoir possibilities may be scarce at

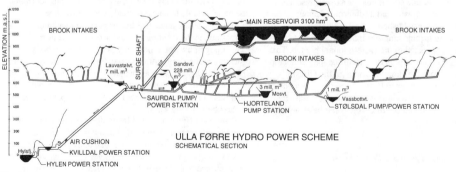

ULLA FØRRE HYDRO POWER SCHEME
SCHEMATICAL SECTION

Fig 5

the level most suitable for run-off collection. Other areas - valleys or lakes even at higher levels - may offer more attractive possibilities, and a pumped storage plant in a two-level layout may be the answer.

Lake piercing

In some cases natural lakes can provide storage, either with moderate dams, or through a lake piercing - or both. The lake piercing, tapping it at a deep point, converts the lake to a reservoir by using the storage volume below the natural water level in the lake. See Fig 7.

Tailoring the scheme

Through the previously described flexibility parameters, a tailoring of the complete tunnel and storage system is possible. The sewing together of different watersheds, lakes and rivers, to an interconnected complex, often comprising pumped storage plants, can offer an optimized utilization of the whole area. This flexibility also facilitates environmentally favourable solutions.

4. COMPONENTS CONTRIBUTING TO ECONOMY

Decreasing Tunnel Costs

The relative cost of hard rock tunnelling has steadily

decreased over the last decades.

Although the hard rock tunnelling technique has a century-old history in our country, a number of radical, cost-saving improvements the last decades have given it an amazingly higher level of performance. Without going into detail, it can be stated that improvements made in all aspects of hard rock tunnelling activities have continuously brought advance rates up and costs down the past 15 years. The improvements have contributed to economize the whole tunnelling process, and higher efficiency has in fact more than balanced the general price increase.

Increasing Tunnel Advance Rates

Fig 8 will show the development of the output of the drilling operation as one main parameter, and Fig 9 gives the total picture of obtainable advance rates.

An ordinary weekly advance rate with the drill-and-blast system in medium good rock - could reach about 100 - 120 m. On a yearly basis, taking into account normal disturbances and tunnel support work, one can estimate average advance rates of approx 2.0 - 3.5 km, somewhat depending on the tunnel cross-sectional area.

In the last decade the full face boring machines, the TBM's. have entered the arena also in hard rocks, and, indeed have come to stay. Some 150 km of tunnels in crystalline rocks have now been bored in Norway, and there are more to come. Weekly advances of 200 m are common with peaks exceeding 350 m. However, as a reasonable estimate, a yearly advance of 6 - 7 km is considered realistic.

Short Construction Period

The total catchment area to be taken into a collecting system will most often be great, and the length of tunnels accordingly. The distance between opposite outskirts of a catchment area to be collected may be 40 - 50 km, and the resulting total tunnel length up to 100 km.

The question often arises, how such extensive tunnelling work can possibly be adopted within reasonable cost and time frames.

The answer is, thanks to the remarkable tunnelling development with increased capacity and decreased relative prices, it really is within reach to intercept even remote catchment areas which previously were beyond economy.

As a value-added effect of higher tunnel- and excavation performance is a shortening of the total timeschedule for the whole design and construction of the plant.

Low Tunnel Cost Impact

It is easy to think that an extreme use of long tunnels will make up a major part of the total costs. However, not surprisingly, the tunnelling normally represents only from 15 to 20% of the total costs.

Even a considerable overrun of the tunnel cost estimate would not have a decisive impact on the final economical result. For example, with the tunnels making up 20 % of total costs, an overrun of, say 30 %, which in this context would be substantial, will result in "only" a 6 % raise of total costs.

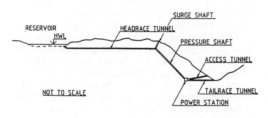

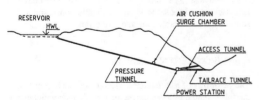

Fig 6

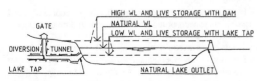

Fig 7

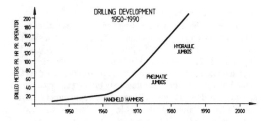

Fig 8

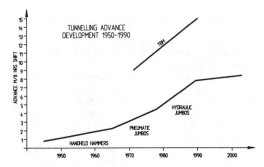

Fig 9

Unlined Tunnels = Low Water Velocities

It should be kept in mind in this context, that the rock tunnels in question are all unlined - in principle. A fully lined tunnel will cost approx. 3 - 4 times as much as an unlined one, and in addition require a substantial extension of construction time. In countries and technical environments where the philosophy still prevails, that even watertunnels in competent rocks need to be fully lined, the flexibility of a tunnelized design is expensive to obtain.

However, it should be mentioned the paradox that an unlined tunnel could imply less headloss than the lined one.
Explanation:
-The generally low price for excavation affords an economically cross sectional area much larger than for a lined tunnel.
-This fact is accentuated by the current development of heavy tunnelling equipment. In many cases the contractor, using his rubber tyred equipment, will offer a lower price for a larger cross sectional area than what may be theoretically required by design criteria.

Providing the extra cross-sectional area is in this way cost-free or marginally attainable, and an oversized water-tunnel can accomodate future expansions to meet the demand for higher power capacity.

5. ENGINEERING GEOLOGY

Parallelling the improvements in tunnelling technique, progress in modern engineering geology is another important success factor in the hydropower design.
This science of "fishing out" the mountains' secrets, can be characterized, although with some lack of respect, by saying that "while the geologist is putting sophisticated names to the problems, the engineering geologist will provide practical solutions to them".

Engineering geology and hydropower development have benefitted from each other. And like hard rock tunnelling methods, engineering geology has been under constant development and improvement over the last decade.

Rather well known is the ability to locate faults and weakness zones on a preliminary stage of planning. On the basis of maps and air photos, one can thus avoid difficulties, advice on adequate measures to be taken, or recommend favourable location and orientation of rock caverns and tunnel alignments.

Less known, perhaps, is the development and sucessfull application of computer based finite-element analysis, giving a representative picture of the internal principle stresses in a given rock mass.

When designing unlined high pressure waterways, which in some cases may be crucial to successful design, it is of vital importance to ensure that the water pressure never will exceed the minor principle stress in the surrounding rock mass. Applying a finite element model in analysing rock stresses gives a powerful design tool. Such analyses have been the basis for steadily new record-breaking, economical and daring designs, with unlined headrace shafts exposed to as much as 1000 m water pressure.

During excavation, and before the final decision on where it is safe to change from lined to unlined design, and thereby expose the rock to full water pressure, it is of vital importance to have the calculated figures verified by actual in-situ internal stress measurements. These measurements, which can be rather tricky, are now executed using reliable methods developed and proven by the engineering geologists.

The importance of estimating internal principle stresses versus maximum water pressure can not be overemphasized. There are examples of unsuccessful designs of unlined high pressure waterways where this obviously was ignored, or rather, was unknown. And with catastrophic results!

Based on such proven technology, the hydropower designers really do have powerful tools available for economically and environmentally favourable design. They will know where to locate the underground facilities, or perhaps even better, where and how not to locate and orientate. They will have the best background for safe and economical lay-outs, and a sound basis for reliable time and cost estimates.

6. ENVIRONMENTAL ASPECTS

The flexibility of a "tunnelized" design approach in utilizing the full potential of a catchment area, is obviously dictated by economy. However, the added value of environmental benefits will be an increasingly important feature of such a flexible design.

Environmental impact considerations have become a crucial issue in all aspects of industrialization, including generation of energy. While hydro electric energy offers superior advantages as an emission-free and fully regenerative source of energy, the environmental consequences of large reservoirs is causing growing concern. The objections to large, artificial reservoirs created in main valleys, are well known and must be taken seriously.

Obviously, a "tunnelized" design that utilizes natural lakes for storage and regulation can help minimize this problem. It gives the designer a greater flexibility in searching for alternative reservoir locations and possibly avoiding large dams. In most cases studied, it has been possible, even with improved economy, to establish environmentally favourable solutions, applying a "tunnelized" design approach.

RESUMEN

*Modern, high level tunnelling technique and engineering geology knowledge open up for new approaches to the Hydro Power Development.

*Intercepting run-offs from wide catchment areas, often remote located, is within reach.

*Reservoirs can be located with great flexibility in an environmentally friendly way.

*The costs and the time frame available offer substantial advantages.

1 Tunnelling and underground works, technical, economic and environmental aspects

Hydropower'92, Broch & Lysne (eds) © 1992 Balkema, Rotterdam. ISBN 90 5410 054 0

Uruguay River basin projects underground works

R. H. Andrzejewski, P. C. F. Correia, O. G. Parente Filho & C. S. Martins
Eletrosul, Centrais Elétricas do Sul do Brasil S.A., Brazil

ABSTRACT: The large scale hydropower development of the Brazilian Uruguay River basin is planned with four major projects within the next decade: Itá - 1620 MW; Campos Novos - 880 MW; Machadinho - 1200 MW and Barra Grande - 920 MW.

Hydropower development is favored by the physical characteristics of the sites, featuring high gradient rivers, narrow valleys and sound rock foundations. Such features, the high runoff and the local topography led to design with high capacity diversion and power tunnels, that are standard for most of the projects. The underground works comprise more than 1.5 million cubic meters of rock excavation for the four projects.

High flow velocities are accepted in the unlined diversion tunnels. The geomechanical conditions of the rock mass are considered for the design of drainage and stabilization measures. Detailed construction planning was carried out to support design studies, cost estimation and bid evaluation.

1. THE URUGUAY RIVER BASIN

The Uruguay River basin in Southern Brazil is the last one with large hydroelectric potential still to be developed, near the major load centers of Brazil.

Twenty-three hydropower projects were inventoried in the upper strech of the basin within Brazilian territory. The 10 most important projects comprise 8,000 MW of installed capacity. Downstream, in the international strech of the basin, three Argentinean-Brazilian projects are planned with 5,000 MW of total installed capacity (figure 1).

Many of the Uruguay River basin projects are planned to be put into operation from the year 1998 to 2005.

The local conditions at the projects sites are such that significant underground works for power and diversion tunnels have been considered in the plants design.

2. LOCAL CONDITIONS

The basin landscape is hilly, with deep v-shaped valleys. The region is part of the geological Paraná Sedimentary basin which comprises areas in Argentina, Uruguay, Paraguay and more than 1,000,000 km2 in Brazil.

The world's largest basaltic volcanic rock occurences are found in the geological Paraná Sedimentary basin. These volcanic rock formations were originated from extensive and sucessive horizontal lava flows with thicknesses of about ten meters each, while the total thickness reaches over 1500 m in some places.

The river course is strongly influenced by subvertical geological discontinuities (fractures and faults) which impose a pattern with rapids, waterfalls, as well as successive changes in river course direction with some turns of more than 270 degrees. The physical river characteristics favor hydropower schemes implementation.

The faults are inactive but some small natural seismic movements have been recorded in the area, as well as some insignificant ones induced by the filling of neighboring reservoirs (Itaipú, Salto Santiago, Foz do Areia and Passo Fundo).

The monthly runoff does not show well defined seasonality, although higher discharges and major floods are likely to occur from May to October, period where the largest floods were recorded. Steep and v-shaped valleys, ground of low permeability soil and significant frontal storms impose fast and high peak floods of short duration. (Table 1).

3. PROJECTS

The large scale development of the Uruguay River basin is planned to start with the implementation of four projects:
. Itá (1620 MW), on the Uruguay River for which preliminary works are under progress,
. Campos Novos (880MW), on the Canoas River with basic design completed,
. Machadinho (1200 MW), on the Pelotas

Table 1 - Hydrology data

Project	Drainage area km^2	Recorded flows (m^3/s)			Flood discharges (m^3/s)		
					Return period (years)		
		mean	max	min	10	50	10.000
Itá	44,500	1,080	29,620	114	19,210	37,790	45,100
Campos Novos	14,200	311	7,853	27	4,290	10,410	14,970
Machadinho	35,800	783	23,000	61	13,950	27,320	38,300
Barra Grande	13,000	281	8,300	46	5,700	12,700	16,530

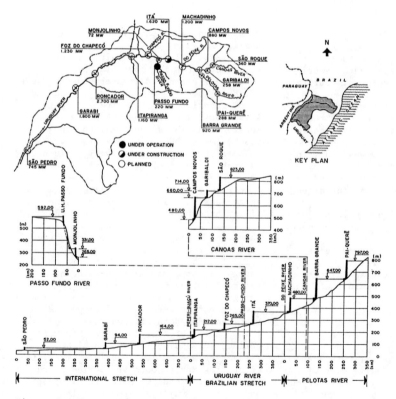

Figure 1. Uruguay River basin planned hydropower plants scheme

30

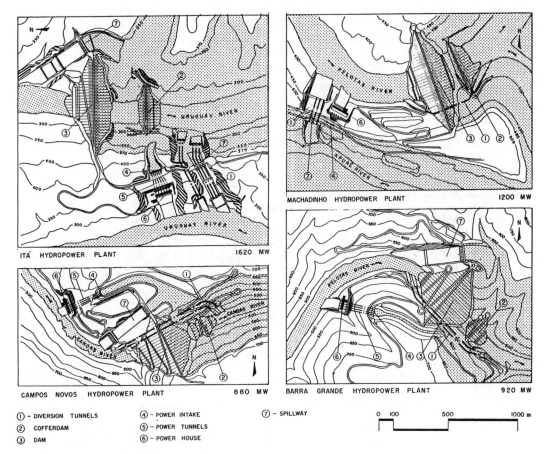

ITÁ HYDROPOWER PLANT   1620 MW

MACHADINHO HYDROPOWER PLANT   1200 MW

CAMPOS NOVOS HYDROPOWER PLANT   880 MW

BARRA GRANDE HYDROPOWER PLANT   920 MW

① - DIVERSION TUNNELS   ④ - POWER INTAKE   ⑦ - SPILLWAY
② COFFERDAM   ⑤ - POWER TUNNELS
③ DAM   ⑥ - POWER HOUSE

0   100   500   1000 m

Figure 2. Project lay-outs

River with basic design completed, and
. Barra Grande (920 MW), on the Pelotas
River with feasibility studies partially
completed.

All project lay-outs are typical narrow
valleys schemes as shown on figure 2.
Table 2 and Table 3 present main project
data.

4. HYDRAULIC DESIGN

Diversion tunnels were designed
considering construction schedule and
seasonal flood occurrence probability.

River diversion is programmed to happen
just after the period of high probability
of floods. The main cofferdams
construction is planned to be fast due to
the possibility of high flows even in low
flow season. The dam construction is
planned to attain the necessary height to
provide protection against 500 years
return period floods before the

forthcoming high risk season.

High resistance of the basalt rock was
verified on similar projects with the same
geological type formation resulting in
acceptance of high flow velocities in the
tunnels for design floods (Table 4).

Power tunnel design considered economic
aspects such as cost and power reduction
due to hydraulic losses. The hydraulic
transients were analysed with help of
mathematical models for normal and extreme
turbine and gate operation. The transient
analysis confirmed the suitability of the
dimensions determined by the economic
design and signed the equipment design and
parameters for gates and power units.

5. GEOMECHANICAL DESIGN

Large tunnels, up to 14 m diameter, have
been successfully built in similar basalt
conditions in Brazil (Foz do Areia, Salto

Table 2. Project data

Project name	Itá	Campos Novos	Machadinho	Barra Grande
Installed power (MW)	1,620	880	1,200	920
Mean power (MW)	802	410	626	361
Garanteed power (MW) *	795	381	577	330
Number of units	6	3	4	4
Turbine type	Francis	Francis	Francis	Francis
Gross head (MW)	105	180	110	167
Reservoir area (km^2)	141	32	266	95
net volume (km^3)	2.67	0.16	6.00	2.87
Spillway capacity (m^3/s)	45,800	14,970	33,680	17,000
Dam volume (103m3)**	8,500,000	11,670,000	10,130,000	12,172,000
Volumes (103m^3):				
Soil excavation	10,054,000	2,201,000	1,740,000	1,180,000
Rock excavation	8,285,000	8,628,000	4,350,000	11,210,000
Quarry excavation	-	1,185,000	5,180,000	-
Underground excavation	568,000	425,000	215,000	334,800
Rock fill	9,900,000	12,354,000	11,200,000	12,905,000
Earth fill	2,450,000	412,000	1,800,000	479,000
Concrete	578,000	308,000	574,000	333,000

* 5% risk garanteed power ** Concrete faced rock fill dams

Table 3. Underground works data

Project	Diversion tunnels				Power tunnels			
	number	max inlet* diameter (m)	mean* diameter (m)	mean length (m)	number	max inlet* diameter (m)	mean** diameter (m)	mean length (m)
Itá	02	18.5/18.5	14/14	520	06	12.5/ 11.4	8.00 c 7.50 s	192
	03	18/17	16/15	510				
Campos Novos	02	25.5/22	16/15	820	04	10.7/ 8.0	7.00 c 5.00 s	365
Machadinho	01	17.5/17.5	14/14	457	-	-	-	-
	03	18/18	16/14	218				
Barra Grande	02	18/18	14/14	890	04	9.3	6.3	321

* Excavation diameter (heigt/width)
** Linning diameter: c Concrete lined section diameter
s Steel lined section diameter

Santiago and Segredo projects).

The general tunneling conditions are good. Design and contract documents consider New Austrian Tunneling Method stabilization support measures for fractured diabase dikes, faulty zones, open lava flow contacts, and wheathered or expansive zones expecially when near tunnel top. Rock bolts of 200 kN capacity and maximun 9 m length, wire meshes and shotcrete with thickness up to 20 cm are specified. Experience obtained on similar projects and carefull field investigations allowed the classification of diversion tunnelling into three classes. Class II where few tension bolts are necessary, shown on Figure 3, is the normal expected pattern.

Table 4. Diversion tunnels- hydraulic data

Project	10 year flow		500 year flow	
	Q (m3/s)	V (m/s)	Q (m3/s)	V (m/s)
Itá	16,400	14.4	23,900	21.0
Machadinho	12,500	16.8	19,140	25.8
Campos Novos	4,192	9.7	9,072	21.0
Barra Grande	4,720	12.1	9,500	24.0

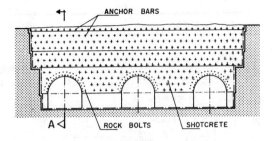

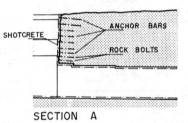

SECTION A

Figure 4. Diversion tunnel portal treatment

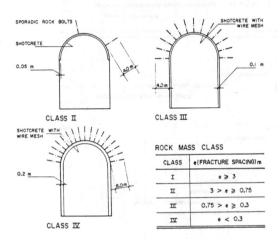

Figure 3. Rock mass classification and expected tunnelling treatment

Every tunnel portal will be protected by wire meshes and anchored with 6 m lenght bars. Around the tunnel mouths, paralel to the tunnel section, 2 rows of tension bars complete the normal consolidation anchoring (figure 4).

The power tunnels will be fully concrete lined up to the dowstream horizontal section near the powerhouse which will be steel lined. The criteria for the length of the steel lining is based on the possibility of rock mass hydraulic fracturing due to internal tunnel water pressure and reduced rock cover.

Rock consolidation measures and a drainage system were designed for whole power tunnels, specially for contact and fractured zones (Figure 5).

During construction, with the aid of blasting seismic monitoring, limits on blast power will be set to avoid damages on neighboring structures and adjacent excavations.

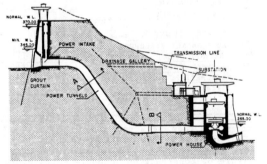

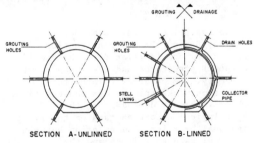

Figure 5. Drainage system Itá hydropower plant

6. CONSTRUCTION PROGRAMMING

The construction of an Uruguay Basin project will take about 5 years from the main civil works commencement to comercial operation start up.

33

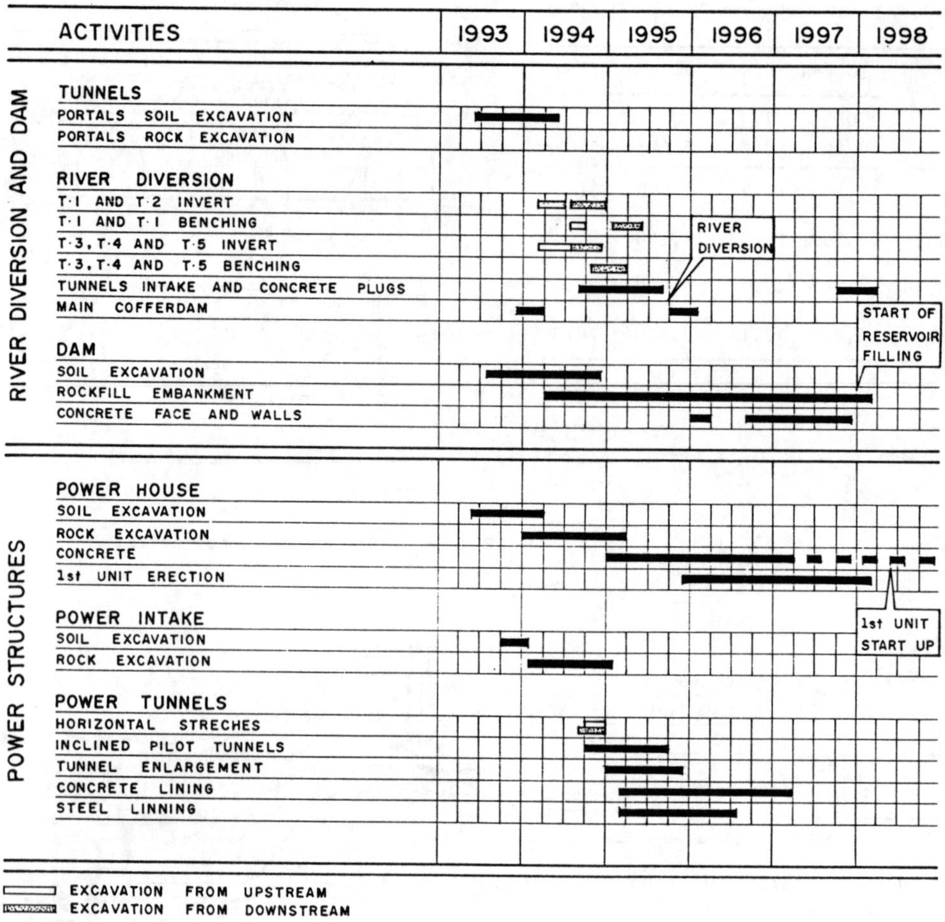

Figure 6. Construction schedule – Itá hydropower plant

Two typical critical paths occur on Uruguay Basin projects construction schedules, one is related to the erection of power structures and their equipment assembly, the other is related to diversion works and the dam construction, which is highly dependent on the hydrological conditions and on the river seasonal behavior. The relation between critical path activities and the construction of other noncritical structures are planned to optmize cost, resources and production on the construction site.

6.1. Itá underground works schedule

Itá Hydroelectric power plant is a typical example of Uruguay River basin projects. At Itá, five tunnels will divert the river through a narrow neck bypassing the local long river loop. With mean lenght of 540 m the 5 diversion tunnels total 500.000 m3 of underground excavation. Three tunnels 16 m high and 15 m wide are located higher than the other two 14 m diameter tunnels.

These higher tunnels, designed to operate during high flows, do not have gated inlet structures. Their ultimate closure is done by concrete plugs constructed during low flood season with cofferdam protection. The lower tunnels steel gates closure, just after plugging of the upper tunnels, starts the reservoir filling.

The construction planning, developed to support design studies, cost estimate, and bid analysis, considers 15 months for diversion works tunneling. With priority for the gated tunnels 1 and 2 the underground work starts after the upstream tunnels portals clearance. Afterwards the underground excavation is shifted to downstream to allow for the inlet concrete structures erection.

The drilling equipment adopted for planning purposes were: Two Atlas Copco H175-38 type jumbo boomers working side by side for heading excavation drilling, Roc-605 type drilling machines for bench drilling, one CAT988 loader, and 35 metric ton dumpers for hauling. The heading works progress is estimated to be 3.3 m per shift of 10 hours.

The six power tunnels, with 180 m length each, are part of the powerhouse intake scheme. The power tunnels are concrete lined, 8 m diameter, except for the final 80 m section that is 7.5 m diameter steel lined.

The power tunnels construction starts upstream with the upper horizontal section excavation and downstream with the horizontal section excavation. Both horizontal sections are linked with pilot tunnels that are excavated upwards and widened to full section downwards.

The main equipment considered in planning the power tunnels execution is the same used in diversion tunnels plus an Alimak type platform, for inclined pilot shaft excavation. and 8 Atlas Copco BBC 16 WTH type drilling machines platform mounted for the tunnel downwards enlargement works.

ACKNOWLEDGEMENTS

The authors acknowledge the support given. By Dirce T. Locatelli for editing the text, Fernando Flores and Paulo R. Nascimento, for drawing the figures.

Hydropower'92, Broch & Lysne (eds) © 1992 Balkema, Rotterdam. ISBN 90 5410 054 0

Design and construction of a high pressure concrete plug at the Torpa Hydropower Plant

Øivind Bøhagen
Berdal Strømme a.s., Partner of Norpower A.S., Norway

ABSTRACT: During the last 7-8 years a new method to reduce permanent leakage from high pressure plugs has been developed in Norway. This new multistage grouting technique includes both cement grouting and chemical grouts (epoxy and polyurethane).The grouts are injected under high pressure (50-100 bar). An access gate plug to retain 455 m of water pressure has been successfully installed at Torpa Hydropower Plant using this technique. After commissoning the total leakage from the plug is less than 1 l/min.

1.0 INTRODUCTION

Torpa Hydropower Plant is located in Oppland in the southern part of Norway. The 9450 m long unlined headrace tunnel was excavated from two access tunnels, one near the intake at Dokkføy and one in the powerhouse area.

To maintain access to the headrace tunnel, the air cushion surge chamber and the sandtrap in front of the penstock for control and maintenance in the future, access gate plugs had to be installed. The plug in the powerhouse area had to be designed for a head of 455 m. Layout including positioning of the plug is shown in Fig. (1)

2.0 DESIGN

2.1 General design conditions and criteria

More than 80 pressure shafts and pressure tunnels with water pressure ranging from 150 m up to 965 m are presently in operation in Norway. Long experience and technical evaluations have produced the design criteria.

Basic loads:
In addition to the direct unbalanced water pressure, the plug is also subjected to pore pressure on the interface concrete/rock, and to (interior) pore pressures between concrete and steel liner.

As direct water pressure shall be taken the maximum static head plus the maximum surge (water hammer). As pore pressure at the wet end shall be taken the maximum static head, decreasing to zero at the opposite end in accordance to the situation at hand, e.g. the depths and locations of cutoff, grout curtains, length of steel liner etc. A typical example of a plug load distribution is shown in Fig.(2).

Rock conditions:
The essential siting criteria for a plug call for detailed logging of the geology in the relevant part of the tunnel to secure optimum abutment conditions, and in situ hydraulic fracturing tests to verify the required minimum rock stress conditions. To avoid hydraulic splitting minimum rock stress must be higher than the water pressure. Norwegian plugs are sited in rocks of great variety (granite, gneiss, sandstone, quartzite, phyllite, greenstone, dolomite and arkose; all of Precambrian or Palaeozoic age.) The rock mass permeabilities are usually in the range of $K = 10^{-7}\text{-}10^{-11}$ m/sec.

It is a must that the tunnel section at the intended plug site is excavated with care so as to minimize overbreak and fracturing. Prior to installing the plug, the adjacent rock face must be meticulously scaled with machine and hand tools.

Concrete permeability criteria:
The permeability coefficient shall not exceed 10^{-11} m/sec, and the water intrusion shall not surpass 25 mm when subjected to unbalanced pressure of 0.3 Mpa, 0.5 Mpa and 0.7 Mpa acting over a three day period, with each pressure level acting one day.

This is regularly achieved when the aggregates contain a fraction of at least 8% passing the 0.25 mm sieve, and the water cement ratio is below 0.50.

Structural function:
The plug must fill two prime functions:

- The entire unbalanced water pressure must be taken by the plug and transferred to the surrounding rock

- Leakages must be minimized

Due to the ragged interface concrete/rock, the failure mechanism along a potential line of failure here will be complex, and include both shear-pressure and shear-tension failure together with bond failure.

According to current practice, shear stresses are considered evenly distributed over the entire interface and the maximal permissible stress is set to 0.5-0.7 MPa for the serviceability limit state.

To handle waterleakage the plug must have a minimum length corresponding to the waterpressure (head). Length of 3-5% of the head is normal. Steel lining in gated plugs is often made shorter than the concrete part; a design that not only saves cost but may also make the following grouting more easy. Steel lining of 1-2% of the head is normal, Fig. (3) shows length of plugs and steel lining versus water pressure at some Norwegian hydropower plants.

Leakage and grouting:
A substantial permanent leakage at a high pressure plug may represent a significant economic loss, and is a potential hazard. Further it is a nuisance and must be counteracted with grouting. Recorded leakages through some Norwegian plugs, plotted against the pressure gradient and the type of grouting treatment, is shown in Fig. (4). It should be noted that most of the leakage (90%) refers to the concrete/rock interface. The diagram demonstrates how the chemical grouting applied to the more recent plugs (later than 1985) has served to reduce leakages. These chemicals (polyurethane, epoxy) are

used in addition to cement grouts and are inserted under high pressures (50-100 bar).

2.2 Design of the Torpa high pressure plug

The plug is designed in accordance with the criteria above.

The plug is given a total length of 20 m. The cross section of the tunnel in the area is 32 m^2 which gives an average shear-stress at the interface rock/concrete of 0.32 MPa. The total length of the plug is 4.4% of the water

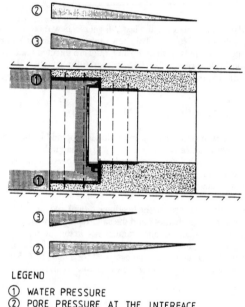

LEGEND
① WATER PRESSURE
② PORE PRESSURE AT THE INTERFACE ROCK/CONCRETE
③ PORE PRESSURE AT THE INTERFACE CONCRETE/STEEL LINING

Figure 2: Plug with typical loads.

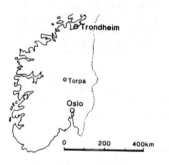

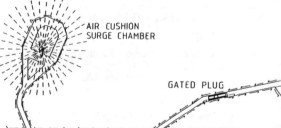

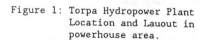

Figure 1: Torpa Hydropower Plant
Location and Lauout in
powerhouse area.

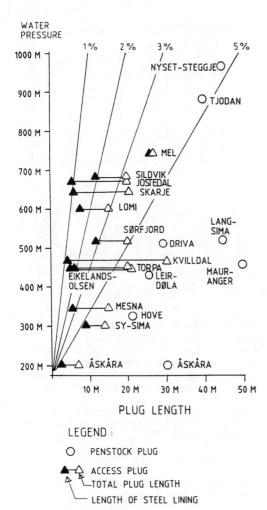

Figure 3: Plug length versus water pressure at some Norwegian hydropower plants.

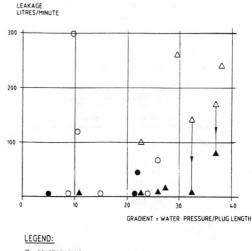

LEGEND:
○ PENSTOCK PLUG
△ ACCESS PLUG
○△ CEMENT GROUTING ONLY
●▲ CEMENT + CHEMICAL GROUTING

Figure 4: Water leakage versus hydraulic Gradient at some Norwegian high pressure plugs (Water heads-higher than 400 metres.

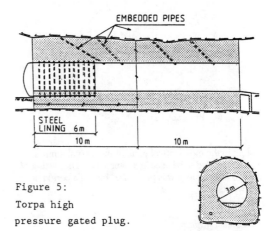

Figure 5:

Torpa high

pressure gated plug.

pressure (head). The access through the plug is circular with a diameter of 3.0 m. The gate is placed at the upstreams end, it has a 40 mm thick and a 6.0 m long steel lining. This length is 1.3% of the water pressure (head). The plug is shown in Fig. (5)

3.0 CONSTRUCTION

3.1 Pregrouting

Prior to concreting of the plug, the surrounding rock was grouted as shown in Fig. (6). A grout curtain including two fans of 8 holes with length 20-30 m were then drilled. Systematic permeability testing with pressure up to 45 bar in the first fan and 60 bar in the next was carried out. Only minor leakages were detected (max.

0.25 Lugeon), and the holes were treated with cement grouting at 60 bar.

3.2 Concrete work

The casting was separated into three sections as shown in Fig. (5), starting with the floor in the steel lining area. This was performed before the steel lining was installed. After installation the two last sections were cast, starting with upstream section including the steel lining and ending with the downstream section. To ensure complete

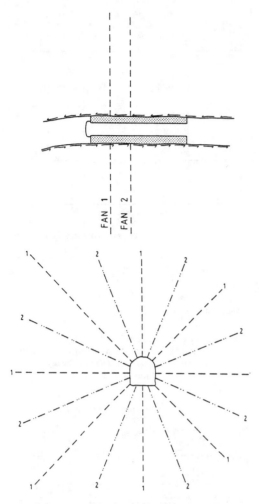

Figure 6: Pregrouting

casting in the most critical parts of the crown, 4 pipes were embedded to be used at a later stage. The casting of the two last sections was performed during a period of two weeks.

3.3 Contact grouting

To avoid leakages along the interfaces rock/concrete, concrete/steel and at the interfaces between the plug sections, grouting were performed. For this purpose special grouting hoses were mounted before casting. The hoses were installed as shown in Fig. (7) The hoses were fixed to the rock surface in 6 positions all around the crossection of the tunnel. At each position the hoses were separated in four parts.

To cover the interfaces concrete/steel, hoses were installed in 4 positions, each separated into two parts. At the vertical construction joint hoses separated in four parts were installed.

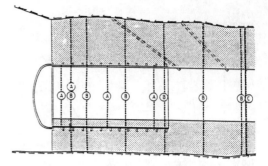

LEGEND
- (A) HOSES AT THE INTERFACE STEEL/CONCRETE
- (B) HOSES AT THE INTERFACE ROCK/CONCRETE
- (C) HOSES AT THE CONSTRUCTION JOINTS

Figure 7: Concat grouting

It is of great importance that the temperature inside the concrete plug is close to what it will be after commissioning when the grouting starts. At Torpa thermistors were installed for monitoring the temperature in the concrete mass. Fig. (8) shows how the temperature varied during the periode after casting.

The contact grouting started approximately 8 weeks after concreting. At this time the average temperature inside the concrete plug was 10°C .

The grouting was performed as follows:

1. Tunnel crown area: Post casting with expanding concrete through embedded pipes.

2. Interface rock/concrete: Upstream and downstream hose position was injected with polyurethane at a pressure of 20 bar to establish barriers for the following grouting.

3. Interface rock/concrete: Remaining hose positions between the upstream and downstream barriers were injected with epoxy at a pressure of 70 bar.

4. Construction joints: All hose positions were injected with epoxy at a pressure of approximately 20 bar.

5. Interface concrete/steel lining: Upstream and downstream hose position were injected with polyurethane at a pressure of 20 bar.

6. Interface concrete/steel lining: Remaining hose position was injected with epoxy at a pressure of 20 bar.

At all positions the grouting started with the floor section, continued with the walls, to end with the top section.

Total grout take was:

Polyurethane 135 kg

Epoxy   1480 kg

In addition to the hose positions mentioned above, an extra hose at the rock surface near the upstream end of the plug, and a fan of embedded pipes to the rock surface at the downstream end of the steel lining had been installed to be used if the water leakages would become higher than acceptable.

3.4 Water leakages

Fig. (9) shows the relation between water leakages and waterpressure during filling of the headrace tunnel.

The tunnel was filled with 0.85 m^3/sec of water which means that the pressure increased with 8.5 m/h at the most.

The first leakage was registered when the waterpressure was approximately 20 bar, and it increased up to 10 l/min at 30 bar and 25 l/min at 40 bar.

Suddenly at a pressure of 41 bar, water poured out of one of the grouting pipes embedded to ensure complete casting of the crown area. The total leakage was 60 l/min at this stage, but it did not increase when the pressure rose to 45.5 bar.

3.5 Post grouting

In an attempt to reduce the waterleakages it was decided to post grout. Polyurethane and expanding cement (NONSET 50) was injected through the pipe with the water leakage.

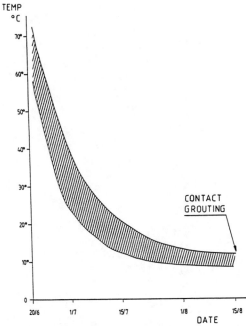

Figure 8: Concrete temperature

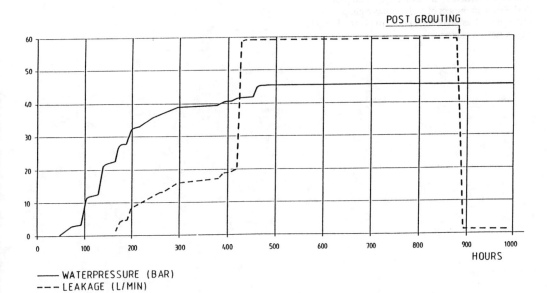

—— WATERPRESSURE (BAR)
--- LEAKAGE (L/MIN)

Figure 9: Leakage versus water pressure

41

Total grout take was:

Polyurethane 175 kg
Expanding cement   1600 kg

The grouting was very successful, and the leakage was reduced to less than 1 l/min.

3.6 Summary

A substantial leakage at a high pressure plug may represent a significant economic loss, and it may imply a potential safety hazard. Further it is a nuisance, and must be counteracted with grouting.

The access gate plug at Torpa is one of a number of high pressure plugs successfully performed in Norway during the last years. They all have been performed by use of this multistage grouting methods, which includes both cement grouting and chemical grouts (epoxy, polyurethane) injected under high pressure (50-100 bar).

A number of plugs treated by this technology are presented in Fig. (4), and they all show insignificant or no leakage. In some cases older leaking plugs have been re-treated with chemical grouts, giving remarkable improvement.

REFERENCES

Bergh-Christensen, J: 20. august 1990.Research project of high pressure plugs. Report no. 3: Access gate plug at Torpa Hydropower Plant (Tverrslagsport ved Torpa Kraftverk).

Bergh-Christensen, J. and Bøhagen, Ø: 20 juni 1991.Research project of high pressure plugs. Report no.7: Design of high pressure plugs. (Prosjektering av høytrykkspropper)

Hydropower'92, Broch & Lysne (eds) © 1992 Balkema, Rotterdam. ISBN 90 5410 054 0

Design of Dongfeng underground power station

Cao Pufa
Mid-South Design Institute, Guitang, Changsha, People's Republic of China

ABSTRACT: This paper describes several problems met in the design work of Dongfeng underground project and the corresponding treatment measures, including layout, NATM,anit-seepage provisions in the Karst region, as well as the surrounding rock monitoring design. So far, the excavation and the treatment measures have proven to be satisfactory in the construction practice.

Dongfeng power station is the second cascade hydroelectric power station on the main stream of Wujiang River in Guizhou Province, China. A 165m high hyperbolic arch dam which formed a reservoir with a capacity of 1.025×10^9 m^3 . The spillway and flood relief tunnel are arranged at the left bank, the diversion tunnel and underground powerhouse system at the right bank. There are three upper orifices and two middle orifices in the dam for flood relief.

The underground powerhouse houses three turbine-generator units with each capacity of 170 MW, and total capacity of 510 MW. The main machine hall cavern is 105.5m× 21.7m×48m (length× width× height) with the major axis striking N7° E. The 66m× 19.5m× 26m main transformer, switch station and draft-gate cavern which are in parallel with the machine hall are arranged in down-stream 31m from the machine hall. The two big caverns are connected by three busbar galleries (31m×6.5m × 11.2m) and one access tunnel (32m×6.6m× 9.0m) to the auxiliary transformer. These tunnels, as well as the access tunnel, three diversion tunnels and shafts, three draft tunnels and tailrace tunnel, adits, air drain galleries of busbar, main exhaust fan room, main air suply fan room, drainage galleries, galleries for impervious curtain grouting, etc. constitute a large underground cavern group.

Up till now, most cavern excavating work has been finished, and the support work for surrounding rock is carrying out followed the excavation work.

1 Layout of the station

On account of the general project layout, the underground powerhouse is arranged at the right bank. After the intake position is decided, the main problem is how to select a proper position for the main machine hall. According to the analysis result of the geologic condition (see Fig 1), there is no big fault through the 250~ 350m wide zone between fault 6 and fault 7, and the rock mass in this area is comparatively sound; in addition, the dissipator on the river bed has a limitation to the tailwater outlet position, so the main caverns such as the main machine hall must be arranged between the F6 and F7. After scheme comparing, the scheme of the underground powerhouse system arranged as semi-ring around the right dam abutment is selected.

As for choosing the major axis (or the powerhouse axis), the first thing to be considered is the smooth arrangement of the water way; and the second, to make the included angle between the major axis and the strike of the rock stratum, or the angle between the axis and the worst NNE joint plane as large as possible (>30°), and at the same time take the included angles between the major axis and several main faults (like F34, Fd2-15, Fd2-16, Fd2-9) into consideration. Thus it can be seen that the axis of the layout requirement of the hydraulic structures.

Conventionally, the principal stress direction of the in-situ stress is also a main factor to determine the major axis. In Dongfeng, the reconnaissance report shows that the principal stress is mainly the horizontal stress (σ =12.2MPa), and the included angle between the major principal stress and the major axis is 77 degrees, nearly to be orthogonal. It is very harmful to the stability as far as this intersection angle is concerned. The conclusion of the horizontal

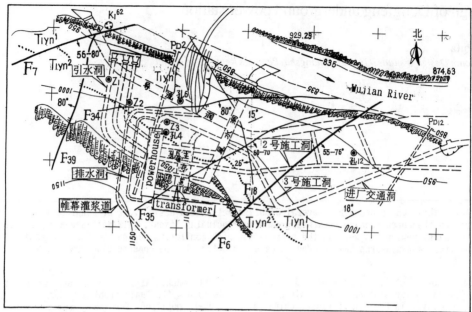

Fig.I: Plan layout of underground power house system of Dongfeng project.

stress as the major in situ stress is proven to be correct from the deformation analysis of the surrounding rock after excavated. Because the magnitude of the in situ stress is not high in respect of the rock strength, the convergency value measured from the end wall and the uplift deformation value of the arch crown are roughly close to the result of the original non-linear finite element analysis. Alough the direction of the in situ stress is unfavourable to the surrounding rock, the situation is not too serious. This fact indicates that the direction of the in situ stress may not be taken as the main control factor of selecting the major axis when the magnitude of the in situ stress is not high and it is difficult to make layout.

The adit arrangement has a great effect on the excavation volume of the underground powerhouse. So consideration of selecting the powerhouse location should be combined with construction method and procedures. With adit No.1 as access tunnel to the powerhouse, adit No.2 as tailwater surge tank, adit No.3 converted to insulation-oil store and air drain, and adit No.4 as main fresh air intake tunnel and jury drainage pump room, almost all adits of Dongfeng project are combined with the permanent project facilities, thus make the adits more usefull and consequently the cost is reduced.

As a result of neglecting ventilation and blow down systems in the preliminary design stage, several galleries and caverns are added in the construction design stage in Dongfeng

project, it affects the economy of the general layout scheme to a certain extent, and also weakens the integrity of the rock pillar between big caverns, increases the cavern cross parts, so the stability of the surrounding rock is affected.

2 Application of NATM

NATM is not only a construction method or design theory, but also a systematic method for design, excavation, support, monitoring and construction management of underground project.

The ultimate aim to use NATM is to make use of the load-carrying capacity of the rock to the fullest extent, and reduce the quantites of lining and supporting. The design and construction works of Dongfeng underground project are carried out on the basis of this principle.

As a kind of flexible support which can adapte to the deformation of surrounding rock, the support by shotcrete and bolting naturally becomes the soul of the NATM. When design this kind of support, the first step is selecting appropriate support parameters by means of analyzing project conditions and geologic conditions; the second is choosing the optimum excavating procedure and excavating method according to the project and geologic conditions, in addition, the reasonable monitoring item and monitoring method of the surrounding rock should also be choosed.

2.1 The Design of Supporting

Up till now, there still has not been a general method for design of support by shotcrete and bolting in engineering field. The supporting paramters of Dongfeng project are choosed mainly based on the following achievements in three aspects.

The first is the classification indicatrix of the geologic characteristics of the surrounding rock. The national standards, ministerial specifications and Barton classification system have recommended some support parameters which can be used for reference. But parameters for big caverns such as Dongfeng underground powerhouse are not specified in the specifications and standards mentioned above for lack of sufficient data. The Q system classification is a conclusion of 212 cases, which includes not only the recommended support parameters, but also the experimental formulas [1] for calculating support pressure.

Table 1 gives the support parameters of Dongfeng project on the basis of Q system.

Another deciding parameters mothod is on the result of analysing the stability of the surrounding rock. The stability analysis work for the surrounding rock of the underground powerhouse is carried out through three different manners: the plane nonlinear finite element analysis; geologic mechanical/physical model test; and linear elastic stability analysis. In analysis, the main structure planes (fault, stratification plane, and main joint group) of the section under consideration are simulated, and the comparison of surrounding rock stabilities under different excavating procedures are made by simulating the cavern excavating procedures. The aralysis result indicat the plastic yield area (or area distributed with cracks due to shearing and tension forces), the stress isogram and deformation situation of the surrounding rock. The depth of

The first is the classification indicatrix surrounding rock yield area is a very meaningful reference value for determining the length of the bolts. According to plasto-elasticity, the surrounding rock of the air face will enter into plastic yield state due to radial stress releasing and tangential stress concentrating near the air face, and this yield area is usually called "relaxation depth". The adjacent surrounding rock out of the yield area should be tangential stress concentrated area, and out of the stress concentrated area is the stress steady area where little disturbance due to excavating can reache. Generally speaking, it is advisable that the bolts should be anchored into the stress stready area. The relaxation depth values obtained from the stability analysis of surrounding rock in Dongfeng underground powerhouse are listed in Table 2.

The last reference of determining support parameters is the specific project which has similar project condition and geologic condition with the project considered. In China, the Lubuge underground powerhouse which was completed is similar in some aspects with Dongfeng underground powerhouse. The design and geologic parameters of these two underground powerhouse are compared in Table 3. It is necessary to indicate that the support parameters chosen in Lubuge projected have been proven to be suitable with small surplus in safty by checking after construction (relaxation area and displacement measurements).

The comparison of support parameters of Dongfeng underground powerhouse with those of foreign countries' underground powerhouses are given in Table 4 [3][4][5].

Besides the analysis mothed in the above three aspects, results calculation and analysis by block theory serve also as one of important bases of reference. There are six potential unstable blocks around the underground powerhouse and the cavern for main transfermer caused by the tectonic planes, Outcrop area of these blocks accounts for 40% of the total air face area at the vault. Analysis for these blocks showed that support by system anchors and shotcrete with mesh can basically maintain stability of the blocks. So no more local anchers were provided for the vault. Beyond, the experience indicates that for the vault, the stability of unstable blocks with even greater height would be significantly improved provided that there will be a surrounding rock 1/3 tp 1/4 of cavern span in thickness making up a stable bearing arch ring.

TABLE 1   SUPPORT PARAMETERS BASED ON Q SYSTEM

Possition	Q vol.	Support Type	Recommended parameters	Bolt length
Arch crown at south end	4.7	20	B(tg)1~2m +s(mr)10~20cm	5.255
Arch crown at north end	2.5	24	B(tg)1~1.5m +s(mr)10~15cm	5.255
Both wall	7.5	23	B(tg)1~1.5m +s(mr)10~15cm	5.255

45

Table 2 Relaxation depth by various analysis

Item		Relaxation Depth (m)
Program (1) (Note 1)		Relaxation depth 5m on average Max. 12m at right arch
Program (2) (Note 2)		3~4m in general Max. 10m at intersections of faults
Linear elastic calculation (1) (Note 3)		1.5~2.0m in general Maximum 6m at walls
Linear elastic calculation (2) (Note 4)		5m in general at vault 10m in general at vault (upstream)
Physical model test (Note 5)	k=1.3	Incipient crecks at right corner of vanlt
	k=1.94	5m at middle of vault 3.5m at both arches Cracks found at wall

Note 1: Two large intercalations (1) and d(40) were calculated, system anchors 1m×1m, L=4m
Note 2: Two large intinalations (1) and d(40) were calcuted, without anchorage support
Note 3: Tensile strength R=0.7MPa, f=1.0, c=1.2MPa
Note 4: Tensile strength R=0.7MPa, f=0.7, c=0.06MPa
Note 5: Initiation of falling of sand grain at cavern periphery when k=1.5 (k--increasing factor for loading on periphery model)

2.2 Design of excavation sequence and excavation method

Due to the non-linear response of the rock, the excavation sequence, would exert greater influense on the distribution of stresses in the surrounding rock. Two alternatives of excavation sequence were compared in stabiliy analysis, that is, excavation of the two large cavers downward in parallel and successive excavation. The calculation indicated that it was more favourable to conduct the excavation of the two caverns downward in parallel. The stress condition was also analysed for simultanes excavation of the top and bottom of the cavern (tairace tunnel). And as the analysis showed, such an excavation sequence was very unfavourable for stability. After the upper and lower portions of the cavern's rock mass was excaveted, the middle portion will experience a very high concentration of stresses. And cut-off of the middle portion corresponds to occurrence of huge stress relief at the middle of the high walls, thus exerting rather unfavourable influence on stability of the walls, it should be prevental.

The smooth blast was used for all the major caverns of the Dongfeng project. The location of cut holes, spacing for blast holes and peripheral holes, charges and coupling coefficient were determined through series of tests. Owing to properly selected parameters, the planimetric line and shaping was among the better cases both at home and abroad, achieving the half-hole rate up to over 90%, locally as high as 95%, under a unfavourable condition when the vault of the powerhouse was affected considerably by horizontally developed beddings.

The appropoxe excavation sequence and the high quality smooth blast have create very favourable condition for stability of the surrounding rock. It is seen from this that the excavation is critical in the NATM. At same project, the poor smooth blast for the tairace tunnel led to quite a few failures of collapse of roof.

2.3 Monitoring

In the monitoring design, apart from selection of appropriate normal sections, consideration was given mainly to surrounding rocks at intersections of caverns and monitoring unstable blocks resulted in by block analysis. The deformation of the rocks and stresses in anchors were the main items for monitoring. A total of 29 sets of multiple point extensometer and 45 sets of anchor stress meter (including rock bolt anchored crane beam) were embedded. In addition, measurement of convegent gauges was conducted.

It was rather difficult to carry out deformation abservation of surface of such large cavern as the main powerhouse with either convergent guage or transit because of its access.

2.4 Adjustment of Support Parameters while Construction

Because geologic conditions were fully exposed after excavation of caverns, a vast amount of monitoring information were obtained, the anchor support parameters for most of caverns were subjected to adjustment repeatedly after these data were studied and analysed. For the main powerhouse, the total amount of anchor was reduced by more than 30% after three times of adjustment, while for some caverns such as some sections of tailrace tunnel and part of sections of air-exhaust tunnel the amount of anchor was somewhat increused.

Temporary supporting anchor was made for zones where cracking of intercation beds and joint planes was visually observed, for locations with exudation of underground water, for intersections of caverns and for location with considerable deformation. The factors such as local improvement of geolgic parameters, smaller deformation of rock, relatively longer stand-up time, etc. served as the main basis for reduction of system anchors.

2.5 Quality Supervision

For NATM, it is necessory to conduct quality supervision over excavation and supporting.

The criteria of overcut and undercut and the criteria of half-hole rate for smooth blast are the principal aspect for excavation. It was prescribed in design not to allow undercut and that the overcut should not exceed 20cm. The practice of excavation showed that it was difficult to keep under the overcut of 20cm, especially for the horizontal rock formations. The practice indicated that it was reasonable to prescribe the relief difference of the excavation configuration.

For system mortor anchors and shotcrete support, it is necessary to strictly control the density of the mortar in the anchor holes and drawing test must be performed for anchors which should be resistant to drawing (e.g. the rock bolt anchored crane beam). For shotcrete layers it is necessary to strictly control the thickness of the layers and the bond between the shotcrete and the rock.

3 Seepage Control Design for Karst Area

The seepage control design for the underground powerhouse was made based on adequate investigation and analysis of hydrogeologic condition in the powerplant area. In respect of the rock mass, it has a weak permeability. Activity of underground water was controlled mainly by intercalations, faults and karst caves and karst erosion. There are two major karst cave systems. One is developed along fracture zone of F7 upstream, another is along F6 fracture zone downstream. Both of the large karst passage system served as intercepts for underground seepage water related to major caverns respectively upstream and downstream. Between the faults F6 and F7, the karst was weakly developed and only single karst caves were developed along some faults.

There are two supply sources for underground water. One is the infiltration of atmospheric precipitation from the surface and the another is the seepage from the reservoir, which infiltrates into the plant area after passing through the groute curtain. Corresponding drainage measures were taken in the design for two sources. A drainage curtain consisting of three drainage alleries was provided between the groute curtain and powerhouse. The infitrating water from the surface is intercepted and drained with the help of the drains in the galleries. In addition, drain holes were provided on the cavern surface according to seepage occurring on the surface after excavation, with dampproof barrier added along the cavern periphery.

4 Treatment measures for several special problems.

The main machine hall cavern and some other caverns are across several large and small faults. With corrosion developed in varying degrees in some large faults, when excavation is across these faults, unstable block appeared in the crown portion of the powerhose due to dissection of a group of well developed horizontal intercalations. In this area, excavation operation had to be stopped when the driving face was just across, or even before across these faults, and go on after bolting some anchor bars along the edges of the faults. For loose rock mass, the ideal temporary support is by shotcrete, which is quick and efficient. However, in Dongfeng project, the temporary anchor bar bolting support was commonly adopted because of allocation of the construction machines and of construction management for shotcrete.

Big limestone karst were not met during excavation of the main caverns. For small crossion areas in the air face of rock, the treatment measure was to fill them by shotcrete or mesh shotcrete after cleaned with high-pressure water, and the filling result is satisfactory due to a certain impact pressure of shotcreting.

In the design of layout for a group of caverns, it is sometimes difficult to avoid cavern intersections. The intersection often has three free faces with stress concentrated, so rock relaxation easy occurs and consequently rock unstability is resulted. For this reason, the excavating and blasting operations had to be controlled strictly, or presplit blasting was adopted, and at the same time many lengthened rock bolts provided. In later stage, some intersections were lined with reinforced concrete when necessary.

Crane beams anchored in rock mass were provided in the main machine cavern and in the cavern of the main transformer. In the crane beam design, the longitudinal stiffness of the concrete beam could not be too high, otherwise, the non-homogeneous deformation occured in different section of rock mass would have harmful effect on the diagonal tension anchor bar and the beam structure. In addition, the crane load was also unfavouable to the stability of the surrounding rock near the crane, therefore, the anchor bars in a certain range of the crance beam elevation had to be increased and lengthened to ensure no local unstability occuring in the surrounding rock. In order to transmit the crane load into the depths of the surrounding rock, the diagonal tension bar had coated with a layer of asphalt in 2m long near the opening, it can both reduce the friction drag between the mortar and the anchor bar, and avoid the bar overhigh local stress developed near the opening. The actually measured stress of the anchor bar after construction had proved that the stress caused by the surrounding rock deformation is better than that of the other similar project.

CONCLUSIONS

It can be concluded that the Dongfeng underground project is successful either in the selection of the position, major axis and support parametrs, or in the application of NATM. On the viewpoint of the regional geology, the Dongfeng project is located in a place where the engineering geologic condition and hydrogeological condition are relatively better. The excavation and support operations of the underground caverns are economical and safety.

REFERENCES

N.Barton, R.Lien and S.Lunde. 1974. Engineering classification of rock mass for the design of tunnel spport

Wang Defu. 1987. Utilization of support of rock bolt and shotcrete at Lubuge project

Wuhan Institute, Reseach report, 1985. The underground powerstation abroad

G.Reik. 1986. Stability control-Cirata Powerhouse cavern

Table 3   Comparison of Supports for Powerhouses of Dongfeng and Lubuge

No	Item of comparison	Lubuge powerhouse	Dongfeng powerhouse
1	Parameters of support	$\Phi25$ @1.5m×1.5m, L=5.0m L=6.0m	$\Phi25$@1.2m×1.3m,L=5.0m L=7.0m alternatively
2	Excavation span	19.0m	21.7m (above crane beam)
3	Lithologic caracter	Breccia limy dolomite Limy dolomite	Thin-bedded limestone Medium-and thick-bedded limestone
4	Compression strength	45~60 MPa	60~80 MPa
5	Tensile strength	2~3 MPa	1~1.3 MPa
6	Elastic modulus	3×10^4~4×10^4 MPa	0.8×10^4~2.5×10^4 MPa
7	Shear strength	c=0.3~0.4MPa Φ=50° ~55°	C=0.3MPa Φ=29° ~35°
8	Class of wall rock	II,III	III(locally II after excavation)
9	Q value	5.0~12.0	2.5~7.5
10	Situ stress	σ =13.5~15MPa, Direction in parallel with powerhouse axis,	σ =12.2MPa,Direction in 77° with axis. Powerhouse intersects with faults F34,fd2-15,fd2-16
11	Faults	Powerhouse intersects with faults f3 and f5	fd2-9 and fd2-3
12		Main powerhouse and main transformer cavern staggered in elevation,being 45.0m apart horizontally	Main powerhouse and main transformer cavern nearly at the same elevation, being 31.0m apart horizontally.

Table 4 Comparison with Several Underground Powerhouses in Respect to Support Parameters

Name of project	Basic conditions	Adopted support parameters
Monger Indonesia	Dimensions of cavern: width 20.1m, height 40.4m, length 112.5m; elliptic in shape; breccia, horizontal joints developed,$E=0.25 \times 10^9$ MPa, $\mu=0.2$, compression strength 10MPa,$Q=1.0 \sim 7.0$ (poor$\sim$ordinary),self weight stress field	1. Prestressed anchorage cable @3.5m×3.5m, L=18m 2. Anchor $\Phi 25$ @1.5m×1.5m, L=4$\sim$6m 3. Shotcrete 15cm with mesh 4. Reinforced concrete lining 1.0m
Cirata Indonsia	Dimensions of cavern:width 28$\sim$35m, height 50m,length 250m,elliptic in shape;eruptive breccia,andesite $E=0.2 \times 10^4$ MPa, $\mu=0.2$,C=2MPa,$\Phi=55°$	1. Shotcrete 25cm in 3 layers with 2 layers of mesh 2. Prestressed anchorage cable 15$\sim$20m long (100t) @2m×4.0m 3. Prestressed anchor 5$\sim$7m long (10t) @1m×2m
Waldeck II German	Shale and clay schist alternating $E=3.0 \times 10^3$ MPa,compression strength 45.6MPa situ stress,confining pressure factor $\lambda=0.4$ Dimensions of caver:33.4m×54m, elliptic in shape	1. Prestressed anchorage cable @3m×4m(17° , 125t) L=23.5m 2. Prestressed anchor @1.33m×1.5m(12t),$\Phi 24$, L=4$\sim$6m 3. Shotcrete 24cm (3 layers of mesh)
Imaichi Japan	Siliceous sandstone,breccia,horny sandstone Section of cavern:33.5m×15m,elliptic shape	1. Prestressed anchorage cable @4m×2m(10t), L=18m 2. Prestressed anchor @1m×2m (15t),L=6$\sim$7m 3. Shotcrete 32cm (with mesh)

Hydropower'92, Broch & Lysne (eds) © 1992 Balkema, Rotterdam. ISBN 90 5410 054 0

Tianshengqiao Hydropower Project

Chen Jialiang & Wang Baile
Guiyang Hydroelectric Investigation & Design Institute, People's Republic of China

ABSTRACT: Tianshengqiao(TSQ) Hydropower Project is on the Nanpan River, south of China. The project consists of a Rcc dam, three long headrace tunnels, three surge shafts, six penstocks and a surface power station. The geological condition is very complicated. The paper mainly describes the engineering treatments upon headrace tunnels.

1 INTRODUCTION

Tianshengqiao(TSQ)Hydroelectric Project is located on Nanpan River, in the up stream of Hongshui River, which forms the boundary between Guizhou Province and Guangxi Autonomous Region. The catchment area above the dam is 50194Km², average annual flow is 19.4 x10⁹m³. The project will mainly serve the function of electric power generation. The total installation capacity is 1320MW, with a annual power output of 8.2 billion KWh. In order to meet a rapid growth of the power demand in Guangdong, Guizhou Province and Guangxi Autonomous Region, the first unit of 220 MW is scheduled to be put on line by the end of 1992. The project is composed of head works, water conveyance system and powerhouse. The water conveyance system includes three circular pressure raised tunnels with an average length approx. 9.5 Km each and diameter of 8.7-9.8m, differential type surge shafts (21M in diameter, 88m in height) with upper chamber(10m in width, 10m in height, 160m in length) and six underground penstocks with a diameter of 5.7 m each, average length 590m each. See fig. 1 and fig. 2.

2 GEOLOGICAL CONDITION ALONG THE WATER CONVEYANCE SYSTEM

From dam site to power house the Nanpan River is 14.5 Km long, the gorges on the both bank are narrow and steep while the valley cuts deeply with a concentrated natural fall of 181m.

From the inlet to Yacha gorge, the tunnels pass through karst medium and high mountain area. Each tunnel is 8100 m in length and is buried 300-760m deep. From Yacha gorge to the power house, the area is formed by sandy shale, while from Yacha gorge to the surge shafts, the tunnels passing through the low hilly area are buried 150-300 m deep and have a length of 1400m each. See fig. 3.

1.1 Stability of the surrounding rock along the tunnels

The tunnel alignment passes through hard and semihard limestone and dolomitic limestone with a length of 8100m each, acconnting for 85% of the total length. The interbedding of sandstone and argillaceous shale made of soft layer and hard layer, has a length of 1400m, occupying 15% of the total length. Most of the hard and semihard limestone and dolomitic limestone is fresh and complete and has a high mechanical strength (with 80-100MPa compressive strength) favouring the stability of surrournding rock. The interbedding of sandstone and argillaceous shale has a low mechanical strength (with 30MPa compressive strength) which doesnt favour the stability of surrounding rock.

1.2 Surge water and outside water pressure

Along the tunnel alignment water penetration has been observed during the excavation. Especially during the flood season surge water has been appeared after strong raining. Counting more than 3 sections along the tunnel No.1, the maximum amount of surge water reaches 10 m³/s or so.

According to the data of the water table

of a few boring holes and the phenomenon during excavation, the outside water pressure can be obtained as 4MPa and used for lining design.

1.3 Rock burst

On the upstream of Yacha gorge, most sections of tunnels are deeply buried. These sections existed a high earth stress recorded with more than 25MPa.

At the sections with hard brittle limestone and dolomitic limestone, during the excavation, duo to the sudden releasing of earth stress in the rock mass, rock burst happened frequently.

1.4 Karstic caves

From inlet to Yacha gorge total 9 large karstic caves have been discovered during excavation. The maximum length of cave containing clay and rock fragment across the tunnel can be 20-30m. Another more than 12 medium and small karstic caves also have been discovered along tunnel No.1. The number of karstic caves along tunnel No.2 and No.3 may be greater. So it takes

excavation and lining design more difficultly.

3 CONSTRUCTION OF HEADRACE TUNNEL NO.1

The construction of the headrace tunnels uses three adits. The tunnel(about 4000m in length) between the starting point of the tunnel and the point 0+4100m was excavated using adit No.1 in the downstream direction by means of the drilling and blasting method. The tunnel(about 3300m in length) between the point 0+4100 m and 0+7437 m was excavated using adit No.2 in the upstream direction by means of TBM with 10.8m diameter. The tunnel(about 2300m in length) between adit No.2(0+7437m) and surge shafts was excavated using adit No.3 by means of the drilling and blasting methed too. The primary support of tunnel for section which was excavated by means of the drilling and blasting method uses rockbolts and shotcrete partly. When the karstic cave was encountered, the measures were taken to install steel rib supports and place conrete on the sidewall besides rockbolts and shotcrete with wire mesh.

After overcoming the difficulties taken by large karstic caves, surge water, rock

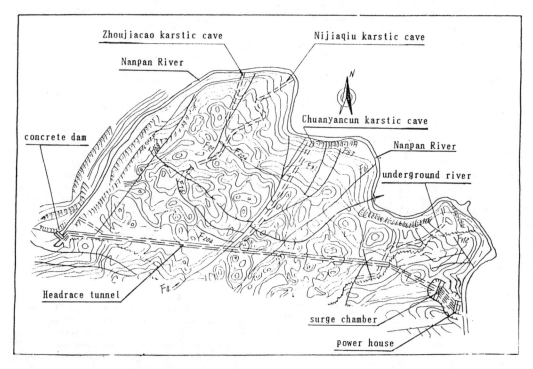

Figure 1: Layout of the key position of Tiangshengqiao secondary stage power station.

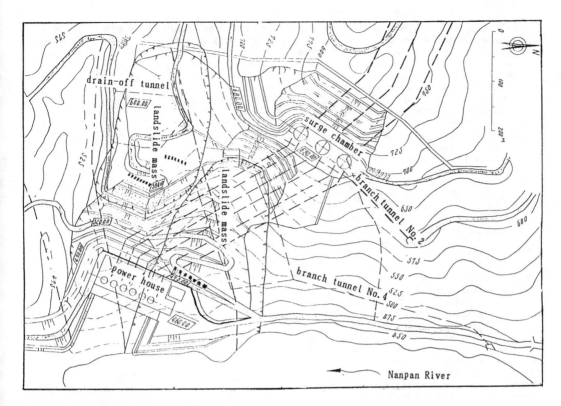

Figure 2: Layout plan of the Tiangshengqiao
secondary stage power station region.

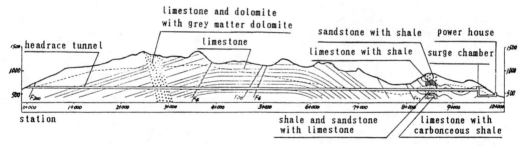

Figure 3: Vertical section of the project.

burst and other disadvantage geological
condition,the excavation of headrace tunnel No.1 had been completed throughout by
July 1991 and the excavation of the headrace tunnel No.2 will have been completed
by May 1992.

4 DESIGN OF TUNNEL LINING

The thicknesses of concrete lining for
most sections of headrace tunnels are 0.4,
0.5 and 0.6m,according to the geological
conditions of surrounding rock.In some
concrete linings there are one or two
rings of circle reinforcement.The diameters of reinforcement are 20 to 25mm with
a spacing of 20cm.

The uncontrolled crack criteria was a-
dopted for concrete lining design. The
thickness and amount of reinforcement in
it is depend on either water pressure inner the tunnels or outer water pressure e-
xerted on the lining of tunnels.

The lining of those sections across karstic caves or falts zones must resist
the disadvantageous fluence induced by the
three factors as following.The first is
the water pressure in the tunnel that reaches more than 0.75MPa,the second is the
high water pressure out of the tunnel that
reaches more than 4MPa,and the third is the
obvious settlement of the concrete lining
under which the foundation is composed with
clay or other similar material which is
too soft and weak to support the weight of
the tunnel.Once the joint between two
blocks of the tunnel would be destroried the
large amount of leakage might happen and
it would be very dangerouse for project.

Based on the analysis above,two measures
were adopted to strengthen the structure of
the tunnel.The first one is to thicken the
lining,i.e.the thickness of the lining is
increased from 0.4 or 0.6m to 0.9-1.75m.
The second is to strengthen the consolidation grouting out of the tunnel.The deep-
ness of grouting hole is increased from
4.5m to 8m and the grouting pressure is
increased from 1.5MPa to 4MPa.

At some sections(about 110m in length)
the extent of consolidation grouting under the foundation of the tunnel is more
expanded.Those measures above are being
taken now.Once the measures were completed,
the lining of the tunnel would coorperate
with the surrounding rock for resistance
inner and outer water pressure and minimize the settlement of the lining more efficiently.

And the necessity of excavation of another drain tunnel with a section of 3.3m
and 7000m in length is being under consideration to reduce the outerwater pressure
of the tunnel.

5 CONCLUSION

The Tianshengqiao Project is one of the
biggest hydropower projects under construction in China.There are three long headrace
tunnels and other underground structures
composed in it. The geological condition
is very complicated and difficult.In construction technique and design theory many
new problems were encountered and now
they are being solved.The numerous karstic
caves,frequent occurrence of rock burst,
exist of hight water table outer tunnels
and huge quantity of underground works
have made much trouble for construction and
design.After investigation in detail for
improvement construction technique and design the project construction goes on
fluently.The concrete lining and grouting
of tunnel No.1 are being taken and it is
expected to complete throughout by Sept.
1992.According to the schedule the first
unit of generator will be put into operation by the end of 1992.

Hydropower'92, Broch & Lysne (eds) © 1992 Balkema, Rotterdam. ISBN 90 5410 054 0

New drilling techniques and the design of adits of long hydro tunnels

L. Da Deppo & C. Datei
University of Padua, Hydraulic Institute 'G. Poleni', Italy

A. Rinaldo
Department of Civil and Environmental Engineering, University of Trient, Mesiano di Povo, Italy

ABSTRACT: The choice of number and length of adits for construction of long hydraulic tunnels has received attention in the past, among others, by the authors (1981) because finding a solution bearing significant factors in mind (like economy of adit lengths, the presence of geologic controls or topographic peculiarities, the need of adequate rock cover, construction timing) for the choice of the most convenient layout defines a difficult optimization problem. New drilling technologies seem to have substantially modified the overall cost scheme thereby altering the optimal tunnel layout and configuration. Upon a review of the progress experienced in this field since the previous paper by the authors appeared, this paper examines some consequences of the new drilling schemes on the optimal layout of the tunnel and adits.

1. INTRODUCTION

The paper discusses the impact of new drilling technologies on the design of layouts of long tunnels, seen in perspective of operations research tools.

Hydraulic tunnels, particularly those associated with hydro projects, frequently have a plane polygonal layout joining the plant's intake works to surge tank or to the forebay. Traditionally this type of layout has been preferred to the rectilinear one, despite the latter's greater possibility for development. Its course is approximately parallel to that of the valley.

Therefore the factors that have a bearing on the design choice, are mainly as follows: the presence of geologically difficult zones; topographic peculiarities; the need for sufficient rock cover; and the need to shorten as much as feasible the distance from possible secondary intakes or from possible adits.

The opportunity of considering tunnel sections of different size, as the layout changes, could perhaps be taken into consideration, provided it is shaped also on productive and economic evaluations. Execution time and scheduled time tables also should be taken into account in the general economy of the work.

Finding a solution which bears these factors in mind for the choice of the most convenient layout yields a difficult optimization problem.

Under such conditions, the problem has not been widely discussed. A starting point was a rather excessive schematization of it, in a paper by Martelli (1895), who intended to minimize the construction time. More recently Conte (1950) has characterized the deviation which the tunnel axis has to undergo in crossing two zones in terms of a different cost of development and indicated where to place the intersection: between tunnels or between tunnels and adits. Indri (1970) has investigated the most convenient number of adits (equal and equidistant) for the construction of a tunnel having a constant unit cost, and has considered some variants to the layout that would bring construction costs to a minimum.

These analyses have been generalized by the authors (1981) who proposed a scheme for design based on: i) operations research tools, specifically mixed integer programming and ii) a model of costs divided into fixed and variable parts according to the distance to the nearest access. The rationale for the choice of model was that the cost per unit length was known to depend of a constant part and a distance-to-access type of cost typically dependent on transportations (materials, manpower, seepage water and air).

The interest in dealing with this problem springs from many considerations. The process can, in fact, be applied at the design stage, to select the "best" from among all the possible layouts and corresponding sections, assuming a builder has average equipment and operating techniques. It can, likewise, be very useful to the builder who studies a tender just on the basis of the available equipment and on the methods imposed by it (and, therefore, on some constraints); or, alternatively, after the bid is concluded, for the purpose of putting forward a building scheme which is different from the one suggested by the tenderer or the consultant. The process offered is equally interesting when determining possible changes in construction which might be necessary because of peculiarities or unexpected situations arising during the work.

This paper examines the foremost features of the method of analysis and critically discusses its potential for the understanding of the impact of new drilling techniques on hydro tunnel design.

2. METHODS

The cost per unit length of a tunnel consists of a constant part (explosives, cleaning, machinery) and a part which varies according to the distance l from the closest access point (for transport of materials and air, bailing out, etc.).

Both the entrance and exit of the tunnel are treated in what follows as any intermediate adits. The unit cost $c_i(l_i)$ of the i-th reach in which the poligonal layout is subdivided has been considered in various manners, and often its functional dependence on distance l_i has been considered linear (e.g. Da Deppo and Rinaldo (1981)).

The cost function of the whole tunnel is constructed as a sum of terms. The costs of eventual adits are to be added to it. Every distance l_i will have to be expressed by a relationship that takes into account the existence or non-existence of the adits. The function is easier to manipulate by using integer variables, thereby yielding a problem of integer programming.

The following example will help explain the manner in which the generic distance l_i can be represented.

Let us consider the scheme of Fig. 1a) where 1, 2, 3, 4 and 5 represent tunnel sections, and [a], [b], [c] three possible adits, being L_1, L_2, L_3, L_4, L_5, and L_a, L_b, L_c long, respectively. Suppose we want to calculate the distance l_2 of the barycentre (a point equidistant from both ends) of section 2 from the nearest access point (such as the entrance or exit of the tunnel, or connection adit). Simple considerations of the layout and of the possible adits allow some layouts to be discarded. Let us suppose that the following inequalities arise:

$$L_1 + L_2/2 < L_2/2 + L_3 + L_4 + L_5$$

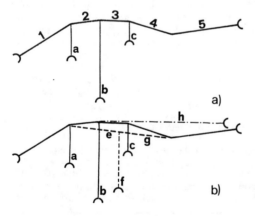

Fig. 1: a) Representation of a typical tunnel with three possible adits ([a], [b] and [c]). The numbers represent various sections of the tunnel; b) Possible alternatives in the planar layout are considered and treated as adits. An integer (either 0 or 1) is assigned also to the adit [f] and the new links [e], [g], [h].

$$L_2/2 + L_b > L_1 + L_2/2 > L_2/2 + L_3 + L_c \qquad (1)$$
$$> L_2/2 + L_a$$

In these conditions, the patterns $(L_2/2 + L_3 + L_4 + L_5)$ and $(L_1/2 + L_b)$ can be discarded, leaving the other three to be considered; the first one, $[L_1 + L_2/2]$, exists and can certainly be built; the others, $[L_2/2 + L_a]$ and $[L_2/2 + L_3 + L_c]$, are shorter and could appear as a result of the optimization process. Let us then associate with each adit, a, b and c, a variable x_i, (i = 1,3), defined in the following manner:

$$x_i = \begin{cases} 1 \text{ if the (possible) adit is taken} \\ 0 \text{ otherwise.} \end{cases}$$

The distance l_2 from the barycentre of section 2 to the nearest access connection must therefore be:

$$l_2 = L_1 + L_2/2 +$$
$$+ [L_2/2 + L_3 + L_c] x_3 - [L_1 + L_2/2] x_3 +$$
$$+ \{L_a + L_2/2\}x_1 - \{L_1 + L_2/2 + [L_2/2 + L_3 + L_c]x_3$$
$$- [L_1 + L_2/2] x_3\} x_1 \qquad (2)$$

It is worthwhile mentioning that no approximation is involved in eq. (2), the distance from the closest access being exactly computed as a function of the existing adits.

The generalization of eq. (2), i.e., its extension to more complex cases, is straightforward and may be obtained by an iterative process. Figure 1b) shows a case in which possible design alternatives for the planar layout are considered, branches and links being treated as viable alternatives to the basic layout. Although with slight complications, the procedure briefly described with reference to Fig. 1a) can be generalized into a general iterative scheme which computes exactly the distance to the nearest access as a function of the existence of predetermined distinct paths.

It is worthwhile mentioning that the method leads to a suitable treatment of layout design in heterogeneous geologic formations, where longer paths might be considered because of lower unit costs of drilling (e.g. higher compactness of the formations or higher likelihood of singularities).

It may be useful, for clarity's sake, to point out that in the example illustrated before, the section (the one indicated by number 2) between two possible adits, has been considered as a single solution. As a matter of fact it is necessary to subdivide the section into short lengths to allow them to be connected with possible adits or exits at the end of the whole section. In fact, if the section were not subdivided into parts, it would be assigned to a single exit. This would probably lead to an incorrect solution because one could not conclude in advance that it would be convenient to attribute one part of the section to the other exit.

The subdivision, therefore, clearly reduces the risk of making a possibly inaccurate evaluation, more easily than using a strict procedure necessary to assign each section of the whole tunnel to exit or adit. The extreme position of two parts having a common exit would therefore define the section to be assigned to the exit itself. Furthermore, inside the section between two adits the subdivision allows the assignment of different values of the unit cost of construction, so as to take into account the different characteristics of the rock, and unusual conditions such as faults, water seepage, gas or a different geometric shape of the finished section.

In certain cases an unusual condition may be considered more easily by putting an adit at either end, each evaluated according to their actual cost. Because of their prohibitive cost, these might not then appear among those considered as possible.

The objective function to bring to a minimum then takes the following form:

$$F(X_n) = \sum_{i=1}^{NP} L_i c_i(l_i) + \sum_{j=1}^{NF} K_{Fj} x_j \qquad (3)$$

where X_n is the set of $x_1, x_2, \ldots x_{NF}$ variables of the problem.

The first sum is extended to all the NP parts, having length L_i, into which the tunnel is subdivided. The term $c_i(l_i)$ is the i-th unit cost of construction computed for a distance l_i and l_i is evaluated by our scheme. The second sum is extended to $x_1, x_3, \ldots x_{NF}$ possible adits; x_j has a global cost K_{Fj}.

The degree of the relationship indicated in eq. (3) is equal to the number of the paths joining the barycentre of the section under consideration with the possible adits, the lengths of which are shorter than the distance between the barycentre and one of the tunnels' terminals.

The constraints to be introduced are:

$$x_j = \begin{cases} 1 \text{ if the (possible) adit is taken} \\ 0 \text{ otherwise.} \end{cases}$$

Other constraints (either equality constraints, i.e. $f_j(X_n) = 0$, or inequality constraints, i.e. $d_j(X_n) \geq 0$) can be adopted to satisfy particular conditions, e.g., to contain the total construction time pre-set within certain limits. The calculation procedure adopted, like that used by other authors for the solution of similar problems, consists of turning the search for the solution of the constrained minimum problem, such as the one illustrated above, into a sequence of searches for a minimum objective unconstrained function by using, in the stages of optimization, functions derived from eq. (3) after suitable modifications. Such modifications take into account the weight of the constrained conditions. while, by operating with appropriate penalties, they lead the search for the minimum towards the values of the variables of the problem which are admissible for the given constraints (Gisvold and Moe (1972)).

The general form of the objective function at the k-th stage of optimization, modified in this way, is given by:

$$P_k(X_n) = F(X_n) + r_k \sum_{j=1}^{n_d} [d_j(X_v)]^{-1} +$$

$$+ (r_k)^{-1/2} \sum_{m=1}^{n_u} f^2_m(X_u) + s_k G_k(X_n), \qquad (4)$$

where:

X_n is the set of the $x_1, x_2, \ldots, x_n$ variables of the problem;

X_u is a sub-set of X_n constituted by u elements selected from among $x_1, x_2, \ldots, x_n$;

X_v is a sub-set of X_n by v elements

$F(X_n)$ is the value of eq. (3) computed on X_n;

r_k, s_k are coefficients computed at the k-th iteration, whose expression is, for instance, used in Da Deppo and Rinaldo (1981);

n_d is the number of inequalities that may appear in the constraints $d_1(X_v), d_2(X_v), \ldots, d_{nd}(X_v)$ led back to the general form $d_i(X_v) \geq 0$;

n_u is the number of equalities appearing in the constraints $f_1(X_u), f_2(X_u) \ldots, f_{nu}(X_u)$ related to the subset X_u;

$G_k(X_n)$ is the value of a penalty function computed on the whole X_n, its form being for instance in Box et al. (1969).

Of all the methods of finding the minimum value of the function (free from constraints because of the artifice (4)) a gradient method is deemed most appropriate (known as Davidon's).

The interesting feature of the above set of equations and of the solution procedure is that no restriction is posed to the particular choice of unit cost of construction $c_i(l_i)$ thus allowing for the treatment of complex geologic environments.

3. NEW DRILLING TECHNIQUES

In the field of civil engineering works, tunnelling - in particular for hydraulic projects - has experienced the most substantial technological progress in construction methods within the last 50 years (Vielmo, 1991). Innovations dealt with research, new materials, advances in engineering mechanics. As a result, materials and methods for tunnel excavation and lining allowed for constructions unfeasible a few years ago because of poor quality of the rock, dimensions, lengths or pressure regimes.

The foremost innovation is the mechanical excavation via boring machines operating on full circular section (TBM) or equipped with a point head cutter. These machines have been reported to operate in rock formations with mono-axial pressure resistance of 100 MPa and presently even up to 200 MPa. Mechanical excavation methods found important applications also in construction of vertical (or inclined) shafts, for which a pilot-bore is first dug followed by a rise-boring.

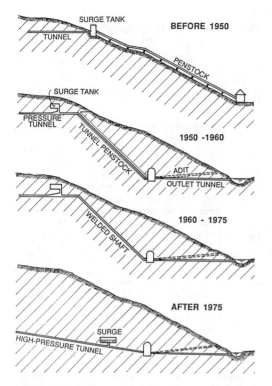

BEFORE 1950

SURGE TANK
TUNNEL
PENSTOCK

SURGE TANK
PRESSURE TUNNEL
TUNNEL PENSTOCK
1950 -1960
ADIT
OUTLET TUNNEL

WELDED SHAFT
1960 - 1975

AFTER 1975
SURGE
HIGH-PRESSURE TUNNEL

Fig. 2: Evolution of power-plant schemes in Norway as a result of new tunnelling techniques (after Broch, 1989; Vielmo, 1991).

Advances in tunnel technology also yielded improvements in the lining protection of the tunnel. Precast elements are lined sometimes even in complete rings.

A most important byproduct of developments is a considerable increase in safety for operators, deeply appreciable knowing the past stability conditions during construction.

A novel realm of problems in tunnelling is related to the so-called environmental impact. Rather than pertaining offenses - quite limited indeed - to landscape sightings, tunnelling impacts may yield overall rock deterioration or seepage and groundwater flow alterations which call for due attention. Nevertheless mechanization allows (beside the unconstrained choice of diameters) for reductions in the overall length of tunnel by reducing the need for adits and for freedom of design choice from external morphological constraints. As such, the design of layout of tunnels has been substantially modified by the introduction of new technologies.

The pilot-boring technique has been adopted for very large tunnels where a TBM excavation of a section of 3.00 m diameter (or more) is enlarged afterwards with traditional methods (Da Deppo et al., 1991). In such conditions sustaining the tunnel (even in difficult geologic conditions) yields much less

troubles than traditional excavation in full section. Furthermore, pilot boring yields a perfect experimental phase in reconnoissance of geologic formations and allows for bailing out and drainage of seepage waters and for easier air supply and tunnel ventilation in the front region.

New technologies, therefore, substantially altered the ratio of fixed and variable unit costs, much favoring the former. As a result, straighter and simpler layouts would be optimally designed in relatively homogeneous rock formations. Nonetheless, the proposed methodology serves well in engineering the best layout in complex formations, where heterogeneity results in different unit costs thereby allowing for many potentially suitable design solutions.

As an example, Figure 2 illustrates the evolution of power-plant schemes in Norway related to widespread mechanization on the site and advanced automatization/instrumentation (after Broch, 1989; Vielmo, 1991). The tendency towards high-pressure tunnels and simpler layouts is clearly shown therein.

4. NUMERICAL APPLICATION

As an example, the method illustrated above has been adopted for the Pieve di Cadore-Soverzene hydroelectric tunnel which belongs to the Piave-Boite-Maè-Vaiont system, Italy.

Fig. 3 shows the layout of the works, a schematic layout and the longitudinal profile.

The adits indicated are the ones actually carried out during the plant's construction (1947-51). Other possible adits have not been examined, because concrete elements for their location were not available and because it seems now that the ones already built are more than enough.

It should be further noted that the constraint of the adit's existence has to be related to the crossing of the Montina, Vaiont, and Gallina tributaries (adit numbers 4, 9, 12 and 12b) by an overhead pipe and to a tube bridge over the Boite and Maè tributaries (adit numbers 2 and 12) being set in.

As no actual figures are available on the adits' cost without any loss adjustment for the development of the application, it has been assumed that the total installation costs of the same adits, along with the progress unit costs, are the same for all the accesses.

The main geometrical features of the tunnel are:
- two intake tunnels 3.50 m in diameter and 446 m long;
- a tunnel section 4.50 m in diameter and 15733 m long;
- a section 7677 m long and 4.70 m in diameter; and,
- a final section 2351 m long, consisting of two tunnels in parallel with a diameter of 5.00 m.

While progress and covering costs depended non only on the diameters but also on the rock features in this application, for the above-mentioned reasons, costs of construction have been assumed for the single stretches of constant diameter that depend only on the design's standard section.

In the first case, the following costs ($) have been

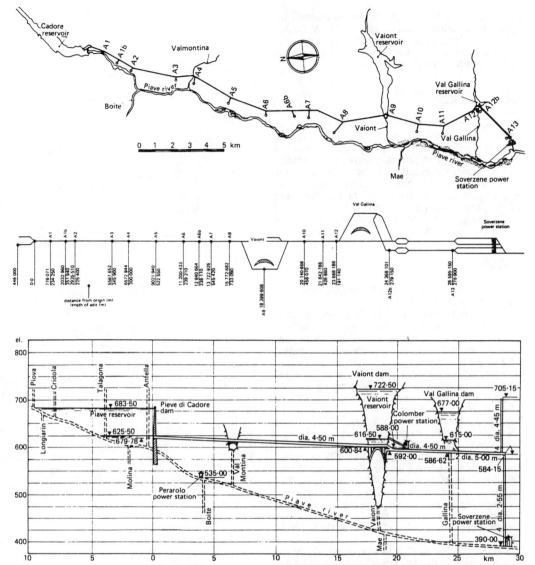

Fig. 3: The Pieve di Cadore-Soverzene tunnel showing its planar layout and a longitudinal profile. Units are in meters (after Da Deppo and Rinaldo, 1981).

assumed as an example (after Da Deppo and Rinaldo (1981)):
- installation cost of one adit = 120000;
- adit's progress cost (per metre) = 460 (1 + 0.000111 l);
- tunnel's progress and lining cost (per metre) diameter (ϕ) = 3.50 m = 744 (1 + 0.000133 l);
- tunnel's progress and lining cost (per metre) diameter (ϕ) = 4.50 m = 972 (1 + 0.000133 l);
- tunnel's progress and lining cost (per metre) diameter (ϕ) = 4.70 m = 1020 (1 + 0.000133 l);

- tunnel's progress and lining cost (per metre) diameter (ϕ) = 5.00 m = 1088 (1 + 0.000133 l).

The coefficients adopted in the example lead to a doubling of the unit cost of the tunnel for every 7500 m of progress and of the adit for every 9000 m.

Taking as a possible starting solution the one anticipating the existence of all adits, the optimum solution reached was one in which adits 1b, 7 and 10 were left out.

To verify the dependence of the solution on the choice of unit cost, it has been assumed, as a second case, that the tunnel's unit cost doubles every 6000 m and the adit's cost every 9000 m. In this example the optimum solution discards adit 1b. Any further imbalance of unit costs - as related to an estimation of the impact of new technology - could reduce the number of adits further.

5. CONCLUSIONS

The application of non-linear programming with integer variables allows a special class of comparatively frequent problems arising in constructing civil engineering works to be treated, that is, those where the unit costs depend on the distance from switchyards, store-houses, etc. Application to the design of layout for long hydrotunnels suggests that the proposed methodology may be a valid tool for large hydropower projects.

REFERENCES

Martelli, G. 1895. Studio sulla distribuzione dei punti d'attacco di una galleria. *Il Politecnico.*

Conte, J. 1950. Etude sur la détermination du tracé économique d'une galerie. *La Houille Blanche.*

Indri, E. 1970. Alcune considerazioni sul tracciamento di gallerie per derivazione di acque. *L'Energia Elettrica* 8.

Gisvold, K.M. and Moe, J. 1972. A Method for Nonlinear Mixed-Integer Programming and its Application to Design Problems. *Journal of Engineering for Industry.*

Box, M.J., Davies, D. and Swann, W.H. 1969. Non-linear optimization techniques. Edinburgh (Scotland): *Oliver and Boyd.*

Da Deppo, L. and Rinaldo, A. 1981. Choosing the number and length of adits for a long tunnel. *Water Power & Dam Construction*, V. 33, No. 9:36-39.

Vielmo, I. 1991. Gallerie idrauliche: interdipendenza fra tecnologie costruttive e progettazione, Keynote lecture, *Proc. of the Congress "I grandi trasferimenti d'acqua"* , Cortina d'Ampezzo (Italy), Vol. 2.

Da Deppo, L., Datei, C. and Furlanetto, G. 1991. The diversion of the Adda river (Northern Italy) for the by-pass of a landslides. In: *"Environmental Hydraulics"* (J.H.W. Lee and Y.K. Cheung editors), Balkema, Rotterdam, Vol. 2:1419-1424.

Broch, E. 1989. Die Entwicklung von reichtausgekleiteden Hochdruck-stollen und geschlossenen Schwallraumen mit Luftpolster in Norwegen. 38 *Salzb. Koll. Geomechanik*: 51-60.

Hydropower'92, Broch & Lysne (eds) © 1992 Balkema, Rotterdam. ISBN 90 5410 054 0

A review of Norwegian high pressure concrete plugs

Tore S. Dahlø
SINTEF Rock and Mineral Engineering, Trondheim, Norway

Jan Bergh-Christensen
Berdal-Strømme A.S., Sandvika, Norway

Einar Broch
Norwegian Institute of Technology, Trondheim, Norway

ABSTRACT: A study on design, construction and operation of high pressure plugs was carried in the period of 1987 to 1991. Details of more than 30 plugs with static pressure head ranging up to about 1,000 meters were collected and analyzed. While the concrete length varies from 2 to 5% of the static water pressure, the steel lining may be as short as 0.4 % of the water pressure head. The technological development since 1980 is documented by reduced hydraulic conductivity of the plugs.

1 INTRODUCTION

A research project has been carried out to document and analyze experience in design, construction and operation of high pressure concrete plugs in Norwegian hydro power plants. The purpose was to evaluate the technology, and develop cost efficient guidelines for concrete plugs which could also be used for planning and construction of plugs for high pressure gas storage caverns.

Information of about 150 concrete plugs was collected. The data base includes about 30 plugs with water heads above 400 meters, constructed in the last 20 years. During the 1980's, many high pressure plugs were constructed, with static head ranging up to 1,000 meters. The study has concentrated on newer plugs, especially because of modern standards and grouting technology that may be expensive, but also efficient in terms of reduced leakage compared to older cement grouted plugs.

2 PLUG TYPES

The two main types of concrete plugs used in hydroelectric power plants are shown in Figure 1.

The penstock plug is located at the upstream end of the steel penstock, at the transition to the unlined pressure tunnel. Access to the unlined tunnel system

Table 1. Key figures for some major plugs.

SITE	WATER HEAD[1] m	YEAR	CROSS SECTION m²	LENGTH CONCRETE m	LENGTH STEEL m	WATER LEAKAGE l/min
NYSET-STEGGJE	964	1987	25	55	Penstock	< 60
TJODAN	880	1984	17	45	Penstock	2
TAFJORD K5	790	1982	18	88	Penstock	50[3]
SKARJE	765	1986	25[2]	20	5.5	< 15[3]
MEL	740	1989	22	27	27	1[3]
SILDVIK	640	1981	26	35	12	<240
JOSTEDALEN	622	1989	35	20	5	6[3]
LOMI	565	1978	20	15	9.5	190
LANG-SIMA	520	1980	30	50	Penstock	120
SØRFJORD	505	1983	20	20	12	10[3]
KVILLDAL	465	1982	31	30	4	4[4]
TORPA	455	1989	32	20	6	< 1[3]
EIKELANDSOSEN	455	1986	20	20	5	8
STEINSLAND	454	1980	20	20	10.2	4[4]
KOLSVIK	449	1979	23	20	10	30
SKIBOTN	445	1979	18	12	7.6	96[3]
LEIRDØLA	441	1978	26	30	Penstock	< 54
SAURDAL	410	1985	49	40	1.5	5[3]
ORMSETFOSS	373	1988	22	22	7	< 3
DIVIDALEN	295	1972	10	13	4.5	<120

1) Max. static head
2) Varies from 20 to 30
3) Remedial grouting at first water filling or later
4) Within accepted limits

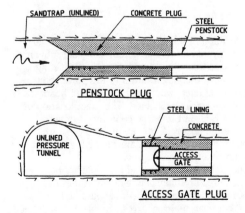

Fig. 1 General layout of penstock plug and access gate plug (from Bergh-Christensen 1988).

is usually provided by an access gate plug located in the construction tunnel adjacent to the pressure tunnel.

Figure 2 shows the trend towards higher water pressure for unlined pressure tunnels in Norway.

The tendency of increasing water head since 1970 is related to more extensive use of unlined pressure tunnels, especially after introduction of the air cushion surge chamber technology, see Gomnæs & al. (1987) and Goodall & al. (1898).

3 DESIGN

There are two fundamental requirements for the design of a concrete plug. Primarily, it must have the static capacity necessary to carry the load from the water pressure. Secondly, specific requirements must be satisfied in terms of leakage. Both in the design and construction, there normally are few problems related to load capacity, although the length and layout of the concrete structure often seem to be a subject for discussion. Less attention seems to be paid to the leakage problems, although one conclusion from this study is that efforts to achieve the optimum water tightness also are very important, both to the functioning of the plug and the total construction costs.

The plug design may vary with respect to the length of both the concrete structure and the steel lining. Figure 3 illustrates the design of two different access plugs, constructed in 1989.

For access plugs, the steel lining is normally shorter than the concrete, and may be located in the upstream, intermediate or downstream part of the plug. The access gate may be located anywhere along the steel lined section. The shape of the plug may be quit simple and uniform, or it may vary along the length axis according to the suggested stress distribution.

3.1 *Plug location*

The location of the plug in relation to the power station layout and tunnel system has great influence on the construction costs. The water conducting system from the plug location to the power house will be steel lined, whereas the upstream headrace system is principally unlined. The exact location of the plug is also of importance in terms of safety, because the host rock for the unlined pressure tunnel must be able to sustain the high water pressure upstream of the plug.

The main rock mechanical principle adopted for unlined pressure tunnels in Norway is that the maximum static water pressure should be less than the minor principal rock stress, reduced by a defined factor of safety. The minor principal rock stress in the plug area must therefore be known before the

location of the plug is finally decided. Traditionally, the rock stresses have been measured by the three dimensional overcoring technique, but during the eighties hydraulic jacking tests have been commonly used as an alternative or supplementary investigation method.

3.2 *Plug length*

It is commonly acknowledged that the plug length should be related to the actual water head. As shown in Figure 4, the length of both the concrete structure and the steel lining (for access plugs) may vary within wide limits, even for the same water head. The steel lining normally is shorter than the concrete, the extreme being the Saurdal access plug with a steel lining of only 1.5 meter at a static head of 410 meter. Still, sometimes the steel lining of the access plug may even be of the same length as the concrete structure (Mel plug).

Figure 4 shows that the length of the concrete

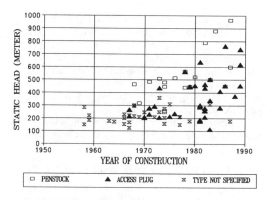

Fig. 2 Max. static pressure at plug vs. year of construction.

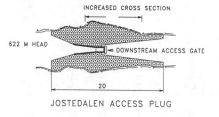

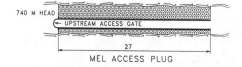

Fig. 3 Sketch of Mel and Jostedalen access plugs.

structure for an access plug ranges from about 2 to 5% of the maximum static water head (in meter). For tunnel cross sections ranging from 8 to 50 m², this represents a maximum shear stress of about 0.4 MPa at the plug circumferential area, assuming a uniform shear distribution in the rock to concrete interface. This also is the maximum shear stress allowed for uniaxial situations according to former standards for concrete structures (for uniaxial concrete strength 25 MPa, i.e. C25).

The maximum linear hydraulic gradient along the plug axis (ratio of water head to concrete length) that may be calculated for a shear stress of 0.4 MPa will be ranging from 20 to 50 for the tunnel cross sections in question. This also complies with a traditional rule of thumb for plug design in Norway, which is based on the assumption that higher gradients may lead to unacceptable high leakage. This gradient criterium may be considered radical. Benson (1989) has for instance suggested that the maximum hydraulic gradient should be as low as 20 for massive, hard and widely jointed rock types.

In reality, the uniform shear distribution supposed in this design principle is not valid. Numerical modelling carried out during the research project has shown that the shear stress will concentrate within the first five meters of the upstream part of the plug (assuming steel gate located upstream so that the water pressure is not acting from inside the plug structure). The shear stresses rapidly decrease further downstream along the plug. Therefore, if one considers the actual stress distribution within the concrete body as calculated by numerical methods, relatively short plug lengths could be allowed. In practical design however, one should also consider the three dimensional water flow regime and the limitations with respect to grouting. In this context, it is the authors' opinion that the minimum plug length for high pressure plugs that are supposed to act as water tight constructions should never be less than five meters.

4 CONSTRUCTION AND OPERATION

There has never been reported any plug malfunction or failure related to overloading. The only kind of "failure" experienced is unacceptable high leakage. Normally, remedial grouting will be carried out during the first water filling or at a later stage. But the criterion for remedial grouting may vary a lot among the plug owners.

4.1 Grouting

A description of the grouting methods for concrete plugs has been presented by Bergh-Christensen (1988).

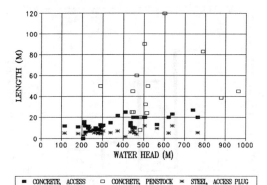

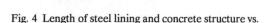

Fig. 4 Length of steel lining and concrete structure vs. static water head.

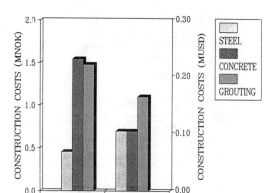

Fig. 5 Construction costs for Mel and Jostedalen access plugs (in million Norwegian kroners and million US dollars).

The quality and extent of rock and concrete grouting is of great importance both for the final construction costs and the leakages at the plug. This is illustrated in Figure 5, in which the construction costs are given for the two plugs shown in Figure 3.

As can be seen, the grouting costs are in the order of about 40 to 45 % of the construction costs for both plugs (both constructed in 1989). Grouting of the rock mass prior to concreting works amounts to about 10% of the total cost. Remedial grouting during or after water filling accounts for about the same amount.

Whereas the steel lining at Mel is 5.5 times longer than for Jostedalen, the lining costs were only about 50 % higher. This is due to the more complicated design at Jostedalen, in which case especially the gate construction is expensive.

The total costs are higher for the Jostedalen plug than for the Mel plug, even though the Mel plug is

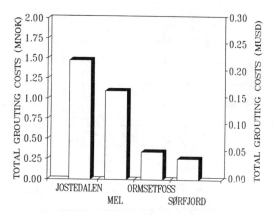

Fig. 6 Grouting costs for some access plugs.

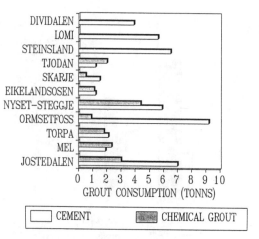

Fig. 7 Grout consumption at some concrete plugs.

longer than the other. The concrete volumes of the plugs are about 600 m³ and 700 m³ for Mel and Jostedalen respectively. The simplicity of the Mel construction as compared to Jostedalen is probably the main reason for the cost differences.

A comparison of grouting costs for several plugs is shown on Figure 6. The costs are actual costs at the year of construction.

At both Jostedalen and Mel, the most modern grouting technique with both polyurethane and epoxy injection at high pressure through grouting hoses has been used. At Ormsetfoss, this was done at a less ambitious extent as much of the grout was injected through boreholes immediately before the first water filling. At Sørfjord, epoxy was not used, and all grouting were done by boreholes.

The final injection to seal the plug is often done during or shortly after the first water filling. The costs paid for this grouting normally ranges in the order of 0.1 to 0.4 MNOK (million Norwegian

kroners). If remedial grouting is required at a stage when the contractor has demobilized, the cost may be in the order of 0.5 MNOK or higher (1990).

The reason for the variations of the grouting costs is probably more related to the quality of the work, including planning and design, than the grout volume needed to achieve an acceptable leakage level.

The consumed grout mass as documented for some plugs is shown in Figure 7. As can be seen, several tonnes of (fine grained) cement is normally injected. Most of the cement mass is used to fill the voids that normally will develop in the contact between rock and concrete at the tunnel roof. If cement grouting is neglected or not thoroughly performed, large quantities of the far more expensive chemicals will be needed.

All the cement consumed at Mel and Tjodan was used to fill voids at the tunnel roof. That also may have been the case for Nyset-Steggje and Eike-landsosen. At Torpa, minimum 1.6 tonnes of the cement grout was used for the same purpose. Whereas the rock mass at Nyset-Steggje was not grouted because of low permeability, a great part of the cement consumed at Ormsetfoss was used for rock grouting. This was also the case at Jostedalen. The rock grouting at Ormsetfoss was done through the same boreholes and in the same operation as the rock to concrete contact shortly before the filling started. At Steinsland and Lomi, the same grouting procedure was followed as for Ormsetfoss, although chemical grout was not used. At Dividalen, the rock mass was grouted prior to the concrete casting works, and the rock contact was grouted at a later stage.

Normally, the plug is constructed at the very latest stage before the power plant is put into operation. The plug construction period should therefor be as short as possible. The cast concrete temperature will often raise to about 60 to 70° during hardening. The plug will cool down gradually, but at a slow and decreasing rate. To be efficient, the grouting should only be done when the concrete temperature has reached an acceptable low level.

Because the construction of the plug is in the critical path of the overall timetable, grouting may be carried out too early, and both the tightness of the plug and the grouting expenses will suffer. Careful planning and control with the concrete temperature is the solution to this problem.

The extent of leakage that may be accepted varies. Some has only been satisfied if the final leakage is less than 10 l/min at a water head of 200 meters, while others have accepted leakages of 200 to 300 l/min at a head of 500 to 600 meters (Lang-Sima and Lomi).

High leakages may give operational problems, they represent loss of energy, and will increase the maintenance costs. If one assumes that the leakages does not increase with time, a decision on remedial grouting may be made by comparing estimated cost of grouting with present value of production losses. This way, it has been reported that leakages of 60 l/min at Sørfjord power plant may be balanced against a production loss of about 100,000 NOK during the working life of the power plant (500 m head, prize level of 1983).

The efficiency of the grouting works is believed to be dependant on the grouting pressure in relation to the water head and the rock stresses. For several plugs, the grouting pressure has been considerably higher than the water pressure.

Figure 8 shows how the grouting pressure for some plugs is related to the static water pressure. At Torpa and Sørfjord, the grouting pressure was higher than the minor principal rock stress as indicated by overcoring measurements. At Torpa, the grouting pressure was even higher than the hydraulic jacking pressure measured at the plug location.

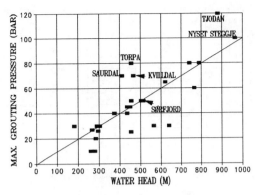

Fig. 8 Grouting pressure vs. static water head.

4.2 Leakages

When relating the water leakages to the water head, there apparently is no connection. In theory, if the concrete plug is homogeneous, the leakage should decrease with decreasing pressure gradient (Darcy). However, linear regression analysis does not correlate the leakage to the hydraulic gradient (Figure 9) Nor has there been found any correlation between leakage and length of the steel lining or linear gradient at the steel lining. To illustrate the latter, the Saurdal access plug, with a steel lining of only 1.5 meter at a water head of 410 m (gradient 273) has a leakage of 15 l/min. In comparison, the Sildvik plug has a leakage of about 240 l/min at a gradient of 53 (water head 640 meter and 12 meter steel lining).

The leakage is best correlated to the year of construction (coefficient of determination R^2 = 0.23). The modern plugs apparently are better sealed than the older ones, as will be discussed later. It also seems that the access plugs are less leaky than the penstock plugs.

The leakage changes with time. Detailed information are given from Saurdal, Tjodan and Tafjord. At Saurdal, the leakage was about 140 l/min after the first filling. Additional grouting by polyurethane at a pressure of 6 MPa (410 m water head) through a curtain of drillholes from the downstream end about two weeks after filling reduced the leakage to about 15 l/min. Later on, the leakage decreased further by 60 to 70 % within the next year.

Even stronger reduction of the leakages occurred at Tjodan (Figure 10). No remedial grout has ever been carried out. The initial leakage after the first water filling was about 50 l/min, which was reduced to about 5 l/min during the first year of operation. In the next four years, the leakages decreased

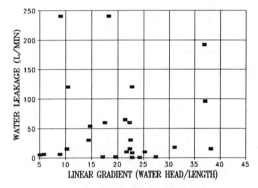

Fig. 9 Leakage vs. linear hydraulic gradient.

further, and was only one per cent of the initial leakage at the beginning of 1990. During the first seven years of operation, the pressure shaft was emptied twice. The owner believes that because of the emptying, suspensions with fine grained materials may have infiltrated the plug and caused the self sealing that have been observed.

Tafjord K 5 shows that the leakages also may increase with time. The power plant was put in operation in January 1982. At that time, the leakage was 160 l/min. The power plant operates with two water reservoirs, at 790 and 675 m head (maximum) respectively. In June 1984, the leakages were about 240 and 190 l/min, dependent on the reservoir. In 1986, the leakages were reduced about 25 to 40% by remedial grouting. However, later the leakages again increased with about 35 l/min per year. During the autumn of 1989, the leakages increased from 260 to 400 l/min.

Pore pressure measurements indicated that leakages through the rock mass were not changed, in which case the concrete plug was the reason for the increased leakage.

Remedial grouting was again carried out. About

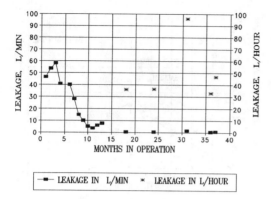

Fig. 10 Leakages at Tjodan (880 meter water head) 1984 - 1987.

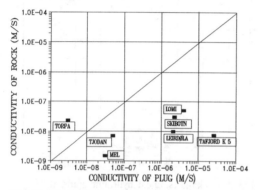

Fig. 11 Hydraulic conductivity of the plug structure vs. the host rock mass. (Partly based on Palmstrøm 1987).

2.8 tonnes of cement, 0.75 tonnes of polyurethane and 0.4 tonnes of epoxy were injected at a pressure of 8 MPa through drillholes from inside the steel lining. The external maximum pressure that could be allowed on the lining was only 5.5 MPa (19 mm thickness). Strain gauges were therefor used to control the pressure load on the lining. Because previous grouting apparently had been unsuccessful, the grouting pressure was increased. An upper pressure limit equal to the minor principal rock stress was suggested. Experience from Torpa indicated that the maximum equivalent external load on the steel lining induced by the grouting process was between 10 to 50% of the grouting pressure. Measurements confirmed that the external (grout induced) pressure on the steel lining during grouting was about 3.2 MPa at the maximum, ie. 40% of the grout pressure applied. Leakages after water filling were reduced by about 70 to 85 % compared with leakages before grouting.

4.3 Plug tightness

Assuming that the leakages are uniformly distributed within a homogeneous concrete cross section perpendicular to the length axis, the hydraulic conductivity of the plug may be calculated according to Darcy's law. The flow area used in the calculations is the difference between the tunnel cross section and the cross section of the steel lining.

There is apparently no correlation between the conductivity and the water pressure. The mean conductivity found is $2.7 \cdot 10^{-6}$ m/s for the access plugs and $4.7 \cdot 10^{-6}$ m/s for the penstock plugs. The reason why the penstock plugs show twice the conductivity of the access plugs may be related to the length of the grouted section. The access plugs are grouted over a longer part of the length than the penstock plugs. If the concrete structure is significantly longer than the grouted section, the ungrouted part does not contribute much to the sealing.

The conductivities indicated above may be regarded as high if one considers that "water tight" concrete (conductivity less than $1 \cdot 10^{-12}$ m/s) normally is prescribed.

The conductivities calculated for most of the access plugs ranges from $1 \cdot 10^{-6}$ to $1 \cdot 10^{-7}$ m/s. These are high figures compared to the conductivity of the host rock measured at the site (Figure 11). However, a commonly used "rule of thumb" suggests that cement grouting is only effective for a hydraulic conductivity greater than one Lugeon, or $1 \cdot 10^{-7}$ m/s.

The plug conductivities calculated relate to the grouting pressure as presented in Figure 12. The figure suggests that the best results are achieved if the grouting pressure is high.

At Torpa, the grouting pressure (8 MPa) was about 10 % higher than the hydraulic jacking pressure. One may also notice that the conductivity of this plug is about one order of magnitude lower than for any other plug.

Plugs with access gates located in the upstream or downstream end of the steel lining may behave differently when subjected to high water pressure. For plugs with gate in the upstream end of the steel lining (Lomi, Kolsvik, Sørfjord, Saurdal, Skarje, Mel and Torpa), the mean conductivity is $9.2 \cdot 10^{-7}$ m/s. For Ormsetfoss, Holen, Jostedalen and Eikeland-sosen, in which case the gate is in the downstream part, the mean conductivity is $2.9 \cdot 10^{-7}$ m/s. Also, the conductivity is fairly constant for the latter.

There seems to be a correlation between conductivity and concrete strength. The mean values are $5.6 \cdot 10^{-6}$ and $7.9 \cdot 10^{-7}$ m/s for C25 (13 plugs) and C35 (6 plugs) respectively, which gives a factor of seven in favour of the higher strength. The reason may be that the concrete design is more carefully evaluated when a higher strength is chosen.

Quality in general, performance of the work, follow up routines during construction etc. may also be more rigorous for higher strength.

The plugs are in different rock types and rock conditions. The majority is located in gneiss and granite, which is believed to represent somewhat unfavourable rock types in terms of permeability. An other main rock group is mica schists and phyllite, which are more incompetent and plastic, and less permeable than the other. But the conductivity does not differ significantly because of rock type. Also, the degree of jointing seems to be of minor importance regarding the conductivity. The reason for this is apparently related to the high conductivity of the plugs as compared to the host rock, which suggests that geological variations normally will have little influence on the total leakage. Also the final plug location is usually selected based on a careful engineering geological mapping of the potential plug area during excavation, in order to find the best possible location for the plug.

4.4 Comparison of modern and older plugs

There has been a significant improvement in the Norwegian grouting technology since the beginning of the 1980's. This is especially related to the increased use of chemical grout and micro cement. When the Tjodan penstock plug was constructed, grouting hoses was introduced for concrete plugs in Norway (see Palmstrøm and Schanche, 1987). The technological development is illustrated in Figure 13. As can be seen, the hydraulic conductivity of modern plugs is about one to two orders of magnitude lower than for the older plugs.

5 CONCLUSION

A study of experience from design, construction and

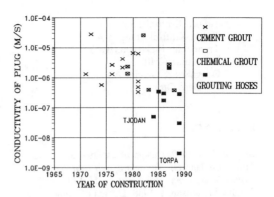

Fig. 13 Conductivity of plug vs. year of construction.

operation of 150 high pressure concrete plugs has shown that the traditional design basis for plugs located in tunnels with cross section ranging up to 50 m² implies a total plug length between 2 to 5% of the static water head.

The final leakage through the plugged tunnel length will strongly depend on the quality of the concrete structure and the grouting works carried out. Normally, most of the leakages occur along the rock to concrete contact, especially in the tunnel roof. The lay out and design of the concrete and the steel lining will influence the plug behaviour and hence the extent of grouting and construction costs.

In the study, some important parameters affecting the plug behaviour have been analyzed. Other factors not included in this study also must be considered. Especially the quality of the excavation works through the plug area will be of great importance (damage to the host rock caused by blasting etc.).

The final leakages may often seem high, and have been found to vary within a wide range. Still, none of the leakages documented seems so high that they should not be accepted, based on a cost/benefit analysis.

ACKNOWLEDGEMENT

The research was carried out as a part of the GUN project ("GasslagerUtvikling i Norden"; Gas Storage Development in Scandinavia). This project is financially supported by Neste OY, Swede Gas AB, Sydkraft AB, Vattenfall and Statkraft. Financial support was also received from The Norwegian Water System Management Association. Important information was received from the following owners and hydro power consultants:

    Berdal Strømme AS
    Bergenshalvøens Kommunale Kraftselskap
    Kristiansand Energiverk
    Lyse Kraft
    A/S Salten Kraftsamband
    Sør-Trøndelag Kraftselskap

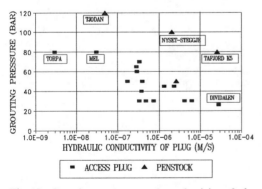

Fig. 12 Grouting pressure and conductivity of plug structure.

Tafjord Kraftselskap
Troms Kraftforsyning
Trondheim Elektrisitetsverk
L/L Tussa Kraft
Ødegaard & Grøner AS
I/S Øvre Otra

REFERENCES

Benson, R.P. 1989. Design of Unlined and Lined
  Pressure Tunnels. *Tunnelling and Underground
  Space Technology,* Vol.4, No.2: 155-170.
Bergh-Christensen, J. 1988. Design of high
  pressure concrete plugs for hydropower projects.
  Int. Symp. on Rock Mechanics and Power Plants,
  Madrid: 261-268.
Broch, E. 1988. Unlined High Pressure Tunnels and
  Air Cushion Surge Chambers. *Proc. Int. Symp.
  Tunnelling for Water Resources and Power Projects.*
  New Delhi, India:10.
Goodall, D.C, Kjørholt, H., Dahlø, T.S. & Broch, E.
  1989. High pressure air cushion surge
  chambers. *Int. Conf. on Progress and Innovation
  in Tunnelling,* Toronto: 337-346.
Gomnæs, P.Chr., Myrset, Ø. & Fleicher, E. 1987.
  The art of Norwegian hydropower design.
  Proc. of the Int. Conf. on Hydropower in Oslo,
  Tapir publishers: 43-54.
Palmstrøm, A., and Schanche, K. 1987. Design
  features at Tjodan save time and money.
  Water Power & Dam Construction, June: 6 p.
Palmstrøm, A. 1987. Norwegian design and
  construction experiences of unlined pressure
  shafts and tunnels. *Proc. of the Int. Conf. on
  Hydropower in Oslo,* Tapir publishers: 87-99.
Selmer-Olsen, R. 1985. Experience gained from
  unlined high pressure tunnels and shafts in
  hydroelectric power stations in Norway.
  *Norwegian Soil and Rock Eng. Association,
  Pub. 3, Norwegian Hydropower Tunnelling*
  Tapir publishers: 31-40-

Hydropower'92, Broch & Lysne (eds) © 1992 Balkema, Rotterdam. ISBN 90 5410 054 0

Underground power house complex of Chamera hydel project

E. Divatia, Brijendra Sharma, M. R. Bandhyopadhyay & Rajeev Sethi
National Hydroelectric Power Corporation Limited, New Delhi, India

ABSTRACT:Layout of the underground power house complex of the 540 MW Chamera hydroelectric project on river Ravi, India is briefly described. Design of rock support for the two caverns of the complex is discussed. Difficulties experienced during excavation of the caverns are stated. Analysis of, and modifications to the rock support required to match the observed behaviour of the surrounding rock mass, midway through the excavation, are discussed. Results of monitoring of performance during and after excavation are presented.

1 INTRODUCTION

The Chamera hydroelectric project is located in Chamba district of Himachal Pradesh, India. The project envisages utilisation of waters of river Ravi over a gross head of 207 m for generation of 540 MW of hydropower. The project has been taken up for construction in 1985 with bilateral agreement between India and Canada.

The project comprises a 140 m high concrete gravity dam across Ravi, a 6400 m long and 9.5 m diameter power tunnel on the right bank of the river, a 25 m diameter power tunnel surge shaft, a 8.5 m diameter partly concrete lined and partly steel lined (embedded in concrete) pressure shaft trifurcating into steel lined penstocks of 5 m diameter each, immediately upstream of an underground power house, which will house 3 units of 180 MW each. A

2400 m long tailrace tunnel crosses over to the left bank under Ravi river to develope additional head. The layout of the project is shown in Figure 1.

The underground power house complex comprises two caverns viz.the main power house cavern, 24.5 m wide 31.6 m high and 112.5 m long and the transformer cavern, 17 m wide 14 m high and 114 m long. Three shafts of 14 m diameter each are provided to take care of tailrace surges. Operation of draft tube gates is also through these shafts opening into the transformer gallery. The main power house cavern and the transformer gallery are conected by three inclined shafts for bus ducts each being 5.2 m x 4.4 m and 57 m long. The machine hall cavern is approached by a 8 m x 7 m and

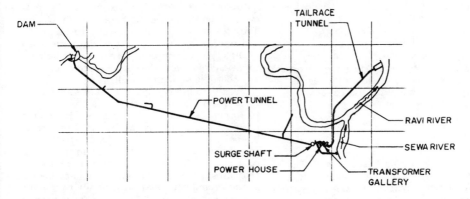

Figure 1 Layout of the project

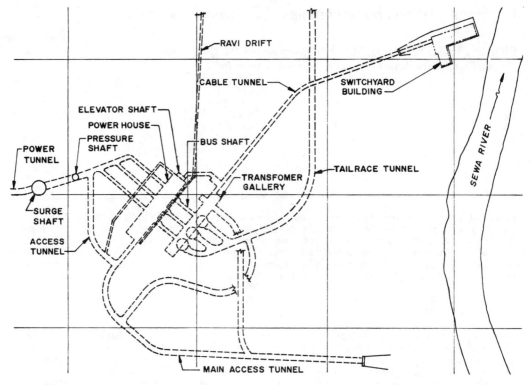

Figure 2 Layout of power house complex

546 m long main access tunnel. A 7.5 m x 7 m and 240 m long access tunnel connects the transformer gallery with the main access tunnel. A cable tunnel of size 4.4 m x 4.9 m and 363 m long carries power and control cables from power house to the gas insulated switchgear housed in a building in the switchyard located outside. The layout of the power house complex is shown in Figure 2.

Excavation of the caverns was taken up in early 1987 after completion of the access tunnel completed in early 1991. The project including the power house complex is in an advanced stage of construction and is likely to go on stream by December 1992.

Initial design of rock support for caverns, behaviour of rock mass surrounding the caverns during excavation, modifications in the support design and the performance of the caverns are briefly described in the following.

2 GEOLOGY

The project area lies within Lesser Himalayas containing a variety of rock types viz.granitic gneiss, carbonaceous phyllite, quartzitic phyllite, limestone, metavolcanics, sandstone and shale.

This sequence is intensely folded. Three major thrusts are present in project area. Jutogh Thrust marks the contact between Jutogh formation and Dhundhiara formation. It is exposed slightly upstream of dam. Shali Thrust separates Dhundhiara formation and Panjal volcanics of permo-carboniferous age in which underground power house complex is located. These are thrusted over Murree formation, giving rise to Murree Thrust exposed in the tail race tunnel near outlet.

Power house complex is located in fine grained metamorphosed andesitic basalt (metavolcanics). The rock mass is variable blocky to foliated and intersected by five sets of discontinuities of different orientations, predominant being foliation joints and shears. Foliation joints are continuous, slightly undulating and generally moderate to closely spaced. Thin interfolial shear seams ranging in thickness from 1 mm to 10 cm, filled with clay and rock fragments traverse this rock mass at 2 m to 4 m intervals. In addition, tight/clay filled joints and shears of variable orientation also exist in the volcanite rock mass. The orientation of power house and transformer caverns has been kept normal to the prominent foliation joint set to minimise the adverse effects of these discontinuities in an otherwise not very favourable rock mass

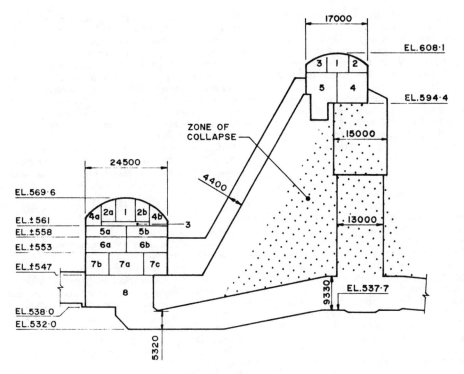

Figure 3 Power house and transformer caverns - excavation sequence

for excavation of caverns of such large span. In a major portion of the cavern, the rock mass is comparatively dry except at some locations where damp to very light drip conditions were encountered.

In addition to surface geological mapping, seismic refraction surveys and diamond core drilling done during investigations 1.5 m x 2 m exploratory drift from Ravi river side was also excavated to a length of 525 m to help locate the power house cavern and other underground components of the power house complex.

General discontinuity data including location, nature and extent of shear zones and shear seams obtained from geological mapping of the drift and the rock mass assessment made was used in finalising the location and orientation of the two caverns and other underground components of the power house complex. In fact the location of the power house cavern was shifted about 150 m further into the hill from the location envisaged earlier because of the presence of shear zones in the drift. Near its end the drift runs 30 m above the crown of the power house cavern along its downstream wall. It was also utilised for conducting insitu rock tests viz flat jack and plate load tests and collecting rock samples for laboratory testing. In addition, instruments like multiple point bore hole extensometers were installed from two cross cuts in

the drift to monitor rock mass deformation during various stages of excavation of the power house cavern. Improved ventilation during excavation of the power house cavern after completion of top heading was an additional advantage of the drift.

3 POWER HOUSE GEOMETRY

The main cavern of the power house is 24.5 m wide and 112.5 m long with a maximum height of 31.6 m (Figure 3). The upstream wall of the cavern has 3 openings of 7.0 m excavated diameter for penstocks which have a rock pillar of 15.5 m between them. The downstream wall has 6 openings, three each for bus ducts and draft tubes. The openings for housing the bus ducts, each 6.2 m wide and 11.0 m high, are horizontal for an initial reach of 9 m connecting to the transformer cavern through 5.5 m wide and 4.4 m high inclined bus shafts. About 10 m underneath each of the bus duct openings are the draft tube openings, each of which is 8.4 m wide and 5.3 m high. The rock cover between the two openings decreases from 10 m to 7 m, along the draft tube. The rock pillar between the bus shafts is about 16 m. The transformer cavern is 17 m wide and 114 m long with a height of 14 m. Three draft tube surge shafts

71

of excavated diameter 13 m/15 m open into the transformer cavern.

The shape and dimensions of the openings of power house, transformer gallery, bus shafts etc. were decided mainly on the requirements of electro-mechanical equipment to be supplied by Canadian firms.

The number and size of the openings downstream of the large power house cavern shown in Figure 3 required careful sequencing of the various stages of excavation of different component structures and immediate installation of the designed rock support.

4 ROCK SUPPORT DESIGN

From the data obtained in the drift, the average rock mass quality for the caverns was assessed to be 'fair' except in a small reach where it was 'poor' due to the presence a 3 m to 4 m thick zone of very closely foliated to crushed rock mass. As there are very few cases of such large caverns in the rock of similar quality, continuous review of rock support during construction, as fresh information on geology encountered becomes available, was considered necessary. Based on geological data from drift and partial excavation of various access tunnels, it was assessed that the rock mass behaviour would be controlled by:
- near surface loosening, especially in and adjacent to the shear zones

- individual block failures, generally in association with the foliation.

As a flexible support system, using a combination of rock bolts, anchors and shotcrete could readily cope with such failure mechanisms, it was considered prudent to adopt these elements of support in design with regular monitoring of performance of the structure by detailed instrumentation. Installation of necessary rock support, soon after excavation, was a prerequisite to this design approach.

Initially, geological data from Ravi drift and access tunnels was projected on the major cavities and rock support requirements determined. Several approaches were considered.
  1. Review of precedent practice
  2. Support recommendations based on various rock mass classification systems
  3. Empirical methods
  4. Stress analysis using two dimensional boundary element method
  5. Block stability analysis

For design of rock support of caverns, prior to start of actual excavation, the support pressures required for stability of the roof were estimated to be 80 kPa to 120 kPa. The zone of overstressing in the roof was estimated to be about 6 m for a fair quality rock (RMR 50).

Accordingly two support options were considered:
  1. 7.5 m long, 25 diameter rock bolts (yield 267 KN) on 1.5 m square grid.
  2. 6.0 m long, 25 diameter rock bolts (yield

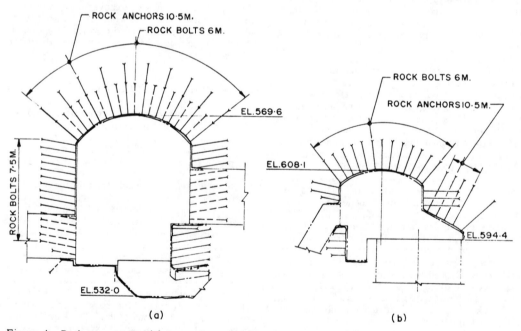

Figure 4   Rock support for (a) power house cavern (b) transformer cavern

204 KN), on 1.5 m square grid as primary support and longer, high capacity anchors (yield 843 KN) on a 4.5 m square grid as secondary support.

Only 25 diameter bolts (204 KN) were readily available in India while the others were to be procured from Canada. Considering the long lead time required for procurement of bolts from Canada, the second option was adopted. 51 diameter hollow core anchors, 10.5 m long, were used for secondary support. The total length was made up using bars in lengths of 3.5 m each. The depth of anchors was sufficient to ensure formation of the required zone of compression. Further, these larger anchors were sufficient to ensure stability of the roof against block failure.

A cavern of this size, supported with only rock bolts and shotcrete, was being constructed for the first time in India in Himalayas and required due care during construction. Also the Ravi adit instrumentation, which was to be installed prior to the start of the central top heading, could not be installed till the completion of central top heading excavation (Figure 3). Further, excavation of the central top heading revealed rock mass conditions somewhat less favourable than anticipated. In view of the above and limited experience of the executing agency in installing flexible rock support, the level of required rock support was reassessed and anchor spacing was decreased to 3.0 m.

During slashing for heading, the geological data encountered was projected on the yet to be excavated walls and block stability analysis performed. The support levels of 80 kPa

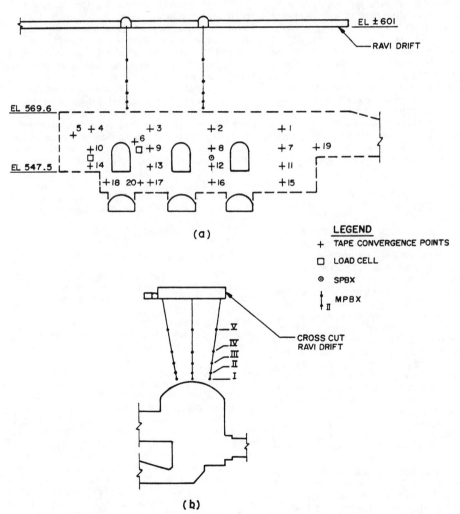

(a)

LEGEND
+ TAPE CONVERGENCE POINTS
□ LOAD CELL
⊙ SPBX
MPBX

(b)

Figure 5 Instrumentation plan for power house cavern(a) longitudinal section (b) cross section

with 7.5 m long rock bolts were considered sufficient (Figure 4). This support had however to be increased later due to cracking in the walls of the cavern and the rock mass ratings being less than anticipated.

Since the geological conditions in the transformer cavern were anticipated to be essentially similar to the ones in power house cavern, the rock support design is similar. For the 17 m wide cavern 6 m long 25 diameter rock bolts(204 KN) on a 1.5 m square grid were provided, both in the roof as well as on the walls.

Additional rock bolts were provided on a suitable pattern at portals of all openings in the walls of the two caverns viz.penstocks, draft tubes, bus shafts and draft tube surge shaft domes.

In addition to the rock bolts, the support of roof and walls, both in the power house as well as in the transformer cavern, also included shotcrete and wiremesh. Thickness of the shotcrete was 100 mm except in the roof of power house cavern where the thickness was 200 mm.

The walls of the power house cavern have been designed to be fully supported by the flexible support system of shotcrete and rock bolts. Further, a gap of 100 mm has been provided between crane beam and columns and the rock wall of the cavern to prevent any transfer of rock pressure to the column - beam frame. Even at locations where crane beam is anchored to rock removable shims have been provided to ensure no load transferance from rock to column - beam frame.

5 INSTRUMENTATION

The design concept for rock support of the cavern was based on regular monitoring of rock mass response to excavation and consequent modification in the degree of support. For this purpose a detailed instrumentation scheme consisting of installation of multiple point bore hole extensometers (MPBXs), single point bore hole extensometers (SPBXs), convergence measurement stations, load cells on bolts and anchors, strain gauges, survey targets etc was evolved.

Two measurement sections representing average and worst geological conditions existing in the exploratory drift were selected for installation of MPBXs from the invert of the drift about 32 m above the crown of the power house cavern. Two cross-cuts were excavated from the drift for this purpose. Each section had three MPBXs with five measurement points at about 1 m, 2 m, 5 m, 10 m and 20 m above the crown of the cavern (Figure 5).

Convergence measurement stations were proposed at three different sections along the length of the power house cavern with each section having measurement points at four different elevations on the walls.

Load cells were provided on anchors and bolts in the roof and walls. Vibrating wire strain - gauges were planned for embedment in shotcrete.

In the transformer cavern also MPBXs and convergence measurement stations were provided. Extensive provision was made for convergence measurement stations in all other underground excavations of the complex viz penstocks, bus shafts, draft tubes including manifolds etc.

6 EXPERIENCE DURING EXCAVATION

6.1 Excavation sequence

Both the power house and transformer cavern were planned to be excavated in stages with specified sequence of excavation steps (Figure 3). For the power house cavern a 6 m wide 7.5 m high central top heading for the whole length was the first stage of excavation (1) followed by slashing of the sides to full width (2a,.......4b). Excavation was planned to be lowered in a number of stages (5a.......8). Similar stages, though less in number were planned for the transformer cavern.

For excavation of power house cavern, the central top heading was planned to be enlarged to full width, in a short length under the extensometer measurement stations established in the drift, so that the deformation in the rock mass due to excavation could be measured and rock support modified, if required. During actual excavation, slashing of the central top heading to full width was done as per above sequence with minor modifications. Installation of secondary support of anchors was however, slightly delayed.

6.2 Cavity in draft tube

Concurrent with the excavation of power house cavern, excavation in other components of the complex viz.transformer cavern, draft tubes, penstocks etc was continuing. Excavation of draft tube no.1 was being done in pilot heading, with the slashing to full width following some distance behind. When the pilot excavation had been done upto about 15 m from the downstream wall of the powerhouse, a cavity, 8 m to 10 m high, developed in the roof of the excavation. Location of the cavity was near the junction with the surge shaft. Excavation of power house heading had also just been completed at about this time. Though efforts at stabilisation of the cavity by grouting etc.continued this could not be effectively stabilised and material kept on falling into the already excavated pilot, access to which was blocked by the collapsed muck. During benching excavation in the power house, the cavity had extended upto about 27 m above the draft tube. Attempts to fill the cavity by cement grout through holes drilled

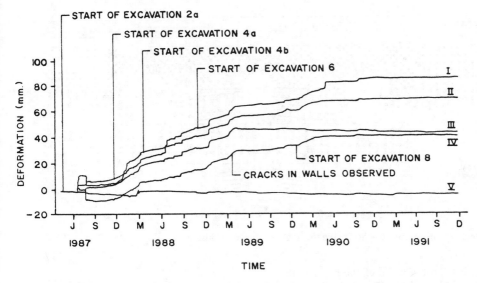

Figure 6   Deformation of power house roof - typical extensometer data

from cavern floor continued. Despite heavy grout takes the cavity finally extended upto transformer cavern in November 1988. Approximate zone of collapse is shown in Figure 3.

6.3 Cracking in the cavern walls

After completion of the central heading of the power house cavern,excavation proceeded in specified steps upto El 553. During excavation of the bench upto El 553 and below, the measurement of deformations in the roof through extensometers continued. The measurement of convergence of the walls could, however, not be done in a continuous manner due to frequent damage to the eye bolts during blasting and other excavation related activities.

During excavation of the central gullet 7a (Figure 3), cracks in the shotcrete on the downstream wall were noticed. Some cracks, though less pronounced, were also observed on the upstream wall. At the time of cracking observed in walls, excavation of the inclined portion of the bus shaft of Unit 1 had reached close to the bottom, leaving only about 8 m length still to be excavated. Excavation in the horizontal portion had also been done to El 553 along with the excavation of power house cavern. Excavation of the other two bus shafts and the draft tubes had been completed earlier.

Extensometer installed in the rock mass above the roof of the cavern showed a maximum deformation of about 65 mm near the crown (Figure 6).

Cracking on the downstream wall was more pronounced in the rock mass between the bus duct openings, particularly around bus duct ope-

ning for Unit 1. This indicated overstressing of rock pillars between these openings. The overstressing was naturally more intense in the rock mass around excavation for bus duct no.1 because of the large collapse in the draft tube. Poor quality of rock mass in this reach, identified as the worst in the whole length of the power house cavern, also contributed to this overstressing and consequent cracking.

7 REVIEW OF DESIGN

With the appearance of cracks in the downstream wall of the power house cavern and access to bus shaft, the excavation work in the area was stopped for some time to review the rock support design andexcavation procedures, particularly in the light of the large collapse in the draft tube no.1.

A finite element analysis, using slightly modified material properties based on the geology actually encountered in excavation, was carried out. computed deformations in the roof broadly matched with the observed deformation data and showed a zone of overstressing in the downstream wall about 8 m deep for a fully excavated cavern with support pressures of 250 kPa. Additional rock bolts were, therefore, provided to increase support pressures from 80 kPa to 250 kPa. The depth of these bolts varied from 10 m to 12 m (Figure 7). Similar reinformcement was provided in the upstream wall of the power house having penstock openings.

In the rock pillars between bus shafts also additional reinforcement was provided in the form of 10 m long rock bolts. These bolts

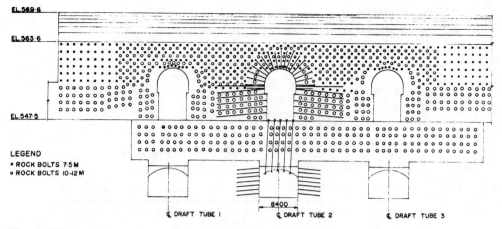

LEGEND
• ROCK BOLTS 7·5 M
∘ ROCK BOLTS 10-12 M

8400

Ȼ DRAFT TUBE I   Ȼ DRAFT TUBE 2   Ȼ DRAFT TUBE 3

Figure 7 Rock reinforcement - downstream wall of power house

were installed from the walls of the horizontal
access to bus shafts. Bolts from two adjacent
openings overlapped to provide uniform rigidity
in the rock mass of pillars. Similar reinforcement was also provided in rock pillars between
adjacent draft tubes on the downstream and
penstocks on the upstream wall. To ensure
stability of roof of draft tubes, rock bolts were
also provided from horizontal bus shaft floor.
For clarity such reinforcement provided from
horizontal bus shaft walls and floor, and from
draft tube walls is shown for Unit 2 only
(Figure 7).

Reinforcement provided in walls after cracking
consisted mainly of 36 diameter rock bolts (both
tensioned and untensioned) instead of 25 diameter provided earlier.

Excavation of bench was resumed after installation of the additional support in the already
excavated portion. Further excavation was
mostly done by ripping with resort to blasting
only when essential. This blasting was associated with vibration monitoring.

Excavation of power house cavern below
El 553 with modified rock support was associated
with more rigorous deformation monitoring of
the roof and walls. Wall convergence with tape
extensometers was measured at different elevations in four measurement sections. Typical
convergence data of one such station with measurement points 9, 13 and 17 is shown in
Figure 8. Similar instrumented sections were
established for all remaining excavation in
other components of the power house complex
and convergence monitored for satisfactory
performance.

Deformation data from MPBXs (Figure 6) and
from tape extensometers (Figure 7) shows
satisfactory performance of the power house

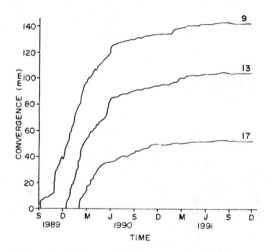

Figure 8 Convergence of power house walls
(typical)

cavern during and after completion of excavation. Excavation for most of the components
of the power house complex has been completed
some time back and installation of the generating machines for the three unis is well advanced.

Hydropower'92, Broch & Lysne (eds) © 1992 Balkema, Rotterdam. ISBN 90 5410 054 0

Modern tunnelling equipment opens for new project solutions

H. Holen, E. D. Johansen & J. Drake
Statkraft SF, Oslo, Norway

ABSTRACT: The planning as well as the construction works for the Svartisen hydroelectric scheme are carried out by Statkraft SF, - the owner of the plant. The first plans for the project were presented in the early 70's. Drill and blast - tunnelling was regarded as the only realistic method, and the technical solutions, operational lay-out and precalculations were based on that assumption. The development of tunnel boring equipment in the following years led to altered operational plans, reduced costs and last but not least, reduced damage to the terrain and improved environmental conditions even for the tunnellers.

About 3 decades have past since the Tunnel Boring Machine method was discussed for the first time in Norway as a possible alternative to the drill and the blast method in tunnelling.

Statkraft, as a Government owned company, has always felt the obligation to play an active part in the development and testing of new equipment and methods for the hydro power construction.

The potential possibilities in the TBM-method were early recognized, but until the beginning of the eighties it was impossible to find a TBM that had a fair chance to cope with the hard, massive rock which is normal in Norway.

A new era for TBM-boring in Norway started in 1981 at the Ulla-Førre site when Statkraft proved that TBM-boring from now on could be

carried out successfully even in hard rock.

Since 1981 and until today Statkraft has bored more than 95 km of tunnels in various parts of Norway, in various geology, diameters from

3.5 m to 8.5 m with TBM's from different suppliers.

During the decade that has passed since we started boring the technical development has continued.

As an example the following table shows the development of series 140 Robbins TBM's over the last 20 years.

What this development actually means can be read out of the following diagram which idealized shows the correspondence between cutter load and penetration rate.

Table 1. Table-series 140 TBM

YEAR	1972-73	1979-80	1989-90
Model	142-145	147-210	1410-252
Diameter	4,27 m	4,32 m	4,3 m
Geology	Limestone	Dolomitic, Limestone	Micaschist, Schist, Granite
Cutterdiam	305 m/m	394 m/m	483 m/m
Cutterthrust	89 KN	178 KN	314 KN
Total cutter head thrust	3 115 KN	5 425 KN	9 160 KN
Cutterhead drive	447 KW	671 KW	2 345 KW
Approx. weight	100 t	113 t	262 t

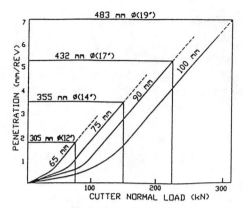

Fig. 1. Idealized force - penetration curved for hard rock. Curved represent different spacing and cutter diameters.

Even after the actual excavating has started we are still looking for possibilities to improve the final result.

It is interesting to observe that when the plans for the Storglomfjord Hydroelectric project, which is a part of Saltfjell - Svartisen Hydroelectric scheme comprising more than 200 km's of tunnels, were presented in 1975, it was wholly based on drill-and blast tunnelling.

The reason is simple. Our experience from TBM tunnelling in Norway was next to none.

In 1976, however, The Division for Construction Engineering at Norwegian Institute of Technology sat down a group of students and scientists with the following tasks:

1. To develop a prognosis model for calculation of costs and penetration rates as functions of geological properties.

2. To find out if and where TBM-boring could be an alternative to drill and blast in this country at the present stage.

3. In case, sort out the relevant factors that should be taken into account.

Their report came late in 1976 and a new group was formed in 1977 to look into The Svartisen plans to find out if the TBM method was applicable.

Their conclusion was, out of 200 km's, 120 km's should preferably be bored.

There were still many years to come before the actual excavating started at The Svartisen Hydroelectric Project, - as a mentioned in 1987.

But Statkraft had other projects, - Ulla - Førre, Kobbelv - and Jostedalen Hydroelectric schemes. The interest had been woken and before starting at The Svartisen site we had TBM bored about 50 km of tunnels at the above projects.

From the very first meter bored there has been a very close cooperation between Statkraft and the Division of Construction Engineering at the Norwegian Institute of Technology in Trondheim and with NGI. The institute has carried out follow-up studies from all the tunnels bored by Statkraft and other contractors in Norway and abroad. The registration of data has been systematized and scientifically adapted.

The prognosis models for calculations of costs and penetration rates have been gradually improved and we have now a very good and practical instrument for planning and calculation on tunnelling.

Generally the pros and cons of using TBM's vs conventional drill and blast methods can be summarized as follows:

Pro:
Faster advance.
Improved hydraulic characteristics allow ca. 40% smaller tunnel areas for water tunnels.
Improved ventilation conditions allow longer headings.
Improved rock stability reduces rock support works and risk for damage caused by rock burst and similar.
Less overbreak

Con:
High investment cost.
Long delivery time and initial set-up.
Less flexibility in some situations where special tunnel supports are required.
Wider tunnel radii.
Need longer headings to justify cost and delivery time of new TBM's.
More sensible to extreme geological properties.

When the TBM-boring turned up as a realistic alternative to drill and blast due to improve technology, we could, as far as the headrace tunnel is concerned, do the following alterations in the operation plan.
See fig.2.

Due to estimated high advanced rate with the TBM we could carry out the tunnelling works at one face only, and still have the works finished within the time schedule.

Due to improved ventilation conditions there

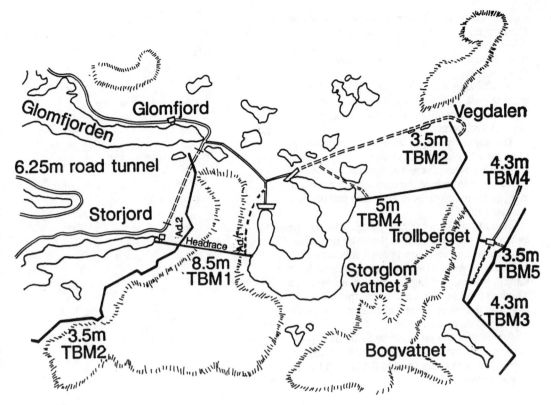

Fig. 2. General lay-out

would be no hazard to the tunnellers health, even at the end of the tunnel. A hazard that would have made drill and blast at one heading impossible. It became unnecessary to open Adit 1, and consequently unnecessary to built up a camp with barracks for the crew, workshops and so on.

Building of a 5 km long road would not be required for the transport of heavy tunnelling machinery, supplies and personnel.

We could also avoid a road that would go through untouched country.

(Even if we do our utmost to be careful we can never avoid leaving traces).

The road would have had to be kept open all the year through. The cost would be considerable, - the damage to the fauna , - unknown.

By boring instead of blasting the head racetunnel the hydraulic properties in the tunnel improved and the area could be reduced from 98 m² to 57 m² keeping the head-loss at the same value. The required tip volume was reduced from about 1000 000 m³ to about 730 000 m³.. and perhaps more important is that all the muck could be placed in Kilvik where it is far easier to hide it than in the

open terrain near the Lake Storglomvatn, where we in the case of drill and blast tunnelling would be forced to hide away 2-300 000 m³.. blasted rock.

The total costs for the two alternatives are estimated to differ with an amount in the range of 20 mill kr in the favour of TBM-boring.

Additionally we have reduced the damage to the terrain substantially and the working conditions for the crew has also been improved. Apart from improvements in the enviromental conditions caused by the fact that there are no nitrous gases from the blasting and even reduced amount of toxic gases from dieselengines, the TBM-boring means highly automatized boring. It also means less hazard to crews health due to rock burst and similar.

A TBM-operator still have a tough job, but the conditions have definitely improved.

Similar results as described above was achieved in Vegdalen, one of the four tunnels in the eastern transfer system.

This transfer system comprises totally more than 40 km's of tunnels. The original intention was to excavate via five crosscuts, - two at Trollberget, one in Vegdalen and one in Beiarndalen, - some km's further up in the valley from the existing crosscut, and finally one near Storglomvatn.

By utilizing a new area in TBM-technology starting just when we were about to start the excavation we could excavate via only one crosscut, - at Trollberget.

The same arguments as described above applies for all the alterations done.

Improved machinery and equipment gives the opportunity to reduce the damage to nature and man, and to improve the total economical result at the same time.

Statkraft is in a special position being the owner and the contractor at the same time. This means that we can chase optimal solutions continuously, - even after the actual works have started, without fearing the large extra invoices which are the private contractors speciality and their usual answer to any alteration in lay-out, time schedule or whatever.

Any improvements made by Statkraft as planner, as contractor, as caretaker of enviromental, as owner is free of charge and to the benefit for the whole nation.

Hydropower'92, Broch & Lysne (eds) © 1992 Balkema, Rotterdam. ISBN 90 5410 054 0

Appraisal of the rock mass interested by underground excavations for the extension of a hydroelectric power plant

Marco Canetta & Alberto Frassoni
ISMES, Bergamo, Italy

Vittorio Maugliani
Aem, Azienda Energetica Municipale, Milano, Italy

Maurizio Tanzini
Elc, Electroconsult, Milano, Italy

ABSTRACT: this paper illustrates the geognostic investigation campaign carried out in the rock mass which will be encountered during the excavation works for the extension of the Premadio (Italy) hydroelectric power plant. These works envisage, in particular, the implementation of an underground powerhouse. The cavern, approx 57x26x38 m, shall be excavated a few meters from the nearby powerhouse, in operation since 1956. The excavation will be carried out in metamorphic rocks characterized by an anisotropic behaviour due to the highly schistose texture and the presence of important dislocation structures with cataclastic rock fillings.

The investigation campaign performed on the site where the powerhouse will be excavated has been carried out from a suitably excavated adit and includes: geostructural investigations, geomechanics investigations, both at site and in the laboratory, and geophysic investigations. These investigations have made it possible to define the geostructural aspect of the site, as well as the rock deformability features and the natural stress condition.

The performed investigations, besides providing the geomechanical parameters necessary for the work design, have also made it possible to acquire data on the vibrations induced by excavation blastings in the adit on structures and equipment of the nearby operating powerhouse. These data will allow the future blastings for the second powerhouse to be sized correctly.

1 INTRODUCTION

Premadio hydroelectric power plant is located in the Upper Valtellina, near Bormio (North East of Milan), in the italian central Alps.

The plant, built in the period 1952-1956, is operated by Aem (the energy utility of Milan municipality) and exploits the water stored and regulated by Cancano reservoir.

The original project envisaged the future expansion of the installed capacity up to 370 MW by enlarging the original power house. For this purpose Aem, after pre-design study led in 1986, assigned in 1988 to Elc Electroconsult of Milan, the design of the enlargement of the power plant (see fig.1).

The main works, needed to attain the full hydroelectric development, are mostly underground, except for the raising of the Cancano dam; this situation required a careful investigation of the geomechanical conditions of the site.

The research carried out by Aem in the technical archives, on the available documentation regarding the existing Premadio power house, have found very little material, probably due to the limited state of the art in the early fifties as far as geomechanical investigation are concerned.

For this reason ISMES of Bergamo in 1990 was entrusted of the implementation of a geognostic investigations campaign, taking advantage of the modern methodologies at present available.

On the basis of the results of these investigations, Elc Electroconsult has elaborated the final design (both civil and electromechanical) and the corresponding bidding documents.

The underground civil works consist of:
- the completion of head-race
- a new power house (57 m x 26 m x 38 m), separated from the existing one by a rock diafragm about 30 m thick,
- a surge shaft
- a surge chamber
- a penstock tunnel,
- the completion of the tail-race

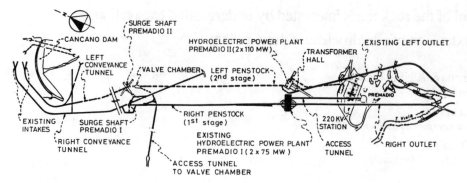

Fig. 1 General scheme of the plant

tunnel.

This paper describes the results of the geognostic investigations for the characterization of the rock mass in wich will be excavated the underground power-house. The geognostic investigations, carried out from an adit excavated along the axis of the designed power house at roof level, include:
- detailed structural survey,
- geognostic drilligs,
- in situ geomechanical tests, including geophysical investigations,
- geomechanical laboratory tests and petrografical analysis of selected rock samples.

2 GEOLOGICAL INVESTIGATIONS

The geological environment of the project area consists of:
- a metamorphic basement (mainly phyllites, with micaschist and gneiss associated and inclusions of amphybolites, prasinites and marbles) where the power house and part of the penstock tunnel are located;
- a dolomitic complex, overthrusted on the basement, where the remaining works are located.

The upthrust fault is striking between WNW-ESE and E-W with a dip ranging from 40° to 65°.

To define the structural set-up of the metamorphic complex, walls and front of the adit, as well as cores and borehole of the geognostic drilling have been investigated as follows:
- lithological identification,
- structural survey,
- core samples stratigraphy,
- T V survey of the borehole.

Moreover, to define the geomechanical set-up of the rock body, the characteristics of joints have been

described. The survey carried out has shown:
- the basic omogeneity of the metamorphic body, consisting mainly of phyllites, with minor gneiss, where the phyllites present poorer geomechanical characteristics than the gneiss;
- the existence of several systems of fractures, shear-planes, faults which affect the rock condition and its geomechanical characteristics; in fig. 2 are represented the diagram of the schistosity and the main faults;
- the occurrence of rock's wedges breaking away along shear planes and/or fracture planes intersecting each other, particularly along "main faults" where the wedges can be as bigger as some m3.

3 IN SITU TESTS

The area where the planned power house will be excavated, has been deeply investigated by means of several different in situ tests, aimed at the evaluation of the structural behaviour of the rock mass as follows:
- first, during the excavation of the adit, convergence measurement and survey of the vibration induced by the blasting;
- core drilling from the adit toward the area of the future power house, with T V survey of the bore hole;
- dilatometric and hydrofracturing tests in the borehole, for the geomechanical characterization of the rock mass;
- flat jack tests on the adit's walls, plate load tests;
- seismic (conventional and micro-rifraction) surveys to determine the propagation velocity of longitudinal (P) waves;
- sonic logs inside the boreholes and sonic tomography between couples of

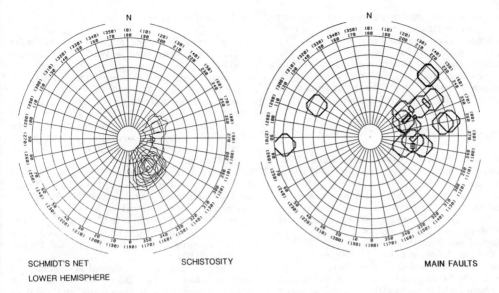

SCHMIDT'S NET   SCHISTOSITY   MAIN FAULTS
LOWER HEMISPHERE

Fig. 2 Polar diagram of the schistosity and the main faults appeared during the excavation of the adit

boreholes to evaluate the distribution of P waves velocity inside the rock mass.

3.1 Vibration measurements

During the blasting of the adit, as the existing power-house had to be kept working, all vibrations induced by the blasting have been recorded in fixed sites, inside the existing underground power house, to assess the effects of the blasting on the existing structures and to adjust the blasting scheme within margins safe enough to prevent dangerous effects.

The recording equipment in the power house has been solidly placed in the rock, in points as close as possible to the blasting front.

The recording has shown that stresses (measured as vibration velocity along the 3 spatial components) induced by blasting, decreased from 7 mm/s (along the trasversal component) at the first blast to values as low as 1 mm/s, in relation with the adjustment of the blasting scheme and the increasing of the blasting distance.

3.2 Convergence measurements

During the excavation of the adit, rock convergence measurements have been carried out in 5 section, installed at the adit contour close to the excavation front, as shown in fig.3, where the diagrams of the measurements carried out is also represented.

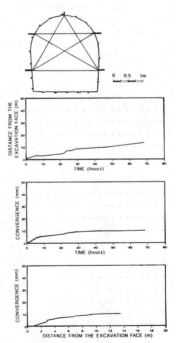

Fig. 3 Convergence measurements: installation scheme and typical diagrams of the measurements

The measured convergence values may vary, according to the sections considered, from one to tens of mm.

These differences depend on the lithology and the shear condition of the rock crossed by the adit. The higher convergency values have been measured as aspected in the main faults zones.

3.3 Flat jack tests

Twelve tests with flat jack have been carried out both in the roof and the walls of the adit.

The tests have shown that stresses are higher in the roof (average values of 4.3 MPa) than in the walls, where values of 3.9 MPa in the right and 3.3 in the left one have been measured.

As far as deformability is concerned, an increase of the modulus have been noticed between the first load phase and the quick load phase, while unloading values are not varying significantly.

It must be pointed out that the load trasmitted on the walls is almost perpendicular to the joints; under these conditions the rock is compacted and therefore shows a deformability higher than in the roof, where the load direction is almost parallel to the joints.

Deformability moduli, measured during loading, range from 3400 to 4600 MPa in the right wall and from 3400 to 5100 in the left one, while in the roof moduli are higher, ranging from 6900 to 16200 MPa.

3.4 Plate load tests

Six plate load tests have been also carried out in the adit, 2 of which parallel to the joints and 4 perpendicular, increasing the load by steps from 4.0 to 8.0 and 12.0 MPa.

Two loading- unloading cycles have been carried out for each load level.

In fig. 4 are represented the deformability moduli measured at different depths from the load surface. It appears the following:
- modulus increases with depth (taking also into account the releasing effect, due to the excavation, in the first 25 cm)
- moduli measured with plates parallel to the joints (tests 1 A-B, 2 A-B, 3 A-B, 4 A-B) are higher than those measured with plates perpendicular to the joints (tests 1 C-D, 2 C-D).
- Test 1 B shows a modulus below the average; that is probably due to the local intense fracturing of the rock, delimiting isolated wedges, easily subject to deformation.

3.5 In-hole dilatometric tests

Dilatometric tests (31) have been carried out in 7 boreholes (ø 101 mm), after selecting with TV camera the most suitable places, to avoid joints and cavity which would invalidate the test.

Tests have been conducted starting from a ground stress of 0.5 MPa, needed to

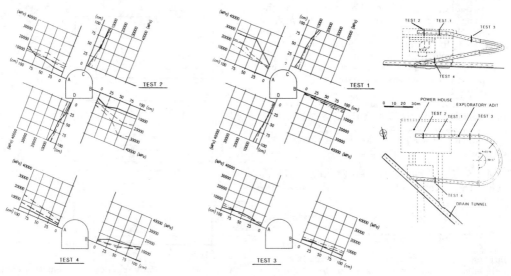

Fig. 4 Plate load test: location of the tests and representation of the deformability moduli

assure the contact between the rubber sheath and the hole contour, up to a maximum load of 9.5 MPa.

3.6 Hydrofracturing tests

Hydrofracturing tests have been carried out in the terminal part of the adit. For this purpose, 4 boreholes (ø 76mm, 20 m long) lying in a vertical plane striking 150°N, have been drilled according to the following scheme: SCF2 hole horizontal, SCF3 vertical downward, SCF1 and SCF4 inclined 45° downward. This geometrical scheme has been selected taking into account the slope morphology over the adit.

The sections for testing have been selected by TV camera as follows: 4 in bore SCF4, 6 in SCF2, 4 in SCF3 and 2 in SCF1.

Test operating conditions were the following:
- 3 pressuring cycles for each test;
- pressuring in the packers and in the test sections increased at the same rate, being the pressure in the packers 2 MPa higher than in the section;
- shut in pressure kept for 10 minutes after the opening of the fracture;
- pressure release in the test section at the end of each cycle and measurement of the pressure build up after closing of the inlet valve;
- taking of the fracture impression for each test.

The pressure versus time diagrams and the fracture traces induced by the tests, show fracturing pattern rather complicated, due to both unparallelism between hole axis and main tension components and rock anysotropy and intense fracturing.

Only 9 out of 16 tests resulted suitable for the processing; for each of these, the parameters determined were:
- the direction cosine of the perpendicular to the fracturing plane in geographical reference frame,
- the values of the stress component s_n normal to the fracture, selected on the basis of a comparative analysis of fracture traces and shut in pressure records.

A computer program for the processing of test data, based on multiple regression analisys, has allowed the determination of the stress tensor, referred to the geographical frame, where x is directed northward, y eastward and z downward. Tensor calculation led from the general tension components to main stresses.

The results are shown in the following table:

Stress	Intensity (MPa)	Dip	Strike
s_1	5.15	66°	172°
s_2	2.27	19°	326°
s_3	1.94	11°	63°

These main stresses are represented in the equiareal Schmidt grid of fig.5, where the direction of the adit is also shown.

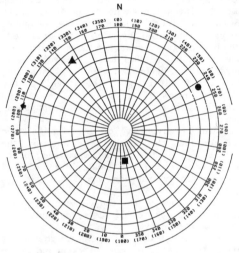

SCHMIDT'S NET
LOWER HEMISPHERE

■ σ_1 = 5.15 MPa

▲ σ_2 = 2.27 MPa

● σ_3 = 1.94 MPa

⚲ axis of the exploratory adit

Fig. 5 Hydrofracturing test: representation of the main stresses in the Schmidt's net

3.7 Geophysical investigations

The geophysical surveys, carried out in the adit and in the drill holes, were aimed at the determination of P waves propagation velocity inside the rock body.

Three different methods have been utilized, namely:
- seismic microrifraction on the adit walls,
- sonic logs inside drill holes,
- sonic tomography between couples of

bore holes.

The operational procedures for the 3 methods were the following:

- For the rifraction, a 12 geophones (spaced 2 m) base line has been utilized, in order to determine the P wave velocity along the rock surrounding the adit, as well as the thickness of the decompressed belt induced by the excavation.

- For the sonic logs it has been utilized a sounding probe (ø 45 mm, designed and built by ISMES) consisting of a trasmitter and a receiver of sonic pulses, both of piezoelectric type, spaced approximately 1 m.

- For the sonic tomography, ultrasonic (or spark type) transmitter and hydrophonic receiver give the running time. To investigate the range of velocity in the sections under test, a computer program elaborated by ISMES is utilized.

Such a program, starting from an initial model, with a S.I.R.T. (Simultaneous Iterative Reconstruction Tecnique) procedure determine iteratively the velocity ranges.

The results of geophysical investigations have shown the following:

- The excavation of the adit has produced in the rocks surrounding the excavated volume, a decompressed belt, generally not thicker than 1 m, with P waves velocity ranging from 2.0 to 4.4 Km/s.

- The velocity in the inner rock are ranging from 4.2 to 5.5 Km/s, depending on fracture intensity.

- The rock body, except for the areas intensely fractured, can be regarded as an anisotropic medium for the wave propagation, with variation up to 10% of the average value of the velocity.

- Two low velocity belts identified, show a good correlation with fault planes.

4 LABORATORY TESTS

More than 70 rock samples from the drilled cores have been tested in the ISMES laboratory. Tests included the following determinations:

- apparent and real volumic mass,
- porosity,
- P waves velocity,
- monoaxial and triaxial compression under deformation control,
- direct tensional strength,
- direct shear strength in the natural joints,
- petrographic analysis.

The results of the tests can be summarized as follows:
- apparent volumic mass: 2.74 g/cm^3

- real volumic mass   2.77 g/m^3
- porosity   1.03 %
- sonic velocity   3290 m/s

In the following tables are shown the compression strength and moduli ratios of samples tested, according to Deere and Miller classification:

Class	Compression	Strength	Samples
	description	value (Mpa)	%
A	very high	> 225.0	0
B	high	225.0-112.5	0
C	average	112.5-56.0	0
D	low	56.0-28.0	40
E	very low	> 28.0	60

Class	Moduli ratio		Samples
	description	value	%
H	high	> 500	80
M	average	200-500	20
L	low	< 200	0

The direct shear tests in the natural joints have indicated the shear strength along rock discontinuity.

On some rock samples representing the litho types of the area investigated, petrographical analysis have been also carried out to define their mineralogical, petrographical and microstructural characteristics.

On these basis the rock mass can be defined as phyllite passing to fine gneiss, with texture varying from sheared to highly sheared and with lepidoblastic structure.

From the mineralogical point of view, the main components are: micas (mainly muscovite and sericite), quarz, feldspars and calcite.

Finally, on few samples of the cataclastic materials filling the main faults encounterd in the adit, laboratory soil tests have been carried out, namely:

- screen analysis,
- Atterberg limits,
- X rays diffractometric analysis of the clay fraction.

These tests have shown that the filling material consist mostly of heterogeneous sand with minor medium to fine gravel, silt and clay.

5 DESIGN OF ROCK SUPPORT

The extensive geological explorations and laboratory and field testing have given an excellent knowledge of the rock mass properties interested by the excavation of the underground plant. Besides, the knowledge of the rock mass was completed by the observed behaviour of the nearby first chamber of similar dimensions.

The assessment of the shear strength and modulus of in situ rock mass was carried out utilizing the results of the site investigations and laboratory tests on intact rock pieces.

In addition the Rock Mass Classification system proposed by Bieniawsky (1974) was adopted. Using the Hoek & Brown (1980) failure criterion in terms of the major (σ_1') and minor (σ_3') principal stresses at failure, the empirical costants m and s of both the intact rock and rock mass have been calculated and the Mohr failure envelope for the estimated values of m and s has been plotted calculating the average friction angle and cohesive stregth of the rock mass. Fig. 6 shows a summary of the plots of σ_1' against σ_3' of the Hoek & Brown failure criterion, adopted in the design of the underground plant.

A first evaluation of stresses and deformations around the openings in the rock was performed considering the anisotropy due to the schistosity of the metamorphic rock mass and the presence of a shear zone traversing the chamber in an asymmetric manner. Finite element (2 dimensional) methods were used. Analyses were made of the cavern cross section under a variety of imposed stresses.

These mathematical models provide the basic informations for subsequent studies on alternative methods and sequences for rock excavation, immediate and long-term support, and final lining.

Both rigid strong linings and flexible linings combined with bolting and anchoring of appropriate lengths will be considered and analysed. In any event, during construction the movements of the rock caused by the different sequences of excavation will be monitored and therefore the stability will be continuosly checked to compare the obtained results with theoretical estimates. As the excavation progresses a better knowledge of the behaviour of the rock will be possible by the monitoring and the design of the rock support measures will be modified where necessary.

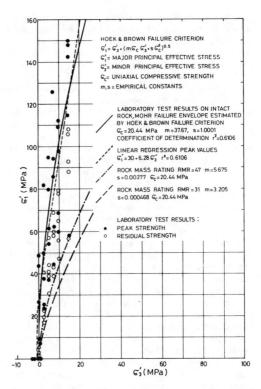

Fig. 6 Hoek and Brown failure envelopes estimated from the laboratory tests and site investigations

6 CONCLUSIONS

For the design of the extension of the Premadio hydroelectric power plant a geognostic investigation has been carried out to satisfy the following criteria.

1. Utilize all the actual reliable in situ and laboratory testing techniques to form a complete picture of the rock mass, in view of optimizing the design of the excavations and supporting structures, with proper safety margins. The investigations included.

- an exploratory adit along the roof of the future machine hall with a geological mapping, an analysis of the geological structure and the monitoring of the behaviour of the rock during the progress of the exploratory excavation.

- flat jack, plate load and dilatometric tests to give a good assessment of the in situ rock modulus;

- some hydrofracturing tests to evaluate the in situ stress field in the area of the cavern, to provide the basis on which to assess observed deformations in the exploratory adit;

- geophysical investigations to determine the distribution of seismic velocity inside and around the future cavern and to study the redistribution of the stresses due to excavation of the exploratory adit;
- laboratory tests on samples to determine density, compressive and tensile strength, modulus of elasticity and Poisson's ratio. In view of the importance of the joints on the behaviour of the rock, shear tests were carried out to determine the shear strength of joints. Normal classification tests were performed on the clayey silt of the cataclastic material filling the main faults.

2. Appraise the geomechanical parameters of the rock mass from the results and interpretation of all the investigations.

Finally the monitoring data about vibrations in the existing power house obtained during the excavation of the adit will allow the best definition of the blasting scheme during the future excavation of the rock mass as far as the safety and at the same time the exercise of existing power house.

Thanks to an extensive and modern geological and geotechnical investigation, an excellent knowledge of the rock mass has been obtained and consequently the geomechanical parameters of the rock mass have been obtained. During the excavation of the caverns a constant control of design assumptions will be carried out monitoring system to control the deformative excavation behavior.

7 REFERENCES

Borsetto, M., Giuseppetti, G., Martinetti, S., Ribacchi, R., Silvestri, T. 1983. Geomechanical investigations, design criteria and observed behaviour of the Timpagrande underground powerhouse, Rock Mechanics and Rock Eng., 16, p. 85-155.

Borsetto, M., Frassoni, A., Rossi, P.P., Garbin, C., Moro, T., 1984. The Anapo pumped-storage power station: geomechanical investigations and design criteria. ISRM Symp. Cambridge.

Forzano, G., Frassoni, A., Moro, T., Rossi, P.P., Vallino, G., 1986. The Edolo underground power station. Int. Symp. on Large Rock Caverns, Helsinki.

Hydropower'92, Broch & Lysne (eds) © 1992 Balkema, Rotterdam. ISBN 90 5410 054 0

Some Chinese hydropower projects adopted Norwegian advanced experience

Gu Zhao-Qi & Pen Shou-Zhou
Department of Hydraulic Engineering, Tsinghua University, Peking, People's Republic of China

ABSTRACT: Lots of Norwegian advanced experiences have been being adopted recently in underground engineering of Chinese Hydropower stations. In Lubuge, Dongfeng, Da-guang Dam, and Guangzhou Pumped Storage power station, rock bolted crane beams were used, some improvements have been made in design and measuring. This technology will also be adopted in some other large power stations. Shafts without steel lining have been adopted in Tian-Hu, Guangzhou pumped storage power station, and so on. The maximum water head of unlined shaft in Tian-Hu station reaches 614 meter. Good benefits have been acquired due to adopting the advanced experiences mentioned above.

1 PREFACE

Until now, in China more than forty underground power stations are under commission. More than twenty big underground power stations are under construction and design.

In the past three decades some new conceptions and new measures were introduced to our country, such as NATM, shot-crete, rockbolt, etc. We developed our code for shot-crete, hydro-tunnel and underground engineering. Several diversion tunnels with low water pressure, less than 60 to 70 meter, were constructed without concrete lining or steel lining.

After 1980, the open policy has been adopted, more and more advanced techniques were introduced into China. The Lubuge underground power station is a very successful example to illustrate some Norwegian experiences.

Today, a lot of Chinese power stations use Norwegian experiences.

2 ROCK BOLTED CRANE BEAM

Lubuge power station is the first station in China using rock bolted crane beam in machine hall, it was put into operation in August, 1988.

It is evident that the rock bolted crane beam has so many benefits, and it is reliable as well. After then, most underground power stations follow Lubuge, see Table 1. And perhaps another ten stations are considering whether using the rock bolted crane beam.

A large number of calculations have been done with numerical method in the design of many of the above-mentioned underground hydropower stations, and it is discovered that the tension stresses of bolts for the rock bolted crane beam increase with the proceeding of excavating, if the horizontal tectonic stresses are quite large, or the elastic modulus of surrounding rock mass is quite small, and or there are unfavorable weak structures in the surrounding rock mass, then, the stresses of bolts will reach relatively large values. At the same time, it is also discovered in calculations that, if the adhension between the rock wall and the crane beam is taken into consideration, after the crane beam is acted by crane's load and weight, stresses of the bolts do not increase much. In general, they vary from 10 to 30 MPa.

In the construction of Lubuge, Dongfeng, and Guandzhou Pomped Storage power station, a number of stress-gauges were embedded along the rock bolts, the observed results were similar to the evaluating results mentioned above. In Lubuge power plant, under general conditions, the maximum observed value reached 95 MPa. In Dongfeng power plant, when the powerhouse was being excavated to the bottom, one of the stress-gauges' read numbers showed the tension stress had increased to 90 MPa. The stresses of bolts for crane beam among one busbar cavern of Lubuge power station, reached 320 MPa in half a month, only until the busbar cavern had been supported by bolts and shotcrete, did the stress values tend to stable. In the case of crane's gravity and load action, the increasement of stresses of bolts was not obvious in all these power stations mentioned above.

All the cases mentioned above show that:

1. It is inadequate to use the static equilibrium method to evaluate the stresses of the bolts for crane beam, because this method does not reflect the force conditions. In the past, after the excavating of a cavern, model tests

Table 1

Name	Capacity of the station (MW)	Capacity of the crane (T)	Max. wheel Load (kN)	Span of crane (M)	Remark
			Load on the beam (KN/M)		
Lubuge	4×150	2 sets, 160	485/262	16.5	under operation
Guangzhou pumped storage power station	4×300	2×200	550/391	19.5	crane beams are finished
Dong-Feng	3×170	2 sets 2×250	670/	20.7	same above
Da-Guang Ba	4×60	2×100	615/	13.0	under construction
Tai-pin-Yi	4×65	200		17.0	same above
Xiao Lang Di	6×300	2 sets 2×250	780/565	23.5	same above
Hong-Jia Du	3×180	2 sets 2×250	680/	19.0	under designing
Zhou-Nin	4×60	2×125	400/	17.5	same above
Long-Tan	9×600	2 sets, 350		25.0	same above

would be carried out with installing bolts in surrounding rock mass to establish experimental model, however, that did not reflect the real force conditions of the rock bolted crane beam. Therefore, the more perfect design is using the finite element method (FEM).

2. Former static equilibrium method did not consider the adhension between the concrete and the rockwall, thereby the stresses of bolts would be very large. If the adhension does not lose, the increasements of stresses of bolts are not obvious after the crane beam being acted by external forces. When using FEM to evaluate, if jointing elements are given between concrete and rockwall, then, before and after the crane beam being acted by loads, the stress values of the bolts will have large variation.

Thereby, we suggest that the following measures should be carried out in order to make the rock bolted crane beam more safe and reliable.

1. On the rockwall nearby the crane beam, more horizontal bolts should be installed for partially consolidating the rockwall and reducing the stress rising of bolts for rock bolted crane beam. Of course, it is also effective by increasing the length and areas of system bolts in this location, the increasement can be decided

by calculation and comparision with FEM.

2. Reduce the eccentric cantilever arm of the crane beam. At present, in rock bolted crane beams designed by China, there are more than 0.5 meter wide gap between the crane ends and the rockwall, the cantilever arm which is the distance of the track center apart from the rockwall reaches 0.75 meter, see Fig 1. We should do our bests to shorten this gap, thus to reduce the cantilever arm. This way will result in that, after the crane beam being acted by loads, the stresses of the bolts do not increase too much.

3. Carefully preserve the integrity of the rockwall nearby the rock bolted crane beam, guarantee a fairly nice adhension between concrete and rockwall, so that the stresses of the bolts do not increase too much after the crane beam being applied loads.

4. In the course of excavating the power-house caverns, the excavating procedure and measures of blasting should be arranged carefully, so that no fairly large displacement occurres at the higher part of the wall. The crossed part of the cavern linked with busbar cavern, access tunnel, and so on, which are near to rock bolted crane beam should be excavated more carefully,

90

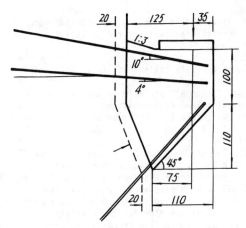

Fig.1 Rock Bolted Crane Beam

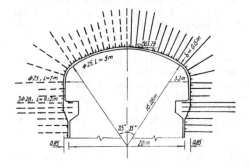

Fig. 2 Roof Supported by Shot-crete and Bolts

so that no fairly large displacement occurres at the rock wall near the crane beam. All the displacements will result in the large-scale rising of the stresses of the bolts for crane beam. It is best to excavate these crossed parts of cavern and permit the accomplishment of large-part displacement of the rock wall in advance, before the bolts for crane beam are installed.

5. Reduce the angle of bolts for rock bolted crane beam, increase their force arm, thus to decrease their stresses.

After all, rock bolted crane beam is one economic and effective method, but it is necessary to understand and master it more deep-going, thus to make it more reliable and safe.

3 ABANDON THE ROOF CONCRETE LINING, ADOPT THIS KIND OF SUPPORTING WITH BOLTS AND SHOT-CRETE INSTEAD

Before 1980's, most of the underground powerplants built by our country, were roof lined by reinforced-concrete, only in several underground power plants, the rock mass conditions were extremely nice, so that the design engineers boldly used the rock bolts and shot-crete for roof supporting.

In recent years, because of the learning of new concepts on surrounding rock mass and the successful experiences of Lubuge power station, more and more projects abandon the reinfoced-concrete roof lining.

Dong-Feng hydropower station on the WuJiang river in Gui Zhou Province is quite a nice example, the underground power house is 105.5 meter long, 20 meter wide, and 48 meter high. The surrounding rock is limestone, the bedding angle is fairly gentle, about 10 deg.. In the limestone, there are several clay seams, and two parallel seams of clay with an interval of 5 to 6 meter and depping about 10 deg., and several little faults on the top of the cavern, the geological condition is not so good, the value of Q varies from 2.5 to 7.5, but the designers yet decide to use the rockbolts and shot-crete for roof supporting, see Fig. 2. The shot-crete is 15 cm thick, and reinforcing fabric is laid inside as well. There are two types of bolt with 25mm in diameter, 5m and 7m in length respectively, which are arranged alternately with an interval of 1.2×1.3m.

During construction, smooth blasting was used, the parameters such distance between any two peripheral holes, total charge and specific charge, and so on, are determined through testing, the explosive forming of the roof was very nice, ratio of half-hole reaches more than 90%, and the highest of 95% in some areas. Nineteen sets of multi-point displacement gauges and thirty-five sets of bolt stress gauges were embeded in the roof rock mass, in addition, the convergency measuring were carried out as well. The deformation and stress were normal During excavation. The maximum value in the stress gauges for bolts was not bigger than 90 MPa, at the beginning, the roof was deformed downward, then upward, the deformation values were 1 to 2 cm, the deformations of upper and down wall were also less than that being estimated originally. Therefore, the number of bolts of the machine hall was reduced correspondingly, the total reduction was more than 30%.

This example of DongFeng power station further affirms that the roof supported by shot-crete and bolts is economic, safe and feasible. At the same time, Guang-Zhou Pumped storage power station, HaiNan Da-Guang Dam underground power station, SiChuan Tai-Ping Yi hydropower station also successfully use this kind of method. In the future, these large size underground power stations such as Er-Tan, Long-Tan, Xiao Lang Di, etc., are planned to use this method, the excavating span of the upper part of the machine hall in these power stations, vary from 25 to 30 m. And the rocks in XiaoLang-Di underground power station are sandstone, fine grain sandstone, which bedding surface dips 8 to 10 deg., the value of Q of the surrounding rock mass varies from 2.0 to 8.0, the geologic conditions are not extremely nice.

It is clear that, using shot-crete and bolts instead of reinforced-concrete lining for roof supporting, will result in not only reducing the excavated volume, but also largely accelerating the constructing. The total reduction of excavated rock volume in Dong Feng hydropower station reaches 65,000 cubic meter. If bolts and shot-crete are adopted instead of concrete lining, the span of the roof will be reduced from 27.0 m to 21.0 m.

4 UNLINED DIVERSION TUNNELS

Before the 1980's, some unlined tunnels had been built in China, but the water heads were quite low, less than 60 to 70 m. At higher water head, the steel lining, at least the reinforced-concrete lining would be used. With opening and reforming, many Chinese engineers and experts have known the foreign advanced experiences of unlined tunnels, of course, including Norweign experiences as well. Therefore, they also boldly use high water unlined tunnels.

4.1 Tian-Hu power station

TianHu power station locates in GuiLin, GuangXi province, with 1100m water head, a generating capacity of 60 MW, and maximum diversion flow of 6.0 cubic meter per second. The surrounding rock mass of the tunnel is granite. The authors had suggested to build an underground power station with unlined tunnels at 1100 m water head. For the reason of the construction, equipment, etc., it was decided that adopting unlined tunnel for upper part of the diversion tunnel and steel pipe for the down part, see Fig. 3. Although this kind of layout was not so reasonable, one section of steel pipe was omitted which construction was very difficult. The maximum internal water pressure upon the unlined tunnel reaches 614 m.

In the near future, the power station will be put into operation. In general, the situation during construction was normal, only in vertival shaft, at the depth of 270 m, on the interface of the fine grain granite intrusive mass, there was a fairly large water leakage, about 2L/s. In the middle of the section which had been planned to be plugged with concrete, there occurred a small fault with a width of 0.5 cm, and including glued crushed-rock seam, the fault dipped about 80 deg., the included angle between the strick line and the long axis of the tunnel was near 90 deg., the leakage flow rate along the fault reached 0.5 L/s. The construction contractor adopted

the corresponding grouting, leakage stoppage, reinforcing measures in terms of design.

4.2 Use concrete instead of steel sheet for

smooth lining At present, in many of Chines high water head diversion tunnels, if the rock mass overburden is thick enough, only concrete is used for smooth lining instead of steel sheet lining.

The famous Guangzhou Pumped storage power station made a precedent in this aspect. The first stage generating capacity was 4×300 MW, with 535 m high water head. Four units were supplied water by one common diversion tunnrl capable of discharging 273 cubic meter per second, which inner diameter decreased gradually from 8.0 m. Both the inclined shaft and main horizontal tunnel were not lined by steel sheet, three very large dimension branch-sections of the tunnel were just lined by concete, only the penstock tunnels about 100 m away from the power house were lined by steel sheet, with diameter of 3.5 m.

Several branch-sections had been analysed by 3-dimension FEM, at these places the maximum internal water pressure reached 725 m, for the sake of reliability, field tests with a scale of 1:2 will be carried out , that costs three million yuan(RMB). Though these measures are by no means nessary , all in all, we at last make the quite important step. Now, extreme large size staions such as Er-Tan, Long-Tan, etc., with quite nice conditions, the diversion tunnels on the upper stream from the grouting curtain, will also adopt the concrete lining instead of steel sheet lining, even though the inner diameters are so big, about 7 to 8 m, the unit generating capacities reach 550 to 600 MW.

4.3 In projects such convey tunnel, diversion tunnel, etc., unlined tunnels are adopted partially in our country.

The Yin-Da-Ru-Qing project in Gan Shu province,

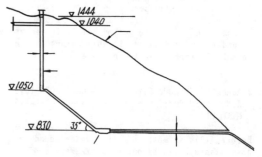

Fig. 3   Tian-Hu Project

has unpressured convey tunnels of more than 70
KM long, the surrounding rock mass in some
sections is fairly sound, unlined tunnels or
some supportings with bolts and shot-crete will
be put into use. After opening, some special
areas, rock surface largely undulates, if they
are lined with concrete, not only too much work
and time are consumed, but also no big discharge
capacity can be acquired. Undulations of rock
wall were measured by the authors, the roughness
was checked by Solvik method, if some
protrusions are excavated by 0.2 to 0.3 m, the
roughness will reduce a lot, without lining, the
discharge capacity can be increased by 10
percent than that of the original design.
The diversion tunnel of Qing Jiang Geheyan
power station in Hu Bei province is more than
600 m long, the dimension of excavated cavern is
15 m wide, 18 m high, the discharge capacity is
3000 cubic meter per second. Rocks at the front
part of the tunnel are limestone, rocks around
the outlet are shales, which beds dip 30 deg..
During construction, a large landslide occurred
at the shale section, caused accidents, resulted
in the delay of the time limit for the project.
The authors took part in the field studing with
some other experts, suggested that in the
sections with fairly sound limestone, only
smooth lining should be used for rockwall and
bottom board supporting, and the roof just should
be supported by bolts and shot-crete. Thus, the
construction schedule was guaranteed again, the
diversion tunnel was finished in one year, at
the same time, the river was cut off. After
then, there have been passed several floods in
flood periods. when the full-flow was formed in
the diversion tunnel, with a velocity of 15 m/s
the operation result showed no problem. This
project was a completely successful precedent
in China.

4.4 Hydraulic splittung test

In the course of designing for the high water
head unlined diversion tunnels mentioned above,
we yet carried out the hydraulic splitting
tests. The capacity of the water pump was just 4
L/min, not the conventional 40 or 50 L/min,
however had quite nice effect, and because of the
small volume and the weight of less than 100 Kg,
this type of pump is quite convenient in moving
and operating.

5 CONCLUSIONS

1. Some advanced Norwegian and other countries'
experiences have been adopted and spread in our
contry, and many excellent benefits have been
acquired as well. Here, we are very grateful to
those foreign experts who have done great
contributions for Chinese construction.

2. Nevertheless, there are still many nics
experiences not to be adopted and spread, such
as the air-cushsion surge chamber, especially the
tailrace air-cushsion surge chamber, underground
power house with quite compacted-layout, etc.,
it is necessary for us to make greater efforts.

Hydropower'92, Broch & Lysne (eds) © 1992 Balkema, Rotterdam. ISBN 90 5410 054 0

Design of rock bolted crane beam

Guan Peiwen
Mid-South Design Institute for Hydro-electric Project, Changsha, People's Republic of China

Han Zhihong
Hydraulic Engineering Department of Tsinghua University, Beijing, People's Republic of China

ABSTRACT: Combining with the rock bolted crane beam in Guangzhou Pumped Storage Underground Plant, the following questions will be discussed in this article: 1. What geologic conditions are demanded when you adopt a rock bolted crane beam and how to deal with the distribution and structural arrangement of rock bolted crane beam, access tunnel and busbar gallery. 2. How to determine the calculation loads of rock bolted crane beam. Suggesting to analyse and calculate crane beam as a system of grid beams or a beam on elastic foundation. 3. How to choose the cross section of rock bolted crane beam and the distribution of anchor rods, including dip angle of rock wall and the main sizes of the beam. 4. Putting forward quanlity standards about construction of rock bolted crane beam.

1 ROCK BOLTED CRANE BEAM AND LAYOUT OF UNDERGROUND PLANT

The stability of rock bolted crane beam mostly depends on rock wall, this involves two aspects. First, rock wall of underground plant should be stable, while, rock wall stability relates to cavern span L and geologic conditions. If L>20m, the classification of adjoining rock should be above class II, if 15m<L<20m, above class III. Second, it should be avoided to layout big tunnel near crane beam at upperstream or downstream wall of underground plant. Because the positions of access tunnel and busbar gallery are high, they affect crane beam greatly.

If access tunnel as high as cut off crane beam, you should let it enter erection bay from end wall of undreground plant. If access tunnel is just near (not cut off) the crane beam, you must put columns at both sides of tunnel entrance.

Besides busbar, low-voltage electrical equipments are usually set in busbar gallery, so busbar gallery is often large. The size of Lu Buge power station's busbar gallery is 7.40×8.68m (b×h), and the distance between busbar gallery top and crane beam bottom is only 4.86m. Although a lot of measures had been taken during excavating busbar gallery, such as excavating guide tunnel, limiting footage, and controlling explosive charge, etc., the stress growth rate of crane beam bolts near the entrance was still 10 MPa/d. Less than two weeks, the bolt stress had been increased to 320 MPa. Until the whole busbar gallery was supported by rock bolts and shotcrete, the bolt stresses were

gradually steady.

Taken warning from Lu Buge, Dong Feng power station adopted a small and low position busbar gallery. In the design of Guang Zhou Pumped storage power station, in order to diminish the size of bus gallery, they had attempted to put low-voltage electrical equipments into main transformer and switch gear room, but this will enlarge the main transformer and switch gear room, and result in costing too much and difficult layout. After a lot of studies, they finally adopt a " small-big busbar gallery", that is in the distance of 8 to 9m from the main house, there are no equipment only busbar in the gallery, size of this section is 5.2×5.5m (b×h), after this small section, the gallery is transited into big section, its size is 8.3×8.8m (b×h), and electrical equipments will be installed into this big section near main transformer and switch gear room. Fig.1 shows the layout of Guang Zhou underground powerplant. Neither affect rock bolted crane beam, nor enlarge main transformer and switch gear room, this " small-big busbar gallery " is a successful method to display electrical equipments.

Anothor thing which needs pay attention to is to make the rock bolted crane beam seamless, because the bending moment of cantilever beam root at seam is two times as big as that of those places far from seam. But construction seams usually need to be set, and bolts must be reinforced at the range of 1 to 1.5m each side of seam.

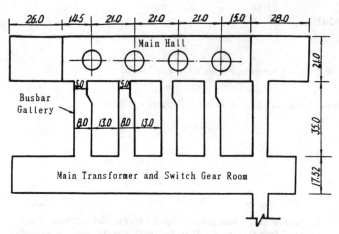

Fig.1　Plane of GuangZhou Pumped Storage Undergroung Plant

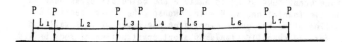

Fig.2

2 CALCULATION LOADS OF ROCK BOLTED CRANE BEAM

The loads acted on rock bolted crane beam
include dead weight of crane beam, weigths of
track and its accessories, these loads can be
converted into unit loads. But the largest
wheel loads of crane are concentrated loads,
while crane beam is a continuous structure, how
to convert these concentrated loads into unit
loads, the following two methods are often used
at presant:
  1. Distribute all these wheel loads along the
distance between the head wheel and the tail
wheel equally.
  2. Distribute the middle two wheel loads along
distance between the two midpoints of near
wheelbases (see Fig.2). This method had been
adopted by LuBuge powerstation and GuangZhou
pumped storage station.

$$q=2p/[L_3+0.5(L_2+L_4)]$$

where, p: the largest wheel load
        q: unit concentrated load
  While author suggests to analyse and calculate
the rock bolted crane beam as a system of grid
beams or a beam on elastic foundation.

2.1 Method of grid beams

Main points of this method is:
  1. Regard the rock bolted crane beam as a grid
structure, consisting of a series of cantilever
beams and a series of continuous beams.
This method is similar to arch cantilever method
of arch dam.
  2. Acted by loads, the displacement of
intersection point of two system axis is equal,
and their interaction forces are equal in value,
opposite in direction. According to these
conditions, enough simultaneous equations can
be set up.
  3. For cantilever beams, crane beam cross-
sections can be used as their calculated
sections. Their deformation matrix can be
calculated as cantilever beams with variable
cross sections.
  4. For continuous beams, they are structures
with variable cross sections, their basic
structures can be simplified as free beams with
unknow bearer moment. This simplify make rigid
matrix regular and convenient for computer
calculation.
  5. The torsion moments between the cantilever
beams and continuous beams is not countered in
calculation.
  According to these hypothesis mentioned above,
a computer program had been designed. A lot of
calculations shows: if the size and loads of
rock bolted crane beam are unchanged, no matter
what grids you divided, the calculation results
about moment and shear at root of cantilever
beam are stable. The moment distribution at
root of cantilever beam see Fig.3 and Fig.4.

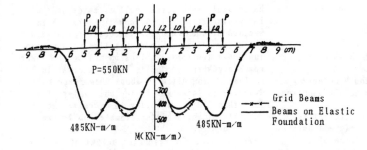

Fig.3 **Moment distribution** at root of cantilever beam **when crane located at the middle of beam**

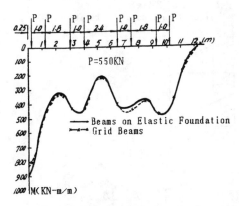

Fig.4 Moment distribution at root of cantilever
beam when crane located at the end of beam

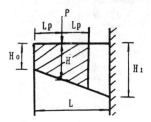

Fig.5

2.2 Method of beam on elastic foundation

Main points of this method is:

1. Regard the rock bolted crane beam as a
series of cantilever beams supporting on elastic
foundation. Obviously, this " elastic foundation "
conform to Winkel hypothesis.

2. The width of beam on elastic foundation is
$B = 2L_p$, (L_p is distance between crane track and
cantilever head, see Fig.5) that means only
take a part of cantilever beam as a beam on
elastic foundation. Error caused by this hypothesis is small, because the deformation, interaction force and force arm at head of cantilever
beam is bigger than that of root.

3. General answers for beam on elastic foundation:

Usually, beams on elastic foundations can be
classified as infinite beams and finit beams,
but crane beam has mobile loads, for some wheels
it may be finite beam, while for others it may be
infinite beam. Especially for a certain wheel, as
its special position, crane beam is infinite for
one side, but it is finite for another side. How
to solve this problem, author think classify is
unnecessary and general answers can be given as
following:

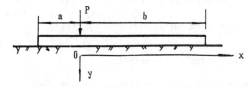

Fig.6 Illustration of a beam on elastic foundation

See Fig.6, the coordinate origin is force
application point, and positive direction is
right. Acted by concentrated load P, the general
answers is:

$$y1 = e^{\lambda x}[c1 Cos \lambda x + c2 Sin \lambda x] + e^{-\lambda x}[c3 Cos \lambda x + c4 Sin \lambda x] \qquad (1)$$

$$(-a \leqslant x \leqslant 0)$$

$$y2 = e^{\lambda x}[c5 Cos \lambda x + c6 Sin \lambda x] + e^{-\lambda x}[c7 Cos \lambda x + c8 Sin \lambda x] \qquad (2)$$

$$(0 \leqslant x \leqslant b)$$

where, $\lambda = \sqrt[4]{K/4EI}$ (3)

     K: foundation coefficient
     E: elastic modulus
     I: section inertial moment of beam on
        elastic foundation

Integral constants $c1 \sim c8$ can be determined
by solve eight equations obtained from the

following conditions:

(a). If x=-a, then M=0, Q=0
(b). If x=b, then M=0, Q=0
(c). If x=0, then y1=y2, θ1=θ2, M1=M2, Q2=Q1-P

After c1~c8 founded, deformation y, cornor θ, moment M, shear Q and reacting force of founction $p_y=k \cdot y$ can be obtained.

4. Calculation of foundation coefficient k:
See Fig.7, coordinate origin is left end point, the height of beam section at distance of x from origin is:

$$h=h_0+(h_1-h_0)x/L \qquad (4)$$

Acted by unit force p=1, the linear deformation along force direction is:

$$r = \int_{Lp}^{L}[(x-Lp)^2/EIx]dx \qquad (5)$$

Where, $Ix=(h_0+cx)^3/12$
then, $r=-12[0.5A(1/h^2-1/h^2)+A1(1/h_1-1/h)+A2\ln(h/h_1)]/EC \qquad (6)$
where,

$$h=h_0+c \cdot Lp \qquad (7)$$
$$A2=1/c^2 \qquad (8)$$
$$A1=-(2Lp+2h_0/c)/c \qquad (9)$$
$$A=Lp^2-h_0 \, A1-h_0^2 \cdot A2 \qquad (10)$$

So, foundation coefficient $k=1/r \qquad (11)$

2.3 Comparison of two methods

According to these two mathematic models mentioned above, two corresponding computer programs had been designed, and a lot of calculations for the rock bolted crane beam acted by a concentrated load or several concentrated loads have been carried out. These calculation results shows: no matter how many concentrated loads appiled, single or several, and which calculation method is adopted, grid beams or beams on elastic foundation, if the calculation conditions are unchanged, the values of calculation results are approximate. Fig.3 and Fig.4 show the comparison of these two methods.
Now that the results of these two methods are approximate, and method of beams on elastic foundation is simpler, author suggests to analyse and calculate rock bolted crane beam by method of beams on elastic foundation.

3 SECTION DESIGN OF ROCK BOLTED CRANE BEAM

3.1 Selection of section figure and section sizes:
Section figure and section sizes can be initially determined according to operation requirement of crane beam, wheel loads,

horizontal brake force, bolt diameter and bolt intervals, etc.. Fig.8 shows the crane beam section of GuangZhou Station. Usually, there are three rows of bolts in rock bolted crane beam, the upper two rows are tension bolts, the lower row are compression bolts. Bolt interval is 70 to 100cm, row pitch is about 50cm, and bolt diameter is 14~36mm. If the bolt interval is less than 100cm, in order to decrease the quantity of bolts, bigger diameter bolts are often adopted.

For initial selection, the height of beam root h_2 (see Fig.9) can be estimate as following:
where, h_2: height of beam root (cm)
P: unit wheel load (KN)
Lp: arm of P (cm)
T: unit horizontal brake force(KN)

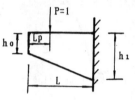

Fig.7

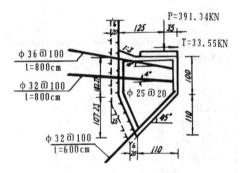

Fig.8   Crean beam of GuangZhou powerstation

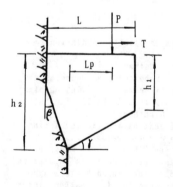

Fig.9

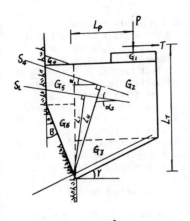

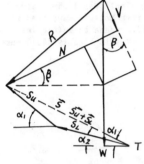

Fig.10 Force Diagram of Rock Bolted Crane Beam

d: bolt diameter (mm)
Notice: formula (12) is presupposed condition of
second grade steel bar and two rows of
tension bolts.

3.2 Determine of bolt parameters

Determine of bolt parameters can be carried out
after section figure and section sizes selected.
According to equilibrium equations of rigid body,
taking the intersection point of press bolt and
rock wall as a rotational centre (see Fig.10),
the forces of tension bolts can be calculated as
following:

Outernal moment is:

$$Mo = \Sigma G_i L_i + P \cdot Lp + T \cdot L_T \qquad (13)$$

Internal moment is:

$$Mi = Su \, Lu + S_1 \cdot L_1 \qquad (14)$$

because,
$S_1 / Su = L_1 / Lu$, and $\Sigma M = 0$, there is

$$Su = Mo / (Lu + L_1 / Lu) \qquad (15)$$
$$S = Su \cdot L_1 / Lu \qquad (16)$$

meanings of these symbols see Fig.10.

vertical join force is: $W = \Sigma G_i + P$ (17)

horizontal join force acted on rock wall is:

$$\overline{R} = \overline{W} + \overline{T} + \overline{Su} + \overline{S}_1 \qquad (18)$$

horizontal component of R is V, vertical
component of R is N.
In order to avoid shear force acted on bolts,
the ratio of V and N should be less than $tg\phi$,
that is

$$f = V/N \leqslant tg\phi \qquad (19)$$

where, ϕ is inner friction angle of rock
Note: in Lu Buge Station $f \leqslant 1.0$, while f
should less than the friction
coefficient of concrete and rock.

According to formula (15)~(19), the bolt
stresses and f are affected by the following
factors:
external loads P, T, etc., they are unvarying
factors;
dip angle of rock wall β, bolt inclinations
α_1 and α_2, over-excavate
value δ, etc., they are varying factors.
Relations between these varying factors and Su,
f is:
1. Bolt stresses will increase obviously along
with the increment of δ, but f will decrease;
2. Bolt stresses will increase slightly along
with the increment of β, because the force arm
will increase sligthly along with the increment
of β, but f will decrease obviously;
3. α_1, α_2 affect bolt
stresses and f as β does, but the changment of
bolt stresses by α is greater than by β, while
the changment of f by α is less than by β .
Overall, the main points need consider when
you determine bolt parameters are:
1. For overexcavate, the smaller the better.
Overexcavate δ should be taken into consider
when you calculate bolt stresses, but it need
not when calculate f.
2. For bolt inclinations, the smaller α is,
the longer force arm will be, so the bolt
stresses will be decreased. If bolts are set
nearly horizontal, not only the construction
error of α has very little influence to the
bolt stress, but also part of reinforcement
in crane beam can be replaced by the horizontal
bolts, so that the total reinforcement can be
decreased. Author suggests if loads are not
heavy and bolt interval is less than 70cm, it
is better to set one row bolts only, and let
these bolts horizontal or nearly horizontal. If
two rows need to be set, it is better to let
$\alpha_1 = 5° \sim 8°$, and $\alpha_2 = 0°$.
3. Whether you want lower bolt stress or want
lower f, you should increase β at first, and
set bolts nearly horizontal. But β can not be
increased without limit, because too big β may

99

widen the plant span. If β<30°, increasing β just widen plant span slightly, so it is better to let β=30°, and then enlarge α to lower f.

For bolted depth of bolts, there are two methods to calculate:

1.　　　$L=L1+35d$　　　　　　　　(20)

where, L1: depth of rock relaxation region
　　　　d: bolt diameter

2.　　　$L=0.15H+2$　　　　　　　(21)

where,　H: height of side wall

Depth of rock relaxation region can be measured at site by sound-ranging. Both Lu Buge station's L1 and Guang Zhou station's L1 is only about 1.5m, L will be too short if L is calculated by formula (20), so formula (21) is is usually adopted. Calculated by formula (21), L of these two station is 8m, but their bolt stresses were too small when the crane load tests were carrying out, and both spread depths were within 1.5m, these results show bolts have reinforce effects. Author suggests bolted depth of crane beam should not less than the length of systematic bolts at same part.

3.3 Comparison

Comparison of results obtained by two methods with different selections of bolt parameters is listed in Table 1.
From table 1, we know that the set tension bolts horizontally and enlarge β slightly is a effective method of selection bolt parameters, this method not only is economic but aslo quicken construction.

4 QUALITY CONTROL STANDERDS ABOUT CONSTRUCTION OF ROCK BOLTED CRANE BEAM

There are two groups of factors about quanlity

controlling, some can not be expressed in formula, such as depth of bolt hole, bolt grout, etc., control method of these factors is strengthen supervision. Others can be expressed in formula, such as bolt tension S, dip angle of rock wall β, bolt inclination α, overexcavation δ, etc.. Besides strengthen supervision, quanlity standards should be put forward for these factors. The presant method of limiting maximum error for a single factor needs improving, so author puts forward a " qualified area " to control these factors. See Fig.11, "qualified area" can express the influence of main factors of β, δ, α on bolt tension S and f, so it is a reasonable standard for quanlity control.

Guang Zhou Station has a 2×200t overhead travelling crane with plant span 19.5m, largest crane load 400t. Its rock bolted crane beam was finished in March, 1991, crane beam load test (static load is 560t, dynamic load is 480t) was

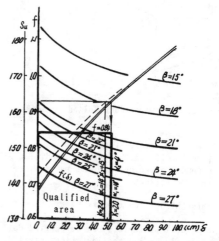

Fig.11　Quality Control Diagram of Rock Bolted
　　　　Crane Beam

Item	Lu Buge		Guang Zhou	
	Old Plan	New Plan	Old Plan	New Plan
top width of crane beam Lu	1.75m	1.75m	1.60m	1.60m
bottom width of crane beam Ld	1.25m	1.25m	1.10m	1.10
head height of crane beam h	1.60m	0.80m	1.80m	1.00m
inclined angle of beam bottom face γ	26° 34′	45°	24° 27′	45°
dip angle of rock wall β	20°	27°	20°	25°
bolt inclination of upper row α	25°	5°	25°	10°
bolt inclination of lower row α	20°	0°	20°	4°
bolt tension of upper row Su	105.9	98.14	169.5	148.83
bolt tension of lower row S	85.9	75.93	134.1	116.95
ratio f=V/N	0.8	0.768	0.73	0.764
sectional area of crane beam	3.512m	3.311m	3.810m	3.261
ratio of saved concrete	0	6%	0	17.1%

carried out in April, 30, 1991. The crane test-
load was increased by steps of 300t, 400t, 480t
and 560t. Supervision of rock bolted crane beam
was done by bolt stress meters and crack meters
which had been installed in advance. Reaction of
these meters in test was very slight, so rock
bolted crane beam is reliable on quality and
works normally.

Hydropower'92, Broch & Lysne (eds) © 1992 Balkema, Rotterdam. ISBN 90 5410 054 0

Optimum design of Fen-He flood discharge tunnel

Han Shanru & Shen Guanda
Hydraulic Resources Ministry of Shanxi Province, People's Republic of China

Gu Zhaoqi & Han Zhihong
Hydraulic Engineering Department of Tsinghua University, Beijing, People's Republic of China

ABSTRACT: In order to save investment and shorten construction period, based on original design, optimum design has been carried out for Fen-He Flood Discharge Tunnel.This article will mainly introduce optimum designs of tunnel lining, intake, and other hydraulic structures, the effects of these optimum designs are also introduced.

1 Brief introduction of Fen-He reservoir and its flood discharge tunnel

Fen-He river is one of the main tributaries of Yellow River. Fen-He reservoir is situated at the distance of 98Km from TaiYuan city, the capital of Shan Xi Province. It is used mainly for flood control, water supply for industry and city, irrigation, and so on. Fen-He dam is earth-dropping-in-water dam with 61.4m high and 0.7 billion m³ storage capacity.

During thirty years, Fen-He reservoir has brought tremendous economic benefit. It was designed according to the flood of 100 years return period, and was checked according to the flood of 1000 years return period. But at present, because of silt accumulation, the silt has occupied 0.345 billion m³ storage capacity, and flood control standard has reduced to the flood of 300 years return period. So, in order to raise the flood control standard, slow down the speed of silting and prolong the life of reservoir, a flood dischange and silt washing-out tunnel with 791m³/s discharge was determined to be built.

The flood discharge tunnel is located in right bank, its total length is 1212m, diameter is 8 m, bottom slope is 1/76, surrounding rock is mica plagioclase gneiss, granite gneiss, quartz-hornblende schist. There are eight faults (F₂, F₃, F₄, etc.) oblique crossing tunnel line, joints can be divided into three groups: N50~70° W/SW<75~84° , N50~70° W/NE<15~40° , N10~15° W/SE<25° . Underground water is crevice water in bedrock. The water absorbing capacity of strong-weathered layer is 0.1~2.0 L/min · m · m, below strong-weathered layer is 0.01~0.1 L/min · m · m .Rock cover between strong-weathered line and tunnel top is thin, the thinnest section is only one times of tunnel diameter.

The construction of Fen-He tunnel was started

in September, 1989, and will be completed after three years. Total investment is 54 million Yuan. Its general layout see Fig. 1. It is asked to influence downstream water supply least during construction, and cost least, but the original design can not meet these demands, so optimum designs for tunnel lining, intake and other hydraulic structures are needed.

2 OPTIMUM DESIGN OF TUNNEL LINING

2.1 Original design of tunnel lining

Original design of tunnel lining was carried out according to the Chinese Code of hydraulic Tunnel. Not considering the primitive stress field of rock, surrounding rock was regarded as outer load for tunnel lining, so tunnel lining was designed thick and heavy, and consume too many reinforcing bars. While, the excavated tunnel section shows the surrounding rock is good, but rock cover is thin. At the range of 100m around 0+900m station, there are only about ten bolts installed in tunnel top, other parts are not surpported completely, while the stability of surrounding rock is fine. All these means surrounding rock can maintain stability by itself, moreover, geologist think most part of surrounding rock is class II or III. According to the Chinese Code, if surrounding rock is class II or III, tunnel lining can be designed by FEM.

2.2 Optimum design of tunnel lining

1. Model and Method
In FEM design, thirteen typical sections is selected along tunnel line (Fig. 2 shows one of these sections), in order to calculate lining stress according to actual cover and loads.

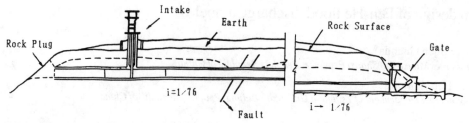

Fig. 1 General Layout of Fen-He Flood Discharge Tunnel.

Calculation region is up to strong-weathered line, rock mass upper strong-weathered line is regarded as outer load acted on boundary. In horizontal direction and beneath tunnel bottom, take five times of tunnel diameter as calculation region.

In order to simulate different thickness of lining and rock cover, A, B, C three types of net is divided, every unit is quadrilateral. Yield condition is Drucker-Prager Yield Criterion. Calculation is divided into three steps. First calculate primitive stress field, second calculate stress state after whole tunnel section excavated, third calculate lining stresses acted by inner and outer water pressure. For safety, take 50% of actual outer water pressure of operating period as calculating outer water pressure. In addition, a extreme condition of inner water pressure maximum, outer water pressure zero is also considered, moreover, the most dangerous condition of inner water pressure zero, outer water pressure maximum adding consolidation grouting pressure and back-fill grouting pressure is also calculated for checking whether lining stresses exceed the allowable compression strength of concrete.

2. Calculation schemes

According to the model and method mentioned above, nine calculation schemes have been selected, see Table 1.

3. Calculation results

Using FEM to calculate every scheme mentioned above, the lining stress can be obtained, so moment M and axle force N of whole cross section can be evaluated. Then according to maximum M and N, reinforcement calculation and check calculation can be carried out.

In the design of tunnel lining, two methods have been used. One is taken from Chinese Code "Reinforced Concrete Hydraulic Structure Design Code SDJ 20-78". This method considers tunnel lining as a eccentric compression member, permits maximum crack 0.25mm wide, that is $[\delta_{fmax}] = 0.25mm$. Calculation results see Table 2.

Another method is based upon the Criterion of Crack Permit Width established by American Concrete Institute No. 224 Committee. Usually the

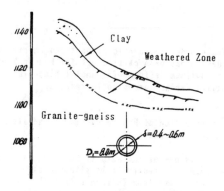

Fig. 2 Profile of Tunnel at 0+830. Um Station.

amount of steel calculated by ACI (224R-80) is far less than that of by SDJ 20-78, but when E=130 MPa, ACI(224R-80)'s calculation result is bigger. For safety, the steel bar of most part of tunnel adopt the calculation results of SDJ 20-78, but for 0+100.2～0+123.5m tunnel section, rock condition is bad, E=130MPa, steel bar adopt ACI's result.

2.3 Comparison of original design and optimum design

The economic benefits of lining optimum design are enormous, it will save tunnel excavated volume 5795.5m³, concrete volume 4631m³, reinforcement amount 738.56t. This shows: according to Code SD134-84 and SDJ20-78, using structural mechanics method to design tunnel lining is a conservative method, it estimates the bearing capability of surrounding rock too low, and regards rock as outer load acted on lining, these concepts are discrepant to real conditions. So new concept of tunnel design should spread vigorously in our country, fully trust and utilize the rock capability of bearing loads. If rock condition is good, concrete lining only take part in a smooth effect to reduce roughness factor, this will save investment

Tabel 1. Calculation Schemes.

scheme No.	lining thickness cm	elastic modulus E MPa	cohesive force C MPa	friction angle φ	station	inner water pressure Pi m	50%outer water pressure Po m
I	40	6000	0.1	40	0+040.2~0+100.2	44.32	22.16
II	80	130	0.01	30	0+100.2~0+123.6	44.62	22.31
III	40	2100	0.1	40	0+123.6~0+315.0	47.14	23.10
IV	40	650	0.01	30	0+315.0~0+330.0	47.34	26.18
V	40	8100	0.2	50	0+330.0~0+600.0	54.05	22.69
VI	40	9000	0.2	50	0+330.0~0+600.0	54.05	0
VII	40	12000	0.25	50	0+600.0~0+815.0	54.05	0
VIII	60	3000	0.1	40	0+815.0~0+845.0	54.05	0
IX	40	2100	0.01	30	Po=31.7m, Pi=0 backfill grouting pressure 20 T/m consolidation grouting pressure 40T/m		

Note: (1) Extreme condition IX is calculated for checking whether lining stress exceed the allowable compression strength of concrete.
(2) VII corresponds to the excavated tunnel section.

greatly and speed up construction. The lining optimum design has been adopted in some tunnel sections with good rock conditions.

3 OPTIMUM DESIGN OF INTAKE

3.1 Original Design
Intake section can be divided into diversion channel section and intake tower section. Diversion channel is 57.65m long and 20m to 30m wide, there is a rock ridge in this section, and a coffer is built on this rock ridge, when whole project completed, coffer will be dismantled by explosion. Intake tower section is 49.6m long, has two orifices 3.2×8.0m, two repairing gates 3.2×8.0m will be installed in gate vault, repairing bench is at 1131.4m elevation, gate house with hoister 2−1×3600 KN is at 1145.6m elevation. The lower part of intake tower is vertical shaft, the upper part is frame tower. See Fig.3 (a).

3.2 Optimum design
A big casing has been used during excavating shaft, because of the fissured surrounding rock, the leakage of shaft is more than 20L/s, and this result in difficult construction. In order to reduce leakage, reinforced concrete lining had

to be used in middle section of shaft.
Excavating division channel is limited by the water level of reservoir in flood season, this means division channel can be excavated only from April, 1 to June, 30 every year, so the completed date of this project will be postponed. The maximum excavating depth is 24m, to complete this excavating task needs drawing off reservoir for four times, so original design not only will increase difficulty of construction, but also will break off water supply for industry, agriculture and city. In optimum design, rock ridge is replaced by rock plug, diversion channel is replaced by tunnel and two orifices replaced by one.

Through optimizing, not only the abandoned water is reduced to minimum and reservoir storage capacity is increased at least 0.1 billion m³ per year, but also the construction is made easy. Intake layout of optimum design:

Upperstream of tower is square transition section with 3.5m wide middle pier and two orifices, each flow section is 3.2×8.0m, two orifices gradual merged into one, flow section is 7.0×8.0m. Before this section is square round transition section, flow section is transited from 7.0×8.0m to φ8m, lining thickness is

105

Tabel 2. Reinforcement Calculation Results
of Tunnel Linning (by SDJ 20-78).

scheme No.	lining thickness	N [T]	M [T-M]	σ lmax [MPa]	[σ lmax] [MPa]	steel bar	δ fmax [mm]
I	40cm	28.44	0.171	0.735	1.599	i Φ16@25	no crack
II	80cm	75.76	6.580	1.458	1.280	b Φ25@25	0.22
III	40cm	56.72	0.575	1.520	1.280	i Φ25@16.7	0.21
IV	40cm	73.28	-0.601	1.878	1.280	i Φ25@25	0.237
V	40cm	33.16	0.201	0.889	1.535	i Φ16@25	no crack
VI	40cm	43.24	0.235	1.169	1.398	i Φ16@25	no crack
VII	40cm	42.02	0.239	1.119	1.415	i Φ16@25	no crack
VIII	60cm	54.08	0.273	1.361	1.280	i Φ25@16.7	0.21

Note: (1) δ fmax is calculated width of maximum crack.
      (2) σ lmax is calculated value of maximum tension, [σ lmax] is allowable
          tension stress of concrete, if [σ lmax]>σ lmax, concrete not crack.
      (3) Mark "i" means reinforcing bars are installed in inner side of lining,
          mark "b" means reinforcing bars are installed in both side.

0.6~1.0m. Silt thickness above centre line of
rock plug is about 10m, the elevation of entrance
bottom after rock plug exploded is asked 1088m.
Intake layout of optimum design see Fig.3 (b).

For surrounding rock of rock plug, the thickness of strong-weathered layer is 4m, weak-
weathered layer is 5m, surrounding rock is
plagioclase-hornblende schist, strike of layer
schistosity is N45° E, nearly vertical, direction
of tunnel axis is N10° W. Surrounding rock is
rather broken, fissures interval is 0.2m, rock
cover of intake tunnel section is about 17m,
underground water level is same as reservoir.

The load component for intake tunnel section
is depended upon construction period. Tunnel
lining is regarded as eccentric compression
member, reinforcing bar is arranged in both side,
reinforcement is Φ16@15cm.

According to geological conditions and structural layout of intake tunnel section, following
measures have been taken in construction.

1. Pre-grouting

The grouting hole line is perpendicular to the
strike of rock schistosity, means hole line is
arranged at an angle of 35° to tunnel line,
grouting holes incline to SE direction, crossing
vertical line at an angle of 15°. The hole depth
is 28~31m, three rows is drilled with hole
interval 6m. From top to bottom, hole is
separated into three sections to grouting,
grouting pressure is 0.4MPa, unit water absorbing
capacity is asked that ω≤0.05L/min·m·m. After
grouting, water leakage is reduced obviously,
when water level is 1113m, leakage of intake

section is less than 3.1L/s.

In order to increase the entire stability of
surrounding rock, besides pre-grouting, two
reinforcing bars of Φ25mm are inserted in
every hole of the middle row from bedrock
surface to tunnel top, these reinforcing bars
play the role of long bolts and reinforced the
top of tunnel. So the effect of per-grouting
and long bolts is remarkable.

2. Partial excavating with upper guide tunnel

Cross section of guide tunnel is 2.5×2.5m,
footage of per cicle is 1~2.5m, advancing bolts
should have been installed before excavating.
Bolts at top of tunnel are Φ25mm reinforcing
bars installed at an angle of 45° to horizontal
line, its length is 3.2m, intreval distance is
1m and row interval is 1.2m. In sectors with
bigger excavated span, steel arch supports are
installed, type of steel support is parabola
arch with three articulations.

The economical and social benefit of intake
optimum design is obvious, it notonly saved rock
excavated volume 45200m³ and concrete 4500m³,
but also avoided drawing off reservoir and
breaking off water supply.

4 OPTIMUM DESIGN OF OTHER STRUCTURES

Repairing gate shaft of intake is also optimized,
because this shaft is a compressed structure
mainly bearing out water pressure, its lining
thickness and reinforcement amount can be
reduced greatly through optimum design.

106

Tabel 3. Comparison of Original Design and
Optimum Design.

station	Original Design			Optimum Design		
	lining thickness (cm)	steel bar (per meter)	steel amount (t)	lining thickness (cm)	steel bar (per meter)	steel amount (t)
0+040.2~0+100.2	60	b 5Φ22	67.39	40	i 6Φ25	39.56
0+100.2~0+123.5	120	b 8Φ25	53.44	80	b 6Φ28	42.24
0+123.5~0+315.0	60	b 5Φ25	276.30	40	i 6Φ25	125.93
0+315.0~0+330.0	80	b 8Φ25		60	i 4Φ25	14.11
0+330.0~0+605.0	60	b 5Φ22	561.76	40	i 4Φ16	
0+605.0~0+815.0	60	b 5Φ22		40	i 4Φ16	83.83
0+815.0~0+845.0	60	b 8Φ25	61.75	60*	i 4Φ25	13.26
0+845.0~0+860.0	80	b 8Φ25	39.24	80*	b 5Φ25	17.66
0+860.0~0+930.0	80	b 8Φ25		60*	b 4Φ25	
0+930.0~0+980.0	60	b 8Φ25	247.00	60*	b 8Φ25	157.54
0+980.0~1+010.0	80	b 8Φ25	93.17	80*	b 8Φ25	93.17

Note: (1) Mark "*" means this tunnel section has been excavated, its lining thickness adopt original design.
   (2) Mark "i" means reinforcing bars are installed in inner side of lining, mark "b" means reinforcing bars are installed in both side.

For exit section, the open-cut volume of original design is 78000m³, but in construction, tunnel was prolonged and exit was advanced, this measure made the open-cut volume reduced to 45000m³, and shortened construction period greatly.

Exit radial gate is 7.0m wide, 6.5m high, maximum thrust is 3200t. The join pattern of bearing beam in original design is difficult for construction, and consumes too much steel. Optimum design changed the original join pattern, made construction easy and safe.

reservoir and breaking off water supply during construction period. Sum total, optimum design will save rock open-cut volume 76000m³, tunnel excavating volume 5800m³, concrete 8000m³, steel and reinforcement amount 900t. Adding up all, about 9.15 billion Yuan(RMB) will be saved, account for 17% of total investment, so benefit of optimum design is evident.

5 CONCLUSION

Using FEM designing tunnel lining is safe and economic, and conform to real load state. If rock condition is good and rock cover is enough, tunnel lining only plays the role of smooth, if rock condition is bad and rock cover is not enough, concrete volume and reinforcement amount also can be saved by fully utilizing bear capability of surrounding rock.

Intake and other structure optimum design not only save investment, but also avoid drawing off

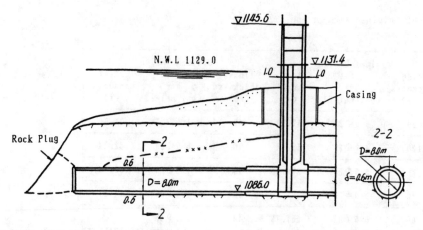

Fig. 3(a) Optimized Intake.

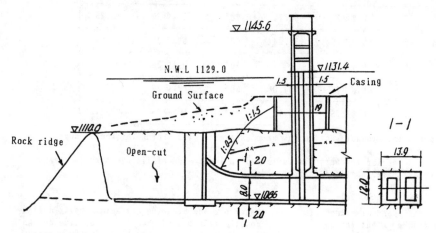

Fig. 3(b) Original Intake.

Hydropower'92, Broch & Lysne (eds) © 1992 Balkema, Rotterdam. ISBN 90 5410 054 0

Reducing uncertainty in underground work for the Snoqualmie Falls Hydroelectric Project

Kim de Rubertis – *Cashmere, Wash., USA*

Michael J. Haynes – *Puget Sound Power & Light Company, Bellevue, Wash., USA*

Robert D. King – *HDR Engineering Incorporated, Bellevue, Wash., USA*

F. Patrick Seabeck – *Converse Consultants NW, Seattle, Wash., USA*

ABSTRACT: The proposed expansion of the Snoqualmie Falls Hydroelectric Project will require excavation of an approximately 22-foot diameter shaft and tunnel in rock. Such excavations often encounter unanticipated geologic or hydrologic conditions which pose significant safety, construction, budget and schedule challenges. Reducing uncertainties and the associated risks for underground work is a goal which must be considered from the earliest stages of design through construction.

1 INTRODUCTION

The proposed expansion of the Snoqualmie Falls Hydroelectric Project (Project) results from the upcoming expiration of an existing federal license. The process required to relicense a project under the Federal Energy Regulatory Commission (FERC) guidelines involves lengthy agency consultation and negotiation, conceptual and preliminary design and assembly of a license application. In late 1990, Puget Sound Power & Light Company (Puget) was ready to develop the preliminary design and a license application. To accomplish these efforts, HDR Engineering, Inc. (HDR) was retained to assist Puget with design and engineering for Plant 1 modifications and Plant 2 expansion. HDR subcontracted with geotechnical consultants to conduct a two phase program to develop the selected preliminary design. The design was completed and a license application submitted in November 1991.

1.1 Project history

The Project is located in Western Washington state, USA on the Snoqualmie River at the 268-foot high Snoqualmie Falls. The existing Project consists of two plants with seven turbines. Plant 1 and Plant 2 went on line in 1898 and 1910, respectively. At the time of construction, Plant 1 was one of the world's first totally underground powerhouses. In 1905, Plant 1 capacity was doubled to meet growing power demands. Plant 2 was enlarged to its present capacity by addition of a second turbine in 1953.

1.2 Existing project features

The existing run-of-the-river Project (42-MW) consists of two plants with the intakes and powerhouses located on opposite banks of the river. All the original principal project features remain in operation today. Diversion for both plants is accomplished by a broad crested concrete dam located immediately upstream of the Falls. Plant 1 (12-MW) was constructed in a rock excavated cavity 280 feet underground and is supplied water via two 8-foot diameter rivetted steel penstocks. Plant 1 houses four 6-wheel, 2-jet/wheel impulse turbines and a single horizontal Francis turbine. The Plant 1 tailrace discharges into the pool located at the base of the Falls.

Plant 2 (30-MW) is an above ground powerhouse. Water is conveyed through a shallow 12-foot diameter concrete lined tunnel, an open gated forebay hewn in rock, a 7-foot diameter rivetted steel penstock and a 10-foot diameter steel penstock (Figure 1). Plant 2 houses one dual discharge horizontal Francis turbine and one vertical Francis turbine. The Plant 2 tailrace discharges into the river about 2,200 feet downstream of the Falls.

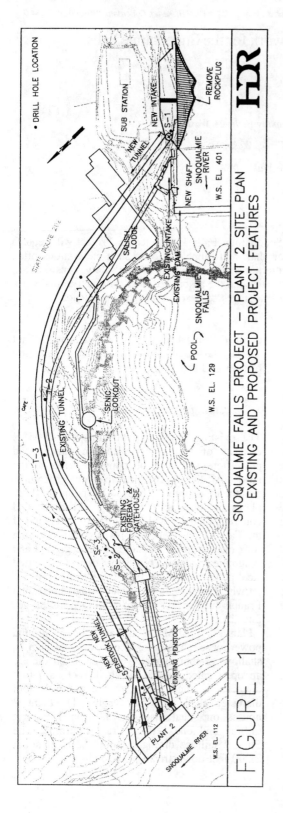

DRILL HOLE LOCATION

HDR

NEW INTAKE

REMOVE
ROCKPLUG

SUB STATION

S-1

NEW
TUNNEL

NEW SHAFT

SNOQUALMIE
RIVER

W.S. EL. 401

SALISH
LODGE

EXISTING INTAKE

STATE ROUTE 202

T-1

EXISTING DAM

POOL

SNOQUALMIE
FALLS

W.S. EL. 129

T-2

EXISTING TUNNEL

SENIC
LOOKOUT

T-3

EXISTING
FOREBAY &
GATEHOUSE

S-3

S-2

NEW
NEW TUNNEL

S-5 PENSTOCK

T-4

EXISTING PENSTOCK

PLANT 2

SNOQUALMIE RIVER

W.S. EL. 112

FIGURE 1

SNOQUALMIE FALLS PROJECT — PLANT 2 SITE PLAN
EXISTING AND PROPOSED PROJECT FEATURES

1.3 Proposed project features

Analyses revealed the optimal arrangement will be to modify Plant 1 by retiring the four oldest turbine units and replacing the horizontal Francis turbine and rivetted steel penstock. Modifications to Plant 2 will be more extensive including: a three bay intake to supply 3,100 cfs, a 20-foot diameter vertical intake shaft 180 feet in height, an 18-foot diameter 1,800-foot long horseshoe shape tunnel, and steel penstocks to convey water to the two existing and one new Francis turbines and a new 2,500 cfs synchronous bypass valve. The proposed Project capacity will increase to 74-MW when complete.

2 CONCEPTUAL DESIGN AND THE GEOTECHNICAL PROGRAM

Geotechnical efforts were initiated at the same time as conceptual design. While conceptual design focused on selecting an optimal flow, turbine size and water conveyance system based on project economics, geotechnical efforts focused on assembling available information, conducting preliminary field reconnaissance and developing siting recommendations for proposed project features.

At this early point in the project, it was important to assemble an initial body of geological data on which to build during preliminary design efforts. Perhaps even more important, was the need to identify those areas critical to upcoming design efforts for which little or no geological information was available and those concerns which might require special expertise to address.

The geotechnical program was conducted to support the conceptual design efforts and was executed in three steps. First, background information was assembled. Available project specific information included original "as-built" drawings, construction records and photographs, inspection and maintenance records, previous geotechnical reports for existing facilities and interviews with knowledgeable Puget staff. Second, detailed surficial site investigations were conducted at specific existing project features and potential construction sites. The purpose of these surface reconnaissance efforts was to identify visible soil, bedrock and groundwater conditions. Third, limited analyses and recommendations were developed for specific proposed project

features for Plant 2 and for general construction issues. During these efforts it was vital for the engineering and geotechnical team to meet regularly and discuss the evolving designs and the growing body of geotechnical knowledge. Anticipated geological conditions influenced the siting and arrangement of proposed conceptual designs and as conceptual designs were relocated or modified, the geological conditions and concerns changed.

As geotechnical efforts progressed, one specific concern relating to tunnel and shaft construction became increasingly important. The Project site is the most heavily frequented recreational site in the state of Washington. Over 1.5 million visitors annually frequent the Project park overlooking the Falls. The park at the Project site has a trail and series of viewing points, including an overhanging lookout located along the cliff precipice. The 800-foot long trail begins at the Salish Lodge and parallels the precipice at a distance of about 20 feet. The Lodge is a premier five-star hotel located immediately beside the Falls on the right bank, set back about 50 feet from the precipice.

In the past, large segments of rock had occasionally broken off the precipice. The concern was the potential impact of conventional blasting and the resulting vibration on precipice stability and guests staying at the Lodge. At this point in the project, it became apparent that specialized expertise in tunneling and blasting would be of great value to Project design efforts to assist in addressing these concerns.

At the conclusion of conceptual design, the design team believed that the proposed features for Plant 2 expansion could be constructed using conventional techniques for an acceptable cost based on the engineering and geotechnical data developed at that time. An exploration program was developed to provide data to answer the questions of rock type and quality, preferred excavation methods, anticipated groundwater conditions, temporary and permanent support requirements and lining requirements. Also, a number of significant uncertainties had been identified. The resolution of which were essential to preliminary design of the Project. Specific uncertainties associated with underground construction which required resolution included:

1 Construction costs. Rock quality, excavation method(s), ground water conditions and seepage potential, support requirements and lining

requirements would greatly influence shaft and tunnel cost estimates.

2 Vibration control. Ensuring vibration could be held to acceptable levels at reasonable cost during construction was both a public safety and guest nuisance concern which had to be resolved.

3 Portal location. Determining the geologic conditions and rock quality at the tunnel portal was needed. Initial siting of the portal had been based on limited topographical and geotechnical information. Limited data was available and the design team was not comfortable with the high level of uncertainty which existed for siting this critical feature.

3 PRELIMINARY DESIGN AND THE GEOTECHNICAL PROGRAM

Preliminary design was executed concurrent with geotechnical explorations. Changes to the preliminary design were made during and after execution of the geotechnical exploration program. Preliminary design efforts focused on refinement of designs and criteria, conducting technical analyses, writing technical reports and developing a complete set of functional drawings. The geotechnical efforts consisted of defining a field exploration program, bidding and award of the drilling contract, employing tunnel and blasting experts, executing and monitoring the field exploration program, developing geotechnical designs and preparing a comprehensive geotechnical report. Throughout the duration of these efforts, the engineering and geotechnical team met regularly to exchange information, modify designs based on acquired data and redirect geotechnical efforts.

3.1 Regional and Project Geology

The Project is located on the main stem of the Snoqualmie River in the Puget Lowland physiographic province. The terrain in the immediate vicinity of the Project is hilly with elevations ranging from about 100 to 500 feet above sea level. To the east and west of the Project, the Snoqualmie River flows over relatively flat floodplain. At Snoqualmie Falls, the river drops 268 feet over a nearly vertical bedrock cliff. Downstream of the Falls the river has incised a steep walled canyon for 1,200 feet.

Glaciation has exerted a profound influence in shaping the present geologic environment of the Puget Lowland. Thick vegetation and the recent glacial deposits obscure bedrock in most areas with the exception of the cliffs immediately adjacent to the river and isolated outcrops.

The Project area is underlain by a complex of volcanic rocks known as the Mount Persis Volcanics which were deposited during Middle to Late Eocene time (about 38 to 47 million years ago) and have an aggregate thickness of up to about 7,500 feet. Bedrock at the Project site consists of a complex interlayered sequence of andesitic to basaltic flows, breccias and pyroclastic deposits. It is believed that these bedrock units were deposited in close proximity to a volcano. Contacts between volcanic units typically are irregular and in many cases identifiable units can be observed to show large changes in thickness over short horizontal distances. Field explorations revealed that andesitic breccias form about 60 percent of the total rock mass, basaltic andesite flows form about 35 percent and the 5 percent balance is comprised of pyroclastic deposits including laharic breccias, lapilli tuffs and welded tuffs.

Bedrock in the Project vicinity has been folded and faulted in response to several episodes of tectonic deformation during the geologic past. Structures in the region resulting from tectonic deformation include joints, shears and faults. Joints are common in each rock unit found at the Project site. Small scale shears related to brecciation during emplacement of volcanic flows and past tectonic deformation are common at the Project site throughout the basaltic andesite. Shears are less common in the andesitic breccia units. A major bedrock shear was encountered downhill of the proposed portal location. However, this shear will lie outside of any proposed underground structures. Faults have not been found or mapped at the Project site.

Groundwater is present in the bedrock discontinuities. The groundwater table in the areas of the proposed shaft and tunnel lies roughly at or within 90 feet above the tunnel crown. This groundwater table elevation is variable because groundwater is generally confined to joint openings, many of which are not well interconnected. The rock itself is essentially impermeable.

Regional seismicity is dominated by earthquake events related to the movement of large segments of the earth's crust along fault zones. Of the six largest earthquakes that have occurred in the region, estimates conclude that three have resulted in Modified Mercalli Intensities of VII at the Project site. A Unified Building Code designation of Seismic Zone 3 was used for preliminary design purposes.

3.2 Summary of explorations

Geologic mapping of the Plant 1 cavity and tailrace and of the general Project area was completed at a scale of 1" = 100'. Eight drill holes were completed, three at the shafts and five along the tunnel alignment. Rock coring was completed with wireline coring equipment. Pneumatic packer pressure tests were performed and standpipe piezometers were installed in Drill Holes S-1, S-3, T-1, T-2, and T-3. Laboratory testing of core samples included point load testing, unconfined uniaxial compression tests, Los Angeles abrasion tests, and specific gravity determinations.

3.3 Rock units

Four distinct rock units were encountered at the site. Light to dark gray or black basaltic andesite was encountered in all of the tunnel and shaft borings. It is present as a densely crystalline, volcanic flow rock which is generally hard and closely to extremely closely jointed. Almost all joints are infilled and rehealed with dark green chlorite which can be commonly broken by hand. Many joint and fracture surfaces are polished and slickensided. Occasional soil infilled joints were observed. Extensive weathered zones were not encountered.

Maroon, brown or gray andesitic breccia was also encountered in all of the tunnel and shaft borings. The andesitic breccia as a whole is usually medium hard and medium to widely jointed. Some joint planes are infilled with chlorite but, many appear to be clean. Sheared and slickensided joint planes are not as common in the andesitic breccia core as they are in the basaltic andesite. Together, the andesitic breccia and the basaltic andesite comprise the majority of the bedrock in the Project area.

Green and red andesite was distinguished from basaltic andesite by coarser and more numerous phenocrysts. The Andesite was also encountered in all of the tunnel and shaft borings. Joints in the andesite were present at a wider spacing and showed fewer highly polished joint surfaces.

Dark colored laharic breccia was present only in the borings located in the downstream end of the tunnel alignment. The laharic breccia is interbedded with pyroclastic rocks including tuff breccias, welded tuffs and lapilli tuffs. The laharic breccia is extremely closely to medium jointed. Some core samples tended to quickly disintegrate into small fragments when placed in water, particularly if mechanically disturbed. The tuff is soft and is closely to medium jointed.

3.4 Rock structure

The rock units occur as interlayered lava flows, volcanic breccias and pyroclastic deposits that have been folded and tilted. In general, the volcanic flows dip upstream. In the field, contacts between adjacent rock units tend to be irregular rather than planer. In the core, contacts are generally sharp and well defined. There are at least four sets of joints, all steeply dipping, that can be observed in the field and in the core. The most prominent joint set is oriented subparallel to the upstream half of the proposed tunnel and normal to the downstream half. Where subparallel, the joint set may control sidewall overbreak. This set of joints is also the most prominent and continuously exposed in the Plant 1 tailrace tunnel.

3.5 Groundwater and seepage potential

Information pertaining to groundwater conditions likely to be encountered in the shaft and tunnel were derived from five sources. These are loss of drill water during explorations, packer testing, installed piezometer readings, observation of seeps from the cliff face and historical seepage in the existing Project features.

During drilling, circulation water was totally lost in five of the eight drill holes, deep in the holes. No circulation loss occurred in only two holes. When a drill hole intersected key open joints, circulation was immediately lost and usually not recovered for the remainder of the drill hole. Five drill holes were water pressure tested. Pressure test results generally reveal that the rock mass exhibits very low to moderate fracture permeability. Most joints appear to be relatively tight. Some high angle joints are open or partially open.

Piezometers consisting of 1-inch-diameter PVC standpipes were installed in five drill holes. Following bailing of the piezometers, water levels were measured. These levels were considered to be at or near static for the time of measurement and are used for preliminary design. Groundwater levels will vary over time in response to precipitation and river level changes. Only a few small seeps were observed in the cliffs along the south side of the proposed tunnel alignment. However, significant unobservable seepage may be exiting the rock through talus and colluvium at the cliff base. Only minor localized seepage was noted in the Plant 1 tailrace tunnel.

It was concluded that groundwater inflows will not seriously affect construction of the tunnel or shaft. A total tunnel inflow of less than 1,000 gpm is anticipated during construction. This preliminary estimate is based on the following:

1 Deep base water levels in the canyon will tend to lower groundwater in adjacent rock masses.

2 Relatively impermeable glacial soils overlie bedrock in most areas near the tunnel and shaft and limit the amount of direct recharge to joints in the bedrock.

3 The explorations and water testing indicate low bedrock permeabilities. The few key joints having high permeabilities are not likely to be hydraulically connected to a steady source of recharge and therefore, will probably show decreased flow rates over time.

4 Historical performance of the Plant 1 cavity and tailrace tunnel, which are essentially dry.

5 Results of the standpipe piezometer measurements. The piezometers indicate hydraulic heads over the tunnel ranging from about 20 to 90 feet.

3.6 Mechanical properties

The best evidence of the strength of the rock masses through which the tunnel and shaft will penetrate is found in the performance of the existing Plant 1 tailrace tunnel and cavity, the natural slopes in the valley walls and the Rock Quality Designation (RQD) of the boring cores.

The performance of the unsupported, unlined Plant 1 tailrace tunnel has been excellent since construction in 1898. The tunnel was excavated chiefly in andesitic breccia, the highest quality rock unit in the project area. Observation of the

tunnel in a "dry" condition reveals only a few localized areas of rockfall from the crown. Some loosened blocks of rock were observed in the tailrace tunnel walls, but the majority of joints are essentially tight or rehealed, showing rock-to-rock contact. These observations indicate that, in the andesitic breccia unit, some loosening of tunnel rock and rockfall in an unlined tunnel would probably occur.

The existing Plant 1 cavity has an unsupported span of approximately 200 feet by 40 feet and has served without any significant problems related to rock instability or water infiltration since its completion. The cavity walls are painted so rock types cannot be directly observed. Based on the jointing patterns, it is believed that andesitic breccia is predominant.

The natural rock slopes of the valley walls range from vertical or slightly overhanging to about 70° from horizontal. Although occasional rock falls occur in these cliffs due to the mechanical actions of freeze thaw cycles and flowing water, the natural slopes exhibit hard and durable rock qualities.

The cores taken at the tunnel, shaft and portal borings revealed average RQD values for each rock unit as follows. (The lengths of the various rock types that will be encountered in the shaft and tunnel are shown in parentheses and will vary from those shown because of the complex interlayering of the volcanic rock. Lengths were estimated for purposes of determining costs associated with excavation, support, and lining.)

	Shaft	Tunnel	Portal
Basaltic Andesite	39 (62)	41 (386)	41
Andesite	73 (64)	81 (217)	N/A
Andesitic Breccia	84 (136)	87 (716)	N/A

3.7 Support Requirements

To determine support requirements, the Q-system was selected as the design basis. As a basis for checking Q values computed for core samples, the Q value for known conditions in the Plant 1 features were evaluated. The rock quality in the existing features is excellent and it should be

114

noted that these features were excavated predominantly in andesitic breccia, which is considered to be the most competent rock at the Project site. Q values were computed assuming likely values for the Q-system parameters easiest to define: $RQD=90$-100, $Jn=4$, $Jw=1$, $SRF=1$-2 and applying a range of values for the term $Jr/Ja=0.9$ to 4 (various combinations of joint roughness and alteration will produce this range of values). Based on the selected variables, Q values of between 10 and 100 were calculated in the andesitic breccia of the existing Plant 1 openings. Q values in this range indicate that little or no support will be required. Since this is in fact, the actual condition in the Plant 1 openings, the use of the Q system for design was employed with confidence.

The mechanical properties of the rock masses which will be encountered in the tunnel and shaft appear to vary with stratigraphy. Values for the Q-system variables for each of the generalized rock units were developed from the core. Selected Q values for permanent support were computed as follows. Basaltic Andesite $Q=1.1$, Andesite $Q=13.4$ and Andesitic Breccia $Q=28.7$. These values were modified for portals, tunnel intersections, and prominent discontinuities. For temporary support, Q was multiplied by 5.

Based on these computed Q values, the basic strategy for ground support consists of temporary support using rock bolts and shotcrete during construction followed by unreinforced, cast-in-place concrete lining the full length of the shaft and tunnel with steel lining for about 200 feet upstream from the portal. Rock bolting will be necessary in limited areas as the excavation proceeds (spot bolting), and as designated reinforcement (pattern bolting). For preliminary design, 1-inch diameter (#8), fully resin encapsulated, untensioned bolts are assumed. Minimal thicknesses are proposed for both unreinforced concrete lining (12-inches thick) and steel liner (3/8-inches thick). Steel liner thickness will be based on handling requirements. Grouting of the liner contact will be required.

3.8 Excavation methods

The tunnel can be driven using conventional drilling, blasting, and mucking techniques. Due to the relatively short tunnel length, it is expected that this is the method that will be employed.

Hydraulic drills should perform satisfactorily. It is anticipated that a four drill jumbo with a separate drill for burn holes will be used for a full heading approach and a two drill jumbo and one track drill combination for the top heading and bench technique. One bypass will probably be required to facilitate the mucking procedure. Since there will be only one opening to the area, there will be no natural ventilation. Therefore, fans, ducts and deflectors will be required. Electric power and water will be required.

As an alternative, a tunnel boring machine (TBM) could be used to drive the tunnel. Employment of a TBM would depend primarily on equipment availability at the time of bidding and other economic factors. There would be certain performance benefits if this method could be used. Vibrations at the cliff face and at the Lodge would be minimized. The tunnel cross section would be more uniform and would require fewer temporary supports. Groundwater drainage will be toward the tunnel portal with no pumping required. Water will be directed into a settling pond and oil/water separator. Water quality testing will be necessary to reduce the potential of contaminants being introduced to the river.

There are four feasible methods of excavating the large diameter shaft: (1) raise boring from the bottom up and then slashing by drill and blast methods to full diameter, (2) vertical crater retreat (VCR) blasting utilizing full depth blast holes, (3) conventional sinking using drill and blast techniques, and (4) conventional raising using a raise climber and drill and blast techniques. The raise bore and slashing technique appears to be the most attractive alternative. The tunnel would have to be driven full length prior to commencement of the shaft construction. Blasting vibrations could be controlled without difficulty. The VCR technique may offer some significant cost advantages, providing the drilling can be accurately controlled. Modifications can be implemented near the surface to avoid unacceptable ground vibration effects. Conventional shaft sinking would not require completion of the tunnel prior to shaft construction. However, the time required and the mucking would be at relatively high cost. Due to fragmentation requirements, vibration levels could be expected to be relatively high. A shaft of this diameter would create problems for full diameter conventional raising but possibly a combination of raising and subsequent slashing might be economically feasible.

3.9 Excavation controls

Ground vibration effects from blasting operations are of concern with respect to the environment in and around the Lodge. Less critical, but still very important, are potential effects on the cliff face. The visitors facilities are not expected to experience any harmful effects. Monitoring will be performed in the existing forebay area when blasting is being first initiated. Monitoring with approved blasting seismographs at appropriate locations will be performed whenever blasting operations are in progress. Recordings taken above the tunnel during its progress upstream will provide background information vital to the design of later blasts accomplished in the Lodge area. A scaled distance versus peak particle velocity plot will be maintained to assist in blast design. The recommended maximum allowable peak particle velocity is 0.50 inch per second under any portion of the Lodge foundation. A 1.0 inch per second limit is recommended for other critical structures.

A general blasting plan will be required of the Contractor for review, comment, and approval. Separate plans will be required for each type of excavation or when blast design changes are contemplated. For each blast, regardless of location or size, a complete report will be submitted on the day of the blast. A single pay line will be established at the desired outside neat line of concrete. Excavation, repairs, and additions beyond this line will be at the Contractor's expense.

3.10 Preliminary design modifications

Based on Phase II geotechnical findings, several modifications were made to the preliminary design. The downstream portal was relocated about 20 feet vertically and 60 feet horizontally. Portal relocation provided increased rock cover above the portal, improved portal access, and potentially avoids the weaker laharic breccia rock unit entirely. The tunnel grade was raised the same 20 feet, thereby shortening the height of the intake shaft.

4 FUTURE GEOTECHNICAL WORK

Before final designs are completed, there are a number of uncertainties that require resolution. Additional exploration will resolve some of the uncertainties, but some conditions may not be completely resolved until construction.

4.1 Unresolved uncertainties

Safe limits for control of drilling and blasting have been proposed, and it is believed that these limits can be met without significant impact to unit prices for rock excavation. Construction control and flexibility in decision making will be required to achieve a proper balance between risk and progress. More information is required to finalize the design of temporary and permanent support for the tunnel, shaft and portal.

4.2 Additional exploration

At the intake shaft, two shallow drill holes are recommended to better define the character of the rock mass. At tunnel grade along the upstream half of the alignment, three deep drill holes are needed to supplement information on the stratigraphy, discontinuities and groundwater conditions. For the downstream half of the tunnel alignment and at the proposed tunnel portal, either an exploratory adit or closely spaced deep drill holes are recommended to gather data on stratigraphy, jointing and rock quality. The present investigation revealed that this area of the tunnel has the poorest quality rock and the least amount of rock cover over the crown of the tunnel. An exploratory adit is preferred because it will provide better information and will be available for inspection by prospective contractors. Alternatively, four or five deep drill holes could provide sufficient information for final design. The additional exploration, testing and analysis will be completed before final design is undertaken.

The objective of future geotechnical work will be to reduce the uncertainties for underground construction to a level which can be managed during construction, a level without major surprises about the conditions encountered and the rock behavior.

4.3 Construction monitoring

Careful monitoring of blasting will be required. Peak particle velocities and overpressures will be measured to insure that safe limits are not exceeded. Liaison will be maintained with the Lodge management to learn what adjustments can be made to the blasting program to reduce any impacts to Lodge operations. Underground construction will be carefully monitored to identify and resolve unforseen conditions and to record geologic conditions.

5 CONCLUSIONS

The program to reduce uncertainty in the Project's underground construction relies on conventional geotechnical engineering to progressively develop an understanding of how the ground will behave during construction and operation. At each step in the design process, additional data has been gathered to form the basis for analysis and application of judgement. The goal has been to reduce uncertainty at the time of construction to a level which does not effect budget and schedule. At Snoqualmie Falls, major problems in underground construction are not anticipated.

ACKNOWLEDGEMENTS

The authors gratefully acknowledge the assistance of Don Harris, blasting controls, and Dennis Lachel, underground excavation.

REFERENCES

de Rubertis, Kim, Seabeck, F. Patrick, 1992. Geotechnical report for Snoqualmie Falls Hydroelectric Project

Hydropower'92, Broch & Lysne (eds) © 1992 Balkema, Rotterdam. ISBN 90 5410 054 0

Design, construction and operational experience at the Torpa air cushion surge chamber

R.S. Kjølberg
Berdal Strømme a.s., Norway

R. Kleiven
Oppland County Energy Board, Norway

ABSTRACT: The Torpa Power Plant, one of the 2 power stations in the Dokka project, features an air cushion surge chamber. The total volume of this chamber is 17,400 m³ and the air volume 10-12,000 m³.

In the surge chamber area, the rock overburden is 225 m while the water head is 450 m.

To prevent air leakages, a water curtain above the chamber has been constructed. The water curtain pressure is between 2 to 5 bar higher than the chamber pressure. One of the main design criteria imposed, was that the curtain pressure must be less than the minimum principle stress in the rock, in order to prevent hydraulic fracturing. Extensive hydraulic fracturing tests have therefore been performed.

The air cushion surge chamber has been built in the shape of a doughnut. The water curtain holes were drilled from a "manifold" chamber above the surge chamber.

The installed air compressor capacity is 940 Nm³/h. Two different control systems for monitoring the water level in the chamber, are also installed. After 2 years in operation, tests indicate that there is no air leakage through the rock mass, so long as the water curtain excess pressure exceeds 2 bar.

INTRUDUCTION

The Oppland County Council Energy Board owns the Dokka power plant. The construction work started in 1986 and was finished by autumn 1989. The map, figure 1, shows the main layout of the project. The Lake Dokkfløy forms the main reservoir for the Dokka development with a storage capasity of 250 million m³.

The dam is built as a 680 m long and 80 m high rock-filled dam.

The headrace tunnel from the reservoir to the power plant is 9.2 km long and the static head is 470 m. The tributary Synna is taken in through a 2 km long tunnel. The lower part of the headrace tunnel is constructed as a pressure tunnel.

The tailrace tunnel from the Torpa power plant is 10 km long with the outlet in the Dokka river. Here a small rock-filled dam is built to form an intake reservoir for the Dokka power plant. The tributary Kjøljua also flows into this reservoir.

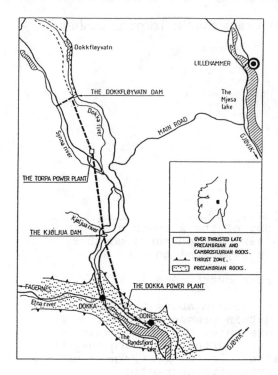

Fig. 1   The Dokka project and the
         main geological features.

The headrace tunnel to the
Dokka power plant is about
10 km long and the head is
130 m. From this power
station a 3.8 km tailrace
tunnel leads the water to
the Lake Randsfjorden.

The installations in the
power stations are:

Torpa power station
    :2x75MW=150 MW
Dokka power station
    :2x22 MW=44 MW

Total   194 MW

The production is 520 GWh in
an average year.

GEOLOGY

The project area is geologically located in the

socoalled "sparagmite-area"
in south eastern Norway. The
rock in this complex overlys
a precambrian gneiss and is
separated from this by a
thrust zone.

The Dokka power plant and
the trailrace tunnel are
located in the gneiss, while
the Torpa power plant and the
air cushion surge chamber is
located in the over thrusted
sparagmite complex, which
consistes of sandstone,
quartzite meta claystone and
limestone.

Due to the overthrusting
the rather hard and brittle
rock is broken and the
complex subjected to folding
and reverse faulting. Blocks
of rock are thrusted over
each other and separated by
tectonic zones. The dip
direction is northwards and
joints parallel to the
bedding, are the most
frequent. The air cushion
surge chamber is completely
located in meta claystone
between two reverse faults.

BACKGROUND FOR THE CHOICE OF
AN AIR CUSHION SURGE CHAMBER

From the cross section figure
2 it can be seen that the
rock overburden along the
lower part of the headrace
tunnel, is below the
pressure line. Above the air
cushion surge chamber, the
overburden is only 50 % of
the water head. An ordinary
surge shaft was therefore
not realistic and a closed
surge system had to be
evaluated.

When the planning of the
surge system started, 9 air
cushion surge chambers were
already in operation in
Norway. Figure 3 shows the
air loss from these chambers
versus the ratio of the
natural pore pressure in the
rock, to the air pressure in
the chamber.

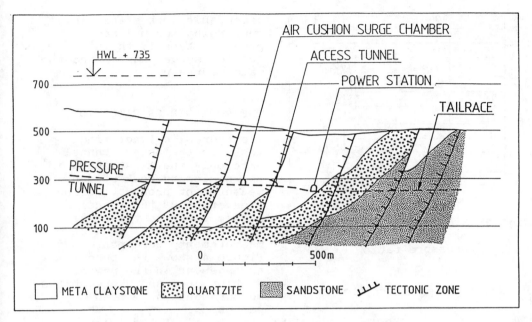

Fig. 2 Geological section along the Torpa tunnels

As can be noticed from the figure, the chamber of the Kvilldal power plant, which has a pore water/air pressure ratio of 0.6, had an unacceptable air loss, but after a water curtain was constructed and put into operation, the air loss was successfully reduced.

At the Torpa power plant systematic and heavy grouting around the chamber was originally planned to reduce and control the air loss. Later, when the positive effect of the water curtain at the Kvilldal chamber was reported, the plan was changed and it was decided to install a water curtain at Torpa as well.

At Torpa the topography limits the pore pressure to maximum 22.5 bar. The chamber pressure is close to 45 bar and hence the pressure ratio (ref. figure 3) is 0.5 or lower. The only existing chamber with such a low pressure ratio, was the Tafjord Chamber, where the air loss was reported to exceed what was acceptable. However, for this chamber no water curtain was installed.

The rock type at Torpa was found to be favourable. The meta claystone, in which the chamber is located, is normally a rock type with low permeability. Besides this, the rock stress situation was of vital importance. To prevent air leakage from the chamber a water excess pressure in the water curtain had to be applied. This excess pressure should not be allowed to exceed the minimum principal stress in the rock. If this was the case, a hydraulic fracturing of the rock would probably take place and cause uncontrollable air and water leakage from the chamber.

At this stage, the first rock stress measurements in the access tunnel to the power station had been performed. The hydraulic

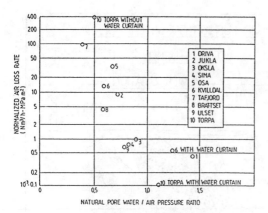

Fig. 3 **Air leakage as a function of pore pressure to air pressure ratio**

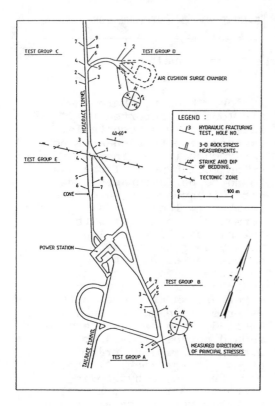

Fig. 4 **Location of stress measurements at the Torpa project**

splitting test results were favourable, showing an

acceptable rock stress level, concerning the unlined pressure tunnel, the air cushion surge chamber and the water curtain.

THE ROCK STRESS SITUATION

The hydraulic fracturing tests in the access tunnel were only the first steps in the necessary rock stress measurement process. During excavation of the tunnel system close to the surge chamber, stepwise rock stress measurements were performed. Figure 4 shows the tunnel arrangement in the power station area and where stress measurements have been performed.

In total 35 hydraulic fracturing tests, in 5 test groups have been performed. Additionally three-dimensional rock stress measurements, by the overcoring method, have also been performed in 2 core drilling holes. Besides the rock stress measurements, permeability tests were of vital interest.

RESULTS

The results from the tests in group A and B were decisive for the location of the power station and the pressure tunnel system connected to it. The shut in pressures measured in these groups varied from 5.2 to 11.9 MPa with a maximum water pressure in the tunnel of 4.5 MPa.

The measurements in test group C were performed before the location of the air cushion surge chamber was determined and founded the basis for the evaluation of the water curtain pressure. In this test group, the shut in pressure varied from 4.3 to 7.3 MPa.

In the chamber tunnel, test group D was performed. Some measurements in very long test-holes failed, but reliable tests showed shut in pressures of 4.5-5.0 MPa.

The measurements in test group E were performed to verify the stress situation in the sand trap area.

The calculated minimum principal stresses based on the 3-D overcoring tests showed considerable deviation from the shut in pressures, meassured in the hydraulic fracturing tests. The overcoring tests, also showed considerable deviation from the regional tendency for measurements of this type. On the other hand the hydraulic fracturing test resembles more the actual situation in the pressure tunnel and tests the rock mass to a significantly larger scale than the overcoring tests. Therefore the 3-D measurements were only regarded as a supplement to the hydraulic fracturing test results.

The lowest shut in pressure value in the pressure tunnel outside the surge chamber area was 4.3 MPa, while the lowest reliable value measured in the chamber was 4.45 MPa. The single values were considered to reflect local variations, while the global average of 5.7 MPa was used as the design basis for the chamber and the water curtain.

The air pressure in the chamber is 4.4 MPa, slightly lower than the lowest reliable shut in pressure value. The water curtain was designed to have a water pressure of 4.9 MPa, which means an excess pressure of 0.5 MPa compared to the air pressure in the surge chamber.

DESIGN OF THE AIR CUSHION SURGE CHAMBER

Figure 5 shows the original design of the chamber. The water curtain holes were planned to be drilled from one chamber, obliquely over the roof of the parallel chamber. A finite element analysis showed however, that this solution would lead to considerable air loss from the outside walls of the chambers and for this reason additional holes in these walls were necessary. The pipeline system required to feed all these holes with water, was found to be too complicated and also too exspensive. Besides, due to the risk of pipe failure (a pipe failure had occurred in the Kvilldal chamber and put the chamber out of operation for a long period), this solution was rejected. Then a solution, shown in figure 6 was chosen. The chamber has a doughnut shape. From a central shaft a pressure chamber was excavated 10 m above the roof of the main surge

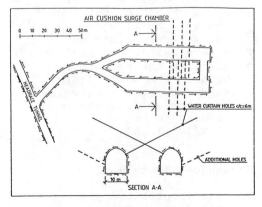

Figure 5. Preliminary surge chamber design

123

chamber. From the pressure
chamber, which has a 6 m
diameter, the water curtain
holes are drilled like the
spokes in a wheel over the
air surge chamber.

The shortest distance from
the chamber roof to the
holes is 7 m and the holes
are drilled down to a level
slightly lower than the
chamber roof.

A concrete plug has been
cast in the shaft to the
pressure chamber and the
water curtain holes are
pressurized with water from
a pipe through the concrete
plug.

The total volume of the
air cushion chamber is
17,400 m³ and the air volume
is between 10-12,000 m³.
During the excavation of the
chamber, systematic test
drilling; water loss
measurement and cement
grouting has been performed.
The water loss measurements
were performed at two
pressure steps, 20 and 50
bar, the latter corresponds
to the water curtain
pressure. Leakages exceeding
1 Lugeon have been cement
grouted.

Water loss measurements
have also been performed in
the water curtain holes. The
water losses were small.

The rock appeared to be
quite impermeable, but some
joints with small water
leakages were registered.
These are mainly orientated
parallel to the bedding.

Based on the water loss
measurements, the mean value
of the permeability was
calculated to be
K = 3 x 10 -15 m².

Mechanical Installation

Two air compressors are
installed for pressurerising
of the chamber. They are

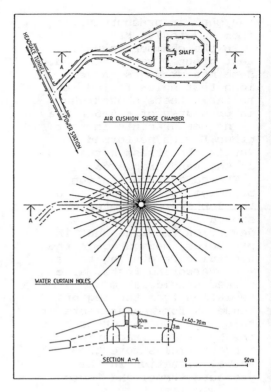

Figure 6. Surge chamber as
built at Torpa

located close to the
powerhouse. Total capacity of
the compressors is 940 Nm³/h.
To fill 10,000 m³ of air to a
pressure of 44.5 bar in the
chamber, a total time of 473
hours (20 days) was needed.
The air is fed into the
chamber through steel pipes
with a diameter of 100 mm.
The distance from the
compressor station to the
surge chamber is approxi-
mately 450 m. The steel pipes
follow the tunnel wall and
are fully embedded in
concrete and end close to the
roof at the highest point of
the chamber. This is due to
emergency discharging of air
through the same pipes. To
discharge air from the
chamber, manually controlled
as well as remotely
controlled valves are
installed. The remote control

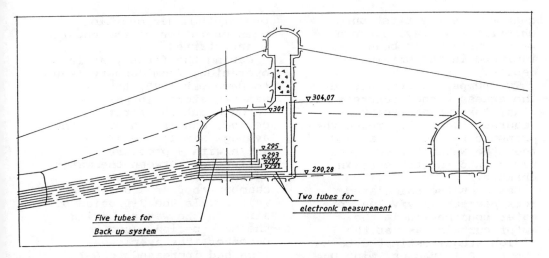

Fig. 7 Measurement system for the chamber water level

valve is installed due to the high noise level, which was measured at 130 dBA, at a distance of 35 m from the valve.

MEASUREMENT SYSTEM

Chamber Water Level

Due to the stability in the water way system and the danger of air escaping from the chamber into the headrace tunnel, control of the water level in the chamber is of vital importance. Hence, two different control systems, an electronic and a manual, are installed. The electronic system, which is based on two pressure transducers, with a precision of 0.01 %, is located outside the concrete plug in the access tunnel. They are connected to the surge chamber, through steel pipes with a diameter of 25 mm. One of the transducers measures the water pressure at the bottom of the chamber and the other the air pressure close to the roof, at the highest point of the

chamber. This pipe is filled with water. A computer gives the altitude of the water level, and gives signals for critical water levels as well as internal defects. These signals will automatically start or stop the air compressors. However, due to very low air losses, the compressors are normally started and stopped manually.

By use of manometers connected to the pressure pipes, the water level in the chamber can also be calculated. In addition to these systems, a very simple manually operated system is installed. Five plastic tubes, which start at different levels in the chamber, end up in manual valves outside the concrete plug in the access tunnel. By successively opening the valves, the water level can roughly be measured, by seeing whether air or water is escaping from the tubes.

Water Curtain

The Torpa air cushion surge

chamber, is the first surge chamber in Norway, where a water curtain has been included in the original design.

Two pumps, located outside the access tunnel concrete plug, take water from the headrace tunnel, increase the pressure by about 4 bar and feed the water through steel pipes into the top curtain chamber.

Based on the calculated rock permeability value, the water consumption in the water curtain was, at the design stage, estimated to be about 350 litres pr. min. Due to uncertainties in a program of complex calculations, water pumps with a capacity of twice the calculated consumption were installed (700 l/min.). So far the actual water consumption, is measured at only about 40 litres pr. min. during operation.

LOAD REJECTION

Rejection of the load creates some oscillation of the surge chamber water level. A load rejection from 2 x 80 MW with the water level at 293 m.a.s.l. induces a raising of the water level to 294 m.a.s.l. and a fall down to 292.5 m.a.s.l. At the same time the penstock pressure is increased by 6.7 %.

AIR LOSS

The air filling of the chamber started when the water filling of the system was finished. Pumping of water into the water curtain started simultaneously. Prediction of air loss had been calculated, but it was obvious that uncertainty about the air loss would be one of the main questions when operation of the power plant started.

During the first year in operation, from January 1990 to December 1990, the air was registered in periods. Through out the first period, January to June, the average air loss was 4.6 Nm^3/h with a pressure difference between the water curtain and the surge chamber roof of 3.0-3.5 bar. This equals the diffusion of air into the water. During the next period, June to December, the average air loss had increased to 6.7 Nm^3/h. In this period the water curtain excess pressure had been reduced to 2.0 bar.

For a couple of days in December 1990, the pressure difference, between the water curtain and the surge chamber roof, was kept at approximately 1.5 bar. The average air loss during this short period was 300 Nm^3/h. To verify this result a new test was carried out in June/July 1991. For this test the water curtain was turned off for a period. The test showed an air loss of 400 Nm^3/h. When started again and the water curtain pressure increased to 2 bar excess pressure, the leakage stopped.

Based on these tests, it seems likely to conclude that as long as the water curtain excess pressure is higher than 2 bar, the air loss at the Torpa air cushion surge chamber is a result of the air diffusion into the water and hence there is no air leakage through the rock mass.

REFERENCES

Kjølberg, R.S.:
   Air cushion surge chamber
   at the Torpa power plant.
   Proceedings of Storage of
   Gases in Rock Caverns,
   Trondheim 1989, Balkema.

Kjørholdt, H.:
   Gas tightness of unlined
   hard rock caverns. Dr. ing.
   thesis
   Department of Geology and
   Mineral Resources
   Engineering University of
   Trondheim, Norway 1991.

Hydropower'92, Broch & Lysne (eds) © 1992 Balkema, Rotterdam. ISBN 90 5410 054 0

Geotechnical design of air cushion surge chambers

Halvor Kjørholt & Tore S. Dahlø
SINTEF Rock and Mineral Engineering, Trondheim, Norway

Einar Broch
Norwegian Institute of Technology, Trondheim, Norway

ABSTRACT: An air cushion surge chamber is an alternative to the traditional open surge shaft in hydro power plants, and has been used in Norway since 1973. The surge chambers have proven to satisfy the hydraulic demands, and have also shown to constitute an economic alternative that gives substantial freedom in the lay-out of the tunnel system, and the siting of the plant. Air leakage prevention is the major challenge when designing and constructing an air cushion. The paper shows how it is possible to handle this and other geotechnical aspects in an efficient way.

1 INTRODUCTION

Air cushion surge chambers are used as an economic alternative to the traditional open surge shaft for damping of headrace tunnel transients from changes in powerplant loading. As illustrated in Figure 1, the air cushion surge chamber is a rock cavern excavated adjacent to the headrace tunnel, in which an air pocket is trapped. The surge chamber is hydraulically connected to the headrace tunnel by a short (< 100 m) tunnel.

The air cushion concept was originally introduced to improve the economy of a hydro-powerplant where a traditional open surge shaft would be an expensive solution due to topographical reasons (Rathe 1975). The air cushion solution also gives substantial freedom in the lay-out of the tunnel system, and the siting of the plant. It is no longer necessary to maintain shallow, nearly horizontal, headrace tunnels for surge shaft economy (Figure 2a). Schemes which have used air cushions have tended to slope the headrace tunnel directly from the reservoir towards the power station as indicated in Figures 1 and 2b.

Air cushions are also favoured where the hydraulic head of the headrace is above ground surface. In such cases, construction of an open surge shaft would require erection of a surge tower, which may be expensive and environmentally undesirable in comparison with an air cushion.

2 HYDRAULIC REQUIREMENTS

The hydraulic design of air cushions follows the

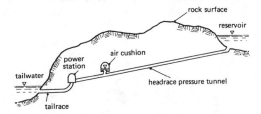

Fig. 1 Concept of powerplant with air cushion surge chamber.

same principles as design of traditional open surge shafts. Pressure surges in an open surge shaft system is according to the changes in water level in the shaft. In a system with an air cushion, pressure surges gives compression and expansion of the cushion according to ideal gas law. One should note that for normal surge periods (periods less than 5 minutes) the air cushion responds adiabatically.

As the hydraulic design of a surge shaft is a question of finding the necessary water surface area in the shaft, the necessary air volume is the key factor for an air cushion. Traditional Norwegian design practice for high head plants, that allows pressure surges up to 10 to 15% above static, is also adopted for the air cushion sites.

To avoid air-escape from the surge chambers, they are located a few meters above the roof of the near-by headrace tunnel as shown in Figure 3. The volume of the water bed should be such that there is a high degree of safety against air escape from the chamber during unfavourable combinations of downsurge and surface wave action in the chamber. However, a major blow-out may be caused by possible

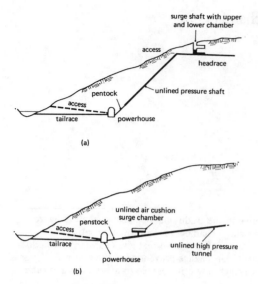

(a)

(b)

Fig. 2 Comparison of layouts for underground plant using: (a) open surge shaft; and, (b) closed air cushion surge chamber.

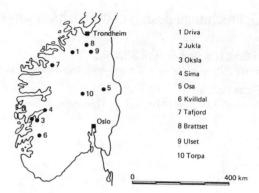

1 Driva
2 Jukla
3 Oksla
4 Sima
5 Osa
6 Kvilldal
7 Tafjord
8 Brattset
9 Ulset
10 Torpa

Fig. 4 Location of powerplants with air cushion surge chambers.

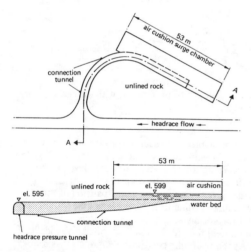

Fig. 3 Plan and profile of the Ulseth air cushion surge chamber.

failure of control equipment, improper gate operation or accidents (upstream blocking of the headrace tunnel). The mechanisms of such an event should be studied, and measures should be taken to ensure that the consequences of a possible blow-out would be tolerable. No blow-out has ever happened in Norwegian air cushion surge chambers.

Normally the total surge chamber volume needed is 50% higher than the necessary air volume. The shape of a surge chamber is not a crucial point, except that the water bed response to surges may be problematic for very long caverns. Long caverns can be avoided by for instance giving the cavern a ring-shape. More details about the hydraulic design of air cushions are provided in Goodall et al (1988).

3 GENERAL FEATURES AND LAYOUT OF EXISTING AIR CUSHION SURGE CHAMBERS

A total of ten air cushions have been commissioned to date, all of them situated in southern Norway, as shown in Figure 4. General features of the air cushion sites are listed in Table 1. The diagrams in Figure 5 show the cavern volumes and internal pressures.

The first air cushion was commissioned at the Driva power plant in 1973, the latest one at the Torpa plant in 1989. As indicated in Table 1, air cushions have been used for power plants with capacity from less than 50 MW to more than 1200 MW.

An air cushion must be located within a limited distance from the turbine due to hydraulic reasons. In practice, distances up to more than 1000 m are found acceptable, at least at some of the sites.

Figure 6 shows two different layouts of the tunnel system in the area from the surge chamber to the power station. Usually the surge chamber consists of one single cavern, but doughnut shaped caverns around a centre pillar have been used at the Kvilldal and Torpa sites (see Figures 7 and 9). The vertical cross-section of the caverns ranges from approximately 90 to 370 m².

The ratio between maximum air cushion head and the minimum rock cover varies extensively from one site to an other. At the first air cushion, Driva, the minimum overburden is twice the air cushion head (in m of water column). At the most extreme site,

130

Table 1. Features of air cushion sites.

Site	Comm-issioning date	Power-plant capacity (MW)	Distance to turbine (m)	Conect. tunnel length (m)	Vertical cavern cross-sect. (m²)	Installed compressor capacity (Nm³/h)	Ratio between max. air cushion head and min. rock cover (m/m)
Driva	1973	140	1300	20	111	425	0.5
Jukla	1974	35	680	40	129	180	0.7
Oksla	1980	206	350	60	235	290	1.0
Sima	1980	500	1300	70	173	270	1.1
Osa	1981	90	1050	80	176	2320	1.3
Kvilldal	1981	1240	600	70	260-370	500	0.8
Tafjord	1982	82	150	50	130	260	1.8
Brattset	1982	80	400	25	89	700	1.6
Ulset	1985	37	360	40	92	360	1.1
Torpa	1989	150	350	70	95	940	2.0

Torpa, the overburden is only half the air cushion head.

The cavern volumes are generally less than 20,000 m³, except for the Kvilldal surge chamber that has a volume of 110,000 m³. As many as six of the air cushions have pressures exceeding 4 MPa. The highest pressure is reached at Tafjord where the maximum operating pressure is 7.7 MPa, equalizing a water head of 780 m . The air cushion itself occupies typically from 40 to 80% of the cavern volume, which corresponds to a water bed thickness in the surge chamber of 2 to 5 m.

Table 2 contains information about the rock type at the air cushion sites. Eight of the sites are located in various types of gneisses and granites, and the other two in phyllite and meta siltstone respectively. All caverns are essentially unlined, although some rock reinforcement, mainly in the form of rock bolts and shotcrete, have been used at a few sites.

Each air cushion is connected to one or more compressors which are located in the access tunnels (Figure 6), or as at one site (Osa), at the ground surface above the air cushion. A system of pipes and cables connects the air cushion to the compressor(s) and monitoring equipment. The compressors serve two purposes: First, to establish the air cushion before commissioning of the plant. Second, to compensate for air loss during operation.

A minimum and maximum air cushion volume is defined as a part of the hydraulic design. The air cushion volume is monitored by measuring the water bed level. All air cushions are equipped with at least two separate water level monitoring devices to safeguard against possible instrument malfunction. Air cushion pressure and temperature are measured directly only at a couple of the sites. The static air cushion pressure can be computed from the height difference between the reservoir and the cavern. The temperature is found to vary only by approximately 5°C on a seasonal base due to changes in

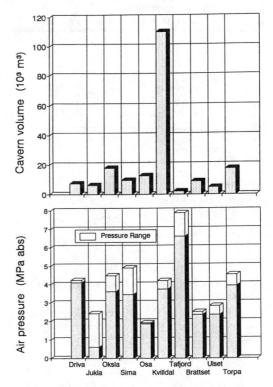

Fig. 5 Volumes and pressures of the air cushion surge chambers.

water bed temperature (which reflects seasonal temperature fluctuations in the reservoir). More details about air cushion monitoring can be found in Goodall et al (1988) and Kjørholt (1991).

131

Table 2. Air cushion fractures.

Site	Rock type	Natural rock permeability* (m^2)	Ratio between air cushion pressure and natural ground water pressure	Air leakage (Nm^3/h)	Air leakage (%/day)
Driva	banded gneiss	no data	0.6 - 0.7	0	0
Jukla	granitic gneiss	$1 \cdot 10^{-17}$	0.2 - 0.7	0	0
Oksla	granitic gneiss	$3 \cdot 10^{-18}$	1.0 - 1.2	< 5	< 0.01
Sima	granitic gneiss	$3 \cdot 10^{-18}$	0.8 - 1.2	< 2	< 0.01
Osa	gneissic granite	$5 \cdot 10^{-15}$	1.3	900/70**	11/1.0
Kvilldal	migmatitic gneiss	$2 \cdot 10^{-16}$	> 1.0	240/0***	0.2/0
Tafjord	banded gneiss	$3 \cdot 10^{-16}$	1.8 - 2.1	150/0***	5/0
Brattset	phyllite	$2 \cdot 10^{-17}$	1.5 - 1.6	11	0.2
Ulset	mica gneiss	no data	1.0 - 1.2	0	0
Torpa	meta siltstone	$5 \cdot 10^{-16}$	1.7 - 2.0	400/0***	2.0/0

* to obtain hydraulic conductivity in m/s, multiply by $\approx 10^7$
** before/after grouting
*** without/with water curtain in operation

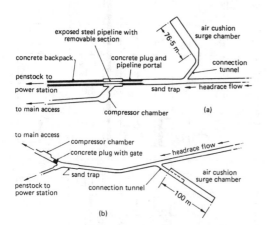

Fig. 6 Air cushion surge chamber arrangements.
(a) Oksla, (b) Brattset.

4 OPERATIONAL EXPERIENCE

4.1 General

A record of operational experience includes:
- Dynamic response of the air cushion as compared to computed behaviour
- Functioning of monitoring equipment
- Compressor operation
- Rock mass stability
- Air loss and air loss prevention

Dynamic response, monitoring equipment and compressor operation are discussed to some extent in Goodall et al (1988). In this article it shall be mentioned that none of these factors have turned out to be major challenges, neither for air cushion design nor operation. It is further important to note that no specific problems related to rock stability have been recorded at the air cushion sites.

The air loss from an air cushion may be due to both air dissolution in the water bed and leakage through the rock mass. The dissolution loss per year ranges from 3 to 10% of the compressed air (depending on surge chamber geometry and pressure). This loss has no practical implication for the plant operation other than a need for supplementary air-filling once or twice a year.

The measured air leakage through the rock is listed in Table 2. Six of the air cushions have a natural air leakage rate that is acceptable. Three air cushions have no air leakage at all through the rock mass. At four air cushions (Osa, Kvilldal, Tafjord and Torpa), the natural leakage rate was too high for a comfortable or economic operation. At these sites remedial work has been carried out to bring the leakage down to an acceptable level, see Table 2. One should also note from Table 2 that these four sites are located in the most permeable rock masses of all ten air cushions. Experience from the leakage prevention work at these sites are discussed below.

4.2 Leakage prevention work at Osa

Osa air cushion is located in the most permeable rock mass of all the ten air cushions, more than thousand times more permeable than the least permeable (Table 2). Although a significant cement and chemical grouting program was undertaken at the time of construction, the air leakage was measured to 900 Nm³/h shortly after startup. After eight months of operation the plant was shut down for further grouting.

The grouting was completed within three months. A total of 36 tons of cement and 5500 l of chemical grout were injected. This brought the leakage down to 70 Nm³/h, which is comfortably managed by the compressor plant, even though the leakage is the highest of all the air cushions.

4.3 Leakage prevention work at Kvilldal

At Kvilldal a major weakness zone passes within 50 m of the cavern periphery and is probably responsible for a low natural ground water pressure in the surge chamber area, and thereby a higher air leakage rate than for a more homogeneous rock mass. Monitoring of the air cushion during the first year of operation showed an air leakage rate of 240 Nm³/h.

A water curtain consisting of 47 boreholes of 51 mm diameter was adopted to reduce the air leakage. These boreholes are kept pressurized with water at a pressure of 1.0 MPa above the air cushion pressure. A plan view of the surge chamber and the over-lying water curtain is shown in Figure 7. The intention of this first water curtain used at an air cushion was to limit the leakage only. However, the water curtain showed to be able to totally eliminate the air leakage through the rock mass.

4.4 Leakage prevention work at Tafjord

As at Kvilldal, the Tafjord air cushion was first commissioned without any leakage preventing measures undertaken. But, even though the leakage at this site was somewhat less than at Kvilldal, the compressors installed to supply the air cushion did not have the sufficient capacity. The surge chamber at Tafjord was therefore out of operation from 1982 to 1990 (i.e the cavern was completely water filled). An attempt to grout a major fracture intersecting the cavern did not improve the leakage condition. Fortunately, the Tafjord air cushion is not crucial to the operation of the Pelton system to which it is connected. The power plant has consequently been able to operate without a surge facility.

A water curtain was installed at Tafjord in 1990, partly as a research project. The curtain consists of 16 core drilled holes (diameter 56 mm) which covers both the roof and the upper part of the cavern walls as illustrated in the plan view in Figure 8. Also at this site the air leakage disappeared when the water curtain was put in operation at the design pressure (0.3 MPa above the air cushion pressure).

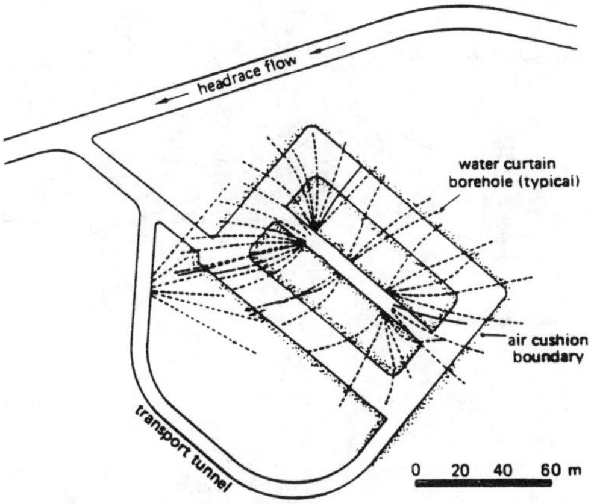

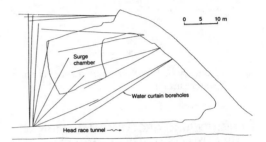

Fig. 7 Plan of Kvilldal air cushion surge chamber with water curtain.

Fig. 8 Plan of Tafjord air cushion surge chamber with water curtain.

4.5 Leakage prevention work at Torpa

The Torpa air cushion is the only one where a water curtain was included in the original design. The water curtain consists of 36 boreholes (64 mm diameter), drilled from an excavated gallery 10 m above the cavern roof (see Figure 9). In addition to the water curtain, grouting was undertaken during construction to improve the rock condition.

As for the two other water curtains, no air leakage has been registered from the air cushion when the water curtain is in operation at design pressure (0.3 MPa above the air cushion pressure). To get an idea of the air leakage potential at Torpa, the water curtain was turned off for two days. This resulted in an "immediate" leakage rate of 400 Nm³/h. The leakage ceased as soon as the water curtain again was put in operation. The measured leakage rate corresponds very well with results from theoretical calculations.

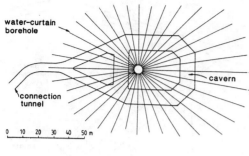

Plan

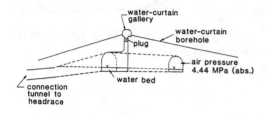

Vertical section

Fig. 9 Geometry of Torpa air cushion surge chamber with water curtain.

5 ENGINEERING GEOLOGICAL DESIGN OF AIR CUSHION SURGE CHAMBERS

5.1 General

The location of a surge chamber is limited by hydraulic constrains to be within approximately one km from the turbine. The challenge for the engineering geologist is to find the best location within this area. The most important factors are avoidance of rock masses with poor stability and high permeability. It is of course also essential to ensure that the rock mass has the sufficient capacity to withstand the internal pressure, in the same way as for the nearby unlined headrace tunnel.

The final location of the surge chambers in Norway is based on mapping and tests carried out from the headrace tunnel. The tests in question are permeability measurements and hydraulic jacking. Core drilling is done to verify the rock quality of a selected site.

Engineering geological mapping includes first of all mapping of:
- Rock type
- Strike, dip and frequency of rock joints, fracture zones and other discontinuities
- Rock mass permeability on the basis of water inflow
- Fracture roughness and infilling

To obtain the best rock stability possible, the long axis of the cavern are normally oriented so that it bisects the angle between the strike of the principal joint sets. More details about the engineering geology related to location and orientation of underground caverns can be found in Broch (1988).

5.2 Air leakage

Air leakage through the rock mass has shown to be the most critical factor for a successful air cushion. The air leakage can be evaluated by use of the following equation presented in Tokheim and Janbu (1982):

$$Q_{gw} = \psi \; \frac{\pi \, K \, L \, P_o}{\mu_g \, G} \; \left(\; \left(\frac{P_g}{P_o} \right)^2 - \left(\frac{P_e}{P_o} \right)^2 \; \right)$$

where Q_{gw} is gas leakage through the rock mass (m^3/s at pressure P_o), K is rock mass permeability (m^2), L is characteristic length of storage, μ_g is dynamic viscosity of gas, G is geometry factor, P_o is reference pressure, (normally atmospheric pressure, Pa abs), P_g is storage pressure (Pa abs), P_e is pressure at a plane isobar away from the surge chamber (normally ground surface with atmospheric pressure), ψ is a factor describing the relative leakage reduction due to the presence of groundwater. The factor ψ varies between one for "dry" rock, and zero if there is a positive groundwater pressure gradient towards the air cushion.

The above equation shows that for a given surge chamber geometry and pressure, the air leakage depends on the surrounding ground water pressure and the rock mass permeability.

To obtain a non-leaking air cushion without introducing leakage preventing measures it is necessary that the natural groundwater pressure at the air cushion site (to be interpreted as the ground water pressure before construction of the surge chamber) is significantly higher than the air cushion pressure. In this way there will be a positive ground water pressure gradient (note pressure gradient) towards the air cushion during operation ($\psi = 0$), which is the criterion to avoid an outward air leakage flow. This is discussed in detail in Kjørholt (1991).

If the natural ground water pressure is insufficient to totally prevent air leakage, an air leakage according to the above equation should be expected. For ratios of air pressure to natural ground water pressure less than unity, both the permeability and the ground water pressure will play a significant role for the leakage rate. In cases where the air pressure exceeds the natural ground water pressure, the permeability will be the dominating factor for the leakage rate.

5.3 *Air leakage prevention*

If the estimated or experienced leakage from an air cushion exceeds the acceptable level, remedial actions can bring the leakage down below this limit. There are mainly two actions in question: Grouting and installation of a water curtain. Our experience is that the cost benefit ratio for a water curtain is considerable more predictable than for grouting.

By grouting, the permeability of the surrounding rock mass will be reduced, and the air leakage will decrease correspondingly. Experience shows that grouting can only reduce the permeability to a certain extent, and can not be used to totally eliminate the leakage. Practical experience indicates a leakage reduction by maximum one order of magnitude if the grouting takes place ahead of excavation. Grouting after excavation is generally less effective than pre-grouting.

The working principle of a water curtain is to increase the ground water pressure around the air cushion artificially. A water curtain consists of an array of boreholes above, and sometimes also along the sides of the air cushion. These holes are connected to a water pump that maintains a permanent water pressure in all these holes, Figures 7, 8 and 9 show the water curtain design at the three air cushions which have such an installation. If the water curtain pressure is high enough to establish a pressure gradient towards the air cushion in all potential leakage paths, no leakage will take place at all, as stated above. Guidelines for design of water curtains to obtain complete air-"tightness" are provided in Kjørholt (1991).

Water curtain operation is a matter of keeping the water pressure in the boreholes at a given level. To obtain this, it is necessary that the water curtain pump operates continuously. Two types of pumps have been used, triple-plunger pumps and centrifugal pumps. Plunger pumps are used if the water has to be pumped from low pressure, while centrifugal pumps have been used in the cases that the headrace have been used as water supply.

At Kvilldal both types of pumps have been used. A plunger pump was used for the first three years until a failure occurred in the supply line for the water curtain. Since it was believed that this failure may have been caused by vibrations introduced by the pump, the plunger pump was replaced by a centrifugal pump. Torpa also uses a centrifugal pump, while Tafjord has a plunger pump mainly because of the high pressure level.

Experience has shown that if the pump stops, an air leakage will develop. How fast the leakage develops can vary significantly, and is a question of overburden, air cushion pressure and the structure of the rock joints. At Kvilldal it takes several months to approach the stationary leakage level experienced before the water curtain was installed. Tests at both Tafjord and Torpa showed that full leakage was reached within few hours. At all sites the leakage stops "immediately" after the water curtain has been put in operation again.

6 CONCLUSIONS

Air cushions have proven to be an economic alternative to the traditional open surge shaft for a number of hydro power plants. Experience shows that the hydraulic design should follow the same principles as for an open surge shaft. The geotechnical design of the air cushion cavern should follow the same basic rules as for other rock caverns.

Air leakage through the rock masses is the major challenge when designing and constructing an air cushion. A certain leakage may for economical reasons be accepted. If, however, the leakage exceeds a given limit, both grouting and the use of water curtains are possible actions. Experience has shown that grouting will reduce the leakage to a certain extent, while a water curtain is able to eliminate the leakage through the rock.

REFERENCES

Broch, E. 1988. General report on underground power plants. *Proc. ISRM Symposium on Rock Mechanics and Power Plants*, Madrid.

Goodall, D.C., H. Kjørholt, T. Tekle, & E. Broch, 1988. Air cushion surge chambers for underground power plants. *Water Power and Dam Construction*, November: 29-34.

Kjørholt, H. 1991. *Gas tightness of unlined hard rock caverns*. Dr. thesis, Norwegian Inst. of Technology, Trondheim, Norway.

Rathe, L. 1975. An innovation in surge chamber design. *Water Power and Dam Construction*, June/July: 244-248.

Tokheim, O. & N. Janbu 1982. Flow rates of air and water from caverns in soil and rock. *Proc. ISRM Symposium*, Aachen, Germany: 1335-1343.

Hydropower'92, Broch & Lysne (eds) © 1992 Balkema, Rotterdam. ISBN 90 5410 054 0

Sanliurfa tunnels system

C. Kurt

State Hydraulic Works 16th Regional Directorate, Sanliurfa, Turkey

ABSTRACT: The Şanlıurfa Tunnels System will be one of the longest irrigation tunnels system in the world when completed in 1993, comprising two concrete lined tunnels each 7,62 m in diameter, 26,4 km in length and discharging water from the Atatürk Dam reservoir at a maximum rate of 328 m³/sec. The supplied water will be conveyed to the Urfa-Mardin plains by an open channels system providing the irrigation of 476 474 hectares of land,of which 327 825 hectares will be by gravity and 148 649 hectares by pumping. Şanlıurfa Tunnels, is the most important component of the South-East Anatolia Irrigation Project, ("GAP") and will be entirely constructed by using Turkey's own resources.

1 PROJECT DESCRIPTION

Fig 1 shows the location of the project, Fig 2 its layout and profile and Table 1 summarises the technical characteristics. The tunnels were excavated by Paurat road-headers and drill and blast methods.

2 GEOLOGY

The geology of the Şanlıurfa area comprises a series of gently folded shallow dipping marls and crystalline or marly limestones of upper Cretaceous to Tertiary age. These are in turn overlain by Upper Tertiary pla-

ŞANLIURFA TUNNELS IRRIGATION SCHEME

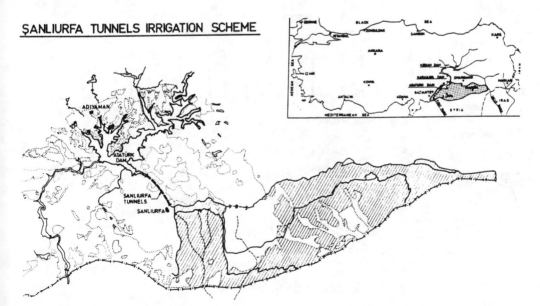

Fig 1 Location of the project in Turkey

teau basalts particularly around the city of Şanlıurfa. The sedimentary units tend to be homogeneous over a wide area. Shallow folding in the late Miocene age was accompanied by regional faulting trending NE-SW and the development of large graben structures along the Turkish - Syrian border.

Şanlıurfa Tunnels are driven almost entirely in upper Cretaceous grey marls (5-15 Mpa). Chalky marls and marly lime stones (30-75 Mpa) of Eocene age are found in the downstream last few kilometers. The tunnel crosses an anticline whose limbs generally dip NE - SW. A few highly weathered water

filled basaltic dykes were also found towards the outlet of the tunnels and caused some excavation and support problems when they were first encountered.

Principal discontinuities in the rock are the bedding and 3 sets of subvertical joints that run parallel to the regional tectonic fold patterns. These joints are commonly smeared with a few millimeters of clay and were often the cause of minor crown or shoulder collapses patricularly before the introduction of rapid acting rock bolts. Figure 3 is a geological map and section of the tunnels.

Table 1. The technical characteristics.

Type	:Circular, reinforced concrete lining
Length of tunnel	:2x26,4 km parallel tunnels
Inclination	:0,00063
Excavation diameter	:Approximately 9,00 m
Lined diameter	:7,62 m
Lining thickness	:0,40 - 0,95 m
Amount of discharge	:328 m³/sec
Geological formation	:Calcareous marl, clayey marl
Hydraulic load	:Approximately 40,0 m
Area to be irrigated	:476 474 ha

3 ACCESS AND CROSS TUNNELS AND SHAFTS

The task of controlling and dispersing ground water was made easier by the presence of 3 access tunnels from the surface and cross connections between main tunnels every 500 m with a ventilation/dewatering shafts every kilometer.

The presence of these cross connections allowed the contractor to drive the tunnels using multiple headings and at one stage twelve headings were operational at the same time. Sections of difficult ground were commonly attacked from both sides (or sometimes from just the safer of the two)

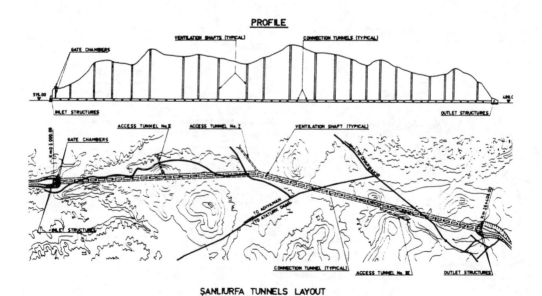

PROFILE

ŞANLIURFA TUNNELS LAYOUT

Fig 2 General layout and profile

a flexibility offered by this regular system of cross tunnels.

4 GROUNDWATER CONTROL

Groundwater was found throughout the tunnel, generally as minor drips percolating through small shear zones along clay filled discontinuities. Its presence in the tunnel created some softening of the grey marl invert which in places required substantial over excavation to sound rock prior to concreting.

Table 2 Stratigraphy

Plio - Quaternary	Gravelly clays
Tertiary Pliocene	Basalt lavas
Miocene	Calcareous clays
	Marls
	Limestone conglomerate
Eocene	Crystalline limestone
	Chalky marls
Upper Cretaceous	Grey marls
	Marly limestone

5 TUNNEL EXCAVATION

Paurat roadheaders were deployed because the tunnel is for the most part in the softer grey marl horizon. Excavation of the remaining section was made principally using conventional drill and blast methods. Excavation was on a traditional heading / bench/ invert method. Figure 4 shows excavation stages.

5.1 Roodheaders

The roadheaders were ideally suited to excavate the grey marl, consistently performing well and producing an excellent shape with little or no overbreak. Softening of the tunnel floors by groundwater caused manoeuvrability problems in places. Advance with contemporaneous support was achieved using a staggered top heading excavation of 2/3 one day, 1/3 the next. This system was developed in preference to the normal full top heading advance when it was found that stress relief was occuring in the crown and leading to minor col-

GEOLOGICAL MAP

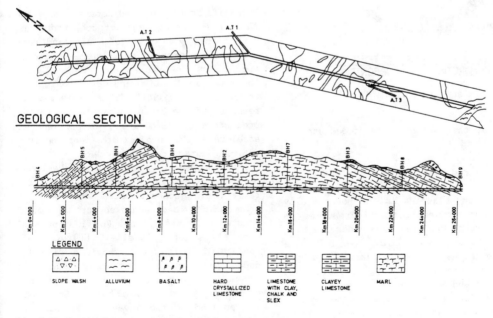

GEOLOGICAL SECTION

LEGEND

SLOPE WASH ALLUVIUM BASALT HARD CRYSTALLIZED LIMESTONE LIMESTONE WITH CLAY, CHALK AND SLEX CLAYEY LIMESTONE MARL

Fig 3 Geological map and geological section

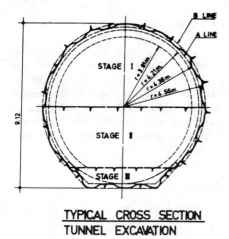

TYPICAL CROSS SECTION
TUNNEL EXCAVATION

Fig 4 Tunnel excavation cross section

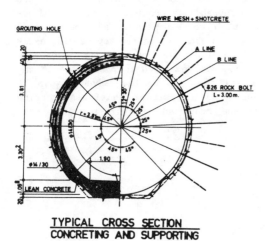

TYPICAL CROSS SECTION
CONCRETING AND SUPPORTING

Fig 5 Rock support components

lapses in the horizontally bedded strata.

5.2 Drill and blast

Drilling and blasting in the horizontally
bedded lithologies frequently resulted in
a flat crown. Profiles were generally difficult to maintain within the A-B line
despite the use of smooth blasting techniques. Drilling and blasting was also used
for bench and invert excavation.

Excavation of the top heading was achieved by using a wedge cut pattern with delay detonators. (Fig 6) to limit the amo-

ount of overbreak it was found beneficial
into incorporate the technique known as
"smooth blasting". Each drill hole was
38 mm diameter and 3 meters long. The number of holes drilled for each cycle was
59, consuming a total of 60 kg of gela-
tinite. By adopting this method the average pull was around 2,75 m which yielded
97 cubic meters of blasted rock. The middle holes were each charged with 2 kg of
explosive whereas contour holes were charged with 0,5 kg per hole. Remaining holes
were charged to suit the prevailing rock
conditions.

During excavation of the second stage
middle bench, perpendicular holes were
drilled to a regular pattern with a spacing of 2,0 meters and 1.5 meters, with a
length of 3,0 meters, see Fig 6. Each hole
was charged with 2.5 kg of explosive. In
this case a typical cycle would be carried
out over a length of 80 to 100 meters of
tunnel bench, achieving a vertical pull ot
2,5 meters. To limit the amount of overbreak the delay charge was increased from
the tunnel longi- tudinal axis to the side
holes.

5.3 Rock Support

Rockbolts, shotcrete and wiremesh were the
three primary rock support components and
were placed according to a standard pattern Cement grouted bolts were discontinued after about 26 km and replaced with
the more rapid acting and safer resin
anchored or "Swellex" bolts. This was in
response to the crown instabilities mentioned above and a desire by the client to
accelerate the project to meet newly
defined contact completion dates.

The response to the change in the rockbolting system was evident almost immediately. Resin anchored bolts were mostly
used and provided secure support within 10
minutes of installation. For instant support Atlas Copco "Swellex" bolts were
used. Both bolts were provided with a
double plate system to tightly retain the
wiremesh. A nominal tension of 5 tonnes
was given to the resin bolts. Shotcrete
was sprayed in two layers 5 cm each to
complete the support shell. Figure 5 shows

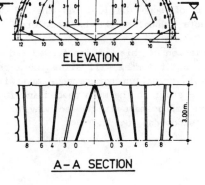

DRILLING AND BLASTING PATTERN FOR STAGE I EXCAVATION

ELEVATION

A-A SECTION

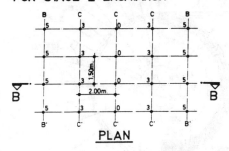

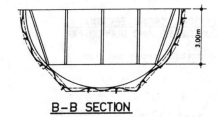

DRILLING AND BLASTING PATTERN FOR STAGE II EXCAVATION

PLAN

B-B SECTION

Fig 6 Drilling and blasting pattern

the rock support components.

6 CONCRETING

Steel shutters 12 m long for casting the concrete invert and arch were designed to allow rapid steel fixing.

After the third excavation stage the invert concrete was cast. The invert forms were self-propelled moving by means of winch and hydraulic jacks.

The arch concrete was poured subsequent to invert concrete by using fully hydraulic special forms which were transported on rails.

During invert and arch concrete pours water leakage insulation was provided by PVC waterproof seal installation on the junction points. Figure 7 shows a typical concrete section.

Reinforcement bars for the arch were adjusted before fixing the arch formwork.

7 GROUTING

Approximately one month after completion of a section of concrete lining contact grouting was undertaken. This was achieved by drilling 8 radial holes, 65 mm in diameter through the lining for a distance of 2 m from the surface of the concrete lining into bed rock. The pattern was repeated every 4 meters along the tunnel with the addition of a vent/check grout hole in the crown of the arch midway between every radial pattern. As a rule injection with neat cement grout was carried out, but where larger openings were encountered primary grouting with a sanded grout was followed by high pressure grouting (up to 5 bars) with neat cement grout to fill the voids left by contraction.

Consolidation grouting was anticipated, but due to the good rock quality (confirmed by water pressure tests), was found to be unnecessary. Figure 8 shows a typical grouting pattern.

8 TUNNEL INLET APPURTENANT EQUIPMENT

It is worth mentioning how water will enter the tunnels from the main approach channel. Both portals are at the base of a 30 m high reinforced concrete retaining wall. These are covered with a moveable trashrack to prevent damage to the downstream control gates which comprise

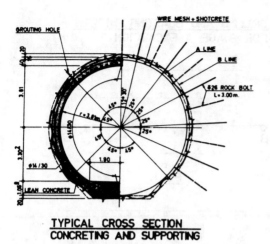

TYPICAL CROSS SECTION
CONCRETING AND SUPPORTING

Fig 7 Concreting cross section

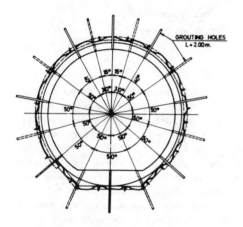

TYPICAL CROSS SECTION
FOR GROUTING

Fig 8 Grouting pattern

stoplogs and a rolling gate. These would
be activated in the unlikely event of a
collapse along the tunnels or hazard on
the downstrem gates.

9 TUNNEL OUTLET APPURTENANT EQUIPMENT

Each tunnel terminal structure is designed
to be fitted with radial gates which is
the most crucial and sensitive control
element of the whole tunnel system as it
has to function continuously for control-

ling water supply. The service gate is
complimented by a guard gate for
inspection and maintenance.

10 CONCLUSION

Despite of some difficulties the 52 km
Şanlıurfa Tunnels will be completed by
July 1993.

Completion of this Turkish funded and
Turkish built tunnel will allow the first
phase of irrigation to commence and will
herald a start to the much needed and long
awaited improvement in the economy of
Turkey's poorest region.

142

Hydropower'92, Broch & Lysne (eds) © 1992 Balkema, Rotterdam. ISBN 90 5410 054 0

The coupling numerical methods for underground engineering

Li Jie
Chengdu Hydroelectric Investigation and Design Institute of MOE.MWR, People's Republic of China

Peng Shou-Zhuo
Tsinghua University, People's Republic of China

ABSTRACT: In this paper the mechanical and numerical models have been established, the three-dimensional non-linear coupling numerical calculation program of finite element and boundary element, of finite element and infinite element have been studied and presented, they are especially appropriate for analysis of displacement, stress, yield zone and stability of surrounding rock mass of caverns. Secondly, the correctness of the programs has been proved by some typical examples. Finally, the program has been applied to three-dimensional non-linear finite element analysis for large underground caverns, the results are consistent with that of prototype test. As a result of effectiveness of the program the calculating results are reasonable, the CPU time, the input data can be saved and boundary conditions can be treated more exactly.

1 INTRODUCTION

Underground plants will comprise about 40% of hydroelectric power stations which are to be built in recent times in China. The character of underground power house is big span, high lateral rock wall and having big caverns such as transformer chamber and surge chamber in the vicinity. The three-dimensional non-linear FEM analysis for stability of cavern rock mass is very complex and spends a lot. This paper presented a three-dimensional nonlinear coupling numerical analysis program of finite element-boundary element, of finite element-infinite element. It is provided with 3-D isoparametric brick element, 3-D joint element, 1-D bar element (simulating rock bolt), 3-D infinite element and 3-D boundary element. The established numerical method and the program NFEMB-3D can save CUP time, reduce input data and treat boundary conditions more exactly. It is flexible, effective, especially appropriate for analysis of stability for surrounding rock mass of underground engineering.

2 CONSTITUTIVE MODEL

Rock mass and its weak formation, or joint plane, are material of little resistance or no-resistance to tension. So plastic yield and tension yield of material were considered when constitutive model was established in the paper.

1. Plastic yield: The "friction model" yield surface which is on the basis of Mohr-Coulomb law is suited to rock and concrete. The singularity of M-C yield surface make plastic matrix calculation difficult, it can be avoided by Drucker's cone yield surface but error is greater. Zienkiewicz-Pande's yield surface which is improvement of M-C yield surface not only avoids calculation difficult of plastic matrix but also approaches behavior of material. The program NFEMB-3D can make use of any one of these three yield surfaces for describing yield of rock mass. Z-P hyperbolic yield surface is recommended in this paper. Fig.1 shows the comparison on π plane for three yield surfaces.

2. Tension failure: When tensile stress is over tensile strength, tension failure will be occurred. The condition can be considered as:

before initial tension fracture,

$$F = \sigma_1 - \sigma_t \leqslant 0 , \tag{1}$$

after initial tension fracture,

$$F = \sigma_1 \leqslant 0 , \tag{2}$$

where σ_i = principal stress in i direction, i=1,2,3, σ_t= tensile strength, F is yield function.

3. Yield criterion for rock joint plane: They include criterion of shear slide, of

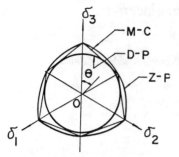

a) yield surface on π plane

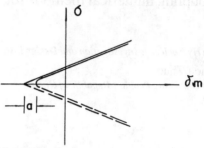

b) hyperbolic on σ- m plane
   for Z-P Criterion

Fig. I Three plastic yield criteria on π plane

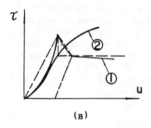

(a)

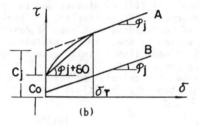

(b)

Fig. 2 The constitutive model of rock joint

tensile fracture and the effectiveness of
bolted joint plane.
   (1) Shear slide along joint plane.
The joint model is shown in Fig.2.(a), the
curve ① is suited to rough joint, curve ②
is appropriate for smooth joint. The shear
strength can be shown in Fig. 2.(b), in
which σ_T is relate to shear dilatation
for rough joint, $\sigma_T = C_J / [tg(\phi_J + \delta_o) - tg\phi_J]$,
peak strength and residual strength are
described by curves A and B, respectively.
According to Mohr-Coulomb law, the shear
slide criterion for joint plane can be
written as follows:
   peak strength,

$$F = |\tau_i| - C + \sigma_n tg\phi = 0, \qquad (3)$$

   residual strength,

$$F = |\tau_i| - C' + \sigma_n tg\phi' = 0, \qquad (4)$$

where i=1,2, c, φ and c', φ' are initial
cohesion, internal friction angle and residual cohesion, residual friction angle
of joint plane, respectively,
σ_n = normal stress, $\sigma_n < 0$, it means when
σ_n is compressive stress, the expressions
(3), (4) have their physical meaning.
   (2) Bolt effect of bolted joint plane.
Cohesion increment C_b on bolted joint

plane can be expressed as follows:

$$C_b = F_d sin(\alpha_b - \alpha_J)/A_r, \qquad (5)$$

in which F_d is the maximum bolt force,
$F_d = 1/\sqrt{3}(\sigma_y A_b) \cdot 1/90(\alpha_b - \alpha_J)$, σ_y = yield
stress of bolt, A_b = area of bolt cross sec-
tion, α_b, α_J are angles at which bolt,
joint cross X axis, respectively, A_r = area
of joint plane.
   Shear strength increment τ_b on bolted
joint plane can be expressed as follows:

$$\tau_{bi} = A_b E \triangle u_i sin(\alpha_{bi} - \alpha_{Ji})/[2A_r(1+\mu)] \qquad (6)$$

in which i=1,2, E = module of elasticity,
μ = poisson's ratio, $\triangle u_i$ = mean relative
displacement of joint plane in tangential
direction.
   (3) Tensile fracture condition on joint
plane.

$$\sigma_n - R_t \geq 0, \qquad (7)$$

in which σ_n = normal stress on joint
plane, tensile stress is positive; R_t =
tensile strength of joint, R_t=0 in general.
After tensile fracture, normal stress on
joint plane becomes zero, i.e. σ_n = 0,
and shear strength on corresponding joint

plane can be expressed as:

$$\tau_1 = \tau_{b1}, \text{ (bolt effect is considered)} \quad (8)$$

$$\tau_1 = 0, \text{ (bolt effect is not considered)} \quad (9)$$

4. Yield condition of bolt element.

$$\sigma < [\sigma_m], \quad (10)$$

in which σ = tensile stress of bolt, $[\sigma_m]$ = permissible tensile strength of bolt material, assuming that compression failure for bolt is not possible.

3 PRINCIPLE OF COUPLING NUMERICAL METHOD

In underground chamber excavation design, the displacement and increment and their total value, which are caused by excavation step by step in the initial stress media, must be calculated. It is not optimum to solve problem by means of general FEM method subdividing domain from the point of view of economic and boundary condition treatment. The coupling calculation of finite element and boundary element or infinite element can make good. It is favourable to treatment of non-linear property, joints, more complex constitutive model for rock mass to simulate region near caverns by means of finite element.

It is advantageous to economy to simulate far region of caverns by means of boundary element or infinite element.

The solution procedures of couple method are similar to those of FEM. It is attentive that when the node equilibrium equations are established for nodes of common boundary between FE region (finite element region) and BE (boundary element region), the effect of FE region and BE region must be all considered. The total contribution to nodal force, which is made by elements surrounding the node and belong to FE region, is expressed by P_1, another total contribution to nodal force, which is made by BE region, is expressed by P_2, the external load applied to node is expressed by P, then we have equilibrium condition:

$$P_1 + P_2 = P, \quad (11)$$

the displacement expression of P_1 and P_2 can be obtained by FEM and direct boundary integral method respectively, so the displacements in (11) can be solved in accordance with FEM solution procedures. It is attentive that the element nodes of FE region must coincide with those of BE region on the common boundary of the two regions, the number of boundary element must be equal to that of finite element

in the vicinity.

The couple method of finite element and boundary element not only has higher exactness but also has less inputdata, decreases number of linear simultaneous equations so it is time saving. On the other hand, we must make the contribution matrix to the assembled stiffness matrix, which is made by BE region, inverse. Matrix inversion is time consuming and inverse of contribution matrix is fully populated; when the displacements in expression (11) is solved, the well banded and symmetric character of the assembled stiffness matrix has been failured, it eliminates the effect of decreasing number of linear equations in coupling method of finite element and boundary element to some extent. Under certain circumstances, we are interested in the displacements and stresses near the hole periphery, it is reasonable to simulate the character of infinite region by means of infinite element, on this condition, the assembled stiffness matrix remains symmetric, well banded and problem has the same solution procedures as those of FEM.

Taking aim at flexibility, the program provides three functions: finite element calculation, couple calculation of finite element and boundary element, couple calculation of finite element and infinite element. Of course, it can do either nonlinear calculation or linear calculation.

4 PROGRAM VERIFICATION AND APPLICATION

1. Verification of finite element - infinite element coupling calculation
In order to verify finite element-infinite element couple program, a circle of infinite region in Fig.3 was taken as an example, plane strain was simulated by three dimension problem, radius=5m, longitudinal thickness=1m, $E=3.2 \cdot 10^4$Mpa, $\mu = 0.22$, internal pressure = 20 Mpa. It was calculated by thirty 8-node 3-D elements and six infinite elements, the results is shown in table 1.

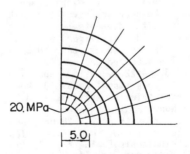

20.MPa

5.0

Fig. 3 Finite-infinite element couple

Table 1					Mpa
Radius(m)	5.83	7.50	9.50	11.83	14.75
σ_r theory	-14.7	-8.89	-5.54	-3.57	-2.30
σ_r calculating	-14.6	-8.82	-5.50	-3.56	-2.30
σ_θ theory	$\sigma_\theta = -\sigma_r$				
σ_θ calculating	14.9	8.88	5.56	3.53	2.28

2. Verification of finite element-boundary element couple calculation.

In order to verify finite element-boundary element couple program, a global hole of infinite region was taken as an example, radius= 2m, internal pressure=1.0Mpa, E=10⁴ Mpa, μ =0.25, the region near the periphery was simulated by twenty four 8-node 3-D elements, the element length along the radial direction=1m, BE and FE region were coupled by twenty four 4 - node boundary elements of BE region, which are in the vicinity of FE region. Comparing the results with those of boundary element method is shown in table 2.

Table 2				0.01Mpa	
Radius(m)	5.0	6.0	7.0	8.0	9.0
σ_r coupling	-4.61	-2.61	-1.63	-1.09	-0.76
σ_r BEM	-4.07	-2.36	-1.48	-0.99	-0.70
σ_θ coupling	2.23	1.28	0.80	0.54	0.38
σ_θ BEM	2.03	1.18	0.74	0.50	0.35

* in accordance with theory, $\sigma_\theta = -\sigma_r/2$

3. Application for a large underground engineering

A large underground power house in China was calculated by two coupling methods provided in the paper, one of the calculated examples is given as follows:

The size of underground power house, bus wire tunnel, tailrace tunnel, transformer chamber and element discretion is shown in Fig.4. The far region whose displacements were assumed to be zero was simulated by 8 -node 3-D infinite elements, some bolt elements were laid along periphery of power house. Fig.5 shows the domain subdividing. The measured ground stress in domain: σ_1 =29.5, σ_2=22.1, σ_3=18.7 Mpa. The parameters of rock mass: E=35·10⁴ Mpa, μ =0.17, internal friction angle ϕ = 60°, cohesion c=5 Mpa, tensile strength R_t=1.7 Mpa, residual strength c' = 1.25 Mpa and ϕ'=54°. The designed four excavation steps were shown in Fig.5. The Z-P plastic yield criterion and tension fracture criterion were applied in the analysis.

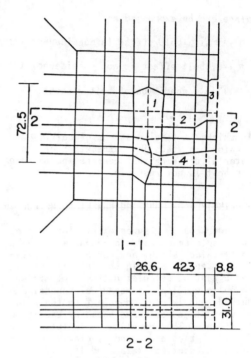

1. power house   2. bus wiretunnel
3. maintransformer chamber   4. tailrace

Fig. 4 The size of caverns snd element discretion

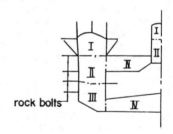

Fig 5 Excavation steps and bolt-layout

The periphery displacement and plastic yield zone after excavation step Ⅲ and Ⅳ are shown in Fig.6. After excavation of tailrace and bus wire tunnel, the first principal σ_1 has increased 25.4 ~ 52.1%, in comparison with the σ_1 of excavation step Ⅲ, which occurs on the periphery of tailrace.

The results indicate that the maximum displacement 4.472 cm occurs on the middle part of upstream rock wall of power house, the excavation of lower part of chamber results in upward displacement, the maximum is 3.745cm on the bottom of powerhouse.

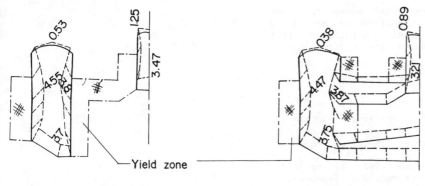

Yield zone

a) after step Ⅲ   b) after step Ⅳ

Fig. 6   penphery displacement (cm) and yield zone

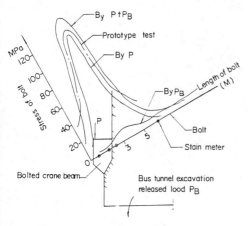

Fig. 7   Bolt stress of bolted crane beam

Table 3   The prototype test results of Bai Mountain bolted crane beam,China

loading phase	load (t)	displacement (mm) horizontal/vertical
design loading	76	0.30/0.39
over-loading	189	1.54/2.15
	196	1.81/2.45

*: loading continues 16-6 min's

The plastic yield runs through the zone between caverns,it indicates that the disturbance of initial ground stress is serious. The excavation of tailrace and bus tunnel makes the plastic yield zone, which is near the cross of power house and bus wire tunnel, expanded. The maximum compressive stress 47 Mpa occurs near the

upstream top corner of power house.

This problem has also been calculated by couple method of finite element and boundary element and 3-D BEM and 2-D FEM, the comparison analysis demonstrated that the results are reasonable and consistent.

Bolted crane beam have been largely used in Norway. There are a few power house having used this type of beam in China,but it has a widely application future. In the paper, bolt stress and displacement of bolted crane beam also have been analyzed. Contact joint between beam and rock wall was simulated by joint element, the crane load P on the beam is 132 t/m, excavation released load of the bus tunnel below the beam is noted as P_B, Fig.7 shows bolt stress of the beam and prototype test result in the similar project. Crane beam has a maximum horizontal displacement 0.67^{mm}, maximum vertical displacement 0.77^{mm}, comparing with prototype test displacement on Table 3, the displacement is reasonable, results indicate that in this powerhouse using bolted crane beam is possible.

5 CONCLUSION

1. The coupling numerical method and the NFEMB-3D program presented in this paper are especially suited to the stability analysis of surrounding rock mass and underground engineering design, their advantages are flexibility, saving time, decreasing input data, exactly treatment of boundary condition, and effectiveness.

2. The correctness of the program has been verified by the well known classic problems.

3. A large underground power house in China has been analyzed by this program for many conditions, one of the calculated

examples is shown in this paper, the comparison analysis, engineering experience and prototype test demonstrated that calculating results are reasonable and consistent.

REFERENCES

Wang Hong-Ru and others 1988. The elestic and plastic analysis for complex rock fundation. Chinese Journal of Geotechnic Engineering.
H.Larsson, T.Olofsson 1983. Bolt action in jointed rock. International Symposium on Rock Bolting Theory and Application in Mining and Underground Construction, Vol.1 P179-191

Hydropower'92, Broch & Lysne (eds) © 1992 Balkema, Rotterdam. ISBN 90 5410 054 0

Underground powerhouse chamber design of ERTAN hydroelectric project

Li Wo-Zhao
Chen du Hydroelectric Investigation and Design Institute of MOE.MOR., People's Republic of China

A. INTRODUCTION

ERTAN hydroelectric project is located on the lower reaches of Yalong river in the southwest part of China's Sichuan province and at a distance approximately 46 km from Panzhihua city. The multi-purpose hydroelectric project is developed mainly for power generation. The main features consist of a 240 m high double curvature concrete arch dam, left bank underground powerhouse complex, log passing facilities, right bank spillway tunnel etc. There are 6 sets of units installed in the powerhouse. The capacity is 550MW for each. The design waterhead is 165m, the unit inflow is 370m³/s and annual output is 17,000GWh. The left bank underground powerhouse complex is comprised of a power intake penstocks, underground powerhouse, main transformer chamber, tailrace surge chambers. The powerhouse size is 280.3x25.5x65.38m (Length x width x height).

The layout of the complex is shown in Fig. 1.

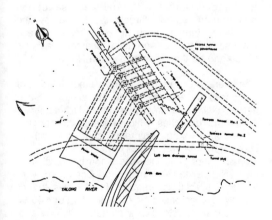

Fig.1

B. THE GEOLOGY CONDITION OF THE POWERHOUSE COMPLEX

The powerhouse is located inside the left bank of the high mountain. The average rock cover is approximately 250~300m. The rock is mainly comprised of syenite, gabbro and meta basalt. The rock strength is high, the rock mass is fresh and intact. The rock is classified as grade A and B, part of them belongs to grade C, according to the classification method of CHIDI (Chen Du Hydroelectric Investigation and Design Institute, Ministry of Energy and Ministry of Water Resources, P.R.C) for ERTAN project.

According to the appearance in exploration adits, only near the endwall of auxiliary powerhouse No. 1 distributed part weathered rock of joint-wall chloritized basalt. It's joint intensive zone and fracture zone and filled with chloritized mineral, belongs to grade E-3. There are no big fault zones in the rock mass. There are some lessly developed fracture zones (The width less than 0.2m) which are distributed in the most abundant rock types occasionally.

The four main joint sets in the rock mass are as following:
    Set 1: Strike N30° ~ 50° E. Dip 60° ~ 80° NW
    Set 2: Strike N40° ~ 60° W. Dip 60° ~ 80° NE
    Set 4: Strike EW   Dip 45° ~ 70° S
    Set 6: Strike EW   Dip 25° ~ 30° S

Joint sets 1,2 and 6 are well-developed in syenite and joint sets 1, 2, 4 and 6 are well-developed in basalt. Most joint surfaces are close and mainly filled with calcareous spar veinlet.

The main parameters of the rock mass are listed in table 1.

Table 1

classification of rock mass quality	R.Q.D. %	E_o Gpa	rock mass				joint surface
			μ	$\text{tg}\phi$	C Mpa	V_p m/s	$\text{tg}\phi$
A	ζ_s	80	35	0.17	1.73	5	3600	
B	$P_\beta F_s$	75	35	0.17	1.73	5	4300	0.65~0.75
C	ζ_s	70	15	0.2	1.43	3.2	3300	
E	$P_\beta F^o$	25	2.5	0.25	0.58	0.8	2300	
fault			0.3~1.0	0.35	0.36~0.5	0.05~0.2		

NOTES: ζ :syenite $P_\beta F$:Beta basalt
$P_\beta F^o$:Joint-wall chloritized basalt

Table 2 The relationship between cave wall and fracture

Orientation	Upstream wall	Downstream wall	Stereogram of fracture
N 6°E			
N 6°W			
N 40°W			

Twenty-four measurement sets for the in-situ rock stresses have performed at the ERANT project area among which the 14 sets are spatial in-situ stresses. The displacement observation inversion for the convergence tunnel of underground powerhouse has been taken. The test result shows that the complex is located in the high in-situ stresses area. The stress is mainly tectonic stress.

The magnitude of the in-situ stress is improved from the upper part of the valley to the river bottom. The stress at river bed bottom reaches 40~60MPa. The result of the 9 sets spatial in-situ stress testing shows that the direction of the Max. principle stress is between N $10° \sim 30°^E$, the dip angle is smaller than 30°, the dipping is to river bed. The magnitudes are stable. In syenite, $\sigma_1=20\sim30$MPa; In basalt, $\sigma_1=30\sim40$MPa. The average σ_2/σ_1, σ_3/σ_1 is 0.53,0.33 respectively.

C. SELECTION OF POWERHOUSE CHAMBER AXIS

The principle shall be followed as a general rule to determine underground hydroelectric powerhouse chamber axis:

1. Consider the general layout of the complex in order to be favourable to operate safely, shorten the construction period and save the investment.

2. Make a larger angle between the orientation of the powerhouse complex and the strike of the main joint sets of the surrounding rock as far as possible.

3. Make a quite small angle between the orientation of the powerhouse complex and the direction of the Max. principal stress.

According to the above principles, the three schemes of the orientation of N6° E,N6° W,N40° W have been studies from the relationship between joint sets, in-situ stress, layout of project and chamber axis.

According to taking the joint sets 1, 2 and 6 to be as the representative structure planes, the stereographic projection analyses done for the

three axes with the orientation of N6° E,N6° W,N40° W, the joint sets 1, 2 and 6 combined with the free surface except the particular condition would not form the unstable sliding wedge for the downstream wall, so the downstream wall is basically stable. The effect of the structure planes on the upstream wall are varied for different orientations. The wall of the chamber with the orientation of N6° E is basically stable as shown in table 2. The orientation of N6° W and N40° W will form the slide wedges sliding along the joint set 2 for the upstream side wall. When the chamber is at the orientation of N40° W, comparing with the axis of N6° W, the unstable rock mass formed by the upstream wall structure planes has large volume and is distributed widely on the upstream wall. Because the strike of the joint set 2 is nearly in parallel to the longitudinal orientation of the chamber, it might imperil the crane rail ledge of the powerhouse to be formed. To sum up, from view point of high side wall stability influenced by the joint sets, the orientation of powerhouse chamber is unsuitable to turn more to the direction of NW. And because of the downstream wall is largely weakened by the transverse tunnels such as bus tunnel, tailrace tunnes, so the powerhouse chamber axis is suitable to be the orientation of the N6° W.

The lateral pressure ratio ($\lambda=\sigma^H/\sigma^V$), see table 3, of the sections perpendicular to the axes, of the 5 sets spatial in-situ stresses measured from the adit 4 and its branches which passes through the powerhouse area illustrates that the stress component lateral ratios ($\lambda=\sigma^H/\sigma^V$) of each measurement point increase when the axis turns from N6° E to NW direction. Compared with average lateral pressure ratio of N6° E axis, that of N6° W axis increases by 7.7% and that of N40 W increases by 33.6%. The average λ of N6° W axis is near to 1. In addition, taking consideration of the average Max. principle stress direction N22° E the angles between this direction and the orientation of

N6° E, N6° W,N40° W are 16°, 28° and 62°
separately. Some documents indicate that stability
of chambers is best when the longitudinal axes of
chamber forms a angle of 15° to 30° with the Max.
principal stress. So in the respect of considering
the in-situ stress, the axes of N6° E and N6° W are
better than that of N40° W. The axia of N6° W is
more convient for the layout of water conduits and
can shorten the length of penstocks. To sum up the
comparison mentioned above, the axis of N6° W
directions has greater advantages in the aspect
such as the rock mass stability, the layout of the
structures, constructions and period etc. The
orientation of N6° W for the powerhouse chamber
is chosen finally.

D LAYOUT FOR THE CHAMBERS

Because of the big and small chambers of
generating system composing a complicated space
complex, the surrounding rock stability is complex
and important. The surrounding rock stability of the
complex firstly depends on the main chamber's
stable condition.

The following three alternatives have been studied
for the layout of the three big chambers
(powerhouse chamber, main transformer chamber,
tailrace surge chamber) of the ERTAN underground
powerhouse.

Sheme 1: The axes of the three big chambers are
parallel shown in Fig.2.

Table 3 Relationship between plane
stress differet underground house
axis and lateral stress factor

Axis of cave / Test point Stress / Test point		N6°E	N6°W	N40°W
σ_{1-1}	σ_{α}	8.82	8.75	16.96
	σ_v	10.88	10.88	10.88
	λ	0.81	0.80	1.56
	η	0	−1.2%	48.1%
σ_{1-2}	σ_H	18.51	20.31	29.68
	σ_v	15.68	15.68	15.68
	λ	1.18	1.33	1.89
	η	0	11.28%	37.51%
σ_{1-3}	σ_α	17.56	19.17	23.63
	σ_v	17.01	17.01	17.01
	λ	1.03	1.13	1.39
	η	0	8.85%	25.9%
σ_{1-4}	σ_η	19.21	19.72	23.22
	σ_v	25.18	25.18	25.18
	λ	0.76	0.78	0.92
	η	0	2.56%	17.39%
σ_{1-5}	σ_η	8.51	8.33	13.99
	σ_v	10.78	10.78	10.78
	λ	0.79	0.86	1.30
	η	0	8.14%	39.23%

Notes: σ_v — vertical stress $\lambda = \sigma_H/\sigma_v$ η——λ. change rate
σ_H — horizontal stress; Unit of stress: M's

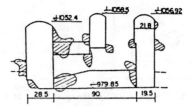

Fig.2 The axes of three chambers are
parallel

Sheme 2: The three big chambers are arranged
as Fig. 3 shown.

Sheme 3: The two chambers are parallel to each
other as Fig. 3 shows. (Arranging the main
transformer on the turbine floor between the units
of powerhouse, the main transformer chamber has
been cancelled.)

The tree alternatives have been analysed as
following from the conditions of station generation
and maintenance:

Sheme 1, the layout of the electric equipment is
symmetrical. It's convient for the line outgoing and
the repairing of the main transformer.

Sheme 2, the location of main transformer is 120m
higher than isolated bus tunnel. The heat emission
is difficult and the losing of electric energy. In the
mean time, the operation and maintainance are
inconvient and the crane equipment for main
transformer is needed.

Sheme 3, the main transformer is near to the units,
the bus is shortest the loss of the energy is little,
but there will be more interference when the main
transformer is checked and repaired. In case of
main transformer fire, it will imperil the powerhouse
safety.

The plane elastic FEM analysis for the three
alternatives of surrounding rock mass stability have
been performed (See Fig. 2, 3, 4). The rock mass
surrounding the chamber contains small range of

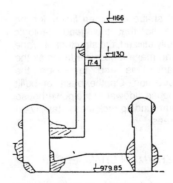

Fig.3 The three chambers are arrange

151

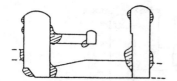

Fig.4 The two chambers are parallel to each other

rupture zones for scheme 2 and 3. For scheme 1, there are comparatively large rupture zones in the rock mas between chambers, but they are discontinous.

The operation conditions of scheme 1 is better than that of scheme 2 and scheme 3, the stability is slightly worse for the scheme 1. Thus, the alternative of parallel layout for three chambers is taken.

E. CHAMBER SHAPE AND IT'S OPTIMIZATION

The shape of the chamber should be studied in order to select the reasonable chamber shape for construction of underground works at the high in-situ stress area.

The straight wall with arch bullit shape, oval and "mushroom shape" have been studied for the main chambers of ERTAN underground powerhouse. According to the calculating results of the finite element, the comparison of the surrounding rock fracture zone of Fig. 4, and Fig. 5 shows that at the same in-situ stress field condition, the periphery stress and the range of the rupture zone are slightly small for the oval shape, but the improvement of them is limited. If the oval shape is chosen, the excavation span of the chamber section will be enlarged. The span of main powerhouse and the main transformer chamber will be more than 34m, 23m separatively. Although the space for chamber has been increased, the utilizing coefficient is low. The oval shape is unsatisfactory at the aspect of layout.

The "mushroom" shape is shosen finally for the main powerhouse so that the good geologic condition can be fully utilized, the structure of crane rail ledge which can make it possible to install the crane in an early time and to fasten the powerhouse construction. On the basis of bullit shape, the powerhouse sidewall of the powerhouse can improve the surrounding rock mass stability on some certain degrees.

F. SPACE BETWEEN MAIN CHAMBERS

The determination of the spaces between the chambers uses the analogue method having taken

the built underground works experience both domestic and abroad into consideration. The initial spaces between chambers have been decided firstly, then finite element method analysis and the model test of geomechanics methods have been used to prove the stability of the rock column between chambers. It's of benefit to operate and lessen the energy losses that the spaces between three big chambers are shorten. The in-situ stress of ERTAN powerhouse area is high, if the spaces between chambers are too small, the large rupture zone of surrounding rock mass will occur to influence the stability of chambers. So it's very important that the spaces between the chambers of the underground powerhouse are taken for the purpose of safety and economics. The statistical materials of rock column thicknesses of 32 large and media underground powerhouse chambers both domestic and abroad illustrate that rock colomn thickness (L) between chambers, the excavation span (B) of the maximum neighbor chamber and the chamber height (H) have the following relations (See Fig. 6):

$$L/B = 0.6 \sim 0.18$$
$$L/H = 0.35 \sim 0.8$$

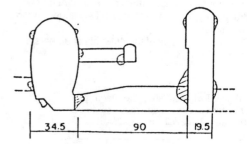

Fig.5 The two chambers with oval shape are parallel to each other

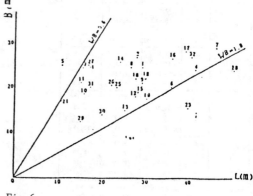

Fig.6

The alternative chamber spaces for scheme A,B of the three big chambers of ERTAN underground powerhouse have been compared during the design stage (See Table 4). The plane, three dimensional linear elastic-plastic finite element analysis simulating the excavation stage (Considering the shotcrete and rock support) have been taken for scheme A and B. For scheme A the plane geomechanics model test has been completed.

The analyses and model test above show:

1. According to the results of the plane and three-dimensional linear elastic analysis using the Mohr-Coulomb criteria, the chamber maximum horizontal displacement for both schemes shall occur at the middle part on the sidewall of the powerhouse chamber and tailrace surge chamber. The displacement is 4 to 5 cm. The difference isn't large. But the difference of surrounding rock mass rupture zone is obvious. The result of three dimensional finite element shows that the rupture zone of the three chambers for scheme A shall be continued as Fig. 7 shows. For the plane analysis the rupture zones shall be discontinued but the max. depth will be 18 to 20 m. (See Fig. 2). The fracture zones for scheme B are lessened obviously as Fig. 8 shown.

Table 4 Relationship between rock column thickness and L/B , L/H

Schem	Powerhouse		L_{1-2}	L_{2-3}		Transformer chamber		L_{3-4}	L_{4-5}		Surge chamber	
	B_1	H_1	L_{1-2}	B_2	H	P_2	H_2		H_3	P_3	H_3
A	28.5	72.55	40	1.40	0.55	20		36	30	0.39	21.8	17.56
B	31.2		57.4	1.92	0.83	17.4			52	0.67		

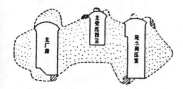

Fig.7 The rupture zone of Schem A– three dimensional liner elastic method analysis result

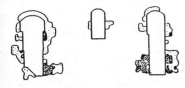

Fig.8 The rupture zone of Schem B – plane liner elastic method analysis result

2. The result of plane elastic-plastic for scheme B shows that the stress shall be improved obviously when the joint support measures of the shotcrete, wiremesh, rock bolts and rock anchors etc. are taken.

3. The plane geomechanics model test for scheme A shows that the opening load is 25.6 Mpa and the failure load is 46.8 Mpa, the overloading coefficient K=46.8/25.6=1.82. The safety margin is a bit small. The overloading coefficient shall be increased by 16.7% when the enhanced shotcrete and rock bolts support are taken.

The design engineer has warried about the spaces between scheme B, but according to following analyses:

1. The surrounding rock for three main chambers is excellent. The compressive strength is over 160 MPa. From the appearance of the adit located at the powerhouse area, there aren't caving in the adit excavated for years. The rock surface is smooth. The rock mass hasn't been loosing obviously.

2. According to the static materials shown Fig. 6, the L/B for about 50% station is 1 to 1.5. The L/B for the adjusted space between chambers is 1.1 to 1.5, and L/H is 0.46 to 0.55 which is near average value. Comparing the Churchill falls station and La Grande LG-2 station which are almost similar in the aspect of geologic condition, the dimensions of the chambers, the unit capacity with the ERTAN power station, the L/B for them is 1.04 to 1.17 and L/H is 0.61 to 0.63. The L/B and L/H are close. The space between chambers of Baishan underground station which was built during 1980's in China and has been operating for years is only 16.5m. The L/B is 0.66. It shows that the adjusted spaces between chambers of ERTAN underground powerhouse is more reasonable.

3. The result of the plane geomechanics model test for scheme A shows that overloading coefficient when the rock treatment measure is taken is increased by 16.7% compared with that of no supporting. The overloading coefficient on the condition of supporting is 2.12. Because the model test condition is limited, the shear strength of the simulating rock material ($\psi=60°$), and the supporting effects of prestressed anchorage haven't taken into consideration for the model test, so the result of the model test has some certain safe margin.

In the mean time, the following measures have been taken for the rock stability of the three big chambers:

1. The prestressed rock anchors have been used on the sidewall of powerhouse and tailrace surge chamber.

2. Move the GIS to the surface in order to lessen the main transformer height for 12m.

3. To adjust the inner layout of the powerhouse, keep part of the rock mass under the turbine floor

remaining. The "mushroom" shape was taken as shown in Fig. 9.

The elastic-plastic FEM analysis indicates that the surrounding rock mass stability for the adjusted powerhouse complex meets the requirement.

G. THE FINAL DESIGN SECTIONS AND TYPICAL SUPPORTS FOR ERTAN UNDERGROUND POWERHOUSE

The optimized transverse sectins and the typical supports after adjusting are shown in Fig. 10 and Fig. 11.

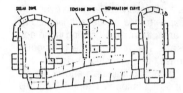

Fig.9 The rupture zone of Schem A after optimization

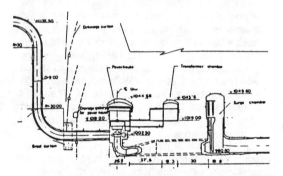

Fig.10 The section of the three biggest chamber
optimization

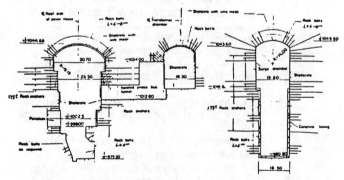

Fig.11 The typical rock treatment of the
three biggest chambers

Hydropower'92, Broch & Lysne (eds) © 1992 Balkema, Rotterdam. ISBN 90 5410 054 0

Unlined high pressure tunnel at Zhouning hydropower station

Liu Wengao, Cai Mingfu & Shi Zhiqun
Fujian Investigation & Design Institute of Water Conservancy & Hydropower, People's Republic of China

ABSTRACT: The paper introduces the geological exploration, ground stresses measurement of a proposed unlined high pressure tunnel at Zhouning hydropower station. Three dimension linear elastic and elastoplastic finite element computation shows that the secondary state and the third order state of stress fields around the rock mass adjacent to the tunnel is of compression in nature, which makes unlined tunneling feasible.

1 INTRODUCTION

Zhouning hydropower station, the second stage of Muyang river cascade development which is now under preparatory construction, is located in the east of Fujian province, having a total installed capacity of 250MW and an annual electric generation of 658Gwh. The water-retaining structure is a 73.4m high RCC dam, provided with 12.4km power tunnel having a diameter of 6.8m, and a maximum diversion flow of 70m³/s, leading to an underground powerhouse on its right bank. The high pressure tunnel is of shaft-type with an internal diameter of 4.7m, a height of 380m and a maximum head of 546m.

2 GEOLOGICAL EXPLORATION

The topographical elevation above the proposed underground powerhoure is between 200m to 777m above sea level. The surrounding land surface is of strongly degraded medium to low hilly morphologic region, with the topographical contour gradually drops eastward. Exposed rock mass can be seen in the huge and undulating hills, which are mainly composed of hard and compact K-feldspar (miarolitic) granite, having a high saturated compressive strength of 125-180MPa.

The high pressure tunnel and the underground powerhouse are to be built in slightly to fresh granitic rock mass. The adjoining rock which has been rated as class I and class II constitutes 94% of the total length, while those as class III and class IV, 6%. The permeability of rock mass is practically negligible except at very few fault and fracture zone. Under the favourable geological condition stated above, it is clear that the feasibility of unlined high pressure tunnel would be mainly determined by the magnitude of ground stresses and the depth of rock overburden.

One of the most important condition in the construction of an unlined pressure tunnel is to ensure that an adequate depth of overlying rock mass is secured so that the minimum of ground stress at any point along the tunnel is greater than the tunnel water pressure at that point. In the stage of high pressure tunnel alignment and preliminary verification, the rule-of-thumb criteria introduced by Norwegian experts and Snowy Mountain Engineering Corporation of Australia were used in determining the depth of overburden, giving a factor of safety between 1.58-1.90, see figure 1.

3 IN SITU GROUND STRESS MEASUREMENT

Hydraulic fracturing method was used in measuring ground stresses. The stress measurement bore holes (ZK604, ZK605) were set at the proposed under-ground powerhouse and the proposed high pressure shaft, having a depth of 342m and 450m respectively. The results are shown in table 1 and figure 2.

The following conclusion can be drawn from the ground stress measurement result.

(1) The ground stresses and the topographical elevation relationship can be described by the following formula:

$$S_H = 20.5484 - 0.0305H$$

$$S_h = 12.8090 - 0.0184H$$

where H: Ground elevation (m)

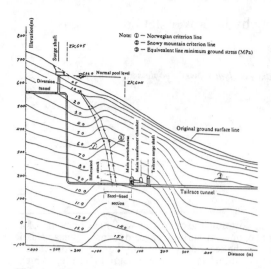

Figure 1

The ground stress regression analysis based on the data obtained from the two boreholes shows fairly high coefficient of correlation, with S_H and H, S_h and H being 0.895 and 0.919 respectively. It demonstrates the in–situ measurement result has fairly good regularity.

(2) The horizontal principal stress, measured in the compressed–fissure zone of perfect K–feldspar miarolitic granite or in that of closed crevasse, increases with depth, indicating tight cementation of the fissures and showing no effect on the principal stresses.

(3) The coefficient of lateral compression in the region is between 1.43–1.70, indicating the absence of stress concentration. The maximum and minimum horizontal principal stresses are relatively uniform, which is quite favourable for the stability of caverns.

(4) The ratio of the minimum horizontal principal stress to the static hydraulic pressure in high pressure tunnel S_h / h_s are 1.68–2.76 in barehole ZK605 and 2.05–2.38 in borehole ZK604, which meets the criterion of minimum principal stress for unlined high pressure tunnel and far exceeds the safety requirement of $S_h / h_s > 25\%$, see figure 3. This proves the design of high pressure tunnel without steel lining is feasible.

(5) The mean orientation of the maximum horizontal ground stress in the region is N40° W, almost perpendicular to the fracture zone with NE line of strike, providing a sound basis for aligning the axis of underground powerhouse caverns.

4 STRUCTURAL DESIGN

To prevent local instability or rock fall around the rock mass adjacent to the high pressure tun-

nel due to internal water pressure fluctuation, which may endanger the safety operation of turbine generator set and to avoid soluble filling material and intercalated clay in faults and joints from being eroded by high water pressure, which may develop into seepage passages, as well as to minimize water head loss and to reduce the diameter of high pressure tunnel, it is decided the high pressure tunnel to be lined with reinforced concrete.

The mathematical model of reinforced concrete lining structure is based on the harmonic condition of displacement in reinforcing steel, concrete and adjacent rock mass, the ratio of tunnel hydraulic pressure supported by reinforced concrete and adjacent rock mass can then be computed. The country rock is to be consolidated by high pressure grouting. The internal hydraulic pressure supported by reinforced concrete has been designed in comply with crack limit criterion, while the thickness of reinforced concrete lining is controlled by the stability against external water pressure.

5 NUMERICAL ANALYSIS

In order to evaluate the secondary state of stress field after the excavation of high pressure tunnel being completed and the stress strain behaviour of the third order state of stress field after the tunnel being put into operation so that the possibility of underwater tunnel piercing may be analysed, three dimensional linear and nonlinear finite element has been applied in the computation. In three dimensional plastoelastic finite element analysis, the yielding model criterion developed by Zienkiewiy–Pandit has been used, while in three dimensional nonlinear iteration computation, deformable plastic stiffness iteration method has been adopted, using a correction method introduced by Sinwandane and Desai for deviatoric stress modification.

The structure of the finite element networks is selected at the shaft's cross section having an elevation between 202–203m, which contains 90 parallelepiped elements of 8 nodes each, see figure 4. Primary ground stresses before excavation are: $\sigma_x = -9.38\text{MPa}$, $\sigma_y = -13.08\text{MPa}$, $\sigma_z = -12.44\text{MPa}$. Full section excavation of tunnel face and static water head of 450m have been assumed in the computation.

The computed secondary state of stress field shows that two circles of finite element networks develop 1m deep plastic zone. The basic nature of stress redistribution due to tunnelling reveals that the radial stress σ_r is released and the tangential stress is increased. The stress variation is gradually diminished along radial direction behind the tunnel wall, approaching the primary ground stress field at a distance of threefold diameter. The stress computation result is shown in table 2.

Table 1. Ground stress measusement result by hydraulic fructuring method

Point of measurement		Depth (m)	Elevation H(m)	Measured ground stresses (MPa)				S_H Orientation
				S_H	S_h	S_v	S_H / S_h	
Borehole ZK605	1	156.0	478.0	6.58	4.25	4.11	1.55	
	2	211.0	423.0	8.32	5.10	5.56	1.63	N28 ° W
	3	259.0	375.0	9.91	5.83	6.83	1.70	N49 ° W
	4	317.0	317.0	10.71	7.16	8.08	1.50	
	5	351.0	282.0	12.54	8.00	8.89	1.57	N50 ° W
	6	408.0	225.0	10.84	6.83	10.44	1.52	
	7	435.0	198.0	13.64	8.5ᶜ	11.50	1.59	N40 ° W
	8	445.0	189.0	13.76	8.70	11.76	1.58	
Borehole ZK604	1	153.0	382.0	7.74	5.41	4.02	1.43	N25 ° W
	2	178.0	357.0	9.09	6.22	4.71	1.46	
	3	251.0	284.0	11.49	8.06	6.62	1.43	
	4	293.0	242.0	15.24	9.03	7.74	1.68	N30 ° W
	5	315.0	220.0	15.49	9.82	8.32	1.58	N38 ° W
	6	334.0	201.0	17.32	10.56	8.82	1.64	

Note: S_H: Maximum horizontal principal stress
S_h: Minimum horizontal principal stress
S_v: Vertical principal stress

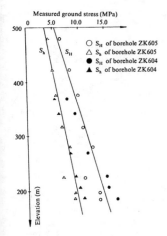

Measured ground stress (MPa)

○ S_H of borehole ZK605
△ S_h of borehole ZK605
● S_H of borehole ZK604
▲ S_h of borehole ZK604

Figure 2

Figure 3

Table 2. Stress computation result of adjacent rock and tunnel lining

Location	Element No.	σ_r(MPa)				σ_θ(MPa)				$\tau_{r\theta}$(MPa)			
		I	II	III	IV	I	II	III	IV	I	II	III	IV
tunnel lining (0.5m thick)	11			−3.72	−4.09			2.95	−0.13			0	0
	13			−3.73	−4.09			2.95	−0.13			0	0
	15			−3.72	−4.09			2.95	−0.13			0	0
	17			−3.72	−4.09			2.95	−0.14			0	0
	19			−3.72	−4.09			2.95	−0.14			0	0
the 1st circle of adjacent rock (0.5m thick)	21	−10.01	−5.50	−8.27	−8.73	−12.45	−27.14	−24.81	−24.37	−2.25	−1.51	−1.77	−1.77
	23	−12.99	−3.27	−5.94	−6.39	−9.47	−16.99	−14.28	−13.85	−1.85	−0.97	−1.00	−0.99
	25	−13.53	−2.60	−5.71	−6.23	−8.93	−14.05	−12.14	−11.64	1.10	0.28	0.39	0.58
	27	−10.89	−3.97	−6.67	−7.12	−11.57	−20.17	−17.81	−17.38	2.53	0.94	0.97	0.97
	29	−8.72	−5.00	−7.72	−8.16	−13.74	−25.28	−22.95	−22.54	4.62	0.01	0.02	0.03
the 2nd circle of adjacent rock (0.5m thick)	31	−10.01	−5.15	−7.43	−7.76	−12.45	−21.79	−20.16	−19.83	−2.25	−2.46	−2.69	−2.69
	33	−12.99	−3.99	−5.98	−6.31	−9.47	−15.31	−13.49	−13.18	−1.85	−2.15	−2.15	−2.15
	35	−13.53	−3.25	−5.21	−5.54	−8.93	−13.01	−10.98	−10.67	1.10	0.66	0.64	0.64
	37	−10.89	−4.33	−6.36	−6.70	−11.57	−17.68	−15.94	−15.62	2.53	2.10	2.13	2.12
	39	−8.72	−4.64	−6.65	−7.00	−13.74	−21.39	−19.60	−19.33	4.62	0.12	0.07	0.09
the 3rd circle of adjacent rock (2.0m thick)	41	−10.01	−7.40	−8.46	−8.69	−12.45	−17.35	−17.12	−16.91	−2.25	−2.64	−2.69	−2.70
	43	−12.99	−7.66	−12.59	−12.39	−9.47	−13.18	−7.97	−8.21	−1.85	−2.70	−2.74	−2.74
	45	−13.53	−7.33	−7.48	−7.71	−8.93	−11.88	−11.04	−10.82	1.10	0.89	0.89	0.89
	47	−10.89	−6.94	−7.55	−7.78	−11.57	−15.09	−14.57	−14.36	2.53	2.72	2.69	2.68
	49	−8.72	−6.61	−7.34	−7.58	−13.74	−17.93	−17.58	−17.34	4.62	0.40	0.28	0.29

Note: I — Represents stress component before rock mass excavation
II — Represents stress component after rock mass excavation
III — Represents stress components of lined body and adjacent rock mass under tunnel water pressure in linear elastic finite element calculation
IV — Represents stress components of lined body and adjacent rock mass under tunnel water pressure in elastoplastic finite element calculation

The third order state of stress field, as the tunnel subjected to high hydraulic pressure, develops tangential stresses $\sigma_\theta = 2.95$MPa (tensile stress)   in linear elastic rock mass and $\sigma_\theta = -0.13$MPa (compressive stress)   in elastoplastic rock mass. This indicates the concrete lining has already cracked and the tunnel hydraulic stress is completely supported by the adjoining rock mass. Comparison of the stress field in the country rock and the excavated tunnel reveals that the radial stress σ_r increases, while the tangential stress σ_θ decreases,   as shown in table 2. The rock mass behind the tun-

Maximum internal hydraulic pressure (MPa)

Minimum ground stress from regression method

Measured min. ground stress

Dynamic hydraulic pressure line

Static hydraulic pressure line

Starting point of steel lining

Length of high pressure tunnel tunnel (m)
(from the centre line of surge shaft to that of underground powerhouse)

Minimum ground stress (MPa)

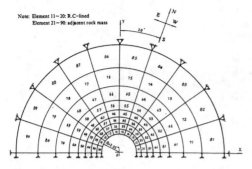

Note: Element 11~20: R.C-lined
Element 21~90: adjacent rock mass

Figure 4

nel is in compression condition, with the minimum tangential stress in the first circle $\sigma_\theta = -11.64$MPa. showing a fairly large margin of safety. Consequently, hydraulic piercing will not occur in the surrounding parent rock mass.

6 CONCLUSION

From the above mentioned criteria of overburden analysis, ground stresses measurement and the relationship of the minimum ground stress and the internal hydraulic pressure. It is clear that the proposed high pressure tunnel has been deeply located in perfect, compact, hard and fresh granite, having a ratio of static hydraulic pressure to rock overburden between 0.90−0.71 and a factor of safety between 1.58−1.90, which is up to the requirement of relevant criteria.

Ground stresses measurement by hydraulic fracturing method gives the ratio of hydraulic pressure to minimum ground stress being 1.68−2.50, which meets the safety requirement of the minimum ground stress should be larger than 25%−30% of the tunnel hydraulic pressure. The secondary state of stress field after tunnel excavation and the third order state of stress field after tunnel operation subjected to water pressure in the neighbouring rock mass are of compressive stresses.

The verification result mentioned above indicates that the high pressure tunnel free of steel lining at Zhouning hydropower station is feasible, without causing possible failure due to hydraulic piercing. however, further study is required to design a completely unlined tunnel during construction in future based on the specific condition of local geology, ground stress and adjoining rock mass.

Hydropower'92, Broch & Lysne (eds) © 1992 Balkema, Rotterdam. ISBN 90 5410 054 0

Prediction for field stress and stability of underground caverns

Liu Ying
Ninth Hydroelectric Engineering Bureau, Gui Yang, People's Republic of China

Peng Shou-Zhou
Hydraulic Department of Tsinghua University, Beijing, People's Republic of China

ABSTRACT: On the basis of measured stress and displacements of six profiles along the periphery of caverns which was excavated in the first step, the field stress was determined. FEM analysis for the critical cross-section of power house were carried out, the key excavation block and other factors which have significant influence upon the stability were pointed out.

1 INTRODUCTION

East Wind power station is located in south-west of China.

In the stage of investigation and design, lots of work had been done, such as the tectonic stress measuring in exploration tunnels, and many times of stress and stability analysis with FEM.

In the stage of construction, large number of surveying instruments had been embedded. After having excavated the roof, many data of deformations were measured. Deducing the tectonic stresses from these data, and predicting the variation of stresses and the stability of the surrounding rockmass in the next several excavating stages, we discovered that, when the cavern had been excavated to the final stage, the most dangerous cross-section would have the problem of stability of surrounding rock mass and need consolidating.

In the real course of excavation, several dangerous cracks filled with clay, were found in the rock wall, and pinched out gradually, thereby it was not necessary adding to additional pre-stressed anchors, and the measured data also showed the excellent stability of the surrounding rock mass, only one stress meter's reading value was too large in the part of assembly bay, but after the measure of consolidating was adopted, the surrounding rock mass tended to stability.

2 CALCULATING CONDITIONS AND THE RE-INFERRING OF THE TECTONIC STRESSES

From the measured results of tectonic stresses in the investigating stage, we learned that there were fairly large horizontal tectonic stresses near the power-house site, horizontal component of stress σx reached 119.4 MPa. After finishing the work in exploration caverns and boreholes, we basically knew the situation of the faults, jointings and the limestone near the power house area. Four faults crossed by the longitudinal axis of the powerhouse, dipped relatively steep. But the limestone beds dipped fairly gentle with some clay seams. The shear strength was very low.

After both the roof of the powerhouse and and the main transformer room had been excavated, a group of values of displacements were measured on the six surveying profiles, it was possible to re-infer the distribution of this area's tectonic stresses from these data.

In order to check the reliability of the data of the tectonic stresses measured originally, even more in order to predidt the stability of the cavern groups during the excavating proceeding, at first we used the FEM to deduce the tectonic stresses, which was then used to analyse the stability of the surrounding rock mass.

Before calculating, it was necessary to choice the most dangerous cross-section in advance. After deeply thinking, we thought the steep dip faults had extremely great influnce upon the rock wall, cross-section which the faults crossed the bottom of the wall, would be relatively dangerous, the determined cross-section is shown as Fig.1. Fig.1 illustrates the stages of excavating and sketch layout of the bolts, one bolt bar in the figure, stands for several bolt bars in real situation.

During the calculation, all the main geologic structures were simulated. Above the roof there were two approximately horizontal clay seams, within the calculated range, there still existed two small fairly steep dip faults.Both the clay seams and faults were standed by jointing elements, the other rock masses were reguarded as anisotropy materials.

Table 1. Several main parameters of stratified rock

Name of stratrum Parameter of material	No.1	No.2	No.3	No.4
E (parallel bed) Mpa	25000	35000	15000	6000
E (perpendicular bed) Mpa	20000	15000	8000	5000
Poisson's ratio μ	0.24	0.24	0.24	0.28
Cohension on bed layer C Mpa	1.000	1.200	0.600	0.25
Inner friction angle tg ϕ	1.0	1.1	0.9	0.6
specific gravity γ (t/m)	2.68	2.68	2.68	2.68

Table 2. Main parameters for clay seams and faults

Name of statrum Parameter	No.1	No.2	No.3	No.4	No.5	No.6
Cohension on bed layer : C (Mpa)	0.04	0.04	0.10	0.20	0.089	0.20
tg ϕ	0.5	0.5	0.6	0.5	0.4	0.5
E (Mpa)	1000	1000	2000	1000	1000	1000
Kt (Mpa)	370	370	740	370	370	370
Kn (Mpa)	3830	3830	7660	3830	3830	3830

Parameters listed in the two tables, were determined large number of field tests and lab tests.

For the clay seams and faults, two-dimension non-linear model was adopted, they were consider- ed as ideal elastoplastic materials. Bolts and prestressed anchors were considered as linear elastic materials, and standed by 1-D rod elements.

The re-inferring results showed that, when the periphery horizontal tectonic stress was 100 Mpa, the periphery shear stress was 7 MPa, the effect of fitting was the best. The theoretical tectonic stresses at the observed points according to the above-mentioned tectonic stresses, are compared with the observed values in Table 3. The theroretical displacements in the stage 2 and stage 3, are listed in table 4. From the latter two tables, the fitting is extremely nice, that shows, the originally measured tectonic stresses are basically reliable, and the inferred periphery stresses can be used to further analyse the stability of the surrounding rock mass.

3 STABILITY ANALYSIS OF SURROUNDING ROCK MASS

In former design stages, FEM analysis had been carried out for many times, and no same calculated cross-sections at each time, there was an analysis, with the faults in the calculated map being shown as what the dotted lines define in Fig. 1, because the faults were near the roof, and both the horizontal tectonic stresses and the normal stresses on the faults were fairly large, neither the roof nor the high wall was the most dangerous. The excavating stages were divided in terms of the real con- struction procedure, when we analysed the excavating of busbar tunnels, pen-stock tunnels, and the tailrace tunnels, the elastic modulus were simulated by proportional reductions respectively. The total excavating period was divided into 5 steps, which were the stages from 2 to 7 in the calculation. In stage 8, concrete in power house was deposited, in stage 9, the bridge crane was being acted by loads in machine hall, comparatively speaking, the loads on the bridge-crane were not so big, therefor, for the stability of surrounding rock mass, the most dangerous stage was the 7, duirng this stage, the excavation had been finished, but the concrete was not deposited yet. Of course, for the bolts of the rock bolted crane beams, there perhaps would occcur quite big stresses, when the crane was being applied by loads.

Table 3

Stress Comparison	σx	σy	σz
Theoretic values	-11.32	-5.93	-0.844
Observed values	-11.94	-5.94	-1.0
Errors	5.2%	0.2%	14.6%

Table 4

Profile	No.point	No.stage	Location			Theoretic value	Measured value
No.1	B 2		main	upstream wall		4.24	2.31
No.2	A 2	Ⅲ		vault		5.01	0.75
No.3	D 2		house	downstream wall		3.73	6.54
No.4	B 5		trans-	upstream wall		2.96	-0.14
No.5	A 5	Ⅱ	former	vault		3.69	3.52
No.6	D 5		house	downstream wall		2.97	0.01
No.4	B 5		trans-	upstream wall		4.65	0.41
No.5	A 5	Ⅲ	former	vault		3.88	4.12
No.6	D 5		house	downstream wall		4.33	6.01

Remarks: 1. the positive sign stands for the direction to the cavern, and, the negative sign is
the opposite.
2. conclusions from the table:
2.1. on the profile of the roof of main transformer room, the observed values are quite
approach to the theroretical values. (profile 5)
2.2. on the profile of the roof of machine hall, the observed values are far less than the
theroretical values(profile 2), the main reason for this phenomenon is that the late
installation of surveying instruments for this profile causes the "losing" of partial
displacement.
2.3. for the downstream wall, the theroretical values are lower than the
observed ones, but, the difference is not big.

From the calculation, we found that, after the
third and forth stage, deformations of walls
were small, there were small increasing in
stresses of bolts, the yield zone of surrounding rock mass and the clay seams in the faults,
was small, and the rock mass was stable, but,
after the fifth , sixth and seventh stage, the
excavation had been finished, several faults on
the rock wall appeared, there was serious
sliding tendence, stresses of system bolts in
upper and down stream wall, surpassed the yield
strength in large scale, so were the stresses of
bolts for rock bolted crane beam, deformations
of roof were small, just 1 to 1.5 cm, but the
deformations of upper and down stream wall were
quite big, more than 4 cm, the biggest reached
5.4 cm. Faults in the deep from the wall and
rock mass around the tailrace tunnel, stepped
into plastic yielding in large range, see Fig.3,
and the slow convergence of deformation showed
the tendence of losing stability. Therefore,
prestressed high strength anchors had to be
installed on the lower half of the walls in
machine hall, with an interval of 1.5 ×1.5 m,
about 14 m in length, and prestress of 40 t per
anchor, after being dealed in generalities,
these anchors are shown as Fig.1. After the
consolidation mentioned above , deformations of
upper and downstream wall had small decrease-
ments, so did the yield zone, but there was quite
large decreasing for the stresses of bolts
installed around the cavern, the convergence was
accelerated in calculation, especially after the
deposition of concrete in power house, there

was a nice stable stituation, and large decreasing of plastic yield zone, even when the crane was applied upon loads, the yield zone was still very small, see Fig.4. So it was suggested that within the influential range by the two faults, the above-mentioned prestressed rock anchors should be installed. Even so, the most dangerous stage was still that when the power house was excavated to the bottom.

4 SITUATIONS IN CONSTRUCTION

In the course of construction, multi-point strain meters, stress-meters for bolts (including bolts of rock bolted crane beams), crack meters, some seepage pressure meters and thermographs, etc., had been embedded in the machine hall, there was still five cross-sections for measuring the convergence of rock bolted crane beams. In the main transformer cavern, multi-point strain meters and stress-meters for

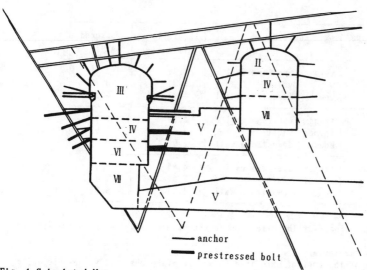

anchor

prestressed bolt

Fig. 1 Calculated Map

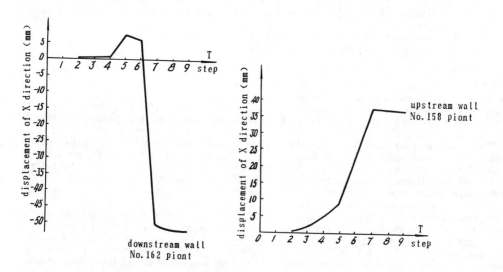

Fig.2 The maximum X-direction displacement of upper and down stream wall

bolts had been embedded as well. The typical locations of various kinds of embedded surveying instruments are shown in Fig.5.

With the excavating step by step, lots of important informations from so many surveying instruments were acquired, and are listed as following:

4.1 Untill the forth stage of excavation (the level of hydraulic turbines and busbar gallery), the measured stress values of each bolt were not

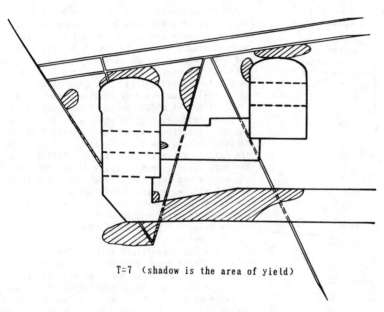

T=7 (shadow is the area of yield)

Fig.3 Yield zone of the periphery rock mass and faults without prestressed anchors, after the full excavation of the powerhouse.

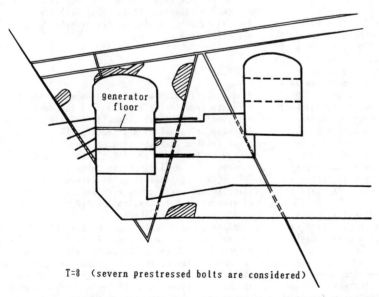

generator floor

T=8 (severn prestressed bolts are considered)

Fig.4 Yield zone of the periphery rock mass and faults with the prestressed anchors being installed, after the the concrete had been deposited.

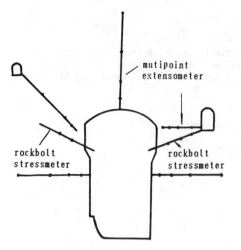

Fig.5 Layout of conveying instruments

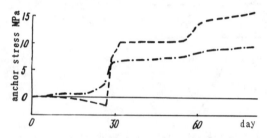

Fig.6 The observed stress of an anchor bar

so big, 60 percent of them varied from 10 MPa to
25 MPa, and 80 percent of that occurred after the
forth excavating stage, see Fig.6. Stresses of
bolts in upper stream wall were larger than that
in downstream wall, stresses of bolts were
greatly relate to blasting, and increased after
each blasting. After the forth stage, on the
downstream wall, the maximum stress in the bolts
for rock bolted crane beams reached 58 MPa.
After the power house had been excavated
thoroughly, the maximum stress values in the
stress-meters of the bolts for crane beams
reached 95 MPa, this stress-meter was embedded
in one side of the asembly bay near the access
tunnel the reasons why its stress was very large
were that:
  1. on the lower half part of walls, there were
two rows of system bolts not being installed.
  2. near the crossed part of access tunnel.
  3. near a small fault.
  Stress values in every stress-meter for bolts
near this calculated cross-section, were less
than 90 MPa, but the calculating results showed
that, when the cavern had been excavated to the
bottom, the tensile of common bolts and the bolts
for rock bolted crane beams reached 150 MPa,

some special ones even reach 200 MPa. The reason
for this diference is that, the faults in the
calculation, pinched out below the crane beams
in real situation, and has good combination and
no clay seams, so the stability of surrounding
rock mass is greatly improved.

4.2 The stretching values in crack-meters
located between the rock bolted beams and the
rock wall, were small, the biggest one is on
the downstream wall, 0.545 mm. That indicated
the basically nice cohesiveness between the roak
wall and the concrete beams, the convergence
value of distance between two crane beams was
also small. After the hydroturbine floor had been
finished, in the middle of the power house along
the long axis, that is thought to be approch to
the most dangerous calculated cross-section, the
maximum convergence value is near 4 mm, less than
the calculated value, and mainly resulted from
the excavation of hydroturbine floor. When the
excavating was near the bottom, the convergence
would have some increasements again.
  Because the geologic situations described
above had changed, and because the stress and
strain values in real conditions were lower than
that estimated, the construction contractor
abandoned the prestressed anchors for the lower
half of the wall, saved about 1 million yuan(RMB).
But, in the assembly bay and some bifurcations of
tunnel, besides supplying the missing bolt bars
in terms of the original design, we added a small
number of rock bolts again. The stability of the
power house as a whole is by far better than
that estimated originally, and the really
installed rock bolts are by far less than that
designed originally. At present, both the
depositing of concrete and the assembly of units
are being done.

5 CONCLUSIONS

It is an effective method to deduce the tectonic
stresses from the deformations of the roof
observed in the first stage. It is needed in the
deducing calculations and choosing the designed
cross-section that the most unfavourable
geologic structures are taken into consideration.
Otherwise, it is difficult to garantee the safty.
This method of predicting the deformations and
stability of surrounding rock mass in the course
of excavating according to the deduced tectonic
stresses, is of feasible.
  Unfavourable structure planes play a
controlling role for the stability of the
surronding rock mass. The faults with sliding
rock blocks facing the free plane, resulted from
the intersection of steep dip and high wall, are
of the most dangerous. And the measure of using
the prestressed anchors to consolidate them ,
thus to avoid losing stability, is of safe and
feasible. But the role of these rock anchors

effecting on reducing the deformations of rock mass are limited.

Under the premise of guaranteeing the stability of the surrounding rock mass as a whole, the operation of the rock bolted crane beams is of safe and feasible, though the distance between two crane beams has the tendence of decreasing convergence, the numerical value is basically small, within the range of adjustable. Although the bolts for the crane beams have born quite large stress during the excavating, the finedesign and adequate system bolts can guarantee the safe operation of the crane beam.

It is more dangerous to excavate the crossed parts of cavern and the last several levels of cavern, that needs to be paid high attention, and it is necessary to carry out prediction in advance and adopt some measures to make sure the safty.

The surveying during construction is of extremely important, adjusting the supporting measures according to the measured data, can guarantee the safty, and save the consruction costs. Therefore, the construction contractor should attach great importance to it, regard it as the necessary work.

Hydropower'92, Broch & Lysne (eds) © 1992 Balkema, Rotterdam. ISBN 90 5410 054 0

The role of monitoring information in the design of underground powerhouses

Lu Jiayou & Du Lihui
Department of Geotechnical Engineering, Institute of Water Conservancy and Hydroelectric Power Research, Beijing, People's Republic of China

Ji Liangjie & Liu Yongxie
Chengdu Prospecting and Design Institute, Ministry of Power Resources, People's Republic of China

Yang Ziwen
Large Dam Safety Supervision Center, MWREP, Hangzhou, People's Republic of China

ABSTRACT: The underground powerhouse of the Lubuge Hydroelectric Power Station is 125 m long, 38.4 m high and 18 m wide. It is built in a Triassic formation consisting of limestones, limy dolomites and dolomitic limesstones etc. The surrounding rock of the powerhouse has been reinforced by rock bolts and thus is rather stable. During the construction of the powerhouse, the deformation of the surrounding rock was monitored. In this paper, the initial stresses in the rockmass before excavation were obtained by back analysis of the deformation data observed in the undrground powerhouse; a failure model which accounts for the elasticity and brittle fracture of the surrounding rock was established by analysing the result of deformation observation, the phenomena in the surrounding rock after excavation and the results of rock mechanics experiments; the stress distribution in the surrounding rock and the reinforcing effects of rock bolts were computed by the FEM and the mechanism of anchorage effects was analysed. The result of the present study suggests that for underground projects carried out under similar conditions in the future, the cost for anchorage supporting can be further lowered by improving the design.

1 INTRODUCTION

In the early stage of underground project design in China, the Prodadianov's theory was generally adopted. Because that theory was outmoded, projects designed by it were of tremendous amount of work, long period of construction and high cost. In the Lubuge hydroelectric project, tunnels in the area of the underground powerhouse are crisscross and the structures are of relatively large scale. The underground workshop is large in size, with an excavation 125 m long, 18 m wide and 38.4 m high. 39.25 m downstream from the main powerhouse, the main thansformer-switch room is arranged in parrallel, for which an excavation 82.5 m long, 12.5 m wide and 25.7 m high was made. 50 m from the downstream sidewall of the main transformer-switch room, there is the tailwater gate chamber. Besides, there are the four bus galleries, transport and ventilating adits, etc. In order to change the backward situation of underground project design, a series of studies were carried out in relation with this project, including laboratory and in-situ tests on the mechanical behaviors of rocks, study on rock classification, in-situ geostress measurements, deformation observation in lab and in the prototype adit, back analysis of geostresses based on the result of deformation observation, physical simulation of

the tunnel deformation and stress distribution, numerical analyses, etc.

The work presented in this paper was carried out in the summarization stage when all the items mentioned above had already been completed. In this study, the stress state in the surrounding rock was analysed by the FEM. From the test results of the mechanical properties of rocks at Lubuge, it has been found that the instantaneous stress-strain relationship of the rock is almost linear and that the time-dependence of deformation values of the powerhouse observed in-situ is not significant. Therefore, an elastic-brittle fracturing-failure model was adopted in computation. The initial rock stresses used for computation were back analysed from the observed deformation values of the powerhouse.

In some of the numerical analyses by the FEM conducted in the early stage of this project, the elastoplastic model was adopted which led to rather large a plastic region. The result of the present computation showed that the integral stability of the powerhouse is good, with only localized fracturing of rock in some individual locations. Such fracturing phenomena were found in some bus galleries, which were quite similar to the computational result. The computational result showed that only the

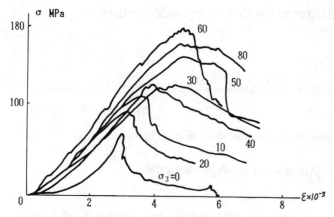

Fig. 1 The stress-strain relationship of brecciated limy dolomite obtained by triaxial test

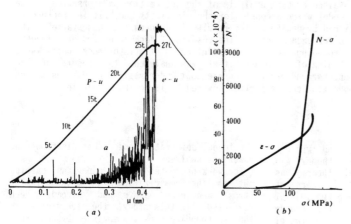

Fig. 2 The relationships of stress versus strain and total AE count versus stress for dolomitic limestone uniaxial compression

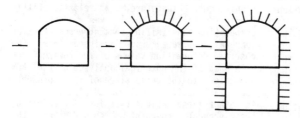

Fig. 3 The flowchart for computing the stresses in the surrounding rock of the undergroundd powerhouse

reinforcement by rock bolts is enough for the surrounding rock of the Lubuge underground powerhouse and that for underground powerhouses built in the future under similar geological conditions, it would be possible to improve the design so that the cost of rock bolts can be lowered further.

2 MECHANICAL MODEL

Rocks in the site of the underground powerhouse of the Lubuge Hydroelectric Station. are limy dolomites and dolomitic limestones. From the results of compression tests on rocks using an MTS servo-controlled stiff testing machine, it was found that the stress-strain relationship of these rocks is essentially linear up to the peak stress. There are four stages of deformation: first, the closure of initial cracks; second, the linear deformation; third, the non-linear deformation which accounts for 8-15% of the total deformation before the peak stress, and finally the unstable fracture propagation after the peak stress(see Fig. 1). Uniaxial tests showed that the rock failed by brittle fracture. The authors have also carried out uniaxial compression tests and acoustic emission(AE) measurements on dolomitic

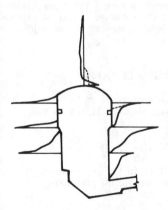

Fig. 4 Deformations of the surrounding rock at an observation section of the underground powerhouse

limestones collected from the site of another project, the stress-strain relationship of which is also nearly linear. However, it can be seen from the results of AE measurement that fracturing in rock starts at about one third of the peak stress(Fig. 2).

We adopted the Griffith criterion to express the fracture initiation in rock and the Coulomb-Navier criterion to describe rock rupture. In this way, the Griffith and Coulomb-Navier criteria express respectively the lower and upper bounds of rock failure while the stress-strain relationship from fracture initiation to rupture still remains linear. This is the basic point of the elastic-brittle fracturing-failure model used in the present computation. The same model has also been used to analyse rockbursts(Lu 1987, Lu et al 1988, 1989). However, for rockbursts to occur, not only the above strength criteria must be satisfied but also the criterion of instability should be fulfilled(Lu et al, 1990). In contrast, for the usual analysis of failure mode, only the strength criterion needs to be satisfied.

3 WAY OF COMPUTATION

Firstly, the upper stage was excavated and the stress state in the surrounding rock of the upper stage was computed; secondly, rock bolts were applied to the upper stage; and finally, the lower stage was excavated to the designed cross-section and stresses in the surrounding rock and in rock bolts were computed. In this final period, rock bolts in the upper stage had affected by the excavation of the lower stage.

In the computation, rock bolts in the lower stage were assumed to have been embedded before the excavation of the lower stage and thus the computational result did not reflect the real stresses in rock bolts but was only an approximation for reference. The flowchart of computation is shown in Fig. 3.

4 INITIAL STRESSES IN THE SURROUNDING ROCK AT THE UNDERGROUND POWERHOUSE SITE

In order to observe the deformations of the

Table 1 Values of geostresses in the underground power house site(MPa)

	σ_x	σ_Y	τ_{XY}
Actually measured	7.20	10.90	3.50
Back analysis for the model adit	5.38	10.00	1.98
Back analysis for the main plant	6.78	10.12	-2.30

surrounding rock, a test adit scaled down to 1/10 of the size of the underground powerhouse and 32 m long was excavated in a prospecting adit which runs in the same direction as the main powerhouse. Geostresses computed by back analysis of the observed deformations are listed in Tab. 1. It can be seen from the table that the values of geostresses obtained by back analysis and by actual observation(Tab. 1) are close to each other. Moreover, the direction of the major principal stress is subparallel to the slope surface of the valley near the powerhouse site. These imply that the compuation results are reasonable.

Three sectios for observation were deployed during the excavation of the underground

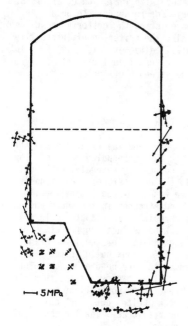

Fig. 5 Range of fracture and stress state in the surrounding rock

powerhouse. Fig. 4 shows the positions of instrumentation and the deformation values on the boundary of the surrounding rock at sections concerned in the present computation. To account for the effects of powerhouse excavation, the observed deformation values were corrected empirically and then used in back analysis by the BEM to give the initial stresses in rockmass(Tab. 1).

It is seen clearly from Tab. 1 that the values of geostresses given by back analysis for the main plant are close to the values obtained by actual measurement and by back analysis based on the deformation of the model adit, except that direction of shear stress is opposite which makes the major principal stress no longer subparllel to the valley surface. The reason may probably lie in the fact that the deformation of the plant is larger on the downstream side than on the upstream side due to the existence of a bus gallery near the downstream side of the section for deformation observation and that there are more excavations on the downstream side. Because of these factors, the geostresses obtained by back analysis and no longer the real initial geostresses. However, the result of direct analysis using these geostresses should correspond to the actually measured deformations and can reflect the actual working state of the surrounding rock. Therefore, the stress values in the rockmass obtained from back analysis of deformations of the main plant were adopted in our computation.

5 MECHANICAL PARAMETERS OF ROCKS

The mechanical parameters of rock and rock bolts used in computation are listed in Tab. 2.

6 RESULTS OF COMPUTATION

6.1 Range of fracture in the surrounding rock and stress state prior to fracturing

Table 2 Mechanical parameters adopted in computation

Material	Volume Weight	Compressive strength	Tensile strength	shear strength		Poisson's ratio	Deformation modulus $E_o \times 10^4$
	γ_D (T/m³)	σ_R (MPa)	σ_T (MPa)	ψ (°)	C (MPa)		(MPa)
Limestone	2.70	74.20	4.90	55.00 65.00 68.00	4.00 7.00 9.533	0.20	3.25
Rock bolt	2.67	320.00	320.00		160.00	0.20	21.00

Fig. 6 Fractures of the surrounding rock in a bus gallery

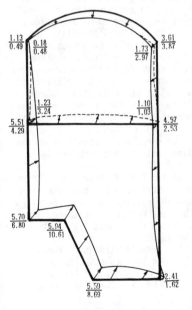

Fig. 7 Deformations of the surrounding rock around the underground powerhouse

The distribution of fractured areas, locations of fractures and the stress state prior to fracturing are shown in Fig. 5. In order that these contents can be expressed clearly in the figure, the stress vectors in the intact rock are omitted. The term "fracturing depth" concerned in the ensuing paragraphs are referred to as the depth of points of fracture initiation. As for the depth of fracture extension, we have been unable to give an estimation in our computation.

After the excavation of the upper stage had been completed, fractures occurred at the corner points on both sides. The fracture plane on the upstream sidewall dipped upstream at an angle of about 74°, while the fracture plane on the dwnstream sidewall dipped downstream at an angle of about 70° on the surface and turned to about 80° at greater depth.

After the excavation of the lower stage, tensile fractures occurred on both the upstream and downstream sidewalls. The fracture points were usually 20 cm deep, but the fracture point on the upstream sidewall reached 1.8 m in depth. Fracture planes on the upstream and downstream sidewalls dipped upstream and downstream respectively with a relatively steeper angle. It is very interesting that in bus galleries normal to the downstream sidewall, fractures visible to naked eyes were found in the surrounding rock in a bus gallery(Fig. 6). The fracture planes were distributed vertically (parallel to the sidewall of the powerhouse). The density of fractures was 2-3 fractures per meter within the depth range of 5 m and then decreased gradually to about one fracture per meter and finally vanished beyond 10 m depth in the surrounding rock. Fractured zones in bus galleries were influenced by the three-dimensional stress concentration at the intersection of the underground powerhouse with the gallery and thus the stress state was the worst. Therefore, more fractures were produced in the rock surrounding the underground powerhouse along radial directions. Because the influence of bus galleries was not taken into consideration in our computation, fractures in our results were distributed in a smaller depth and non-parallel to the peripheric surface of the powerhoues. Nevertheless, the angle between the fracture and the powerhouse surface was very small. Considering that the initial stresses, mechanical parameters and the FEM procedure used in our computation all include some errors, the fracturing in the surrounding rock obtained from our computation could be considered as fairly close to the phenomena actually observed, showing that the mechanical model adopted in computation is correct. In contrast, the result obtained in the early stage by use of the elastoplastic model gave a

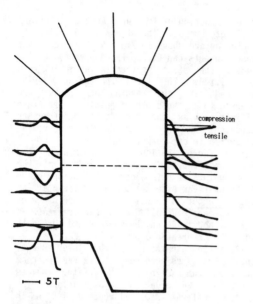

Fig. 8 Stresses acting on the rock bolt

large area of plastic region which was far apart from the realtities. A fractured zone amounting to 6 m in depth occurred on the upstream floor. It dipped dwnstream at an angle of 20-30°. In the depth range of 2-4 m, the fractured zone was shear in neture. On the intermediate and downstream floor, a zone of tensile fracture occurred. The upstream fracture plane dipped downstream while fracture plane on the downstream side dipped upstream. Fracture planes in the intermediate location were distributed crisscross with gentle dip angles.

6.2 Deformation of the surrounding rock

In Fig. 7, dashed lines denote the deformation after the excavation of the upper stage while solid lines the deformation after the excavation of the lower stage; the numerator and denominator denote the horizontal and vertical deformations respectivly. The sense of the deformation vector indicates the trend of deformation of the surrounding rock after stress release. The direction of deformation of the surrounding rock usually points to the excavation except a few corner points. The computed deformation values were smaller than the observed but the rate of deformation attenuation in the outward radial direction was smaller too, usually 0.125 mm/m. The actually measured deformation of the surrounding rock dropped abruptly at a depth greater than 3 m from the surface. All of these showed that the deformation values were affected by the blast-

loosening of the surficial layer of the surrounding rock which cannot be reflected in our computation.

6.3 Effect of rock bolts and their stress state

Since the time-dependence of deformations of the surrounding rock in the Lubuge project was not significant, rock bolts installed after the excavation of the upper stage began to work during the exacavation of the lower stage, while rock bolts installed in the lower stage corresponded to pre-installed ones in the computation. Therefore, the stress state obtained by computation was on the larger side. In some segments of the rock bolt, the stress was compressive. This was due to the fact that the rock bolt was embedded and thus its deformation was consistent with that in the surrounding rock which was non-uniform in the radial direction. Fig. 8 shows the stress distribution on the rock bolt.

7 CONCLUSIONS

From the computational results, it can be expected that local fracture and block-falling of rocks on sidewalls of the underground powerhouse may occur. Fractured zones of this kind and fracture belts on the floor are loosened regions in the common sense. In fact, the actual ruptured region is usually larger than the computed one owing to the influences of blasting and discontinuities of rock. Provided shotcrete reinforcement is given to such regions, the integral stability of the surrounding rock can readily be guaranteed.

Based on the fact that ruptured regions in the upper stage after excavation only occurred at some corner points, it can be inferred that the ruptured region on sidewalls should be somewhat related with the number of steps divided in excavation. It is suggested that the number of steps of excavation should be as less as possible. The computation result shows that the surrounding rock at arch supports is in a good state and thus is favorable to the construction of angle-table type crane beams. However, caution must be taken that the first step of stepwise excavation must be large enough so that the corner point where stress concentration takes place does fall into the position of the angle-table.

Computational results indicate that the stress in the rock bolt is not high and thus there is still some potential for lowering the cost of rock bolts by the inprovement in design.

ACKNOWLEDGEMENT
The authors are grateful to Mr. Yang Guohua of
the Research Department, Beijing Designing
Institute, Ministry of Power Resources, who
made the back analysis of geostresses in the
powerhouse site using the BEM.

REFERENCE

Lu Jiayou, 1987. Study on mechanism of
    Rockburst in a headrace tunnel. Proc. of
    Inter. Conf. on Hydropower. Oslo, Norwây
Lu Jiayou, Wang Changming, Huai Jun, 1988.
    FEM analysis for rockburst and its back
    analysis for determining in-sitn stress.
    Proc. of 6TH Inter. Conf. on Num. Methods in
    Geomechanics.Innsbruck, Austria
Lu Jiayou et al, 1989. Brittle failure of rock
    around the underground opening. Proc. of the
    Inter. Symp. on Rock at Great Depth PAU,
    Franch
Lu Jiayou, Du Lihui, 1990. Application of
    numerical method in the predict and controll
    on rockburst, Proc. of Symp. on the
    Application of Numerical Method in
    Engineering. Press of Tongji University.
    Shanghai, China

Hydropower'92, Broch & Lysne (eds) © 1992 Balkema, Rotterdam. ISBN 90 5410 054 0

Pressure tunnel in the Urugua-i dam (Misiones, Argentina)

Juan C. Malecki, Carlos A. Prato & Raul A. Lescano
Inconas S.R.L., Professional Engineering Consulting Services, Córdoba, Argentina

ABSTRACT: This paper includes, in a synthetic manner, the experience acquired from the design and subsequent loading operation of the pressure tunnel in the Urugua-i hydro-development, with scarce rock overburden on the lower section. This is a description of the approach used to check, in an economic and rapid manner, that at least within the range of pressures tested, the hydrofracturation phenomenom exists. The design of the lining is centered on satisfying the requirements of durability and control of water flow, related to the technology of the material used and the surrounding rock mass. Reference is made to the behaviour after the first pressure tests and the influence on the drainage system of the rock mass, which provide interesting orientative data, since the lower section of the tunnel is a media not free from the risk of hydrofracturation, although structurally significant for the resistance to internal pressures.

INTRODUCTION

The Urugua-i hydroelectric scheme is composed by an RCC gravity dam, with a zoned material embankment dam on each side. The powerhouse has an installed capacity of 120 MW, fed through a 900 m long and 7 m diameter tunnel.

The project was built on the Urugua-i creek, 6 km upstream from the creek discharge into the Paraná river, in the NE corner of the northern province of Misiones. The owner is the provincial government-owned electrical company, EMSA. The contractor was Consorcio Urugua-i, and the engineering was provided by Bechtel Int. The management of the project was performed by Inconas SRL.

There was a panel of experts and its members were Dr. Don U. Deere, Dr. Giovanni Lombardi and Dr. Flavio Lyra.

GEOLOGICAL FEATURES

The tunnel is within a series of nearly horizontal basaltic layers. For engineering purposes, the basaltic rock was classified into two principal types:

1. Microcrystalline basalt: very dense, massive with very small size voids, dark gray and very fractured.

2. Amigdaloid or vesicular basalt: They show a very porous texture, the colour is reddish and is slightly fractured. The amigdaloid basalt voids are filled with other minerals (silica, calcite) and the vesicular basalt has its pores free from any material.

The thickness of the layers is variable, most frequently less than 10 m. The strength of these rocks is very high, generally above 50 and below 180 MPa. The most noticeable discontinuities are the subhorizontal ones, and the majority correspond to the contact between layers, even though there are some subhorizontal discontinuities within the same layer. One of these discontinuities, of highly altered rock, was found between marks 175 and 268 m, measured along the longitudinal axis of reference. It was a zone 1,50 to 2 m thick, sloping towards the upstream side. The sloping configuration of the altered zone meant that the affected length of the tunnel was shorter than 50 m.

GEOMECHANICAL CLASSIFICATION OF THE ROCK

The two types of rock possess different mechanical properties and the differences are very noticeable. The microcrystalline-basalt is very fractured. The typical block is a prism of an average size of

0,50 m. There are three sets of well defined fractures, two vertical at almost
90° from each other and a third horizontal one. The vesicular and amigdaloid basalt has few fractures, mainly subhorizontal, and of very rough surfaces.

Table N° 1. Summary of the most important
geomechanical parameters.

Parameters	Dense Basalt	Amigdaloid Basalt
UW (t/m3)	2,9	2,6
Q	19,50	22,50
RMR	67	63
σ c (MPa)	170	54,0
$\varnothing$	58°	36°
C (MPa)	21,6	6,7
Longit.wave veloc.(km/s)	4,25	4,0
Shear wave veloc. (km/s)	2,50	2,22
Modul.dynamic deform.(MPa)	40300	32900
Modul.static deform.(MPa)	22250	21850
Poisson coefficient	0,23	0,28

where:
- Q and RMR: Classification parameters
  (1,2)
- The values of the strength shown are
  the result of tests on specimens obtained through drillings
- The value of the static modules, corresponds to correlation between shear
  wave frequency and in situ deformation
  measurements (3) in brazilian projects
  located in the same geological environment.

RELATIONSHIP BETWEEN INTERNAL PRESSURE,
ROCK COVER AND THE NEED FOR LINING

A great number of failures in pressure
tunnels are related to zones in which the
internal pressure was higher due to the
equivalent pressure resulting from the
overburden (4,6,8). In this case there
are two locations in which the above mentioned condition exists. One of them, under the dam, and a stretch somewhat beyond the toe of the dam, the other between mark 750 m and the downstream end
of the tunnel. It is in these locations
where a possibility exists that hydraulic
fractures takes place, due to water passing through the lining (11).
What is conditioning the risk of hydrofracturing is the stress pattern within
the rock mass, which it is necessary to
evaluate (7) as well as the possibilities
of dissipation of the water pressure. A
complete evaluation implies the measurement of the natural stress pattern in the
rock, evaluating the magnitude and directions of the principal stresses. If
the smaller of the principal stresses is
larger than the internal pressure of the
tunnel, there is no risk of hydrofracturing. If this is not so, a lining is necessary to prevent the passage of water.
When the lining is partially pervious, as
in this case, it has to be considered
that cracks in the lining might exist,
allowing the transfer of the internal
pressure to the rock, therefore an adequate drainage system to relieve the
pressures has to be provided. In the Urugua-i tunnel, pressure relief boreholes
were performed.
To evaluate the risk of hydraulic fracture in the rock due to water losses
through the lining, a simple and practical method was used. Boreholes were performed from the inside of the tunnel at
different locations, and water pressure
tests were performed, locating the pressurized segments of the borehole my means
of plugs. The preselected segments had to
contain preexisting fractures (5).
A total of five horizontal boreholes
were performed, and they were aimed at
intercepting the principal patterns of
the crack. The pressures were within the
range of 1,6 and 5,0 MPa.
The diameter of the boreholes was 3"
with continuous extraction of borehole
cores. The inspection of these cores allowed the selection of the areas to be
tested that contained only one crack.
Each test was performed in two cycles,
the first (Fig.1) until the existing
crack began to open up, which was evi-
denced when a marked increase in the absorption of water ocurred, marking an
abrupt change in the slope of the pressure vs. water flow diagram; when these
conditions were reached, the water pressure was reduced to zero, and a second
cycle started, during which the crack re-
opened at lower pressures and it was under these conditions that only the rock
stresses acted.
During the first cycle the stresses
produced by the bond existing between the
two blocks was destroyed. This bond is
due to solidified salty solutions or residual hydrothermal liquids, vestiges of
which are commonly found in basalts of
this area.

In Fig. 1 where the two cycles were plotted (water absorption vs. water pressure), it can be seen that both results show a tendency towards the same level of pressure of the crack, and this is the joint in which the water pressure equals the stresses across the plane of the joint.

The results obtained indicate that for pressures twice the internal pressure of the tunnel under normal operating conditions, no hydraulic fracturing occurred. That applies to almost any crack of the Urugua-i tunnel, since the tests were performed on cracks at different positions, except horizontal.
Results are shown in Table 2.

Table N° 2.

Bore Hole	Overburden(m) and Rock Pressure (MPa)	Section of Test (m)	Re-opening Pressures (MPa)
PHF1		7,23- 7,73	1,6*
	25 m = 0,67	11,40-11,90	3,0*
	(pi = 0,60)	13,25-13,75	2,2
PHF2		7,81- 8,31	3,0*
	25 m = 0,67	10,37-10,87	3,0*
	(pi = 0,60)	13,85-14,35	2,8*
PHF4		9,25- 9,75	2,4
	51 m = 1,40	10,80-11,30	2,6
	(pi = 0,62)	11,65-12,15	2,4
PHF5		13,48-13,98	3,2
		17,25-17,75	1,8
	51 m = 1,40	15,48-15,98	2,0
	(pi = 0,62)	21,90-22,40	4,8*
		23,90-24,49	1,8
PHF7		8,50- 9,00	2,8*
	25 m = 0,67	13,81-14,31	4,0
	(pi = 0,70)	18,02-18,52	4,8

pi: Internal pressures

* up to this value without re-opening

These tests do not reveal the magnitude nor the direction of the stresses, and measurements performed in different parts of the world (9) have shown that the horizontal stresses are larger than the vertical ones, in zones close to the surface.

These considerations suggest that, in this case, the horizontal stresses are larger than the vertical ones, and it could be concluded that the minor principal stress is vertical.

FIGURE 1

BOREHOLE PHF 5

MARK 392.00

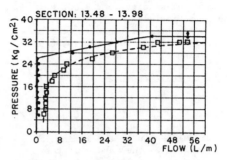

SECTION: 13.48 - 13.98

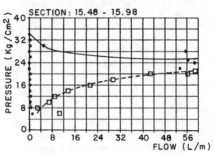

SECTION: 15.48 - 15.98

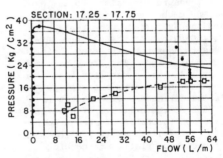

SECTION: 17.25 - 17.75

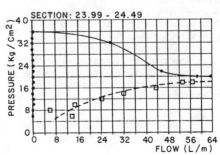

SECTION: 23.99 - 24.49

REFERENCES:
. 1ST. CYCLE
□ 2ND. CYCLE

177

DESIGN PARAMETERS

- Adopted for the rock mass
  Modulus of elasticity   20.000 MPa
  Unit weight   2,70 m^3
- Adopted for the concrete
  Modulus of elasticity   27.000 MPa
  Poisson coeficcient   0,25
- Internal diameter of tunnel   7 m
- Thickness of lining concrete   0,30 m
- Steel lining (16 mm thickness)
  .Shaft below intake
  .End of tunnel (length 20 m)

STRUCTURAL CALCULATIONS

The structural design of the lining is directly related to the nature of the rock mass and the internal and external pressures. In this case, under these particular conditions of pressure and overburden, there are two zones along the overall length of the tunnel:
i) Zone between the toe of the RCC dam (mark 180 m) and the surge chamber with a length of 493 m.
ii) Zones between mark 750 m and the downstream end (mark 930 m), 180 m long, and under the dam, 80 m long.

Along the first zone, the maximum static pressure within the tunnel is lower than the pressure equivalent to the overburden. The mechanical properties of the rock and the hydrofracturing tests performed, indicate that the maximum internal pressure does not represent a risk to the surrounding rock. The above circumstance, in addition to the fact that along this segment the maximum internal pressure is lower than the lithostatic pressure, leads to the criteria of checking the lining on the basis of strain compatibility of the lining with the rock. The specified characteristic strength obtained was 23 MPa at 250 days. The static modulus of elasticity of the concrete is 28800 MPa using the practical formula Eh = 6000 f'c (MPa) and, considering the assumed elastic modulus of the rock Er = 20000 MPa, the result is that the maximum stress in the concrete lining due to internal pressure, using the formula for thin shells is:

$$\int t = pi \frac{Eh}{Er}$$, pi being the internal pressure within the tunnel

The maximum value of pi in this stretch is 68 m of water head. Thus:

$$\int t = 0,68 \times \frac{28800}{20000} = 0,98 \text{ MPa}$$

This value is lower than the average tensile strength of the concrete (which is 2,3 MPa), and therefore, it is not necessary to perform the crack control, at least related to internal pressure.

Minimum reinforcements were provided for cracks due to thermal effects and contraction.

In zone ii), the water hammer condition increases the value of internal pressure to take into account in the structural verifications. Here, the maximum value of internal pressure is pi = 0,97 MPa, thus resulting in:

$$pi > \tilde{\gamma}_r \; h$$ (where $\tilde{\gamma}_r$ is the specific weight of the rock and h the depth)

Thus, in addition to the verification of the stresses in the lining and the rock, considering the compatibility of deformations of rock and concrete, it is necessary to guarantee the structural strength of the concrete lining to comply with the above mentioned requirements.

The criteria of compatibility of deformations leads to a maximum stress in the concrete of:

$$\int t = 0,97 \times \frac{28800}{20000} = 1,4 \text{ MPa, which is lower than the concrete strength (2,3 MPa)}$$

To check the lining as a structure, a fictitious internal pressure was adopted $p = pi - \tilde{\gamma}_r \, h$, calculating therefore the tangential forces as being N = p.r (r = radius of the tunnel). The reinforcement was designed to limit the cracks to a maximum of 0,2 mm under load N.

The criteria used for the cracking calculations is that used for non-constrained reinforced concrete structures, therefore not taking into account the surrounding rock as indicated by C.E.B.

It was admitted however, that using thin shell formulas in this case, it is only orientative for the design and distribution of the reinforcement, since the cracking of the rock could affect the validity of the assumptions made in a significant manner, because in addition to the deformations due to the internal pressure, there are deformations due to "hydraulic jack" effect in the cracks of the rock, should water pass beyond the lining.

To cover all possible circumstances of behaviour of the lining and the rock, it is necessary to ensure full contact between the lining and the rock. For that purpose, grouting was performed to fill cavities, to provide contact in the boundary lining-rock and to consolidate the rock within a 5 m thick ring.

It must be pointed out that the grouting pressures used were relatively low, around 7 kg/cm^2, and the amount of boreholes could have been scarce, therefore the consolidation of the surrounding rock and the contact lining-rock were not totally adequate to give full validity to the fictitious internal pressure used in the design calculations. Under these conditions, the pressure relief drainage system used for the rock in the zones of shallower overburden is relevant.

Between marks 180 m and 220 m (near the dam) and between mark 750 and the downstream end, where the overburden was not providing an equivalent value to counterbalance the internal pressure, vertical and inclined drains (30° towards the upstream end of the tunnel) were drilled from the surface, reaching 8 m below the tunnel invert.

The vertical drains were drilled 7,50 m apart, along each side of the tunnel and at 5 m from the sides of the tunnel. Inclined drains were drilled between the vertical drains, but 1 m closer to the tunnel, therefore the resulting mean spacing of the boreholes measured at ground surface was around 3,8 m. All drains reached 8 m below the invert and their diameter was of 3". Drains were also drilled in a natural lower zone 80 m to the left of the tunnel.

At the downstream end, and from the almost vertical face of the rock surrounding the upper part of the tunnel, 6 drains were drilled. Four of them were almost horizontal and 30 m long. The other two were sloping 50° until reaching 8 m below the invert (10).

BEHAVIOUR UNDER LOAD

During the successive tests performed after the first filling, related to the powerhouse and to the tunnel itself, water losses were detected and the measurements indicated 1300 litres per minute. Nearly 50% of that amount came from the drains.

Inspections performed revealed the existence of numerous cracks along the entire length of the tunnel. In the upstream section and up to the surge chamber, the cracks were mainly vertical and horizontal, rarely inclined. From the surge chamber to the downstream end (there is a steel lined portion along the last 20 m) there were many cracks inclined 30°, downwards towards the downstream end. The cracks were located on both sides and on the upper half of the lining. Spacing was about 1 m and the width was between 0,2 and 0,3 mm, and very few between 1 and 1,5 mm. After emptying the tunnel, the access of water into the tunnel was noticeable and the flow was higher on the downstream side.

To reduce the water losses, the most conspicuous cracks received treatment (about 10% of the total). First, the water access was prevented with rapid setting cement and afterwards a 7 cm wide masking plastic tape was placed on the crack. Following this, the masking tape was covered with an epoxi bituminous paint to a width 3 times the width of the masking tape.

The tunnel has been full for a year and a half (from mid-July, 1990), several measurements were performed (with all gates closed, the water losses are calculated measuring the water level descent either at the intake structure or at the surge chamber) and they are stabilized at around 900 to 1000 litres per minute, and about one half comes through the drains.

Water losses are minimized at present and the record shows less than 100 litres per minute in the drains.

Movements of the rock on the surface were also controlled. A 1 mm upward movement was verified, as regards the initial condition prior to the first filling.

REFERENCES

1) Barton, N. et al. Engineering Classification of rock masses for the design of tunnel supports. Rock Mechanics, Dec. 1974

2) Bieniavsky, Z.T. The Geomechanics classification in rock engineering applications. Proc. 4th Congress of I.S.R.M., Montreux 1979.

3) Bieniavsky,Z.T. The "Petite Sismique" technique, a review of current developments. Proc. 2nd Conference on Acoustic Emission Microseismic Activity in Geologic Structures and Materials. Trans Tech Pub., Rockport, 1979.

4) Blind, H. and Schwarz, J. Limits for pressure tunnels without steel linings. Water Power, July 1987.

5) Brekke, T.L. and Ripley, B.D. Geotechnical Engineering challenges in the design of pressure tunnels and shafts. The art and science of geotechnical engineering. Prentice Hall, 1988.

6) Broch, E. The development of unlined pressure shafts and tunnels in Norway. Proc. I.S.R.M. Symposium on Rock Mechanics: Caverns and pressure shafts, Aachen, Vol. 2, 1982.

7) Deere, D. U. and Lombardi, G. Lining of pressure tunnels and hydrofracturing potential. Victor De Mello Volume. Edit. Edgard Blücher Ltda., Brasilia, 1989.

8) Hendron, A.J., Fernández, G. Jr., Lenzini, P.A. and Hendron, M.A. Design of pressure tunnels. The art and science of geotechnical engineering. Prentice Hall, 1988.

9) Hoek, E. and Brown, E. T. Underground excavations in rock. The Institute of Mining and Metallurgy. London, 1980.

10) Lombardi, G. and Deere, D. U. Personal Communication.

11) Schleiss, A.J. Design of pervious pressure tunnels. Water Power, May 1986.

Hydropower'92, Broch & Lysne (eds) © 1992 Balkema, Rotterdam. ISBN 90 5410 054 0

Underwater tunnel piercing, state of art and recent experiences

Øivind Solvik
SINTEF NHL, Norway

ABSTRACT: A short survey of the most frequently used design of submerged piercing of tunnels is given. The paper deals with at least three dominating methods and some varieties and points out hazardous consequences if the design is incorrect or if the constructional work deviates from preconditions which are vital to the physical processes involved in the blasting.

1 INTRODUCTION

The great number of natural lakes in Norway are frequently used for water storage, mainly for hydro power production. A storage can be provided using a dam to raise the water level or lowering the waterlevel by excavating a tunnel under the lake and finally blast a hole in the lake bed making lake tap possible. Since Norway started developing hydropower in the beginning of this century many hundred lake taps have been done. Most of the blastings carried out before 1960 were based on thumb rools, good thinking and practical experiences. Since 1960 comprehensive research has been done to be applied in the design and execution of underwater tunnel piercing. This has most certainly lead to a better understanding of the physical processes. It is justified to say that we now are capable to find a safe design for any situation involving a submerged piercing.

2 THE OPEN SYSTEM

A general view of a lake tap is shown on Figure 1. There are many ways to do a lake tap blasting. The tunnel plug may be left open to atmospheric pressure through the gate shaft. Water filling is necessary in such a case to prevent violent upsurge and transport of debris from the plug that may result in damage to the gate. To prevent destructive shock wave it is essential to cover the face of the plug with an air cushion. The volume of this airbag must be sufficient to receive the gas from the explosive keeping the gas-pressure within acceptable limits. If the air bag volume is too small, the resulting gas pressure

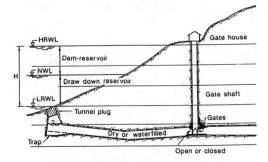

Figure 1. General view of a lake tap.

may be to high and if the volume is too big the resulting pressure from the explosive may be less than the external waterpressure. In such a situation the inflow of water will cause a postcompression which may be unacceptable. An inflow into a small chamber may be close to frictionless converting the inflow energy into compression-energy. Figure 2 shows how dangerous this high pressure may be.

The additional pressure in the air pocket caused by the explosive is depending on the amount of explosive, the volume and pressure in the air pocket before blasting. Figure 3 shows the resulting presure crused by the explosive.

Precompression of the air pocket is often an adequate and necessary way to reduce the max-pressure. But one should keep in mind that precompression higher than the external water column may result in loosing the air pocket which one cannot do without. The reason for that is the distructing shock waves that arise in

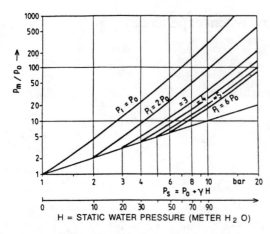

Figure 2. Pressure rise by frictionless inflow.

P_0 = Atmospheric pressure.
P_s = Abs. static pressure at the gate level.
P_m = Max. pressure in the pocket.
P_1 = Air pocket pressure before blasting.

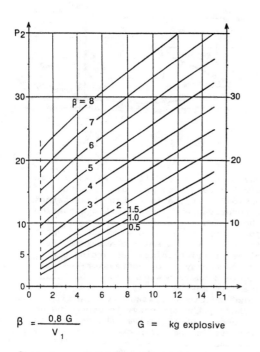

$$\beta = \frac{0.8\ G}{V_1} \qquad G = \text{kg explosive}$$

P_1 = Precompression in chamber

V_1 = Volume of precompressed air at pressure P_1

P_2 = Resulting max. pressure

Figure 3. Resulting gas pressure caused by explosive.

water with no damping air bag.

The precompression of the air bag in an open system is fixed by the waterfilling. The upsurge in the shaft is further a function of the inside and outside waterlevel besides the friction and turbulence losses.

To execute a lake tap using the open, waterfilled method, it is required to have completely knowledge of this multilateral process. But if so, the method is highly flexible and can be adapted to most of the existing cases. During the preparation work and execution it is necessary to controll waterlevel, airpocket volume and pressure.

3 THE CLOSED METHOD

The closed method is defined as a situation where the tunnel and the plug is closed to atmosphere by a gate at the upstream side of the gate shaft. The pressure in the tunnel before blasting is usually equal to the athmospheric pressure. But precompression is possible and may sometimes be necessary. For long tunnels and limited water depth on the plug, the max-pressure will be just a little more than static water head. For a short tunnel and a deep located plug, the max-pressure after the blasting may be unacceptably high. Figure 4 shows how the length/depth-ratio influences the max pressure. If the length-depth ratio is less than 2, the max-pressure increases rapidly and ratio less than 2 therefore is generally not acceptable.

The compression after the blasting is approximately adiabatic. The energyloss caused by friction and turbulence reduces the final max-pressure. That is the reason for why long tunnels give lower max-pressure than short tunnels. Likewise, a smooth concrete lined tunnel will offer a much higher max pressure than an unlined blasted tunnel. Figure 4 shows roughly this effect for a 40 m^2. tunnel.

If the precalculation proves too high pressure, precompression of the total air volume should be considered. For the calculation of the pressure using this closed method a numerical model has been worked and used for some years. This model is good and reliable for pressure calcula-tions, but not satisfactory to describe the transport of debris from the plug. This flushing process should be studied in a physical model which can be adapted to model-laws by means of a special perfor-mance of the model taking into account the fact that the atmospheric pressure is equal in model and prototype. This is an obstacle to the use of an Froudian model, but we have solved the problem by a special model technique.

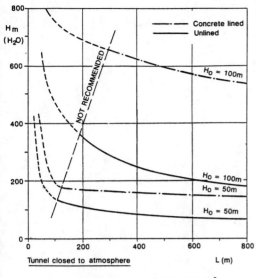

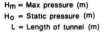

Hm = Max pressure (m)   Tunnel area : 40m^2
H$_0$ = Static pressure (m)   Singularloss :c = 1.0
L = Length of tunnel (m)

Figure 4. Max. pressure, closed system.

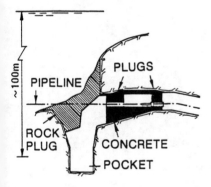

Figure 5. Pipeline shore appear. (Completed.)

4 CLOSED OR OPEN METHOD?

The discussion about the preference of the open or closed method, waterfilled or dry tunnel is some inadequate. The important thing is to select the method which is the best in any case. That requires a complete understanding of the different physical processes involved. We would like to emphasize the importance of precompressing a closed chamber. If that is done properly a submerged blasting can be done at very high head against small closed chamber. An example is the shore approach at Hjartøy of an oil pipeline from the North Sea, see Figure 5. The recommended pre-compression was here 5 bar and the blasting was after-

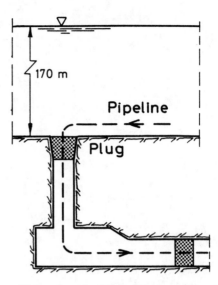

Figure 6. Pipeline. Shore approach. (Not completed.)

words judged to be 100% successful. There is really no limit to the depth where a piercing can take place. For the time being a 170-200 m deep pipeline piercing is under planning for A/S Shell Norge, see Figure 6. The limits are rather geological conditions and rock quality than lak of knowledge of the physical processes. In this case a precompression of the chamber is essential and is recommended together with some waterfilling to limit the transport of debris.

5 FINAL COMMENTS

In this paper has only been discussed the design criteria of a submerged piercing. It should be emphasized that thorough investigation of the geological and rock mechanical conditions are absolutely necessary. The numerical model is an adequate and time-saving way to examine any closed system. The model has been tested against a number of done blastings and therefore it is reliable. The transport of the debris is some more complicated to handle. The only way to test the transport or trap-system is still to use a physical model. One should not believe in good thinking and apparently good ideas if you deal with inflow velocities 10-20 m/sec. In such a case the debris is all mixed with the inflowing water forming a complete suspension passing through any kind of practical trap. That is a situation where the physical model still is superior.

Design criteria for asphalt and concrete pavement in unlined tunnels

Øivind Solvik
SINTEF NHL, Norway

ABSTRACT: The paper deals with dimensioning criteria for pavement lining in generally unlined tunnels. Pulsating forces set up by the roughness elements are described as dimensioning lifting forces in mild sloping tunnels. In steeper tunnels the pore pressure comming up during emptying the tunnels may be dimensioning for the lining.

1 INTRODUCTION

Due to good rock quality and not to forget confidence in the selfsupporting ability of the rock, hydropower tunnels in Norway generally are left unlined. Only 4-5% in average of the total tunnel lengths estimated to 3000 km are lined all around with conrete. Criteria for finishing of the tunnel floor have varied. Very often the tunnel spoil has been left open with additional material for the maintenance of the roadway during the excavation period. This procedure has very often lead to turbine damage, especially if hard minerals as quartz and feldspar are present. In this case expensive sandtraps are recommended. Removal of tunnel spoil to flush with the blasted rock face peaks will usually offer cheap additional tunnel area. This is therefore more frequently recommended than the above mentioned method. Turbine damage may still occur, and sandtrap still is considered to be necessary. These problems can be avoided if the tunnel spoil is removed completely from the tunnel floor. This is a time consuming and expensive method and the cross-sectional area gained will not make up for the increased roughness which means increased headloss. Sandtrap should not be necessary in this case, but the roadway for inspection is lost and would have to be rebuilt if substantial maintenance should be required in the future.

Recently we have introduced asphalt floor lining as an alternative to concrete as a method to reduce headloss, eliminate the need for sandtrap and save the roadway for later routine inspection or maintenance work. Asphalt linings have long been in common use as lining of canals and dikes and as cores in rockfill dams. Asphalt used as tunnel floor lining is the latest area of application. For the design of asphalt floor linings it has been necessary to carry out some research regarding friction and roughness factors which are needed to calculate the gain in power production due to reduced head loss. The research has also shown that the design criteria for the thickness or weight of the floor lining are given by pulsating forces set up by the tunnel roughness. In steep tunnels, however, dimensioning pore pressure may arise in the base material under the lining during emptying the tunnel.

2 ASPHALT FLOOR LINING MEANS INCREASED POWER PRODUCTION

We can describe the hydraulic roughness of an asphalt lining by means of the socalled Manning factor. This factor roughness coefficient n, in this paper replaced by the resiprocal value M = 1/n, has been determined by using a model flume for 5 different types of asphalt. It is obvious that the roughness factor is heavily dependent upon the finishing treatment of the asphalt surface, the grain size distribution and the amount and type of bitumen. The test results are shown in figure 1.

For the smoothest asphalt a Manning factor of 75-80 should be possible. We assume M = 75 to be a reasonable value and have used this value to show by an example how asphalt lining can reduce the headloss and thereby increase the power production. In figure 2 we have compared the head loss pr km length for a 40 m² tunnel for four different methods of floor treatment.

The annual benefit for the asphalt lined tunnel compared to complete removal of the tunnel spoil, depends very much on the dura-

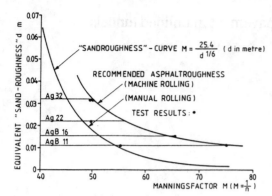

Figure 1. Test results, Manning factor for asphalt.

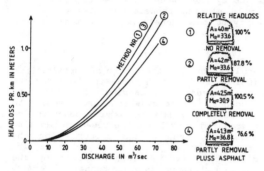

Figure 2. Headloss per km for 4 different treatments.

tion history of the load. It may make up for the asphalt cost. In addition comes the extra benefit represented by the permanent access road for inspection and maintenance work. If asphalt could be placed during the excavation period, constructional advantages in transporting could be added.

3 DESIGN CRITERIA FOR THE ASPHALT

Asphalt linings in water tunnels have to be made to resist strains different from those encountered when used as roadway lining. Some of them can be taken into consideration by constructional precautions. Extra uplift caused by trapped air after a filling up procedure could easily be avoided by arranging evacuation openings in the lining at suitable locations. Such openings should be narrow enough to prevent erosion of the base material. The movement of water will cause friction loss which means a sloping hydraulic gradient. Inhomogenous base material and cross-sectional area variation may give differences between the grade line on top of the lining and in base material and set up a lifting force. To solve this

problem we recommend as a general measure to arrange continous openings in the lining on at least one side. Abrupt constrictions in the tunnel will reduce the water pressure on top of the lining, but not necessarily to the same extent underneath the lining...At the upstream end of full-circumference lining, we therefore recommend to increase the weight of the asphalt lining by 50% over a 10-15 m length. The same procedure is usually followed at the downstream end of niches.

The design criteria for the lining has turned out to be given by pulsating forces set up by the roughness of the tunnel. These forces are acting on top of the lining as well as in the permeable base under the lining, giving rise to a force imbalance which varies in magnitude as well as direction. When the difference force is upwards directed we have a lifting force which is equal to the design force. Prototype measurements and model tests have shown that these forces are proposional to the velocity-head and are also a function of the absolute roughness or equivalent "sand roughness" of the tunnel.

The previous history of this research programme is interesting, but let us not hope typical of Norwegian hydraulic research: An asphalt lining with a minimum thickness was done in a head-race tunnel. During the first running season with maximum water velocity the lining was partly destroyed. The percentage distruction varied from 0-100%. This means that we had unintendently a full scale test with a safety factor of the lining ±1.0. We could study how the breaching process had developed step by step. It was obvious that the weakest part of the lining had been lifted slightly every time the pulsation force had the necessary magnitude. Gradually the lining was lifted and prevented of going completely back to its basic position because of sandgrain that also had moved. The lifting procedure could be compared with the use of a jack. At last big bubbles in the asphalt were formed. They probably burst when they reached a hight of 10-15 cm because that was the maximum hight to be observed.

The result of this event was that the lining was repaired by increasing the thickness. It also gave us an unique opportunity to research on the problem as we had observed part of the answer. Pressure transducers were therefore installed in the repaired and reinforced asphalt. In this case we could vary the velocity, but of course, not the roughness of the tunnel. A free surface flow was not possible in this tunnel and all this shortcomings lead to an extension of the research program.

First we did a model-study where both velocity and roughness could be varied. Then one more full-flow tunnel and two free flow tunnels were involved in the research. We found that both velocity, roughness and the

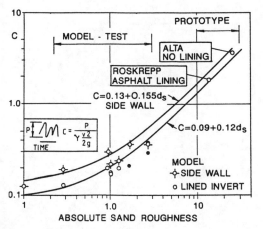

Figure 3.

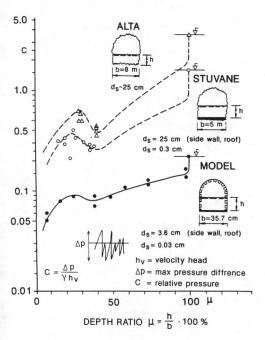

Figure 4.

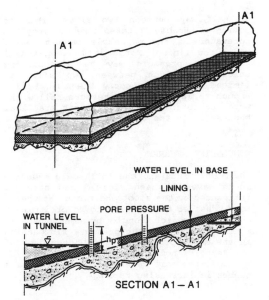

Figure 5.

higher pressure pulsations.

A free surface flow in a tunnel gives much lower pulsations. Compared to a full flow tunnel the pulsation is reduced by roughly 50%. But the reduction is depending on the degree of fillup. This is shown on figure 4.

The dimensioning lifting force is less than the maximum pulsation difference. From the first full scale measurement in a full-flowing tunnel we have found the dimensioning lifting force to be 40% of the maximum pulsation difference. For the time being we conclude with the following formula for a lifting force acting on an asphalt or concrete floor-lining with the rest of the periphery unlined:

$$h = (0.04 + 0.05 \ d_s) \ v^2/2g \qquad /2/$$

d_s = equivalent sandroughness in cm
h = water column, same dimension as $v^2/2g$

LIFTING CRITERIA DUE TO PORE PRESSURE IN THE BASE MATERIAL

The asphalt or concrete floor lining should be foreseen with small continous openings at the connection with the tunnelwall. This is to prevent overpressure caused by the hydraulic headleos that may accumulate under the lining, see figure 5. For tunnels wider than 6 m it is also recommended to arrange a small slit in the middle of the tunnel.

Special attention must be taken if the tunnel has a steep slope, 1:20 or steeper. When emptying the water in the tunnel, which should not be too timeconsuming, dangerous pore pressure arises in middle of the tunnel

filling degree were the dominating factors effecting the pressure pulsation.

For a full tunnel flow the maximum pressure difference h in water columns could be expressed by:

$$h = (0.09 + 0.12 \ d_s) \ v^2/2g \qquad /1/$$

(d_s = equivalent sand roughness in cm)

This equation is shown in figure 3 and is valid for a floor lined tunnel. Recordings from an unlined tunnel all around show some

or half way between two slits. This is coming up because it takes time to empty the porewater in the permeable base material under the lining. For a 1:10 steep tunnel this pore pressure may be dimensioning unless the width between two slits is limited to 3 m. To empty the tunnel slow enough to let the pore pressure follow the water table in the tunnel is not practical because it is too time consuming.

CONCLUDING REMARKS

The model tests and the full scale measurements have all been limited and strongly influenced by the site conditions. We therefore consider the results to be preliminary and the research should be continued. We would mention that the pressure fluctuation in the model provided to be undependent of the Reynolds number (Re) for Re $>3 \cdot 10^5$.

This is the same value where the Strouhals number is increasing rapidly.

The 'Collect-and-Transfer' system in hydropower developments – A cost effective and environment saving solution

J. Kristiansen & O. Stokkebø
Norconsult International, Nesbru, Norway

ABSTRACT

During more than 100 years of hydropower development, Norwegian engineers have introduced several noteworthy methods and techniques which have been put to worldwide use becoming known as Norwegian specialities. The overriding idea has been to use rock as a structural element and to develop and improve cost-saving solutions in tunnels and caverns. By going underground greater freedom is gained to arrange reservoirs, waterways and power stations in a more cost-effective way than placing them stepwise along the river course. The Collect-and-Transfer System was a result of the underground development. These solutions of underground development also usually have less impact on the environment. The surplus excavated rock masses are mostly used to restore the landscape.

COLLECT AND TRANSFER SYSTEM

The basic idea is to reduce the number of reservoirs and power stations by collecting water from several rivers and streams into an interconnected and integrated system by means of tunnels and shafts. In this way the natural head can be utilised without constructing high dams. Reservoirs can be located at natural mountain lakes or where damsites are favourable regarding costs and environment. The traditional step-wise hydropower development along a main river course, where a series of high dams have to be constructed, often has a serious impact on the environment and can cause great local resistance if populated areas or woodland are inundated.

Obviously, the collect and transfer system cannot be looked upon as a blue-print which can be used anywhere with the same advantages. Topographical, geological, hydrological and environmental conditions are decisive factors when assessing hydropower potential in a given area.

GENERAL

The topographical conditions and the high precipitation in Norway are favourable for hydropower developments. This, together with the fastly increasing demand for electricity, made it necessary to construct a great number of hydropower plants.

Today Norway is the world's greatest consumer of electric energy per capita, and hydroelectric generation represents 99 % of the total electricity generation.

UNDERGROUND WORKS

Geological conditions in an area are among the decisive factors when choosing project layout, excavation and tunnelling methods.
The rock quality in Norway led to the use of rock as a structural element in caverns, shafts and tunnels. Hard rock tunnelling techniques were developed, and cost-saving, unlined high and low pressure tunnels and shafts are used in a great number of hydropower projects.

The successful introduction of full face boring machines in hard rock also led to an even more extended use of diversion of water over long distances.

RESERVOIRS

Due to these new tunnelling techniques a greater degree of freedom in location of reservoirs was achieved. Natural mountain lakes or other beneficial damsites as regards costs and environment could be chosen.
When topographical conditions are favourable, the reservoir can be located at a high level, and from lower levels, water can be transferred by pumping into the reservoir to utilize the obtained high head in the electric generation.
Furthermore, it is possible to transfer water to the main reservoir from other lakes or rivers located on approximately the same altitude or higher, sometimes combined with a local power station.

COLLECTING TUNNELS

The main tunnel in a hydropower project is the headrace tunnel which usually consists of a low pressure part, an unlined pressure shaft or tunnel, and finally a steel lining.
The headrace tunnel leads the water from the reservoir to the power generating units, and the alignment should be adjusted to collect any streams on the way. If the system contains a surge shaft, this can also be utilized as a stream intake shaft.

Cost effective shaft excavation methods made it possible to establish separate collecting tunnels to transfer water from groups of streams or single rivers. The collecting tunnels are excavated on a high level to reduce the length of the intake shafts.

INTAKE SHAFTS

When introducing stream intake shafts in a hydropower headrace system, it is particularly important to make computations of hydraulic stability and oscillations.

In Norway much effort has been concentrated on developing new shaft boring technology. Improved boring methods, in combination with the use of helicopters for construction of intake structures, led to cost effective stream intakes. Some of the more daring engineering solutions led to problems with air supersaturation and air blowouts, however. This in term led to more experience and after intensive research on the matter at the Norwegian Hydrotechnical Laboratory and other institutions, design recommendations could be introduced.

PUMPED STORAGE

Pumping of water to higher levels is a well known method in hydropower planning, and the method is very appropriate in the Norwegian collect and transfer system as pumping introduces a new degree of freedom in the planning solutions.
Pumping for storage is mostly economic in periods with surplus water and supply of cheap off-peak electricity. The value of the stored water is then higher than the value of the water used for surplus power generation, plus the pumping costs.

The generating sets are usually of the reversible type but may also be separate pumping plants if economical justifiable.

COLLECT AND TRANSFER IN PRACTICE

In the following one of the largest hydropower projects in Norway is used as to illustrate the collect and transfer method. The drawing shows a simplified longitudinal section.
The project utilizes the water in three hydropower stations on various levels. An extensive

system of tunnels collect and transfer the water to the different power stations.

The main reservoir is located on a level above the tree line where the environmental impact is limited. The natural conditions were favourable to create a large and inexpensive perennial reservoir. The main reservoir can be filled up from its own catchment area, by gravitational transfer from neighbouring catchment areas (A) and also by pumping.

The purpose of the large reservoir is to increase the firm power and meet the seasonal variations of the power demand. Because of the large storage volume, it will also give considerable national security in dry years. By making use of pumping, other reservoirs can be held to a limited size and for short term regulation only.

From the main reservoir it is possible to utilize the head in the first power station (I). This station, however, is also supplied with reversible turbines and pumps water into the main reservoir in periods with surplus water and with supply of surplus electricity.

The figure shows that the first power station (I) is located on a middle level. This level contains many interesting potential water resources. A neighbouring catchment area (D), however, was situated on a lower level. It was found feasible to collect, transfer and pump this water to the reservoir (a) on the middle level. It was also possible to collect streams and lakes and transfer them direct into the same diversion tunnel (C) without pumping.
Due to appropriate levels, another catchment area (B) could be collected and transferred direct to the reservoir (b) without pumping.

These two reservoirs on the middle level (a and b) have equalised regulated water levels, and are also intake reservoirs for the second power plant (II) situated on a lower level in the system. This leads to a complicated but effective optimal use and regulation of the water on the middle level:

● From one or both reservoirs (a,b) the water can be pumped into the main reservoir through the reversible turbines in the first power plant (I).

● From one or both reservoirs (a,b) the water can be utilized in the second power plant (II).

● The discharge from the first power plant (I) can be directly utilized as turbine inflow to the second power plant (II).
● The turbine inflow to the second power plant (II) can be a combination of discharge from power plant (I) and stored water from the reservoirs (a,b).

The tailwater from the second power plant (II) discharges into a regulated lake about seventy metres above sea level. From this reservoir the water is finally utilized in the third power plant (III) at sea level. This power plant, however, has to share the water with the salmon in the river course. This is achieved by closing the third power plant during a couple of months in the summer.

HYDROPOWER ENGINEERING

Norwegian hydropower engineering skills are backed by almost a century of involvement with both small and large scale schemes in the domestic development programme. This engineering skill embraces every element of schemes, including all types of concrete and fill dams, waterways, tunnels, canals, shafts, penstocks, power plant, switchyards, transmission lines and substations.
The understanding of the environmental importance in a hydropower development was also early recognised by the Norwegian engineers, and environmental conditions in combination with hydrological, topographical and geological conditions are decisive factors when assessing hydropower potential. Going underground has also become an important element in the environmental evaluation.
A combination of all these factors has made it possible to develop the typical underground collect and transfer system. It is obvious, however, that each hydropower development needs planning by experienced engineers because of its complexity. Therefore all hydropower schemes will, due to natural conditions, be so different that the collect and transfer system should not be considered as a blue-print which can be used everywhere with the same advantage. It can, however, represent an economic and environmentally friendly alternative in many cases.

LONGITUDINAL SECTION OF A COLLECT AND TRANSFER SYSTEM

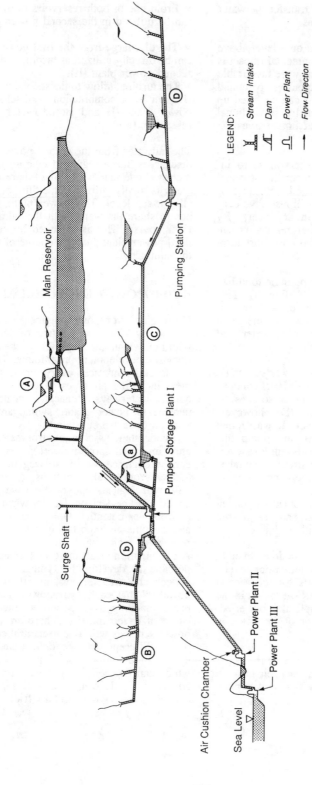

Main Reservoir

Surge Shaft

Pumping Station

Pumped Storage Plant I

Air Cushion Chamber

Power Plant II

Power Plant III

Sea Level

LEGEND:

⅄   Stream Intake

Ⴠ   Dam

⨅   Power Plant

→   Flow Direction

Norconsult ❖   Feb. 92

Hydropower'92, Broch & Lysne (eds) © 1992 Balkema, Rotterdam. ISBN 90 5410 054 0

Long term stability in Norwegian hydro tunnels

Alf Thidemann
SINTEF Rock and Mineral Engineering, Trondheim, Norway

Amund Bruland
Norwegian institute of Technology, Trondheim, Norway

ABSTRACT: The paper presents the results of a five year project on stability inspections of hydropower tunnels in Norway. Observed block falls and other stability problems are described by frequency, volume and reason. A model for calculation of costs related to block falls are given. Finally, guidelines for rock support of hydropower tunnels are suggested.

1 INTRODUCTION

A project on "Support of Hydro Tunnels" was completed in July 1991 and gives the basis of this paper. The client was the Water System Management Association.

The aim of the project was:
 - To clarify if the long term stability was in accordance with the presupposed safety and economy.

Another goal for the project was:
 - To give suggested guidelines for the rock support work by analysis of the observations, considering production safety, energy saving and economy.

The total length of Norwegian hydro tunnels is now nearly 3500 km. About 95% of the tunnel length is excavated in the last 40 years, as shown in Figure 1.

To illustrate the significance of the subject it can be mentioned that the yearly costs of tunnel support are estimated to exceed NOK 100 million (1USD = 6NOK). It was presupposed that some of the tunnels or tunnel stretches could be "oversupported". On the other hand some weakness zones could be underestimated.

The most expensive and difficult case of unstability noticed during the project is from a rock fall blocking the whole cross section of a tailrace tunnel. The repair and support costs were in the order of NOK 10 million, excluding production losses.

Stability problems are reported from all parts of the world. Many of them have more serious consequences than any of the examples described in our study.

The project is based on inspections of accessible hydro tunnels and reports from earlier inspections. A total of 35 tunnels corresponding to a length of 330 km has been studied. The tunnels have cross sections from 6 m^2 to 70 m^2 with operation ages from 1 to 65 years. The investigations are supplied with data from all known and reported greater failures in Norwegian hydro tunnels.

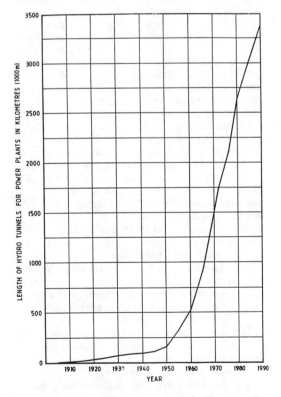

Fig. 1 Total length of hydro tunnels in Norway.

2 NORWEGIAN SUPPORT PHILOSOPHY AND COSTS

Rock reinforcement is a very wide and complex topic that includes main elements as geological engineering, rock mechanics and construction engineering. The basic material is the rock mass. In Norwegian support philosophy the rock mass is not regarded as a material needing support, rather as a material capable of resisting considerable loads, even in jointed condition. The engineering geologist has to take advantage of the self-supporting capabilities of the rock mass and the positive aspects of rock as a construction material.

The acknowledgement that has formed the basis of design, excavation and support is the need of flexibility in the management of the actual rock conditions and their local variations. The rock mass itself with the variations in discontinuities, stress situation and water content, will be the dominating factor. In addition comes the manner in which we treat the rock mass during blasting, and in eventual supporting works.

Medium jointed rock mass and even heavier jointed rock may carry high loads under relatively small deformation, provided that the joint surfaces have a satisfactory friction. When meeting clay-infected joints or joints covered with weak minerals, a stability problem may occur.

The greatest problems are normally connected to crushed zones or other weakness zones. These zones very often have a high degree of jointing and contain clay minerals. The stand-up time can be very close to zero if the zone contains swelling clay. This means that if there is a combination of swelling clay and access to water, there will be a very short time between blasting and a major rockfall.

Other problems are rock burst and spalling caused by high rock stress. Furthermore, the stability can be influenced by high water pressure and leakages.

The costs of tunnelling and tunnel reinforcement have decreased relatively with time taking the cost index in account. Improvement in equipment, material, organization and rock blasting technique are the main explanations.

An analysis of tunnel excavation costs and tunnel support costs shows a relative reduction in the support costs during the last decade.

One can separate between the following demands for support:
- Safety precaution (or preliminary support).
- Permanent support (on or behind the working face).

The practical difference are not always present, because a support for safety may constitute the permanent support or part of this support.

A review of the typical support methods in Norwegian hydro tunnels might be covered by the following:

- Scaling.
- Rock bolts.
- Shotcrete.
- Concrete lining.
- Pregrouting.

In the following we shall concentrate on three methods: Rock bolts, shotcrete and concrete lining.

In Figure 2 the average cost of rock support is related to the excavation cost of the tunnel. Tunnels with a cross section of about 40 m^2 are chosen as reference.

From Figure 2 can be seen that it sometimes may be a cost saving alternative to use a light support at the working face followed by a heavier support behind the face. Though the tendency seems to be a reduction of the cost difference between works at the face and behind the face, the increase in advance by a combination may be considerable.

Figure 3 and Figure 4 give the rock reinforcement costs as a function of tunnel cross section.

3 TUNNEL OBSERVATIONS

The data from the observations are presented in two separate tables. For every power plant the first table includes: Type of tunnel (headrace tunnel, etc.), year of construction/completion, year of inspection, cross section, length, laboratory analyses, rock support in per cent of the tunnel length for systematic bolting, shotcrete and concrete lining. For each support method is shown the main reason, like: Weak rock masses (0), jointing (1), rock stress (2), water inflow (3), crushed zones - inactive (4) and crushed zones - active (5).

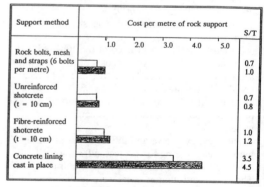

Fig. 2 Relative costs of rock support and tunnel excavation (Average values).

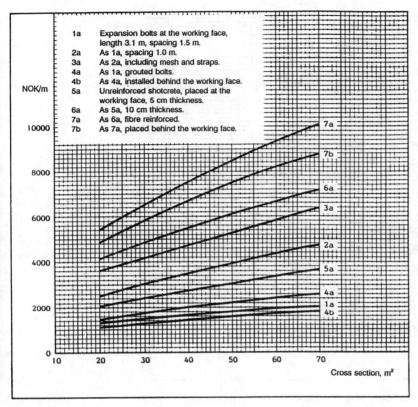

Fig. 3 Rock reinforcement costs for bolts and shotcrete as a function of tunnel cross section. (Price level 1991-01-01).

The other table gives the data from observations during inspections of the tunnels. The observations are divided into groups: Block falls or rock falls with volume less than 5 m³ including the number of blocks, the total volume, main reason (like 0 to 5 above for all groups), open area above the rock fall both in m² and in per cent of the tunnel cross section. For rock falls exceeding 5 m³ the same information is presented except for the number of blocks.

The table also includes damage on rock reinforcement. For bolts, shotcrete and cast in place concrete, the number and type of damage are shown. The types are: Bolt overloaded (1), bolt - poor quality of work (2), bolt - corrosion (3), shotcrete - minor cracks or outfall (4), shotcrete - overloaded (5), shotcrete - poor quality of work (6), concrete lining - minor cracks or weaknesses (7) and concrete lining - collapse or major damage (8).

4 ROCK FALLS

A summary of the observations shows:
- By far, the predominant part of the hydro tunnel length are stable, especially if minor block falls are excluded (less than 0.05 m³).
- Major stability problems are mostly observed at weakness zones.
- Block falls and minor rock falls are registrated in an average number of 3.5 per km tunnel. The volume varies between 0.1 and 3.5 m³ with an average value of 0.8 m³. The main reasons are jointing, high rock stress and small weakness zones with inactive material. Block falls are mostly from unsupported stretches.
- Rock falls exceeding 5 m³ are found to be one for every 5.5 km as an average. The medium free area for the water flow above the rock falls is about 80%. See Figure 5. Rock falls blocking the tunnel are not included in the figure.

In Figure 5 the rock falls are distributed after their relative magnitude.

Two rock falls blocking the whole tunnel profile have been found during the inspections. All known rock falls with a total blocking of tunnel have a number of 10. This figure are believed to cover all Norwegian hydro tunnels with a total length of nearly 3500 km.

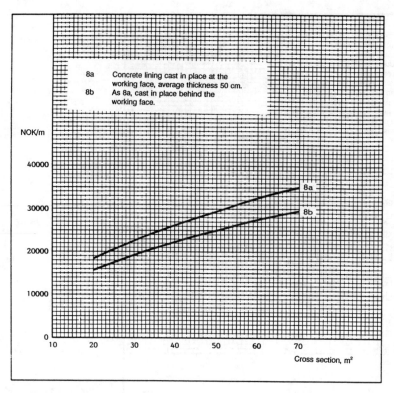

Fig. 4 Rock reinforcement costs of cast in place concrete as a function of tunnel cross section. (Price level 1991-01-01).

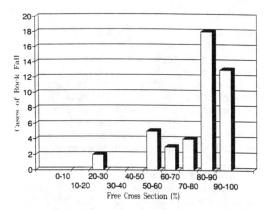

Fig. 5 Number of rock falls and the free area in per cent of the original cross section.

5 DAMAGE ON ROCK SUPPORT

All registered defects on support installations are shown in Figure 6. 20 cases of damage on rock bolts are observed. Instability of bolted blocks are for about 80% caused by poor quality of work, a specially due to lack of grouting at the whole bolt length. About 20% of the instable bolted blocks are overloaded. No bolts are observed to have failed because of corrosion. This goes for the ungrouted expansion bolts for preliminary support as well.

Tunnel support with shotcrete and cast in place concrete has mostly damages of a minor scale like cracks, etc., without importance for the total stability. A total of 39 defects on shotcrete and 17 defects on concrete lining are observed.

The actual time periode from a tunnel is taken into use until a major failure occurs, may vary from a few weeks to years. Figure 7 shows the main reasons for and type of supporting works at 49 examples of major failures. Some examples from earlier inspections with detailed and reliable observations are included.

The inspections have confirmed that major failures in most cases are due to weakness zones containing swelling clay. The zones are mostly not secured at all or are shotcreted. The main reasons why the zones are underestimated during the excavation time are:
- The zones were dry and apparently stable.
- The zones were well consolidated.
- The zones had a favourable orientation.

196

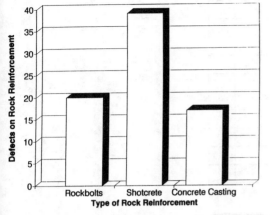

Fig. 6 Defects on different types of rock reinforcement.

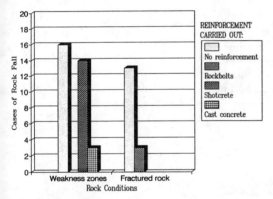

Fig. 7 Registered major failures with main cause and types of support.

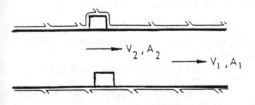

Fig. 8 Block fall in a hydro tunnel.

- The swelling clay was not activated before water filling.

6 HEAD LOSSES AND COSTS

Formulae are developed to calculate head losses and costs caused by rock falls. The basis is general for-mulae for head losses and production in hydro power plants.

A commonly used formula for production is :

$$E = 8.4(Q \cdot H \cdot T - Q_{max} \cdot \Delta H \cdot T_c)$$

where E = yearly production (kWh)
Q = yearly average waterflow (m³/s)
H = average gross head (m)
T = number of hours in a year = 8760h
Q_{max} = maximum waterflow (m³/s)
ΔH = sum of head losses (m)
T_c = number of hours (with Q_{max}) corresponding to an equivalent cubic average value of the waterflow during a year (h)

The second link in the formula expresses the year-ly loss in production caused by the head loss:

$$\Delta E = 8.4 \cdot Q_{max} \cdot \Delta H \cdot T_c$$

A case of block fall is illustrated in Figure 8, with a velocity and free cross section at the block of v_2 and A_2 (cavity in the roof is not included). Velocity and tunnel cross section before and after the block are v_1 and A_1. The relation between the two cross sections is:

$$\alpha = A_2/A_1, \text{ and } v_2 \cdot A_2 = v_1 \cdot A_1$$

which gives:

$$v_2 = v_1/\alpha$$

Head losses with respectively cross section reduction (H_r) and expansion (H_e) are calculated by:

$$H_r = K_r(v_2)^2/2g$$

$$H_e = K_e(v_2)^2/2g$$

where K_r and K_e are the head loss coefficients, which can be expressed as follows:

$$K_r = 0.5(1-\alpha)$$

$$K_e = (1-\alpha^2)$$

Adding these head loss coefficients will give the coefficient for a singular head loss. Because of the irregularity of single blocks it is assumed that the singular head loss by minor block falls may be doubled. Major rock falls will get a more regular shape from the waterflow.

The coefficient for singular head loss by minor block falls is:

$$K_s = 2(K_r + K_e)$$

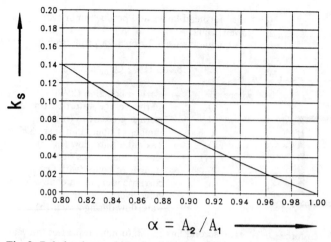

Fig. 9 Relation between the head loss coefficient K_s and $\alpha = A_2/A_1$.

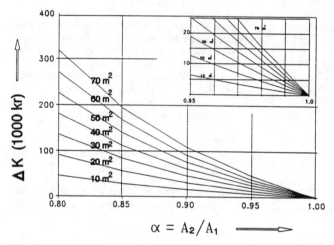

Fig. 10 Capitalized energy loss (ΔK) by minor block falls as a function of cross section and α.

and for major rock falls:

$$K_s = K_r + K_e$$

Singular head loss:

$$H_s = K_s(v_2)^2/2g = K_s(v_1)^2/\alpha^2 \cdot 2g$$

Figure 9 shows the variation of the coefficient K_s by minor block falls with the relations between the cross sections α within the observed magnitudes of α.

The energy loss and the related capitalized costs may now be calculated. Figure 10 illustrates the capitalized cost of a minor block fall as a function of α and tunnel cross section. Here are presupposed: velocity $v_1 = 1.4$m/s, $T_c = 2500$h and the economic

limit value of investment in hydro energy is NOK 3.50 per kWh.

Corresponding costs for major rock falls are shown in Figure 11.

For differing values of velocity, etc. the following correction factors are used:

Velocity-cor. factor : $K_1 = (v/1.4)^3$
T_c "   "   : $K_2 = (T_c/2500)$
Limit inv."   "   : $K_3 = $(Investment cost in NOK per kWh/NOK 3.50)

The observed size and frequency of the block falls (3.5 block falls per km with a medium size of 0.8 m^3) will give a capitalized energy loss of about NOK 18,000 per km tunnel (40 m^2). To avoid single block falls - if possible - costs are supposed to be higher.

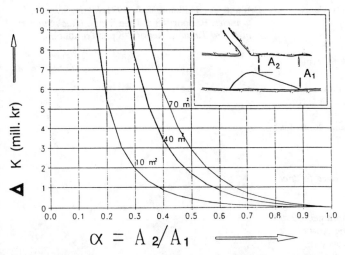

Fig. 11 Capitalized energy loss (ΔK) by major rock falls as a function of cross section and α.

CONTROL ADHESION

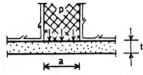

$$p \cdot a = 2 \sigma_{ad} \cdot \delta$$

$$\sigma_{ad} = \frac{p \cdot a}{2 \delta}$$

CONTROL FLEXURAL STRENGTH

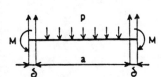

$$\sigma_b = \frac{M}{W} = \frac{\frac{p \cdot a^2}{12}}{\frac{1 \cdot t^2}{6}} = \frac{p \cdot a^2}{2 t^2}$$

Fig.12 Cross section of weakness zone. Control of support.

The observed size of major rock falls ($\alpha = 0.8$), will give a capitalized energy loss in a 40 m³ tunnel of about NOK 170,000.

If the rock fall had continued and filled half of the tunnel, the energy loss would have increased to about NOK 1500,000.

7 CONCLUSIONS AND SUGGESTED GUIDELINES FOR ECONOMICAL TUNNEL SUPPORT

With an improved final scaling it is believed that a considerable number of the block falls would have been eliminated.

Tunnel inspections have confirmed that the most important stability problems are related to local faults and other weakness zones, a specially if the zones contain swelling clay.

In connection with the use of shotcrete, the following points are very important:

At a weakness zone the adhesion between concrete and rock is vital to the ability of the shotcrete to withstand load before primary failure, as shown in Figure 12. When punch loaded, a narrow stripe (δ) will transfer load to the siderock. (δ is measured to about 3 centimetres in large scale tests). Fibre-reinforcement adds considerably to the flexural strength of the shotcrete. Procedures are developed to control the use of shotcrete based on laboratory swelling pressure tests of the clay and field tests of the concrete-rock adhesion.

7.1 Conclusion for minor block falls:

- With some improvement in working quality and in scaling at the end of the excavation period, the

199

Norwegian support concept will be in accordance with the required stability and economy.

7.2 The main conclusions for major rock falls are:

- The study has confirmed in general that the Norwegian support concept is, when used, in accordance with the required safety and economy.
- The significance of an adequate quality control is verified.

7.3 Following superior guidelines are suggested for the support of hydro tunnels:

1. The tunnels have to be supported in a necessary degree to prevent injury of the tunnelling crew or damage to the equipment during the excavation period.

2. A thorough final inspection shall be carried out at the end of the excavation period to identify any potential unstable part of the tunnel. The inspection has to be managed by personnel with engineering geology competence.

3. Systematical support to avoid any single blockfall is not economically recommended. Systematical replacement of preliminary support - like ungrouted rockbolts - is not supposed to be necessary.

4. Emptying and inspection of the tunnels after a few years to remove blockfall of importance to the energy losses is recommended.

5. The engineering geologist shall give his special attention to the weakness zones and carry out laboratory and field tests to verify the decision of supporting method.

8 ACKNOWLEDGEMENT

The authors wish to thank the Water System Management Association for the support of the project and their kind permission to publish the results.

REFERENCES

Thidemann, A. 1981. *Langtidsstabilitet i vanntunneler.* Dr.ing.-avhandling. Geologisk institutt, NTH, Trondheim, 306 s. (*Long Term Stability in Hydro Tunnels*. Ph.D. Thesis, 306 pp).

Vassdragsregulantenes Forening. 1985. *Falltap i kraftverkstunneler*. Rapport 127 s. (Water System Management Association. *Head Losses in Hydropower Tunnels*. Report 127 pp).

Thidemann, A. 1987. Long Term Stability in Norwegian Hydropower Tunnels. *Proc. of the Int. Conf. Underground Hydropower Plants*. Vol.2: 843-855. Oslo.

Bruland, A., Thidemann, A. 1991. *Sikring av vanntunneler*. Rapport 88 s. Trondheim. (*Rock Support in Hydro Tunnels*. Report 88 pp. Trondheim.

Hydropower'92, Broch & Lysne (eds) © 1992 Balkema, Rotterdam. ISBN 90 5410 054 0

Comparison on design methods of temporary supports using Sanliurfa tunnels

H. Tosun
State Hydraulic Works, Ankara, Turkey

ABSTRACT: In this paper comparative analyses on various tunnel design methods of temporary supports are given with applications to the Şanlıurfa tunnels. The empirical method consisting of conventional analysis, geomechanics classification and Q–system, and the analysis based on the convergence behaviour of tunnel, namely convergence – confinement method are compared by considering the rock conditions and the support systems. Comparison of various methods about temporary support is based upon the support systems, rock loads, construction sequences and cost analysis.

1. INTRODUCTION

Tunnel construction is expensive endeavor, when compared with other civil engineering structures. According to recent surveys, the support costs range from 30 to 50 percent of the total cost and may even go upto 70 percent. Therefore, it is desirable to provide a cost reduction in underground construction. It seems possible that the cost may be recuded by increasing the efficiency of support design and construction.

An efficient and economical tunnel design practice dictates selection of supports that suit best to the conditions of rocks encountered during tunneling. Thus, selection of the most suitable tunnel design method appreciably reduces the cost of tunneling.

The Şanlıurfa tunnel is one of the most important aspect of the Southeastern Anatolian Project. The tunnel is located between Sanlıurfa city and the Atatürk dam (Figure 1). It is designed to convey water from the reservoir of the Atatürk dam to the Urfa (Harran) plain and also to generate electricity (50 MW) at its outlet. It is a pressurized tunnel with an internal water pressure of approximately 400 kPa. Its excavation was completed at mid of 1991. When entirely completed, it will be the longest water conveyance tunnel of the world consisting of two concrete–lined (double tube) tunnels each 26.4 km long and 7.62 m in diameter. Its excavated diameter is aproximately 9.0 m. The tunnel is designed to convey 328 cu.meter per second of water at an average flow rate of 3.5 m/s which will be utilized for the irrigation of $4.6x10^5$ hectares of land in the Sanlıurfa and Mardin plains.

Shotcrete and rock bolts are used as temporary support systems. Shotcrete is applied in double layers (15 cm thick) with wire mesh combined with rock bolts of 3 m long and 2x2 m spacing. Cast–in–place concrete lining of varying thicknesses (minimum 40 cm) provide the permanent support of the tunnels. Excavation is performed in three steps by tunnel boring machines (TBM) and locally by blasting. The rate of advance of the TBM is 6 m per day.

2. SITE GEOLOGY

Geological investigation along the alignment of Şanlıurfa tunnel is detailly conducted by State Hydraulic Works (DSİ,1977). Germav formation (limestone, marl, and shale alternations –Late Cretaceous); Gercüş formation (clayey and chalky limestones containing chert nodulus– Paleocene–early Eocene); Midyat formation (crystalline limestone and marls–Eocene); basaltic lava flows (Pliocene); and Quaternary deposits (alluviums, colluvium, alluvial cone deposits) constitute the dominant lithology of the tunnel alignment. Approximately 85 percent of its total length is located within the marls of the Germav formation. The Germav formation starts with marl, mudstone and shale alternations and continues with limestone,

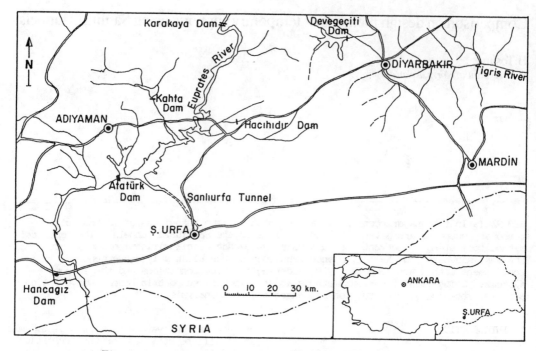

Figure 1. Map showing the location of Şanlıurfa tunnels.

argillaceous limestone and marl. The thickness of the lower sequence exceeds 500 m and constitutes the major bedrock along most of the tunnel length.

The grayish to cream marls generally dominate the lithology of the lower sequence. The individual layers are 50–60 cm thick and lie horizantally. The hydraulic conductivity of the marls is very low (0–2 lugeon units). Thus, they are very poor water bearers.

The upper sequence is characterized by a lower clay and a higher carbonate content. The limestones locally contain chert nodulus. Its individual layers are 40–50 cm thick. Approximately 4500 m long section of the tunnel follows the upper sequence (Figure 2).

The Gercüs formation is easily differentiated from the Germav formation by its white color and highly friable nature. It consists of limestone, marl and shale alternations. The limestones contain abundant chert nodulus. The individual layers are 80–150 cm thick.

Basaltic lava flows are observed in the form of local exposures which lie approximately 100 m to 200 m above the tunnel elevation. They are related with the Pliocene activity of Karacadağ volcano. The Quaternary deposits consist of alluviums, colluviums, and alluvial fan

deposits of vary thicknesses. Clays and silts are rather dominant within the alluviums.

3.ENGINEERING GEOLOGY

3.1. Engineering geological charecteristics of Germav formation

Along the tunnel alignment the bedrock possesses rather uniform and/or relatively homogeneous lithological characteristics. Thus a special emphasis is given to the types and characteristics of the discontinuties encountered within the bedrock.

The tunnel alignment is intersected by numerous faults and/or shear zones. The faults generally dip 40 to 90 degrees and their displacements range between 6 to 80 cm for the 1200 m section considered in this study. The width of the fault and/or shear zones is rather narrow and infrequent overbreaks may be expected where they intersect the tunnel at low angles.

The joint surveys conducted between stations 11+200 and 15+000 indicate that the dominant set strikes northeast and dips at angles greater than 40 degrees. Generally one or two major sets plus random are encountered within the bedrock. Spacing of

202

the joints are highly variable. Their apertures are narrow to tight with smooth, planer, and rarely polished and/or slickensided surfaces. Infilling materials are generally calcite and clay. Iron oxide staining of joint surfaces is rather common.

3.2. Geomechanical properties of the rock masses surrounding the tunnels

Various laboratory and in-situ tests were conducted to determine the mechanical behaviour of the bedrock. The plate loading tests conducted within the adits GT2 and YT1 have yielded the modulus of deformation values ranging between 570 MPa and 1825 MPa within the pressure range 0 to 3.5 MPa, respectively. On the samples obtained from borehole SK 21 the average values of the poisson's ratio, the secant modulus of elasticity, and the uniaxial compressive strength were determined as 0.16, 1500 MPa, and 8.5 MPa, respectively.

In the tunnel three stations were selected to determine the mechanical properties of the rocks and two stations to monitor the convergence behaviour of the tunnel. At station 25+500, the plate loading test has yielded an average deformation modulus of 6315 MPa within the pressure range of 0 to 3.5 MPa, whereas by Rocha dilatometer an average value of 1500 MPa was determined within the pressure range of 0 to 5.0 MPa.

At station 24+500 three deformation moduli tests were conducted. The pressumeter, Goodman jack and plate loading tests have yielded average deformation values of 1312, 1186 and 3958 MPa, respectively.

At station 13+700 plate loading and laboratory rock mechanics tests were conducted. The modulus of deformation ranges between 2520 MPa and 11100 MPa, averaging 6255 MPa within the pressure range of 0 to 3.5 MPa. The compressive strength, dry density and poisson's ratio averages 24.0 MPa, 2010 kg/m^3 and 0.20, respectively. The triaxial test results yielded cohesion values within the range of 0.4 to 1.0 MPa and the angle of internal friction value of 25 degrees.

4. COMPARATIVE ANALYSES

In this study comparative analyses were carried out using empirical methods consisting of conventional analysis, geomechanical classification and Q-system, and convergenge-confinement method. Analyses were carried out on the bases of rock loads, support systems, construction sequences and cost analysis.

4.1. Rock loads

In the empirical design, load acting on the support element is estimated by means of rock mass classifications. For the conventional analysis, rock load is determined as a function of the height of loosened rock, referring to Terzaghi's rock load concept (Terzaghi, 1946), was estimated at the range of 0.35(B+H$_t$) to 0; where B is the width of tunnel. An external support system was considered in the design of Şanlıurfa tunnel with the conventional analysis (Tosun, 1985). For a rock unit weight of 20 kN/m^3 the height of loosened rock gives 130 kPa of rock pressure directly acting on the steel ribs.

Geomechanical classification introduces a rating obtained from the detailed enginering geological investigation (Bieniawski, 1974). An empirical equation [h$_t$ = ((100−RMR)/100)xB] has been developed to estimate the rock load acting on the support system (Unal, 1983). From this equation it is estimated that the rock load ranges between 50 and 100 kPa for a rock unit weight of 20 kN/m^3. It has also been determined that rock mass rating (RMR) ranges between 45 and 74, that is, from fair to good rock class. The RMR values suggest that the range of cohesion and internal friction angle of the rock mass are equivalent to 200−400 kPa and 25−45 degrees, respectively.

The quality Q calculated for the 1200 m length of the tunnel between station 11+200 and 15+000 varies 1.13 to 80. For the same portion, rock mass rating (RMR) values were determined and the results were correlated with quality Q. The empirical relationship derived between RMR and Q values by Bieniawski (1979) and Rutledge (1978) were confirmed with the results obtained through the analyses of the 1200 m portion of the tunnel. The correlation of RMR and Q values for the Şanlıurfa tunnels is presented in Figure 3.

An empirical equation relating the permanent support pressure and the rock mass quality (Q) has been developed for the roof and the walls of a tunnel (Barton et al, 1974 and 1980). The permenant support pressure can be calculated as a function of quality Q and joint roughness number and joint set number. The 160 kPa of permenant support pressure was calculated for the roof of tunnel, as based on the worst conditions of tunnel.

In the convergence-confinement method rock load is defined as a function of

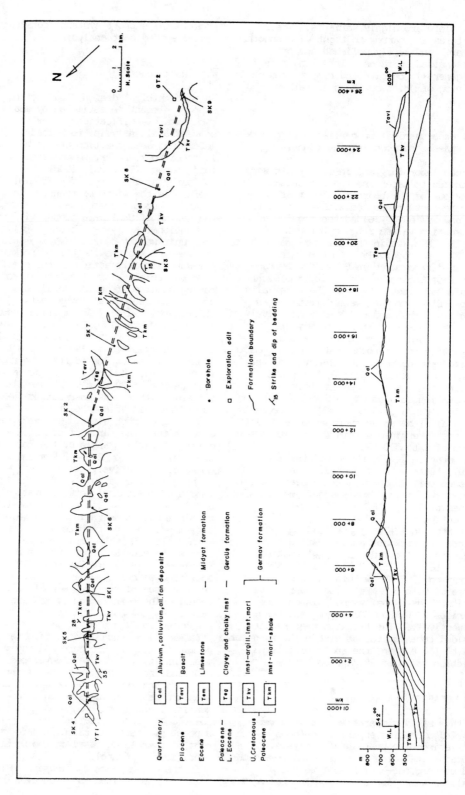

Figure 2. Geological cross section of Sanliurfa tunnels.

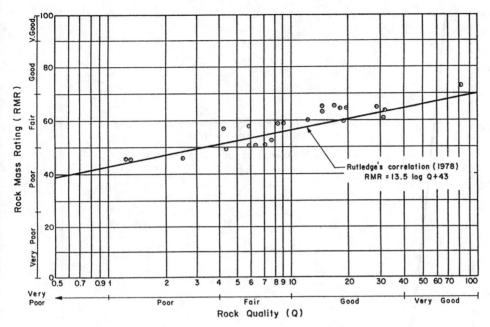

Figure 3. Comparison of the results obtained by geomechanics
classification and Q-system between stations 11+750 - 12+205
and 14+635 - 14+985.

primary stress conditions (Ladanyi, 1974 ;
Hoek and Brown, 1980). The stress-
deformation curve of surrounding rock is
plotted for estimating the limit pressure
to be supported (Figure 4). From the rock-
support interaction analysis, the limit
pressure have been determined as 180 kPa
for flexible support system.

4.2. Support systems

Selection of support systems are based on
the most unfavorable conditions of the rock
mass and the limitations of the selected
design methods. In the conventional
analysis an external support system,
depending on the recommendation of
Terzaghi's rock load concept, was selected.
Three steel ribs having semi-circular shape
(8W40, 8W28 and 8W20) were considered as a
temporary support for Şanlıurfa tunnel.
The analysis suggests that the 8W40 steel
rib is sufficient for a factor of safety of
1.2.

Geomechanics classification requires
systematic rock bolts having 20 mm
diameter, 4 m length and 1.5–2.0 m spacing,
and shotcrete of 5–10 cm thick at the crown
with wire mesh and 3 cm at the walls.

The supports suggested by the Q-system
consist of rock bolts and thin layer of
shotcrete. Based on the quality Q and the
effective span or diameter of excavation,
the ground category of rock mass and its
support requirement may be changed. For
the worst conditions encountered, as at
stations 13+300 and 14+700, Q-system
suggests rock bolts (untensioned and
grouted) at 1.0 spacing and shotcrete of
2.5–5 cm thick.

The convergence–confinement method
introduces an analysis that widely utilizes
different support systems. For the case of
Şanlıurfa tunnel two temporary support
systems were selected: (a) shotcrete (6 cm
thick) combined with 26 mm diameter rock
bolts spaced at 2 m intervals, (b) light
steel rib (8W20) as an external support
system. Based on field measurements and on
the engineering judgement the initial
deformation for flexible and rigid support
systems was assumed as 2.0 and 3.0 cm,
respectively.

4.3. Construction sequence

The conventional analysis of empirical
design methods do not assume a construction
sequence. Geomechanics classication
requires an excavation as heading and one
bench with drilling and blasting (Table 1).
Advancing the top heading between 1.5 and

205

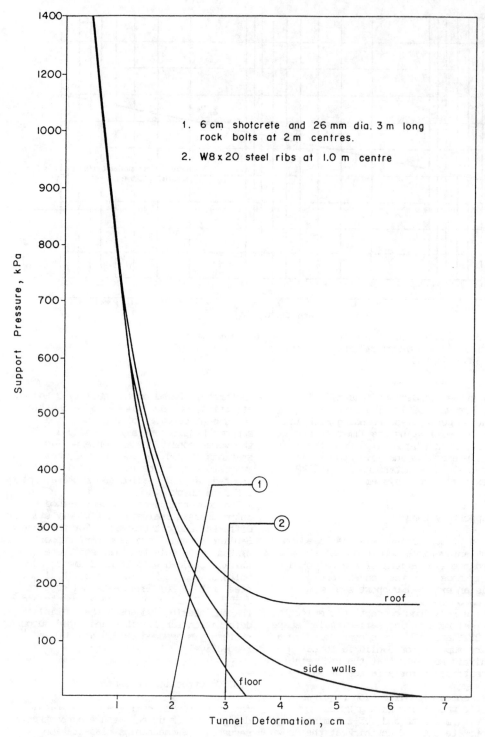

Figure 4. Rock – support interaction analysis of convergence-confinement method with two different support systems for the Şanlıurfa tunnel.

Table 1. Summary of the comparative analyses.

Method	Rock Load (kPa)	Support System	Construction Sequence	Tunneling Cost TL/m ($/m)
Conventional Analysis	130	W8x40 type steel rib with semi-circular shape,space at 1.2 m.	—	210 873.16 (496.17)
Geomechanics Classification	100 *	Systematic rock bolts, having 20 mm dia. 4 m length and 1.5-2 m spacing. Shotcrete, having 5-10 cm in crown with wire mesh, 3 cm on walls.	Top heading and bench by drilling and blasting. 1.5 - 3.0 m advance in top heading. Commencing the support after each blasting and completing support 10 m from face	110 453.99 (259.89)
Q- system	160	Systematic rock bolts (grouted and untensioned) with 1 m spacing and shotcrete 2.5-5 cm thick	—	108 508.39 (255.31)
Convergence Confinement Method	180 **	6 cm thick shotcrete combined with 26 mm dia. and 3 m long rock bolts spaced at 2 m	—	83 611.71 (196.73)
		Light steel rib (WX20) spaced at 1.0 m interval		131 035.62 (308.32)

* Support system suggestions are independent of rock load height consideration.
** Rock load is limit pressures obtained from the rock-support interaction analysis.

3.0 m, commencing support after each blast and completing support 10 m behind the face are the recommended steps in the construction sequence.

The stand-up time of 80 hours, for quality Q and RMR greater than 0.55 and 55 respectively, is obtained by means of Q-system and geomechanics classification. If the situation described above is not valid an immediate collapse may occur.

4.4. Cost analysis

The cost analysis of support system for each design method is carried out throughout the study. The cost level is based upon the local employment conditions and the 1984 prices. The cost of excavation is not included in the analyses. Because payment per cubic meter of excavation, given in the specifications of State Hydraulic Works, does not change with the method of excavation adopted such as blasting and TBM.

The analysis is based upon the actual conditions and support requirements for the Şanlıurfa tunnel. The cost of temporary flexible supports required for the convergence confinement Method (83 611.71 TL or $ 196.73) is the lowest cost of all. The highest cost of temporary support per lineer meter of the tunnel (210 873.16 TL or $496.17) was calculated for the rigid steel ribs of conventional analysis. When these two extreme values are compared, a ratio of 2.5 can be found. The cost of external rigid support system of convergence-confinement method exceeds all the others except the rigid steel ribs of conventional analysis. The cost analysis of support systems with other subjects of comparative analyses is summarized at Table 1.

5. CONCLUSIONS

"Fixed load" assumption seems to be the major shortcoming of the design methods. The load acting on the support elements varies with deformation allowed to occur in the rock, when the tunnel supports are defined as indeterminate structures. Even though all design methods used in the comparative analyses consider the rock support interaction with different degree. The ones which are directly based on it present less costly and more effective support system. In this study, it has been shown that the heavy rib (8W40) of conventional analysis with fixed load assumption, spaced at 1.20 m, is equivalent to the light rib (8W20) of convergence-confinement method. This conclusion arises

from the consideration of ground-structure interaction.

A flexible support system has an economical superiority, when compared with the external rigid support system (Table 1). By a flexible support system, it is possible to mobilize the ground resistance to its optimum extent to favorably redistribute stresses around opening and to improve rock properties. It has been shown that the cost of shotcrete-rock reinforcement system for the temporary supports is approximately a half of the rigid steel ribs of the conventional analysis.

REFERENCES

Barton,N., Lien,R. and Lunde,J. 1974. Engineering classification of rock masses for the design of tunnel support. Rock Mechanics 4 :189-236.

Barton,N., Iosed,E., Lien,R. and Lunde,J. 1980. Application of Q-system in design decisions concerning dimensions and appropriate support for underground installations. Subsurface Space 2 :553-567.

Bieniawski, Z.T. 1974. Geomechanics classification of rock masses and its application in tunnelings. Proc. 3rd Int. Cong.on Rock Mech. ISRM. Denver. V.2A.

Bieniawski, Z.T. 1979. The geomechanics classification in rock engineering applications. Proc. 4th Int.Con. on rock mech. and foundation Eng. Mexico.

DSİ. 1977. Geological report of Urfa tunnel of lower Euprates project. General Directorate of State Hydraulic Works. (Unpub.)

Hoek, E. and Brown, E.T. 1980. Underground excavation in rock. The Institution of Mining and Metallurgy. London. England.

Ladanyi, B. 1974. Use of long term strength concept in the determination of ground pressure on tunnel lining. Proc. 3rd Cong. on Rock Mech. ISRM. V.11. Part B. Denver. Colorado.

Rudledge, T.C. and Preston, R.L. 1978. New Zealand experience with eng. classifications of rock for the production of tunnel support. Proc.Int. Tun. Sym. Tokyo.

Terzaghi, K. 1946. Introduction to tunnel geology: In rock tunneling with steel supports (ed.by Proctor and White). Commercial sheaping co. Youngstown. Ohio. p:17-99.

Tosun, H. 1985. Comparison of tunnel design methods and their application to Urfa Aqueduct tunnel. Ms.D. thesis at the Middle East Tech. University. Ankara. 141p.

Ünal,E. 1983. Development of design guidelines and roof-control standards for cool-mines roofs. Ph.D. Thesis. The Pennsylvania State University. 355 p.

Hydropower'92, Broch & Lysne (eds) © 1992 Balkema, Rotterdam. ISBN 90 5410 054 0

A study on the effect of shotcrete and rock bolting support in underground opening

Yang Shuqing
Wuhan University of Hydraulic and Electric Engineering, People's Republic of China

ABSTRACT: The effects of shotcrete and rock bolting support on the power house of Lubuge Hydropower Station are studied in this paper by model tests and FEM calculation. The results indicate that the support has sign nificant effects on improving working condition, raising load-bearing ca- pacity and increasing overall stiffness of surrounding rock. When the sup- port is utilized the failure load of model will be increased by about 50% and the maximum absolute displacement of power house will be decreased by about 20%.

1 INTRODUCTION OF HYDROPOWER STATION

Lubuge Hydropower Station is located at Southwest China and its installed capacity is 450 MW. The underground power house with dimension of 125X 17.5X39.4M (lengthXwidthXheight) stands where the rock stratum mainly consists of limestone and at depth 30M below the power house exist two faults F_{209} and F_{313} . The rock mass and faults has following mechanical parameters: E=31.9GPa, $\mathcal{M}$=0.22, C= 6.9MPa, φ=65° for rock and E=7.84GPa, $\mathcal{M}$=0.3, φ'=55°, f=0.3-0.4 for faults. The initial stress of rockmass is considered mainly gravitational stress. Owing to the influence of the valley topography, the direction of the maximum principal stress in- tersects the vertical axis of the cross section of power house at 25° (Fig. 1).

2 MODEL TEST AND METHOD CALCULATION

The stress and failure mode of sur- rounding rock are studied by plane strain geomechanical model tests, whereas the absolute displacement due to excavation is analysed by the laser speckle method. According to theory of similarity we have simu- lated the main mechanical parameters of rock, faults and shobcrete rock bolting support. Fig. 2 and Fig. 3

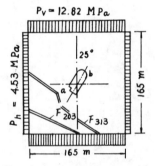

Fig.1 Simulated scope and load of model

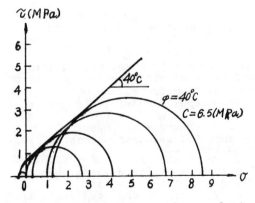

Fig.2 Mohr's envelope of equvalent material

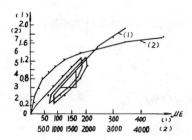

Fig.3 Uniaxial (1) and triaxial (2) stress-strain curves of equivalent material

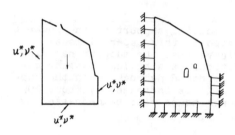

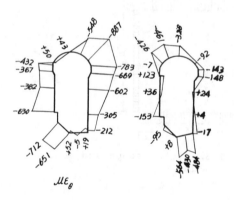

$\mu\varepsilon_\theta$

Fig.5 The secondary strain distribution of surrounding rock

show the Mohr's envelope. uniaxial and triaxial stress-strain curves of equivalent material.

The steps of model tests are as follows:

1. To excavate the opening and measure the secondary stress of surrounding rock;

2. To do shotcrete and rock blotting support of the opening then to measure the stress again;

3. To carry on the load until the model is exhausted.

The calculating chart of FEM is

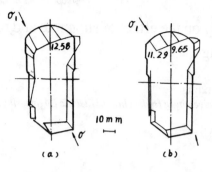

Fig.6 Distribution of the absolute periphery displacement of opening (a) Non-supported (b) Supported

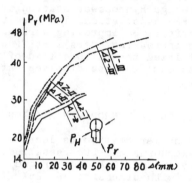

Fig.7 Opposite displacement of opening
I-homogeneous model non-support
II-homogeneous model with support
III-nonhomogeneous model with support and faults

shown as Fig. 4, so the initial stress field and the secondary stress of surrounding rock due to excavation are callculated.

3 ANALYSIS RESULTS OF MODEL TEST AND EFFECT OF SUPPORT

3.1 Analyses of strain and displacement

Fig. 5 is the secondary strain distribution of surrounding rock. The maximum of concentrations of tangential strain are developed on right arch support and bottom of left wall, and they are 2.7 and 2.9 times their initial strain.

Fig.6 shows the absolute periphery displacement of opening measured by laser speckle. Comparing Fig.6 (a) with Fig.6 (b). it is clear that both of them are similar. Since the displacements of surrounding rock are dominated by the initial earth stress field, so the maximum displacements are developed at the same place whether the opening was supported or not. Meanwhile the direction of maximum displacement is identical with the initial maximum principal stress. Of course, the value of the maximum displacement of supported opening has been decreased by about 20%.

Fig.7 shows the opposite displacement of opening during failure test. Because the shotcrete and rock bolt can raise the overall stiffness of surrounding rock, p-Δ curve of supported opening has more steep slope than the curve of non-supported opening. It means that, at the same load, the supported opening has less displacement. And the higher the load goes, the larger the difference becomes between the supported and the non-supported openings, which manifests the significant effect of support especially as the load is comparative strong.

3.2 Failure mode and failure mechanism

As shown in Fig. 8 whether the opening is supported or not, the initial failure place the failure process and the final failure mode all are similar, it is mentioned before at right arch support and the bottom of left wall exist the concentrations of tangential strain, so from which the failures have started. With the rising of load along the opposite side walls, one path of the failure develops from right arch support down to the bottom and another from the bottom of left wall up to the arch support. It can be seen that on each model have found a certain amount of macro crackes parallel to the maximum principal earth stress and some even stretch into the inner of surrounding rock. At same time the both walls formed two cuneiforms and finally fell down. On the bases of failure mode and process as described above, we deduce that the failure is resulting from high tangential strain and is accelerated by function of compression-

shearing.

Tab.1 shows the results of tests, where the initial failure load corresponds with the appearance or observable microcrack and the final failure load with the loss of load-bearing capacity. As it is shown, the final failure load of the supported opening is about 50% higher than that of the non-supported.

4 EFFECT OF THE SUPPORT

If the support is constracted well, it can work together with surrounding rock. In this case the support not only bears the acting pressure of surrounding rock, but also gives

Tab.1 The initial and final failure load for model

Load(MPa) Model	Initial	Final
I Nonsupported	2.62	4.05
II Supported	3.00	5.94
III Supported	2.80	5.94

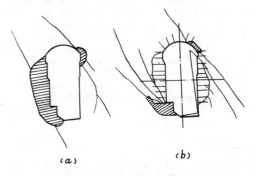

(a)   (b)

Fig.8 The failure mode of different model (a) non-supported (b) Supported

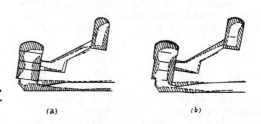

(a)   (b)

Fig.9 Vector diagram of displacements of opening

Tab. 2 Comparing data of maximum displacement of opening (mm)

Displacement / Model and place	Δv		Δh		Δt	
	FEM	Las.	FEM	Las.	FEM	Las.
I Left arch	12.68	12.17	6.27	4.5	14.10	13.90
Support					$\alpha = 64°$	$\alpha = 61°$
III Left arch	11.4	9.23	6.61	5.31	13.0	11.23
Support					$\alpha = 59.4°$	$\alpha = 58°$

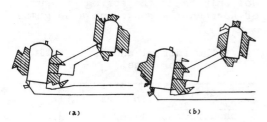

(a)   (b)

Fig. 10 The failure scopes of opening

rock the reacting force. Thus working condition of rock will be improved. If the shotcrete layer is very thin and all wall of opening is lining, the shotcrete layer can adjust to the deformation of surrounding rock as a soft support. And the rock bolting can raise the load-bearing capacity of surrounding rock.

5 RESULTS OF FEM

The vector diagram of displacements is shown as Fig. 9 and the failure scopes of opening are as Fig. 10. Tab.2 lists the comparing data of maximum displacements of opening.
Comparing the obtained results by model tests with that of FEM, it found that both are corresponding well.

6 CONCLUSION

6.1 The characteristic of stress distributions, the failure place and failure mode of opening mainly depend on the initial earth stress and the shape and the size of ex-

avated part. Obviously the shotcrete and rock bolting can improve working condition raise load-bearing capacity and decrease displacement of surrounding rock, but it can not change the characteristic of stress distribution and failure mode of opening.

6.2 Results obtained by model test correspond well that of FEM

6.3 Model test of shotcrete and rock bolting in laboratory is very difficult. This paper has not only studied the secondary stress, displacement of opening and effect of shotcrete and rock bolting support but also analysed the failure mode and failure mechanism, which undoubtedly presents the model test a new practical way to the engineering application.

REFERENCES

Brady, H.G. & Brown, E.T. 1985, Rock Mechanics for Underground Mining. GEORGE ALLEN & UNWIN.
Jaeger, Charles. 1979. Rock Mechanic and Engineering. New York, Cambridge University Press.
Yu Xiefu & Zhen Yingren. 1983. Study on Stability of Surrounding Rock of Opening. Press of Coalindustry, Beijing.

Hydropower'92, Broch & Lysne (eds) © 1992 Balkema, Rotterdam. ISBN 90 5410 054 0

Design of reinforced concrete high pressure tunnel of Guangzhou pumped-storage power station

Yao Lian Hua
Guangdong Provincial Design Institute of Water Conservancy and Electric Power, People's Republic of China

ABSTRACT: This article discussed about the design and construction of high pressure large diameter tunnel in Guangzhou Pumped-storage Power Station ------the first project of pumped-storage power station in China------which is under constructing at Guangdong Province. The high pressure tunnel is 8.5m in diameter and 1059m in length. It is embeded deep under granite rock mass and consists of two 50 degree inclined tunnels and three horizontal tunnels. This article presents the summarized results of a study carried out by the author on the design of high pressure large diameter tunnel with reinforced concrete lining by the means of different theory and method. The theory and method of pervious pressure tunnel is also discussed.

The high pressure tunnel in the project of Guangzhou pumped-storage power station consists of two sections of inclined tunnel and is connected with upper, middle and lower section of horizontal tunnel, they are D8.5m and D8.0m in diameter and 1059.068m in total length. The detail lengths are respectively 14.146m, 346.86m, 90.37m, 288.117m and 215.037m, while the total lengths of 4 bends are 4*29 671m.

It is known that at the middle stage of Yanshan Mountain the rock properties consists mainly of coarse biotite granite, and the high pressure tunnel is located in the thick wall mass of granite. Except for alteration in varying degrees occuring in the rock

Fig. 1. Plan view of high pressure tunnel

mass arround the structural fissure, rock features are comparatively uniform and integral. The longitudinal section of high pressure tunnel is embeded deeply under granite rock mass and meet with two important faults (F145 and F2) at the site of upper and lower section of inclined tunnels.

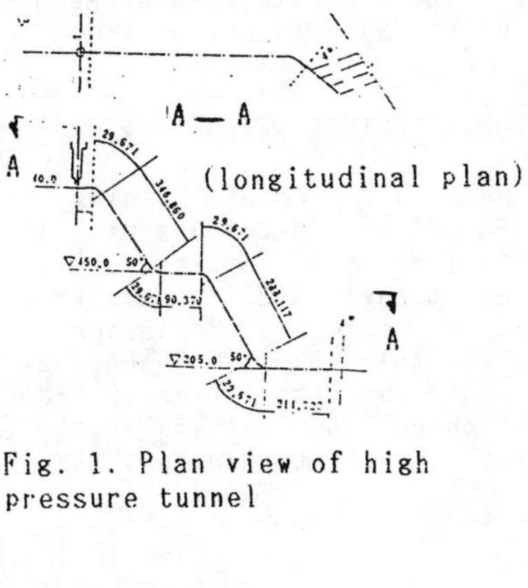

Plan view of delivery water system and the geological condition of this project is respectively shown as Fig.1.and Fig.2.

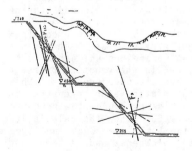

Fig. 2. Geological condition of high pressure tunnel

CROSS SECTION OF TUNNEL & TYPE OF LINING

With regard to the use of rock cover as a criterian for adequate confinement, the high pressure tunnel in both

longitudinal and cross profile can be checked by the formula of Snowy Mountain and Norwegian Criterian for Confinement:

1)Snowy Mountain Criterian

$Crv=hs \div Rw/Rr$

$Crh=2 \div Crv$

2)Norwegian Criterian

$Crm=(hs \div Rw \div F)/(Rr \div Cosb)$

The result of these two criterian are shown in Table 1 and Fig. 3.

Table 1 Ratio of h/Ho

No	Hx (m)	Ho (m)	h (m)	b (o)	h/HO	Crm (m)
A	740	76.8	100	35	1.30	58
B	450	366.8	290	35	0.79	231
C	450	366.8	230	38	0.63	240
D	205	611.8	450	36	0.74	390
E	205	611.8	490	36	0.80	390

Both high pressure grouting prestressed concrete lining and reinforced concrete lining have been designed to meet the need of suitable lining of the tunnel.

High pressure grouting prestressed concrete lining is one of the constructure which can bear the internal pressure and all outer loads action. It is assumed that the rock mass, concrete and the cement liquid are informed as a good quality of lining which processes of the forces acting at the tunnel. In this respect, the design will present an optimal condition of tunnel during operating,concrete of lining and the boundary rock mass will prevent from cracking and perviousness. There are two important parameters which has to be determined, one is the grouting pressure and the other is pressure loss. When 1.2 times the static water pressure is given as grouting pressure, the pressure loss will be 0.65. Then the calculating results of the tunnel are 60cm and 120cm in concrete thickness and 4.5Mpa and 7.3Mpa in grouting pressure.

Due to the fact that the diameter of this high pressure tunnel is fairly large, and all the design datum of the prestressed concrete lining and grouting pressure have to be determined by a large number of laboratory test works, especially the constructing period of

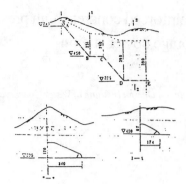

Fig. 3. Criterian for confinement

this project is so short that it has not enough time and design schedule to fulfil these works. However, both the reinforced concrete lining and the high pressure grouting have to be discussed.

The design interal water pressure (Homax-m) and the outer water pressure (Ho'-m) of the tunnel that had pointed out in Fig.1 with the symbol A, Ao, B, Bo, C, Co, D and E are Homax=75.7, 260.8, 430.8, 435.4, 439.9, 573.8, 707.7, 715.1; Ho'=20, 36, 58, 55, 46, 64, 90, 98. Table 2 gives the characteristic of the rock.

Table 2 Characteristic of rock

rock class	f (Mpa/cm)	Ko	Er (Mpa)	l	Rr (T/M)	Kr (cm/s)
2	8-9	80-100	$25.6 \div 10^3$	0.2	2.6<$1.0 \div 10^5$	
3	5-6	40-50	15.0 "	0.25	2.5	2.683"
4	3-1	13-5	9.7 "	0.28	2.4	4.919"

In this paper, there are three different methods are used to analyze this project, they are Mechanics Of Structure Method, Finite Element Method and Pervious Pressure Tunnel Method.

MECHANICS OF STRUCTURE METHOD

The thickness of concrete lining is assumed as 60cm. According to the design criterian, the results of the high pressure tunnel in 29 cross section are shown in Table 3. By means of increasing the thickness of concrete and improving the characteristic of rock mass, the comparison results of those section which is situated in weak rock mass are seperately given in Table 4,5.

Moreover, using the computer program, the results of a large number of sections that is situated in good rock mass are satisfactory to meet the need of minimum width (0.2mm) of rock. However, when there are faults or weak rock mass crossing through the tunnel, this method will give more or less unsatisfactory results.For exemple, when the elasticity modulus of rock mass is about 9.7∗1000Mpa, the reinforced concrete can support 210m of internal water pressure only, the area of steel bar in per meter of tunnel is so large that it is unable to arrange steel bars into the concrete. For this reason, it is necessary to discuss the calculation in the site of weak rock mass and fault. Therefore not only by calculation,but also the construction countermeasure of weak rock mass and fault have to be determined.

Table 3 Results of mechanics of structure method

No	Ho (m)	Ho'rock (m)class	Hc (cm)	Zg (Mpa)	Ag (cm2)	W (mm)
1	97	20 2	60	26.8	20.36	<0
2	135	23 3	60	42.2	39.27	<0
3	158	24 3	60	54.4	39.27	<0
4	183	25 4	60	198.0	116.3	0.16
5	212	28 2	60	24.7	39.27	<0
6	220	28 4	60	198.0	193.19	0.16
7	232	31 2	60	30.2	39.27	<0
8	240	32 4	60	198.0	234.71	0.16
9	328	42 3	60	144.2	39.27	0.09
10	415	58 2	60	79.0	39.27	0.014
11	420	55 4	60	198.0	608.4	0.16
12	435	52 3	60	17.9	39.27	0.134
13	553	62 3	60	16.9	56.55	0.143
14	586	66 4	60	98.0	953.0	0.213
15	636	76 4	60	198.0	1056.8	0.213
16	724	90 2	60	165.8	56.55	0.107

Table 4 The comparision results of improving the characteristic of rock mass

NO	Hc (m)	KO (Mpa/cm)	Zg (Mpa)	Ag (cm2)	W (mm)
14	60	130	198	608	0.16
	60	195	187	479	0.18
24	60	195	187	929	0.18
	60	260	181	803	0.16

Table 5 The comparision results of increasing the thickness of concrete

NO	Hc (m)	KO (Mpa/cm)	Zg (Mpa)	Ag (cm2)	W (mm)
14	80	260	187	354	0.16
22	80	260	187	702	0.16
24	80	260	187	802	0.16

FINITE ELEMENT METHOD

Along the lower inclined tunnel, the transverse section of tunnel that crossing through fault F2 in the top and middle position are shown in Fig.4. By using the SHP5 finite element computer program, in the case of tunnel which take loads of internal water pressure and rock stress, deformations and stresses of the concrete lining, steel bars requirement and boundary of rock mass can be estimated. Then following the results of reinforced stresses, the allowed limitation width of concrete crack can be checked.

By means of finite element method varying assumption of design condition can also be put into the calculation datum to performed various schemes for further analyzing and discussing. By doing so, the difficult problem of high pressure tunnel that crossing through F2 fault could be solved with the aid of the optimal results of the method mentioned above.

Table 6 shows the result of concrete lining in different thickness, 21different requirement of steel bars, and also the rock mass when take into account the consolidation grouting to improve its characteristic. From Table 6, it gives the opinion that high pressure tunnel which crossing through weak rock can be solved by increasing the thickness of concrete and improving the characteristic of rock mass with high pressure grouting

The final results of this part of tunnel are adopted as bellow: the thickness of concrete must be increased to 80cm, on the inner layer of concrete two sets of circular steel bars of 2∗8D25@12.5cm in per meter tunnel , and on the outer layer there is one set of circular steel bar of D25@12.5cm, while along the circular steel bar, there are 3 sets of longitudal steel bar of D25@12.5cm,in addition behind the outer part of concrete, some part of F2 fault with the deep of 40cm have to excavated.The width of this expanded range is the same as the fault width. The design of consolidation grouting of this part of tunnel is 10 holes per section, space between holes are 2m, hole depth 6m, grouting pressure is 6.1Mpa.

Table 6 Main results of finite element in fault section

SECTION   111

No	Hc (cm)	Er (1÷1000Mpa)	steel bar	Zg (Mpa)	w (mm)
1	60	3.0	8D28	282.3	0.49
2	60	3.0	10D28	278.5	0.48
3	60	3.0	2÷8D28	257.1	0.37
4	60	4.5	8D28	221.4	0.24
5	60	4.5+3.0	8D28	227.3	0.27
6	60	4.5+3.0	10D28	222.0	0.29
7	80	3.0	8D28	266.6	0.31
8	80	4.5+3.0	8D28	213.1	0.03

SECTION   222

1	60	3.0	8D28	243.0	0.33
2	60	4.5+3.0	8D28	200.2	0.16
3	80	3.0	8D28	242.8	0.18
4	80	4.5+3.0	8D28	199.8	<0

SECTION   111

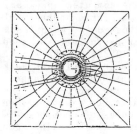

SECTION   222

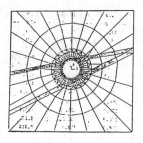

Fig. 4 Mathematic model of finite element

DISCUSSION OF THE DESIGN OF PERVIOUS TUNNEL

This method was published in WATER POWER & DAM CONSTRUCTION on May 1986 by A.J.Schluss. The author present a new design procedure, which takes into account the seepage pressure and the influence of fracture deformation on secondary permeability in rock mass.

According to the current criterion of pressure tunnel of China, it is pointed out that the forces applied to the concrete of tunnel is transmitted along the lines of seepage as a body force. If the design condition of tunnel calculated can also be performed with this method, and take into account the distribution of the body force which action the concrete line and rock mass.

Nowadays, more and more high pressure with large diameter tunnel will be designed and constructed in China, therefore many engineers and professors are making great effort to study hard on new design method, with their scientific efforts and achievements, the design of pervious tunnel of Guangzhou pumped-storage power station has also been discussed. Table 7 shows the main results of this method,it is pointed out

Table 7 Results of pervious tunnel method

NO	R.HO C.(m)	Kr (÷1)	Er (10÷Mpa)	n	y (%)	Zg (Mpa)	Ag (cm2)
1	2 97	1.789	25.6	186	97.8	144.5	6.7
2	2 205	1.789	25.6	179	98.4	139.6	7.9
3	3 158	2.683	15.0	179	97.6	139.6	12.4
4	4 183	4.919	9.7	179	95.7	139.6	25.8
5	2 212	1.789	25.6	179	97.7	139.6	15.9
6	4 224	4.919	9.7	179	98.4	139.6	31.0
7	2 233	1.789	25.6	179	98.4	139.6	12.2
8	4 240	4.919	9.7	179	95.7	139.6	33.8
9	3 328	2.683	15.0	179	97.6	139.6	25.7
10	2 415	1.789	25.6	179	98.4	139.6	21.9
11	4 420	4.919	9.7	179	95.7	139.6	59.2
12	2 520	1.789	25.6	174	98.4	136.8	28.3
13	4 586	4.919	9.7	174	95.5	135.4	87.1
14	2 620	1.789	25.6	174	98.3	135.4	34.5
15	4 636	4.919	9.7	174	95.6	135.4	94.5
16	2 715	1.789	25.6	174	98.3	135.4	39.8
17	2 724	1.789	25.6	174	98.3	135.4	40.3

that the effort of fracture in concrete lining and rock changed the amount of internal water pressure. If the force is assumed as a body force, the rock mass will support a large percentage of the internal water pressure.

The most important thing of this method is how to decide the main parameter of concrete and rock permeability (Darcy) factor. Pay much attention to the high pressure grouting and prestressing by rock grouting or gap grouting is one of the criterian for pervious pressure tunnel.

By means of the design of this method, the problem that cannot be done by mechanics method of structure will be solved.

CONCLUSION

Design of high pressure with large diameter tunnel by means of traditional statics method, when it is embeded deep enough in good rock mass, all the method that has discussed above can present a satisfactory results of lining and reinforcedment. However, in the case of weak rock and faults are crossing through tunnel section, design of finite element and pervious pressure tunnel will be used to determined the optimal thickness of concrete and steel bar as well as the excavation requirement of weak rock and fault.

It is known that the water delivery system of pumped storage power station always arranged with large diameter and high pressure tunnel. The design of this kind of tunnel not only in the theory but also in the calculation method, designer must pay much attention and need much of an effort to handle it.

The final results of high pressure tunnel in this project are presented as follows:

.The thickness of concrete except the fault section described above is 60cm.

.In good rock, there are only one set of circular steel bar of 8D16, 8D20,8D28,8D30 in per meter long of tunnel.

.The longitudinal steel bar along the circular section is D16@12.5cm.

.All the length of tunnel are designed with consolidation grouting of 10 holes per section, and arranged at internals of 2.5m,the depth of holes are 5m, grouting pressure are 3.0-6.1Mpa.

.In class 4 rock section along the upper and lower inclined tunnel the arrangement of steel bar will be increased to 2 or 3 sets of circular steel bars, namely 2*8D25@12.5cm and 3*8D25@12.5cm, the longitudinal steel bars are D25@12.5cm.

.In class 4 rock section, the parameter of grouting is the same as the other section, only the amount of

grouting pressure at the upper inclined tunnel and middle horizontal tunnel is increased up to 4.5Mpa.

Notation

Crv=vertical rock cover
Crh=horizontal rock cover
Crm=minimum rock cover
hs=static head
Rw=unit weigth of water
Rr=unit weigth of rock
b=slope angle
F=safety factor
h=depth of rock from outer lining to ground surface
HOmax=maximum water pressure
HO'=outer water pressure
f=solid coefficient
KO=unit modulus of elasticity
Er=modulus of elasticity of rock
l=poison's ratio
Kr=permeability (Darcy)
Zg=stress of reinforcedment
Ag=steel bar area
W=crack width in concrete
Hs=thickness of concrete
y=percentage of water pressure support by rock
D=diameter of steel bar
*1=1*100000 m/s

Hydropower'92, Broch & Lysne (eds) © 1992 Balkema, Rotterdam. ISBN 90 5410 054 0

Some technical problems in the construction of a tunnel in soft rock

Guangqian Zhang, Yusheng Zhang & Zhiyuan Nie
Yindaruqin Engineering Construction Management Bureau, People's Republic of China

Kaouxe Liu
Shaanxi Institute of Mechanical Engineering, Xian, People's Republic of China

ABSTRACT: Tunnelling in soft rock had long been a challenge to engineers. Technical problems met during the construction of the 15.723 Km Pandaoling (PDL) Tunnel in the Northwest of China, may have been one of the most challenging. Stability, tunnelling under adverse geological conditions, cracking of the final concrete lining and other problems, required detailed analysis and cautious consideration. This paper is a brief summary of some problems encountered, tunnelling methods, some innovative analytical approaches and their application during construction of PDL Tunnel.

1 INTRODUCTION

Pandaoling Tunnel is a 15.723 Km. long diversion tunnel, presently the longest tunnel in China. It is the 37th Tunnel of the Yindaruqin Irrigation Project which is a large scale irrigation project partially funded by the World Bank in Gansu Province, in the northwest of China. The maximum depth of overburden is 404 metres. For 2.64 Km overburden depth is greater than 300 metres and 3.91 Km is 200 and 300 metres. The remaining length has less than 200 metres of overburden.

The lithology encountered within the tunnel was Tertiary formations for 12.84 Km length with the remainder Cretaceous Period. The rock types encountered during excavation were sandstone, conglomerate and other silty rocks such as sandy mudstone, muddy sandstone. Most of the rocks were loosely cemented. The quality was poor to very poor. In-situ density varied from 19.5–24.2 KN per cubic metre. Uniaxial compressive strength Rp = 0.01–45 Mpa. Laboratory testing of Young's modulus of saturated samples gave results between 165–1220 Mpa.

Although ground water was encountered in about 4 Km of the tunnel length, the worst conditions were where ground water was in combination with weak rock. In some places, ground water was highly aggressive.

Due to the poor quality, low strength, low cohesion and other unfavourable geological conditions, the stability of the tunnel, both during and after excavation, was one of the most critical factors which needed to be adequately studied and carefully evaluated.

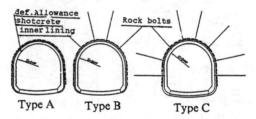

Type A　　　Type B　　　Type C

Figure 1.

In the design and the construction of the tunnel, many numerical analysis, experiments, in–situ measurement and monitoring were carried out. Most of the suggested approaches including the back–analysis and deformation forecasting were successfully tried in the construction of the tunnel.

2 DESIGN AND RESEARCHES

The typical cross sections of the tunnel are shown in Fig. 1. Three types of supporting system (as in A,B C) were designed to adjust to the variation of geological conditions encountered.

2.1 Shape of the cross–section

For tunnelling in soft rock, a structurally rational shape should be circular or horse shoe shape with invert. Construction convenience dictated the above shape. Numerical analysis, in–situ measurement and cracks found on the inner lining and many other difficulties met in the tunnel-

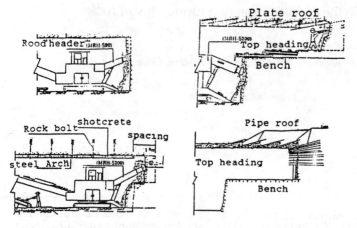

Figure 2. Methods of excavation

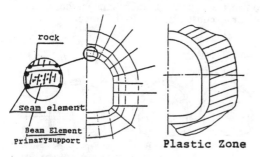

Fig. 3 FEM Mesh

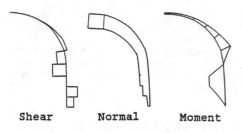

Shear   Normal   Moment

Fig. 4

ling have reflected the disadvantages of the shape.

2.2 Construction

"NATM" was adopted in the design and construction and specified in the tender documents to adjust to variations in geological conditions.

Roadheaders of type MRH–S200 were selected for drifting. In ordinary rocks, full face drifting was carried out by roadheader. For zones where loose sandstone was encountered, the top heading was usually excavated manually, and the bench was excavated roadheader. Forepoling with plates or pipes for temporary support of the crown was commonly used for the top heading. Steel arch and / or steel mesh was used for more than 13 Km with a spacing of 0.8 − 1.3 metres where serious geological conditions were encountered. [1]

2.3 Design load and analytical models

The design load (especially the ground pressure) on the supporting structure had long been a very complicated problem for designers. Although there were suggestions of analytical approaches, numerical models and the selection of parameters in some Standard Design Codes, most of the determinations and / or selections of parameters required much more experience than was available. No Code can provide solutions to every problem or include every condition that may be encountered in a specific tunnel. Maybe, this was one reason that many studies should be made in the design of a tunnel.

Both Bedded Beam Models and Finite− Element Models were suggested in China's Standard Design Code. Each has its advantages and disadvantages.

3 STABILITY DURING AND POST EXCAVATION OF THE TUNNEL

3.1 General

Since the geological conditions were poor in PDL Tunnel, the stability of the tunnel, both during and after excavation was of utmost importance.

The physical properties of the exposed rocks exhibited a high degree of variation. This variation caused numerous difficulties for the selection of appropriate parameters.

The Back—analysis Method made choosing of appropriate parameters more reliable.

Elasto—Visco—Plastic Finite Element Method analysis was used for estimating stresses, deformation and the stability of surrounding rock, together with the structural safety of lining support .[2] [6]

The FEM Mesh is shown in Fig. 3

In the computer program, pseudonodes were designed to make the deformation (rotation) in Beam element compatible with the deformation in linear elements. The rotation of the beam was stored in the pseudonode. The rotational continuity was guaranteed through a common pseudonode in adjacent beam elements .[3]

The use of beam elements in FEM programs found the common point between FEM Models and the classical Bedded Beam Model. This program also gives the forces (moment, normal, shear) in lining structure which are more convenient for the use of structural designs.

With the consideration of both "spatial effect" and "time effect (viscosity)", The numerical simulation gave good correlation between calculated deformation and in—situ measurement.

Fig. 4 gives some results of numerical analyses.

4 MEASUREMENTS BACK—ANALYSIS

Field measurements which are an integral part of "NATM" were required by the Contract Specification. At first, the convergence measurements were used to determine appropriate time for installation of the final lining.

As the excavation advanced, the geological condition deteriorated. Some cave—ins and ruptures occurred. Both the Contractor and Management Bureau became more aware of the importance of interpretation of field measurements. Thereafter, the in—situ measurements were also used to gain a better insight into the overall behaviour of surrounding rock and the tunnel supporting systems.[2]

The now widely accepted Back— analysis Method was adopted to the interpretation.

Numerical simulation also suggested the possible insufficiency of the designed primary support in zones where rock was very soft ($\varphi < 30$,C <0.1 Mpa for example) and the depth of overburden was greater than 300 metres.

A simple solution to the possibility was to design a stronger reinforced final lining for example.

In practice, this problem never occurred. The main reason was that the rock encountered was stronger than predicated and no ground water was encountered in zones with high overburdens. Much research was carried out on back—analysis

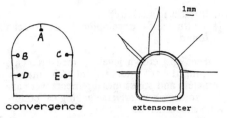

convergence   extensometer

Fig.5 Deformation instrumentation and measurements

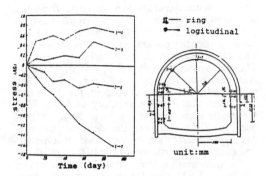

Fig.6 Stress instrumentation and measurements

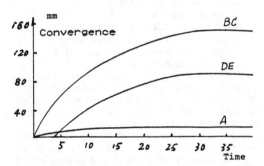

Fig.7 shows a typical convergence measurement result and the result of back—analysis.

and monitoring. One of the outstanding achievements was the development of a method for determination of site stresses and the visco—elastic parameters of surrounding rock through Back—analysis. [5]

The measurements which were carried out in the construction of the PDL Tunnel are as listed :

(1) Convergence and crown levelling
(2) Stresses (or forces) in rock bolts
(3) Pressure on shotcrete
(4) Pressure on final lining
(5) Extensometer

(6) Stresses in steel arch
(7) Width and development of cracks found.

Convergence and crown levelling measurements were arranged at intervals of 20 metres. Extensometers and stress measurements were arranged at zones where adverse geological conditions were encountered.

Typical instrumentation is shown in Fig.5 and Fig.6.

Fig.6 also shows the stress development in the inner lining.

The Mechanical parameters obtained from the measured convergence by the Back–analysis Method were: $E = 50$–100 Mpa, $\mu = 0.3$ (Assumed), viscosity parameter $\eta = 10E$ Mpa $*$ day. The site stress was assumed to be $\sigma_y = -\gamma H$, The confinement coefficient K_o was obtained by Back–analysis. $K_o = 0.7$ – 1.0, i.e $\sigma_x = K_o \sigma_y$. The results indicated a greater strength for the in situ rock than that from laboratory samples. With the mechanical parameters from Back–analysis used in FEM analysis, the deformation forecasting and safety alarm were tried in some sections during construction of the PDL Tunnel.

For deformation forecasting, both determined Model (i.e FEM with the input parameters from previous back–analysis) and an uncertainty Model—Grey system predication Model—GM(1,1) Model was tried.

The determined Model of deformation forecasting was well suited for predicting the future deformation of the point under consideration. The method was no more than a multi–parameter regression and extrapolation.

The Grey Model was used to predicate the possible deformation at the section which was to be excavated. The Model GM(1,1) was set up according to the measured deformation of previously excavated sections.

Although the accuracy of the GM(1,1) Model was not as expected, it did provide an acceptable way to "guess" what was going to happen before excavation. Fig.9 shows one of the forecastings and the correlation after excavation.

Grey Model can also be used for the forecasting of mechanical parameters like E, G, C, Rp etc. and geological parameters like RQD.

The safety alarm was based on the deformation forecasting. Also, the alarm system could not be absolutely reliable since unpredictable abrupt changes in geological conditions could occur within a few meters.

With less reliability and inaccuracy, deformation forecasting and structural safety alarming assisted a lot in the monitoring.

5 SOME PROBLEMS MET DURING THE CONSTRUCTION

The extreme complexity of geological conditions

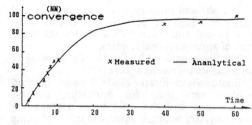

Fig.8 Convergence and backanalysis result

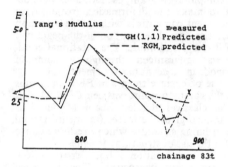

Fig.9 The Grey Model Prediction

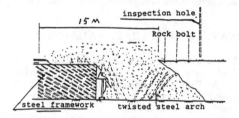

Fig. 10   Illustration of cave–in

in PDL Tunnel hampered the Contractor and Site Engineer during the construction.

5.1 Stability Safety

One of the most frequent and disturbing problems was stability during the excavation phase.

The typical types of instability could be classified into cave–in and rupture. Cave–ins occurred frequently both in zones of water ingress and zones of cohesionless sand.

One large cave–in which resulted in three months delay, occurred in Sept.1987, when the cross–section was being enlarged because of insufficient allowance for the inner lining and the replacement of the roadheader.

Fig. 10 shows an illustration of the cave–in.

Rupture of the primary lining and outflow of working face also occurred in the construction.

Fortunately, during the tunnelling process, there was no loss of life.

5.2 Deformation and Control Measures

The maximum convergence was about 240 mm. There were also some sections where deformation could not stabilize until measures were taken.

The general measures for controlling deformation were:
    (1) Closure of the invert
    (2) Local rock bolts
    (3) Steel arch and mesh
    (4) Secondary application of shotcrete
The practice proved theeffectiveness of these measures.

5.3 Cracks found on inner lining

Cracks were found in zones comprising of 3 km of the inner lining, within a month or so after the placement of the concrete.

Also many analyses and measurement experiments were made on this problem.

FEM analysis and experimental study suggested that the ring cracks were less harmful and were probably due to shrinkage and thermal effects. Longitudinal cracks which were harmful to the stability of lining structure, were probably caused by insufficient structural strength, greater ground pressure than expected, and the combination with thermal stresses.

Most of the longitudinal cracks were found in the vicinity of the arch foot or the top of side walls, where large tensile stresses were measured and were indicated by analysis, as shown in Fig.11.

5.4 Long term stability under operation

The Hydrologic properties of the rocks encountered were very poor. Some sandy rock became cohesionless when saturated, mudstone and other rocks also lost much of their strength. This may cause problems of long term stability of the substructure when the tunnel becomes operational. Seepage from the tunnel will affect the mechanical properties of the surrounding rock and consequently jeopardize the stability of the tunnel.

6 CONCLUSION

The PDL Tunnel is one of the longest tunnels driven in very soft rocks. Numerous difficulties and problems were met during excavation and construction. Other problems or difficulties may arise in future when the tunnel is operational.

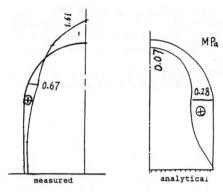

measured   analytical

Fig. 11 Stress distribution

Regardless of these problems, the completion of excavation was accomplished by Jan. 1992, with no major incidents. This successful completion was no doubt due to the concerted efforts of the Contractor (Kumagai Gumi of Japan) and the Yindaruqin Engineering Management Bureau.

Construction responsibility, tunnelling experience, numerical simulation, monitoring and other engineering measures also contributed to make the execution as safe as possible.

7 ACKNOWLEDGEMENTS

The authors here are in debt of Mr.G Palmer of Snowy Mountains Engineering Corp. for his help in preparing this paper, and are grateful for the support of Leaders of the Commanding Office of Yindaruqin Irrigation Project, and many thanks to Mr.Zhang Deli, Ms. Bao Jianwu, Ms. Wang Zhuqing for typing and some drawings.

8 REFERENCES

[1] Zhang Guangqian Nie Zhiyuan "Construction of PDL Tunnel in Soft Rock", National symposium on rapid tunnelling method with TBM. Aug. 1991.
[2] Liu Kaoxue, Zhang Guangqian, Zhang Yusheng "Some Technical Problems in the Construction of PDL Tunnel", Research Report.
[3] Liu Kaoxue, Zhang Guangqian "The Visco–elosto–plastic FEM analysis for PDL Tunnel" Research Report, 1989.
[4] Liu Kaoxue, Zhang Guangqian "The Application of Grey System Prediction in the Construction of PDL Tunnel" Proceeding of 4th National Conference on Numerical and Analytical Method in Geotechnics, April,1991.

[5] Liu Kaoxue, Zhang Guangqian "Visco–elastic Back–analysis for a Tunnel Driven with Roadheader", proceeding of 1st National Conference of Rock mechanics for Youth. November,1991.
[6] Zhang Wugong "Elasto–Visco– plastic FEM analysis for PDL Tunnel" Research report, 1990.

Hydropower'92, Broch & Lysne (eds) © 1992 Balkema, Rotterdam. ISBN 90 5410 054 0

Study on stability in surrounding rocks of an underground powerhouse under complex strata

Zhao Zhen-ying & Ye Yong
Wuhan University of Hydraulic & Electric Engineering, People's Republic of China

ABSTRACT: By means of some geomechanical model tests, the excavation steps of underground openings during the construction are imitated for a large underground hydro-power station in China, and the field of secondary stresses and displacements around the openings produced during the excavation is seriously studied by the authors in this paper. On the basis of the work mentioned above, the stability of rock masses around openings is appraised and some suggestions to strengthed this project are put forword.

1 INTRODUCTION

It is a conspicuous problem that the stability of rock masses around large openings in complex geological conditions during the construction of underground projects. It must be on the basis of the state of stress and deformation in surrounding rocks to reinforce project. But there are many factors, including excavation steps, opening type, the characteristics of primary stress field, opening arrangement, rock behaviour, geological structure and so on, having effect on the state, which makes the problem of the surrounding rock stability more complicated. Therefore, only by theory caculation is it not enough to make a faithful comment for opening design, and it is necessary that the deeper cognition on the distribution of stress and displacement around openings under the condition of the many influence factors is intensified by using model tests in laboratory in order to provide a more referential scientific basis for design bulunches, and a comparision for similar projects as well. For the stability of surrounding rocks about a large underground hydroplant in China, the authors carried out several sets of geomechanical model tests to study the distribution of stress and displacement in the course of excavations and the corresponding measurement to reinforce the project.

2 GENERAL SITUATIONS OF THE PROJECT

The hydroplant, composed of a cluster of crisscross openings including a powerhouse, a main transformer hall, tailwater tunnels, and so on (Fig. 1), is located in right bank of dam site in the calcareous rock formations. There exist three faults

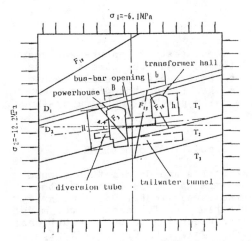

Fig. 1 Geological structures, boundary loads and opening dimension. H=48.6m, B=21.7m, h=25.9m, b=19.5m.

(F_3 , F_{15} and F_{16}) mear or throuth the openings and two mud interlayers (D_1 and D_2) above the roofs of openings, especially, one of which-D_2-is only 3-4m away by the arch crown of powerhouse. In addition,there exists a large fault, F_{18} , about 50m away from the left arch abutment of power-house. Because these structural planes cut apart the rock mass into several parts, to be sure, the integrity of surrounding rock is greatly destroied and the stress state in rock masses is deteriorated.

The tectonic stress is main in the natural rock mass by means of site exploration: the direction of the maximum principle primary stress σ_2 is nearly horizontal, and the minimum principle one σ_1 is near the gravitational stress of overlying strata. On the model test section, the parameters of geostresses are: σ_1 =-6.1MPa; σ_2=-12.2MPa, the angle between and horizontal plane is -4.4 , the corresponding lateral compression coefficient λ =2. The mechanics parameters in rock masses are as follows:

Tab.1 Mechanical parameters

Rock stratum	T_1	T_2	T_3
Elasticity modulus E(10^3MPa)	25.0	11.5	5.5
Poisson's ratio μ	0.24	0.25	0.28

place	D_1	D_2	F_{11}	F_3	F_{13}	F_{16}	T_1 / T_2	T_2/ T_3
tgφ	0.38	0.38	0.70	0.40	0.40	0.46	0.60	0.35

3 MODEL TEST

The model testing measurement adopted two methods: the stress was measured by the electrometric method-- Model No. I, and the displacement by the laser speckle photography--Model No. II.

3.1 Simulation conditions

The stress and deformation around openings was investigated by the plane-stress model. The similarity criterion was followed in the model tests according to below equations: $C_\varepsilon \cdot C_E /C_\sigma =1$, $C_\varepsilon C_l /C_\delta =1$. $C_\mu =1$. $C_f =1$. The friction factors of the faults, interlayers and other interfaces were simulated by putting some paper into every structural plane in the model according to $C_f =1$, and the elasticity moduli were done according to different rock strata. In terms of many comprehensive factors including simulation range, measurement precision and equipment conditions, the similarity constant and other data were elected as following in Tab. 2:

Tab. 2

Model Nos.	Similarity Constant					Model Dimension (cm)	Simulation Range (m)
	C_ε	C_σ	C_E	C_δ	C_L		
Mod.I	2	13.16	6.58	400	200	100×110	220×220
Mod.II	1	6.58	6.58	400	200	60×60	240×240

The boundary loads were exerted upon the model boundaries, or principle planes, according to the magnitudes and directions of principle stress in order to avoid the difficulty that shear stresses were exerted upon the boundaries. In addition, the model gravity weight was neglected in the course of the measurement.

3.2 Test equipments

The loading system was composed of two rectangular rigid frames (140cm x140cm, 80cmx80cm), oil pressure cells, oil pumps as well as an automat of stabilizing oil pressure. The models were vertical type, and the loading was on four sides of the models. In addition, polytetra=fluoroethylene was adopted for antifriction on the side surfaces. The strain measurement system, including a strain meter of static resistance with high-volume measuring points and a minicomputer, can achieve automatic measurement, calculation and print while the model tests are being carried out.

4 KEY TECHNOLOGY

The key points in the model tests involve two problems:

The first one is how to carry out the test procedure. At first, before excavation, the loads, kept constant in the course of whole test, are exerted upon the model to form an expected initial stress field. Then, after the first stage of excavation according to real steps of construction, the measurement is carried out to obtain the data of the stress and deformation of redistributions at this stage till the beginning of the second stage. For other further stages, the test will be carried out in the same way of the first stage until the end of the test. It is obvious that these data obtained in the tests are able to fit in with those in actual situation of site construction only according to this procedure.

The second one is that the displacement is measured by the laser speckle photography, which has the great advantages of high accuracy without damage and restraint to the measured object; in addition, whole displacement field can be gained so that the maximum displacement and its corresponding position is obtained accurately.

5 ANALYSIS OF RESULTS

The excavation of openings is divided into ten stages. Now, some diagrams on the distributions of stress and displacement, only to certain partial stages among whole those, are given here for shortness' sake (as shown Fig. 2—5).

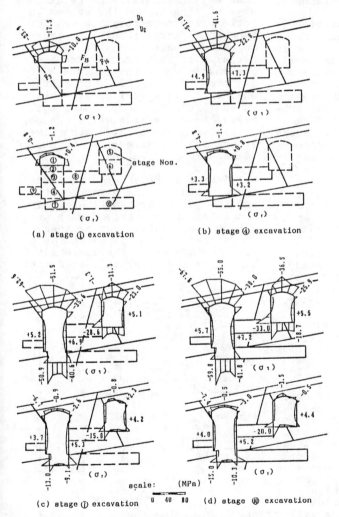

(a) stage ① excavation

(b) stage ④ excavation

(c) stage ⑦ excavation

scale: (MPa)
0 40 80

(d) stage ⑩ excavation

Fig.2 Stress distribution along boundaries (σ_τ:tangential stress;σ_τ:radial ones) and the Nos. of different stages (①—⑩) of excavation.

227

5.1 The stress distribution around openings (Fig. 2)

5.1.1 After the excavation of the 1st stage, the arch face of powerhouse is formed. The opening, whose depth-span ratio is 0.38, has a type of large-span and low-side wall. The tangential stresses σ_t on the arch are increased by a big margin in comparison with the initial ones at corresponding positions. The value of maximum compresion stress among those is about twice that of initial one, which occurs at left arch abutment. However, the radial stresses σ_r on it are universally released and meanwhile there has been a few tensile ones at right arch abutment.

5.1.2 In the course of excavation from stage ② - ④, the height of side-wall of powerhouse is on the increase with the fall of floor level. At the end of stage ④, the powerhouse, whose depth-span ratio has been 2.0, has a type of high side wall. The manner of stress redistribution is that the values of σ_t on the arch are on the increase, whereas those of σ_r on it, on the decrease because of the arch action during this course; at the end of stage ④, the maximum value of compressive stresses, up to -51MPa, is almostly close to the rock compressive strength, which still occurs at left arch abutment; and the stress distribution on the arch is non-symmetry, especially at the right arch abutment where the value of σ_t is only about half that of σ_t at left arch one; in addition, the stresses on the side wall are released greatly, where tensile stresses appear universally. The reasons for those mentioned above are that, firstly, the side wall of powerhouse is very high, about 38m, and the horizontal loads are great; secondly, the integrity of surrounding rocks is destroied as a result of the cutting of geological structure planes on the rock mass, which makes the load bearing capacity of surrounding rock weakened.

In fact, the authors had completed a set of homogeneous model test by the beginning of the non-homogeneous ones introduced in this paper for comparison between both results. Its results showed that no tensile stresses appeared on the side wall and the stress distribution was also symmetry in the main, for the surrounding rock kept intact.

5.1.3 From stage ⑤ — ⑥, main transformer hall is formed. Its manner of stress variation is vary similar to that of the powerhouse. Tensile stresses universally appear on the side wall of the hall in consequence of the effects of both the horizontal loads and the fault cutting.

5.1.4 After the 7th stage, both the powerhouse and the hall have come up to their design dimension. Comparing with the former stage, the character of stress distribution is still no too much change, that is to say, the side walls are still the relaxation area of stress, and the arches and bottoms the concentration area of tangential stress; but the stress magnitudes have some variation, especially at left arch abutment of powerhouse where the maximum σ_t is 5.3 times the magnitude of initial one, which is a very great variation; in addition, the surrounding rock masses near the side walls are universally in the state of two dimensional tensile stress, which should be paid much attention to.

5.1.5 From stage ⑧ — ⑩ all openings, including the diversion tube and the tailwater tunnel etc., are completed. In comparison with stage ⑦, the excavation during these stages has few influence on the major openings—the powerhouse and the hall: the stress amplitude of most

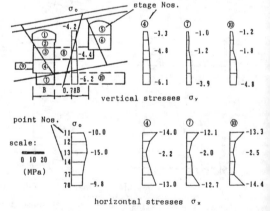

Fig.3 Stress distritution on the centre axi of rock pillar (σ_o—initial stress).

positions is not beyond the scope of 10%. Only at the corners of openings, the phenomena of stress concentration are of a litter serious.

5.2 The stresses in the rock pillar (Fig. 3)

It can be shown clearly from Fig. 3 that after the excavation of stage ④, the variation of vertical stresses on the centre axis of the pillar is no more obvious in comarison with initial stress field;whereas that of horizontal ones on it is remarkable greatly: the stresses of every measuring point near ½ height of the side wall are released obviously, especially, the stress at point No.13 (-2.2MPa) is only 0.14 the value of initial one; in addition, in consequence of the cutting of faults, some phenomena of stress concentration occur at point No.11 and No.78, and the maximum stress among those reaches 1.4 the magnitude of the initial stress. As mentioned above, the excavation of the powerhouse results in the stress relaxation in high side wall, the influence of which obviously spreads to the centre of the pillar.

After the excavation of stage ⑦, the vertical stresses in rock pillar decrease universally, which is very obvious at the upper half parts of the pillar—in particular, the value of stress at point No.11 has decreased to 0.25 that of initial one;the horizontal stresses are of continuous decrease. The reason for those is that the thickness of the surrounding rock at left side wall and arch of the hall is lessened because of the cutting of structure planes including D_2, F_{15} and F_{16}, which results in the deformation towards the inside of the hall (as shown Fig. 5) at left arch abutment and further bring about the stress relaxation in the pillar. On the other hand, the stress distribution in the pillar is non-continuous, for the pillar is divided into two parts by F . What is more, the excavation of crisscross openings further weakens the integrity in the pillar, which is also unfavourable to the stability of pillar.

5.3 The influential region due to excavation (Fig. 4)

5.3.1 There are some tensile stresses within the centain region of the surrounding rock near the left side wall of powerhouse as a result of excavation, but in rock mass at about once the wide of powerhouse, the stress nearly maintaints the state of the initial stress without disturbance by excavation. Within the surrounding rock near the right side wall of powerhouse, the influential region of tensile stress due to excavation is about 1/4 the wide of the pillar, which makes the stress state in pillar more deteriorated. The influential region near the right side wall of the hall is similar to that near the left one of the powerhouse.

5.3.2 As the lateral compression coefficient $\lambda=2$, the horizontal loads play a control role at the arches and bottoms of openings. The values of the tangential stresses on the arch boundary of powerhouse are about 5 times those of the initial

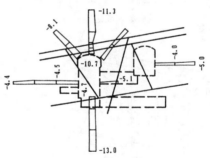

(a) initial stress field (σ_1)

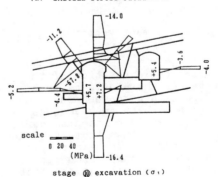

stage ⑩ excavation (σ_1)

Fig. 4 Stress distribution at the hydroplant region.

ones; therefore, the influential region near the arch due to excavation is deeper than that near the walls so that the stress near the arch crown, in the rock mass at twice the wide of powerhouse, has still a certain fluctuation in comparison with the initial one. The region near the bottom and arch abutments of powerhouse is similar to that near the arch crown. On the other hand, because there exist several faults and interlayers within rock masses (especially near arches), the continuity of stress distribution is not very good.

5.4 The displacements around openings (Fig. 5)

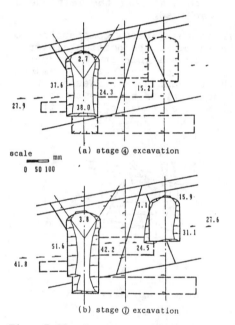

scale ___ mm
0 50 100

(a) stage ④ excavation

(b) stage ⑦ excavation

Fig. 5 Displacement vector diagram

5.4.1 After the excavation of the 4th stage, at the middle parts of both side-walls of powerhouse, the displacements which are convengence ones in the nearly horizontal direction are greater than those at other positions; and the maximum displacement is about 37.6mm, which occurs at the left side wall. At the foot of powerhouse, there are also some convengence displacements. In addition, owing to the effect of hori-zontal loads, there are some upheave displacements at the arch.

5.4.2 After stage ⑦, the character of displacement distribution of powerhouse is nearly without variation, but the maximum displacement which occurs at the left side wall of powerhouse has reached 51.6mm that is a great value.

After the transformer hall is formed, some convergence displacements appear at its right side wall, right lower corner and left arch abutment; whereas some displacements towards the outside of the hall do at its left lower corner and crown. It is obvious that this tendency is coincident with the behaviour of the stress distribution, i.e., there are some great tensile stresses at the left arch abutment and the right lower corner, whereas some great compressive ones at the left lower corner.

5.4.3 After stage ⑩, the excavation of transverse crisscross openings has some influence on the displacement distribution of major openings (the powerhouse and the hall): the maximum displacement has reached 57.7mm which occurs at the left side wall of powerhouse (its Figure is omitted).

As judaged by the tendency of the displacement direction in whole hydroplant region, the displacements of powerhouse due to excavation play a control role on the displacement field: for example, in consequence of the great convergence displacements at the right side wall of powerhouse, there are some displacements towards the powerhouse in the rock pillar, and also some great ones towards the outside of the hall at its left lower corner.

6 DISCUSSION

6.1 The character of the stress distribution after excavation is that the arches and the bottoms are the concentrition area of tangential compressive stress, where the maximum compressive stress is about 5 times the value of the initial stress at the same position; whereas the side wall are the relaxation area of tensile stress. This character makes the surrounding rock masses

230

in the dangerous state. It has been proved by the over-load failure model tests that the phenomena of compression and shear failure appear at both the arches and the bottoms to different degrees, and there is also a certain dangerous condition at the side walls. It is not difficult to understand that the condition mentioned above is closely related to both the character of the geostress field and the type of openings with high side wall.

6.2 The character of secondary stress distribution around openings depends on the rock behaviour, the initial stress field and the type of opening, but the complex geological structures decide the deterioration degree of stress state, the positions of local failure and the possibility of instability.

6.3 The displacement due to the excavation of powerhouse has a great influence on the displacement distribution in the rock pillar as well as around the hall, meanwhile decides the character of displacement distribution in whole hydroplant area.

6.4 The rock pillar is separated into two irregular triangular bodies by the cutting of the faults and interlayers, especially at the body near powerhouse, if the arch of powerhouse is destroied, the side walls of it can also be no stable. As authors think, the load bearing capacity of the pillar can be raised if the pillar is strengthened by mean of anchorage and grouting to improve its integrity, which may be more effective and economical than thickening the rock pillar.

6.5 It has outstanding advantages to apply the laser speckle method to underground structural model test: firstly, this discontiguous measurement method has high accuracy without damage and restraint to measured object; secondly, whole displacement field can be gained so that the maximum displacement and its corresponding position are obtained accurately. Therefore, this is a very good test technology deserving of recommendation and development in the field of underground structural model test.

REFERENCES

Zhao Zhen-ying & Ye Yong 1989. Model tests on surrounding rock stability in underground openings of D.F. hydroplant. Wuhan Univ. of Hydraulic & Electric Eng., China.

Zhao Zhen-ying & Ye Yong 1988. Surrounding rock stability of underground openings. Proc. ISTWRPP: p. 225. New Delhi, India.

Ye Yong & Zhao Zhen-ying 1988. The analysis on surrounding rock stresses in underground openings of D.F. hydroplant. Proc. of WUHEE: p. 21. China.

Hydropower'92, Broch & Lysne (eds) © 1992 Balkema, Rotterdam. ISBN 90 5410 054 0

Engineering geological features and rock support – Guangzhou pumped storage power station cavern

Zhu Yunzhong
Supervision Board, GPSPS, MSDI, Wuhan Hydroelectric College, People's Republic of China

Liu Xueshan
Supervision Board, GPSPS, People's Republic of China

Xiao Ming & Yu Yutai
Wuhan Hydroelectric College, People's Republic of China

ABSTRACT : This paper mainly presents the engineering geological features and the support consideration on GuangZhou Pumped Storage Power Station Cavern.

1 INTRODUCTION

GuangZhou Pumped Storage Power Station (GPSPS) is located in GuangDong province of China. The installed capacity of the first stage is 1200 MW (4x300 MW), and also has 1200 MW at the planned second stage. The total capacity is 2400 MW, which is the largest Pumped storage power station in China and the world to date. The project with a head of 530 m is mainly composed of upper and lower reservoirs, underground water delivery system, and power cavern system. The power cavern is 146.5 m long, 22 m span, 44.5 m high and has an overburden of 350 m to 450 m. Both the excavation and the support in power cavern have been finished. Monitoring has indicated that support is both reasonable and efficient.

2. ENGINEERING GEOLOGICAL FEATURES

The rock in power cavern area is a coarse biototic granite which was formed at the middle stage of Yanshan Movement, and it is weakly weathered to fresh granite, The rock at both sides of geological fractures has been altered by middle to low temperature hydrothermal action. Recommended index for physical and mechanical properties of altered rock is shown in table 1.

The power cavern is sited in rock massif where geological structures are relatively simple, which is between fault F4 and fault f7012 (see Fig. 1). The orientation of the main fracture varies from NE-NNE to NW-NNW, which intersect the longitudinal axis of power cavern with large favourable angles. In power cavern, there are four important faults and alterations with an accumulative width of 2.7 m, which is 1.9% of power cavern length (see Fig. 1, Table 2). Among the four big fractures, because of widely influenced deeply altered, fault F4 and fault F7012 are outmost important. The others are slightly or weakly altered with a relatively small width. In addition, there develop a few gentle dip angle joints with a length of 1m - 3m.

The in-situ stress measurements in power cavern are indicating that the maximum average principle stress is 12.2 MPa, with a WE orientation almost in a horizontal plane, while the minimum average principle stress is 6.4 MPa. Close to the working

Table 1 Physical and mechanical index of altered granite.

Degree of alteration	Unit weight (g/cm 3)	Elastic modulus (GPa)	Deformation modulus (GPa)	Poission ratio	C (MPa)	∅ (°)
Weakly altered	2.56	6-10	1.5-2.5	0.25	0.4	31
Slightly altered	2.56	22	8	0.22	0.65	38
Fresh	2.61	40	20	0.20	1.5	48

Table 2. General features of the main fractures in power cavern .

Number	Occurence	Width (m)	Intersect angle with the power cavern axis	General features
f7027	N25°-30° E/ SE<85°	0.2-0.5	52°	Slightly altered, Composed of three alterations
f184	N40°W / SW<85°	0.2-0.3	60°	Slightly altered , clayed
f182	N50°-60°W / SW<85°	0.8-1.0	45°	Slight to middle altered clayed
f7012	N35°-40° W / NE<50°-65°	1.0-1.3	62°	Middle altered, fractured, clayed

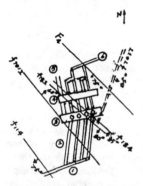

Fig. 1 Cavern section at El. 240 m
⑦Manifold ② Penstock③Power cavern
④Transform hall⑤ Access tunnel⑥Tailrace tunnel

face which is about 10m to the fault f7012, some rockburst took place (See Fig. 2), resulting in total volume of more than 10 cubic meters. A rough analysis shows that f7012 close to the face and local stress concentration are all attributable to this geological surprise.

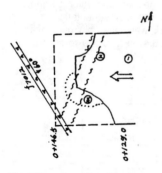

Fig. 2 Location of rockburst
① Power cavern②Alteration
③ Rockbursting area

In the power cavern area, granite is almost impervious, there is fissure water only in fracture zones. All together, there are about 10 seepage points spotted all over the cavern roof where the flow totals about 2-3 l/min.

Alteration is the biggest feature of the engineering geological condition of this project, most of them distribute at the both sides of NE, NW and NNW fractures. After excavation, due to the formation of free surface, if an alteration is exposed to water (of moisture from air), its physical and mechanical properties degrade either rapidly or slowly, depending on montmorillonite content (See Table 3, Fig. 3). When the content is more than 5%, the alteration will swell and disintegrate rapidly if it is exposed to water; When the content is less than 5%, after excavation, at first, behaviour of the alteration is sound rock, but after a period, it will also swell and disintegrate. After excavation, if it is covered in time protecting it from exposing to water or maintaining the surrounding rock mass pressure, the alteration will be of a relative high strength without swelling and disintegrating.

Three-dimensional elastoplastic FEM analysis has indicated that power cavern stability is mainly controlled by the geological structures.

In the light of Surrounding Rock Classification in China Hydroelectric System, in power cavern, ratio of class 1 (very good rock) is 43.5%; class 2 (good rock) is 36,5%; class 3 (fair) is 17%; class 4 (poor) is 1%. Engineering condition in power cavern is good, and suitable for priority of a flexible support scheme.

3 SUPPORT CONSIDERATION

Rock support in power cavern includes grouted

Table 3 : Swelling test results

Number	Degree of alteratin	Testing method	M-Content (%)	Swelling pressure (MPa)	Swelling displacement (mm)	Hours (h)
Bp-2	Middle	Free swelling	15.6		8.88	238
Bp-6	Middle	Free swelling	7.5		0.045	44
Bp-3	Middle	Swelling pressure	15.2	0.48		223
Bp-4	Slight	Swelling pressure	4.3	0.072		255

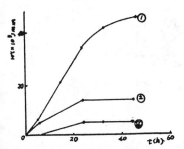

Fig. 3 In-situ swelling test
① Surface ②10cm depth③15cm depth

bolting, wire meshing and shotcreting. Based on the supporting place, rock mass characters and relaxation zone of surrounding rock mass, supporting parameters have been decided by using engineering comparison method, monitoring control method and supplementary calculation. There are three characters with power cavern rock mass: the first is that the space between steep dip angle joints is large, the second is gentle dip angle joints are underdeveloped, the third is rock mass is intact with a relatively high strength. So, in the support consideration, the requirement for bolting is to reinforce the big faults, alterations, local unstable rock blocks, relaxation zone and surface fissures.

The amount of alteration in power cavern is not big, and also the swelling test has indicated that swelling displacement is limited, rockbolting and wiremeshing and shotcreting are efficient measurements to reinforce the alteration. However, weak material of faults and alterations have to be careful scaled and cleaned followed by in-time shotcrete covering.

1. Roof support
There are three kinds unfavourable factors in power cavern roof, the first is low strength faults and

alteration, the second local gentle dip angle joints and the third is relaxation zone of surrounding rock. Low strength faults and alterations are reinforced by spot bolts. Distance form bolt hole to the relevant discontinuity is 1.5 m. Make sure that spot bolts must be in a stitching and splitting pattern, cut through the discontinuity to be reinforced, with 2 m at least on both ends resting in sounding rock (See Fig. 4).
Gentle dip angle joints and relaxation zone are reinforced by pattern bolts. Relaxation zone in sound rock is 0.7 m - 1.0 m, in fractured rock is about 1.5 m. Pattern bolts should cut through the zone with 2m past. So, the length of pattern bolts is increased to 5 m. Between the roof and haunches, 5.3 m and 7.0 m long pattern bolts have been used to offset stress concentration and large deformation (See Fig. 5). Diameter of pattern bolts is 25 mm, the spacing is 1.5 X 1.5 m.

Because there are alterations in the power cavern, all the roof has been covered by wiremesh and shotcrete (See Fig. 6). Before shotcrete applying, rock surface has been thoroughly scaled and cleaned. In seepage area, drainage measures have been adopted, such as drill holes and connected with water pipes.

2. Side walls support
The height of power cavern side walls is up to 44.5 m. Because there are relaxation zone, unfavourable geological fractures and the possibility of rock mass failure result from secondary stress, side walls supporting is needed.

Spot bolts is used to reinforced the low strength faults and alterations, the supporting principle is the same as in the roof. Stability analysis have been done to the wedges formed by two or more than two sets of discontinuities. Take into consider the pattern bolts reinforcement effect, spot bolts are used to reinforce the wedges whose stability safety factor is still less than 1.2 (See Fig. 7).

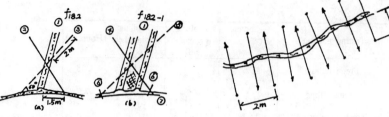

Section   Plan

Fig. 4 Spot bolting in power cavern roof ①Fault ;   Alteration ;② L=6m ø25 ;
③ L=8m , ø 25 ;   ④ L=9m , ø 25 ;   ⑤ L=10m , ø 25 ;   ⑥ Shot bolts : L=2m , ø 25;
⑦Reinforcing steel bar :   L = 5 m , ø 25

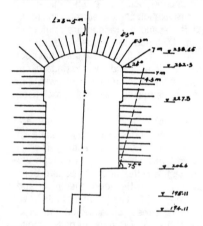

Fig. 5 Pattern bolting in power cavern .

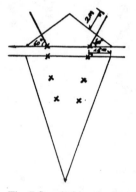

Fig. 7 Spot bolting for unstable
wedge in the side walls

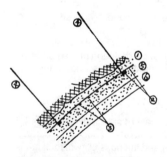

①, The first layer, Min 5 cm
② Reinforcing bar , ø8 , @ 1.5 m x 1.5 m
③ Wiremeshing ø6, @ 200 mm x 200 mm
④ Pattern bolting , ø 25, @ 1.5 m x 1.5 m
⑤The second layer, Min 5 cm
⑥The third layer, Min 5 cm

Fig. 6 Support detail of power cavern roof

According to the NGI rockmass classification, the general Q value of side walls is 50-100 support pressure is 10-45 KPa (See Fig. 8). If the diameter of pattern bolts is 25 mm and the horizontal space is 2 m, the vertical space should be 1.5 m. In power cavern, the average dip angle of the main geological fractures is 75 degree. From the point with an elevation of 206.6 m at the side wall, a line with a plunge of 75 degree is drawn, pattern bolts of side wall should cut through this line and with at least 2 m at the inside (See Fig. 5). The part below wlevation 206.6 will be backfilled with concrete, there is about 1 year from the completion of excavation to the completion of concrete placing, temporary support is needed.

Based on the upper analysis, the parameters of pattern bolts have been decided as follows: The diameter is 25 mm; The horizontal space is 2 m; The vertical space is 1.5 m and there are two kinds of length in a stagger pattern: the first one is 7 m, the second is 4.3 m. In the parts below elevation 206.6 m, the length of all the pattern bolts is 4.3 m

236

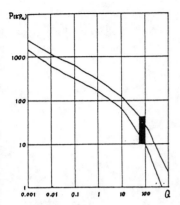

Fig. 8 Curve between Q value
and support pressure .

in the upper stream wall, and no pattern bolts in the down stream wall (See Fig. 5). In the unstable wedges place, the diameter of the pattern bolts is increased to 28 mm, horizontal space is decreased to 1.5 m, the length of 4.3 m long bolts is increased to 7 m.

The part above elevation 206.6, the principle of wire mesh and shotcrete is the same as in the roof. The part below elevation 206.6 m, 7 cm thick shotcrete has been applied on alterations area or poor geological condition area.

3. Intersection support
In the power cavern side walls, complex three-dimensional stress has formed in the intersections with some other tunnels such as busbar galleries. Blasting, secondary in-situ stress and geological structure interacted one another, this would result in rock mass failure. Prebolting is used as indicated in Fig. before the excavation. In the excavation, careful blasting and in-time bolting is adopted with a shot progress of 1 m - 1.5 m for each round.

4. Conclusion
In order to oversee surrounding rock behaviour, three monitoring sections have been arranged during the excavation. And now, both the excavation and the support in power cavern have been finished. Monitoring has indicated that the support is both reasonable and efficient.

Reference
1. Yu Yutai, Xiao Ming, The surrounding rock stability analysis for the opening of the underground power hose, Proceedings of the First International Conference on Hydropower, June, 1987, Norway.
2. Zhu Yunzhong, Lu Xuchu, Liu Xueshan, Siting & Supporting Optimization of Power Cavern, Proceedings of International Symposium for Pumped Storage Development, October, 1990, BeiJing, China.
3. Nick Barton, Cavern Design for HongKong Rock, Norwegioan Geotechnical Institute, 1988, Oslo, Norway.

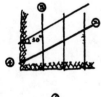

① Busbar gallery
② Patter bolting , L=2.5 m
   ø25, @ 1.5mx1.5m
③Prebolting , L=3m , ø 25
④Power cavern

Fig. 9 Intersection support.

237

2 Reservoirs and environmental aspects in water resources management

Hydropower'92, Broch & Lysne (eds) © 1992 Balkema, Rotterdam. ISBN 90 5410 054 0

Restoration of the generating capacity of the Guatapé hydroelectric powerplant

A. Amaya, M. Gómez & G. Wolff
Porce II Division, Empresas Públicas de Medellín, Colombia

ABSTRACT : Since 1978, when the second stage of the Guatape Hydroelectric Powerplant (560 MW), owned by Empresas Públicas de Medellín, started operating, some problems in the performance of the refrigeration system of the units were noticed. It was found, after having devoted a long time to investigations, that the water used for refrigeration, taken from the reservoir, gave rise to scales in the heat exchanger tubes. That entailed a rapid decreasing of the tube capacity, and compelled the operations personnel to frequently shut down the units to clean the exchangers, with serious technical and economical consequences.

After having identified the problem, different solutions have been attempted, such as physical and chemical treatments in the reservoir, the waterways, the pump pits, the refrigeration system, and even in the tailrace tunnels. It was finally decided, after having analyzed different alternatives, to redesign the system and change the water source, as the only solution guaranteeing a reliable operation of a public facility fundamental for the electric system of the country.

1 INTRODUCTION

Among the different factors that may unfavorably influence the technical and economical operation of the hydroelectric powerplants, and also affect their reliability within an interconnected system, mention must be made of the water quality of the reservoirs. Reservoir pollution is due to diverse causes such as decomposition under anaerobic conditions of vegetal and animal organic material, the characteristics of the impounded area, the nature of the soil materials of the surrounding land, the possible residential and industrial wastes, an even the materials used for the construction of the dams.

In some hydro facilities, the effects are noticed just after the plant starts operating. The main effects are presence of stinking gases in the powerhouses, corrosion of the electric equipment and the metal structures, excesive heating of some parts of the units, and alteration of protective equipment performance.

Problems like these ones appeared in the Guatapé Hydroelectric powerplant, and gave rise to a detailed investigation of the causes and the framing of alternative solutions. That is the subject of this paper.

2 DESCRIPTION OF THE GUATAPE HYDROELECTRIC POWER PLANT.

The Guatapé hydroelectric powerplant is located, in the Andean region (Northwestern Colombia), in the department of Antioquia, 50 km east of Medellín City. Is was built in two stages, each of 280 MW; the first stage was completed in 1971 and the second one in 1978.

The most important works of the plant and their main characteristics are the following :

A reservoir, with a capacity of 1240 million cubic meters, up to date the biggest one in the country, that covers 6240 hectares and is fed by the Negro-Nare river with a flow rate of 50 m3/s.

An earth and rock fill dam, that has a maximum height of 59.5 m, a length at the crown of 900 m, and a total fill volume of 3.4 million cubic meters.

Two intake towers, each 58 m high, 10 m in diameter and with eight intakes.

Two shafts, each 87 m deep and 3.5 m in diameter, convey the water from the intake towers down to the headrace tunnels. The tunnels, each 4895 m long and 3.5 m in diameter, and separated by a distance of 111 m, are parallel and communicate with the atmosphere through two 150-m high surge tanks.

Two 874-m long, 48° inclined penstocks that convey the water to the distributors, The gross and net heads are 824,28 m and 703,28 m, respectively.

The powerhouse is located 650 m below the surface (it is one of the deepest in the world) and consists of four parallel caverns, two for eight turbine-generator units, and the two other to house six single-phase transformers plus a spare transformer in each one.

The following is the main equipment in the powerplant : eight 4-jet, vertical shaft, 114,832 hp, 514.3 rpm, Pelton turbines with their corresponding speed governors; eight three-phase, vertical shaft, 774 MW, 13.8 kV generators; and eight 1-m diameter spherical valves.

The refrigeration system, which is especially important in this analysis, is completely open, without recirculation. For this reason, the water used for refrigeration can not be treated, making difficult the control and solution of the problems mentioned above. The water for this system passes first through the turbines, and then is stored in two pits, each one for four generating units. The storage is possible due to the level difference between the pump pits and the tailrace tunnels to which the pits are connected. Water is pumped to the refrigeration system for units and trans- formers by means of four main pumps with a capacity of 5 200 l/min. The refrig- eration system is made up of radiators and oils for the turbine guide bearings and water-oil heat exchangers for the transformers.

The access to the powerhouse is through a 1965-m long road tunnel with an 11% downgrade, which is also used for the high voltage and control cables leading to the surface substation. The water passing through the turbines and the re- frigeration system is conveyed to the surface by means of two 4568-m long, 30- m2 section, parallel, tailrace tunnels.

3 CHARACTERISTICS OF THE PROBLEMS IN THE GUATAPE HYDROELECTRIC POWERPLANT.

The main effects noticed, which have affected the performance of the Guatapé powerplant, are the following :

3.1 Malfunctioning of the refrigeration system.

Although the problems were noticed for the first time approximately two months after

having began to impound the second stage reservoir on May 23, 1978, the first stage equipment was affected in first place, due to obstruction of the refrigeration system and other causes.

From the above mentioned date on, every day the temperature rise was more critical to the point that the units ran out of service. The problem was limited not only to the piping obstructions but spread to the generator bearings and radiators.

By examining the encrusted material affecting the refrigeration system, it was found that it consisted mainly of vegetal organic material and mineral iron compounds.

To solve the problem, it was necessary to clean the units more often. Due to the continuous coupling and uncoupling of the equipment, wear increased, bringing about a limitation of the availability of the units during the system operation.

3.2 Damage to electric equipment and fittings.

Parallel to the refrigeration problems, there appeared in the powerhouse odors characteristic of the hydrogen sulfide (H_2S) as well as corrosion of the exposed metal surfaces. In addition to this gas, the presence of sulfur dioxide (SO_2) was also detected. In the powerhouse, it was common to find many of the effects caused by H_2S and SO_2, such as voltage drops in the brush contacts of the generators, blackening of the relay contacts, weakening of the tin and lead weldings, failures in the coils of the auxiliary relays, corro- sion of the copper busbars, etc., especia- lly in the period when the causes of the problem were not completely known and no effective control measures were enforced.

3.3 Effects on the operations personnel.

Though the hydrogen sulfide concentrations in the powerhouse did not reach critical levels, in accordance with the analyses carried out, the personnel working in the plant underwent some troubles as headaches, throat irritation, and cough, which made necessary to adapt the operators shifts, due to the existing environmental condi - tions.

4 CAUSES OF THE POWERHOUSE PROBLEMS

As mentioned earlier, the problems in the Guatapé powerplant are directly related

to the water quality in the El Peñol
reservoir, mainly because of the vegetation submerged in the area when the
second stage started to operate.

During the first stage of the reservoir
1300 hectares were impounded, and the
vegetation to be submerged was not removed, mainly because the area was occupied by pasture and stubbles, remaining
after the cultures were harvested before
impounding.

Most trees and bush, but no grass, were
stripped of the area to be impounded for
the second stage. However, part of the
trees and bush existing in the intake
area could not be removed, being the main
cause of reservoir pollution, though the
presence of the aquatic plant Elodea sp.,
over 2 hectares in the intake area, also
contributed to pollution.

4.1 Decomposition of organic matter and pollution of the reservoir.

The organic matter of vegetal origin is
chemically made up of proteines, polypep-
tides, nucleic acid, carbohydrates
(monosaccharides, oligosaccharides,
polisaccharides), lignin, resins, facts,
waxes, and other compounds of minor
importance.   The elements entering into
these compounds are hydrogen, carbon,
oxigen, nitrogen, phosphorous, potassium,
calcium, magnesium, sulfur, iron, manga-
nesium, etc., many of which, when
decomposition occurs, forms salts, oxides,
hydroxides, acids, etc., or remain in
ionic form.

The decomposition of the organic matter
is mainly carried out by microorganisms
(bacteria and fungi).   In such process,
the aerobic bacteria are the most important
ones, but under impounding conditions the
decomposition is performed by the anaerobic
bacteria, that require no oxigen for their
activity, producing an increase of carbon
monoxide (CO_2), a decrease of oxigen due
to the high demand of the microorganisms
that carry out the degradation of vegetal
wastes, reduction of pH, formation of
acids and other organic and inorganic
compounds such as methane and the hydrogen
sulfide (H_2S) mentioned earlier.

4.2 Sulfur, iron and their relation to the problem.

The simple compounds of these two elements
are the most important ones, from the
point of view of the subject being
discussed.

4.2.1 Sulfur.

In the sulfur mineralization, the proteines, the peptides, and other organic
sulfur compounds are polimerized up to
their aminoacid state and, continuing
with the mineralization process, the
sulfur in the aminoacids, under anaerobic
conditions, like those prevailing in the
reservoir, is reduced to H_2S by two genera
of bacteria, desulfuvibrium and desulfuma-
culum.

The reduced sulfur (H_2S) remains
dissolved in the water of the lower reservoir layers (hypolimnium), and escapes to
the atmosphere only when destratification
occurs.   As the water entrances of the
intake towers are very close to the
bottom, the water being conveyed to the
generation units drags hydrogen sulfide
in solution.   Due to the turbulence in
the Pelton wheel, the hydrogen sulfide is
liberated, and causes galvanic corrosion
and the other harmful effects already
mentioned.

4.2.2 Iron.

Different from sulfur, that comes mainly
from the organic matter of vegetal origin,
the iron source is the mineral material
arriving at the reservoir as sediments in
suspension, produced by erosion and
transported by the streams and the runoff.
Iron also enters the reservoir from the
internal landslides and from the earth
mass movements in the banks due to the
waves produced by the motorboats.

Part of the iron is also associated
with organic compounds present in ferns.
In a manner similar to sulfur, iron is
reduced in anaerobic conditions to ferrous
forms producing a dark brown coloration of
the water, corrosion and scales in the
refrigeration system, when transported in
solution in the water moving the turbines;
due to aeration, iron can be precipitated in
ferric form, and may also produce the
precipitation of other ions, which
explains the presence of corrosion films,
hardened salts, clays and microbiological
mud, and especially bacteria such as
bacillus ferroxidans and B thioparus,
whose colonies can grow up to the point
of obstructing the pipes, besides the
corrosive effect, which was the problem
in the Guatapé hydroelectric powerplant.

4.3 Other causes.

Other causes that can contribute to the

water pollution of the El Peñol reservoir although not up to a critical level, are the waste water discharged without treatment from the sewarage systems of El Peñol and Guatapé towns, and the residential and industrial waste waters from the Rionegro valley.

5. EXTENT OF THE PROBLEM IN OPERATION AND GENERATION.

5.1 Problems during operation.

Regarding the problems brought about by H_2S, their control was effective from the beginning; for this reason few difficulties were experience in the powerplant operation due to this gas. The main problems were related to the refrigeration system, which had to be frequently subject to maintenance and physical and chemical treatments of its different parts, such as bearings of the turbines and generators exchangers, radiators and piping.

At the beginning, the problems were limited to the first stage units and refrigeration system. From February, 1979 on, problems in the second stage units began to appear.

5.2 Problems during generation.

The effect of the problems, as regard generation, was thought at the beginning not to be meaningful, given the volume of the reservoir that would permit to store the water corresponding to the unsold power.

Nevertheless, as time passed, the generation capacity was noticed to decrease, since full load operation was not possible, due to the overheating of the units as a consequence of the piping obstructions.

It is important to mention that the uncertainty experienced by the operations personnel, unable to guarantee the reliability of the system and to keep the service operating efficiently, was an important aspect difficult to estimate.

6. ACTIONS TENDING TO CONTROL THE PROBLEM.

When the environmental problem appeared in 1978, Empresas Públicas de Medellín saved no effort in the search of appropriate solutions to guarantee the service. For this, a task force was created with the participation of its technical personnel, directly or indirectly related to the problem.

The first thing foun by this task force was the total ignorance about the nature, the causes and the solutions of the problem. For this reason, consults were made abroad, which let to know a similar case, by then under control, occurred in the Capiravi-Cachoeira hydroelectric powerplant, located in the Paraná state, Brazil.

To take advantage of that experience, a technical commission was sent to the site to get information on control solutions applicable to the Guatapé case.

The Brazilian experience contributed to the better knowledge of the causes of the problem and the solutions to be implemented.

As the characteristics of the problem did not allow unique solutions, it was tackled in different places (reservoir, powerhouse, tailrace tunnels). furthermore a procedure was established that included both practical and research actions.

6.1 Practical actions.

The practical actions included physical and chemical treatments in the powerhouse, which were changed or left in accordance with their effectiveness.

6.1.1 Treatments in the powerhouse

By the time the piping obstructions were notice for the first time, some chemical treatments with Drewsperse were designed, to be used in the maintenance periods, recirculating the compound in solution through the bearings of the turbines by means of a pump. During the annual maintenance, a complete physical washing was carried out, for which the bearings were dismantled. The exchangers were washbored with water to eliminate obstructions. By the beginning of August, 1978, the refrigeration problem reached a critical point, and the treatments previously used were not effective any more. For this reason, it was necessary to resort to more effective physical and chemical treatments.

PHYSICAL TREATMENTS

- Washboring of each part of the system and later washing with water.
- Application of water under pressure to the obstructed parts.

- Application of hot air to dry deposited sediments and later application of water to remove sediments.
- Periodic injection of air to the refrigeration system.

CHEMICAL TREATMENT

- Treatment with Drewsperse.
- Injection of different chemical solutions, such as calcium hypochlorite, sodium hypochlorite and trisodic phosphate, to the radiators, the pump pits and the bearings, respectively.

The most successful treatment, of those listed above, happened to be the one including dismantling of the equipment, and washboring and later washing with water of the parts exhibiting problems. The effectiveness of the other physical and chemical treatments was, in general, low.   For this reason, it was necessary to resort to new solutions.

From mid-1979 on, the following treatment were designed:
- Washboring of the affected parts with later washing with water.
- Injection of air to the system at short time intervals, with the refrigeration system operating, to create turbulence and force sediments to be transported.
- Washing with a solution of sulfuric acid, with concentrations between 1% and 2%, using a pump to recirculate the fluid through each of the affected parts, and later washing with a basic solution of sodium hydroxide, to neutralize the residual action of the acid.
- In addition to this, a great number of treatments were designed with application of chemical solutions, for which the differents parts of the refrigeration system were adapted to permit the recirculation of the chemical compounds to be used.

The most successful chemical treatment, as regards the control of the problem, was the application of sulfuric acid solutions, with concentrations between 1% and 2%, and later washing with a basic solution.   The main characteristics of this washing is that the sulfuric acid is capable of disolving the iron present in the sediments as trivalent iron in the form of acid or hydroxide.

Of the physical treatments, the washboring of the affected parts have been decreased, due to the long time required for execution and the wear and maladjustment of the equipment.

The air injection to the system, at short time intervals, is periodically made as a preventive measure.

6.1.2 Treatments of the reservoir.

To solve the problems brought about by hydrogen sulfide and iron, it was decided to inject air by means of compressors to the reservoir botton in the intake area. The purpose of such injection is to accelerate the anaerobic decomposition, to avoid the formation of H_2S, or to oxidize it to prevent its corrosive action, and to oxidize iron to cause its precipitation and later sedimentation, preventing with this process the corrosion and obstruction of piping.

6.1.3 Treatment of the tailrace tunnels.

To decrease the effect of the hydrogen sulfide (H_2S), an air extractor was installed at the portal of the first stage tailrace tunnel, without noticeable results.

6.2 Research activities.

In addition to all the actions previously described, and to keep the problem under strict control, a research plan was devised with the support of a chemical laboratory located in the Guatapé powerplant.   As a result of the studies carried out in the laboratory, it was found that exists a cyclic variation of the reservoir water quality, which is optimum during the rainy periods and becomes worse during the dry periods.

With the purpose of completing the research activities, the services of the University od Lund (Sweden) were retained to carry out complete limnological studies of the reservoir; these were made by mid-1980.

From the different investigations, it could be concluded that the reservoir tended to a recovery process, showed no eutrophication, and that the pollution elements and the phytoplankton were at low levels.

7.   ANALYSIS OF ALTERNATIVES FOR A FINAL SOLUTION.

- Treatment of the refrigeration water: an analysis was made of the technical feasibility of precipitating the iron present in the refrigeration water using a binder like the alluminum sulfate, to later remove the precipitate by continuous blow off. However, the environmental characteristics of such water (low pH, organic matter in solution, high concentration of sulfides and low concentration of dissolved oxygen) do not permit its treatment by means of the processes commonly used.

Nevertheless, although the water conditions were favourable, the process would require the construction of big structures for continuous aeration, for which no room is available in the existing cavern.

- Placement of filtering fabrics : this procedure retains flocules but blocks the water flow. When a test was carried out the pump pit was dried in two hours, preventing the operation of the refrigeration system.

- Installation of electromagnetic filters : there exist electromagnetic filters used in industry to treat waters containing iron impurities. After consulting several manufacturers, it was concluded that these filters do no operate when the chemical composition of the iron in the water is $Fe(OH)_3$. The Physico-chemical investigations carried out and the sampling of the reservoir water indicated that the iron content is 60% Fe^{++} and 40% Fe^{+++}. Consequently, this was not an alternative for the final solution of the problem.

- Closed circuit refrigeration system : an analysis was made of the possibily of using a closed refrigeration circuit, with very good quality water as primary circulating medium; the heat would be dissipated in the exchangers to be located in the tailrace channels that receive the water from the turbine pits.

This alternative was not found feasible, due to the operating restrictions on the turbines, caused by the installation of the exchangers in the tailrace channels, since one of the functions of the Guatapé powerplant generators is to compensate the reactive power during the low charge hours; for this, the generators take power from the network with the turbines rotating without water. In this mode, there would be no water flow through the tailrace channels to cool the primary refrigeration circuit and, therefore, the units could not operate.

- Outside water sources : after analyzing and discarding the former alternatives, it was concluded that the only feasible and safe solution was to take water from outside unpolluted sources to be conveyed down to the powerhouse.

As regard the supply of good quality water, among the several possible alternatives, two were analyzed to select the safest and most economical one.

- Pumping from the Bizcocho river : this alternative considered the intake to be located close to the powerplant and the water to ve coveyed through the second stage tailrace tunnel. This options was discarded, due to the difficulties for construction and, mainly, because a hydroelectric powerplant so important for the national electric system does not admit the restrictions that would be imposed.

- Pumping from the Guatapé river : an analysis was made of the possibility of taking water from La Clara creek (750 l/s), at its confluence with the Guatapé river by using a spillway wall. There would be a pumping station and a piping system along the access tunnel to convey the water to the cavern. For this purpose, several studies were conducted on water quality, flow rate reliability in dry and wet periods, costs, ease of construction, maintenance during operation, reliability and possible environmental impacts. After these analyses, it was concluded that this solution will permit to increase the generating capacity of the powerplant in 5% (28 000 kW), and to eliminate the frequent dismantling of pipes to be washed. Furthermore, the alternative was deemed to be the most feasible and safest one, and to present the minimum interferences during construction. Therefore, it was chosen as the final solution for the problem of bad quality refrigeration water, and presently is under construction.

8. MAIN BIBLIOGRAPHIC REFERENCES

Bjork, S. and Gelin, C. (1980).Limnological Management of the El Peñol Reservoir, Colombia - Preliminary Report. Sweden : Institute of Limnology, University of Lund.

Coral, L., Lora, M. and Medina, H. (1978) Report on the Technical Visit to the Capiravi-Cachoeira Hydroelectric powerplant Brazil (in Spanish). Medellín : EE.PP.MM.

Medina, H. and Rivera, P. (1983). Operation Problems of Environmental Origin in the Guatapé Hydroelectric Powerplant (in Spanish). Journal of Empresas Públicas de Medellín, 5 (1), 53-77.

División Playas. Iron Effects on the Refrigeration System of the Guatapé Powerplant. EE.PP.MM. January, 1988.

Hydropower'92, Broch & Lysne (eds) © 1992 Balkema, Rotterdam. ISBN 90 5410 054 0

The experience of CEMIG in studies of hydroelectric plants

Arthur J. F. Braz, Osvaldo C. Ramos & Raimundo P. Batista Neto
CEMIG, Companhia Energetica de Minas Gerais, Brazil

ABSTRACT: Hydroelectric production plays a very important role in Brazil owing to its great potential. This works presents the experience of CEMIG in the feasibility and river basin studies of hydroelectric plants for future expansion of its generating system. Emphasis is given to the methodological aspects and the problems found in the application of the Brazilian methodology.

1 INTRODUCTION

Since the Brazilian electric production system is composed predominantly of hydroelectric power plants (more than 90% of the total installed capacity is hydro) and there is still a great underdeveloped potential, equivalent to 255 GW (corresponding to yearly production of 1,200 TWh or about 7 million barrels of oil/day), and only 30% is in operation, the studies for the dimensioning of these plants play a very important role in order to reach a technical and economic optimization of the energetic potential of a site, taking into account the possible social and environmental constraints.

CEMIG (Companhia Energetica de Minas Gerais), the utility of the state of Minas Gerais, was created in 1952 and has presently an installed capacity of about 4,400 MW and 18,000 employees. It has the greatest hydroelectric potential of the South-Southeastern interconnected system (the most important in Brazil and responsible for more than 70 % of the installed capacity), with important rivers flowing through the state and a potential for more than one hundred medium and large size plants, to be in operation until the year 2010, according to the expansion planning made to supply the forecast demand.

The studies are carried out at two levels. The first corresponds to river basin development, where whole basins are analyzed and the best sites and schemes for plant construction are evaluated and selected (survey studies). The second level corresponds to feasibility studies, where a specific site is analyzed.

This work reports the experience of CEMIG in river basin development studies and feasibility studies, according to the methodology established in 1982 by the GCPS (Coordinating Group for the System Planning). This methodology is presented with analysis and discussion of case studies. Finally, a new methodology that incorporates probabilistic criteria is presented.

2 RIVER BASIN DEVELOPMENT STUDIES

The main goal of river basin development studies is to find the best set of possible sites for plant construction, taking into account the whole basin. A scheme of sites for hydroelectric plants corresponds to a possible way for dividing the natural available head of the river through the definition of the maximum water level and tailwater level of each site. Theoretically, the number of possible schemes for such a river basin is enormous but the number is initially reduced by selecting good topographical sites or eliminating sites that do not fulfill the pre-requisites for minimum installed capacity, reservoir fill-up duration, etc (Pereira 1984a).

The number of alternatives that still remain to be studied can be very high (many hundreds or even thousands) so a simple methodology and fast computational models to rank them according to a cost-benefit ratio has to be used. After this preliminary selection the best alternatives are studied in greater detail.

The benefit criterion is met through the evaluation of the additional firm energy that a new plant brings into the system. The firm (or guaranteed) energy of a system is the maximum continuous load that this system can meet without any energy shortage in the critical period. The critical period, shown in Figure 1, is the period of time in which all reservoirs of the system are depleted to meet the load demand and is associated with the worst drought recorded in the past. For the Brazilian system, the critical period extends from June/1949 to November/1956 (90 months), lasting so long because of the many multi-annual storage reservoirs. The firm energy of a plant is its mean production in the critical period.

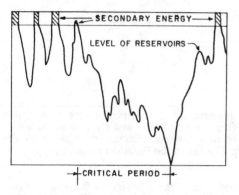

Figure 1 The critical period

Note that in these studies the benefits associated to peak demand and secondary energy are not explicitly evaluated, being a common practice to adopt a generalized capacity factor for all plants.

The costs of a project involved in the analysis are the total capital investment cost and operation and maintenance costs.

Hence, the cost-benefit index (or merit index) is found by:

$$CB = \frac{INV.CRF + OMC}{FE} \quad (1)$$

where:

CB = cost-benefit or merit index, in US$/MWh;
INV = investment cost of the project, in US$;
CRF = capital recovery factor, considering the project's life cycle of 50 years and a discount rate of 10% a year;
OMC = yearly operations and maintenance cost, in US$;
FE = firm energy of the project, taking into account the benefits downstream, in MWh/year.

The economic evaluation of each site is done by comparing its merit index with a limit index, called reference index (RI), which can be considered as the cost of substituting its firm energy by another available generation source. Thus, a project is competitive only if its merit

index is lower than the reference index; otherwise, it is eliminated. In order to consider the environmental aspects, the projects that present low indexes (energy production/flooded area) are also eliminated. It should be noted that the process of elimination is iterative, given the cascade operation of the hydroplants.

Since each possible scheme generally presents a different energy production, it is necessary to homogenize them through complementation up to the alternative of greatest production, by the reference index. In this way, the merit index of each alternative is given by

$$CBA = \frac{INV.CRF + OMC + (FE^* - FE).RI}{FE^*} \quad (2)$$

where:

CBA = cost-benefit or merit index of the alternative, in US$/MWh;
INV = total investment cost of the alternative, in US$;
CRF = capital recovery factor, as defined in (1);
OMC = yearly operations and maintenance cost of the alternative, in US$;
FE = firm energy of the alternative, in MWh/year.
FE^* = firm energy of the alternative of greatest production, in MWh/year.

CEMIG has recently finished two river basin studies, namely the Jequitinhonha river basin (CEMIG 1989a) and the Sao Francisco river basin (CEMIG 1989b). The final scheme for the Jequitinhonha river basin was composed by 14 sites and 6.5 TWh of firm energy. The study of Sao Francisco river basin was of greater complexity, involving 10 rivers and 42 possible sites. The number of complete schemes for that basin could reach as many as 8,000, disregarding any social or environmental limitation. After taking into account these factors, the 8 best alternatives were studied in greater detail. The selection of the best alternative was based on the environmental and economic aspects, leading to a scheme with 21 sites and 13.1 TWh of firm energy.

CEMIG and ELETROBRAS (the federal power agency) are now carrying on the studies of Doce river basin. For these studies a methodology that takes into account with greater detail the environmental aspects was developed (ELETROBRAS 1989). The preliminary studies have already finished and CEMIG will conduct the final studies in 1992.

3 FEASIBILITY STUDIES

Feasibility studies correspond to an optimization of the energetic potential of a specific site in terms of technical and economical aspects and taking into account the possible social and environmental aspects.

The optimization procedure consists of a cost-benefit comparison, seeking a maximization of the energetic benefits and minimization of the costs. The main parameters of a plant are determined, i.e., the maximum and minimum water levels, the power capacity and the characteristics and number of turbines and generators.

In order to proceed with the economic analysis, the energetic benefits are valorized by reference costs that come from expansion planning programs and are divided into costs for energy (US$/MWh) and cost for peak capacity (US$/kW/year).

Therefore, the process for dimensioning and establishing a given parameter (for instance, the minimum water level) is:

. for the technical and feasible range the energetic benefits are evaluated by computational production models that simulate the system's operation. These benefits are the firm energy (the average production in the critical period) and the peak capacity (using a simplified load duration curve);

. the firm energy and the peak capacity are multiplied by the reference costs to obtain the economic equivalent benefit, as:

$$EB = FE \cdot MCE + PC \cdot MCP \quad (3)$$

where:

EB = yearly total economic benefit, in US$;
FE = firm energy, in MWh;
MCE = marginal cost of energy, in US$/MWh;
PC = power capacity, in MW;
MCP = marginal cost of peak capacity, in US$/kW/year.

If there are benefits of secondary energy they are evaluated by comparison of fuel costs savings.

The increases in investments are evaluated in parallel. The cost/benefit ratio is obtained and the process iterates until the investment increases are greater than the increases of energetic benefits.

The experience of CEMIG with the application of this methodology will be described.

3.1 Evaluation of the minimum water level

The minimum water level has to be determined for each alternative of maximum water level. This is done by assuming that the plant will be, firstly, a run-of-river plant. For each value of minimum water level, the energetic benefits are evaluated.

The main problem found in this step is the relative position of the plant in the cascade. Since the Brazilian system is multi-owned, even on the same river the plants are owned by some utilities. Because of this, sometimes there is a conflict of interest between CEMIG and the system as a whole.

When the plant is situated upstream from the plants of CEMIG and other utilities, there are no drawbacks. In these cases, the minimum water level is limited mainly by technical constraints than by economic ones. This happened, for instance, in the feasibility studies of Nova Ponte (CEMIG 1983) and Bocaina power plants (CEMIG 1987).

It is known that a certain volume of water has much more value upstream than downstream and the construction of any reservoir upstream will increase the firm energy of the plants downstream. But a problem may arise when the increases of the head in a plant are more important than increases of storage.

Capim Branco (CEMIG 1986), in CEMIG's point of view, is an end-of-river plant (there are two plants upstream, Nova Ponte and Miranda, and only one downstream, Sao Simao). However, from the whole system's point of view, the storage in Capim Branco is very important (there are 7 plants downstream). Since the utilities which own plants do not participate in the investments for building the plant, a compromise solution was found, limiting the storage capacity of Capim Branco to a value acceptable to both CEMIG and the system.

Notice that this problem also arose in the studies of Miranda (CEMIG 1985) and Formoso power plants (CEMIG 1989c). New rules for determining the minimum water value which take into account the benefits of reservoirs are being researched and CEMIG is now studying a new methodology to quantify these benefits.

3.2 Evaluation of the maximum water level

The evaluation of the maximum water level is done through the same procedure used to determine the minimum water level, i. e., a cost-benefit comparison.

During this step the investments are detailed and the social and environmental constraints are rigorously analyzed.

The economic analysis for the choice of the maximum water level must be completed with evaluations of the expected initial times for filling the reservoir and to refill it after a drawdown (the critical period). Since there is no

249

methodology for evaluating these times (the only recommendation is that the re-filling time does not exceed three hydrological cycles), what is done is to simulate the plant with synthetic streamflows and compare these expected times with those from existing plants. This was the procedure adopted in the studies of Nova Ponte and Bocaina.

The main problem here is that according to the GCPS' methodology, all reservoirs have full storage in the beginning of the critical period. Hence, increases in the storage are very economic to the system. The solution to this problem is the study of various sceneries, using synthetic streamflows, when the reservoirs may not have their maximum storage. These aspects will be further discussed.

3.3 Power capacity evaluation

Conceptually, the evaluation of the power capacity is the same as the one adopted for the parameters already discussed. The main difference is that machines may be installed in the future and the scheduling of the investments to build the structure must be taken into account. In this step the projected heads of the turbines are also calculated.

In order to evaluate the variability of the efficiency of the turbine with the inflows and the head, a computational model was developed, therefore helping the choice of the number of machines.

4 A NEW METHODOLOGY FOR FEASIBILITY STUDIES

As stated before, the energy values are taken from computational simulations using the historical record of streamflows, assuming that the worst drought will repeat itself and all the resources of the system will be available (the critical period).

Although the current methodology presents a solid economic basis, and a very intuitive and plausible supply criterion (guaranty of supply under the worst of the conditions), problems have arisen with the consideration of just one scenery (the critical period). For the system as a whole, different basins present different hydrological cycles, some with high inflows in the critical period. This fact conducts to a wrong choice for the power capacity of a plant and the machines tend to be idle in a great part of the time. This way, it is said that the current methodology is a simplification of a probabilistic process. Moreover, the worst drought registered in the past will not be exactly the same in the future and the risk of energy shortages is not measured.

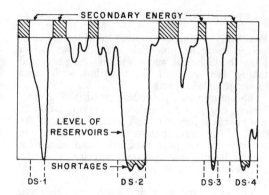

Figure 2 Typical dry spells

The proposed methodology intends to be a natural extension of the current one, maintaining the economic concepts. It is based on the concept of guaranteed energy at a preset level of risk, assuming that the system will not be able to meet the whole load and that there is a specific risk of shortage (in the Brazilian case, the supply criterion is 5% annual risk of energy shortage).

The main principle is to measure in different situations (dry spells), using synthetic streamflows, the performance of the plants in the system and using these results estimate the energy availability of the plants. Figure 2 shows typical dry spells (DS):

To obtain the system's guaranteed energy, simulations are done for a given target load. The number of years with any shortage divided by the total simulated years gives the annual risk of shortage for that load. The process is iterative and finishes when the risk is equal to 5%. The final load will be the system's guaranteed energy. The guaranteed energy production of each plant is obtained through the identification of the several droughts in the simulation of each plant in many periods (sceneries).

This methodology has been developed since 1988 (Braz, 1988) and the results obtained up to now are very satisfactory.

5 CONCLUSION

This paper presented a methodological approach to evaluate the energetic and economic benefits associated with a project, questioning some aspects of the current methodology for the feasibility studies and proposing a new one.

As a function of its position in the cascade and its storage capacity, the problems found in the dimensioning of hydroelectric power plants are more or less complicated. The social and environmental aspects have also to be evaluated.

In the Minas Gerais State, there is still a great

number of sites located in places with very low populational indexes. In those places, the hydroelectric projects would have no considerable environmental impact, being, in many cases, a decisive factor in regional development.

It is very important to note that associated with the presented new methodology of probabilistic energetic benefits (according to the criterion of 5% of annual risk), there is already a methodology to consider social and environmental benefits.

REFERENCES

Batista Neto, R. P. et alii. 1989. The experience of CEMIG in feasibility studies of hydroelectric power plants, (Portuguese). X SNPTEE, Group I. Curitiba, Brazil.

Braz, A.J.F. 1988. Hydroelectric power plants dimensioning with probabilistic energy supply criteria, (Portuguese). MSc thesis, UFMG. Belo Horizonte, Brazil.

CEMIG - 11.134-PN/GR2-068/83. 1983. Feasibility studies of the Nova Ponte project, (Portuguese). Belo Horizonte, Brazil.

CEMIG - 11.158-EM/GR1-001/85. 1985. The dimensioning of the reservoir of the Miranda project, (Portuguese). Belo Horizonte, Brazil.

CEMIG - 11.130-EM/GR1-008/86. 1986. Feasibility studies of the Capim Branco project, (Portuguese). Belo Horizonte, Brazil.

CEMIG - 11.163-DE/GR1-001/87. 1987. Feasibility studies of the Bocaina project, (Portuguese). Belo Horizonte, Brazil.

CEMIG - 99.825-GM/GR1-008/89. 1989a. Energetic and economic studies of the Jequitinhonha river basin, (Portuguese). Belo Horizonte, Brazil.

CEMIG - 99.821-GM/GR1-012/89. 1989b. Energetic and economic studies of the Sao Francisco river basin, (Portuguese). Belo Horizonte, Brazil.

CEMIG - 11.141-GM/GR1-008/89. 1989c. Feasibility studies of the Formoso project, (Portuguese). Belo Horizonte, Brazil.

ELETROBRAS. 1989. Final report of the preliminary studies of the Doce river basin, (Portuguese). Rio de Janeiro, Brazil.

ELETROBRAS. 1984. River basin development studies reference book, (Portuguese). Rio de Janeiro, Brazil.

GCPS. 1982. Criteria for the dimensioning of hydroelectric power plants, (Portuguese). Rio de Janeiro, Brazil.

Pereira, M.V.F. 1984a. Hydroelectric system planning. IAEA Technical report series 241:303-336. Vienna, Austria.

Pereira, M.V.F. 1984b. Stochastic streamflow models for hydroelectric systems. Water resources research Vol 20,3:379-390.

Hydropower'92, Broch & Lysne (eds) © 1992 Balkema, Rotterdam. ISBN 90 5410 054 0

Repercussions of hydroelectric developments in Northern Québec, Canada: The case of the La Grande Complex in James Bay

N. Chartrand & C. Demers
Environment Branch, Hydro-Québec, Que., Canada

ABSTRACT: This paper focuses on the evolution of newly created environments as a result of the construction of the La Grande hydroelectric Complex in the James Bay region and their effects on wildlife and habitats. It will also examine the modifications to land use created by the project and the economic, social and cultural repercussions on the Cree Communities.

1 INTRODUCTION

The hydroelectric Complex on the La Grande Rivière, known as the La Grande Complex is located in the James Bay territory in Northern Québec, Canada, some 1000 km north of Montréal around the 54th parallel (figure 1). The La Grande Complex watershed covers an area of 176,800 km^2. The source of the La Grande Rivière is at an altitude of approximately 1,100 meters. Most of the river travels over a vast plateau of rolling hills and countless lakes created by glaciers. The 150 km coastal plain is marked by peat bogs and clay deposits.

The cold continental climate is characterized by very contrasting seasons, with the annual average temperature being -4°C, and the frost-free period lasting only 80 days. At the start of the project in the early seventies some 7,000 Native Peoples made use of the wildlife resources for subsistence purposes.

Phase I of the development of the hydraulic resources of the La Grande Rivière began in 1973 and was completed in 1985. It includes three very large power stations, La Grande 2, La Grande 3 and La Grande 4 and two major diversions, Caniapiscau and Eastmain-Opinaca, for a total capacity of 10,282 MW. Phase II, which is scheduled for completion in 1996, comprises La Grande 1, La Grande 2A, Laforge 1 and 2, Brisay and Eastmain 1

generating stations for a total capacity of 5,437 MW.

Once completed, the bodies of water created by the La Grande Complex will cover some 15,873 km^2 of the 390,000 used by the Cree for their traditional activities.

2 ENVIRONMENTAL MONITORING PROGRAM

When the project to harness the hydroelectric potential of James Bay in Northern Québec was announced in 1971, very little was known about the environment of the territory and questions were raised about its effect on the environment and the Native Peoples. A federal-provincial task force recommended accelerating the acquisition of knowledge by using the La Grande Complex region as a vast natural laboratory.

This recommendation was the impetus for the 1973-1977 study of terrestrial and aquatic ecosystems and the development, in 1978, of an environmental monitoring network designed to meet the following objectives:

-use the data gathered to rationalize remedial measures (fish and wildlife habitat improvement, bank stabilization) and reservoir management;

-take advantage of the knowledge acquired to improve impact prediction methods and mitigative measures for future projects;

-evaluate, using a scientific method, the physical, chemical and biological changes in reservoirs likely to affect resources closely associated with traditional activities of the Native Peoples.

Other studies were aimed at tracking the evolution of all the biophysical components of the environment and the studies on the caribou and waterfowl continue today.

The economic and social impact of the La Grande Complex and of the presence of methylmercury in the reservoirs was also examined. Other studies monitored changes in native use of the territory. More recently, studies were undertaken to determine the impact on wildlife harvesting of opening the territory to the general public.

This paper will focus on the evolution of newly created environments and their effects on wildlife and habitats. It will also examine the modifications to land use created by the project and the economic, social and cultural repercussions on the Cree communities.

In another paper written by the same authors, the emphasis will be put on the mitigative measure put forward to ensure the quality and biological productivity of the milieux affected in order to permit Native Peoples to carry on their traditional activities. It will also discuss how the lessons learned over the last seventeen years help put into perspective foreseeable repercussions of the future projects in Northern Québec such as the Great Whale River Project (cf.C. Demers, N. Chartrand, "Lessons from 17 years of environmental monitoring at the La Grande Project, Québec, Canada).

3 ENVIRONMENTAL MONITORING NETWORK

Since the area affected by the La Grande Complex is very large, the network was concentrated in the La Grande 2, Opinaca and Caniapiscau regions. All the other regions were either comparable to one of these or underwent basically the same changes.

At each of these reservoirs, and downstream from control structures, powerhouses or spillways, sampling stations were set up for water quality, phytoplancton, zooplancton, benthos, fish and mercury. Also, a number of stations were located in undisturbed habitats, to determine whether a variation was due to natural agents or the hydroelectric development.

The monitoring process began two years before development got underway. Still in operation, this is one of the most important monitoring networks in the world due to its scope, its duration and the number of variables measured.

3.1 *Water quality and biological productivity*

During the monitoring period, water quality has always been adequate for aquatic organisms. In response to prolonged water residence time, nutrient increases and availability of organic material, phytoplanktonic and zooplankton biomass increased in most of the affected waterbodies. Fish monitoring revealed no food shortages. Stomach content surveys indicated that benthic organisms were relatively abundant in the new habitats. In fact, most fish species showed increased biomass, growth rates and condition factors in most of the surveyed waterbodies.

Biological activities peaked a few years after impoundment and are now gradually declining to pre-impoundment levels.

3.2 *Estuaries*

Diverting the Eastmain and Opinaca rivers into the La Grande Rivière has reduced the flow of fresh water into the Eastmain River estuary by 90%. In the La Grande Rivière estuary, the freshwater flow in winter has increased eight fold, while some conditions have remained much the same.

The Eastmain River estuary is shallow and extends 27 km. Whereas it was almost entirely filled with fresh water under natural conditions, it now contains brackish water over its first 10 km. In addition, the estuary is now subject to tides for its entire length. A drop in water level has also been noted, along with greater turbidity caused by increased inflow of sediments. These sediments now settle in the estuary and contribute to the formation of a substrate more favourable to establishing a benthic community. Marine fish species venture further into the estuary, while freshwater species are pushed upstream. However, no major changes have been detected in the balance of these populations, and fish yields have remained the same.

In the La Grande Rivière estuary, the considerable increase in flow has brought about effects that are generally the opposite of those observed in the Eastmain River estuary. In winter, for example, the La Grande Rivière's plume (the five-metre-thick surface layer of fresh and brackish water floating on the denser salt water) is three times larger than under natural conditions. The summer plume is comparable to what it was previously.

After 10 years, it may be concluded that these two extreme, opposite ecological changes have not caused any perceptible impact on the fish populations using these estuaries. Condition factors of the different fish species, distribution of age groups among their respective populations, and yields from experimental fish catches are comparable to those under natural conditions. It even seems that eliminating spring flooding does not affect the biological productivity of a northern oceanic habitat. Actually, northern rivers contribute very few nutrients to their estuaries, and the species present have adapted to great fluctuations in salinity.

3.3 Mercury levels in fish

The mercury content in fish in the James Bay basin, as in the rest of the Canadian Shield, is naturally high. The majority of predatory fish in natural lakes not affected by hydro developments show mercury concentrations well above the Canadian federal marketing standard of 0.5 ppm and often above the U.S. federal standard of 1 ppm.

Impoundment of the La Grande Complex reservoirs resulted in an increase in the mercury levels in fish flesh. However, the evolution in the mercury content over time depends primarily on the species of fish and the type of reservoir.

Five years after impoundment, the mercury content in the non-predatory species 400 mm in lenght (lake whitefish) in the La Grande 2 reservoir increased by a factor of 4 (0.16 mg/kg to 0.57 mg/kg). Afterwards, the mercury levels started to drop and in 1990 the average mercury content in lake whitefish was 0.45 mg/kg.

This evolution is more gradual in fish-eating species. Eleven years after impoundment, the mercury concentration in a 700 mm northern pike at the La Grande 2 reservoir increased from 0.61 mg/kg to 3.43 mg/kg. Among young fish, however, a drop in mercury levels has been noted since 1986. Once the older fish are eliminated from the system and replaced by younger ones, a marked decrease in the concentrations at standard lenghts should be observed.

The mercury concentration in non-predatory species is significantly higher in river sections located downstream of the reservoirs. In 1988, lake whitefish showed a concentration of 1.17 mg/kg downstream of the La Grande 2 station compared to 0.45 mg/kg upstream. This bioaccumulation downstream can be explained by a change in feeding habits. The lake whitefish are consuming fish that come from the La Grande 2 reservoir.

Also, downstream from the power station, the effect of mercury export is limited to the fresh water plume of the La Grande Rivière. Therefore all the fish along the James Bay coast outside the influence of the river plume are not affected by methylmercury.

The study of several reservoirs of the Canadian Shield, which were created between 10 and 70 years ago, suggests that it could take between 20 and 30 years before a return to natural conditions.

3.4 Waterfowl

The forming of the reservoirs in Phase I of the La Grande Complex resulted in the disapearance of 6,142 km^2 of habitat with variable waterfowl potential, and an estimated 7,000 to about 9,450 breeding pairs may have been consequently displaced.

However, these habitat losses, which total less than 6% of the overall surface areas of the watershed cannot have any significant impact on the total numbers of waterfowl migrating through Québec, which amount to 1.1 million breeding pairs.

The La Grande development also involved the diversion of the Eastmain-Opinaca rivers.

255

Reduces current velocity and wider herbacious and shrub-like habitats have created favorable nesting habitat for waterfowl. Although intensity of utilization by waterfowl varies considerably depending on the month and the area in question, data indicates that waterfowl use these rivers more now than under natural conditions. These increases, however, are too small to have any effect on the overall reproduction in Québec.

Habitats along the east coast of James Bay are important areas for waterfowl during spring and fall migration. The Crees harvest a large part of their subsistence resources in those seasons.

Studies conducted since 1975 have not shown any change in ecological factors capable of producing measurable changes in habitats since the modification of the freshwater flow conditions by the development of the La Grande Rivière. There was, in particular, concern about the effect that the change in salinity would have on the aquatic plants, especially the eel-grass, which are food for migrating waterfowl. The enlargement of the La Grande Rivière freshwater influence area has not had any negative effects on the coastal habitats that are either protected by the ice cover or naturally subjected to seasonal variations in the salinity, since just as large plumes existed under natural conditions during the flood of all the tributaries. As yet, there are no signs of modification in distribution or density of marine-plant communities and waterfowl continues to stop there in the same numbers as ever.

3.5 *Caribou*

Since 1975, Hydro-Québec, its subsidiary the Société d'énergie de la Baie James and the Department of Recreation, Fish and Game have carried out inventories and behavioural studies of the Caribou on the territory concerned by this project. The first phase of these studies made it possible to estimate the size of the population of the George River caribou herd which grew from 200,000 in 1976 to 680,000 in the period 1984-1988. It is believed that the herd has now stabilized or started to decline slightly, but no surveys have been done since 1988. The Leaf River herd is believed to be increasing and was last surveyed in 1991. However the data are note available yet.

The environmental monitoring revealed the positive role played by the frozen surface of the reservoirs in the movements and feeding of the caribou. In fact, the banks of the reservoirs and of the islands within them are to a large extent covered by spruce stands which, before the land was flooded, were too far from the frozen bodies of water to be used by the caribou. Also, there are peat bogs and shrub and lichen-covered areas near these banks which can also serve as calving and feeding habitats for the caribou.

It should also be mentioned that government inquiries and scientific research have shown that the September 1984 drowning of 10,000 caribou 400 km downstream of the Caniapiscau reservoir in a 22 m high waterfall on Caniapiscau River was due to natural causes.

4   Cree communities

After the announcement by the Government of Québec of its intention to develop the river systems draining into James Bay, the Québec Association of Indians, applied in fall 1972 to the Québec Superior Court for an injunction to stop all work in the area. An injunction was delivered one year after but was stayed pending the outcome of the appeal lodged by the Société d'énergie de la Baie James or until the court decided otherwise.

At the same time, the Cree and Inuit of Québec, the Québec Government and other Québec parties, and the Canadian Government began to negotiate and an agreement in principle was reached in November 1974. The Grand Council of the Crees of Québec dropped all legal proceedings, recognized Québec's jurisdiction over the territory and gave up its land rights. In return, the government confirmed the Cree's right to hunt, fish and trap on all of the line at all times and granted exclusive rights on approximately 75,000 km^2 of the territory used by the Cree for their traditionnal activities (figure 1). Finally, the James Bay and Northern Québec Agreement was signed in November 1975 (JBNQA). Under this agreement and subsequent ones the Native Peoples received more than $500 millions CAN in financial compensation.

When studying the impacts on the human environment of the La Grande Complex, it is difficult to distinguish impacts of the Complex from the effects brought about by the James Bay and Northern Québec Agreement or the changes in Cree society which would have occurred even without the project. Nevertheless, follow-up studies allow an initial assessment of both the project and the agreement's effect on Cree communities.

4.1 *Way of life and economic development*

The land submerged by the reservoirs represents 3.4 % of the total area of the hunting territories used by the Cree for their traditional activities.

Various remedial measures such as clearing selectively and building access ramps, landing areas and snowmobile trails facilitated both summer and winter use of the reservoirs for travelling to the adjoining or nearby traplines. However the increasing mercury levels in fish have significantly limited the use of reservoirs and prompted trappers to concentrate their activities within their traplines.

Whereas, in 1971, 600 families regularly took part in hunting, fishing and trapping activities, the total has held steady at about 1,200 (or one-third of the Cree population) in recent years due to the Cree Hunters and Trappers Income Security Program and Board as defined in the JBNQA. This program yields an average of $10,000 CAN per family in addition to the food equivalent and the profits from fur sales.

This program has made it possible to preserve a way of life that had appeared doomed because of the drop in the market price of furs and the increase in cost of practicing traditional activities. The income provided by the program has also permitted more widespread use of higher-performance hunting and trapping technologies, such as snowmobiles.

Furthermore, the salaried job market now represents an attractive alternative to traditional activities. The significant growth in salaries, a sign of a stronger job market, is no doubt the most important impact of the project and the Agreement. The jobs previously offered were precarious and seasonal. A large number of these jobs are now permanent and much better paid. Generally provided by the Cree Regional Authority (CRA) or by one of its many affiliated organizations, they are concentrated primarily in the public service sector, which has been developed regionnaly as a result of the transfer of government services to the CRA. The average annual income obtained by this type of employment increased from $1,580 CAN to $15,000 CAN between 1971 and 1981.

The very rapid population growth, which rose from 7,000 individuals in the early seventies to 10,300 in 1989, has brought about a growing need for housing and infrastructures. This factor, combined with contracts issued for the implementation of Phase II of the La Grande Complex, has favored the expansion of the construction industry. The Cree worked on contracts totalling $30 million CAN between the middle of 1988 and the middle of 1989. Also, Hydro-Québec has initiated a program in the James Bay area to train and hire Cree for permanent positions with the utility. Its aim is to have 150 Cree in permanent employment by 1996.

Finally, interest payments resulting from the massive indemnities paid out under the various agreements have been invested in the start-up of new entreprises, notably in an airline company: Air Creebec.

4.2 *Opening up of the territory by access roads*

Development of the La Grande Complex also led to the construction of an extensive road network. The main Matagami-La Grande 2 road, which is 620 km long, links the Complex facilities to the road network in southern Québec. A second, east-west axis, 580 km in length, joins up with the main road and provides access to facilities located further east. The community of Chisasibi is now also connected to the road network established for construction of the Complex. The other coastal communities of Wemindji, Eastmain and Waskaganish are linked to this network by winter roads or snowmobile trails. Opening up the Cree communities led to a considerable drop in the prices of certain products previously brought in by boat or plane, and increased economic and cultural exchanges with cities in the south.

Establishing this road network also facilitated use of the region's wildlife resources. While portages previously determined the siting of camps, these are now most often located near roads and reservoirs.

Road access and snowmobile use have further enabled the Crees to travel much more frequently between camps and villages. It has thus become easier for trappers to get fresh supplies, when needed, and to send fish or game to the village on a regular basis.

The presence of these roads and trails has promoted recreational hunting and fishing, both by salaried Crees who live year-round in the villages and by white populations who fish and hunt mainly in the southern part of the James Bay territory and increasingly to the east of the La Grande 2 facilities. Conflict with the personnel who manage the road system and with the Cree can also arise when the hunting regulations are not successfully monitored, which can happen since the territory is so large and difficult to supervise.

4.3 *Health and social services*

Considerable progress can be observed in health care and social services. For instance, infant mortality among the James Bay Cree dropped by half between the mid 1970s and the mid 1980s.

Every Cree village now has a dispensary providing first aid care, home visits and community health services. The Chisasibi Hospital is equipped to provide the James Bay communities with general care, obstetrics, pediatrics and chronic care services. There is also an air ambulance service.

4.4 *Education*

The Cree School Board, created in 1975 under the JBNQA, is run entirely by the elected representatives of the Cree communities. This board, which took over the existing schools, has all the power conferred on other Québec school boards as well as a number of special decision-making powers

over school calendars, text books, teaching materials, and the number of Native and non-Native persons required to teach in each school.

Cree youth can now do their primary education and much of their secondary education in their villages. Two villages, one of which Chisasibi, even have comprehensive high schools. It should also be noted that Cree is the language of education up to the third grade in all Cree schools.

5 CONCLUSION

The preceding brief description of the results of monitoring of the positive and negative impacts of the hydroelectric development of the La Grande Rivière certainly does not do justice to all the studies that have been conducted in that region.

After 17 years of studies and follow-up of the repercussions, the results confirm that there have not been any ecological catastrophes that can be attributed to the La Grande Complex. On the other hand, Hydro-Québec recognizes that there have been major upheavals in the physical and hydrological milieux. Terrestrial ecosystems have been replaced by aquatic ecosystems, but it is noticed that the interdependence of the physical, physico-chemical and biological elements is now in equilibrium or is about to become so.

The development of the La Grande Complex and the implementation of the James Bay and Northern Québec Agreement have made it possible for the Cree to gain access and participation in the mainstream economy of the country while preserving their hunting, fishing and trapping way of life.

In its long-term development plan, Hydro-Québec now intends to complete Phase II of the La Grande Complex and to develop the Grande Baleine Complex some 150 km north of the La Grande Complex. Once again, Hydro-Québec will continue to seek new approaches and solutions to develop and operate its facilities in harmony with the environment. While allowing Québec to develop its resources for the benefit of the entire population of Québec, this project should ensure once again the economic self-sufficiency and autonomy of the Cree and other Native Peoples.

REFERENCE

The lessons of the La Grande Complex (Phase I), Hydro-Québec. Proceedings of a conference in the general program of the 59th Annual Meeting of Association canadienne-française pour l'avancement des sciences, Sherbrooke, Québec, Canada, May 1991.

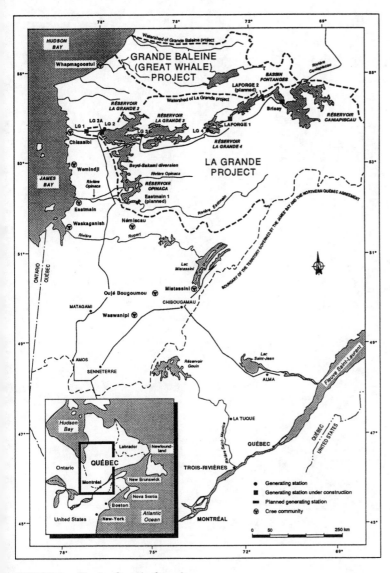

FIGURE 1 : The La Grande Complex

259

Hydropower'92, Broch & Lysne (eds) © 1992 Balkema, Rotterdam. ISBN 90 5410 054 0

Caverns design for Hong-Jia Du project and its reservoir influence upon environment

Wang Bo-Le & Chen De-Chuan
Gui Yang Institute of Water Resources & Hydropower Investigation and Design, People's Republic of China

Liu Tian-Xiong
Tsinghua University, Beijing, People's Republic of China

ABSTRACT: Three units with total generating capacity of 540 MW are installed in Hong-Jia Du hydro-power station. The reservoir storage capacity is 5.0 GM³. Except the rock fill dam with reinforced concrete face, the flood discharge system, draw-off system, diversion system and genera-ting house are located in left mountains, Many questions, such as election of the tunnels' axis, layout and supporting for the underground cavern group, rock burst, etc., are discussed in this paper. The background investigation, variation and prediction of 36 environmental factors, and the compre-hensive evaluation on the reservoir influence upon environment are discussed as well. At the same time, the second pollution of the water mercury is studied in detail.

1 INTRODUCTION

The Wu Jiang River is a main tribtuary of the Yangtze River, with abundant water and high water head. It is planned ot be explored by 11 stages, with a total generating capacity of 8795 MW (Fig.1 & Fig.2).

Hong-Jia Du power station is the first stage and locates on the Liu Chang River, north tributary of the Wu-Jiang River in Qin Xin county of Gui Zhou province. Besides generation, Hong-Jia Du power station can increase the electricity and guarantee output of the downstream stations. It also does good to flood control, water supply, irrigation, obstructing sand, breeding, tourism, and promote the development of mining and other industries around the reservoir area.

2 TOPOGRAPHY, GEOLOGY AND LAYOUT OF PROJECT

Strata of dam site are limestone, clay shale and sand shale. The dam site is an unsymmetrily V-shaped valley. The right side slope of it is fairly gentle, and the left bank is a two-storey cliff with about 100m in height. the main faults in dam site are F_8, F_{13}, F_{15}. Hydraulic structures except the rock fill dam with reinforced concrete are arranged in the left mountains, forming a large cavern group (see fig.3). These tunnels penetrate the rock strata mentioned above at big angles. That is good for safty of the tunnels.

fig.1. Cascade Development of the Wu Jiang River

1. Hong-Jia Du (540MWH)   2. Dong Feng (510MWH)
3. Suo-Feng Ying (420MWH) 4. Wu-Jiang Du (1050MWH)
5. Gou-Pi Tan (2000MWH)   6. Si Lin (840MWH)
7. Sha Tua (800MWH)   8. Peng Shui (1200MWH)
9. Da-Xi Kou (1200MWH)   1'. Pu Ding (75MWH)
2'. Yin-Zi Du (160MWH)   3'. The Mao-Tiao He River
                              cascade project

3 DESIGN OF THE UNDERGROUND CAVERN GROUP

3.1 Flood discharge system

1. Tunnel-type spillway
The enter port is a open-kind overfall weir with a size of 2×12m, and with a radia gate on it.

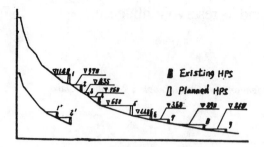

fig.2. Longitudinal view of the Wu Jiang River

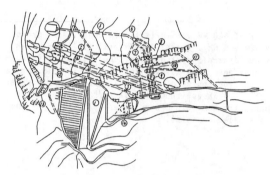

fig.3. General layout of the Hong-Jia Du hydroe-
       lectric project

1. Dam   8. Underground power house
2. Spillway tunnel   9. Tailrace tunnel
3. Discharge tunnel   10. Tailrace surge tank (shaft)
4. Draw-off tunnel   11. Power plant entrance tunnel
5. Power tunnel   12. Diversion tunnel
6. Surge tank(shaft)   13. Upperstream coffer dam
7. Penstock   14. Downstream coffer dam

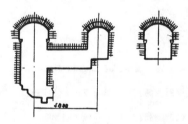

fig.4. Cavern for underground power house

There is an unpressured tunnel behind the weir.
The cross-section of the tunnel is rectangular
with upper arch. Its discharge capacity is
5400 m³/s, V_{max} =41m/s. There are three air-
mixed trough along the tunnel. Depth of air-mixed
water is from 9.6 to 15.0 m. Energy dissipation
of ski-jump type with side diffusion is adopted
at the outlet. The test biggest pit is 13m deep.
The spillway penetrates limestone, clay shale

and several faults. In limestone region, it is
excavated by full section. In the region of clay
shale, fault and poor quality rock mass, step
excavation is adopted.
   2. Discharge tunnel
It discharges the flood when the water level
reaches the checking level PMT. It also plays
the role of draw-off reservoir. The discharge
tunnel is rectangular with upper arch with a
slope of 9‰ , and it is 699 meters long. It is
unpressured, with a discharge capacity of
1640m³/s. There are three air-mixed troughs
along the tunnel, the water depth is from 7.56m
.to 8.82m. Energy dissipation of ski-jump type
with a narrow stitch is adopted at the exit.
   3. Draw-off tunnel
The diversion tunnel No.1 is restructed to be
a draw-off tunnel. It takes 70 or 100 days to
empty the reservoir.

3.2 Headrace system and tailrace system

   1. Upperstream surge chamber. It is a circula
reflected surge chamber with throtle located in
limestone rock.   It is 106.44 meters high with
a diameter of 23m. There are three safty gates
in it.
   2. Tailrace surge chamber. The results of
calculation show that tailrace surge chamber
is needed. The arrangement is that, machine hall,
main transformer room and tailrace surge chamber
are placed parallely. The lower part of tailrace
surge chamber is three independent shafts and the
upper part is a linked type chamber. Computation
shows that no resonance will occur in upperstream
and downstream surge chambers.

3.3 Underground power house system(see fig.4
    and fig.5)

   1. Arrangement of power house. Underground
power house is arranged in the hard limestone
between clay shale and fault F8. The reservoir
and underground power house is separated by
clay shales. So the leakproof and moisture-proof
conditions are nice. In order to keep the
stability of the side wall of the underground
power house with a big span, the axis of the
power house is arranged parallel to the direction
of the maximum ground stress. It crosses the main
faults(such as F13) at a big angle.
   2. Measures of reducing the machine hall span.
tubular butterfly valve is adopted instead of
the traditional butterfly valve. In the lots of
region, Bolts and shotcrete are adopted as
permanent supporting. And bolt crane beam is
adopted (see fig.4).
   3. Drainage and ventilation in underground
power house. There are two rows of galleries for
drainage. Drainage holes are drilled in the
vault and the wall. The drainage ditch and

moisture-proof partition are placed in the power house. In order to form a natural vetilation system, two passages to power house are laid. Ventilation can also be attained by manual work or by machinery. The control room is located on the ground and the imformation is transmitted by ray cables.

4. Construction. The excavation depth of machine hall is about 52.6m. It is done by four different floors. Two branch tunnels are excavated for construction. During excavation, bolting must be finished timely for supporting.

5. Rock burst and prevention. The ground stress in the power house site is very high, from 25 Mpa to 30 Mpa. Split rock core has been found in the bore hole. The rock split rapidly when it had been taken out. It shows that rock burst maybe happen during construction.

Improve the stress of surrounding rock mass:
Arrange the longitudinal axis parallel to the the maximun ground stress; Short chamber footage excavation should be adopted. Release the stress by predrill holes.

Reinforce the surrounding rock mass: Prebolt the work face with anchors or steel wires.

Other protecting measures: Shelter from the bursting rock by steel wires or nylon nets; safty nets are hung above the machines.

Besides the methods mentioned above, measuring on spot and forecasting may be also adopted .

4 RESERVOIR INFLUENCE ON ENVIRONMENT AND POLLUTION OF MERCURY

The storage capacity of Hong Jia Du reservoir is 5.0 Gm^3, the area is 80.5 Km^2. According to the project scale, the valley and social environment, 36 environmental factors are selected for research. Background investigation of these factors are carried out, evaluated them by quality or by ration. Then synthetical evaluation is carried out by quality quota. The result shows that the quota of environment increases 0.13977(23.6%) in which the quota of natural environment increases 0.05134, the quota of social environment increases 0.08843. It shows that the project can improve the environment , and the improvement of social environment is greater than that of natural

environment. Among the 36 factors, 23of them (63.9%) get the better change, 10 of them(27.8%) get the worse change. But the worse change can be reduced, even can be eliminated by some measures. 3 of them get no change. So far as the environment is concerned, Hong-Jia Du project is feasible. It does more good than harm to the environment.

4.1 Evaluation on background of the liquidoid mercury in the water of Hong-Jia Du river

1. Form of the liquidoid mercury and its circulation.
2. Background of liquidoid mercury. Table 1. shows the ten sections measured data during 1986 -1987.

The reason why liquidoid mercury is more produced especially during flood period is that the produced mercury of indurstry and

Table 2. Measured values of mercury in bottom clay (PPM)

| Hong-Jia Du bottom clay | 0.11 ~ 0.15 |
| Wu-Jiang Du bottom clay | 0.095 ~ 0.25 |

Table 3. Background value and measured value (PPM)

deposit in Rhineian River	0.2	yellow clay of Gui Zhou	0.262
averge content in shale	0.2	red clay of Kun Ming	0.286
edge region of the world	0.35	bottom clay in Hong-Jia Du	0.11~ 0.15
deposit of 87 lakes	0.186	bottom clay in Wu-Jiang Du	0.25

Table 1. Measured values of liquidoid mercury (PPM)

position	time	flood period	average water period	dry period	annual mean value
Hong - Jia Du	average measured value	0.00179	0.00022	0.00052	0.00084
Wu - Jiang Du	average measured value				0 ~0.0005

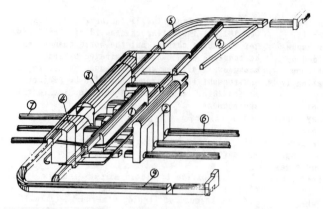

fig.5. Section for the underground power house project

① Main transformer chamber ② Tailrace surge tank(chamber)
③ Foundermental power house ④ Secondary power house
⑤ Access tunnel for transformer chamber and tailrace surge chamber
⑥ Tailrace tunnel   ⑦ Penstock
⑧ Exhaust ventilation and cable tunnel
⑨ Access tunnel to power house(also for exhaust ventilation and placing cables)

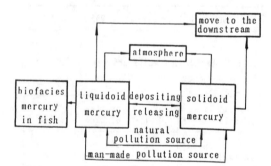

fig.6.   Circulation of liquidoid mercury

mining leads to mercury pollution. The origion value is high, forest cover rate is low. Soil erosion leads to the natrual pollution. And the suspended mercury in turbulent flow deposits uncompletely.

3. Solidoid mercury. Table 2. shows the mercury values.

The forms of mercury depositing in bottom clay are complexed: accumulation and release; depositing and resuspending; importing and moving out; oxidation and redution, etc(see fig.6). They affect each other and lead to the dynamic equilibrium. The background value and measured value in some areas are listed in table 3.

Compared to background value (0.262PPM) of Gui Zhou yellow clay, content of mercury in bottom clay of Hong Jia Du and Wu Jiang Du is low. It is not thought to be polluted.

4. Background of mercury in fish. Table 4. shows the measured value of mercury in fish.

Table 4. shows that fishes in Wu-Jiang Du and Hong-Jia Du river are not polluted. The total mercury and methyl mercury in fish are lower than Chinese standard(0.3PPM). Chinese scholars regard 6 percent of the total mercury(0.2PPM) as the standard of methyl mercury in fish.

4.2 Forecast of the liquidoid mercury of Hong-Jia Du reservoir

Wu-Jiang Du reservoir which is 168.8 Km downstream far from Hong-Jia Du is consulted for research on mercury. Wu-Jiang Du reservoir area is 47.5Km² with the storage capacity of 2.14Gm³. Between these two reservoirs there are San-Cha He river and Mao-Tiao He river in which content of mercury is high.

Table 5 shows the forecast values according to the runoff of three periods: flood year, average water year and dry year. It is based on the Japanese method:
$$C_t = C_o + (1 - e^{-t/T})(R C_i - C_o)$$

Table 5. shows that the average value of liquidoid mercury for years increases little each year. It keeps Stable five years later. The requirement of national standard is satisfied. No pollution will happen after the reservoir has been built.

The source and background of mercury of Hong Jia Du are lower than that of Wu-Jiang Du. After Hong-Jia Du reservoir is built, the deposited mercury and liquidoid mercury are

Table 4. average content of mercury in fish

collecting region	meat		liver		gill	
	total	methyl mercury	total	methyl mercury	total	methyl mercury
Hong-Jia Du	0.139	0.115	0.09	0.058	0.071	0.051
Wu-Jiang Du	0.153	0.134	0.086	0.086	0.080	0.012

Table 5. The forecast values of liquidoid mercury for Hong-Jia Du ($\times 10^{-4}$PPM)

year	1	2	3	4	5	10	20
average year	6.84	7.35	7.52	7.55	7.56	7.58	7.58

not higher than that of Wu-Jiang Du. The methyl mercury in fish is low also. It does not lead to any pollution.

4.3 Summarization

1. The liquidoid mercury reaches dynamic equilibrium after Hong-Jia Du reservoir has been built for four years. The mercury in fish and deposited mercury increase a little. But it is lower than the standard. The liquidoid mercury is improved meeting the national standard II, without any harm to human beings and fish production.
2. In the valley of Hong-Jia Du reservoir, if some measures are adopted(such as increasing the rate of plant cover, transforming the teraced fields, controlling industrial pollution and the farm chemicals containing mercury), the mercury is imported to the reservoir through natural system. The mercury imported to the reservoir by soil erosion will keep the equilibrium.
When the deposited mercury reaches balance, it is still lower than the natural standard. Thus water will not be polluted by the depositing mercury.
3. No methyl mercury is drained off by the chemical industry. The pollution caused by the methyl mercury formed by organisms is a little, and the reservoir conditions are unfavourable for producing the methyl mercury.

5 CONCLUSION

Large cavern group is the focus of this project. It needs careful design and construction. The project influences the natural and social environment greatly. It will produce good economical and social effects and lead to better influence on environment comparing with the bad influence. The project is economically reasonable and feasible. It does good to the environment. Related departments of Chinese government are getting ready to start this project presently.

Hydropower'92, Broch & Lysne (eds) © 1992 Balkema, Rotterdam. ISBN 90 5410 054 0

Erosion in regulated reservoirs

Tor E. Dahl
SINTEF NHL, Trondheim, Norway

ABSTRACT: Due to extensive regulation for hydro power, serious erosion along the shores at some reservoirs has been observed. The degradation of the tributaries was particuarly dramatic. A programme to monitor the development of the erosion of three reservoirs was established. The programme consists of: - Aearial photography - Regular photo recording of interesting spots - Measurement of suspended material in the reservoirs.

The programme was more intensive the first years after 1973, but is still being followed up at intervals of years.

Among the findings so far is

a. The rate of degradation and eroded areas at the outlets of the tributaries is related to - drainage area - soil composition - time.

b. The development of new quasistable shores is dependent on - wave activity - soil composition - ice conditions - time.

c. The sediment concentration varies regularly with seasonal and hydrological conditions such as - waterlevels - wave activity - ice conditions - waterdischarge in tributaries.

For some of these relations it has been possible to formulate general expressions.

Findings under (a) will be reported in this paper.

BACKGROUND

Several natural lakes in Norway are used as reservoirs for purpose of hydro power production.

To increase the volume of the reservoirs the water levels have sometimes been raised by damming, but the most characteristic feature is the utilization of volumes below the natural water level. Some of the reservoirs are located in glacial deposits.

The annual regulation and especially the annual lowering of the water level below the natural level has caused erosion which is attributed to four mechanisms:
- Slides resulting from loss of slope stability during drawdown.
- Stream degradation at outlets to the reservoir during drawdown.
- Wave action on new sections of the shore profiles which are not protected by armouring layer of coarse sediments.
- Ground water seepage.

A project for investigation of the erosion process caused by the regulation was organized from 1973, financed by The fund of licence fees (Konsesjonsavgiftsfondet).

Fig. 1. Slide released during the first year after the water level was lowered below the natural waterlevel.

Fig. 2. Stream degradation in fluvial deposits.

Fig. 3. Wave erosion of shore.

Monitoring of actual shoreline erosion at selected spots, tributari degradation of the outlets to the reservoir and monitoring of suspended sediment load were parts of the project. The main part of the project took place between 1973 and 1980, but irregular control observations are still being carried out as follow-up action.

The project has mainly been reported in Norwegian. Some details have also been referred in international fora. (Carstens and Solvik, 1975, Nielsen, 1980.

SITE SELECTION AN CHOICE OF MONITORING PROGRAMME

Important criteria for choice of reservoirs were:

- Recent regulation, such that as much as possible of the initiation process could be covered.
- Glacial deposits along a major part of the lake shore.
- Seasonal drawdown, well below the natural lake level.

- Accessability, particularly during the snowmelt period.
- Locally available operators for regular sediment sampling.

Figures 4-6 show the three lakes that were chosen, all recently regulated.

THE SITES AND THE FIELD OBSERVATIONS

Gjevilvannet

The lake is long and narrow, with numerous tributaries as shown on Fig. 4. Large bank sections consist of fine sand, extremely vulnerable to erosion, while the contents of fine, easily suspendable sediments are moderate.

A narrow strip of the bed along the natural shoreline was covered by a stable armour layer of sorted gravel. The regulation, 0.8 m up and 14.2 m down, exposed large areas of unprotected bottom below the armour layer to wave action and groundwater erosion.

Rapid degradation of the tibutaries started right after the first drawdown. Some of the major tributaries were then permanently protected against further erosion, while some more amateurish attempts to stop erosion in minor creeks were futile, Fig. 8. A gradual flattening of the exposed shoreline has occurred, and the minor creeks have now developed a rather stable profile, too.

Regular slides occurred during the first drawdown season at some places as marked on Fig. 4.

The field programme comprised bed samplings, profiling of beach cross sections, and visual inspections at irregular intervals. Regular samplings of suspended material, visibility measurements by Secci disk, and temperature recordings were made from boat or from the ice near the outlet by local personnel. Water level recordings were supplied by the power company.

Málvatna

The reservoir is formed by damming together three lakes, marked I, II and II on Fig. 5. The surrounding terrain above normal water level is mainly exposed rock, while underwater deposits are probably of postglacial origin, consisting mainly of fine silts with mean grain size about 0.04 mm. During drawdown, the original linking creeks between the lakes function as draining channels for the two smaller lakes into lake III. This has caused drastic degration of the passage between the lakes. Together with the degradation of tributaries and shoreline erosion from waves and seepage, this has resulted in suspended sediment loads so heavy that they are otherwise rarely found in Norwegian watercourses.

No measures have been taken to control the erosion. The field programme comprised bed samplings and irregular visits for inspection and photos. A shoreline profile was surveyed in 1975, but the bench marks were

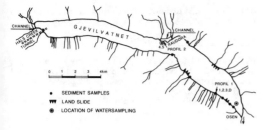

Fig. 4. Lake Gjevilvannet, with interesting spots.

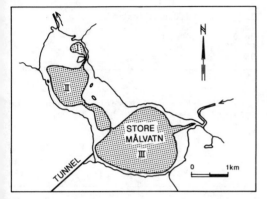

Fig. 5. The lakes Målvatna, with HRWL and LRWL.

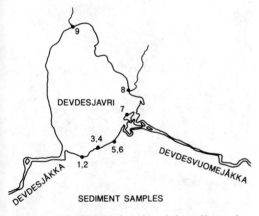

SEDIMENT SAMPLES

Fig. 6. Lake Devdesjavri, with tributaries.

later destroyed by severe erosion.

Regular samplings of suspended sediments were made from the power station discharge. Some samples from the lake surface have also been taken during inspection visits in 1974 and 1975.

Table 1. Typical data for Devdesjavri.

Map, Fig no	5
Natural WL	410,5-
Regulated WL	413-380,5
Length km	3
Width km	2
Grain size mm	0.06-08
Damages	Erosion of stream Beach erosion
Measurements	Air photos Photos Sediment Consentration
Observed	1973-77-86

Devdesjavri

The lake has a pear-shaped outline, with one major and seven minor creek tributaries, Fig. 6. It is located in a bed of glacial deposits of varying composition. The range 0.06-0.8 mm refer to nine samples along the eastern side of the lake where the deposits are particularly heavy. This lake also had a naturally stabilized shoreline. Regulation 3 m up and 30 m down resulted in very intensive erosion in all the tributaries. No measures have been taken to stop the erosion.

The field programme comprised bed samplings and irregular visits for observation and photographing, but no shore profile has been taken. The most intensive erosion occurred in the creek deltas. This development was studied by use of air photos from 1972, 1973, 1976, 1982, 1986.

DEGRATION OF THE STREAM DELTAS

When the reservoirs were lowered for the first time below the natural wter level, the unprotected deposits outside the tributaries in the reservoirs were exposed to erosion.

The erosion which started in the fine fluvial deposits propagated backward and destroyed the previously stable streambed. Before this process came to a balanced situation the backward erosion in most cases had propagated far beyond the natural shore line.

The rate of scour development is very much dependent on the type of sediments. Fig. 7 below shows the erosion caused by a small stream where the fine sediment layer was relatively thin with coarser materials below. The degradation therefore came to a stop after a short period.

Fig. 8 shows an unsuccessful attempt to duct a small stream through the unprotected area of fine sediments below the natural line.

In all three reservoirs the scour was considerable at the outlets of the tributaries.

Fig. 7.

Fig. 8. Stream degradation and unsuccessful repair attempt.

However, the rate of scour in Gjevilvatnet was limited compared to Store Målvatn and Devdesjavri. The reason for this was that measures had been taken in Gjevilvatnet to protect the outlets of the tributaries against erosion. Fig. 8 shows one of the unsuccessful ways of doing it, but several inlets were protected by the building of artificial thresholds which were very effective and stopped the scour from propagating backward beyond the natural shore line.

In the store Målvatn and Devdesjavri no measures were taken to limit the erosion at the outlet of the tributaries.

The scour development in these two reservoirs were quite simular. By using air photos we have tried to quantify the volume of eroded sediments at the outlet of four tributaries in Devdesjavri. The development over time and the eroded volume compared with the catchment area and discharge for the four streams have been studied.

About the data

By utilizing air photography we have estimated the volume of eroded sediments at the outlet of the four major tributaries to Devdesjavri.

Five images were taken; the first one before the regulation of the reservoir.

Table 2. Air photography.

Date	Water Level
72	410,5 m (natural WL)
73-07-07	405,5 m
76-07-11	408,0 m
82-07-03	399.5 m
86-05-25	396.0 m

The lowest annual waterlevels since the regulation started have been as follows.

Table 3.

year	73	74	75	76	77	78	79	80
L.W.L.	?	380	381	396	382	387	383	389

year	81	82	83	84	85	86	87
LWL	385	382	394	389	386	390	

Fig. 9. Air photo 1973.

270

Fig. 10. Sector of Fig. 9 stream no 4.

The drainage areas of the four streams in question are quite different in size.

Table 4.

Stream no	Drainage area k m²
3	2.4
4	210.0
5	20.0
6	6.1

Result of eroded area/volume, estimates.

Table 5.

Stream	1973		1976	
	area	volume	area	volume
	m²	m³	m²	m³
3	7800	114000	10000	130000
4	216000	$3.2 \cdot 10^6$	248000	$3.9 \cdot 10^6$
5	45500	440000	50000	520000
6	10050	201000	14750	229000

Stream	1982		1986	
	area	volume	area	volume
	m²	m³	m²	m³
3	10000	150000	10000	150000
4	293000	$4.95 \cdot 10^6$	303000	$5.2 \cdot 10^6$
5	72500	716000	72500	716000
6	14750	300000	14750	300000

It has not been possible to measure the eroded volume exactly because the air photos have not been taken when the waterlevel was lowest. The reason is that the waterlevel is

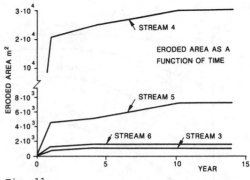

Fig. 11.

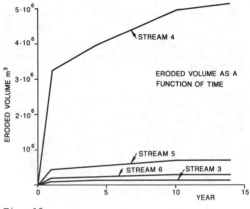

Fig. 12.

lowest in late winter when the whole land-scape is covered by snow and ice, and the elevation of the WL starts when the snow starts melting.

As can be observed from Table 5 a major part of the erosion took place the first year. The scour development is dependent on water discharge and time Fig. 11 shows the development for the four streams over a period of 14 years.

As mentioned, most of the erosion at the stream deltas took place the first year.

Records giving the elevation of the reservoir and the water discharge out of the reservoir through the powerstation for the period 73-05-20 to 73-07-07 are available. The latter date coincides with the day we took the first air photos after the first lowering of the reservoir.

These records gives us the opportunity to compare the inflow to the reservoir with the eroded volume in the delta of the four tributaries. The four tributaries together drains 95.4% of the catchment area.

As can be seen from Fig. 14 a unit volume of water transported to the reservoir by a stream which is draining a small catchment area, carries more sediments than the same unit volume of water coming from a larger catchment area.

Several factors such as geomorphological

271

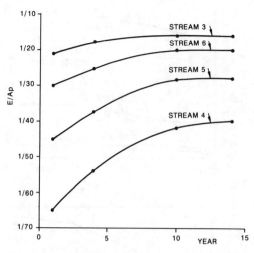

Fig. 13. The relationship between eroded material and drainage area as a function of time. $E/Ap = m^3/m^2$

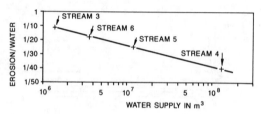

Fig. 14.

Table 6. The data of water supply and the amount of erosion in the four deltas.

Stream no	Water volume mil.m³	Average discharge m³/s	Erosion m³	Erosion water v.
3	1,4	0,3	114000	~ 1:12
4	124,0	28,7	3200000	~ 1:39
5	11,8	2,7	440000	~ 1:27
6	3,6	0,8	201000	~ 1:18

The relationship between eroded materials and conveyed water is given for each stream in Fig. 14.

and hydralical differences can explain this. But in this particular case we are of the opinion that the small catchments area respond more rapidly to the change in day and night temperature.

The peak discharge during daytime in the small streams are relatively higher than the peak discharge in the larger streams draining larger areas.

As the ability of streams and rivers to carry sediments is closely related to the water discharge (Q^n) and the exponent is

allways greater than 1 the effect we see in Fig. 14 is not surprising.

SUMMARY

In the reservoir Devdesjavri we have closely followed the degradation of the tributaries at the outlet to the reservoir.

The annual lowering of the waterlevel to 30 m below the natural level exposed fluvial deposits in the reservoir to erosion.

The erosion started in the fine deposits in the reservoir and propagated backwards beyond the natural shore line.

The eroded areas were measured from air photos, and the volume of erosion was also estimated by using air photos.

The development shows that the erosion was very intensive the first year the reservoir was lowered.

The erosion and degradation culminated first for the streams draining small catchments and later for the bigger streams.

REFERENCES

Carstens, T., Solvik, Ø. 1975. Reservoir eosion. Proc. 2nd World Congress on Water Resources, New Delhi.

Nielsen, S.A. 1980. Erosion in reservoarer som følge av årsregulering. (Erosion in reservoirs caused by annual regulation). Nordic hydrology meeting, Vemdalen.

Hydropower'92, Broch & Lysne (eds) © 1992 Balkema, Rotterdam. ISBN 90 5410 054 0

Lessons from 17 years of environmental monitoring at the La Grande Project, Québec, Canada

C. Demers & N. Chartrand
Environment Branch, Hydro-Québec, Que., Canada

ABSTRACT: This paper will focus on two points; first, the lessons from the mitigative measures and, secondly, the lessons from the environmental monitoring program. The mitigative measures must be designed in accordance with the expressed needs of the Native peoples who use the territory and their pertinence judged on a cost-benefit basis. Other measures could be introduced in locations other than the sites directly affected by the project. We also give our definition of environmental monitoring, its place among other environmental activities and its application for future projects.

1 INTRODUCTION

The hydroelectric project on the La Grande River is located in the James Bay territory on the 54th parallel some 1,000 km north of Montreal, Quebec, Canada (Figure 1). The La Grande project watershed covers an area of 176,800 km^2.

We started our environmental studies in 1973 at the same time as the construction work began. Very little was known about this territory, as it could only be accessed by air. The governments of Canada and Quebec and Hydro-Québec, directly or through its subsidiary, the James Bay Energy Corporation, implemented the most imposing program of environmental studies in Canada. The region was to be considered a vast natural laboratory where multidisciplinary studies and research would be conducted in order to determine how the ecological processes would be changed by the hydroelectric development projects.

We now have seventeen years of knowledge on this territory and this paper gives the lessons we have learned from this work. These are particularly useful to us for the impact assessment study of the Grande Baleine (Great Whale) project located only 150 km north of the La Grande project which we will soon be completing (Figure 2).

In another paper by the same authors, the emphasis will be put on the evolution of newly created environments and their effects on wildlife and habitats. It will also examine the changes in land use created by the project and the economic, social and cultural repercussions on the Cree communities (cf. N. Chartrand, C. Demers, "Repercussions of hydroelectric developments in Northern Québec, Canada: The case of the La Grande Complex in James Bay").

2 LESSONS FROM THE MITIGATIVE MEASURES PROGRAM

2.1 *Defining the objectives*

The development of mitigative measures consists in providing tools for managing these impacts adequately. The nature and scope of the impacts therefore must first be determined. Next, before implementing any measure, it is important to properly identify the issues, define the objectives and set priorities. It is then a matter of finding the best measure for mitigating the impacts. Finally, the scope of the residual impact must be assessed.

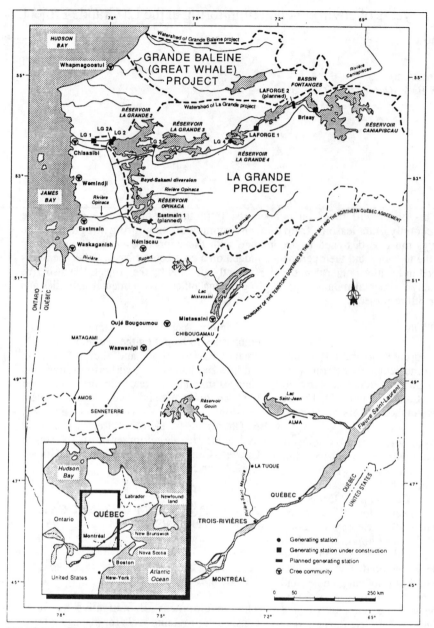

FIGURE 1 : The La Grande Complex

274

PROJECT	Watershed Area sq.km	Construction Schedule	Number of Generating Stations	Capacity MW	Reservoir Area sq.km	Land Flooded By Reservoir sq.km
La Grande Phase 1		1971-1985	3	10,282	13,520	10,400
La Grande Phase 2		1987-1996	6	5,437	2,093	1,105
La Grande Total	176,800	1971-1996	9	15,719	15,873	11,505
Grande Baleine	57,200		3	3,168	3,576	1,786

Figure 2: Technical characteristics of the hydroelectric development projects in Northern Quebec

From this perspective, the balance sheet of losses and gains (in terms of habitats, abundance of wildlife, access to the territory, potential, etc.) must be drawn up and it all assessed in terms of the objectives and cost of the mitigative measure. The La Grande project experience suggests that the remedial facilities would be more profitable if they were built in unaffected areas with high wildlife potential, rather than limiting the measures to the affected areas (which often have low potential such as, for example, the banks of reservoirs).

2.2 *Dialogue*

Many measures applied to the La Grande project were developed in accordance with the situation ten to fifteen years ago. They are no longer necessarily appropriate to the current or future context. Notably, the situation of the main users of the area, the Cree, has changed a lot.

The discussion mechanisms established during Phase I of the La Grande project revealed how important it was to include Native users in the mitigative measure development process. These measures must be developed in consultation with the Native people, especially if they have repercussions on the use of the land. This concept can be broadened to include all types of measures.

Consultations with Native people could make it possible to identify which orientations, wildlife elements and areas should be given preference. Knowing these elements would make it possible to build facilities that would be more useful to them

and that they would make use of as long as the habitats have good wildlife potential.

It is however very important that the dialogue mechanism not be perceived as another bargaining table and that it not aggravate existing or latent social conflicts that may exist in the Native communities.

2.3 *Planning and design*

The ecology of the target resources, equipment operating constraints, financial, technical and practical (maintenance) factors, environmental conditions and contingencies must all be taken into account in the planning and designing of the mitigative measures.

The problems involved in operating hydroelectric facilities must be incorporated in the planning of mitigative measures. For example, it must be ensured that remedial facilities built in sections of the main valley containing a spillway will withstand spills that can occur. In fact, the facilities could be destroyed and the appropriateness of the measure questioned.

Climatic conditions will affect the success of mitigative measures. Assessment of the environmental factors must therefore be included in the planning so that the appropriate specific choices can be made.

The schedule of the impacts must also be included in the planning and design of the program of mitigative measures. In fact, to the extent that it is possible and relevant to do so, the works must be operational when the impacts occur so that they are ready to mitigate them immediately. The construction calendar must also be taken into account so that the construction equipment can be used to implement the mitigative works.

From a practical point of view, works that are simple, technically easy to build and manage and require little or no maintenance should be planned as much as possible. Accessibility is another element that must also be taken into consideration.

Finally, the planning must take into account the fact that a host of unforeseen technical, environmental, financial or human circumstances may occur which can profoundly alter the program of measures.

The planning of these measures should take into account the particular characteristics of each case, since no general purpose method exists.

2.4 Construction and Monitoring[1]

When the remedial works are being built, costs can be substantially reduced by taking advantage of the machinery already in the area. The distance and accessibility of the locations of the works from the construction sites and the construction calendar of the various components of the complex then become very important factors to take into consideration.

Moreover, the new facilities will require maintenance to ensure that they remain functional. For example, the development of a habitat is a form of specialization of the environment and it is important to monitor the evolution of the developed site. In this context, the accessibility of the developed sites is important in order to maintain them, whenever necessary. Maintenance of the works may be prohibitively costly and the cost increases when the distances are great.

It should therefore come as no surprise that the maintenance of the facilities is generally not a priority among the tasks that have to be performed by the Hydro-Québec operating units, since these facilities are often located far from the electricity generating sites. An organization specially designated to carry out this work should see to the operation of these mitigative facilities in accordance with

---

[1] *Monitoring here consists of the activities related to the maintenance of the mitigative works. It does not mean the monitoring of their biological efficiency.*

specific operating instructions. In certain cases, the monitoring and maintenance could be entrusted to the users, which would promote local employment.

A specific financial reserve should also be allocated. These funds should be guaranteed so that any required adjustments can be made, principally those requested by the Native people. This reserve should be protected for this specific purpose and planned for when the facilities are first designed. The planning of their long-term maintenance therefore must take into account these practical elements.

Nevertheless, the fact still remains that the best facilities are those that require the least amount of maintenance. However, if the facilities require maintenance, they must be accessible and simple enough in design so that anyone can take care of them. Ideally, the users of the area.

2.5 Specific lessons

1. Seeding and Planting in Exposed Rivers
The principal objective of seeding in exposed rivers was to speed up the process of reconstitution of riparian habitats and control surface erosion.

In general, the seeding of species in fine deposits proved relevant in the short term but irrelevant in the medium and long term, since natural colonization produces as significant a plant cover in three to five years. The work should be restricted to sectors with a serious risk of erosion. In such cases, action must be taken very quickly after exposure and seed mixes used that are the least disruptive to indigenous flora. Planting shrubs does not accelerate the reconstitution of the riparian ecotone in exposed areas. This type of work does not have any particular advantages over natural reconstitution.

Bulrush could be planted to control the erosion of the banks through sapping as long as the conditions for planting (substratum, water level, etc.) the species are adequate.

2. Planting in Borrow Pits
The Environment Quality Act regulation on quarries and sandpits requires plant restoration in borrow pits. The problems are different from those of exposed rivers, since the soil is often poorer. Moreover, it is

difficult to preserve the thin layer of organic matter and this slows down the recolonization process. The object of the restoration work is essentially to restore the plant cover and the esthetic appearance of the site. The work therefore consists in stabilizing and conditioning the soil, as well as starting the recolonization process. Promoting wildlife is not a recognized objective.

A variety of techniques have been tested to attempt to encourage new plant growth such as the use of wood chips, inoculation and rototilled hay which have not produced the desired results. Fertilizer sticks combined with existing organic matter and a fairly high groundwater level seems an interesting technique. Circular planting, which protects the plants in the centre, is advantageous from an esthetic point of view. On the other hand, since one of the objectives is to recondition the soil, green alder is the species most often planted. Jack pine and willow are companion species. However, the height of the groundwater is often a limiting factor and cuts down on the number of favourable sites for planting willow which is used more by the fauna. Green alder is the species that has fared the best. New growth is difficult on very well sorted material such as fine sand where drainage is excessive.

3. Tree Clearing

Tree clearing in reservoirs does not affect the fish. It should instead be oriented towards promoting fishing. In this case, the cleared areas should be selected in terms of the anticipated harvest, their accessibility and their use. Experience on the La Grande project shows that, before the mercury problem was discovered, fishing sites were intensively used by the Cree. Since certain species can be harvested in the short term after the flooding, clearing for fishing in the reservoirs is always possible. Nevertheless, the mercury problem must be taken into account before recommending any action concerning the reservoirs. On the other hand, certain strategic locations can always be cleared for esthetic reasons.

The clearing of one to five metres along the banks of the reservoirs promotes and accelerates the establishment of vegetation. However, the restored riparian habitat is narrow (under two metres) and is composed of opportunistic plant species that only hold their own because of the clearing made. The environments created by the tree clearing along the banks of the reservoirs are transitional habitats that do not have the qualities of those lost through the flooding of the existing riparian areas. Consequently, given the scale and quality of the flooded habitats, clearing only slightly mitigates the impacts.

The reconstitution of the riparian environment along the banks of the reservoirs therefore appears to be an unsolvable problem. It would be more profitable to clear elsewhere than along the immediate periphery of the reservoirs, in areas with higher wildlife and plant potential in order to create a more significant beneficial effect.

4. Creating of Spawning Grounds

The development of spawning grounds may prove to be profitable locally. However, it is important to clearly determine whether this type of facility is required in terms of the impacts, potential of the site, and requirements of the fish and users. The mercury problem and operating constraints must be incorporated in the planning of this measure. The spawning grounds must therefore be considered in the long term, especially as it is difficult to measure their efficiency and few experiments have been done in Quebec, particularly in the case of lake trout.

5. Fish Ladder

This type of measure has only been implemented in one location in the La Grande project, namely as part of the experimental facility, dike BA-02, southeast of the La Grande 2 reservoir. In this particular case, experience has shown that this ladder is indeed used by certain species, mainly northern pike and black sucker and, to a lesser extent, yellow pickerel. Brook trout and lake trout have barely used it at all. The installation of the ladder however has not had any significant impact on the fish populations in the reservoirs.

The fish ladder is an expensive, very specific facility which may prove to be useful locally. However, previous experiments show that use of it by fish may vary greatly. Its design and siting depend mainly on the choice of which species to promote and the potential of the breeding sites to which it provides access. Finally, it should be noted that this type of facility requires a certain amount of monitoring to ensure that debris do not impede the fishes' ascending or descending of the river.

6. Building of Weirs

In the la Grande project, the water level of two rivers (the Eastmain and Opinaca), exhibiting high faunal potential and playing a very important role in the use of the resources by the Cree, dropped significantly. Four weirs and a complementary structure were built to restore the water level on part of these rivers to their pre-cutoff average.

The objectives of the weirs were many. A combination of circumstances, in fact, led to the building of these works. This type of measure was included in the James Bay and Northern Quebec Agreement to permit restoral of the bodies of water that were exposed. Local users who had their access to the territory greatly reduced when the bodies of water were exposed also had to be taken into account. Other biological, experimental and esthetic considerations affected the choice to build the weirs.

The erection of weirs can mitigate impacts on several components of the environment. Consequently, it is important to take into account all aspects before implementing such a measure. The impacts must, first of all, be assessed correctly and then the physical, ecological and technical constraints considered.

Technically speaking, concrete or rockfill seems to be the best material for constructing weirs. The weir should be stationary. Variable level structures are too complex and difficult to maintain. The weir preferably should contain a pass for low-water flow and for the water to run over the weir during periods of runoff. Ice action must also be anticipated. Designs like that of weir No. 9 (variable level) therefore do not seem to be recommended, due to the operational work required and the difficulty to access them due to their remote location, among other things.

7. Creating of Ponds and Other Waterfowl-Related Measures

In the La Grande project, most waterfowl-related work have been aimed at attempting to improve harvesting conditions.

Three important aspects of the waterfowl-related measures have been studied: 1) improvement of the quality of the harvesting conditions; 2) development possibilities to promote this resource in northern environments; 3) development possibilities in southern environments.

The purpose of creating waterfowl hunting ponds is to promote harvesting of this resource. However, it is important to be prudent and to also consider the situation from the resource's point of view. Also, the development criteria (technology, choice of sites, etc.) and cumulative effects of creating a large number of ponds must be carefully determined. Otherwise, this measure could have negative repercussions on the resource and its habitat.

The most realistic way of mitigating the impacts on the waterfowl in northern regions seems to be to exclude sectors with exceptional qualities in the affected areas (protective dikes). The creating of ponds further south can be considered but only after all possibilities in the north have been exhausted and the North American management context taken into account.

8. Intensive Trapping and Relocation of Beaver

Beaver were intensively trapped in the La Grande project reservoirs following a request by the Cree who wanted to recover this resource before these areas were flooded.

From a biological point of view, the intensive trapping does not seem to have been a relevant measure, since beaver move about and relocate themselves. These conclusions can also be applied to the relocation of beaver unless required otherwise by the Native people.

9. La Grande Project Remedial Works Corporation (SOTRAC)

Under the James Bay and Northern Quebec Agreement (JBNQA), the purpose of SOTRAC was to mitigate the repercussions of the projects on the ways and places the Cree hunt, fish and trap. In practice, SOTRAC mainly provided compensation. In fact, the remedial measures were not aimed at obtaining minor residual impacts after the impacts were identified in the field. SOTRAC's role was rather to meet the expectations and needs of the users and to interface with the Native community.

SOTRAC was involved in over forty different main projects in various ways in the communities affected by the La Grande project. These main programs were divided into three areas, namely the environment, community projects and assisting trappers.

SOTRAC was also particularly useful in helping start up projects and businesses. It also made it possible to maintain ongoing dialogue with the Cree allowing them to express their ideas and to develop projects that seemed to them the most promising.

However some experiments were possibly disappointing. SOTRAC's projects were carried out at the instigation of the Cree for the Cree and with the involvement of the people in the area in their organization. As a result of the La Grande (1986) Agreement, SOTRAC's records and assets were transferred to the Eeyou Corporation of James Bay.

2.6 Conclusion of the lessons of the mitigative measures program

One of the principal lessons learned is not to restrict the measures to the affected sites alone. Implementation of measures in unaffected areas which have the best potential and which are also accessible should be advocated. The construction of simple structures that require the least amount of maintenance and monitoring possible should also be promoted. In addition, implementation of the mitigative measures must be flexible and incorporate the temporal aspect of the impacts and works. This implies that the measures must be carried out early enough so that they are functional when the impacts occur. Finally, dialogue with the users is a key element in the development of a program of mitigative measures.

3 LESSONS FROM THE ENVIRON-MENTAL MONITORING PROGRAM FOR FUTURE PROJECTS

Hydro-Québec have now been gathering data on the La Grande project territory for seventeen years. This makes it one of the most studied areas in Canada. The studies were carried out by government bodies, Hydro-Québec and its subsidiaries and a number of universities. The studies are open to the public and available.

The environmental monitoring network is part of one of the most important environmental monitoring programs in the world in the terms of duration, size of the area studied, number of stations, measurement parameters and variables observed. This program is being carried out by Hydro-Québec which is under no obligation to do so. The construction of a number of additional structures in Phase 2 of the La Grande project brings this program until the end of the century.

For us, environmental monitoring is a research activity, the duration of which will exceed the period of construction of the infrastructures. Long-term environmental monitoring is therefore required but should be optimized over time.

The monitoring and impact studies must be concrete, pragmatic and especially place more emphasis on the needs expressed by the Native users of the land. In addition, the environmental specialists must know more about the sources of impact when a hydroelectric facility is built.

The environmental monitoring and impact studies done for our other projects must take into account the knowledge acquired during the development of Phase 1 of the La Grande project. For example, when the GB 1 reservoir of the Grande Baleine complex, located 150 km further north, is flooded around the year 2000, we will already have over twenty years of data from the environmental monitoring of the La Grande complex, a similar environment with only a few exceptions.

No set recipe exists for studies related to a hydroelectric development project in Northern Quebec. However, the geology, geomorphology, vegetation, climate and animal species vary little. The areas which will require special attention are areas that were covered by seas or glacial lakes and where clay is found. Certain estuarine and coastal environments will require more detailed studies.

As regards the evolution of the reservoirs, the results of the environmental monitoring show that, with the exception of mercury, there have been no surprises. The same evolution pattern as for similar reservoirs elsewhere in the world applies. Changes in the quality of the water occur in the first few years (three to five years) after flooding. There is then a return to conditions similar to the surrounding natural environment. The changes in the water quality have always been adequate for the survival of the aquatic organisms, including the fish. The banks of the reservoirs, after the initial erosion of the fine surface materials, stabilize very rapidly.

The choice of the laboratory to perform the testing is very important. Very strict criteria must be applied to the selection of a laboratory and the various quality control programs. Any change in the laboratory and/or testing methods must be rigorously validated and verified. One laboratory is preferable to many as this makes it possible to better ensure the quality of the results and thus make better spatial-temporal comparisons of the results.

The estuaries and the reduced or increased flow rivers, even those subjected to major hydrological changes, have exhibited no major biological changes. In reality, the biological productivity and diversity of the reduced flow rivers increase as a result of a reduction of certain ecological constraints. The major change in the freshwater plume along the eastern coast of James Bay in the winter has not led, ten years later, to any significant changes in the coastal habitats, such as the banks of marine eelgrass and, consequently, no changes in the resources that depend on them.

A major disproportion has existed between the abundant data and knowledge available on the natural environment and that on the human environment. It was obviously easier to study the animals and plants. Moreover, twenty years ago, the definition of environ- ment was often limited only to the natural environment.

It seems important to determine the appropriateness of the results of the programs of studies and environmental monitoring and the compensation paid to third parties. Such a determination requires the carrying out of adequate sociocultural and economic studies in order to leave as little as possible to conjecture.

Moreover, it is very difficult to separate the repercussions of the development of the La Grande complex from those related to the James Bay and Northern Quebec Agreement and the inexorable, although probably slower, evolution of the Native people. Although it is theoretically possible to study these changes by comparing the evolution of the societies that are affected or unaffected by one or more of these measures, the sociopolitical context of the Native claims increasingly present in Canada is a significant constraint.

The opening up of the territory through the road system (1,500 km of new roads) was a major repercussion of the La Grande project. This completely changed the movements of the Cree and therefore their habits.

The creation of the reservoirs and the presence of the east-west road provide the caribou with an ideal winter corridor. The caribou population has expanded greatly forcing them to move about extensively. The situation is independent of the development of the La Grande Rivière. However, the presence of caribou west of the territory is an additional source of food for the Native people.

Mercury remains the most significant residual impact, although we feel it can be managed by working closely with the people concerned.

We are currently making up an environmental balance sheet of Phase 1 of the La Grande project. Sectorial summaries are already completed. A comprehensive presentation of results is currently being prepared.

REFERENCE

The lessons of the La Grande Complex (Phase I), Hydro-Québec. Proceedings of a conference in the general program of the 59th Annual Meeting of Association canadienne-française pour l'avancement des sciences, Sherbrooke, Québec, Canada, May 1991.

Hydropower'92, Broch & Lysne (eds) © 1992 Balkema, Rotterdam. ISBN 90 5410 054 0

Physical consequences of the NBR hydropower project on the Rupert Bay area: Description of the various interrelated studies

P. Desroches, N. V. Tran, T. J. Nzakimuena & T. T. Quach
Hydro-Québec, Montréal, Canada

ABSTRACT: An important part of the feasibility studies of the Nottaway-Broadback-Rupert hydropower system is devoted to the assessment of the environmental impact in the Rupert Bay area. A vast program of data collection is carried out in the field as to tides, currents, waves, salinity, bathymetry, marine bed characteristics, sediment flow data, river water temperatures and discharges, climatic parameters, ice cover behaviour, etc. These data will serve as input in many studies including a numerical model to simulate hydrodynamic, sedimentologic, salinity and winter flow conditions of a vast fluvial and marine area extending to the James Bay. Moreover, the Broadback river estuary immediately downstream of the first site of the proposed power generating stations is replicated in a reduced scale physical model to assess erosion and ice problems. Subsequent analysis of both observed and simulated data will thus assure an adequate coverage of the physical impact of the NBR project on this area.

1 INTRODUCTION

This paper mainly describes the physical environmental studies that are carried out presently on the Rupert Bay, one of the most important natural area of the Nottaway-Broadback-Rupert (NBR) hydropower system. It starts out with a general description of the NBR complex and the proposed development, then focuses on the Rupert Bay with a description of its natural environment and program of field studies and concludes on the subsequent analysis to be carried out in the future.

2 THE NBR PROJECT

The NBR complex is located on the north westerly part of the province of Québec, the largest of ten Canadian provinces. Its hydrologic basin is part of the James Bay territory, a vast region targeted for development by the government of Québec.

Geographically, the NBR hydrographic basin covers some 130 000 km^2 between the 49th and 52nd northern parallels, immediately south of the La Grande complex. The climate of the territory is of a northern type characterized by long winters (November through March) with temperatures as low as -40°C followed by a rather cool spring time and a short summer (July-August).

This hydropower system has been the subject of numerous studies for the purpose of electrical generation since 1964. Various possible development schemes have already been evaluated. Among these is the development concept proposed in 1982 by the Société d'Energie de la Baie James, recently updated with minor changes and an innovative four-stage development scheme in order to respond to electrical demand in a flexible and cost effective manner.

Because of the fluctuating economy, Hydro-Québec wishes to remain as flexible as possible with regard to its decision to build the complex. Therefore, the project is planned to spread over four successive stages as follows:

Phase 1: 2 505 MW
Phase 2: 3 042 MW
Phase 3: 1 485 MW
Phase 4: 1 721 MW

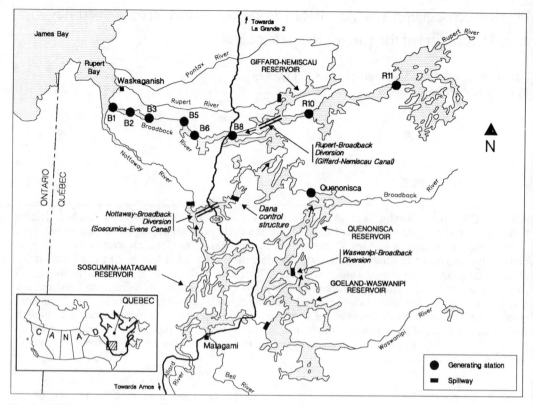

Fig. 1   NBR complex

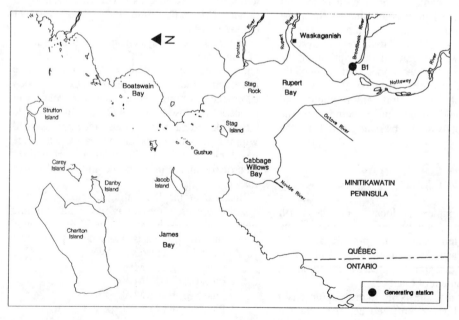

Fig. 2   Rupert Bay

The duration between these stages could vary, therefore giving the utility the latitude it needs to adapt the development of the complex to the changes in energy demand. The first generating stations in the complex could be commissioned around the year 2012, in a scenario still to be determined by the long term development plan.

The concept involves diverting the Nottaway and Rupert rivers into the Broadback river and the expansion of most of the generating stations on the lower section of the Broadback river. In total, six reservoirs would be created and nine power stations will generate an annual output of 46 terawatthours (TWh) and peak power of over 8 700 megawatts (MW).

The six reservoirs of the complex would have a total useful reserve of 35 billion m^3 in the three catchment basins. The bodies of water thus created would cover a total surface of 6 500 km^2, or about 5% of the surface of the natural basins of the three rivers, namely 3% of the land environment and 2% of the aquatic environment.

Development of the Nottaway catchment basin would include two steps: creation of the Goeland-Waswanipi reservoir which would make it possible to divert the upper Nottaway to the Broadback catchment basin, by way of the Quenonisca reservoir, and creation of the Soscumica-Matagami reservoir which would make it possible to divert the lower Nottaway to the same basin by way of the Soscumica-Evans canal up by Dana lake. Development of the Broadback basin would include creation of the Quenonisca reservoir and part of the Giffard-Nemiscau reservoir. It would also include the reaches of generating stations Broadback 1 to Broadback 8. Development of the Rupert basin entails creation of reservoirs Rupert 11 and Rupert 10 and a portion of the Giffard-Nemiscau reservoir and the diversion into the Broadback by way of the Nemiscau-Giffard canal.

On the other hand, environmental studies are carried out to determine the repercussions of the project on the human, biological and physical environments. As regards to the human environment, the studies aim to evaluate the exploitation of the wildlife resources, the forestry and mining operations, the archeological backgrounds, the use of the territory by the native people and other occupants together with the exploitable resources in the territory. As regards to the natural environment, the evaluation focuses, among other things, on the impacts on the climate, river banks, runoff, sedimentation, water quality, vegetation, land fauna, birds and fish. Increase of mercury in wildlife organisms in the reservoirs is also studied.

A great deal of the preceding efforts is devoted to the Rupert Bay area where the discharges of all the power stations find their way to the sea. Studies carried out over this particular area are described in the following sections.

3 THE RUPERT BAY

3.1 The natural environment

The Rupert Bay constitutes a common and vast estuary for the four most important rivers of the region: the Nottaway, the Broadback, the Rupert and the Pontax rivers. It stretches over 50 km long and widens from 3 km at the rivers' confluence to 20 km as the waters reach the James Bay to the north-west, thus covering a total area of 850 km^2. Its depth varies from a mere 2 m to about 6 m towards the widest area.

Geologically, the rivers' bottom in this region is made of relatively recent and soft deposits from the last glaciation periods. In the immediate area of the bay, these deposits are composed mainly of compacted marine clay of the former Sea of Tyrrell. To this day, signs of the presence of the sea are left visible on the land to an altitude of 290 m. The entire Hudson and James Bays were effectively covered by this sea at the time where glaciers retreated about 8000 years ago. The Rupert Bay's bottom and shores are thus mainly constituted of this layer of clay which may reach up to 8 m thick in some places. Other types of deposits such as till, sand and gravel from former sea beaches as well as former glacier blocks and more recent layers of clay, silt and sand make up the complete geological picture of the area. The present shape of the Rupert Bay is thus the result of the combination of the rivers' confluence and an equilibrium between isostatic lift estimated to an average rate of 3 mm yearly, erosion action by the currents, waves and ice, and probably a slight increase in the average sea level.

Between the bay's mouth and the estuary of the Nottaway river, three zones can be distinguished hydrodynamically: the reach between the

Nottaway and Stag Rock near the Pontax river is a fresh water area due to the rivers' discharge; the zone farther downstream from Stag Rock is a mixed water area which extension varies with currents, tide, wind and ice presence; the third area constituted of shores' zones around the bay which during the low tide leaves largely uncovered beaches such as Cabbage Willows and Boatswain Bays. These constitute particularly rich ecosystems for various waterfowl.

Due to the rivers' discharge, a gradually narrowing shore-line, the relatively shallow depth and gradual positive sloping of the bay bottom, tidal waves can dampen from an amplitude of 3 m at high high tide in open sea to a weaker one and phase themselves out as they move up the Rupert Bay to the rivers' first rapids. Contrarily to the mixing zone where a current's inversion is observed, no such phenomenon is noticed in fresh water area between Stag Rock and the three rivers' estuaries. In the winter, the Rupert Bay as well as the James Bay are covered by ice. Various physical properties can be observed for the ice cover which is made of salted water at the bay's mouth and of fresh water farther upstream. Its thickness is 1 m in average, but can reach up to 2 m in some places. Some areas are covered by a rather smooth layer whereas others such as the shores are made up of hummocked ice, which is mainly formed of broken plates of ice created by repetitive tidal action. Although the process of ice formation and breakup is rather complex due to various combinations of weather conditions, flow and tide action, a seemingly consistency can be observed in its chronology, namely the fresh water ice always forms and breaks away ahead of the salt water one.

On the biological side, the Rupert Bay contains a very diverse animal population. The phytoplankton represents the primary production and its presence follows closely the salinity distribution and the type of exchanges between the fresh, the mixing and the salt water zones: the abundance varies from rich to poor when moving from the fresh water to the salt water areas. The zooplankton, on the other hand, although rich as a secondary production source in the fresh water zone, is much less diversified in that zone. Its diversity increases gradually from the mixed to the salt water regions. Also, the fish population is very varied and spatially distributed according to water turbidity and salinity; thus the mixing zone which is the most turbid is also the less divers in fish species. Many marine mammals such as belugas and various types of seals have been noted in the bay's mouth. The emerging land strips due to isostatic lift around the bay are large briny tide-zone marshes of which rich coastal grassy areas are sought after by some 70 species of migrating birds such as Canada geese, snow geese, black ducks and some other twenty waterfowl types.

As to human aspects, the Waskaganish community is the only settlement in a 100-km radius area where close to 1 500 native people have their permanent homes. This native community makes their living partly out of hunting and fishing. In the winter season, transportation by snowmobiles is often made on the ice cover. Fishing catch includes species such as salmonids (brook trouts, ciscoes, yellow walleyes) and northern pikes. Hunting consists of two seasons a year for waterfowl.

3.2 Primary change in physical factors and the potential effects

The main change in physical factors is the modification of the hydrologic regime. As such the NBR project will modify the annual and spatial discharge picture of the Nottaway, Broadback and Rupert rivers which make up 90% of total fresh water flow in the area.

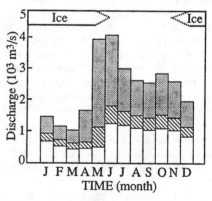

Fig. 3   Monthly natural discharges

RIVERS: Nottaway
Broadback
Rupert

284

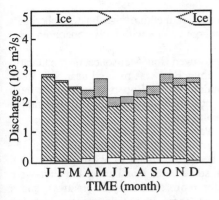

Fig. 4   Monthly discharges after
          development

The discharge patterns before and after the development are depicted in figures 3 and 4. These changes will induce alterations of rivers' ice regime as well as of fluvial sediment transportation to the bay. The sediment contribution by these three rivers is estimated at one million tons per year in present conditions. As a consequece of the change in fluvial regime, varied effects may be expected on the bay's physical environment such as hydrodynamics (salinity, flow velocity and sediment distribution), ice cover and sediment dynamics on the riparian zones.

It is with the objective of defining the relative importance of these repercussions on every environment aspects (physical, bilogical as well as human) that Hydro-Québec had initiated various biophysical studies these last fifteen years. Whenever possible, some of the results have been directly useful in design studies of the B 1 power station located on the first rapids of the Broadback river.

3.3 *Studies of physical characteristics*

The scientific objectives of these studies are three-fold:

1. to define the natural environment dynamics by analyzing the factors controlling the environment evolution;

2. to predict future conditions which goal is achieved by way of a mathematical model and a reduced scale model;

3. to find mitigating measures if needed.

The first environmental studies were initiated before 1981. Oceanographic observations were done at the bay site. One reduced scale model of the Rupert Bay and one of the Rupert river estuary, as well as a finite element and a finite difference mathematical models were used for the interpretation of the collected data.

That programme of studies was reactivated in 1989 when the new concept of four-stage development was put under study. Those evaluations will resume at the end of 1992. They include: an important oceanographic campaign in the Rupert Bay and the adjacent surrounding in the James Bay, the analysis and interpretation of the informations, the development of a finite difference numerical model of that water body, the operation of a reduced scale model of the Broadback river estuary, the analysis of the present and future conditions of the ice regime in the bay area and in the Broadback river and the satellite imagery photointerpretation in order to visualize the global circulation in the bay with respect to water temperature and the suspended sediments.

The following section describes the field observations that were planned in order to collect the input data for all mentioned physical environment studies as well as many other ecological studies.

4 OCEANOGRAPHIC OBSERVATIONS

Field observations were necessary to complete the data defining the physical characteristics of Rupert Bay in the following fields: hydrodynamics (currents, tides, waves, weather, water temperatures and salinities, river discharges); sedimentology (marine and alluvial sources, deposition and erosion dynamics); ice regime (formation process, accumulation zones, effects on the currents and on the erosion potential, thawing process); marine bed bathymetry and geology. Particular attention was given to the most productive bays for waterfowls as well as certain beaches, for which the present and future sedimentary regimes can only be determined by analytical studies and correlation of the data gathered on the above zones.

4.1 Zone and periods of observation

The main observation zone covers the Rupert Bay water body, including the mouth of the Nottaway, Broadback, Rupert and Pontax rivers up to their first rapids, the Cabbage Willows and Boatswain Bays, as well as an external zone or pass in James Bay, located between Charlton and Strutton islands and the entrance of the Rupert Bay. Some water temperature and flow recording stations were in operation on the Nottaway, Broadback and Rupert rivers, upstream from their first rapids.

In the late summer of 1990, the observation period was of the order of forty five days, depending of the parts of the bay. From the beginning of December 1990 to first days of January 1991, there was a daily flight by helicopter over the bay in order to draw descriptive maps of the ice cover formation. There was a winter campaign of one month at the beginning of 1991, where lines of currentmeters were moored under the ice cover to measure direction and velocity of the currents and salinity and temperature of water.

In 1991, the cruise take place during two and a half months, beginning in mid June; the moorings were recording from late June to mid October.

Some of the land based tide and climatological recording stations were operational from the beginning of the first campaign to March 1992, with some periods of interruption.

4.2 Logistics

The Rupert Bay and the of Waskaganish native village are accessible only by water-route during a period extending from the melting of ice cover in James Bay in mid-june to the beginning of the freezing period in mid-november. During the winter season, ending with the melting of the snow, access to the community is done via a "winter road". So most of the transportation is done by airplanes or helicopters.

Considering the dimensions of the zone to survey, in the more intensive period of the campaign, a team of ten to fifteen technicians, specialists and crew members were working in the bay, using up to two helicopters, a twenty meters "lab" vessel named Septentrion provided with launching facilities and lodging for up to six persons. A medium size boat, rubberboats and two large canoes with native operators were also used.

A barge, towed from Moosonee, on the Ontario side of James Bay, was moored near Stag Rock in the middle of the Rupert Bay, to stock fresh water, fuel and part of the oceanographic instrumentation.

There was a network of radio communication between helicopters, the Septentrion vessel, the smaller boats, the village of Waskaganish (from which at some times members of the team were distant for more than sixty kilometres), and a inland base camp (km 257) one hundred and fifty kilometres away from that village.

The localizations of the moorings, of the sediment samples and of the bathymetric and geophysic soundings were done partly by radiometry with a Tresponder system using a set of three antennas and by satellites using Trimble Pathfinder global positioning devices (GPS).

4.3 Hydrodynamics observations

To study the hydrodynamic conditions in this zone, Hydro-Québec moored between ten to sixteen lines equipped with two currentmeters (usually one Aanderra RCM-7 currentmeter in the upper part of the water column and one InterOcean S4 currentmeter in the lower layer).

Three moorings were made on the boundary of the south west entrance of the pass enclosed between the Charlton island and the opening of the Rupert Bay, three other on the boundary of the north east entrance of the same pass, characterised by relatively high currents, and one line in each of the three main channels giving way to the entrance of the Rupert Bay itself. Five to six other lines, depending of the period, were moored in the interior of the Rupert Bay, some of which were moved from one site to another every twenty eight or thirty days in order to increase the spatial density of the points of information in an economical way. The direction and velocity of the currents as well as the temperature and salinity of water were recorded continuously by these currentmeters.

Some of the S4 currentmeters in the shallow zones were also set to record the waves amplitude by pressure measurements. An additional Datawell accelerometric wave meter buoy was launched in a very central position, approximately half way from Charlton island and the entrance of the Rupert Bay.

During the data gathering periods, manual and punctual measurements were done at many points on verticals in various parts of the bay, during complete cycles of different tide conditions, in a way to describe the extension of the fresh water discharge in the salt water body or the saline penetration inside the bay.

4.4 Sedimentology

Each of the two currentmeters of some of the mooring lines were coupled with a D&A Instrument OBS turbidity gauge, linked to a common Cambell data acquisition system concealed in a cylinder. This arrangement permits a continuous following of the sediment concentration variations in relation to tide and meteorological events.

Once again, in order to increase the spatial distribution of points at which this parameter was measured, manual sampling of water were done at different depths on many verticals and during different tide cycles. These samples were analysed with Coulter Counter analyser, model TA2, to determine concentration and fraction of the suspended sediments, and with the laser model LS100, to determine the granulometric distribution by percentage. These data were also used to calibrate the signal obtain from the OBS gauges.

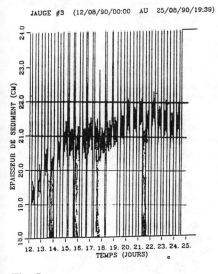

JAUGE #3   (12/08/90/00:00   AU   25/08/90/19:39)

Fig. 5

Sedimentologic JNT gauges, developed by the Institut National de la Recherche Scientifique-Océanologie, of Rimouski, were installed on the intertidal strands, each one coupled to a S4 currentmeters. These gauges are made up of a CS137 source is sheltered underneath the surface of the sediment and of a probe fixed to a pole. This probe is used to read fluctuations of gamma ray attenuation caused by the variations of sediment thickness between the source and the probe. As the net rate of deposit or erosion in the zones of bays like Cabbage Willows and Boatswain is very small, the sensitivity and the continuity of recording of these devices (see figure 5) made it possible to understand the influence of individual events, like storms or high high tides on the sedimentary cycles.

During the cruise period, in conjunction with the bathymetric and geophysic soundings the field team proceeded to sampling of bottom sediments, with a Shipeck grab and a vibro-corer, in different zones of the bay. The INRS-Océanologie proceeded to conventional analysis on these samples to determine the nature and the granulometry of the bottom sediments and used this information for the calibration and interpretation of the subbottom profiling and the side scan sonar records. Collected cores were submitted to Rx scanner so to determine the fine three-dimensional structure of their layering with their true, their compaction degree and their log. These information associated with the subsurface profilers and the side scan records will be interpreted to define the dynamics of the spatial and temporal natural rates of erosion and deposition of sediments in the bay.

Some quality analysis were also performed on sediment samples and other samples collected at the mouth of the most important rivers flowing in the Rupert Bay as well in the James Bay, in order to know the eventual presence of natural level of contamination. Efforts are also made to identify tracers that would help to determine the relative contribution of the marine waters to the sediment budget of the bay and more specifically to dedicated zones like Cabbage Willows bay.

4.5 Bathymetric and geological surveys

More than 1 500 kilometres of sounding were performed, with variable equidistance between

287

the lines depending of the part of the bay, the irregularity of the geometry and the variation in nature and characteristics of the sediment layers.

In conjunction with conventional bathymetric echo sounding with a Raytheon 719, a Mesotec side scan sonar, a Raytheon RTT1000 boomer and an Endo-Western subsurface profiler were used to draw the map of the bottom morphology, the superficial distribution of the sediments, as well as the layers' thicknesses and the bed-rock position. In shallow water, the newly design SEISTEC IKB subsurface profiler was successfully used and gave a very high resolution.

For inter-tidal strands and in the shallow bays, with a very mild slope, in order to better coverage with sufficient precision and at an acceptable cost, an helicopter was used to follow the position of the water line on the shore at different stage of the tide, recording the horizontal coordinates of those lines with a GPS device. The vertical coordinates were obtained with a tide gage.

4.6 Recording stations located on the shore

In addition to all the moorings, recording stations located on islands and on the shore line were operated from June 1990 to March 1992, with periods of interruption gages A tide gauge station was installed on both side of the south west entrance of the pass between Charlton island and the mouth of the Rupert Bay; two other ones on both sides of the north entrance, were the depth is lowering and the amplitude of the tide much smaller than that from the south entrance. Another station was located at the mouth of the bay, on Gushue island, an other on Stag Rock in the upper part of the bay and in 1991, an other station on the south bank of Nottaway river, at the confluence of Broadback river estuary.

Almost from the beginning of the 1990 campaign up to March 1992, there was a ometeorological station on Charlton island, recording wind speed and direction, air humidity and temperature, atmospheric pressure and radiation. Similar station for shorter periods of time were installed on Strutton island and on Stag Rock.

The fresh water flow from the river was monitored at two hydrologic stations, one each on the Nottaway and the Broadback rivers. The temperature of that fresh water entering the bay was simultaneously recorded at a station upstream of the first rapids of the Nottaway, Broadback and Rupert rivers.

5 CONCLUSIONS

Hydro-Québec needed to understand the natural dynamics of the Rupert Bay environment in order to foresee the potential effects of it's project and, if necessary, to implement mitigation measures.

A review of studies of the seventies was a precious guide for the planning of the program of these complementary studies.

This program includes numerical and reduced scale modeling and field observations that were needed to understand natural mechanisms as well to provide calibration data for the models.

The bay under study has rather wide dimensions and the factors having an effect on the natural quality of this water body are very diversified. So an optimization had to be done to determine density of the sounding lines, position and number of moorings and of shore-line recording stations, as well as for manual and punctual samplings which has to be done during different tide, meteorological and fluvial water flow conditions.

The effort that was deployed in the field had been successful and generated a relatively great amount of data. These have now to be analyzed, validated and interpreted in conjunction with the model studies before further action.

REFERENCES

LeVan, D. 1991. Complexe NBR. Avant-projet phase I. Rapport d'étape. Choix préliminaire de la variante à privilégier. Rapport interne. Hydro-Québec.

Vice-présidence Environnement. 1991. Complexe Nottaway-Broadback-Rupert. Programme d'études biophysiques de la baie de Rupert. Enjeu: La modification éventuelle du milieu naturel de la baie de Rupert. Rapport interne. Hydro-Québec

Long, B.F. and Hudier, E. 1987. Nuclear gauges in sedimentary dynamic.One example in the wreaking zone measurement. *IAEA Editor: Isotope Techniques in Water Resource development:* 647-661. Vienna.

Hydropower'92, Broch & Lysne (eds) © 1992 Balkema, Rotterdam. ISBN 90 5410 054 0

Environmental considerations and remedial measures in Norwegian hydropower schemes

Jon Arne Eie & John E. Brittain
Environmental Section, Norwegian Water Resources and Energy Administration (NVE), Oslo, Norway

ABSTRACT: There have been increasing efforts in Norway to incorporate environmental considerations into the planning process and to develop and test new ways of reducing the effects of future hydropower development, while at the same time rehabilitating the environment in and around existing regulation schemes. The aim is to develop measures which maintain biological diversity or encourage changes which reduce the deleterious effects of hydropower on the environment. Landscape and biological considerations are an important aspect of such remedial measures. The most useful measures that have been developed and tested are weirs, substrate improvement and landscaping.

1 INTRODUCTION

In a global contex Norway is in a unique position, 99.8 % of its electricity is produced by hydropower. This means that a high percentage of Norwegian watercourses are affected to various degrees by regulation projects. In an attempt to reduce the adverse effects of hydropower developments increased efforts have been put into environmental and landscape rehabilitation over the last 20 years.

2 PLANNING, ASSESSMENT AND LICENCING OF NEW HYDROPOWER SCHEMES

The Norwegian Water Resources and Energy Administration is the state body responsible for the overall evaluation of hydropower licensing applications before making a recommendation to the government. Hydropower developments are assessed according to the Watercourse Act as well as certain other related legal acts.

The actual assessment and evaluation process can be divided into the following phases: planning, application processing, overall evaluation and decision-making, construction and subsequent site clearance and finally rehabilitation and remedial measures.

2.1 The planning phase

Major environmental encroachments, such as hydropower development require a licence granted by the Norwegian authorities. During the planning phase an active cooperation between the developer and the affected interest groups is established, ensuring among other things, that requests for the necessary environmental impact reports can be made. A provisional notice of the proposed development is given to the interested parties and the planning process can then start. The interested parties may put forward their views at this stage. With regard to environmental considerations it is important that landscape and ecological aspects are drawn into the planning such that the environmental consequences are reduced to a minimum, at the same time as expensive post-regulation remedial measures can be avoided.

2.2 Processing of the project application

Once the licensing application is complete, it is sent by the developer to the Norwegian Water Resources and Energy Administration, where it is checked to ensure that all the required information, documents, reports and so on are included. Considerable effort and expense is used to obtain good impact analyses of the consequences for the environment. These include recreation, fish and game, conservation and cultural heritage, in addition to the more technical and economic aspects. These analyses are an important basis for the decision makers. When complete, the application is circulated to government agencies as well as municipal and county administrations and voluntary organizations for comments.

During this phase, environmental questions are often at the fore, and here the Environmental Section plays a central role in the assesment of possible consequences for the environment. In this process the developer has already to a certain extent taken account of the environment on the basis of economic considerations. In most cases the developer removes the more controversial parts of the plan, for which a licence is unlikely to be granted.

2.3 Overall evaluation and the decision phase

When the application has been thoroughly discussed and supplemented with the evaluations of the various interested organizations, NVE's final evaluation commences. This entails considerable emphasis on environmental considerations and obtaining a balance between the advantages and disadvantages in relation to the legal constraints. When complete, NVE's recommendation is sent to the Ministry of Petroleum and Energy, where an overall evaluation is made. For applications involving a major regulation, the final decision is made by the Norwegian parliament (Stortinget).

All this is a time-consuming and lengthy process. One must, however, take into account the fact that large hydropower projects have been the cause of some of the greatest controversies in post-war Norwegian history, where developers and environmentalists have been pitted against one other. A long and detailed process guarantees that all interests are thoroughly considered before the final decision is reached.
The outcome can vary from full approval to denial of a licence because the environmental impact would be too great in relation to the amount of power gained.

2.4 Construction and site clearing

All licences are granted on specific conditions and there are generally many particular remedial measures which the developer must carry out. These may for example involve landscaping, scientific studies to determine the effects of regulation, fish stocking and measures to increase game.

Landscaping is largely carried out to make sure that the power plants are built in such a way that they fit into the total landscape as well as possible and that unnecessary scars on the landscape are avoided. There will also be need for measures which maintain natural diversity and create a dynamic cultural landscape.

Modern power plants are now usually located underground and are only visable by their entrance gates and switchyards for transmission lines. This building technique has many advantages, both from the point of view of security and the landscape. However, large amounts of rock are a troublesome by-product.

Efforts are made to incorporate this waste rock into other parts of the same power project or to use it in other ways. Great care is taken to find a suitable location, to adapt any rock tips to the surrounding terrain, to cover the surface with a thin layer of soil and if necessary apply fertilizers and seeds or even plant trees over the tip. Left alone, any pile of rock will be covered by vegetation sooner or later. However, on a human time-scale, this proceeds painfully slowly, and much can be gained by the use of fertilizers. In our experience it its very important to establish a plant cover as soon as possible, thus creating a protetctive surface which prevents uncontrolled erosion by surface runoff, as well as providing a foothold for the plant species native to the area.

Large dams are constructions which will endure for very many years. Although they are in themselves large and obvious structures which cannot be hidden in the surrounding terrain, one has chosen to let them be seen as they are, a man-made structure. Efforts, are however, made to blend them into the landscape by giving the dam a curved form "around" the reservoir. Nevertheless, economic considerations have the final word. Emphasis is given to keeping the area below the dam as much as possible free from man-made enchroachments or ensuring that the terrain is landscaped such that minor enchroachments disturb the overall picture as little as possible. In order to build dams it is usually necessary to quarry large amounts of rock, stones and filling materials.

Attempts are made to site quarries and sites for extraction of moraine material in areas which will subsequently be covered by reservoirs and thus not visable after regulation. When this is not possible extensive efforts are made to develop the plans such that the quarry does not appear to be an obvious scar in the landscape. The sharp contours of the quarry can be smoothed out and the waste material laid in the bottom so that the fissures do not appear as sharp edges, but form a softer profile which fits in with the more natural lines. There are also examples where parts of the quarry have been filled with water, giving a water surface within the total landscape.

3 REMEDIAL MEASURES

3.1 Flow management

The greatest environmental effect of hydropower development is the reduction in flow or change in the annual discharge regime. This can produce severe damage to fisheries and to aquatic life in general. In modern developments predetermined minimum flows contribute to reduce the severe effects on the aquatic environment and the landscape.

In order to provide the best possible conditions for fish and aquatic life in general emphasis is given to creating a flow regime which is both flexible and as near as possible to the original regime. This means that under certain conditions artificial floods are created which promote or make possible salmon runs. Such floods can also reduce undesirable algal and moss growth. Care must also be taken not to reduce flows rapidly in order to avoid fish stranding.

Increasing flows above such minimum flows is usually an expensive option and efforts have therefore been made to find and remedial measures which can contribute towards providing suitable conditions for fish and other aquatic organisms, at the same time as landscape and power interests are satisfied to a reasonable extent.

3.2 Weirs

Weirs have been constructed as a remedial measure for more than 20 years and more than 1,000 have been built. Their main purpose is to create a suitable habitat for fish and maintain a certain water level in the river. They can be built of wood, concrete or moraine material. In low discharge rivers they are usually built as a single straight weir, but in high discharge rivers or salmon rivers "Syvde" weirs (V-form) are built. This concentrates the flow and creates a pool below the weir, providing suitable conditions for fish to negotiate the weir. There is a sorting of bottom material below the weir which also provides opportunities for spawning. The pools that are formed, in addition to the weir basin itself, provide suitable overwintering habitats for fish when flows are low.

In order to follow the long-term effects of weirs and to develop and test other remedial measures, a major research programme, the Biotope Adjustment Programme, is in progress. The biological effects of weirs have been monitored for over 15 years. Fish, benthos and bird life have all been studied. Results so far show that weirs have a positive effect. Ducks and waders benefit from the increase in food supply, particularly in the weir basin.

Over a long period the benthos of the weirbasin undergoes a succession towards life forms adapted to slow flowing waters. Fish production also increases and can be used as a basis for reservoir stocking programmes.

3.3 Substrate improvement

In channelized rivers where sand and other fine material dominate the bottom, artificial areas of rocks and stones have been placed in the river. Such areas or "islands" provide improved conditions for fish, by creating a greater diversity in substrate, flow conditions and water depth, as well as providing cover. Large boulders have also been placed in rivers to provide suitable habitat, especially for larger fish.

In Norway, hydropower development often results in lower water temperatures during summer and slightly higher winter temperatures because of releases from mountain reservoirs. In order to improve growth conditions for salmon fry, experiments are being conducted with rearing channels in which additional nutrititional input in the form of grain is provided to increase the production of fish-food organisms.

3.4 Fish stocking

Within the scope of the hydropower licence there are possibilities to impose fish stocking programmes paid for by the developer in order to maintain fish stocks. However, with the restrictions imposed to prevent the spread of fish diseases and the mixing of genetically distinct populations, increasing emphasis is being given to physical measures aimed at improving natural recruitment; thus reducing the need for stocking.

3.5 Wildlife

In connection with hydropower development wetlands can be either drained or flooded, rendering them unavailable for waders and other wetland birds. Amg the remedial measures tested and evaluated are artifical islands constructed to recreate delta areas and the building of weirs in arms of reservoirs in order to maintain a suitable wetland habitat despite low

reservoir levels. This also provides
improved conditions for fish-food organisms.

The maintenance of corridors of woodland and other vegetation along watercourses is of major importance for
biological production in rivers and
streams, as well as being of strategic
importance for birds and game. Such
boundary zones, or ecotones, are areas
of high biological diversity and have
an important function in reducing ero-
sional runoff from agriculture and
industry, but they are often threatened by a wide range of human activities in addition to hydropower.

3.6 Revegetation in reservoirs

The regulation zone of hydropower
reservoirs is very susceptible to erosion and also appears unsightly when
the reservoir levels are lowered.
Experimental planting trials have
been carried out to determine the most
suitable species for this zone characterized by alternate desiccation and
flooding. In large new reservoirs,
planned to be filled over several
years, and which have still retained
considerable amounts of finer material, conditions are more favourable for
the establishment of vegetation. Fert-
ilization trials have also been conducted.

4 CONCLUSIONS

The increasing awareness of environmental issues has given rise to a lic-
encing procedure which is both exhaustive and time-consuming, but which
places emphasis on environmental needs
at the same time as consideration is
given to the different user groups. As
a result of many major research programmes several different types of
remedial measures are being developed
and evaluated in order to reduce the
effects of hydropower schemes on the
landscape and aquatic ecosystems. In
the first instance these measures include different types of weir, substrate improvement, rentention of water in
small bays of reservoirs and the construction of artificial islands.
Even when carried out afterwards, the
extra costs involved in such remedial
measures constitute a very small part
of the total development costs. If
such measures can be incorporated into
the planning phase costs can be reduced even further.

REFERENCES

Brittain, J.E. & Eie, J.A.
   (red.) 1991. Biotopjusteringspro
   grammet - status 1990. Rapport
NVE-Vassdragsavdelingen 12-1991 50 p
Brittain, J.E., Eie, J.A., Brabrand,
   Å., Saltveit, S.J. & Heggenes, J. In
   press. Habitat modification measures
   in Norwegian hydropower schemes.
   Regul. Rivers
Hillestad, K.O. 1989. Landskap i ut
   vikling. (Our Developing Landscape).
   Kraft og miljø 18. Norwegian Water
   Resources and Energy Admin. 117 pp.
Mellquist, P. 1986. Life in Regulated
   Streams. The Weir Project. Norwegian
   Water Resources and Energy Admin. 58
   pp.
Schjetne, S. 1990. Wassekraft und
   landschaftsplanung. Garten u. Land
   schaft 8/90: 19-21.

Hydropower'92, Broch & Lysne (eds) © 1992 Balkema, Rotterdam. ISBN 90 5410 054 0

Study of river bed lowering downstream of a dam using a mathematical model

P.Jehanno & T.Ulrich
Hydraulics and Structural Design Division, SOGREAH, Grenoble, France

ABSTRACT:

This paper deals with a study of river bed lowering downstream of a completed dam.
The dam is of the arch type (height: 155 m, crest length: 170 m, spillway discharge: 1500 m³/s), its objectives being power generation and irrigation.
Due to the fact that sediment is trapped in the reservoir, the river bed downstream of the dam is being lowered.
The riverine forest, which constitutes the major evergreen area of the plain located downstream of the dam, providing produce and fodder for the population and animals during the dry season, may be partly affected in case of excessive lowering of the river bed.
Bed lowering and the effects of remedial measures are studied out with the help of a mathematical model of long-term river bed evolution. This mathematical model takes into account the main characteristics of the sediment, the river and the hydraulic conditions.

1 INTRODUCTION

The aim of the dam is power generation and irrigation. The region where the dam is built is arid. A few kilometres downstream of the dam, which is located in narrow gorges, the river enters a plain, where the bed is relatively steep (slope 2.2 m/km in the upstream reach) considering the fine sandy granulometry of the river bed. Trapping of the larger particles carried by the river within the reservoir might lead to bed lowering downstream. This river depletion might affect the present conditions in the following ways:

.   Destruction of the riverine forest in the plain, which is at present well developed, by lowering of the water table. This riverine forest is of great importance: its constitutes the major evergreen area of the region, providing produce and fodder for the population and animals during the dry season.

.   Effect on the yield of existing intakes supplying irrigation water.

.   Effect on projected new irrigation intakes, intended for the future extension of the irrigated areas.

The purpose of the study was to assess the effects of the dam on the river and to propose remedial measures to keep this impact within acceptable limits.

A mathematical model was used as a basis for drawing conclusions.

2   DESCRIPTION OF THE RIVER

The river is studied over a reach of 200 km downstream of the dam. The river receives its main tributary some 15 km from the dam. The river and its tributary carry a significant discharge all year long. The 2-year return period peak flood of the river before construction of the dam is some 1000 m³/s, the peak of its tributary being close to 500 m³/s.

Downstream of the confluence, some intermittent streams are met, which generally flow for a few days each year.

It is assumed in this study that, after construction of the dam, the power house would be operating at 34 m³/s discharge during periods representing half of the time, and nil discharge the rest of the time.

The river bed, from the dam down to the confluence, is braided and meandering. Its width is 80 to 100 m for limited discharges (i.e. discharges exceeding some 15 m³/s). It can exceed 200 m for large floods. A small sandy cliff (1 m to 2 m) is generally observed at the limit of the riverine forest. The river bed migrates from one year to the other within the limits of the riverine forest. The general slope of the valley at present is close to 2.6 m/km. The real slope for the river is around 2.2 m/km, i.e. 1.2 times less, if meandering is taken into account.

In the last few kilometres upstream of the confluence, the tributary is narrow and meandering (35 to 50 m before the side cliffs are encountered). The general slope of the valley is close to 2 m per kilometre; the slope of the river bed proper is probably 1.7 times gentler. The river capacity before overbank flow takes place is some 500 m³/s. There is rather thick vegetation along the banks.

Downstream of the confluence, the river bed is braided and meandering, except in a few places where some natural controls apparently exist (due not to rock outcrops but presumably to bank cohesion or resistance). The overall width of the river increases downstream.

The average slope of the river is estimated at 1.5 m/km downstream of the confluence.

3   CRITERIA FOR FOREST PROTECTION

Degradation of the river bed in the upper section of the river, from the dam to the confluence, could cause both the river bottom and the alluvial aquifer to fall by several metres.

Thus, a major impact could be a severe alteration of the water supply to the alluvial forest.

Considering that the riverine forest along the river must be preserved by any means, it is necessary to keep this phenomenon of lowering within acceptable limits.

It is considered that the mitigative measures proposed allow a maximum bed lowering of 1.5 m. Indeed, this value is roughly the seasonal fluctuation of the aquifer, and thus it may be considered that the ecology of the species, and more particularly their root system, is adapted to this potential change.

4   THE MODEL

4.1 THE MATHEMATICAL MODEL

The mathematical model (CHAR2) developed by SOGREAH, solves a set of non-linear differential equations of river flow and sediment movement using Preissmann's semi-implicit finite difference scheme. The following assumptions are made:

(a) flow is one-dimensional in accordance with the De Saint-Venant hypotheses,

(b) liquid wave celerities are high compared to bottom perturbation celerities, which justifies the assumption of constant discharge all along a reach of river,

(c) flow resistance is described using the Manning formula with the Manning coefficient constant in time,

(d) bed material is homogeneous or can be represented by a specific diameter of sediment.

The above assumptions lead to the following system of differential equations:

$$\frac{\delta}{\delta x}[\frac{Q^2}{2A^2} + gy] + \frac{g\,Q|Q|}{D^2} = 0$$

$$(1-n)\frac{\delta Z}{\delta t} + \frac{1}{b}\frac{\delta G}{\delta x} = 0$$

$$G = F(y, z, Q, d, ...)$$

Where :

$y(x,t)$ = water stage
$z(x,t)$ = river bed elevation
$G(x,t)$ = sediment discharge
D = conveyance
x = longitudinal coordinate
n = porosity of deposited sediment
$b(y,x)$ = width of river bed affected by solid transport
$A(y,z)$ = wetted area
d = sediment diameter
$Q(x,t)$ = liquid discharge
t = time

4.2 CONSTRUCTION OF THE MODEL

The model represents about 210 km of the river and 10 km of the tributary, using existing maps and topography.

108 nodes were used for the model, generally representing one river cross-section. Averaged cross-sections were used on the model without taking into account local variations in bed morphology encountered in the field at some places.

4.3 CALIBRATION OF SEDIMENT INFLOW TO THE MODEL

Precise calibration of a model requires long iterative steps to adjust a number of parameters on the basis of accurate data concerning the actual morphology and hydraulic conditions along the rivers. As most of these parameters were not available, the following procedure was adopted.

The initial model, with a longitudinal profile obtained from topographical data, was run with the natural hydrograph and sediment loads entering the river and in the tributary adjusted to ensure stability for 10 years. The slope of the tributary was adjusted to obtain more stable conditions downstream of the confluence. The river bed longitudinal profile obtained after 10 years was taken as representative of the "initial state with natural conditions".

As this situation is in fact not exactly stable in this so-called natural state (some aggradation appears downstream of the confluence), the impacts of the dam indicated in this paper were obtained by comparison between the evolution of the bed profile with natural conditions run for another 10 years, and modified conditions also run for 10 years starting from the same "initial state with natural conditions".

Particles of less than 0.15 mm diameter, forming the "wash-load", are not to be taken into account when studying the evolution of the river bed profile.

For calculations made after construction of the dam, sediment inflow is nil in the main river (downstream of the dam) and sediment inflow in the tributary is taken as per the natural conditions resulting from calibration of the model.

According to sediment size, the Engelund-Hansen formula was adapted for calculation of the solid transport.

4.4 MAIN CALCULATIONS

A simple calculation shows that river bed degradation from the dam to the confluence will be far more than 1.5 m (the limit defined for forest protection). Downstream of the confluence, river bed degradation is limited because this reach receives sediment from the tributary and from degradation of the upper reach of the river.

The first calculation was carried out for 10 years with 3 weirs built in the reach located from the dam to the tributary, in order to limit river bed degradation.

The 3 weirs were located respectively at 2, 6 and 10 km downstream of the dam.

The solid and liquid hydrographs used were as follows:

.   natural conditions for the tributary,

.   liquid discharge controlled by the dam and nil solid inflow downstream of the dam.

The results in the upper reach of the river show that many more than 3 weirs were to be built.

The calculation results are summarized in table 1 (see end of the text).

The degradation effect resulting from the dam is observed from the beginning of the reach and extends progressively downstream.

From the dam to the first weir, degradation is nearly stabilized after 10 years, reaching 3.30 m just downstream of the dam. The modified river bed is nearly horizontal at some 40 cm under the level of the first weir. Downstream of the dam, degradation reaches 1.50 m within the first year.

Between the first and second weirs, degradation is still in progress after 10 years. Degradation downstream of the first weir reaches 1.50 m within the first year and 7.40 m after 10 years, still progressing at a rate of 20 cm/year.

Between the second and third weirs, degradation is still in progress after 10 years. Degradation downstream of the second weir reaches 1.50 m within 3 years and 6.30 m after 10 years, still progressing at a rate of 40 cm/year.

Downstream of the third weir, degradation reaches 1.50 m within 3 years and 4.70 m after 10 years, still progressing at a rate of 40 cm/year.

At existing intake no. 1 (located 18.6 km D/S from the dam), the situation is nearly stable. The bed reaches 40 cm aggradation after 5 years and then is lowered again to reach the initial level after 10 years with annual degradation of 11 cm/year. This is due to the fact that the erosion rate between weirs is significantly reduced only after 5 years, leading to reduced sediment load in the upper branch of the river.

5   REMEDIAL MEASURES AND CONCLUSIONS

The above results, even if considered as an order of magnitude, call for urgent remedial works, starting from the upper

Table 1 - Impact of the dam with 3 weirs implemented in the upper reach

Location	Distance km from the dam	Situation after 10 years	
		Absolute aggradation (m)	Rate of aggradation (cm/year)
The dam	0	-3.39	-3
Weir KP 2 U/S	2	-0.41	-2
Weir KP 2 D/S	2	-7.38	-20
Weir KP 6 U/S	6	-0.18	-1
Weir KP 6 D/S	6	-6.27	-39
Weir KP 10 U/S	10	-0.30	-1
Weir KP 10 D/S	10	-4.70	-41
Confluence	15.2	+0.13	+3
Existing intake no. 1	18.6	+0.58	+2
Existing intake no. 2	22.3	+0.21	+1
KP 50	50.0	-1.47	-12

very limited (± 60 cm) from the confluence to the existing ford downstream of existing intake no. 1. Further downstream, the impact calculated is degradation progressing regularly to reach 1 to 1.50 m after 10 years.

part of the river as a priority.

The protection works are designed to limit river bed degradation to a maximum of 1.50 m.

Several other calculations were carried out in order to define the necessary protection works.

These calculations lead to 2 alternatives for protection of the river bed from the dam to the confluence:

. building 22 weirs, with an average spacing of 700 m, able to withstand 1.5 m head,

. providing a "continuous" protection, made up of 54 weirs with an average spacing of 270 m, able to withstand 60 cm head.

Complementary calculations were carried out to study the stability of the lower reach of the river (i.e. D/S of the confluence).

The first conclusion is that the impact of the dam on the lower reach is not significantly modified when the type of protection in the upper reach changes. This is due to the fact that the sediment load downstream comes mainly from the tributary.

The impact calculated in the lower reach is

Hydropower'92, Broch & Lysne (eds) © 1992 Balkema, Rotterdam. ISBN 90 5410 054 0

Alta – Norway's most controversial hydro scheme

L. E. Karlsen
Norwegian State Power Board, Norway

ABSTRACT: Never have the environmental questions connected with a Norwegian hydropower project aroused such public interest - at home and abroad - as with the Alta scheme in North Norway. The project was put before the Norwegian Parliament three times. And on top of that, the matter was then taken to the Supreme Court. Actions from demonstrators influenced the government to halt construction work for one year. However, favoured with a good owner/contractor relation, the work was speeded up, thus reducing construction time from 6 to 5 years and enabling the project to be commissioned in 1987. As to expenditures, 7.5 per cent of the total cost has gone to environmental matters and an amount equivalent to 10 per cent of the total cost was used on a sizeable police force to get construction work started. Among several ecological aspects, the superb salmon fishing in the Alta river was the most important. Therefore no other Norwegian power plant has had such strict flow regulations imposed on it. As far as sociological aspects go, there was in connection with the Alta scheme a great focus on the Sami culture - not only the ancient history, but perhaps even more on the Sami's situation today and in the future. Before and after construction, several multi-disciplinary scientific studies concerning the environment have been carried out.

1 THE RIVER AND ITS CATCHMENT AREA

1.1 Location and landscape

The Alta watercourse is located in the county of Finnmark, in the northernmost part of Norway. (See Figure 1). The river, which has its origin at the Finnish border and its outlet by the small town of Alta at the bottom of Altafjord, is 220 km long, with a watershed of 7500 km². It runs through the largely rural municipalities of Kautokeino and Alta.

Most of the catchment area is located on the so-called «Finnmark plateau», a flat plain lying 350 - 600 meters above sea level. In between smooth ridges it is covered with a network of lakes and marshlands. In the northern part the river has dug its way in, making steep terraces in the soil deposit, covering the bedrock. Further south where the bedrock is covered with layers of sandstone and clay-slate, narrow valleys have been shaped including its canyon, a rarity in Northern Europe.

1.2 Ecological and sociological characteristics

The Alta district and Alta valley can be described as a luxuriant area. One can mention pine forest, which is not common at such high latitudes. Near the power station area, there is also a small aspen forest, which is under legal protection and was not to be harmed by the construction work. The Masimelt, Oxytropis deflexa ssp. deflexa, is only found growing in this area and had therefore to be protected during the construction period.

Above all, the Alta River is famous for its superb salmon fishing.

The district, part of the larger area known as Lapland, is famed for its domestic reindeer herds, the traditional mainstay of Sami (Lapp) life. The herds stay in the inland area during the winter-time. In spring they are driven to areas near the coast and in autumn they return to the inland. In the course of these yearly migrations, some of the herds will pass near the power plant and the access road.

Artifacts from the Sami culture are protected by law. Every construction operation has, according to this law, to be cleared with the cultural authorities. Every construction area has to be investigated, first by registration and later on, if special cultural artifacts or relics are found, an investigation programme has to be carried out before construction work can start. In connection with the Alta scheme there was a great focus on the Sami culture - not only the ancient history, but perhaps even more on the Sami's situation today and in the future. In a way, the Alta case provided the push needed for a more clear identification of the Sami society. In recognition of this, Norway's new Sami Parliament was established in October 1989.

2   THE PLANNING OF THE SCHEME

2.1 Planning history

Over the years, several plans for developing the Alta River were evaluated, and as late as in 1968 a rough plan showed possibilities for a development corresponding to an electricity production of 1400 GWh. This plan presupposed the use of a series of tributaries in the area, among them (See Fig. 2) the Iesjåkka, Tverrelva, Joatkajåkka and Iesjavri. With the high regulated water level for the reservoir in the main river as indicated on this original plan, the Sami village of Masi would have been drowned.

For environmental reasons these earlier plans were revised. When the Norwegian State Power Board - Statkraft - presented their concession application, it was therefore a question of a considerably reduced scheme, which left Masi and several of the tributaries outside the influence of regulation.

The licensing process revealed that the two municipalities involved - Alta and Kautokeino - were against such a hydro power development. Finnmark county, however, sanctioned a smaller development. This local resistance was partly a matter of the expected environmental impact, and partly a political question concerning the rights and future of the Sami to their traditional lands.

The final plan approved by the Storting - the Norwegian Parlament - in 1978 left also the Joatkajåkka free, and moved the power station outlet as far upstream in the main river as the upper limit of the salmon migration. Thus, calculated annual electricity production was reduced to 625 GWh.

Today's Alta Power Plant is owned 60% by the Norwegian government and 40% by Finnmark county. Statkraft is responsible for the planning, development and operation of the plant.

2.2 The plan adopted

The Alta Power Plant is a run-of-the-river plant. The catchment area is approx. 6 000 km^2 with an annual run-off of some 2100 million m^3. The dam - a 140 m high concrete arch dam with a volume of 135000 m^3 - has been built in a narrow gorge 5 km downstream from the outlet of Lake Virdnejavri, which has raised the waterlevel of the lake 15 m. The reservoir obtained is 135 million m^3, or about 6% of the annual run-off. With maximum turbine capacity 1.5 times the mean river discharge, the spillwater   accounts for more than 30% of the total annual run-off. Of an average annual production of 625 GWh, some 40% is produced in winter.

Like almost all of Norway's hydropower plants, Alta is built underground. With a 2 km long headrace tunnel and a short tailrace tunnel, a mean head of 175 m is gained. (See Fig. 3).

Although the maximum rise of Lake Virdnejavri is only 15 m, the part of the reservoir close to the dam can be regulated 65 m, due to a fall of 100 m on the 5 km stretch between lake outlet and dam.

Precautions have been taken at the headrace tunnel intake concerning the salmon and its environment: In order to keep the temperature in the downstream water as close as possible to the «natural» level, two intakes have been made - one near the HWL and another below LWL - thus making it possible to mix water from the different temperature layers. Model tests were carried out to find the best design for the upper intake to avoid air entrainment which may cause air-supersaturated water with possibility of salmon death. The temperature as well as the air situation is continually monitored.

For flood discharge there are, besides the dam crest, 2 tunnels, 2 gates in the dam 60 m below HWL and a diversion tunnel. Capacities are 1200, 800 and 100 m^3/s, respectively. As the riverbed has rather heavy overburden of loose materials (gravel), the flood is best conducted through the flood tunnels.

The power station has two Francis turbines, 50 and 100 MW, and arrangements have been made for a third one. In winter there is enough water only for the 50 MW unit. In order to meet environmental requirements, the station is equipped with a special type of bypass valve with the same discharge capacity as the 50 MW unit, 30 m³/s. When a shutdown occurs, the bypass valve opens automatically. In this way the calculated water discharge through the station remains unaltered.

All functions at the power plant are remote monitored and controlled from the Operations Centre. The bypass valve in the power station is controlled by a local computer, PLS (Programmable Logical Control), which operates completely independent from the remote control equipment. This installation secures a minimum flow of water downstream from the power plant, even if errors should occur in the remote control.

Such a complex type of construction was considered necessary, in view of the strict requirements for an even discharge. Continuous manning of the power plant would have been necessary, were it not for the extensive remote control and local automatic functions.

The temperature as well as the air saturation in the water-stream passing through the plant are continually monitored.

Total cost was 1100 million NOK exclusive of interest during the construction period, i.e. 1.80 NOK per annually produced KWh. Compared with other Norwegian hydropower plants built at the same time, this is rather inexpensive power.

3   MANOEUVRING REGULATIONS

To reduce the environmental impact - especially on salmon fishing - to a minimum, very strict restrictions have been imposed on running the plant.

Except for wintertime when the reservoir will be emptied, discharge is to be very close to natural conditions. Manoeuvring regulations are as follows:

15 Dec. - 1 Apr.: Discharge limited to 30 m³/s.

1 Apr. - 1 May: Discharge can gradually be increased to 50 m³/s by 25 April and to 96 m³/s by 1 May.

31 Aug.: From the time when the reservoir is filled up, discharge is not to deviate more than + 10% from natural water flow.

1 Oct. - 15 Des.: Discharge shall under normal conditions gradually decrease from 85 m³/s to maximum 30 m³/s by 15 December.

After the river downstream from the power plant has been covered with ice (usually by November), the water flow may not be increased beyond the flow at the time of freezing.

Appointed experts on fish and ice conditions can put forward even stricter restrictions if special river conditions so necessitate.

In addition, the reservoir is to be filled in such a way that flood rise will be smooth and the flood maximum less than unregulated, to minimize erosion along the riverbed. The reservoir will also be used gradually from the first ice-cover until the flood period starts, to avoid a sudden drop in the discharge in springtime just before the flood season. A sudden drop in the discharge at this time, with low natural discharge, might cause problems for the salmon fry.

Figure 4 shows the natural and regulated discharge through the year 1988.

The previous mentioned bypass valve in the station levels out variations in discharge from the plant in winter. During the summer, further water is supplied by the flood or diversion tunnels in the dam area. The delay of water from the dam area to the tailrace of the power plant is approx. 20 minutes, which causes a brief reduction in the flow of water just downstream from the power plant.

The additional restriction applied in the winter during freezing - i.e. no increase in the flow of water after the river is covered with ice - has entailed much work concerning production prognoses in order to enable water outlet from the reservoir in the winter within the limits set.

To monitor the hydrology in the river, three measuring stations have been established, two upstream and one downstream from the reservoir. Hourly discharge data from these stations are automatically transmitted to the Operations Centre.

The extremely strict maneouvring regulations place great demands on the functioning of the equipment installed, not least the automatic functions. Particularly the local automatic functions and the remote control have proven highly reliable to date.

4 LANDSCAPING

In planning the Alta Power Plant, great effort was made to achieve a plant which could blend smoothly into the terrain formations. Environmental considerations were paramount.

The 28 km long access road was given special treatment to prevent any sharp contrasts in the very open terrain.

From a spiral tunnel in the right side bank there are accesses to the various dam levels.

The power station is linked to the grid via a SF_6 underground switchgear, thus limiting outdoor features to a minimum. For permanent power supply to plant equipment, underground cables have been used.

The two tunnel tips were given a shape as similar as possible to the contours of the surrounding area. Removed gravel close to the riverside used for concrete works has been replaced by excavated rock.

All wounds and interference in the land have been treated carefully with fertilizer and seed.

Strict restrictions were placed on the contractor's work. Continuous pollution control in the river confirmed that the construction work was not causing any damage.

5 DEMONSTRATIONS

As a reaction to plans to develop the Alta River, environmentalists and others started a protest campaign. The movement had as its main goal to prevent an Act of the Storting for hydropower development of the Alta River.

When the Act passed, opponents decided to do their outmost to prevent it from being put into effect. In Sept. 1979 construction work on the access road had to be cancelled due to demonstrators placing obstacles in the way of the contractor.

After two additional Acts of the Storting, road construction was resumed 15 January 1981. Approximately 600 policemen were required to remove demonstrators who had been barricading the line.

There were some other actions up to January 1982. Then the opponents decided to put an end to their activity. Construction work on the power plant itself was accomplished without any confrontations.

After the demonstrations had stopped and the road was finished, it was possible to show the local people what was really to be done on the 10 km long stretch of the river. The government gave permission to have «open plant» four days a year, particularly intended for the local population. This proved a great success, and it appeared that the need for direct information was enormous. On one weekend more than 2000 visitors called at the plant.

Gradually the local staff from Statkraft felt that the once-considerable objections against the scheme were disappearing.

6 LAW SUITS

After an application for an assessment of compensation had been lodged with Alta District Court, several protests were made against the assessment being carried out, because of procedural errors. One main reason was that rules of international law relating to the protection of an aboriginal population and ethnic minorities restricted the Norwegian authorities' right to interfere with the course of Sami livelihood and culture. Moreover, reference was made to defects in the development petition itself which led to the decision being made on an erroneous basis. Construction works, for the commencement of which prior permission had been given, were postponed to await a judicial decision on this point.

The Norwegian Supreme Court, in plenary setting, delivered a judgment that unanimously rejected all grounds for appeal.

Simultaneously with the assessment cases, and especially in the beginning, a number of other both civil and criminal cases were brought before the courts in connection with the development resolution and its implementation.

With so many disturbances in the construction schedule, it should be noted that all disputes between owner and contractors were solved without any court proceedings, thanks to good Norwegian contractual practice.

7 THE OWNER'S IMPOSED CONTRIBUTION

The two municipalities involved have received financial contributions as well as direct support for certain projects.

A 10 mill. NOK concession fund was set up. In addition the municipality of Alta has received, through a cooperative agreement, 4 mill. NOK as contributions to a community hall, health clinic and roads.

The owner was also required to build two bridges across the Alta River at a cost of 17 mill NOK, and a 30 km long power-line (22 kV) to the Sami village of Masi.

To improve the possibility of fisheries research generally, and the Alta River salmon specifically, the owner agreed to build a hatchery plant at a cost of 14 mill. NOK.

Finally the owner was required to contribute 5 mill. NOK in order to prevent further erosion from the banks in the lower part of the river. This requirement was imposed even though it was stated that the power plant would not lead to any increase in erosion.

8   PUBLIC RELATIONS

8.1 Construction period

During the construction period, Statkraft maintained good communication with the local community through regular meetings with representatives from Alta municipality. At these meetings, any matters related to the project - especially environmental problems - could be taken up.

There were also a special committee dealing with matters connected to Sami and their activity in the plant area. This cooperation lead to excellent relations between the Sami in the district and the staff at the plant.

65 % of the labour force was recruited from the local district.

8.2 Today

Even if demonstrations against the development halted early in the construction period, the plant is still much in focus in the local community. Operation of the power plant, and especially the discharge, is of great concern to local inhabitants. The activity of the State Power Board is followed with keen interest, but in a positive way as seen by those working at the plant.

Our impression is that the political and administrative leadership of Alta Municipality has acknowledged the positive impacts the plant has had, and will have, on the municipality.

The power plant has become an attraction for visitors to Alta. Many organisations hold their meetings in Alta, partially to show the partici-

pants the Alta Power Plant. A visitor's information centre, placed in a cavern close to the dam, will be ready for use in summer -92.

9   ACCESS ROAD

The 28 km long access road has in many ways been a bone of contention. It was closed to public traffic during the construction period, out of consideration for reindeer husbandry and wildlife. On the other hand, the road runs to the most popular resort areas in Alta, both summer and winter. The local population and Alta Municipality therefore have pressed strongly to have the road opened to the public. An open road would also provide possibilities for developing tourism in the area, where the power plant and boat trips on the reservoir will be important features.

However, those who earn a living from reindeer husbandry have consistently opposed the development of tourism in this area. Seen from the environmental viewpoint, there has also been much scepticism towards more tourism, mainly because of the stocks of wild animals.

In order to meet the local community's request for an open road, which at the same time acting in the interests of reindeer husbandry and environmental conservation, Ministry of Petroleum and Energy resolved in 1989 that a 18 km stretch of the road is to be opened to the public for parts of the year, for a trial period of three years. The road is to be closed when reindeer are in the area.

10 STUDIES OF VARIOUS
ENVIRONMENTAL ASPECTS

Before and after construction, several multi-disciplinary scientific studies have been carried out. Main study areas are: Climatical differences, hydrology, ice formation, erosion, sedimentation, fishing and fish biology, reindeer husbandry, archaeology and ethnography, wildlife biology, and botany. The studies are still in progress. Final reports are expected to be published within this year. So far it seems that the plant has had only minor impact on the environment.

11 CLOSURE

This article has been prepared by members of the Committee for Implementation of the Alta Project. L.E. Karlsen - chairman of the committee - has had the overall responsibility for the paper.

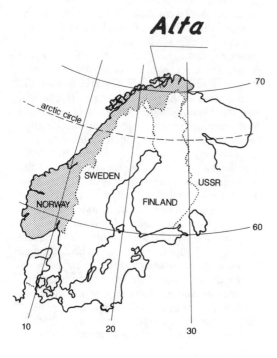

FIG. 1.

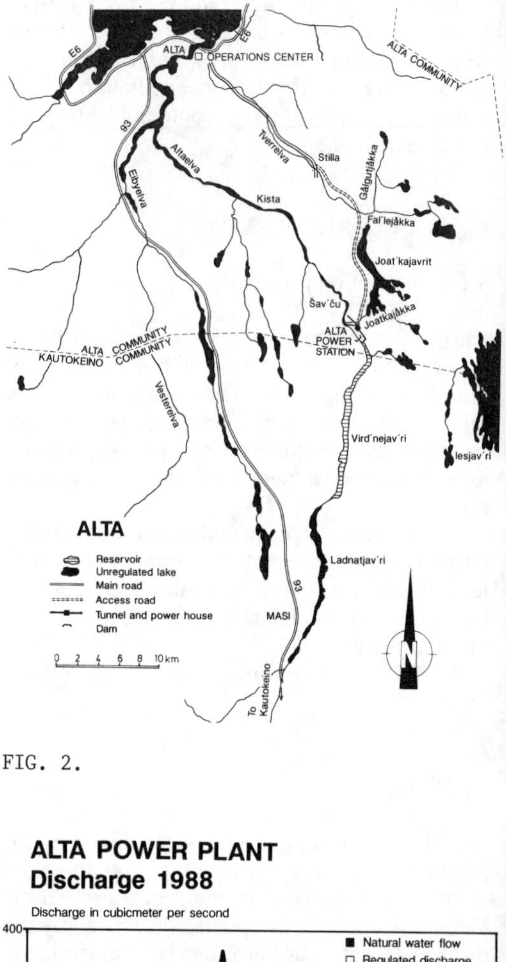

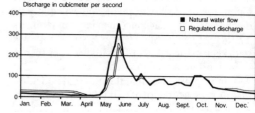

FIG. 2.

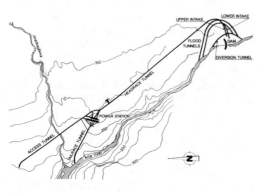

FIG. 3.

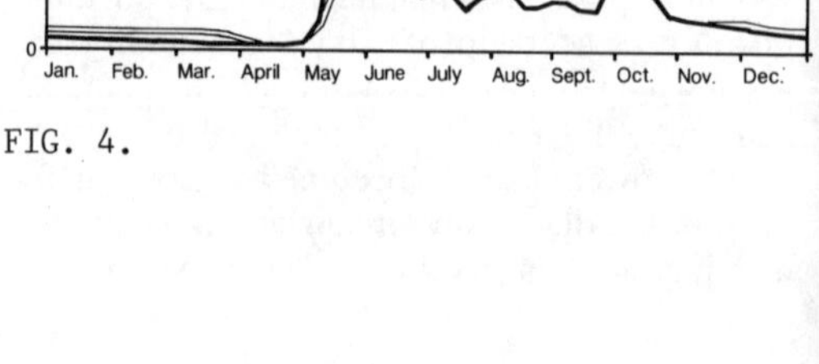

FIG. 4.

Hydropower'92, Broch & Lysne (eds) © 1992 Balkema, Rotterdam. ISBN 90 5410 054 0

Resettlement work of Pak Mun Hydroelectric Project

V. Meevasana
Electricity Generating Authority of Thailand, Nonthaburi, Thailand

ABSTRACT: The Pak Mun Hydroelectric Project is under construction by the Electricity Generating Authority of Thailand (EGAT) across the mouth of Mun river which flows into the Mekong river on the border between Thailand and Laos. The project is a run-of-river type comprising a regulating dam of 17 m high and a powerhouse of four 34 MW bulb turbine generating units. Development of this project is rather controversial. The reservoir upstream of the dam covers about 60 sq km of which about 52 sq km is the original riverine and 8 sq km is the inundated farmplots. There are 248 households who have to be resettled from the affected area. The Government of Thailand has set up two main committee comprising representatives from various concerned agencies to assist the affected villagers. The first is the committee for compensation of land rights and properties. The second is the committee for resettlement. There are many problems during performing the resettlement work. Some problems and solutions are mentioned in this article as case studies.

1 INTRODUCTION

In the past, major problems of water resources development had been concerned with the technical aspects. Nowadays, the problems have been changing to the social aspects. One of the main problems is the involuntory resettlement of people from the project area. It is becoming more apparent that resettlement work is often the largest and most complex matter related to water resources development. The resettlement problems are so complicated that technical knowledge alone can not solve. For resettlement work, the human adaptation is particularly a complex process of individual adjustment. It demands long time, a lot of effort, more expenditure and a considerable support from many parties concerned.

With no exception, the Pak Mun Hydroelectric Project which is under construction in Thailand has to face the same problems.

2 PROJECT FEATURES

The Pak Mun Hydroelectric Project is located on the Mun river, in the northeastern part of Thailand, about 5.5

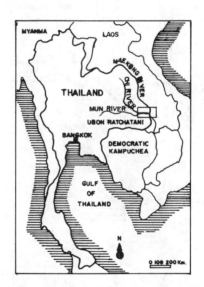

Fig. 1 Project location map

km upstream of the confluence with the Mekong river on the border between Thailand and Laos. The project is a run-of-river type which, unlikes a storage dam project, does not impound a large stagnant

305

mass of water. The dam is a roller-compacted concrete type of only 17 m high. It creates a reservoir of 225 million cu m. The reservoir area of 60 sq km dividing into 52 sq km of original riverine and only 8 sq km of inundated farmplots. The dam is provided with eight automatically controlled gated spillways designed to discharge the highest probable flood that could occur in a thousand years. These gates lift completely off the river bed and when fully open can let the river flow through practically unrestricted. With four 34 MW bulb-turbine generating sets, the project would generate 136 MW, 280 GWh per annum of power to meet the peak demand for electricity in the Northeastern Region and would create the potential for pumping irrigation about 250 sq km of land.

The project was approved for implementation by the Thai Government on May 15, 1990. The project construction is being performed by the Electricity Generating Authority of Thailand (EGAT) with the schedule to be completed by December 1994.

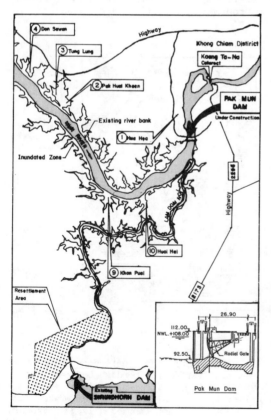

Fig. 2 Pak Mun dam site

3 AFFECTED AREA

There are 13 villages along both river banks affected by the construction of Pak Mun Hydroelectric Project. Since the impounding will utilize only the original capacity of the river channel, the affected houses are located in groups along both banks of the Mun river. The maximum water retention level in front of the Pak Mun regulating dam will be only about 2 m above the original river bank level after the project completion.

The affected areas are divided into two zones i.e. the reservoir zone and the freeboard zone. In the reservoir zone, there are 142 households and 1100 land plots fully and partly inundated, which depend on their sizes. In the freeboard zone, 95 households is considered to be partially affected.

4 SOCIO-ECONOMIC CONDITION IN THE AFFECTED AREA

The affected area is in rural. The average family size is is of 7 persons. The proportion of the inhabitant in the working age is 56 percent. The main occupation is rice farming. The supplementary occupations are off-farm employment, trading and fisheries.

With respect to migration, only 26 percent of households regarding themselves as immigrants. There are several traditional religious rites which are still performed actively. The main

Fig. 3 An existing house in the affected area

306

organizers of these activities are usually the village head men. Most of the household heads have demonstrated that they have some knowledge about politics.

Out of the total 1100 affected land plots, only 42 plots are fully affected. Most of land holding is less than 10 rai (6.25 rai = 1 hectare) of which about 23 percent have land titles. The average land price at beginning of the project implementation was about 2000 Thai Baht per rai. Agricultural practices are still the traditional ones depending on traditional farm tools,human and animal labour. Crop yields are very low.

The average annual household gross income was 14845 Thai Baht. The average annual household expenditure was 12965 Thai Baht, most of which were off-farm expenses. The average annual household net income was 1880 Thai Baht.

5 PROJECT MANAGEMENT

The villagers affected by implementation of the Pak Mun Hydroelectric Project are regarded as the devotees who have to leave their original residences and farmlands so that the area could be fully developed for maximum benefit to the country as a whole. Consequently, the Electricity Generating Authority of Thailand considers providing them in respectful response with a high compensation and appropriate resettlement to ensure the better livelihood. To achieve this objective, two sets of committees were appointed by the Thai Government on May 15, 1990 to be responsible for the project management.

5.1 The committee for compensation of land rights and properties

This committee is in charge of appraising the properties values, providing compensations to the resettlers,appointing sub-committees to carry out their assignments, i.e. the sub-committee for land and properties investigation,the sub-committee for properties compensation,etc.

5.2 The committee for resettlement

This committee is in charge of providing the resettlement area, allotting land for homeplots and farmplots promoting appropriate agriculture and other occupations, relocating the villagers into the provided areas, managing and constructing public facilities and

appointing sub-committees to perform different assignment, i.e. sub-committee for land allotment, sub-committee for agricultural extension and occupation promotion, etc.

Chairman of both committees is the Governor of Ubonratchathani Province where the project is located. The committees composes of experts from various agencies such as land department, forestry department, agricultural extension department and etc. with secretaries from the Electricity Generating Authority of Thailand. The sub-committees have also included the village head men. The representatives of affected villagers are also invited to join the sub-committees or working committees for resettlement works. The management has been closely advised by the Deputy Permanent Secretary to the Prime Minister.

6 COMPENSATION OF LAND RIGHTS AND PROPERTIES

The compensation costs to be paid for the properties within the reservoir area cover two main categories according to the nature of assets to be inundated,i.e., the immovable properties of which land and tree crops are the main components, and the movable properties which include the built-up properties such as private dwellings, buildings, rice storages, fences, schools, temples, and others. The study of the compensation aspect for Pak Mun Hydroelectric Project involved the following steps: (i) collection and review of related information; (ii) field investigations and properties inventory; (iii) analysis of compensation rates for different types of assets and (iv) the estimation of the total cost for compensation of each category properties.

6.1 Compensation of immovable properties

Land values were assessed based on various parameters including factors such as land ownership, land potentiality, land scarcity,local selling prices and official land prices. The local selling prices of lands were collected by means of direct interviews with the villagers and from local land offices where selling prices were reported by the villagers.

The compensation to be paid for the tree crops would be varied with type of the tree crops. These values would be added to the land development cost. The value of each of tree crops in Pak Mun Hydroelectric Project was determined based on the rates previously paid for other reservoir projects, with modifications to fit local conditions.

Payment of compensation for land and tree crops has been a controversial matter between the affected villagers and the implementing agencies relating to the past water resource development projects. Experience gained from other projects reveals that large numbers of rural people have not had legal land ownership. Due to rapid growth of population and dwindling of cultivated areas reserve areas have been continuously illegally encroached upon.

These factors are the main causes of complication in the process of land and tree crop payment, because, by law, lands without proper holding certificates would not entitle to compensation payment. However, to enable the project to be successfully implemented, compromising measures are to be established in order that the illegally occupied lands which have been cultivated for a justifiable period of time could be compensated for.

6.2 Compensation of the movable properties

Movable properties recognized in the compensation are divided into 2 main categories according to its ownership i.e. privately-owned properties and government or community owned properties. These two categories are subjected to different process of compensation payment. For the privately-own properties, the compensation cost is to be paid directly to their owners at the rates previously set up by the authorized committee, while the compensation costs for the government or community structures are normally based on their potential uses and the compensation costs, are to be negotiated case by case with agencies incharge.

Compensations covering for land, tree crops and buildings are paid to the resettlers and land owners at the rates set up by the sub-committee for appraisal of the values of lands and properties as follows:

1. Land compensation rates are set up by the capital account for ownership registration and juristic act.

2. Tree crops compensation is paid on the crop values basis as assessed in the report made by the sub-committee for land and properties investigation and assessment. The kinds of crops determined for compensation in the said report fall into 168 items of perennial plants and 46 items of biennial ones.

Compensations for tree crops are based on the following conditions:

   a. full rate for fruitful plants
   b. half rate for fruitless plants
   c. fair rate for biennial plants only which could not be harvested when evacuated.

3. Houses and structures compensation is based on the following factors
   - removal cost of existing structures
   - relocation cost of existing structures
   - cost of wasted materials
   - transportation cost of removal materials to the new location

6.3 Actual compensation

The value of land and tree crops as surveyed before the project implementation showed that the compensation rate of land with tree crops in Khong Chiam district which is the major affected area should be about 2000 Thai Baht per rai (1 hectare = 6.25 rai). The actual compensation rate of land as approved by the committee for compensation of land right and properties is 35000 Thai Baht per rai or more than 10 times of the proposed figure. In the same manner, the average actual payment of houses and structures is 52000 Thai Baht per household which is about 10 times of the estimated 5300 Thai Baht per household. (See Table 1.)

Table 1. Compensation rates for Pak Mun resettlers.

Item	Description	Affected number	Values of properties as surveyed prior to project implementation	Actual average compensation rate
1	Land	5700 rai	2000 ฿/rai	35700 ฿/rai
2	Houses and structures	248 households	5300 ฿/household	52000 ฿/household

Remark: 1 hectare = 6.25 rai   1 US $ = 25.6 ฿

308

7 RESETTLEMENT

The main principle of rehabilitation is widely recognized as a dynamic approach to assist the reservoir resettlers. It stresses on the concept that enable the people to earn a new living comparable to what they would have got their lands benefited in the same way as the lands of others in the project area. This concept is accepted and regarded as the main direction for planning of the resettlement for Pak Mun reservoir resettlers.

The study methodology comprises the following activities:

1. Collection, and examination of information concerning various resettlement projects which have already been implemented.

2. Synthesis of such information into the form that can be the basis of further planning.

3. Field investigation and survey of the possible resettlement sites and schemes, evaluation and site selection.

4. Collection of data needed for physical planning of selected resettlement site and formulation of resettlement development program.

5. Preliminary layout and planning of infrastructures.

6. Cost estimation for resettlement program and scheduling which includes the recommendations for organization and management structure.

7. Evaluation of probable affects related to the resettlement program.

According to the agreement of the committee for resettlement, the principles for provision of assistance to the resettlers after payment of compensation are divided into many alternatives to match the requirement of the affected villagers as follows:

1. For one who recieves the compensation for house and is owner of the resident area.

Case 1. For one who wishes to move into the new resettlement area, one homeplot together with a new house will be provided.

Case 2. For one who wishes to live in the old village.
1. if still resides in their original place, the ground will be backfilled and his house will be elevated or renovated or rebuilt.
2. if moves into other area in the old village, one homeplot together with a new house will be provided.

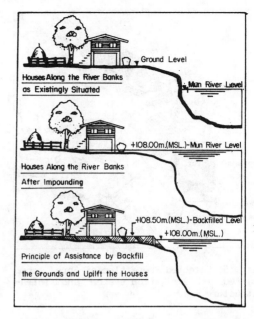

Fig. 4 Assistance of the to be inundated households who wish to reside in their original ground

Case 3. For one who does not want the assistance in Case 1 & 2 and intend to manage the resettlement by oneself, an extra assistant money of 135000 Thai Baht will be paid.

2. For one who receives the compensation for house and is the owner of the resident area and also receives the compensation for farm land.

Case 1. For one who wishes to resettle into the new provided resettlement area outside the old village, one homeplot and one farmplot will be provided with irrigation system and other needed infrastructures.

Case 2. For one who wishes to live in the old villages.
1. if still resides in his original place, the ground will be backfilled and his house will be lifted, or renovated, or rebuilt, and one farmplot in the resettlement area will be provided.
2. if the original land cannot be backfilled and the house cannot be lifted, one homeplot together with a new house will be provided in the old village and one

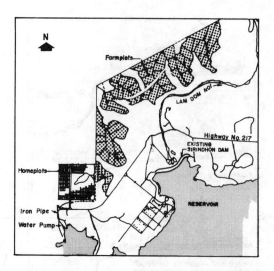

Fig. 5 Selected resettlement area completed with irrigation system and infrastructure

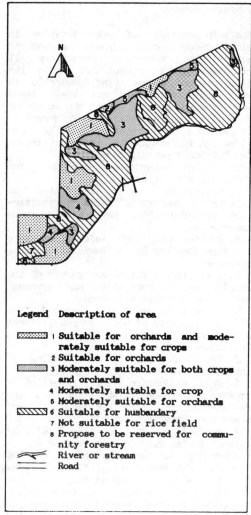

Fig. 6 Survey of land suitability of the selected resettlement area

farmplot will be provided in the resettlement area.

Case 3. For one who does not want the assistance in Case 1 & 2, an extra money of 135000 Thai Baht will be provided for him to manage the resettlement by his own.

3. For one who receive compensation for farm land only, one farmplot in the resettlement area will be provided.

4. For one who receive compensation for house only, one homeplot together with a new house will be provided in the old village.

5. For houses located in the reservoir freeboard zone, the homeplot will be backfilled and the houses will be lifted or rebuilt up to the maximum freeboard level and improvement of some facilities as needed will also be undertaken. In case any damages occur due to the improvement, compensation for those affected will be provided as well.

8 OCCUPATION PROMOTION

Occupation promotion is considered an important task to undertake in parallel with other relevant activities to achieve a better livelihood of the resettlers. The occupation promotion as carried out in former projects often varied in forms. However, one of the distinct patterns of occupation promotion in long term is to encourage an establishment of an agricultural institute like "the Agricultural Cooperative".

For old villages, all infrastructures such as access roads, electricity system, water deep wells,and etc. will be improved and the occupation promotion will be introduced to the villagers.

For new resettlement area, the farmplots will be provided with irrigation system. The soil properties were investigated and confirmed that it is suitable for growing crops.

Table 2. Summary of the resettlement scheme

Category of ownership in the affected area	Alternative and provision of assistance

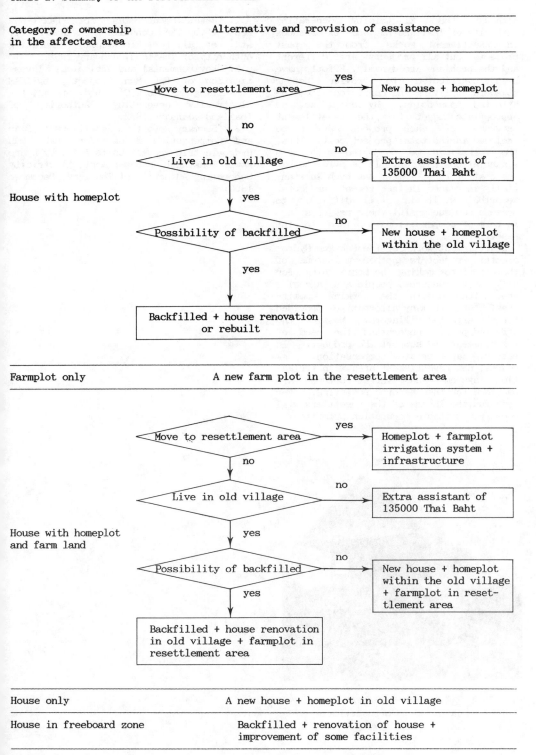

House with homeplot	Move to resettlement area — yes → New house + homeplot
	Live in old village — no → Extra assistant of 135000 Thai Baht
	Possibility of backfilled — no → New house + homeplot within the old village
	Backfilled + house renovation or rebuilt

Farmplot only	A new farm plot in the resettlement area

House with homeplot and farm land	Move to resettlement area — yes → Homeplot + farmplot irrigation system + infrastructure
	Live in old village — no → Extra assistant of 135000 Thai Baht
	Possibility of backfilled — no → New house + homeplot within the old village + farmplot in resettlement area
	Backfilled + house renovation in old village + farmplot in resettlement area

House only	A new house + homeplot in old village
House in freeboard zone	Backfilled + renovation of house + improvement of some facilities

311

9 CONCLUSION

Although the Electricity Generating Authority of Thailand has some experiences in resettlement works from the past projects, but all projects are different and the problems are dynamic. Efforts have been contributed to provide better standard of living and conditions to the affected households. By doing so, an appropriate budget for the resettlement programe of each project has to be included in the total project cost. Thus, the works can be conducted with less financial obstruction. It is evident that the cost of resettlement has been substantially increased in the recent projects. Nevertheless, it is still difficult to measure how successful these projects are in term of compensations provided to the resettlers.

Resettlement works of the Pak Mun Hydroelectric Project is another milestone of the effort for making the best thing for all. Every concerned people and agencies are invited to join the working committees. There are many alternatives provided for the affected villagers. Most of the affected people understand the need of resettlement and support the project. In return, appropriate compensation, new houses, new resettlement area with irrigation system and infrastructures are provided for them before resetting. For long run, the income of the resettlers will come from suitable occupation promotion.

REFERENCES

1. "Resettlement Plan for Pak Mun Project", by TEAM Consulting Engineers Co., Ltd. et. al. for Electricity Generating Authority of Thailand, January 1982.
2. "Environmental and Ecological Investigation of Pak Mun Project", by TEAM Consulting Engineers Co., Ltd. et. al. for Electricity Generating Authority of Thailand, January 1982.
3. "Summary Report on Resettlement Plan and Compensation Rates for Pak Mun Hydroelectric Project Units 1-4", by Hydro Power Construction Department, Electricity Generating Authority of Thailand, December 1990.

Fig. 7 A typical new house, 4 x 8 m, 2 storey, as provided to a resettler

Hydropower'92, Broch & Lysne (eds) © 1992 Balkema, Rotterdam. ISBN 90 5410 054 0

Reservoirs and developing countries

Obeng Michael Adjei
Architectural and Engineering Services Corporation, (AESC) Accra, Ghana

ABSTRACT: Throughout the world, a number of large man-made lakes have been constructed mainly to serve as a source of hydro-power generation. This lecture outlines some of the resulting effects that large reservoirs, particularly in the developing countries, have caused in the field of public health through diseases of immeasurable dimensions which are destructing and diminishing the population of these areas in a gradual but painful manner.

1 INTRODUCTION

The above topic is certainly a broad one, hence in talking about reservoirs and developing countries, I am solely referring more precisely to large man-made lakes that have resulted from the construction of dams, of whatever magnitude, but occupy very vast surface areas.

Despite the good intentions envisaged in the planning and siting of such dams in these developing countries, most of them, if not all are already wreaking as much social and environmental devastation as a major conventional war, only the rate of their destruction is quieter and prolonged.

Table 1.WORLDS LARGEST CAPACITY RESERVOIRS

DEVELOPING COUNTRIES:

Name	Country	Capacity $10^9 \ m^3$
Owen Falls	Uganda	2700
Aswam	Egypt	169
Kariba	Zimbabwe	160
Akosombo	Ghana	148
Guri	Venezuela	138

REMAINING WORLD:

Kahkovskaya	Russia	182
Bratsk	Russia	169
Daniel Johnson	Canada	142

From the Table 1, it is evident that of the largest capacity reservoirs in the world, about 65% of the largest 8 are located in the developing countries with about 50% of these surprisingly in Africa. The technological advancement of these portions of our global formation is inadequate compared to the developed countries and hence even attempts to eradicate the consequences resulting from the lakes become national and ultimately global issues. Global in the mere sense that the same financial institutions that, by virtue of granting of loans to undertake such constructions are compelled to adopt measures financially to support governments to fight the resulting social effects, particularly in public health consequences.

The International Commission on large Dam is blueprinting in Vienna, the destruction of the planet's last free-flowing rivers within a generation, at a rate of 200 dams a year.

2 OBSERVATION

Large dams are the world's largest man-made disasters. The vast majority of large dams are being filled up with silt before they will be paid for. It needs to be emphasised that firms that rely on dam building contracts, funders recoiling and glued into their furnished offices without accountability of any sort, and politicians who view large dams as the opportunity of a lifetime do not possess any monitoring of the negative effects as the uncontrollable mortality rate resulting from diseases.

Diseases that have sprung up/ springing up from these accumulated reservoirs create so much casualties on the rural folks to such an extent that, it is the same financiers who eventually have to cope with funds to undertake researches and find solutions to these environmental catastrophes.

In developing countries, health workers, ecologists, economists, engineers etc., - everyone who is abreast with vivid information on such reservoirs - could testify to the environmental and social upheaval that these dams cause, and how even only a few people actually benefit from them. The electric pylons fly overhead villages, minor towns etc., without the inhabitants possessing any knowledge as to when "electrical visibility" could reach them.

Future planned gigantic projects under the auspices of the World Bank and other lending agencies should be reviewed, terminated and abandoned. The very few tropical rain forests that environmentalists are striving hard to preserve would eventually be engulfed by any of the projected future huge dams that are located from Algeria to Angola and China to Chile.

3 PLANNED EXAMPLES OF LARGE
  RESERVOIRS:

3.1 In Chile, the construction of the projected six dams on the Bio River would prune the Andean Wildlife closer to extinction, and exhilarate one of the most exciting wild rivers on earth.

3.2 In India, the famous Narmada river targeted for two dams would flood areas of considerable proportions not commenting on the 150,000 people that would be displaced from farms and towns. Plans eventually call for 30 major dams costing $20 billion and displacing over 1.5 million people.

3.3 On the Xingu River in Brazil, two dams would create the world's largest artificial lake, drawing 7,200 sq. km. of rainforest, habitats for thousands of Indians. It is therefore no wonder that the Kayapo Indians of the Amazon are leading an international campaign against the Altamina Hydroelectric project. It is estimated that, by the year 2010, the project will flood an area of the Amazon equivalent to the size of the United Kingdom.

3.4 In China, the Yangtze River could give way to a 600-km² reservoir by a dam in the three Gorges (Hubei Province). Displacement of humans alone is targeted over one million and the environmental damage from Shanghai to Chongqing would never be assessed completely anyway.

3.5 The destruction of the most important floodplain forest in Central Europe would be fulfilled if the targeted Gabcikovo Dam on the Danube River in Czechoslovakia is carried out.

3.6 The $45 billion James Bay Project might endanger the traditional Cree Indian Society as well as the Artic wildlife but the environmental consequences would not be the burden of the host country Canada alone, but a world-wide affair.

Nevertheless, the storage of hydraulic energy is still the most practical way of retaining large quantities of energy in an economic way for immediate availability. In order to ensure the role of hydro-power production of energy for all types of communities, efforts must be ensured to exploit the possibilities of operating big run-of-the-river type of power stations with moderate storage reservoirs that would not compound the health hazards of the communities through water-borne diseases as we experience in developing countries.

With the financial support of the UNDP special fund, the Food and Agricultural Organisation, FAO of the United Nations, in collaboration with the corresponding government's concerned, some developing countries as Ghana, Nigeria and Egypt have for some period embarked on research programmes to solving selected secondary problems emanating from the creation of their corresponding man-made lakes. Despite some positive merits derived from these researches in the fields of assisting food production, economic development and in the technical training of nationals, the most fundamental objective of these researches is epidemiological studies, in co-operation with the appropriate public health bodies on public health problems arising from the major ecological changes resulting from the creation of the reservoirs; studies of the vectors of schistosomiasis, malaria, trypanosomiasis and onchocerciasis.

The inhabitants of these areas have no other choices to their places of settlement, since some compensations have eluded most of them, and hence have imbibed superstitious beliefs as the causes of such diseases. Surprisingly, such vectors of viruses causing the outlined diseases exist only in the tropical areas of these developing countries but not in the reservoirs of the Developed World's.

The question that agitates one's mental faculty is: why then at all should the planning of such huge reservoirs be sited in the developing countries, where the resulting effects CANNOT be handled from all angles by Governments and Financial donors?

4 SOME MAJOR PUBLIC HEALTH PROBLEMS

The construction of man-made lakes changes existing ways of life and patterns of disease, and creates conditions in which the risk of explosive outbreaks of infections may be high.

Prevention or control of the many infections which might be associated with man-made lakes requires considerable effort. Each project will present its problems, and it should therefore be a cardinal rule that advisors on public health, biology and sanitation are associated at an early stage with each new reservoir project to provide guidance on studies which should be made and to anticipate those infections which might break out.

4.1 Snail-borne infections

With the construction of reservoirs and their associated irrigation systems in the tropics and subtropics, an extension of the freshwater snail populations and an increased intensity of schistosomiasis can be foreseen. The problems are therefore widely recognised but, unfortunately, a generally applicable solution does not exist.

4.2 Arthropod-borne infections

The most important and destructive among these infections are malaria, onchocerciasis, trypanosomiasis, filariasis and virus infections. Malaria and yellow fever, however, might appear as sudden outbreaks, sweeping through a population and disrupting all aspects of community life. Filariasis infections are less obvious since the parasites requires several months to develop in the human host and the full effects may not be seen for several years.

4.3 Malaria

Most diseases in the tropics are the direct/indirect effects of the malaria parasite. With nearly 100 different species of mosquito throughout the world (with the deadliest located in the tropical zones of developing countries) ready to act as malaria vectors, each with a different ecology and each with facts of its life that are still unknown, there can be no simple or universal control method.

4.4 Onchocerciasis

River blindness, as the infection is mostly called, is very common in large areas of Africa, especially in West and Central Africa, and Central America. Since spillways from dams provide ideal breeding sites for the vector devices such as automatic siphonic spillways and submerged pass pipes should be used wherever possible to prevent breeding and save the health hazards of the communities of developing countries.

4.5 Trypanosomiasis

The foci of African sleeping sickness are especially related to the geography and the vegetation pattern of the countryside. The vector, the tsetse-fly is associated in West Africa with the lakeside and riverside, more specifically with the light forest which often fringes the water's edge. The vegetation pattern around man-made lakes, particularly in the tropics will change and evolve during the years following dam construction, and it will be incumbent upon all concerned to recognise the probable growth of new tsetse habitats.

5 THE TECHNOLOGY

The simplest form of small hydro-power plant/station is the run-of-the-river plant, which has virtually no storage dam. This arrangement, from hydro-power technology, diverts water from the river into a turbine located downstream and back into the river. The mode of deliverance of water into the turbines could adopt any technological approach, but the ultimate gain is to eradicate the creation of these almost stagnant voluminous waters.

Typical examples of the havoc that such lakes have caused are well vexed in WHO and other affiliated publications.

Using this run-of-the-river approach, and embarking on the use of other major waterfalls for small hydro-power plants, the public health standards of inhabitants could never be impaired to such magnitudes as are being observed at present.

Various cost analysis have proved that such small power plants are more expensive to construct than large plants and dams; this could certainly never be true if the human eradication factor with relation to resulting effects from public health diseases are taken into consideration. The destruction is prolonged and could stay with any affected society for any continuous period of human existence.

6 CONCLUSION

Notwithstanding the socio-economic and politico-religious gains that the construction of large dams could offer, it would be quite appropriate for all environmentalists as well as peace lovers to contribute in the elimination of the global catastrophes that large reservoirs create for developing nations. It is no need to be adamant anymore and thereby allow the mass destruction masquerade as development. Everywhere nowadays, it is evident that large dams cost as much as any war, not only in the third world countries, but in the whole world. There is the need to adopt a global approach towards the widespread of various water borne diseases by halting the planned construction of gigantic reservoirs in developing countries and make the life of the third world child a seemingly bearable one in future.

Project proposals should normally include budgetary provisions for health studies as a preliminary step to the development of subsequent health protective measures. It is essential that this advice be utilised at an early stage of each project, even before engineering works commence.

Despite the current concentration of hydro-power in developed countries, the World Bank estimates that 50 per cent of the world's total hydro-power potential lies in developing nations with only 10 per cent hitherto exploited. Consequently, it is in these countries that the world must expect the most rapid expansion in the future, and as such efforts must be made to ensure that the future developments of these hydro-power sources should reduce to eliminate the brutal consequences with regards to Public Health through diseases of incalculable dimensions, particularly in developing countries.

REFERENCES

Karl F. Lager, Man-Made lakes - Planning and Development, UNDP & FAO of United Nations, Rome, 1969, 3:19-25

Mandeline Scott, Water for Power - Some economic aspects. Article in Developing World Water. edited by WEDC, publ. Grosvenor Press International, pp. 420 - 422.

Mosonyi. E. Water Power Development Vol.I, II, English Editions.

Hydropower'92, Broch & Lysne (eds) © 1992 Balkema, Rotterdam. ISBN 90 5410 054 0

Numerical modeling of Rupert Bay – NBR Hydro-electric Complex development

Peter Sheng
University of Florida, Gainesville, Fla., USA

Peter Ko
SNC-Shawinigan, Montreal, Que., Canada

Tung T. Quach, Tonino Nzakimuena & P. Desroches
Hydro-Québec, Montréal, Que., Canada

ABSTRACT:Field data collected during 1990 and 1991 are used for the preliminary calibration of a 3-Dimensional numerical model that permits simulations of the hydrodynamics and sediment transport in the Rupert bay. The numerical model includes the effect of wave action on the resuspension of sediments and an empirical formulation of the flocculation phenomena.The wetting and drying scheme is also included.

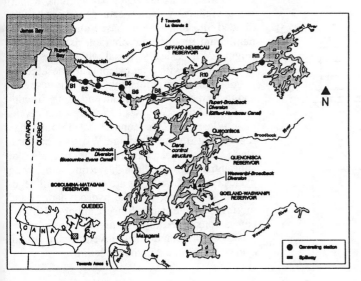

Fig. 1: NBR complex.

1. INTRODUCTION

In 1982 ,the hydroelectric scheme Nottaway-Broadback-Rupert (NBR) Complex (Hydro-Québec, 1990, Desroches, et al., 1992) (Figure 1.) was proposed by Hydro-Québec for study. For this scheme,flows from Nottaway and Rupert rivers would be diverted into Broadback river at various locations.Although the total flow volume will remain unchanged over the year, discharges into Rupert bay will be mainly through Broad-

back estuary. As part of the environmental studies of the Rupert bay due to the development of the NBR Complex, numerical modeling of the entire Rupert bay and part of the James bay was conducted, which is shown in figure 2.

2. MODELING AREA

The study area can be divided into two regions, the entire Rupert bay and part of James bay (be-

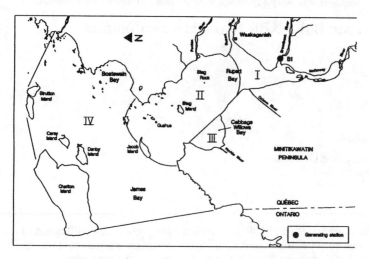

Fig. 2: Rupert Bay.

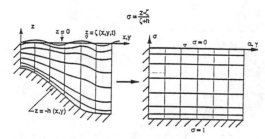

Fig. 3 The vertically stretched grid

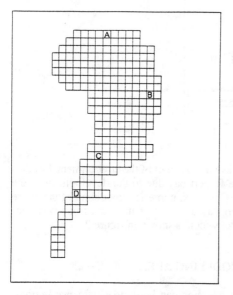

Fig. 4 Grid for zones I, II and III only

ing referred to as the outer bay region). Rupert bay is a relatively shallow estuary and can be subdivided into four regions based on the hydrodynamic and topographic characteristics. Zone I (Figure 2) is basically fluvial while Zone II is a mixing zone, with water depths in these zones ranging from approximately 2m at the upstream end to 10m at the downstream end. Zone III consists of Cabbage Willows bay with water depths mostly less than 2m. Zone IV is the outer bay region.Within the Cabbage Willows bay and along some coastal areas in the outer bay, areas of substantial sizes go through wetting and drying cycles.Part of these areas are covered with vegetation, which may alter the hydrodynamics and mixing characteristics. Since Cabbage Willows bay is considered to have distinctive sediment processes, simulation of the wetting and drying cycles was included in the modeling.

3. NUMERICAL MODELING

Numerical modeling was conducted with the use of the ESHM3D code developed by Sheng(1983,1991). The development of ESHM3D, a 3-Dimensional Estuary and Lake Hydrodynamic and Sediment Model, was based on the following assumptions:

-hydrostatic pressure distribution;
-turbulence time scale is much less than the mean flow time scale;
-turbulence quantities vary little over

318

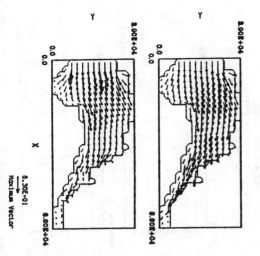

Fig. 5 uv velocity at time of 72,5 hours and sigma of -0,100 and -0,900

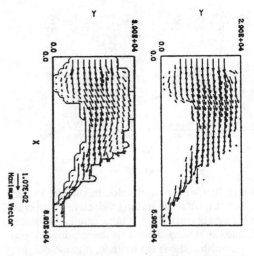

Fig. 7 uv velocity at time of 120,0 hours and sigma of -0,100 and -0,900

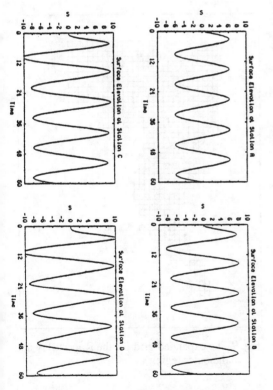

Fig. 6 Surface elevation at station A, B, C and D

the length scale of vertical turbulence eddies, such that vertical turbulent mixing is in near equilibrium condition at all times;

-fine sediment particles (d< 100 m) are sufficiently homogeneous in their sizes such that the suspended sediment dynamics can be represented by the concentration of a single size group;

-volumetric concentration of the fine sediment particles is sufficiently small such that all particles follow the turbulent eddy motions;

-suspended sediment concentration is sufficiently low(< 1000 mg/l) such that non Newtonian behaviour can be neglected;

-measurements of such model parameters as settling velocity, mineralogical composition of sediments, and bed structure are available; and

-wind-wave and wave-induced bottom stresses are available from either measurements and/or wave modeling.

The modeling equations in Cartesian coordinates are:

$$\frac{\partial u}{\partial x} + \frac{\partial v}{\partial y} + \frac{\partial w}{\partial z} = 0$$

$$\frac{\partial u}{\partial t} + \frac{\partial u^2}{\partial x} + \frac{\partial uv}{\partial y} + \frac{\partial uw}{\partial z} = fv - \frac{1}{\rho_o}\frac{\partial p}{\partial x} + \frac{\partial}{\partial x}\left(A_H \frac{\partial u}{\partial x}\right)$$
$$+ \frac{\partial}{\partial y}\left(A_H \frac{\partial u}{\partial y}\right) + \frac{\partial}{\partial z}\left(A_v \frac{\partial u}{\partial z}\right)$$

$$\frac{\partial v}{\partial t} + \frac{\partial uv}{\partial x} + \frac{\partial v^2}{\partial y} + \frac{\partial vw}{\partial z} = -fu - \frac{1}{\rho_o}\frac{\partial p}{\partial y} + \frac{\partial}{\partial z}\left(A_H \frac{\partial v}{\partial z}\right)$$
$$+ \frac{\partial}{\partial y}\left(A_H \frac{\partial v}{\partial y}\right) + \frac{\partial}{\partial z}\left(A_v \frac{\partial v}{\partial z}\right)$$

$$\frac{\partial p}{\partial z} = -\rho g$$

319

$$\frac{\partial T}{\partial t} + \frac{\partial uT}{\partial x} + \frac{\partial vT}{\partial y} + \frac{\partial wT}{\partial z} = \frac{\partial}{\partial x}\left(K_H \frac{\partial T}{\partial x}\right) + \frac{\partial}{\partial y}\left(K_H \frac{\partial T}{\partial y}\right)$$
$$+ \frac{\partial}{\partial z}\left(K_v \frac{\partial T}{\partial z}\right)$$

$$\frac{\partial S}{\partial t} + \frac{\partial uS}{\partial x} + \frac{\partial vS}{\partial y} + \frac{\partial wS}{\partial z} = \frac{\partial}{\partial x}\left(D_H \frac{\partial S}{\partial x}\right) + \frac{\partial}{\partial y}\left(D_H \frac{\partial S}{\partial y}\right)$$
$$+ \frac{\partial}{\partial z}\left(D_v \frac{\partial S}{\partial z}\right)$$

$$\frac{\partial c}{\partial t} + \frac{\partial uc}{\partial x} + \frac{\partial vc}{\partial y} + \frac{\partial (w+w_s)c}{\partial z} = \frac{\partial}{\partial x}\left(D_H \frac{\partial c}{\partial x}\right) + \frac{\partial}{\partial y}\left(D_H \frac{\partial c}{\partial y}\right)$$
$$+ \frac{\partial}{\partial z}\left(D_v \frac{\partial c}{\partial z}\right)$$

$$\rho = \rho(T, S)$$

where (u,v,w) are fluid velocities in the (x,y,z) directions, ws is the settling velocity of suspended sediment particles, f is the Coriolis parameter defined as $2\Omega\sin\phi$ where Ω is the rotational speed of the earth and ϕ is the latitude, ρ is the density, p is pressure, T is temperature, S is salinity, c is suspended sediment concentration, (A_H, K_H, D_H) are horizontal turbulent eddy coefficients, and (A_V, K_V, D_V) are vertical turbulent eddy coefficients. Variable eddy coefficients are implemented in the model. The equation of state is based on the equations given by Eckart(1958).

Vertically stretching(orσ-stretching) is used in conjunction with the Cartesian grids in the horizontal direction. The -stretching introduces extra terms to the original equations of motion. However, most of these terms appear in the horizontal diffusion terms (Sheng and Lick, 1979, Sheng, 1983). (Figure 3)

In the horizontal plane, both uniform and non-uniform Cartesian grids are implemented. A computation grid of Rupert bay is shown in Figure 4.

The model employs a mode-splitting scheme to treat the external mode (vertically integrated variables) separately from the internal mode (3-D variables). This allows the use of different time-steps for the external and internal modes to minimize the computational time. An implicit finite difference scheme is used for the numerical computations of the external mode.

4. BOUNDARY AND INITIAL CONDITIONS

At the outer open boundaries, tidal elevations or currents can be imposed with vertically varying salinity and temperature. Flow boundary conditions are imposed at the river entrances to repre-

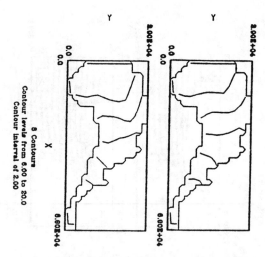

Fig. 8 Salinity at time of 120,0 hours and sigma of -0,100 and -0,900

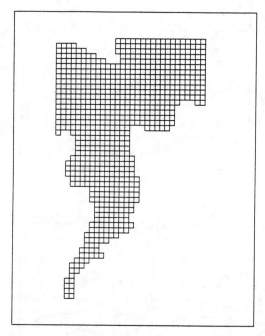

Fig. 9 The expanded grid for the Rupert Bay and the outer region

sent discharges from the rivers. Using a separate wave model, the bottom stress due to wind wave can be computed and used as input. Wind stresses are imposed on the free surface. Variable sediment characteristics are imposed initially on the

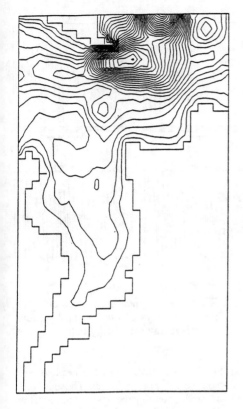

Fig. 10 The bathymetry contours

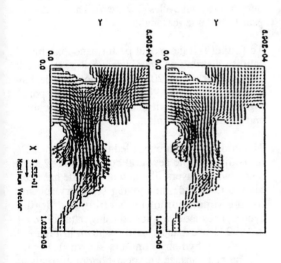

Fig. 11 uv velocity at time of 24,0 hours and sigma of -0,100 and -0,900

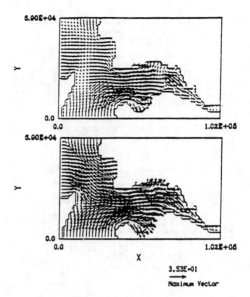

Fig. 12 uv velocity at time of 48,0 hours and sigma of -0,100 and -0,900

bottom. The erosion rates are calculated based on these characteristics.

5. TURBULENCE MIXING

Turbulence affects the erosion/resuspension, deposition, floculation and settling, and vertical mixing of sediment in the water column. In EHSM3D three turbulent models are implemented, namely, constant eddy coefficients, Munk-Anderson type eddy coefficients, and a simplified second-order closure model of turbulent transport(Sheng 1985; Sheng & Chiu, 1986; Sheng et al., 1989a). The simplified second-order closure model is based on the assumption of local equilibrium and is significantly simpler than the complete second-order closure model(the Reynolds stress model). However, the simplified second-order closure has been shown to give good results with little or no tuning of model coefficients(Sheng et al., 1989b & c; Johnson et al.,1989).

6. MODELING RESULTS

Using the preliminary grid as shown in Figure 4 and bathymetrie data collected in the early 1980's, preliminary simulations of Rupert bay have been

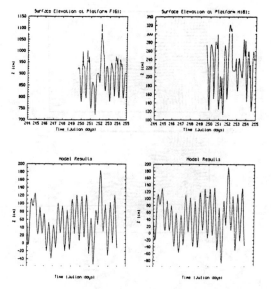

Fig. 13a Simulated and measured water level during Julian days 244 to 254 at 2 stations in Rupert Bay New bathymetry

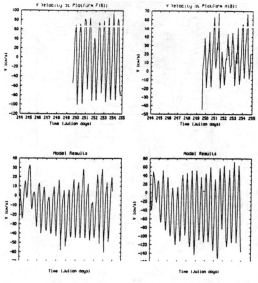

Fig. 13c Simulated and measured v-velocity during Julian days 244 to 254 at 2 stations in Rupert Bay New bathymetry

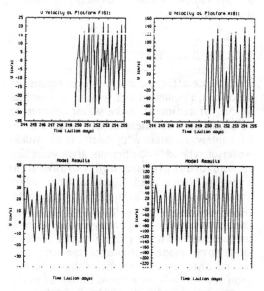

Fig. 13b Simulated and measured u-velocity during Julian days 244 to 254 at 2 stations in Rupert Bay New bathymetry

conducted first for tidal circulation in the Rupert bay with a M2 tide of 50 cm range and zero salinity. The results indicated that strong currents can be generated in the very shallow eastern

region. Stable simulations can be generated with time step on the order of 5 minutes. These include the vector plots showing the instantaneous flow field at 72.5 hours (Figure 5), the time series of the surface displacement at Stations A,B,C, and D during five tidal cycles (Figure 6). The results appear to be quite reasonable.

To further test the model performance, an initial salinity distribution was introduced next, which varies linearly from 20 ppt at the mouth of the bay to 4 ppt at the eastern boundary. The corresponding results are shown in Figure 7 for velocities and Figure 8 for salinity.

The grid was futher expanded to include zone IV in the model. The new grid consisted of 52 x 31 points with 5 layers. The computer code has been revised to include the moving boundary scheme for the simulation of the wetting and drying cycles. The model was tested using data collected in 1990. The expanded grid is shown in Figure 9 with the bathymetry contours shown in Figure 10.The near-surface and near-bottom currents at 24 hours after being forced by a semi-diurnal tide of 1 m amplitude are shown in Figure 11 with the corresponding currents shown in Figure 12.

For the September 1990 simulation,wind forcing

322

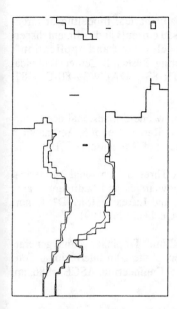

Fig. 14a Location of shoreline at 108 hours

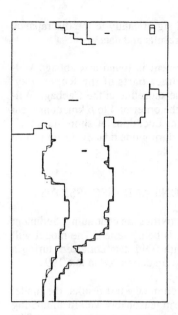

Fig. 14c Location of shoreline at 114 hours

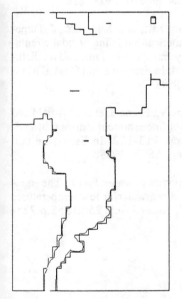

Fig. 14b Location of shoreline at 111 hours

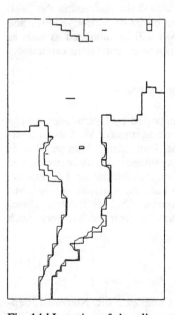

Fig. 14d Location of shoreline at 117 hours

on the free surface was introduced together with the tide and salinity field. The model was used to simulate a period of one month. The results are shown in Figure 13a, 13b, 13c. It appears that relatively large transient currents are generated during the first 10 days of model simulation, which contributes to the discrepancy between simulated and measured currents. After the first 10 days, agreement between simulated and observed currents are much improved.However, the relatively coarse resolution of the bathymetric data in the vicinity of the current meter stations also contributed significantly to the discrepancy between model results and data.

Shoreline movement is found at Cabbage Willows bay and other parts of the Rupert bay. Movement of the shoreline in the Cabbage Willows area is on the order of 4 to 6 km, consistent with observation. Locations of shoreline at selected durations during one tidal cycle are shown in Figure 14a to 14e.

7. CONCLUSIONS AND DISCUSSIONS

The modeling activities are continuing. Sediment transport model is being tested. The model will be calibrated with field data collected during a recent field program conducted in late 1991.

Based on the preliminary test results, the model appears to perform reasonably well in simulating the free surface elevations and currents. The shoreline movements on the order of 4 to 6 km compared well with those observed in the field. The effect of ice cover on hydrodynamics and sediment transport will be simulated as soon as the cases with open water conditions calibrated.

ACKNOWLEDGEMENTS

This study is supported by Hydro-Québec. The supports and encouragement of Mr. Géatan Guertin, Directeur, Mr.Jean-Étienne Klimpt, Chef de Service, Vice presidence Environnement, and Mr. Dinh Le Van, Administrateur d'Ingénierie, Mr. Jean-Pierre Lardeau, Chef de Service, Mr. Jacques Prud'homme, Chef de division, Direction Aménagement de centrales, are very much appreciated .

REFERENCES

Hydro-Québec, Complex NBR, Numéro 1, Septembre 1990.

Desroches P., Bergeron D.et al., " Physical consequences of the NBR project on the Rupert bay area: description of the various interrelated studies", Proceedings of the 2nd International Conference on Hydropower, Lillehammer, Norway, June 16-18, 1992.

Sheng, Y.P., Mathematical modeling of three-dimensional coastal currents and sediment dispersion: model development and application", Coastal Engineering Research Center Technical Report CERC-TR-83-2 DACW39-80-C-0087, 1983, 297p.

Sheng Y.P., "A new one-dimensional ocean current model", Tech. Report No. 566, Aeronautical Research Associates of Princeton, N.J.,1985.

Sheng Y.P., " A Three-dimensional Numerical Model of Hydrodynamics and Sediment Transport in Estuaries and Lakes: EHSM3D", Report (draft) to Hydro-Québec, May 1991.

Sheng Y.P. and Chiu," Tropical cyclone generated currents", Proc. of the 20th International Conference on Coastal Engineering, ASCE, 1986, pp. 737-751.

Sheng Y.P. and Lick W., " The transport and resuspension of sediments in shallow lake", J. Geophys. Res., 84,1979, p. 1809-1826.

Sheng Y.P., Choi J.K., and Kuo A.Y., " Three-dimensional numerical modeling of tidal circulation and salinity transport in James River Estuery", pp. 200-208, In Estuarine and Coastal Modelling, ASCE, 1989a.

Sheng Y.P., Cook V., Peene S. et al., " A field and modeling study of fine sediment transport in shallow waters" , pp. 113-122. In Estuarine and Coastal Modeling, ASCE, 1989b.

Eckart C., " Properties of water, Part II. The equation of water and sea water at low temperatures and pressures", Amer.J.Sci. 256;1958,p.225-240.

Hydropower'92, Broch & Lysne (eds) © 1992 Balkema, Rotterdam. ISBN 90 5410 054 0

Ice formation and breakup in Rupert Bay

B. Michel
Université Laval, Que., Canada

T.-T. Quach, T.-J. Nzakimuena & D. Bergeron
Hydro-Québec, Que., Canada

ABSTRACT: The ice conditions during the formation, growth and decay of ice in Rupert Bay were measured for two winters and, with satellite data, were interpreted in function of climatic and hydrological conditions. The analysis made possible the quantification of the average and extreme ice conditions that are needed in order to predict by computation the effects of modified river discharges on ice conditions when the Nottaway-Broadback-Rupert hydroelectric complex will be built.

1 INTRODUCTION

Hudson Bay and James Bay are the largest bodies of water in the world that seasonally freeze over each winter and become ice-free each summer (Figure 1).

At the south east end of James Bay, Rupert Bay receives the discharges of large rivers, three of which (Nottaway, Broadback and Rupert) might be harnessed by Hydro-Québec for power production. This development would change the timing and rate of flow of fresh water in the bay, which might in turn modify the nature and duration of the ice cover.

In order to assess the environmental changes on the ice cover, Hydro-Québec has sponsored many studies to understand the natural conditions of ice formation, growth and breakup of ice in the bay, so that the changes caused by the development could be evaluated (Michel and Doyon, 1991). The natural conditions of ice in Rupert Bay are the object of this paper.

2 ICE FORMATION

The first day of winter ice regime in Rupert Bay starts when the ambiant temperature

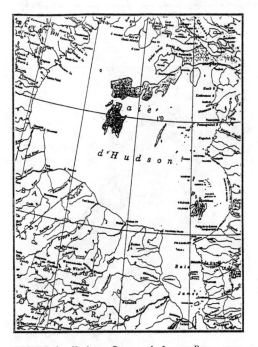

FIGURE 1: Hudson Bay and James Bay.

descends and remains below zero. This was computed from the air temperature data for a period of 58 years and found to occur on the average on the fifth of November.

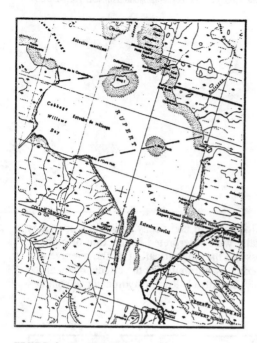

FIGURE 2: Rupert Bay.

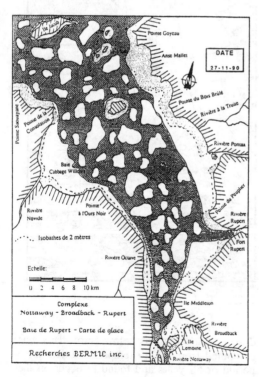

FIGURE 3.

After this date, shore ice and frazil appears almost immediately at the mouths of the Pontax and Nottaway rivers, while frazil appears slightly later at the mouths of the Broadback and Rupert rivers. A map of Rupert Bay is shown on Figure 2. Ice on tidal flats generally forms in mid-November along the islands and coast, the largest amount being seen along the eastern shore of Rupert Bay. During the third week of Novembre on the average, the opposite shores of the Broadback and Pontax estuaries are bridged with ice. Large sheets of thin drift ice form and cover partially the surface of the bay, their movements being dominated by the action of tides, currents and the wind (Figure 3).

Usually, the first continuous cover of thin, smooth ice forms up to Deer Rock, at the beginning of December, by sintering of large plates of ice (Figure 4). However, if the river discharge on this date is high, as was the case in the autumn of 1990, complete bridging by ice the upper part of the bay may be delayed for up to two weeks.

Deer Rock behaves in some ways as a sort of demarcation point for the glaciological phenomena in Rupert Bay. It is the approximate upstream limit for the progression of the saline wedge in winter. The ice that forms upstream of this point is relatively smooth, uniform and saline free, while in the lower reaches of the bay, the shallow channels accumulate ice that is hummocked and slightly saline.

The leading edge of the continuous ice field oscillates for some time about Deer Island. In general, it is only by the third week of December that the large ice floes join together to form a solid continuous cover in the three principal channels at the entrance to Rupert Bay (Figure 5).

At the end of December, the bay is completely covered up to Jacob's Island in an area where large ice ridges have been formed at the entrance to the bay (Figure 6).

3   WINTER CONDITIONS

For an average year, the severity of the

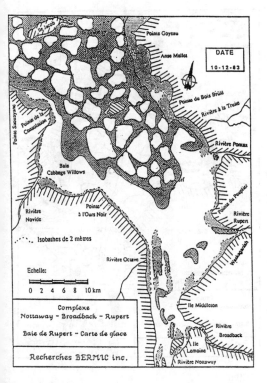

FIGURE 4.

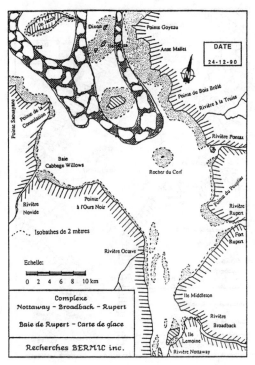

FIGURE 5.

winters is 2334 C degree days. This corresponds to the maximum value of the freezing index With the modified Stefan formula (Michel, 1971) and local measurements of ice thicknesses, the ice thicknesses can be computed for any value of the freezing index. For the average winter this gives a maximum ice thickness between 1,0 and 1,1 meters in the upper part of the bay from Deer Rock. Far from the shoreline, the ice is partially rafted and consolidated with a thickness that can attain between 1,2 and 1,4 meters. The piled-up unconsolidated floes in ridges can reach a thickness of several meters.

The area of zones in which the ice is piled-up and anchored to the bottom is relatively important along the coast, especially on the side to which the dominant winds drive the drift ice at the beginning of winter. This shorefast ice also accumulates around islands and can even form new islands composed entirely of ice that has piled-up on points of higher bed elevation. Cabbage Willow Bay is

completely covered with ice that extends vertically throughout its depth except for a few small deeper channels.

During winter, there are only two channels at the entrance to the bay with relatively smooth ice. The whole surface of the smooth ice in the bay is windswept, consequently the snow accumulation is only of the order of a few centimeters. On the other hand, snow accumulation is more significant (of the order of 30 cm.) in the troughs of hummocked ice.

Hanging frazil dams form at the upper extremities of the Rupert and Broadback estuaries. They are fed by frazil generated in the upstream rapids of these rivers, these rapids remaining open throughout the winter.

The tidal amplitudes in the Hudson and Rupert bays are strongly damped in winter due to the additional energy loses caused by the extensive ice cover. This is particularly true of Rupert Bay where the average tidal amplitude of 2 m in summer is reduced to 0.7

327

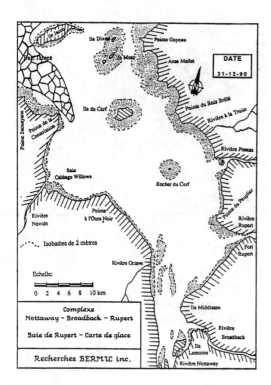

FIGURE 6.

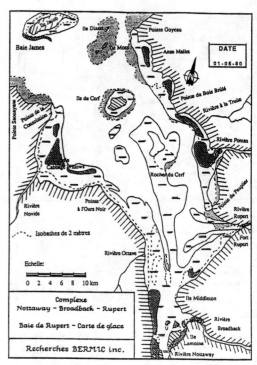

FIGURE 7.

m in winter. The saline intrusion appears to remain stationary during the whole year at a point in close proximity to Deer Rock. An interesting phenomenon manifests itself in winter as well as in summer : the propagation of two consecutive flood tides without an ebb flow caused by longitudinal surface gradients of atmospheric pressure. The tides form due to the passage of strong zones of atmospheric depression across Hudson Bay which are transmitted to the water body through a moving ice pack that is not continuous and solid on this bay. The pressure wave thus produced propagates under the solid and continuous ice cover of Rupert Bay and makes water jets to squirt out from fissures in the cover.

4   ICE BREAKUP IN RUPERT BAY

The melting and breakup of ice in Rupert Bay generally occurs during a relatively short period of about a month around the end of April or beginning of May.

It is initiated at the river mouths where the water is relatively warm due to the southern location of these rivers and the transport of heat from the zones containing rapids. The first open areas in the ice covering the estuaries appear at the end of April, and as in the formation period, the estuary of the Pontax distinguishes itself by being the first to be ice free.

The ice cover is melted both by the warmer water coming from the rivers and the surface melt by thermal exchanges with the atmosphere. In 1978, computations and measurements showed that at "Pointe des Peupliers" 29 % of the ice was melted by the river flows, but that further offshore at Deer Island, the effect of the rivers became negligible compared to the surface melt.

The melting of the surface ice becomes increasingly important during the first two weeks of May. The water initially floods the ice formed on shallow tidal flats, shoreline channels open and the leading edge of the ice

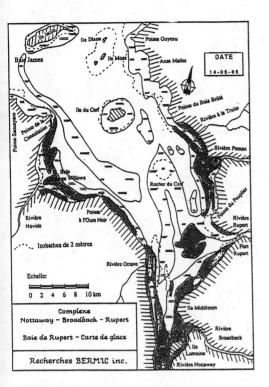

FIGURE 8.

cover retreats rapidly towards the open water from the river estuaries (Figure 7). By mid-May, the majority of the tidal flats and Cabbage Willow Bay are free of ice and pools of water may be seen on the remaining stretches of ice cover in Rupert Bay. The edge of the ice cover has now retreated towards open sea area close to Deer Roch (Figure 8).

If a heat wave occurs, the ice cover continues to retreat rapidly beyond Deer Rock towards the open sea. In general, this cover oscillates for several days between Deer Island and the entrance to the Bay until the end of May (Figure 9). During all the years of study, no ice was observed in the bay after the 10th of June.

5   CONCLUSION

The detailed observations made during 1990 and 1991 (Hydro-Québec, 1991) coupled with those already accomplished during 1978 (Michel and Purves, 1978) as well as satellite

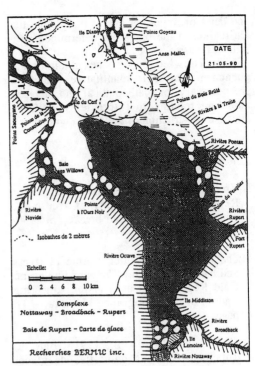

MAP LEGEND

 Land

 Open water

 Uniform smooth ice

 Ice islands

 Slush, brash and ice blocks
(concentration 1-5)

 Slush, brash and ice blocks
(concentration 5-9)

 Small ice floes 10 m - 100 m

 Medium ice floes 100 m - 500 m

FIGURE 9.

photographs allow the ice conditions in Rupert Bay to be relatively well documented. With the present study, the physical characteristics and dates of important events that define the ice regime in the bay were determined on a statistical basis.

The knowledge of the glaciological phenomena and their quantitative evaluation are necessary for mathematical and physical modeling relating to future ice conditions after the construction of the NBR complex.

REFERENCES

Hydro-Québec, 1991. Hydrometric Division : Reports and photographs of freeze-up, winter ice conditions and breakup in 1990-91.

Michel, B., 1971. Winter regime of rivers and lakes. U.S. Army Corps of Engineers, Cold Regions Research and Engineering Laboratory, Monograph 111-B1a, 131 pages.

Michel, B. and Doyon, B., 1991. Baie de Rupert, Conditions naturelles de glace. Report to Hydro-Québec, 66 pages and 7 appendices.

Michel, B. and Purves, W., 1978. Conditions de glace dans la baie de Rupert - 1978. Report to Société d'Énergie de la Baie James.

Hydropower'92, Broch & Lysne (eds) © 1992 Balkema, Rotterdam. ISBN 90 5410 054 0

Croatian environmental regulations regarding hydrotechnical projects

J.Rupči

Elektroprojekt Consulting Engineers, Zagreb, Croatia

ABSTRACT: Since 1984, the obligation has been casted by the Croatian regulations to prepare an Environmental Study for particular projects, including the hydrotechnical. The article presents the basic concepts of the Study development, acceptance and results application. The Kosinj Reservoir has been used as an example, with total capacity of $561x10^6$ m3 and the surface area of $16x10^6$ m2. The major negative and positive effects of this structure on environment have been described, as well as the measures to be undertaken prior to construction and during exploitation of the reservoir in order to mitigate eliminate the negative and achieve the planned positive effects.

1 GENERAL

Each hydrotechnical project is characteristic for its close relationship with nature, namely environment it has been set in. The relation is so intensive that it could be said that the functionally interrelated natural elements and human activity form a hydrotechnical project. Therefore, it is necessary to investigate the natural ambient the project is to be set in, and define with maximum precision possible impact of the project on environment, or the impact of the environment on the project, since they are mutual, already during the planning stage.

Since 1984, the obligation has been casted by the Croatian regulations to prepare an Environmental Study for construction or reconstruction of projects, including the hydrotechnical projects, aimed at environment protection and improvement, conservation of natural resources and man-made structures, and provision of favorable conditions for living, residing and work activities of population. The Study should provide for ecological, spatial, technological and economical bases for determination of Special Environmental Requirements included in Site Planning Requirements issued by the Ministry of Environmental Protection, Physical Planning and Construction for each particular project, which are one of the basic documents for obtaining the plant construction permit.

The Environmental Study is developed in two stages:
1. Preliminary Environmental Study, and
2. Final Environmental Study.

The Preliminary Environmental Study is developed simultaneously with the generation of the Preliminary Study for the project, on the basis of existing data. The Study shall include the review of possible project environmental impact, available measures for mitigation and elimination of damages, and the draft investigation program on environmental impact of the project, as required for preparation of a Final Environmental Study.

The Final Environmental Study is developed simultaneously with development of the Feasibility Study, and it is based on the data obtained by the investigation works both those carried out to define the environmental impact of the project, and those performed for the project itself. The Study shall include the review of possible environmental impact, available measures for mitigation and elimination of damages in the project surroundings, environmental monitoring program to be implemented during the project construction and operation. Additionally, the Final Study shall include the costs incurring from the environmental damage mitigation and elimination, which are the part of the total project construction and operation cost.

The Study development is being monitored and evaluated by a Board consisting of experts in particular fields, assigned by the Ministry.

When the Study is completed, the public

hearings are to be carried out in the project region. All the comments and suggestions resulting from these hearings are to be evaluated by the Study Development Monitoring and Evaluation Board and incorporated in the Study at their discretion. The final Study evaluation related to acceptance of its results is given by the Board.

The present article shall give a presentation of the Environmental Study development procedure and final results, using the Kosinj Reservoir as the example.

2 BASIC RESERVOIR CHARACTERISTICS

The reservoir is the part of the Senj Hydroelectric System, and it regulates regimes of the Lika and Gacka Rivers. The water power potential of these rivers is used in power generation Lika Highlands – Adriatic Sea power system. The reservoir is formed in the karst region of Croatia, by construction of three dams: Bakovac Dam of rockfill zoned type with central clayey core, 55 m high, 529.18 m long in crest, with volume of 1.43×10^6 m3; another dam located in saddle, also of rockfill type, with central clayey core, 24 m high, 516.3 m long in crest, with volume of 0.24×10^6 m3, while the third is the main Kosinj Dam on the Lika River, Marcello type, 74 m high, 413 m long in the crest, concrete volume 247,000 m3. The Kosinj Dam includes the spillway controlled with radial gates, sized to the transformed maximum flow of the 10,000-year return period flood wave of 1.140 m3/sec, the bottom outlet with the discharge capacity at maximum operating water level in the reservoir of 318 m3/sec and the hydroelectric power plant adjacent to the dam with the installed discharge of 90 m3/sec and rated power of 35 MW. The total reservoir capacity is 485×10^6 m3, the effective capacity is 433×10^6 m3, and the surface is 16×10^6 m2. The reservoir seepage losses shall be prevented by the grout curtain, total lenght about 4.5 km and maximum depth 150 m.

3 PRELIMINARY ENVIRONMENTAL STUDY

The Preliminary Study is based on investigations and analyses, and the results of the Optimum Use of the Gacka & Lika Rivers Water Project. This Project has systematically considered the problems related to this area, including the forecast of the social and economic development, use of space, water and land, water quality, and flood control, based on its social, climatological, hydrological, geological, hydrological, geological, hydrogeological and pedological characteristics.

One of the results of this project, regarding the water use, is the construction of the Kosinj Reservoir within the Senj Hydroelectric System, storing the spilling water of the Lika River from the constructed Kruščica Reservoir which is also included in the System. Since the project had comprehensively covered the problems related to the subject region, it presented a sound bases for defining possible impact of the Kosinj Reservoir on the environment and a review of possible measures for mitigation and elimination of damages in the structure vicinity.

The possible impact on population, space planning and utilization, transport communications, municipal infrastructure, monuments, microclimatic changes, groundwater, water quality, flood control, power generation, flora and fauna, fishing and hunting, and aesthetical changes have been determined.

The measures for mitigation or elimination of damages are generally as follows: compensation to the inhabitants for flooded land and structures, construction of new roads and infrastructure, adequate protection of monuments, adequate measures in the area of the future reservoar as concern prevention of the reservoir eutrophication process and water quality preservation, the reservoir management that shall allow for the maximum flood protection of the downstream area, and define the "zero" status regarding the climatic characteristics and water quality, as well as the groundwater status.

4 FINAL ENVIRONMENTAL STUDY

In order to achieve the best quantity definition of the environmental impacts, and undertake the measures for their mitigation or elimination, special investigations and analyses within the Feasibility Study development activities have been carried out, and their results have been entered into the Final Environmental Study. The major studies covering particular effect areas have been listed in the References.

The accent shall be placed on the most important impact of the Kosinj Reservoir, the one it has on environment, while other impacts covered by the Study are not elaborated due to the limited length of the article.

4.1 Impact on the population, space planning and utilization

The major impact of the project is with no doubt the impact it has on the population, space planning and utilization, transportation communications and municipal infrastructure. It is due to the fact that the

Kosinj micro site housing the Kosinj Reservoir is an integrally arranged space with an area of approx. 234 km2 with six major an a number of smaller settlements. The central settlement is Gornji Kosinj, with school, church, community center, medical center, shops, post office, which have generally been used by the complete region. The settlement shall be flooded, and thus the affected zone shall exceed the reservoir area approx. 15 times. The area population counts 3,026, 873 of which are directly affected by the reservoir, either because it shall flood their residential or farm buildings with the arable land, or only the arable land which is the source of income for majority of the population in this region. The people live in three major settlements arranged on the surface area of 57 km2, thus the directly affected zone is 3.5 times larger than the reservoir area. In this area, 389 residential, 195 farm building, and 1314 ha of land shall be bought off. Of 873 people to leave their homes 66% is above 50 years of age, which is one of difficulties related to moving, since the people of this age do not easily adapt to the new way of life.

Having this in mind, the opinion was surveyed in all 426 households to be moved in order to obtain the exact insight into their approach to the problem, so it could be solved in the best possible way. Out of the total number, 77% wants cash remmuneration, 16% construction of new buildings accompanied by arable land on the site located within 10-20 km from the present, and 2% would accept the appartment and a job. The Study concludes that all these requirements should be met.

Regarding flooding of the central settlement, part of the road and infrastructure, the conclusion of the Study is that, on the new site selected by the population of this region during the public hearings, the new local center should be constructed with all necessary buildings, communications between all the settlements, and the new infrastructure.

4.2 Impact on the cultural monuments

The investigations aimed at preservation of cultural monuments have established the continuity of human existence in this region since prehistory. The oldest finds are from the Neolithic, Late Bronze and Iron Age, extending through the Antic and Roman Age to the late Middle Ages.

All the archeological finds and the monuments have been assessed, and on this basis the protection requirements have been set in the form of preservation, detailed land and photographic surveying or removal to the other site. Thus, the construction of a memorial park with the museum exhibiting archeological and ethnological collection, and architecturally valuable structures and comodities of the Kosinj region has been planned.

Even prior to these investigations it has been known that the medieval Kosinj had its printing house where, in 1483, the first Croatian book was printed, the liturgical book in old-Croatian alphabeth. The remains of the old town of Kosinj are located above the maximum reservoir pool, and should be preserved.

4.3 Impact on the water quality

The Kosinj Reservoir effect next in importance should be its impact on water quality. Preservation of reservoir water quality on a particular level is important not only because the water is one of the basic life-supporting elements, but also because in this case it shall be used for water supply of the surrounding settlements and the complete Croatian coast belt.

Examination of the current physical, chemical and biochemical characteristics of the Lika River water indicates that it is generally of the II cathegory, which in categorization means that the untreated water may be used for bathing, recreation and fishery, while the treated water is potable. The forecasts of the water quality in the future reservoir, carried out against the Vollenweider and Dillon model, indicate that gradual eutrophication should already be expected with the present water quality status, so the required precautions should be taken to prevent trophization of the reservoir water.

The major causes of the Lika River water pollution are the waste waters of the town of Gospić and the settlements by the river. Since one of the ways of preventing the water eutrophication is to provide that the water on the reservoir intake is of the higher quality regarding physical, chemical and biochemical characteristics, the Study specifies treatment of the Gospić waste waters prior to their discharge into the Lika River. Additionally, certain limitations are required regarding planning of new industrial plants within the catchment area, and particular attention should be paid to their impact on the Kosinj Reservoir water quality.

During the reservoir construction all the organic matter should be removed from the location to be flooded, and control of torrents and their catchment areas by technical and biotechnical measures in order to prevent erosive processes and entrance of the eroded material into the reservoir carried out. Further, solid waste depositing should be allowed only on the landfills developed specially for that purpose, and construction of landfills for detrimental and hazardous waste should be forbidden.

The process of eutrophication may be decelerated with maximally intensive exchange of all water layers in the reservoir, so this should be accounted for in reservoir management. During the reservoir exploitation, the water quality should be regularly checked, particularly regarding the trophi development, particularly vegetation, and when necessary apply the biological method of water plants by using fish.

Application of the specified protective measures during the Kosinj reservoir construction and exploitation could cause such a condition of the new ecosystem which shall meet the stipulated criteria for water supply, and shall be maintenable with the normal biological protection measures.

4.4 Impact on the microclimatic conditions

The first step in investigation of the microclimatic conditions caused by construction of the Kosinj Reservoir is determining of the "zero" status, to which purpose three weather stations monitoring all relevant climatological features shall be situated in the area with the 20 km radius. Linking these data by mathematical statistics to the Gospić Weather Station with a 60-year monitoring period shall enable precise definition of the "zero" status, and changes in particular climatological features due to the Kosinj Reservoir construction, after its completion.

4.5 The positive effect of the Kosinj Reservoir construction

The construction of the Kosinj Reservoir has been initiated on the basis of the spilling losses of the Lika River water from the Kruščica Reservoir, which is included in the Senj Hydroelectric System. During the 17-year period of exploitation the average annual spilling losses have amounted to $380x10^6$ m3, which makes 42% of the mean annual inflow of the Lika River, and 27% of the mean annual inflow of the Lika and Gacka Rivers, the water of which is used in the Senj Hydroelectric System. The reason lies in torrential characteristics of the Lika River flow which is best

characterized by the variation coefficient of the mean decade flows of 1.35, and insufficient volume of Kruščica Reservoir, with operating capacity of $128x10^6$ m3, which is only 14% of the mean annual inflow of the Lika River, namely 9% of the mean annual inflow of the Lika and Gacka Rivers, since this reservoir is also indirectly used in training of the Gacka River waters for which, due to the topographic conditions, it is not possible to construct the reservoir space. Construction of the Kosinj Reservoir with the $433x10^6$ m3 operating capacity increases the effective storage capacity of the reservoir in this hydroelectric system to $561x10^6$ m3, which makes approx. 63% of the mean annual inflow volume of the Lika River, which is about 40% of the mean annual inflow of the Lika and Gacka river joint inflow, representing the storage space which, provided the management is correct, enables very high level of the water regime training for these water courses.

This assumption has also been confirmed by the analyses of the Kruščica and Kosinj Reservoirs management, carried out on a mathematical model within the Feasibility Study. The analyses resulted in spilling loses reduction from $380x10^6$ m3 to not more than $26x10^6$ m3 a year in average, namely it was reduced approximately 15 times. This has been achieved by determining the most favorable reservoir water levels for particular decades of a years. Since these results, both regarding the spilling quantities and the most favorable water levels have been obtained by the analysis of the time series of the natural flows in a 60-year period, which has been considered sufficiently typical, it should be expected that the obtained results shall be confirmed in the structure operation.

Therefore, by construction of the Kosinj Reservoir, utilization of the water quantities in the Senj Hydroelectric System has been increased from 73% to 98% ; therefore, almost total utilization has been achieved, which reflected in increased power generation of the Senj Hydroelectric System for approx. 380,000 MWh, since the power equivalent of this system is approx. 1 kWh/m3 of water.

Additionally, almost complete rearrangement of the water quantity from the rainy into the dry season would be achieved, which reflects not only on the electric power quality related to generation of peak power, but also on preservation of water quality since during the summer season the reservoir water levels are high, which obstructs the eutrophication processes due to rise in water temperature.

The positive effect of the Kosinj Reservoir construction is reduction, both in magnitude and frequency, of the downstream Lipovo Polje flooding risks. This enables development of intensive agricultural production, which is one of the basic activities of the population in the region.

The Study concludes that the adverse effects the Kosinj Reservoir has on the environment could practically be eliminated would the required protective measures be undertaken; therefore, as regards power generation and water supply, realization of this project would have positive effects.

5 CONCLUSION

The development of the Environmental Study has enabled the interdisciplinary approach to assessment of an individual project resulting in a comprehensive determination of both positive and negative effects of a structure on its environment in the widest sense of the word, and in defining the required measures for elimination or mitigation of negative effects and achievement of positive effects of the planned project in exploitation.

REFERENCES

Climatological and Hydrological Study of the Kosinj Reservoir Site, Croatian Hydrometeorological Institute, Zagreb, 1986

Drechsler R., Archeologic Finds in the Water Project Sites, Zagreb, 1983

Evaluation of the Present Situation Effect and Possibilities for Reservoir Construction in the Gospić and Otočac Municipalities Region, Croatian City-Planning Authority, Zagreb, 1983

Geological and Hydrogeological Investigations, Institut of Geology, Zagreb, 1988

Habeković D., Ichtiological Research in the Kosinj Reservoir Site, College of Agriculture, University of Zagreb, Zagreb, 1988

Historical Preservation in the Kosinj Reservoir Site, Historical Preservation Authority, Zagreb, 1987

Kovačević A., Study on Employment Possibilities in Connection with the Gospić Municipality Development, Croatian City - Planning Authority, Zagreb, 1987

Optimum Use of the Gacka and Lika Rivers Water Project, Elektroprojekt Consulting Engineers, Zagreb, 1984

Plans of Regional Development of the Gacka and Lika Rivers Catchment Area by the Year 2025, Civil Engineering College, University of Rijeka, Rijeka, 1983

Predicting Social and Economic Development of the Lika and Gacka Rivers Catchment Areas for the Period 1980 - 2025, College of Economy, University of Rijeka, Rijeka, 1983

Seismological, Seismotectonic and Neotectonic Investigations, College of Natural and Mathematical Sciences, University of Zagreb, 1988

Štrcaj V., Environmental and Sanitary Study for the Kosinj Reservoir Site, Public Health Protection Authority, Rijeka, 1986

The Kosinj Reservoir and Kosinj Hydroelectric Power Plant, Feasibility Study, Elektroprojekt Consulting Engineers, Zagreb, 1989

Tutek V., Pejnović D., Lay V., A Study on Regional and Social Aspects of the Kosinj Reservoir Effects, Croatian City-Planning Authority, Zagreb, 1987

Hydropower'92, Broch & Lysne (eds) © 1992 Balkema, Rotterdam. ISBN 90 5410 054 0

Balancing of hydropower and non-power resources values

Rodney G. Sakrison
Washington Department of Ecology, Olympia, Wash., USA

ABSTRACT: Federal and state agencies, Indian tribes, hydropower developers, and public interest groups each have a different perspective on the role of balancing in hydropower licensing. The balancing of developmental and non-developmental values is inherent in the states' water allocation policies and Federal Energy Regulatory Commission's (FERC) licensing decisions. The role of balancing in the state and FERC's decision making is changing. This is one of the most significant areas of potential overlap and conflict between the states and the FERC.

1 BALANCING RESOURCES IN HYDROPOWER LICENSING

FERC has always had broad powers in Section 10(a)(1) of the Federal Power Act (FPA) to balance competing interests in reaching a licensing decision. It has traditionally been charged with ensuring that the project selected to be licensed was "[b]est adapted to a comprehensive plan for improving or developing a waterway." The 1989 Amendments to the Federal Power Act (ECPA) specifically required FERC to consider other environmental resources and competing uses: "adequate protection, mitigation of damages to, and enhancement of fish and wildlife (including spawning grounds and habitat), irrigation, flood control, water supply, and other purposes."

1.1 Comprehensive Planning In FERC Licensing

Previous descriptions of FERC's balancing process under ECPA have been, for the most part, reiterations of the statutory directives given to FERC to expand its existing authorities for comprehensive development to encompass broader considerations in licensing

hydropower projects. FERC's exclusive authority under Section 10(a)(1) was not diminished by the 1986 Amendments. ECPA invited FERC to "explore all relevant issues of the public interest", and encouraged FERC further federalize water resources development.

FERC's comprehensive planning responsibility under Section 10(a)(1) of the FPA is taken very seriously by the states, given the sweeping effects that the authority can create. The tendency of FERC and the applicants is to focus on maximizing the power uses of the waterway and meeting the statutory requirements to protect fish, wildlife, recreation, and other uses identified in the pre-ECPA Sections 4(e) and 10(a)(1) of the FPA. The final balancing in FERC's comprehensive planning authority stays mostly in the background during the pre-license consultation phase, and is usually not revealed until FERC issues an environmental assessment and order granting a license.

Despite FERC's allegations, the individual licensing proceeding is not an adequate forum for balancing all the competing uses of a waterway. To the extent that FERC ever attempts to look broadly at a whole basin, through pre-licensing environmental impact statements, it has been constrained to only address resources or uses directly affected by the pending licensing decisions and not consider all uses or potential users. FERC has not issued any formal guidance on the prerequisites for balancing. FERC has put forth its responsibility for centralized comprehensive planning authority in Section 10(a) as proof of its comprehensiveness.

ECPA provided incentives for states and other federal agencies to submit comprehensive plans for FERC's consideration in licensing decisions, under the authority of Section 10(a)(2). Rules governing state and federal comprehensive plans were adopted by FERC, which could signal a new importance of comprehensive plans at FERC. Under FERC's final rule, virtually any plan for any resource that contained data and criteria constituted a plan for their consideration. FERC weights these plans on the basis of the number of beneficial uses of a waterway it considers and balances, and the depth of the data and analysis. This did not instill much confidence among the states that FERC really cared about their comprehensive plans, and has made it difficult for states to commit scarce planning resources for the potentially dubious purpose of gaining recognition at FERC.

1.2 Equal Consideration of Power and Non-Power Values

Equal consideration of power and non-power resources under Sections 4(e) or Section 10(a)(1) has not received a great deal of attention from the states, following the passage of ECPA. Outside of fish and wildlife concerns, the balancing power and non-power values in hydropower licensing was not a big issues for several years following the passage of ECPA in 1986.

FERC recently circulated a paper that addresses in theory how resource values associated with competing beneficial uses should be evaluated (Fargo 1991). FERC

views this as a start toward a more objective evaluation of the "project best adapted to the comprehensive development of the waterway." This approach proposes a technical solution to the balancing question. This fits comfortably with FERC's perspective on licensing, which is removed, by their own admission from the local issues and preferences that give unique values to resource conflicts.

Recently, the hydropower industry has been developing techniques to articulate the balancing issues in hydropower licensing. It is approaching the evaluation of competing beneficial resources from a technical perspective (Carter 1989). Methodologies are being used that evaluate hydropower licensing alternatives by evaluating the effects on resource utilization. Cost-benefit analysis is used to evaluate the alternatives and advanced techniques such as decision theory are employed to appropriately guide the analysis (Decision Focus 1991).

The vulnerability of this approach is the assignment of values to competing uses is very subjective, particularly with inadequate input from the affected public. While technical evaluation of alternatives is necessary and much work needs to be done to familiarize ourselves with these methods, they alone are not sufficient to support the decisions necessary to resolve the resource conflicts inherent in the balancing process.

What can states offer at this time? It has long been held by all states that we are in the best position to fairly weight competing beneficial uses and articulate the public interest issues.

FERC often lacks the necessary information to determine how hydropower projects can be integrated into existing framework of law and water rights. FERC licensing often conflicts with the state balancing of competing uses in water rights and water allocations. Under current practice, the role of state water rights administration is reduced to merely determining the existence of prior vested rights, so that these rights might be struck down, or condemned in a licensing proceeding.

FERC lacks adequate information on the history of water management and water allocations in these areas, or sufficient information on existing, pending and reserved water rights. To do a credible job of balancing the competing uses FERC needs accurate information on stream hydrology, water conservation and reuse, fisheries, recreation, aquatic habitat, as well as the potential to enhance streamflows through trust water rights.

Given the diverse mix of benefits which are the subject of water allocation discussions throughout the West, the record on which FERC's bases its licensing proceedings is inadequate and is a threat to state efforts to allocate scarce water resources among competing uses.

Balancing at the state level is moving away from a technical or economic approach, to one that is more concerned with the political process, and in that sense possibly more representative, acceptable and enduring. These are the issues that have become the backbone of the state, local and tribal government discussions ongoing in the Washington.

1.3 Balancing Competing Uses in State Water Policy

The decision-making process for selecting among future water use alternatives is changing in the 1990's. Competition for scarce water resources has heightened concerns for protection of instream resources and constrained water use options. Decisions regarding appropriate future water uses and among competing water users are extremely complex.

Water resources planning inevitably runs up against the question: "What is the value of water if left for instream uses." This raises fundamental questions as to how environmental and other difficult-to-measure economic considerations (i.e. all intangible values) can be brought into the analysis. No use of water can be increased without reducing, in some way, the availability of water for other uses.

In the State of Washington, the Water Resources Act of 1971 stipulates that the allocation of public waters should attain "maximum net benefits", where benefits exceed costs, "including opportunities lost." Although this is a clear economic efficiency policy, it has yet to be implemented (Shupe 1988). Absent any clear guidance on how to incorporate economic, environmental and social costs in water resources planning, water and power utilities in Washington have attempted to address these issues on their own, with mixed results.

The challenge to power and water planning is to bring the consideration of environmental costs into the decision on balancing between competing users. Water resources planning has a long legacy of proposed techniques for assigning a monetary value to wilderness, recreation, wildlife, aesthetics, and other values hard to quantify. The current practice of evaluating competing uses based on the benefits and costs of a specific project proposal seldom results in consensus among different interest groups.

The identification of the value of water in alternative uses, particularly instream uses, is controversial and always arguable. It is more likely that agreement among interested parties on the value of water left instream can only occur through a regional water planning process that is more broadly based and objective than project specific planning. The future direction of water resources planning in the State of Washington is moving away from economics as the decision tool. The shift to mediation and consensus-building presents a new water resources planning model and introduces additional uncertainties.

Consensus-building at the local and state level can resolve long-standing water disputes in Washington that have forestalled the development of multiple-purpose projects and also left unresolved the extent to which other options, and other uses of the waterway, may be pursued.

2 BALANCING RESOURCES IN WASHINGTON STATE

The State of Washington is blessed with the water resources that have allowed substantial development of hydropower generation, probably more so than any other state. Like of the rest of the country, we are running out of good hydropower sites. New hydropower generation will have to be carefully sited with attention to high value environmental resources.

In Washington State, there are currently 92 operating hydropower projects in Washington with the capacity of generating 26,270 megawatts. That's two dozen nuclear plants. Washington, incidently enjoys the highest percent of hydropower utilization of any state. We currently provide almost 90% of our energy needs from hydropower. This is a significant portion of the state's water resources that are already managed and operated primarily as hydropower rivers.

Currently there are about 30 pending small hydropower projects, as well, being considered in Western Washington, most less than five megawatts. Development of cost-effective new hydropower resources in Washington is dependent on an understanding of the comprehensive hydropower plans that have been prepared for the region. New projects also need to be aware of the water resources planning context and water allocation policy issues currently being debated between state, local and tribal governments.

2.1 Northwest Power Planning Council

The Pacific Northwest Electrical Power Planning and Conservation Act (Pub. L. 96-501) established the Northwest Power Planning Council, and directed the Council to develop a 20-year electric power plan for the region. Through its periodic updates of the Power Plan, protected areas designation, and Columbia River Basin Fish and Wildlife program, the Council balances hydropower development with environmental protection. In accordance with the Act, the Council selects resources that are cost-effective and gives first priority to conservation resources. The preference for conservation resources by the Council recognizes that conservation resources are believed to have less adverse environmental effects than new generating resources.

Section 3(4)(B) of the Northwest Power Act defines system cost to include direct costs, indirect or future costs and such "quantifiable environmental costs and benefits as the (Bonneville Power) Administrator determines, on the basis of a methodology developed by the Council as part of the plan...are directly attributable to such measure or resource." The differences in environmental costs imposed by various resource options could influence decisions about the resources that comprise the resource portfolio.

The Council has identified three reasons why incorporating environmental costs are important in electricity planning. First, if external costs are not included in the cost estimates for resources, then the full costs to society to build a resource has not been fully considered. Second, including environmental costs could significantly alter the order in which resources are developed to achieve a least-cost energy future. The environmental costs of an otherwise relatively inexpensive resource might be large enough to make it more expensive than other resources with smaller environmental costs. Finally, the cost of large, centralized generating resources, such as potential new coal plants, are used as the basis for determining avoided cost and least cost alternatives, and often significantly determine how much conservation is determined to be cost-effective. If generating resources, such as coal, which are nearly cost-effective themselves, do not accurately reflect the environmental costs of the resource, then cost-effective investments in environmentally benign resources such as conservation may be missed, and never pursued (Northwest Power Planning Council 1989).

2.2 Washington State Comprehensive Hydropower Plan

In 1987 a Task Force on Hydropower Development and Resource Protection recommended to the Washington State Legislature that a state comprehensive hydropower plan should be developed in concert with appropriate interests. In 1989 the Washington State Legislature passed hydropower planning legislation and an appropriation to reconvene the Task Force to develop such a plan.
The 1989 Task Force represented the following interests: state agencies and legislative committees, Indian tribes, environmental and recreational groups, utilities, hydropower developers and other interested parties. The executive agencies have relied on their existing statutory authority to develop a plan, building upon the work done by the Task Force (Triangle 1991).

The purpose of the State Comprehensive Hydropower Development and Resources Protection Plan is to serve the broad public interest regarding development of cost-effective electricity and conservation of river-related environmental resources. The plan will do this by guiding new hydropower development toward locations where environmental impact can be minimized and away from areas where development might be incompatible with continued protection of significant natural or recreational resources.

The principle means to achieve this end is the identification of "resource agreement areas." The following three categories are identified:

- Resource Protection areas where hydropower development will likely conflict with significant environmental values, and

- Sensitive areas/hydropower opportunity and Less Sensitive/hydropower opportunity areas where development will not conflict with or may enhance environmental values.

The foundation of the plan is a base of consistent and reliable resource information. The State of Washington currently possesses the nation's most exhaustive base of river information and the most sophisticated computerized data retrieval and analysis capability. The plan incorporates definitions of mitigation and exceptions are provided for certain categories of hydropower development opportunities, such as projects proposed at existing dams.

The State Legislature determined that future development of hydropower and protection of river-related resources shall be guided by policies and programs which:

1) Create opportunities for balanced development of cost-effective and environmentally sound hydropower projects by a range of development interests;

2) Protect significant values associated with the state's rivers, including fish and wildlife populations and habitats, water quality and quantity, unique physical and botanical features, archaeological sites, and scenic and recreational resources;

3) Protect the interests of the citizens of the state regarding river-related economic development, municipal water supply, supply of electric energy, flood

control, recreational opportunity, and environmental integrity; and

4) Fully utilize the state's authority in the federal hydropower licensing process.

Using the authority given by the State Legislature, the executive agencies have used their existing statutory authority to develop a draft plan. The agencies have provided information and technical support to develop the Washington State Hydropower Data Base, the computerized basis for the draft plan.

The draft plan includes the following elements:

- An analysis of natural resources and public uses associated with the state's rivers and streams.

- An analysis of the potential conflicts between additional hydropower development and the protection of significant natural values.

- An analysis of the status of hydropower in the state of Washington, including both existing and potential generation.

- The identification of a process to reach an implementation strategy.

Although the State of Washington process for developing a comprehensive hydropower plan may have been a model for other states, it has been difficult. The hope of reaching a consensus with hydropower developers and environmental groups has been fraught with problems. It has yet to be determined whether the final plan must be formally adopted by the state Legislature. This is for the most part a political determination since the agencies had proposed using their existing authority to submit the plan to FERC for consideration during licensing. The agencies have also proposed to send the state comprehensive plan to the Northwest Power Planning Council for integration with their protected areas plan.

2.3 The Chelan Agreement

The Chelan Agreement, so named for the resort at which it was drafted in October, 1990, is a tri-party agreement bringing together the state, tribes and local government. It seeks to resolve the long-standing dispute between the state and Indians over protection of fisheries habitat, encompassing the provision of adequate instream flows and the allocation of remaining flows to competing beneficial uses. Conservation and improvements in state water resource data management are key to the agreement.

The Chelan Agreement sets out two pilot planning basins where the special interests will seek to mediate their competing positions. There are seven caucuses represented in the Chelan Agreement. The caucuses are: state government, the Indian tribes, local government, and interest caucuses representing fisheries and recreation, environmentalists, business, and agriculture.

Among the most far reaching implications of the Chelan Agreement is the movement to mediation and consensus as the foundation for water resource plan development and away from technical, economic and heavily bureaucratic planning. This is epitomized in the statement that "because of its cooperative nature, the results of this planning process will maximize the net benefits to the citizens of the state." This firmly precludes any higher review at the state level and solves for the present the maximum net benefits riddle found in the Water Resources Act of 1971.

Will this help or impede utilities in determining the extent to which a project is environmentally acceptable and cost-effective? Is this an appropriate solution to the difficulty in incorporating environmental values in balancing resources?

By combining dispute resolution and establishing environmental protection standards, the Chelan Agreement should result in firm allocations of water for competing purposes. The establishment of instream protection levels and out-of-stream allocations will give potential users firm estimates of the yields and costs of their development alternatives. This indirectly establishes a value of the water left instream and sets a price on the water available for allocation. Although there may be insufficient economic efficiency rationale in this method, it results in a readily acceptable balance of competing uses. Moreover, it meets at least one optimization test....it can't be improved upon.

3 CONCLUSIONS

This paper has explored the current balance of powers between the states and FERC. Comprehensive planning and the role of balancing in hydropower licensing are being widely discussed by participants in the FERC licensing process. The role of balancing in the state and FERC's decision making is changing. States' water allocation policies, which inherently balance competing uses, often conflict with FERC's licensing decisions.

REFERENCES

Carter, E.F. & Trouille 1989, Balancing Power and Other Instream Resources, *Hydro Review*, August, 1989, pp. 96-108.

Decision Focus, Inc. 1990, *Evaluating Hydro Relicensing Alternatives: Impacts on Power and Nonpower Values of Water Resources*, Electric Power Research Institute, Palo Alto, CA.

Fargo, J.M. 1991, *Evaluating Relicense Proposals at the Federal Energy Regulatory Commission, FERC Office of Hydropower Licensing*, Paper No. DPR-2.

Northwest Power Planning Council 1989. *Resource Evaluation Methodology*. Staff Briefing Paper, Publication #89-27, Portland,OR.

Northwest Power Planning Council 1989. *Accounting for the Environmental Consequences of Electricity Resources During the Power Planning Process*. Staff Issue Paper, Publication #89-7, Portland, OR.

Shupe, S. & Sherk 1988. *Washington's Water Future: The Report of the Independent Fact Finder to the Joint Select Committee on Water Resources Policy*, Olympia, WA.

Triangle Associates July 1991. *Washington State Hydropower Development/Resources Protection Draft Plan*. Seattle, WA

Hydropower'92, Broch & Lysne (eds) © 1992 Balkema, Rotterdam. ISBN 90 5410 054 0

Dams and reservoirs as an answer for a better water resource management in high dense populated areas of developing countries

Soejoedi Soerachmad
State Electricity Corporation, Indonesia

ABSTRACT: Water is the main and most important thing for human life, but the problem men are facing is that water is not always available every time and in every place. In the rainy season we have plenty and excess of water, but in the dry season we have shortage of water. Construction of dam and reservoirs will help to harness the water flow to the ocean during rainy season and providing water store to the people to be used during dry season, or in an other word giving people a better water resource management possibility. In this paper a comparison between the negative and positive impact of dam and reservoir construction and forest in line with the green policy will be discribed and tabulated which leading to the conclusion that people should not hesitate to construct big dams and reservoirs, even they have to sacrifice land or forest for the inundated area for the reservoirs.

1 INTRODUCTION

We are aware that due to GOD ALIMICHTY HANDS, water plays the most important role in human life. Water sustains the continuity of human life, and shortage of water will cause serious problems, and in extreme cases can cease living **or causing** chaos, dramatic death toll etc.

Water is also very important for plantation, forest animals etc. which in the other hands are also important for the human life.

This is what people feel that ecology and enviroment should be maintained and keep fungsioning as well. Besides used to sustain human living, water is also used to improve prosperity, by means for : irrigation, electricity, industries, transportation, recreation etc.

While water is very important for human life, but the problem men are facing is that water is not always available suficiently every time and in every place. In rainy season we have plenty and excess of water, but in the dry season we have shortage of water.

A relatively constant water availability is more or less required.

Big reservoirs and dams is one of the option to keep relatively constant water flow in a certain water shed.

By constructing reservoirs which are able to keep a huge amount of water in the rainy season and release it in the dry season can help to reduce the problem due to shortage of water in the dry season.

2 THE RESERVOIR

Reservoir is a volume area formed caused by constructing a dam harnessing a river, with the objective to collect water.

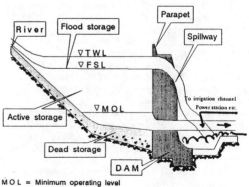

M O L = Minimum operating level
F S L = Full supply level
T W L = Top water level

345

3 THE PROBLEM OF WATER RESOURCE MANAGEMENT IN JAVA ISLAND

Java island is the most dense populated island in Indonesia with the population of ca 840 people per square kilometer.

As a developing country and around 70 % o of the Indonesian people living in Java island, still based their living from agriculture.

Even the Indonesia Government try to suppress the forestation but it will hard to succeed due to the people's rapid growth of more than 2% annually. The agriculture people still got "Hunger of land". The increasing number of agriculture people still need increasing land for agriculture.

In such situation the fungsion of forest as water regulator in a water shed will be continuously and rapidly reduced, and as a cosequence the water shed will be in a critical condition with the characteristics like.
- an extremely low discharge river in dry season, but
- uncontrolled flood in the rainy season
- high degradation of fertilelity
- people in the water shed are not able to plan and implement their income from agriculture
- more difficult to get drinking water
- poverty as a consequence
- chaos and last but not least Dramatical death toll
- etc.

Big reservoir is one of the options to help solving the non constant water dicharge of a river in a water shed to a certain extend. But a series of big reservoirs will contribute significantly managing water discharge of a river in a water shed.

Besides utilized as water canteen for several purposes as : water for irrigation, drickwater, electricity, sanitation, downstream water transportation, water in the reservoir can be utilized for sweet water fish cultivation, water sport etc.

Water in the reservoir also improving the ground water level on the surrounding and downstream area.

Anyhow besides the positive impact mentioned above, construction of big reservoirs has also their negative impact as what always been raised by the Green Activist.

So far noticed big reservoir always disturbing the ecology because constructing big reservoirs always force to move a lot numbers of people from their home land and inundate a significant area in replace of a reservoir.

Now people is coming to dillema in making a choice between the threat of water shortage in an extreme case especially in dense populated developing countries and the need ecologically not to be disturbed as what the Green Activist always strive on.

4 COMPARATIVE PROBLEM ANALYSIS

It is really understood that construction of big reservoirs always disturbing the eco logy.

But people must also realize that population growth in dense populated developing countries like Java island/Indonesia where 80% population still based their living from agriculture, in another manner is disturbing also the ecology continuously.

In this case the disturbing process act gradually but continuously with an amazing effect after a certain period if no other precaution can stop the process.

In the other hand it is hard to prevent the "Land hunger people" to stop deforestation in seeking new or additional land for agriculture.

In such condition although with an amazing effort we are lucky if we can manage the "Land hunger people" to slow down deforestation, but more impossible is to reduce people's land for forestation with the objective to restore a critical water shed.

For that to my opinion, in such condition construction of big dams and reservoirs is very important to obtain the maximum benefit from water resources in a water shed.

We need time, may be 30, 40, or 50 year to modernize agriculture structure population to a more balance industrial and modern agriculture structure, and also restoring the critical water shed.

The critical water shed need to be restored. If we are not able to restore the critical water shed the threat of chaos and dramatical death toll is waiting.

To come to a more objective answer the following approach is recommended.

5 TABULATION OF ENVIRONMENTAL AND ECONOMIC COMPARISON

No.	Construction of big reservoir	Not constructing reservoir
	ENVIRONMENT	
1.	Have to move people in average 5 people per Ha. (average condition in Java island)	Not moving people but population still increase and gradually pressing the environment.
2.	Sacrifice land for inundated area in replace of the reservoir	Not sacrificing land but the "Land hunger people" still a threat for the existing forest
3.	Reservoir improve water management and control water requirement in the downstream area	No water management at all
4.	Reservcir improving the ground water level in the urrounding and down stream area	Not improving ground water level
5.	Reducing flood during rainy season	Bigger and bigger flood during rainy season
6.	Winning time to restore the environment and developing the agriculture people to became a balance modern agriculture and industrial people	Drowned in time by doing nothing
	ECONOMIC	
1.	Need to invest for Dam and reservoir ca $ 150,000. per Ha	No capital cost for investment
2.	1 Ha. reservoir can be used for irrigating 5 Ha. rice field with the addional crop value of $ 2,000 per Ha. per year, or 1 Ha. reservoir gives additional crop value by irrigation of $ 10,000 per year.	No additional value
3.	1 Ha. reservoir can produce electricity average 600,000 KWH/year or $ 45,000 per year	No electricity produced
4.	In addition 1 Ha reservoir can produce sweet water fish ca 10.4 ton/year or $ 400 per year	None

From the above tabulation it can be easily understood that in general cases mostly dam construction is economically viable and environmentally better than doing nothing.

Tabulation for comparison between forest

No.	FOREST	RESERVOIR
1.	Retarding surface run off and increasing water per colation	Storing rain water in the reservoir and better management in down stream area
2.	Improving ground water level	Improving ground water level

No.	FOREST	RESERVOIR
3.	Reducing flood a little bit	Reducing flood much more than forest, because canteen capacity per unit area is much more than forest
4.	Preventing erosion	No erosion occur in the reservoir area, and especally when the reservoir is equipped with green belt
5.	Very beneficial for cleaning air by producing oxygen by the trees.	Plankton and micro organism at the reservoir produce oxygen
6.	Weathering leaves and organic materials fertilzing the soil	Reservoir water used for irrigation act also as fertilizer
7.	A suitable place for the living of several kind of fauna	A suitable place where several kind of fish can be cultivated

From this tabulation it is clearly understood that reservoir has the better environmental aspect per square area compared with forest.

6. CONCLUSION

From the above brief paper some conclusion can be taken.

1. In dense populated developing countries where mostly people are "Land hunger", they have critical water shed which needs to be restored for the better water resource management.

2. Restoring water shed by forestation or increasing forest area is impossible. With maximum effort and strong measures by the Government, we can only hope to slow down continuous destruction process of water shed. While developing countries need time for suppressing their population growth and making a more balance structure of modern agriculture and industrial population.

3. Big reservoir is an answer for such condition since :
- environmental and economically is better than doing nothing, while on the other hand forestation program is impossible.
- Reservoir has a more effective and efficient environment and water resource management capability persquare area.

7 REFERENCES

Kehutanan Indonesia No. 5, Th.V
Lembaga Ekologi UNPAD, 1985
· Laporan penelitian akuakultur dalam pemukiman kembali penduduk di proyek PLTA MRICA
Soejoedi Soerachmad, 1985
Bendungan dan waduk besar untuk penyelamatan DAS kritis di Pulau Jawa.

Hydropower'92, Broch & Lysne (eds) © 1992 Balkema, Rotterdam. ISBN 90 5410 054 0

Mathematical modelling of groundwater flow near semipervious reservoirs

Heinz Brunold
Steirische Wasserkraft- und Elektrizitäts-AG, Graz, Austria

Johann Stampler
TDV, Technische Datenverarbeitung Ges.m.b.H., Graz, Austria

ABSTRACT: This article presents a mathematical groundwater flow model. The model was intended to find answers to questions raised in connection with a planned power project in upper Styria involving the impoundment of the river Mur. Unusual difficulties arose in the preparation of the model due to the extremely heterogeneous conditions in the project area. As the river Mur in its original condition has no direct hydraulic connection with the water table over large parts of the backwater area and replenishment is only by percolation through an unsaturated zone, careful determination of the parameters describing water exchange was a necessity in the calibration of the model. The main result of our studies was the decision to omit any impervious elements from the project for the benefit of an increased groundwater replenishment.

1 THE PROJECT

The Fisching power station is designed as a combined storage and diversion-type facility on the river Mur in a traditional industrial region between the towns of Judenburg and Zeltweg in upper Styria. The project site offers both geological and topographical advantages. The dam will raise the water level of the river by 11 m. Power water will be conveyed to the powerhouse by a 1.1-km long headrace. Turbines will work under a head of 22 m. The main power station will generate an annual 66 million kW, with another 6 million kWh being supplied by the dam turbine installed to cater for compensation water (see Fig. 1). Above the dam, the river will backwater over a length of 4.9 km.

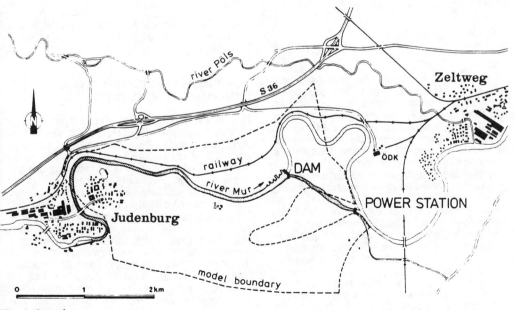

Fig. 1. Location map

2 GEOLOGICAL EXPLORATION AND SURVEY

A large depression of 2000 m maximum depth in the basement of crystalline schists is filled with late Tertiary sediments. These consist of marls, coarse-sandy clays and fine-grained conglomerates which on the whole may be regarded as being of low perviousness. The eroded and very irregular relief of the Tertiary surface is buried under unconsolidated Quaternary sediments of varying, but mainly good permeability. As at the peak of the Würm glacial stage the Mur glacier extended to a point slightly to the west of the town of Judenburg, the loose sediments in the valley floor of the project site are considered to be stream deposits that accumulated in front of the glacier and were then carved into several terraces by erosion.

The material primarily consists of fairly coarse-grained gravel-sand mixtures with conglomerate intercalations varying in thickness and frequency, which are responsible for the steep banks (Fig. 2).

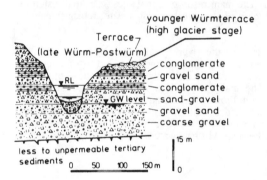

Fig. 2. Geological profile of the valley (after Becker 1983)

More than 40 exploration drillings have been sunk since 1983, most of which have been provided with groundwater measuring instruments. Apart from the engineering geological exploration of the dam and powerhouse sites, studies focussed on the collection and interpretation of hydrographic data as well as pumping and tracer tests for the preparation of a mathematical model.

The great depths to the water table and to the top of the Tertiary formation caused substantial drilling cost.

Locating the bottom of the aquifer, which is mainly characterised by the extremely irregular relief of the Tertiary surface, called for supplementary geoelectrical tests (Fig. 3). The results obtained suggest the presence of two pronounced buried channels separated by a ridge in the headrace and powerhouse area. Differences in level are as much as 40 m in places. The southern channel is 1 to 2 km wide and about 15 m deeper than that to

the north. The ridge in the Tertiary surface combined with the low perviousness of the Quaternary material dams up the groundwater stream and makes the general flow direction parallel to the valley separate so as to continue in the nothern and southern buried channels. In periods of low to medium-high groundwater level, the peak of the Tertiary ridge is not submerged.

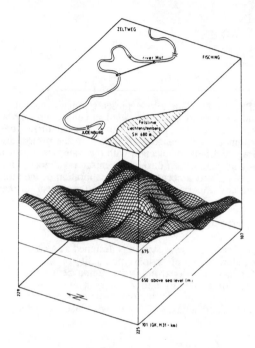

Fig. 3. Three-dimensional model of aquifer bottom approximate outline of tertiary surface (after Walach 1988)

3 GROUNDWATER CONDITIONS AND THE PURPOSES OF MODELLING

Due to the deeply incised river, depth to water table is substantial. Whereas in the westernmost part, i.e. in the future backwater area, the water table is 2 m on average below the bottom of the Mur river bed, with no direct hydraulic communication between river and groundwater, there is some communication between the two downstream of the envisaged dam site. In the section extending 1.5 km upstream of the powerhouse, where the river has cut up to 4 m deep into the Tertiary deposits, groundwater emerges as strata springs from a limited catchment area. As the surface of the Tertiary deposits drops off steeply south of the powerhouse site, the groundwater flows off into the southern buried channel. Over the alignment of the 1-km long tailrace, which will have to be excavated some 2 m deeper, there is again some communication between river and groundwater.

At present, the river feeds the groundwater body over the entire backwater area, especially during floods and mainly through percolation (in an unsaturated zone) through the slopes of the river bed. (Fig. 4). Groundwater level variations range between 3.8 m in the uppermost river section and 1.7 m downstream of the powerhouse.

An area permeability map prepared for the aquifer from the results of pumping tests, grain-size distribution curves and tracer tests has shown that the smallest groundwater thicknesses are all associated with the lowest kf values, i.e. around 10^{-5} m/s (dam and headrace), whereas the two buried channels give relatively good permeabilities, with kf values ranging around 10^{-3} m/s.

Flow velocities as checked by tracer tests vary between 5 and 18 m/d.

Due to the great depths to water table, there are few points where groundwater is used for drinking water supply. The only well of any regional importance in the area under consideration and likely to be influenced is that of the town of Judenburg, which is situated at a distance of 250 m south of the river Mur.

The purpose of the mathematical model was to help hydrologists answer the following questions:

1. What are the effects of groundwater being dammed up within the Quaternary deposits with respect to
   - temporary and lasting rise of the water table?
   - change in flow direction and residence times?
   - inflow to the groundwater body and the resulting water losses from the reservoir associated with an increase in groundwater capacity?
2. What is the effectiveness of the planned provision of impervious elements for the reservoir slopes in lengthening seepage paths and protecting the Judenburg well?
3. What is the scatter of the results for a variation of the input parameters (sensibility analysis)?
4. What is the required length and the expected effectiveness of cutoff walls in continuation of the dam.

The computations carried out to answer the above questions comprised three parts:
1. Area model - large horizontal groundwater model for the whole aquifer concerned;
2. Detailed model of the dam region;
3. Plane vertical models for special structural problems.

In the following we will confine ourselves to the discussion of the large horizontal groundwater model.

4 PREPARATION OF THE MODEL

The first and most delicate task in preparing a model is to define the limits of the region to be investigated. Apart from that, the following parameters are needed:
- top of the aquifuge,
- piezometric head of the water table at the instrument locations at different moments,
- permeability coefficients of the region,
- leakage parameters to describe exchange with surface waters,
- boundary conditions (subsurface inflow to and outflow from the area covered by the model, sources and sinks, groundwater replenishment by precipitation, etc).

For all these parameters, at least isolated values should be available from measurements and exploratory work. To answer these requirements, the following questions should be considered before defining the boundaries of the area under investigation:
- What is the region potentially affected by the project (worstcase assumption)?
- What are the regions that are covered by adequate information from measurements?
- What are the regions where additional measuring schemes are needed to supplement the available information?
- What are the model limits that allow adequate boundary conditions to be determined both for calibration and forecast.

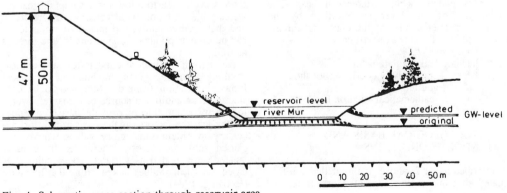

Fig. 4. Schematic cross-section through reservoir area

Much attention was given to these questions in the preparation of the model for the Fisching power project.

A drastic compromise had to be accepted between information requirements and reasonable expense as it turned out to be impossible to cover the whole area potentially influenced by the power project by a dense enough grid of instrument stations. This would have included the major part of the Aichfeld plain, the area extending to the mountain ranges to the south and north, part of the Pöls valley to the west, and the mouths of the Pöls and Granitzen streams to the east.

For this reason, we had to choose model limits that would allow changes in boundary conditions to be predicted with sufficient accuracy and upper and lower limits to be determined for the potential effects from the impoundment up of the river.

5 COMPUTATION METHOD

Among the computation methods available for dealing with mathematical flow models, the Finite Element Method (FEM) has been the most widely accepted over the past decade. This method was also applied in the case under consideration, using the MISES3 program system. This software package, developed by TDV, is a general three-dimensional Finite Element system for solving structural mechanics and field problems with special regard to geotechnics.

Isoparametric 8-noded elements with curved boundaries were used for the model. Quadratic shape functions allow a much coarser mesh to be established than is possible by means of the widely-used linear triangles. In spite of that, we had to divide the modelled domain into more than 500 elements to reproduce the heterogeneous conditions with sufficient accuracy. This corresponds to some 5000 linear triangles for the same accuracy of the results.

Exchanges with surface waters are simulated according to the Leaky Aquifer principle in MISES3. This assumes a semipermeable layer, or leakage layer, to be present between aquifer and surface water, through which water seeps into or out of the aquifer, depending on the difference in head. The conductivity of this layer is described by the leakage parameter lamda, or the mathematical siltation head Dk. This may be written as

$$\text{lamda} = kf / Dk$$

where: lamda... leakage parameter
      Kf....... conductivity coefficient of the aquifer
      Dk...... mathematical siltation head

6 HYDROGEOLOGICAL BASIS AND BOUNDARY CONDITIONS

As described under 2. above, the modelled aquifer, situated in front of the alluvial Mur glacier, is extremely heterogeneous in structure. Underground inflow into the modelled area mainly comes from the Mur valley. Another underground stream, coming from the Pöls valley, generally follows a buried channel situated further north. It is assumed that there is some interrelation between the two streams, but it is definitely the groundwater from the Mur valley that dominates the underground flow system.

Hydrological instrument readings from the bend of the river Mur near Judenburg indicated substantial differences in head between the gauges situated close to the river on the two banks. It was in fact not possible to find a satisfactory explanation for this phenomenon. We assumed local inhomogeneities to be the cause. However, as this area lies at the uppermost end of the backwater area, it was omitted from the model.

To the south and north, model boundaries were located so as to coincide with the streamlines along the centrelines of the buried channels. This choice was based on the assumption that these streamlines would not be affected very much by the impoundment of the river. This idea involved a boundary line running relatively close to the reservoir area, especially to the north. But as the assumptions underlying the model refer to streamline directions rather than the absolute potentials along these lines, this drawback was thought to be acceptable, with the model still being reliable enough to provide satisfactory answers to the chief questions.

The main modelling problems arose in the eastern part because of the presence of the partially submerged ridge in the aquifuge. As the model had to be calibrated for three different groundwater levels with three different extents of submersion, different model limits had to be chosen for each of these conditions.

As MISES3 offers the possiblity of activating and deactivating individual elements, the areas not submerged during low-flow and medium-flow periods were considered in the construction of the mesh, so as to allow all the conditions to be calculated simply by deactivating the elements concerned without changing the mesh itself (Fig. 5.).

To the north and south of the buried ridge, the eastern limit of the model is sufficiently far from the reservoir area to allow the assumption that the water table will not be affected by the reservoir and that, consequently, the measured water levels can be used as boundary conditions for the forecast as well.

Apart from the underground inflow and outflow resulting from the given potentials at the model limits, infiltration from the Mur had to be considered as an additional source of recharge. Hydrologists were not able to provide measured values for the leakage factors, as the measuring effort necessary to obtain satisfactory values would have caused substantial costs. Therefore, these parameters were treated as variables to be calibrated, based on mean values taken from the literature (Herrling, 1984).

Among the few local wells, only the one supplying the town of Judenburg with a mean pumping rate of 35 l/s had to be considered. Comparison of the magnitude of recharge through precipitation with the total flux in the aquifer allowed us to neglect this influence.

7 CALIBRATION OF THE MODEL

The aim of calibration is to determine the parameters controlling groundwater flow in the whole project area. The main parameter is transmissivity, the product of aquifer thickness and the hydraulic conductivity coefficient, kf. For the Fisching groundwater model, another important parameter to be determined was the mathematical siltation head for the rate of infiltration from the river Mur.

The initial values used for the calibration are usually measured values obtained at certain instrument locations. Permeability development between the stations is assumed to be continuous. For the Fisching project, we were provided with a contour map of assumed permeabilities based on the results of hydrogeological and geoelectrical investigations (Harum, 1988; Walach, 1988). Starting from this first estimate, parameters were varied until the measured groundwater levels approached the values obtained from the calculation.

On our model, this iteration process was particularly difficult and time-consuming, as leakage factors had to be determined apart from the permeabilities, and as the heterogeneous field conditions were likely to cause substantial deviations from the initial values.

On the other hand, however, this large depth was an advantage in that no extraordinary accuracy was required for the calibration.

As mentioned earlier, calibration was performed using 3 different quasi-stationary conditions. As a first step, we made a rough estimate of the influence of infiltration from the Mur on the water table. By taking into account this influence, we then determined conductivity coefficients on the assumption that no infiltration occurs. Leakage factors were then varied until the computed water levels agreed with the measured values within the accepted tolerance.

One of the desirable side-effects of this trial and error method of calibration is that engineers in charge of the work get a very good idea of the flow system, its behaviour under different conditions and its sensitivity to variations in the different parameters.

In the course of calibration it was seen that the permeability coefficients had to be modified only little as compared with the initial values. This did not apply, however, to the values of infiltration from the river Mur. Many of the leakage factors determined by computation differed substantially from the values obtained from the literature. On the other hand, however, this confirmed our main observation that infiltration in the western part of the reservoir is much larger than that in the dam area (see Fig. 6).

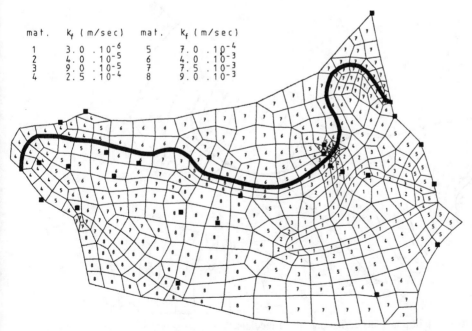

mat.	k_f (m/sec)	mat.	k_f (m/sec)
1	$3.0 \cdot 10^{-6}$	5	$7.0 \cdot 10^{-4}$
2	$4.0 \cdot 10^{-5}$	6	$4.0 \cdot 10^{-3}$
3	$9.0 \cdot 10^{-5}$	7	$7.5 \cdot 10^{-3}$
4	$2.5 \cdot 10^{-4}$	8	$9.0 \cdot 10^{-3}$

Fig. 5. Finite Element mesh showing material numbers

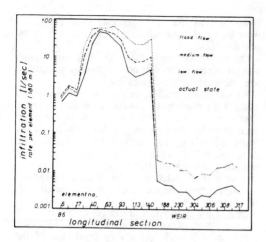

Fig. 6. Curve of infilration rates along the reservoir

Table 1. Actual flow values.

l/s	Q inflow	Q Mur	Q outflow	Q spring
Low flow	530	180	670	1.4
Medium flow	540	260	760	2.2
Flood flow	580	410	950	3.5

Some interesting questions relating to the project were already answered by the calibration computations. Thus, it was seen that replenishment from infiltration was much larger than previously assumed. As can be seen from Table 1, water infiltrating into the aquifer in the reservoir area accounts for more than 40 percent of the total flow through the modelled area during floods, and for as much as 25 percent during low-flow periods.

8 FORECAST - CONDITION AFTER REALISATION OF THE POWER PROJECT

The main goal of the study was to find out whether it was necessary and economical to take any structural measures to ensure watertightness in the reservoir area. Knowledge of the importance of infiltration for the replenishment of the groundwater system made continuous groundwater recharge and the resulting enhancement of the exploitation possibilities appear extremely desirable.

A first computation variant supplied information on the increase in infiltration resulting from the impounding of the river in case no impervious element was provided. The leakage factors determined for the calibration were adjusted on the basis of values taken from the literature. Two extreme cases were studied:

A. increasing the leakage parameter by a factor of 5 in the slopes and

B. same leakage factors as in the calibration.

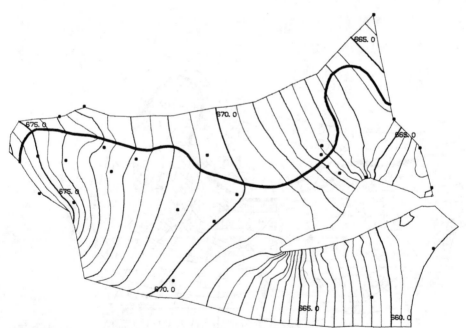

Fig. 7. Contour map of potentials during periods of medium flow

Case A essentially simulates the condition shortly after the impounding of the river with no self-sealing having taken place so far. This alternative yields the maximum possible groundwater level rise. Case B considers a condition following a major period of operation. The low rate of sediment transport of the Mur allowed the assumption that the silt layer accumulating in the reservoir would not be much more impervious than that now present in the river bed, the more so as the transition from percolation to direct infiltration tends to involve an increase in conductivity. Therefore, this assumption will provide information on the expected minimum rise of the water table.

As this computation gave very high infiltration rates of between 1000 and 1300 l/s (for a lowest recorded flow of the Mur of about 8000 l/s), provision of impervious elements on the reservoir slopes was seriously considered. Complete treatment of the slopes was immediately excluded from consideration as the cost involved would have been prohibitive. Simulation of various treatment alternatives showed, however, that only complete treatment would lead to any significant reduction in infiltration. Thus, it was decided that no impervious elements be provided at all, in spite of the large water losses expected. This decision is compatible with the requirements of groundwater management, which will profit by the replenishment with non-polluted water.

The mathematical model has yielded the following main forecasts:

- Within the entire area situated between the ridge in the Tertiary deposits and the town of Judenburg, the water table will rise by an amount of between 7 and 3 m.
- The great differences in permeability and the complex boundary conditions will result in greatly varying inflows.
- The minimum depths to water table will still be more than 7 m even in the lowest portion of the terrace.
- The increase in infiltration will result in a further increase in groundwater capacity (at the beginning an additional 1000 l/s or more will be likely to find its way from the reservoir into the groundwater).
- The reservoir area forms the watershed in the groundwater body. There will be direct infiltration and a corresponding change in flow directions.
- Around the Judenburg well, the relatively regularly raised water table will flatten, which will tend to lower the flow velocities.
- (The water table is not expected to return to the original level. In addition, the low rate of sediment transport and regular flushings will prevent any major silting in the upper half of the reservoir.)

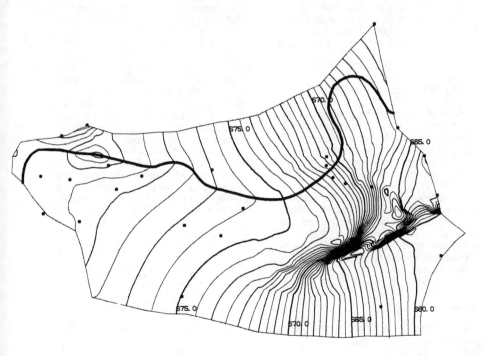

Fig. 8. Contour map of potentials (forecast)

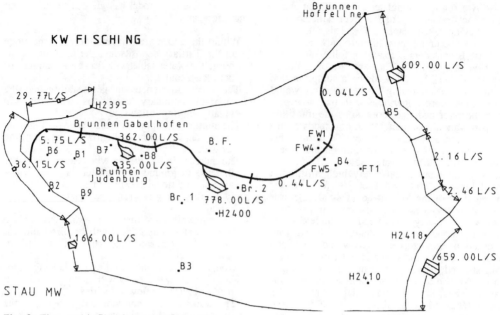

Fig. 9. Flow and infiltration rates (forecast)

- There is no need for the planned impervious elements in the backwater area. The effect of any such measures would at any rate be unsatisfactory, while groundwater replenishment with high-quality water from the reservoir is a desirable effect.

CONCLUDING REMARKS

By way of summary, it should be pointed out that mathematical modelling of the groundwater stream has proved an adequate and low-cost means of answering many different questions relating to the construction of reservoirs in waterbearing soils.

Due to delays in the construction schedule, it has not been possible so far to compare the forecast with measured values. Construction of the power project was commenced at the end of 1991. Work on the model is planned to be resumed when new measurements allow an improved calibration to answer readily and at any time questions arising during operation with respect to potential effects on the groundwater system.

REFERENCES:

Bear, J. and Verruijt, A. 1987. Modelling Groundwater Flow and Pollution. D. Reidel Publ. Comp.

Becker, L.P. 1982. Geologischer Vorbericht, Geologie und Grundwasserverhältnisse im Bereich der Brunnenanlage bei Murdorf/Judenburg KW Farrach, unveröffentlichte Berichte. Graz.

Harum, T. and Fank, J. 1988. KW-Fisching. Hydrogeologische Untersuchungen. Endbericht. Forschungsgesellschaft Joanneum, Graz.

Herrling, B. 1984. Bericht d. 6. Fortbildungslehrgangs Grundwasser des DVWK. Institut für Hydromechanik, Universität Karlsruhe.

Mörth, W. 1988. Geologischer Bericht KW-Fisching. Forschungsgesellschaft Joanneum, Graz.

Stiny, J. 1943. Zur Kenntnis der Quellen und ihrer Schürfleistung. Geologie und Bauwesen, Heft 3.

Walach, G. 1988. Bericht über die geoelektrische Erkundung des geologischen Untergrunds im Gebiet des KW-Fisching der Steweag. Institut für Geophysik, Montanuniv. Leoben.

Worsch, E. 1963. Geologie und Hydrologie des Aichfeldes. Mitt. Mus., Bergbau,Geologie und Technik, Landesmuseum Joanneum, 25, Graz.

Ziekiewicz, O.C. 1977. The Finite Element Method. Mc.Graw-Hill.

Water and sediment resources in reservoirs

H. Støle
SINTEF NHL, Trondheim, Norway

ABSTRACT: Reservoirs play an important role in hydropower development. It is, however, difficult to maintain water reservoirs in river basins with sever sediment transport. Up to now, large dam projects have been the only alternative which provides required regulation capacity in sediment loaded rivers. River transported sediments should not only be regarded as a problem, sediments are a resource which could be harvested in connection with river regulation projects. New approaches to hydropower development in river basins with sever sediment load must facilitate sediment bypass of the dam. Presuming appropriate means for sediment bypass are developed, there is a considerable potential for developing environmental and economical sound hydropower projects which will provide desired regulation capacity within power systems.

1 INTRODUCTION

This paper approaches hydropower development in Himalayan river basins in general based on the authors experience from water resources development studies and projects in Bhutan and Nepal since 1980. Research work carried out over the last 5 years has focused on sediment transport patterns in Himalayan rivers and matters related to sediment control and sediment handling facilities at water intakes and in reservoirs.

The objective of this paper, however, is not to dig deep into hydrological, sedimentological, technological or environmental aspects of regulation of water flows in sediment yielding river basins. The objective is to approach some of the basic principals of hydropower development in rivers where sediment related issues give some of the premises for power system inception and design, and where sediment transport remains a major issue form the conceptual stage through planning and design to operation of any power plant within the power system.

2 WATER RESOURCES IN RESERVOIRS

The ABC of hydropower says that where water flow and head are available, there is a potential for generating hydro-electric power. In a natural flowing river, all potential energy is dissipated along the course of the river. A hydropower plant traps, however, the major part of the potential energy in the withdrawn water flow and converts it to electric power by means of turbines and generators.

Water flow in a river is a natural phenomenon which is extremely time dependant. River discharge might vary with several orders of magnitude from extreme floods to periods of lean flow. Power demand is, however, dependant on the consumers needs for electricity. Human behaviour is also time dependant, but to some extent predictable with respect to energy consumption.

An efficient utilization of power potentials in a

river basin requires normally storage of energy in order to bridge the time gap between natural river flow and power demand. Energy consumers can depend on regular power supply due to energy storage capabilities in the power system. Water reservoirs serve as energy stores in hydro-electric dominated power systems.

2.1 Regulation frequency

Reservoirs play different roles in a power system depending on the required duration of water storage. Reservoirs for short term regulation stores water on daily or weekly basis. Daily peaking reservoirs store water flowing into the reservoir during periods of the day when power consumption is low. Stored water is, however, used sometimes later within the next 24 hours when the demand grows higher than the corresponding river flow at that time.

Reservoirs for long term regulation stores water on seasonal or yearly basis. Water is collected during wet season and used during lean season, or water is stored during wet years and consumed during dry years.

2.2 Primary and secondary reservoirs

Primary reservoirs are located in the river basin where water is halted and stored for some time in water basins behind manmade dams or in lakes where the outlet is controlled by man. Minor water stores like ponds or forebays for short term regulation purposes are sometimes built somewhere along the waterways between the intake structure in the river and the pressurized part of the waterway like shafts or penstocks.

Secondary reservoirs are not an integrated part of the water-ways in the river basin. They serve as "large batteries" which are charged during periods of surplus electric power, i.e. periods of high river flow and low power demand. Pump storage schemes falls into this category. The energy might also be stored as heat and tapped by use of heat pumps. Electric battery technology stores energy by chemical means. These technologies are, however, suffering from constraints related to limited capacity and great losses and they will therefore most probably remain insignificant means for storage of energy in large power systems.

Reservoirs are today an integrated part of almost all hydropower systems. High head gives a high energy content in every cubic-meter stored water. It is therefore favourable to locate reservoirs if possible in the upper reaches of a river basin where high head is available. It is fairly straight forward to compute costs and benefits of reservoirs within the framework of a hydropower system (Gunnes).

Regulation of river flow is also required for other purposes like irrigation, domestic and industrial water supply as well as downstream flood control. All water users must be involved during the planning phase of river regulation projects. This paper deals, however, only with the hydropower aspects of water resources development.

3 HYDROPOWER DEVELOPMENT IN SEDIMENT YIELDING BASINS

Water resources are natures gift to man in the Himalayan region. The extreme sediment loads in the rivers, mainly during the monsoon season, is one of the major problems related to mans use of Himalayas water resources. Protection of soil from erosive forces and improved land use patterns are essential in order to protect land and water resources and enable the people to use these resources for their own development purposes. But sediment transport is and will remain a natural phenomenon in the Himalayas. It is important to protect forests, and it is nice to plant threes which will reduce sediment yield in most cases, but it will not stop this geomorphological process (Carson and Bruijnzeel & Bremmer). Reliable and efficient systems for sediment control and removal of sediments from withdrawn water will therefore always be one out of many preconditions for a successful utilization of water resources in the Himalayas and other similar regions.

The possibilities of regulating river flow in basins where the river carries considerable amounts of sediments are limited. In a geomorphological context it might be observed that sedimentation processes have already neutralized natural reservoirs like lakes and ponds within river basins. Man made reservoirs will sooner or later also face the consequences of reservoir sedimentation and be transformed from water stores to stores of sediments, containing huge volumes of gravel, sand, silt and clay.

In basins where there are no lakes and ponds, the only natural "reservoirs" which regulate river flow to some extent are glaciers, snow covered areas and forests. Man can reduce these natural regulation capabilities through poor land use practices like sever deforestation. Methods for observing the amount of water stored in snow and predict the resulting run-off pattern are developed (Sand), but man has no means available to control and operate these reservoirs in connection with hydropower and irrigation projects. River discharge responds rapidly to heavy monsoon rainfall.

> The need for storage capabilities within a power system and the value of obtained regulation capacity increases with increased sediment transport in the region.

Hydropower engineers tend to group hydropower projects according to the size of their installed capacity. For some purposes it is convenient to use the labels micro-, small-, medium- and large-hydro. In Himalayan rivers where the sediment transport patterns are complex and the sediment loads are high it seams convenient to also categorize hydropower projects according to how sediment transport is taken care of.

> Run-of-river projects and storage projects with large dams are today the only feasible hydropower projects in river basins with sever sediment transport.

The reservoir sedimentation phenomenon has excluded the possibility of building medium sized reservoirs for long term regulation requirements.

3.1 Run-of-River Projects

A run-of-river project (ROR) diverts normally a minor part of the annual river flow for power generation since the installed capacity is normally limited by lean season flow for reliable power supply. Design rule no 1 with respect to sediments is:

> Run-of-river project shall allow 100% of the sediment load in the river to pass by the intake dam.

There is no room for permanent storage of sediments upstream of the dam (diversion structure). Short term regulation might, however, be obtained by maintaining a pond for daily peaking by flushing of the pond through sluice gates in the diversion dam or by rising the crest of the dam during low flow season. This might be achieved by means of flap gates, a rubber dam or flash-boards.

Many ROR projects are economically and technically sound projects, specially in an early face of hydropower development within a power system. A power system based entirely on ROR projects will however be limited to utilize only a minor part of the available water resources and the system will suffer from limited regulation capacity. ROR projects supported by storage projects for peaking purposes will, however, in most cases produce reliable power to a reasonable cost.

Construction of ROR projects will also contribute positively to a power system dominated by thermal power plants. ROR projects will produce base load during wet seasons and serve peaking purposes during lean season. A system combining thermal power with ROR projects will facilitate considerable savings on fuel consumption and reduce the environmental costs drastically compared to a thermal based power system (Lysne).

3.2   Large Dam Projects

A Large Dam Project (LD) facilitates long term regulation of the river flow. They are normally multipurpose plants where regulation contributes to ensure reliable power supply as well as it serves irrigation and flood control purposes. LD projects are based on the following philosophy:

> Large dam projects shall allow all sediments trapped in the reservoir to settle there and still maintain the required regulation capabilities throughout the economic lifetime of the project.

The project might sooner or later be transformed gradually to a ROR project with respect to regulation capabilities as well as sediment bypass necessities.

LD projects major contribution to a power system is the desired regulation capabilities which ensures regular and reliable power supply. But large dams have been and will remain controversial projects. Some of the reasons for the controversies over large dam projects are listed below.

a   Resettlement of many people
b   A site specific, but long list of environmental and ecological impacts
c   Losses of fertile agriculture land and productive forests
d   Climatological impacts
e   Financial constraints for other development activities within the country due to the heavy burden of huge loans
f   Dependence on foreign expertise and technology
g   Socio-economical impacts of a klondyke-like construction period
h   National and international politics

Many volumes have been written on these matters and many topics could have been added. Despite this, it will most probably still be required to build some large dams. The list of impacts may, however, motivate everybody involved in hydropower planning to search for new approaches to hydropower development in sediment yielding river basins.

Hopefully there should be feasible alternatives or supplements to the two extremes: Run-of-river projects on one hand and large dam projects on the other.

4 SEDIMENT RESOURCES IN RESERVOIRS

Within the context of water resources development, reservoir sedimentation has been considered a cost factor only. Sediments transported by rivers are, however, playing an important role within several fields.

4.1 Sand and gravel resources

Gravel and sand are important construction materials used in big quantities in connection with most civil works. Natural gravel and sand are in most countries found only in or close to rivers. Water reservoirs are also sand and gravel reservoirs. Particles are sorted according to size in the reservoir due to the relation between settling velocity and particle size. Appropriate techniques for controlled removal of sediments from reservoirs will facilitate "harvesting" of sediments from reservoirs. It would be possible to be selective with respect to size of withdrawn particles.

4.2 Land development

In a geomorphological context it might be observed that a considerable part of the earths available agriculture land is built up by river sedimentation processes. Mans utilization of these land resources has, however, in many areas required flood protection works. Land reclamation works and flood protection works require construction materials for earth walls and levelling of bigger areas. Systems for withdrawal of river transported sediments would reduce sediment related problems in rivers and reservoirs and provide necessary material for land development programmes.

4.3 Environmental benefits

One undesired impact of large dams are the low content of sediments in the water released from the dam to the downstream reach of the river. Downstream degradation might in some cases cause sever problems with respect to bank stability and lowered ground water level. Sever ecological impacts are (or should be) considered as heavy costs in the overall cost-benefit analysis of large dam projects. The Grand Canyon large dam in USA is a sediment trap which has caused great concerns with respect to ecological changes within the downstream national park and wildlife reserve. In both these cases, systems for maintaining sediment transport in the river by bypassing some of the sediment load will be a gain in the overall cost of a project.

5 MEANS FOR SEDIMENT BYPASS AND SEDIMENT REMOVAL

Chapter 2 dealt with the basic necessity of storing water in reservoirs. The problems related to obtaining water reservoirs in rivers with sever sediment transport were highlighted in chapter 3 and possible utilization of sediment resources in reservoirs was briefly reviewed in chapter 4.

The first phase of hydropower development in a power system is dominated by small and medium hydropower projects of run-of-river type. After some years, however, it will be required to obtain long term regulation capacity within the power system in order to utilize a bigger share of the water resources potential for power generation as well as other purposes.

Project concepts intending to evolve in the space between run-off-river projects and large dam projects, must facilitate a bypass of the reservoir for sediments or reliable means for removal of sediments settling in the reservoir. To develop means for this purpose might be the greatest challenge for hydropower development in river basins with sever sediment transport in the near future.

These new techniques will also provide the opportunity to utilize sediment resources as a spin-off effect. If means for removal of settled particles are developed further, existing reservoirs suffering from reservoir sedimentation could be rehabilitated.

Sediment bypass and removal of sediments from reservoirs requires development of new approaches to operation of hydropower projects as well as new technology. Operation of reservoir and headworks is an important field of hydropower development in river basins with sever sediment transport. Hydraulic and civil engineers play normally an important role during planning, design and construction of power projects. In the best cases they leave a maintenance manual behind before they move on for another project after the last inauguration speech, just when the important period of production, earnings and repay of loans starts. Operation of hydropower projects have been carried out by electrical engineers with some assistance on the mechanical side. Knowledge about river hydraulics and sediment transport should, however, be present within the operation crew, at least during the first years of operation.

We do not have adequate means ready-made for maintaining water reservoirs in rivers with sever sediment transport today. The author do not intend to advocate for sophisticated and fancy solutions for headwork operation. Appropriate technology is required for hydropower development in order to obtain reliable power systems with a minimum of down-time due to mall-operation of unnecessary high-tech components. But it seems somewhat inconsistent to allow use of advanced technology in components related to turbines and generators on one side, but on the other side limiting technology related to headworks and intake to those things which could be monitored and operated by an uneducated gate operator at the damsite. The headworks are after all facing a wild river while the turbines are only exposed to a tamed and controlled water flow.

5.1 Framework for sediment bypass at reservoir projects

Particles settle in a reservoir due to the reduced velocity in the river flow entering the reservoir. Small reservoirs at run-of-river projects are normally flushed after the monsoon in order to regenerate some storage capacity for daily peaking purposes during the lean season. It is required to close down the power plant during flushing in order to generate a swift flowing current in the reservoir area. The efficiency of reservoir flushing is dependant on size and shape of the reservoir and properties of the settled particles. Reservoir flushing will, however, only be feasible for small reservoirs for short term regulation purposes.

The alternative to sediment bypass systems for maintaining regulation capabilities within a hydropower system is large dam projects. High dams are required to take care of the sediment load and to provide power head. A considerable share of dam costs at large dam projects could be spent on sediment bypass facilities at alternative projects with smaller reservoirs located where natural head is more easy available. These less controversial projects would therefore still be able to compete with large dam projects also from an economical point of view.

The difference in water levels upstream and downstream of a dam provides the energy potential required to generate a current with the necessary erosive force to remove and transport settled particles in somewhat bigger reservoirs. A small sized flushing conduit will have a high transport capacity when sediments are continuously feed into the upstream end of the flushing conduit and thus bypassed the reservoir.

> Sediment bypass could either be obtained by sediment diversion systems or by sediment sluicing systems.

The driving force for both types of systems is the hydrostatic pressure difference provided by the dam. They will therefore be operated while the reservoir is operational and the power plant is in operation. The amount of water used for flushing will be controlled and surplus water will in most cases be utilized for this purpose.

Sediment diversion systems withdraw sediment load (mainly bed load) from the river in the upstream end of the reservoir and divert it through a diversion tunnel before it is flushed

back to the river downstream of the dam or to schemes for extracting sediment resources.

Sediment sluicing systems draws the required energy from the reservoir as well. But sediment sluicing systems operates inside the reservoir and removes settled particles from the reservoir. These systems may be compared with dredging systems, but they do not require external power supply.

5.2 Development activities

Considering the large potentials for sediment bypass facilities, several new concepts will most probably be developed during the years to come.

There has been a close cooperation between water development institutions in Nepal and Norway over the last five years. Several promising concepts for sediment control at intakes and for removal of settled particles in sedimentation basins as well as reservoirs have been developed (Bakken, Lysne and Støle). Butwal Power Company (Nepal), The Norwegian Institute of Technology, Norwegian Hydrotechnical Laboratory and Norwegian State Power Board are at present involved in research and development activities concerning sediment handling facilities at hydropower projects.

Laboratory model studies have been carried out for several concepts. Field studies are, however, required before these concepts are ready for practical engineering use.

5.3 Motivation for further development

The list of impacts given in Sec. 3.2 should hopefully motivate everybody involved in development of water resources in river basins with sever sediment transport to search for new ways to obtain regulation capacity. Many professional groups are involved in hydropower development, not only engineers. Contributions are needed from all groups, and also from political and financial authorities, users including the general public and media. Development banks, donors and others who provide funding for hydropower projects will play a crucial role. Their needs for references tends to strain new approaches within all fields of water resources development.

>If the financial institutions do not allow new means for sediment bypass to develop, large dam projects will remain the only feasible energy storage projects in river basins with sever sediment transport.

The following statements will hopefully motivate people involved in hydropower development in rivers with sever sediment load to search for means providing sediment bypass of reservoirs.

The evolution of sediment bypass and reservoir maintenance systems will produce feasible hydropower project which provides regulation capacity in reservoirs of smaller size than large dam projects which have been built up to now for regulation of water flow in rivers with sever sediment load. These projects will be competitive to large dam projects because they will be more manageable with respect to: finance, organization and applied technology. From a development point of view it seems wise to spend less money during construction and rather spend some more money on the operation side later on when the project generates income. This will provide job opportunities locally on long term basis, and it will reduce the klondyke-like peak during construction period as well as the need for capital at year 0. The environmental impacts of hydropower development will be reduced and the exploitable water resources potential is expected to be increased within the region.

6 CONCLUSIONS

>small might be beautiful
large might be impressive
something in between might be difficult
but still - it might be wise

REFERENCES

Bakken, T. 1991, Recovering sediments from intakes and reservoirs in sediment carrying rivers.
Bruijnzeel, L.A. with Bremmer, C.N. 1989, ICIMOD occasional paper no 11. Highland-lowland interaction in the Ganges Brahmaputra river basin
Carson, B. 1985, ICIMOD occasional paper no. 1. Erosion and sedimentation processes in the Nepalese Himalaya.
Eshete, S. and Habtemariam, Y. 1992, Model study of sediment flushing system
Gunnes, O. 1978, Technical economical analysis in hydropower planning
Lysne, D.K. 1991, Personal communication
Sand, K. 1990, Modelling snow-melt runoff processes in temperate and arctic environments
Støle, H. - NORPOWER 1992, Kali Gandaki 'A' hydroelectric project - Detailed feasibility study

Hydropower'92, Broch & Lysne (eds) © 1992 Balkema, Rotterdam. ISBN 90 5410 054 0

Impact of HPP Djerdap-1 reservoir on environment and protection of agricultural areas

M. Vukovic, L. Vajda, M. Miloradov, M. Sretenovic, A. Soro & B. Katic
WRD Institute J.Cerni, Energoprojekt, HPP Djerdap, Belgrade, Yugoslavia (Serbia)

ABSTRACT: The paper deals with the protection against noxious effects of the backwater and describes the system of measurements and monitoring of the backwater regime increasing in time from 68.00 to 69.50. MaxNWL. There is also a description of the water regime and quality, as well as of changes induced by backwater on ice and in the river bed (morphology), and on the Danube river banks. The paper presents the groundwater regime and protection under natural and back-water conditions at the agricultural areas of app. 150,000 ha being protected by 300 km long dikes. It deals also with the changes of the ichthyology. Finally the paper treats the problem of relocation and protection of towns, villages, communications, archaeological and historical monuments in the zone of the existing backwater effect of the HPP Djerdap-1 reservoir.

1. GENERAL

The Danube river is the second biggest river in Europe, with a catchment area of approx. 817,000 sq.km. It is with a course of 2192 km in length, being an international waterway which connects eight European countries with the Black Sea.

The common Yugoslav-Rumanian sector of the Danube river of approx. 230 km in length within which the Danube flows through the Djerdap gorge disposes of significant hydroelectric potential which is characterized with large discharges and relatively great heads. At this sector there are characteristic discharges of the Danube Qmax 1% = 16,000 cms; Qmin 1% = 1,950 cms; Qav = 5,650 cms. Also, at this sector of the Danube, the Djerdap HPP-1 and HPP-2 (fig. 1) are constructed. The basic power characteristics of the main structure i.e. Djerdap HPP-1, (two hydropower plants, each of 6 units, i.e. Yugoslav and Rumanian, two navigation locks 310 m x 34 m, concrete spillweir dam with fourteen bays each of 25 m, and two earthfill dams and other appurtenant structures) are:
- the maximum head of HPP Hmax = 35 m;

- the minimum head of HPP Hmin = 21 m;
- installed capacity Ni = 2,100 MW;
- output 12 TWh/yr
- live storage 500 Mcm; reservoir of app. 400 km in length.

Djerdap-1 HPP falls into the category of the ten biggest plants in the world. Low banks upstream of the Djerdap gorge, a great number of settlements, a few larger towns and industry, woods in unprotected zones, over 300 km of dikes and over 150,000 ha of agricultural areas, means a particular technical challenge for the hinterland protection solutions.

1.1. The basic problems of the reservoir

As a result of the HPP Djerdap-1 reservoir, the following technical problems have been arisen:
- water level elevation and decrease of flow velocity,
- sedimentation of suspended and retention of bed load sediments,
- icebergs gathering and estuaries of tributaries and ice barriers formation,

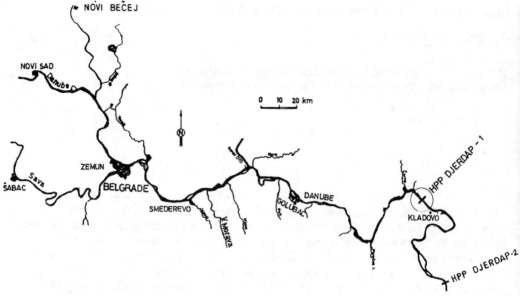

Fig. 1 - Layout of Danube and HPP Djerdap-1

followed by possible consequences,
- perishing and decrease of woods growth,
- eutrophication and endangering the reservoir ecological system,
- endangering the general and seepage stability of the existing dikes in back-water conditions,
- over-moistening and salinization of soil and endangering agricultural production within the protected hinterland,
- endangering all inhabited settlements, by groundwater as well as public and traffic structures and revetments within the reservoir area,
- endangering the stability of the high banks and
- changes of species and scope of ichthyo-fauna in backwater zone and dilemma about the fish ladders construction within the final design.

1.2. Backwater Regime

Headworks of the Djerdap-1 HPP have been constructed at kmD 943. Along the reservoir extending over 400 km, a number of minor and greater tributaries including the Velika Morava river, the Sava and the Tisza rivers among the others empty into the Danube river. Therefore, the Djerdap-1 HPP reservoir has been created

over the complex river system, and it is characterized with the water regime defined through hydrological conditions and the hydropower plant operation regime.

The dam has been constructed for the max. normal water level at 69.50 m.

Regarding the character of the reservoir which is mostly protected by dikes, the backwater regime is inverse, with minimum water level at 63.0 m a.s.l. at the dam (for larger discharges Q > 10,000 mcs).

During the past period the following operation regimes were realized: 68.00 regime experimentally up to 1977 and finally 69.50 regime.

At the longitudinal profile of the Danube within the reservoir zone (Fig.2) the natural water levels and the backwater levels during the flood flows p = 1% and the backwater levels during the minimum flows have been given for 69.50 regime. Changes of backwater regime at the dam have caused the change in length of the reservoir in dependence on the discharge and it is inversely proportional.

In the regime of 69.50 the backwater is extending for the same Qmin on the Danube as far as Novi Sad, while it extends up to Sabac on

366

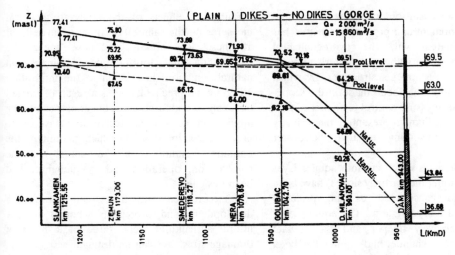

Fig.2 - Longitudinal section of natural and backwater regimes

the Sava river and up to N.Becej on the Tisza river.

Change of water flowing conditions along the reservoir has caused modification of geometric and hydraulic characteristics, which then resulted in variations of transport process and the sedimentation of the suspended material.

2. PROTECTION OF AGRICULTURAL AREAS

Creation of the HPP Djerdap-1 reservoir has caused general rise of groundwater levels over the protected agricultural areas within the reservoir hinterlands (former inundation areas). In the given conditions, demand for execution of works precaution measures intended against high groundwater levels and control of water regime in the hinterland of the Danube and tributaries backwater effects was manifested. Prior to HPP Djerdap-1 dam construction a stretch of low littoral areas had been protected with dykes against flooding in the period of high water levels, while at the most extensive stretch of protected areas there were drainage systems for groundwater arisen by precipitation.

Low agricultural areas in the hinterland of the Djerdap-1 HPP, which represent former inundation of the Danube and the tributaries, could be from hydrogeological

and hydrodynamic aspect schematized as a two-bedded porous medium with aquifer which is covered with a layer of low permeability.

The basic aim of low agricultural areas protection against the backwater effect was to prevent water logging and over-moistening of soil, and indirectly, also, salinization of large complex of hinterland areas.

Such an approach was established with the additional demand for research workers and designers to insure conditions for complex control of water regime within the protected area.

For needs of studying and forecasting groundwater and salt regimes extensive studies were realized. They were supported with numerous hydrodynamic calculations and prognoses of groundwater and salt regimes in the littoral (hinterland) zone of the Djerdap-1 HPP, which were based on the results and the conclusions of previous investigation.

Results of hydrodynamic calculations had proved,that in given conditions, at the design reservoir regime, the area of influence of Djerdap-1 HPP backwater on groundwater of the low hinterland areas was limited to a relatively narrow zone which did not exceed 3 to 4 km.

As to provide optimum conditions for agricultural production development, the criterion for groundwater level maintenance in the surface bed at 0.8 to 1 m depth was established.

Concept of technical solution of the endangered regions development and protection schemes has been coordinated with the hydrogeological conditions prevailing within the hinterland zone. Thus, drainage system, i.e. structures, by which direct interventions has been made in the scope of groundwater pressure control in aquifer of two-layer porous medium, represented the primary measure for preventing negative backwater effects.

Drainage channels which cut into aquifer layer with their bottom (and partially slopes) have been designed and constructed as the principal structures of drainage systems. The aquifer layer is located relatively deeper and the contact between drainage channel and aquifer has been achieved through flowing wells, or rarely calculation of BWR 68.0 with old river channel by the material replacement at the bottom of the channel. On the figure 3, there are characteristic results of comparative hydrodynamic computations for one of about 100 analyzed design profiles.

Parallel to computations and prognosis of groundwater regime the extensive hydrodynamic computations had been made for the purpose of defining the effects of the analyzed technical measures on the salinization process carried out at these low areas. On those areas the evaporation and transpiration dominated in natural conditions in relation to the infiltration arisen by the precipitation. As a result of carried out analyses and forecasts of the effects caused by application of various technical measures to littoral areas, the solutions which guaranteed tne optimum conditions in the course of the construction, operation and maintenance, had been designed.

In accordance with the stated concept of development and protection schemes of low littoral agricultural areas, three basic types of drainage structures can be distinguished:
- protection which is represented by the open drainage channels network,
- protection which is provided by the open drainage channels in combination with the old branches of the Danube and the parts of other waterstreams and
- protection which is ensured by the open drainage network in combination with 960 flowing drainage wells.

Presented characteristics types of applied

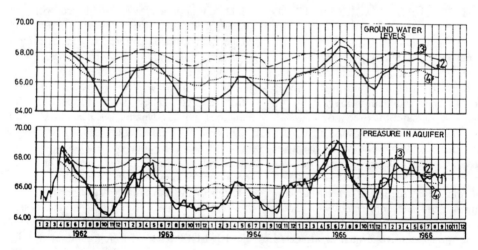

Fig.3 - Results of GWL hydrodynamic calculations

    (1) Recorded piezometric level oscillations
    (2) Natural ground water level (GWL) regime
    (3) GWL calculation of BWR 68.0 without drainage
    (4) GWL

protection and development measures are given on the Fig. 4.

Through systematic following-up of the drainage systems operation effects and periodic interpretation of the obtained data, the operation problems could be noticed and eliminated by reconstruction or enlargement of the drainage systems or the change of the pumping station operation regime. In addition to that, the obtained results and recognition of the regime and operation effects of the drainage systems enable maintenance, appendage and eventual operation regime change to be utmost rationally carried out. Within the scope of drainage systems maintenance in the littoral zone of the Djerdap-1 HPP, the basic problems are associated with ageing process and degradation of flowing wells in function of properties modification (chemical composition) of groundwater.

3. PROTECTION OF HINTERLAND (LITTORAL ZONE) AGAINST EXTERNAL WATERS

A considerable stretch of inundated areas within the littoral zone of the HPP Djerdap-1 reservoir was not protected from internal waters, therefore, these areas were almost every year flooded. Extension in duration of the high water levels in the reservoir and the changed conditions in usage of low hinterland areas imposed the demand for protection of these areas against external water.

Considering the set conditions in preparation of the and designs, technical solutions of protection (dykes) were outlined in such a way as to include:
- protection against overflowing,
- protection of the upstream (unprotected) slope from wave erosion and
- insuring the full filtration (as well as static) stability of the dikes.

Woods degradation would occur in the conditions of HPP Djerdap-1 reservoir operation at a number of sectors in forelands. Thus protection of the external dike slope at these sections has been achieved using lining.

Dimensioning of the dike body and protection of internal (protected) toe is based on the results of filtration stability computation of surface low permeable soil against fluidisation arisen by piezometric pressure in the aquifer.

Increase of filtration stability degree of dike toe has been achieved by construction of ballast made of re-filled sand (fig. 5).

Within the scope of training works intended for protection of low littoral areas from the impact of external waters and reconstruction of the existing dykes, the following works have been carried out:
- construction of new dikes extending 61 km,
- reconstruction of the existing dikes, 252 km long,
- lining of external slope, 2.23 millions m²,
- ballast in the dike toe, 209 km long (13.6 millions m3), and
- revetment construction 26.6 km long.

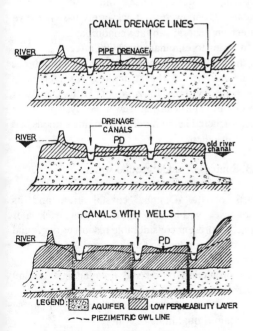

Fig.4 - Types of adopted drainage systems

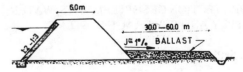

Fig. 5 - Cross section of dike with ballast

4. CHANGES IN THE REGIME OF SEDIMENTS

Change of water regime along the reservoir has caused the change in sedimentation regime due to decrease of waterstream transportability. The consequences of these changes are complex: sedimentation
has occurred, and storage volume decrease. For the purpose of studying the sedimentation process, investigations of deposit regime in the natural conditions have been carried out, as well as the prognoses of the reservoir siltation process for the period of 25 years. During the operation, researches of backwater effect on deposit regime were organized applying survey of permanent cross-sections and balance of sedimented deposit amount at inlet and outlet sections.

Deposit balance has been continually measured since 1974 at inlet sections of the Danube, Sava, Tisza and Velika Morava rivers, as well as at the outlet section, of the Danube, Sava, Tisza and Velika Morava rivers, as well as at the outlet section, i.e. through the dam itself (through turbines and over spillway).

It has been ascertained that a considerable percentage of the total deposit amount has been deposited in the reservoir. In addition to checking of deposit balance prognosis, available observations made possible the analysis of accumulated deposit arrangement along the reservoir.

Analysis of the morphological changes in the river bed during HPP operation has proved that the river bed is mostly filled with sediments at kmD 1,077 downstream (in the gorge part) which caused no restriction as far as navigation is concerned. Comparison of forecasted and recorded suspended sediments movements[*] has proved generally satisfactory.

5. PROTECTION OF SETTLEMENTS, WATER SUPPLY AND SEWERAGE SYSTEMS

5.1. Settlements

Certain number of settlements and industrial areas in vicinity of HPP Djerdap-1 were protected by dikes against high water levels. Change of water level regime resulted in prolonged duration of high groundwaters levels in settlements and industrial zones. Therefore protection measures and regulation of groundwater levels had to be undertaken in those parts of hinterlands.

Having faced to complexity of the problem, a sanitary and ecological conditions, the following criteria for regulation the underground waters have been accepted:
- for town and industrial areas at 3 m of depth,
- for smaller settlements at 2 m of depths.

For control of the groundwaters and maintenance of fixed criteria required for needed conditions on settlements and industrial areas either drainage wells, equipped with pumps or closed drainage collectors were used.

5.2. Sewerage system

In favorable effect of HPP Djerdap-1 as far as sewerage is concerned mainly reflected in duration of flooding outlet structures among present sewerage systems.
Those always mean higher exploitation and maintenance costs. In settlements without public sewerage systems, unless they are protected against groundwaters, backwater has negative effect on general sanitary conditions.

In order to eliminate negative effects of the backwater of the Danube and its tributaries some number of the structures was erected in sewerage systems in zone of HPP Djerdap-1 reservoir. The existing sewerage systems were reconstructed i.e. number of pumping stations had been installed.

5.3. Sources of underground water

Parts of the Danube coastal area and its tributaries within the zone of the backwater effects represent considerable resources of ground waters. Groundwater intakes are used for two purposes:
- as protection of coastal area against ground water and
- as structures of water intake systems for supply of required quantities of water.

The decrease of the velocities of Danube and

tributaries flows has both positive and negative effects on groundwater resources. Backwater effect of the river and formation of accumulation in the river bed do effect the groundwater resources, i.e.:
- due to the higher level in the river capacity of groundwater enlarges;
- riverbed colmatage is larger, i.e. hydraulic losses at the contact of the river flow and aquifer are higher and
- transfer from backwater control to
complex reclamation of the groundwater hinterland (coastal areas).

It is quite certain that at the beginning of accumulation capacity of groundwater sources increases, while, later on, it gradually decreases. The decrease is usually more intensive if compared to the one before backwater effect. During the whole process of groundwater sources capacity changes we faced the ageing problem of the intake structures which usually could not be explicitly quantified.

6. ICE PROBLEMS

Problem of ice in the Danube river had been well known for decades due to very serious movement of ice.

Under natural conditions, in a section which corresponds to back water zone 1, ice problem is critical at three sectors which have unfavorable hydraulic-morphological characteristics. The most critical were the sectors in the gorge (kmD 965-974) and at the gorge entrance (kmD 1,035-1,038), and on the mouth of Velika Morava river kmD 1,180.

Erection of HPP Djerdap-1 has caused hydraulic conditions changes in accumulation zone and in the zone kmD 965-974 there is no condition for making ice blocking. In the second and third zones they could have appeared but only during the flood period.

Ice regime between the first and second sector has been completely changed. In former period ice flotations have appeared only shortly. This sector is situated in lake zone and therefore in severe winter periods ice cover may be expected for a longer period.

Period after reservoir formation (1971-1989) had

less cold winters and was followed by less frequency of severe winters compared to the past years. During last three coldest winters (1972/73, 1984/85 and 1986/87) favorable coincidence of meteorological and hydrological conditions caused normal formation of ice.

7. WATER QUALITY

New water flow has caused changes of reaeration, development of plankton and other biochemical processes.

In order to estimate quality of accumulated water the last few years researches have been carried out from (kmD 1,116.2) to (kmD 926). Researches have covered all physical and chemical characteristics.

Hydrobiological investigations have covered quality and quantity data of plankton, biogenesis of the plankton and estimation of water quality, while radio-activity tests have been accomplished for measurement of total α and β-activity.

Researches results have provided that the input organic load is eliminated by effects, which are getting stronger with longer retention period, i.e. with slower water flow in the reservoir.

Process of reaeration mainly compensate deficit of soluble oxygen in running reservoir water. Efficacy of such a process does not allow favorable oxygen balance, and therefore it has larger deficit along the total flow.

Exception from the established conditions existing with flood period when oxygen deficit is decreased in downstream sector.

Decrease of diversity and abundance of the plankton along testing sector results in less production in lower part of reservoir. Although content of macro and micro nutrition enables high level of plankton production, in particular effects of mixing with changed geometric characteristics, present limitation of production of the plankton in reservoir.
Current lab tests should explain and quantify such a process.

8. CHANGES IN ICHTHYOLOGY

Construction of the HPP Djerdap-1 has caused a series of changes in composition and structure of

371

ichthyology. In accordance with geographical ecological fish spread, the Yugoslav section of the Danube river is divided into two regions, defined by the previous boundary at Sip. The Black Sea migration fish species, which prior to construction come mainly to the zone of the HPP Djerdap-1 to breed, now, are disabled to arrive at their spawning places to leave descendants due to the HPP Djerdap-1 and the HPP Djerdap-2 constructed dams. Now their migration path is reduced to only 18 km, from the Timok estuary to the HPP Djerdap-2. Great water level oscillations conditioned by the operation of the Djerdap-1 HPP, also, make impossible the normal spring spawning of the Danube fish, thus in dependence on the water level height during the period from February to June and water level oscillations the fish habitat within the Yugoslav section of the Danube can be renewed. It means that mainly the natural fish spawning at this section of the Danube is endangered.

9. RELOCATION OF SETTLEMENTS, COMMUNICATIONS, CULTURAL MONUMENTS AND ARCHEOLOGICAL WORKS

9.1. Relocation of settlements

Construction of the Djerdap-1 HPP required relocation of a few settlements, population, as well as relocation of roads and railway, and cultural monuments, and carrying out of significant archeological works.

Seven colonies have been relocated at the Yugoslav (Serbian) side, i.e. 8,500 inhabitants in total. Also, 1,600 residential buildings with over 10,000 m² areas have been relocated. All the remaining colonies located over the area extending from the dam to Beograd are protected by embankments, drainage systems and other technical measures adapted to urbanization and development scheme of these regions.

On the Rumanian side there was all relocated except Orsava and a few colonies with totally 14,000 inhabitants.

9.2. Relocation of communications, cultural monuments and archeological works

Relocation of roads on both banks and railway (12 km) on the left bank only (the Rumanian side) has been carried out along the first 135 km of the reservoir.

Among the cultural monuments within the reservoir area, the most important ones are the Trajan's Table and Lepenski Vir.

The road stretching through the Djerdap gorge cut into the right stone bank dated from 104 A.D. The Trajan's Table spoke of the fact that construction of the same road was completed in Trajan's time. It was moved beyond the reservoir impact at the same location.

Archeological works were intensified by the HPP Djerdap-1 construction on both banks. In such a way Lepenski Vir, prehistorical settlement, almost 8,000 years old was found out at kmD 1,005 on the right bank.

It represents one of the most important monuments of prehistoric Era and it has universal significance, although the Djerdap gorge has been practically "terra incognita" by the sixties of this century.

With the exception of the stated localities which have been explored, about 80 localities have been submerged. Numerous undiscovered fields relating to significant number of prehistoric, classical and medieval necropolis whose existence is assumed with surety on the basis of the discovered settlements, have disappeared, as well. Also, it might be assumed that a great number of prehistoric localities has remained submerged under the reservoir.

The parts of almost all discovered medieval settlements and necropolis have been submerged to a great extent, taking into consideration the fact that, mainly, among them only those parts located above the classical and Early Byzantium fortresses have been explored.

10. CONCLUSIONS

Djerdap-1 HPP construction and the reservoir creation in the Danube river stream and its tributaries have caused a series of changes and troubles. More favorable conditions in relation to the preceding ones have been generally created by their solving, i.e. by means of:

- reconstruction and dike consolidation which have resulted in their general and seepage stability increase;
- drainage systems construction at the agricultural areas and at settlements have provided the conditions for complex development and protection schemes against ground water and salinization;
- water level regime changes in the reservoir zone has made possible better conditions for navigation over this sector of the Danube;
- back water has caused regime modification of deposit which silted in the middle part of the Djerdap gorge;
- the reservoir creation has contributed to betterment of ice regime, particularly, in the gorge stretch over which, in natural conditions, it has significantly amassed;
- the Danube river damming up by Djerdap-1 and Djerdap-2 dams has blocked the natural fish pass for the selected fish species which spawned in the gorge stretch. Studying of subsequent construction of fish ladders which are omitted, are in progress;
- relocation of submerged settlements, accompanied with construction of modern type buildings has enabled more favorable living conditions;
- construction of Djerdap-1 HPP has intensified the archeological works which provided for a prehistoric settlement discovery, i.e. Lepenski Vir about 8000 years old. A significant number of archeological localities, prehistoric, classical and medieval necropolis have been submerged by the reservoir.

During Djerdap-1 HPP operation a great number of parameters and occurrences are monitored. They are important for normal (function) utilization of the reservoir and protection from backwater. Also, secondary effects created by the reservoir formation (water levels, sedimentation, ice, woods degradation, water quality, waves, ground water at hinterlands, etc.) are monitored.

REFERENCES

Numerous studies, investigations and designs which represent a synthesis of long year results and the effects presented at the Symposium held in March, 1990 in Donji Milanovac, (Serbia) have been used: BACKWATER EFFECTS OF HPP DJERDAP-1 ON REGIME AND HINTERLANDS. This symposium was attended by 58 authors who presented 42 papers.

Hydropower'92, Broch & Lysne (eds) © 1992 Balkema, Rotterdam. ISBN 90 5410 054 0

Environment evaluation problems of hydropower projects

Zhang Jia Min
Lanzhou Hydroelectric Investigation & Design Institute, MWREP, People's Republic of China

Yang Feng Wu
Gansu Academy of Sciences, People's Republic of China

ABSTRACT: The authors present three examples of different kinds of the projects to introduce the characteristic of the regional environmental evaluation and the individual emphasis problems to be solved for each project.

The environmental protection work has made considerable achievement in China in the past ten years. The people have been gradually concerned with and paying attention to the environmental protection work. At present, it has been proved by the use of many hydropower projects in China that there is an inseparate interrelation between the developing water resources and environmental aspect.

The environmental evaluation of a hydropower project concerns many aspects and is a comprehensive science. According to the respective natural conditions and characteristics of various areas, as well as the purpose of constructing project, the environmental evaluation of a project must aim at some emphasis problems. In order to provide a reliable basis for engineering feasibility of a construction project, it should get the correct evaluation on the important problems, which seriously influence the environment of the areas.

The authors present three examples of different kinds of projects by first hand experience to introduce the characteristic of regional environmental evaluation and the individual emphasis problems to be solved for each project. The first one is Cai Jia Xia hydropower project, located in the suburb upstream the Yellow River from Lanzhou city. It has low dam and small reservior. A dense industrialized area is nearby. The second one is Hong Jia Du hydropower project on the Wu River, where there are high muntains and deep cut valleys. It has high dam and big reservior. The third one is Jiu Dian Xia hydropower project at the Tao River in the Gansu province. Corre-

sponding to the policy of "promoting irrigation by hydropower", the big irrigation district will be constructed by stages after the hydropower station project is completed. Therefore, it must consider this feature for the environmental evaluation of Jiu Dian Xia project. The characteristics of environment evaluation of the above three projects are different. Each of them has it's emphasis problems to be solved.

After doing concrete work, we recognize that, to do environmental evaluation work generally and specifically is not only the correct method of work, but also the correct way to solve the contrasts and the problems.

1 A brief introduction of three hydropower projects

1.1 The Cai Jia Xia hydropower project on the upper reach of Yellow River.

The Cai Jia Xia hydropower project is located in the Xi-Gu district of Lanzhou city, 34 km west of the Lanzhou railway station and 17.7 km downstream from the Ba Pan Xia hydropower station. There are good communication and transport conditions in the region. The Lan-Xin railway and Gan-Qing highway pass through on the right bank of the dam site. A forest of factories and mills rises around the reservior shores. (figure 1)

The Cai Jia Zia project is a low-head run-of river power station. The normal pool level is 1550 m and the corresponding storage capacity is 15.8 million m³.

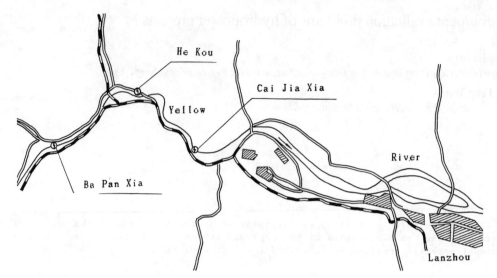

Figure 1   The location of Cai Jia Xia project

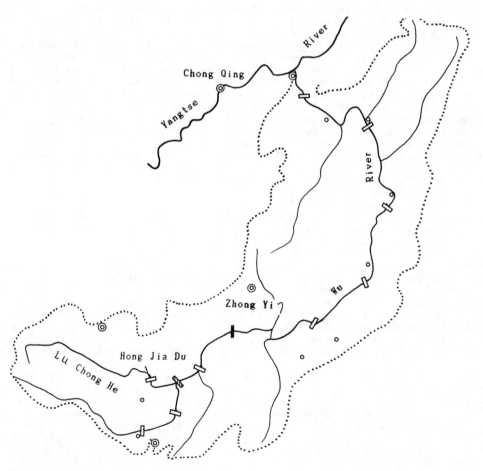

Figure 2   The location of Hong Jia Du project

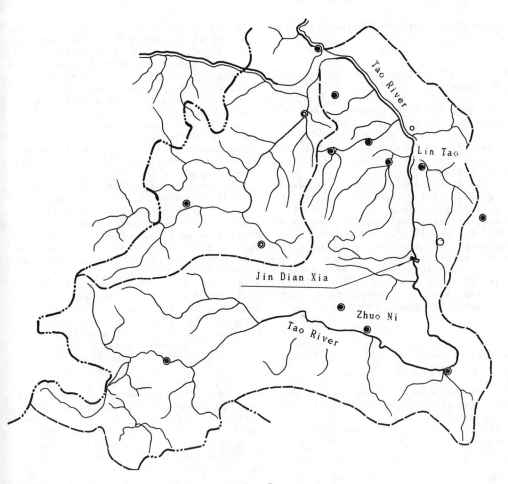

Figure 3　The water catchment area of Tao River

The main structures of the project are comprised of a retaining dam, a gated spillway and a power station, which has an average net effective head(flood season) of 6.8 m, an installed capacity of 96 MW and an annual energy output of 494 GWH/year.

530 mu farmlands, 26 pumping stations and water wheels will be inundated after completion of the project. No population need be resettled.

1.2 Hong Jia Du hydropower project of Wu River

The Hong Jia Du hydropower project is located on the Lu-Chong river, which is an upper distributary of Wu River. The project is 165 km along the highway from Guiyang city. It is the first one of the cascade stations on the Wu River, known as a "reservoir on a dragon's head". The dam will head up the water level of 165 m. The normal pool level is 1140 m. The reservior is a narrow valley type with the function of perennial requlation. It has a storage capacity of 4.589 billion m³ and has a back curve 86 km long. The installed capacity is 540 MW and the annual energy output is 1572 GWH/year. The project's main purpose is generation, but also has a multiple purpose of flood control, irrigation, fishing, tourism and water supply.

A population of 32,700 will be resettled, 21600 mu of farmland, some houses, buildings and mills will be inundated.

The multinational people are crowded together in this region. There are densely populated, but the farmland is less plentiful.

1.3 Jin Dian Xia hydropower project of Tao River

The Jin Dian Xia hydropower project is located in the middle reaches of Tao River, which is a large distributary in the upper reach of Yellow River. The Tao River is 673 km long and has a drainage area of 25527 km^2.

The Jin Dian Xia project is considered a largescale multiple purpose project of irrigation, generation, flood control, log pass, aquaculture, etc. It's designed normal pool level is 2200 m and the corresponding storage capacity is 900 million m^3. The installed capacity is 240 MW. After completing the project, it will create favorable conditions for gravity irrigation, thus 5×10^5 mu farmlands will be irrigated by progressive development.

2 The focal points of the enviromental evaluation work for the three hydropower projects

Based on the general situation of the above three projects, the characteristics and the focal points of the enviromental evaluation are stated respectively as follow:

2.1 The evaluation area of the Cai Jia Xia hydropower project

The area is the outer suburbs of Lanzhou, which is the primary industrial base of Lanzhou. There are more than 300 industrial enterprises of large, middle and small scale in the area, which also has advanced natural conditions. The source of water is rich, the cultivated lands are concentrated and connected together and the soil is fertile. It is the one of the main vegetable producing bases of Lanzhou. The area has a population of 5000 and of which 3000 are townsmen.

The present situation of the area shows us the following characteristics:

a. There are many factories, mills and buildings in the area.

b. The output of the industrial products is very high.

c. The water supply problem appears very important because of the dense po-

pulation in the area.

d. The industrial use of water and the draining of industrial pollutants.

e. The cultivated lands are less, but the utilization ratio is very high.

According to the above features of the area, first of all, we paid attention to minimize the inundation loss of the farmlands. We have changed the development plan of this river reach. The Dual cascades development has replaced the single cascade development of the previons plan. Because there are many factories and buildings in the area, Lan Xin railway and Gan-Qing highway on the right bank of the damsite, the protection against erosion becomes a secondary emphsis problem to be solved. For this reason, a protection wall of 8.5 km in length will be constructed around the reservior. The water pollution and silt control of Yellow River are also important problems. Especially, it is necessary to make clear the relationship between surface water pollution contents of mud flow on the river bed. The water quality of drainage water is considered an emphasis problem because of some large factories near the reservior.

Aiming at above problems, after analysis and evaluation, we have worked out rigorous measures of environmental protection. If these measures can be achieved, the unfavoable influences of constructing Cai Jia Xia project on the natural environment will be quite small. The remarkable economical and social beneficial results will be fully brought into play. Therefore, it will bring benefit to society.

2.2 The evaluation area of Hong Jia Du project

The reservior of Hong Jia Du project will inundate a large area and many people will need to be relocated. From the statistical data, the loss of farmland and resulting grain production per person are great. In addition, the vegetation may be destroyed by the reservior inundation, the moving of populations and the reclamation of the wastelands.

After analysis of the enviromental oharecteristic of this area, we have attached importance to study following problems:

a. Evaluating and planning present land-use situation of the reservior area and the reservior shore.

b. The influence of the reservior inundation and immigrant arrangement on the region's environment.

c. The influence of the project on the terrestrial biology.

Under the composite influences of the natural factors and man-made factors, the present land-use situation of the area has the features that, the farmland is more plentiful, but the forest and herbage are less. There is a higher population, but less farmland. The land utilization ratio is small.

The forest and pasture areas have been destroyed over a long period of time and the cultivated lands have been increased blindly, as a result, the soil erosion has been intensified day by day. The available land for agriculture, forestry and animal husbandry are decreasing. The ecological environment becomes worse with each passing day.

For the purpose of solving the working and living problems of the immigrant in the future, the previous management pattern, which relied on the farming, must be changed. The reasonable agricultural production structure must be built in light of the local conditions. The development of agriculture should be based on the multiple development, which includes farming, forestry, animal husbandry, side-line production and fishery. To improve the ecological enviroment of the reservoir shore, a good circle will be achieved between the ecology and the economy.

2.3 The evaluation area of the Jin Dia Xia hydropower project

According to the police of "promuting irregation by hydropower" the big irrigation district will be constructed by stages after the hydropower station completed. Therefore, the objective of the evaluation should be based on the Jin Dia Xia hydropower project and considering the connection with irrigation district. The comprison and evaluation are carried out by the profit and loss analysis of the environmental economy.

The influence of the project on the ecological enviroment is the main problem of the evaluation. The following problems are evaluated as emphsis:

a. The influence of changing hydrological condition and developing irrigation

area on the downstream irrigation and on the agriculture ecological enviroment.

b. The influence of changing hydrological condition on the water biology.

c. The influence of storing the reservoir on the transpotating bed load of the down stream of the river.

d. The inflnence of the water temperature on the agriculture production of the down stream of the river.

From the above mentioned, the furtures and degrees of the influence of the hydropower projects on the nature environment not only depend on the scale of the project, but also on the classification of the project, as well as on the environment background. In association with the influence on enviroment, the hydropower project may be divided into the hydropower station with big storage reservior, the run-off-river hydropower station and the project, which has main purpose for irrigation.

After introducing above three diffirence projects, we can see that, each project has it's particular and emphsis problems of the environmental evaluation to be solved. The contents and conclution show that, aiming at the important problems to do evaluation work is a best way and a best method of the work.

3 The advance of the hydropower project from the view of the the point of the environment

To build the hydropower project is very beneficial, as long as we have a well-conceived plan and careful work. The key attraction of the water power is it's regenerate ability as a source of the energy. It is proved by the practice of many projects in China. For the purpose of developing water power to the maximum limit, first of all, the economic benefits of the water power need to be correctly understood. For example, doing comparison between thermal and water power, it is unfair, if the comparison is only done by the construction cost of them. If adding the cost of the coal and the transport into the cost of the thermal plant, the comparison will be reasonable. As a result, the cost of the two will be not diffirence. In regard to the long-period operating, the water power has more advantage of the reliability and the long-life performance.

From the point of view of the environmental protection, water power has not pollutions and has some benefit. The firing of coal produces acid rain, which seriously affects the environment and human life. For example, a thermal power plant have fired coal of 3,900,000 t from 1990 and excreted silt coal of 870,000 t.The silt coal have took up the land of 4900 mu. The around environment has polluted seriously by this plant.

In one word, we should make effort to devolop and utilize the water power, to built hydropower station. It is an embodiment of remaking the world by human. However, when developing the hydropower project, must pay attention to the overall planning and do well the enviromental evaluation work. Making the contradiction to be minimum between the hydropower project and the enviroument, the advantage of the developing hydropower project to the benifit for the people will be fully brought into play.

3 Dam safety

Hydropower'92, Broch & Lysne (eds) © 1992 Balkema, Rotterdam. ISBN 90 5410 054 0

Using geo-radar to check dam core quality

T. L. By & F. N. Kong
Norwegian Geotechnical Institute, Oslo, Norway

ABSTRACT: Geo-radar principles and the technical analysis of radar ability in the application of ground investigations are presented in this paper. Field test results for three Norwegian dams are presented to show the potential in the use of geo-radar to check dam core quality.

1 INTRODUCTION

NGI started a project using geo-radar for environmental and geotechnical investigations at the beginning of 1989. With three years of richly varied experience, and also with the technical development of the radar system, NGI is now the leading centre in Norway in the area of geo-radar for technical geology.

The main distinguishing feature of the geo-radar developed at NGI is the use of frequency synthesized signals, where most commercially available systems use impulse signals.[1] The advantage of using frequency synthesized signals is that the bandwidth of the radar signal can be varied by software. This, together with the experience at NGI in the design of wide bandwidth antennas, ensures radar detection of short distance targets (e.g., 5cm asphalt layer), as well as detection of long distance targets (e.g., penetrating a 50m rock layer).

Exploring new areas of application has always been a main goal of the NGI geo-radar project.[2] From June 1989 to October 1991, 46 consultancy jobs have been performed by NGI using geo-radar and the experience obtained in environmental and engineering geology has been rich, including:

- Location of buried objects, such as pipes, cables, waste containers, etc.
- Mapping of waste disposal areas.
- Rock quality control, such as the evaluation of rock quality for industrial building purposes, evaluation of rock quality after blasting and mapping fracture zones in bedrock.
- Dam safety control, detecting dam core depth and quality.
- Mapping topography and stratigraphy of bedrock and soil for the design of geotechnical sampling surveys, and for determining the ground water flow path where there is a concern for hazardous waste contamination.

- Detecting voids in foundation rock or under pavement.
- Determination of road asphalt layer thickness.
- Quality control of concrete: detection of cracks and steel bars, etc.
- Mapping archaeological sites.

2 GEO-RADAR PRINCIPLE

First we discuss the principle of geo-radar in the application of ground investigation.[3] Fig. 1 shows the principle of using radar to detect the height of a dam core top. The radar shown in the figure consists of the following main units: transmitter, receiver, antennas (transmitting antenna and receiving antenna) and data acquisition computer.

The electromagnetic wave, generated by the transmitter and radiated by the transmitting antenna, propagates in the sub-surface medium. When the wave hits the target surface, the target interface will reflect the wave which can be received by the receiving antenna. The ground surface reflection can be considered as the time origin, and measuring the time delay between the ground reflection and the target reflection thus gives the depth D of the target to the ground surface:

$$D = \frac{1}{2} \Delta\tau \, V \qquad (1)$$

where V is the electromagnetic wave speed in the medium, and $\Delta\tau$ is the travelling time.

In using geo-radar, the main difficulty arises from the fact that the medium may strongly attenuate the EM wave. Table 1 shows the EM wave behaviour in common ground material.

As shown in Table 1, we can see that geo-radar can detect sub-surface features at distances of several metres to several tens of metres in low attenuation material such as sand, gravel, rock and fresh water. The distance may decrease to a few

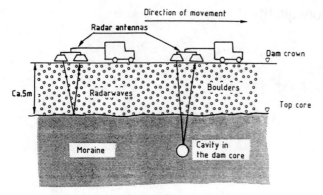

Fig. 1   Principle diagram of radar
detection of dam core

Table 1   Typical attenuation and
estima-tion of detection distance
in common ground material at 100
MHz.

Material	Attenuation (dB/m)	Wave speed (cm/ns)	detection distance (m)
Fresh water	0.1	3.3	100m
Sea water	10^3	3.3	0.1m
Dry sand	0.01	15	100m
Gravel	0.02	14	50m
Saturated sand	0.03-0.3	6	50m
Limestone	0.4-1	12	30m
Shales	1-100	9	<10m
Silts	1-100	7	<10m
Clays	1-300	6	<10m
Granite	0.01-1	12	50m
Dry salt	0.01-1	13	50m
Ice	0.01	16	100m

metres in high attenuation materials such
as clay, etc. With these detection distances in mind, geo-radar will certainly
become an important tool not only for geotechnical investigations, but also for
environmental and archaeological investigations, etc.

In NGI's radar system, a network analyser (HP 8753) is used as the signal generator and receiver. During measurement,
201 frequency samples which cover the
desired signal bandwidth are generated in
sequence. The magnitudes and phases of
the echo at those frequency samples are
received, also sequentially, by the network analyser, and recorded by the data
acquisition computer. The software is
written in ASYST programming language. It
is computationally efficient in performing
signal processing (such as Fast Fourier
Transform, etc.) and in displaying
results.

The power level generated by the network analyser is from -5 to 20 dB mW and
the receiver noise level is -90 dB mW.
Considering the maximum input to the
receiver is 20 dB mW, the receiver has a
dynamic range of 110 dB.

3   USING RADAR TO CHECK DAM CORE QUALITY

A typical Norwegian dam contains a moraine
core which is covered by a thin filter
layer (sand material), layers of gravel
transition zone, and a layer of 1 to 3m
large, blasted rock blocks. The distance
from the moraine core top to the dam crown
is normally 2-4m. Fig. 2 shows the cross
section figure of the Tunhovddammen.

One of the important problems involved
in dam quality control is to check the
absolute height of the dam core top. The
most economical method of doing this is to
perform radar reflection tests on top of
the dam crown, to detect the distance from
the crown top to the dam core top. The
moraine core material is similar to clay
material. Hence, referring to Table 1,
the wave speed in moraine is close to
6cm/ns (ns= 10^{-9} second).

The wave speeds in granite rock and in
gravel material are all faster than 10
cm/ns. One can thus expect a reasonably
large reflection coefficient at the
interface between moraine and filling
material.

Another method of locating the core top
is to use transmission mode measurement
(Fig. 3), where the transmitter antenna
and the receiver antenna are placed on
opposite sides of the crown wall to
measure the received power and the wave
arrival time. Table 2 shows the result
measured at Norway's Oddatjørn Dam.

In Table 2, we can see that when h1 and
h2 reach about 4 metres (Test 3), the
transmission attenuation increases
dramatically. This suggests that at these
antenna locations, the EM wave beam between the two antennas has already met the
dam core since the core material has a
much higher attenuation to EM wave than
the rock does.

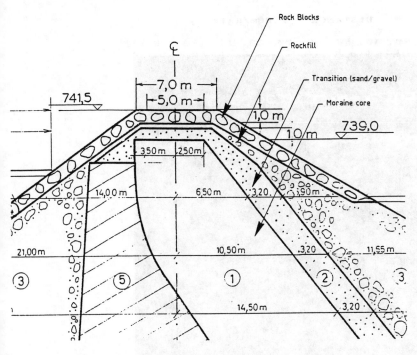

Fig. 2 Tunhovddammen structure (Norway)

Table 2. Cross crown measurement

Test No.	Antenna Position to crown top (h1, h2)	Antenna Distance L	Time delay ΔT	Speed $v = L/\Delta T$	Attenuation
1	2m, 2m	12.7m	83 ns	15 cm/ns	26 dB
2	2.2m, 2m	13.0m	89 ns	15 cm/ns	26 dB
3	4m, 3.5m	17.7m	105 ns	17 cm/ns	50 dB
4	5m, 5m	21.3m	123 ns	17 cm/ns	55 dB

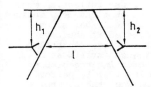

Fig. 3 Cross crown measurements

The distinctive variance of the transmission attenuation while changing the antenna heights h1 and h2 to meet the core top, suggests that it is an effective method to use the cross-crown measurement to detect the dam core location. Since crown walls are normally very irregular, it is not easy either to measure the exact coordinates of the antenna, or to move the antennas. The reflection mode method is a far more convenient test method.

Fig. 4 shows the radar reflection measurement results for one section (120m length) of Tunhovddammen. In the figure, the first reflection is from the interface between the large block rock filling and the gravel transition zone. The second reflection is from the moraine core. In this dam section, the reflection from the core is, in general, strong and stable, and the core is at about 2.5m depth to the crown top. However, we can also see that

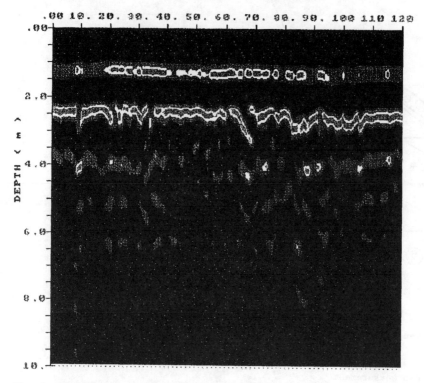

DISTANCE   1.00m/unit

Fig. 4 Measurement at Tunhovddammen. The
second reflection at 2.5m depth is the
reflection from the dam.

the core reflection from unit No. 60 to
No. 110 is not as stable as the reflection
from other distances. This is an example
of good core reflections.

We have also done testing at
Askjelldalsvatn Dam. At the beginning
section at Askjelldalsvatn Dam, we could
hardly point out the core top depth from
the radar result. This dam was opened
after the radar test was performed. Even
experts, after seeing the site, could not
tell exactly where the core top interface
was since it was immersed in the
transition zone and the rock filling
layer. However, at some other sections of
the dam, the reflection from the core top
becomes stable and the core top is at
about 1.5m depth to the dam crown top.

Fig. 5 is the result of the test from
the beginning section of the Nesjen Dam.
From the figure, we can see that the core
reflection is not very stable in this
section. The better part is from 95m to
120m and the worst part is from 38m to
94m. The worst locations are at 45m and
85m.

To summarise the above, we say that
radar measurements can locate the moraine
core top position by using the reflection
mode measurement. Radar measurements can
also tell where there is not a sharp

interface between the core and the gravel
transition zone.

Compared with other types of ground
investigation problems, dam core detection
is amongst the most difficult. The main
difficulty is that the rock filling layer
is hardly homogeneous. Air pockets
amongst large block stones also reflect
high frequency signals. The variance of
the air volume in the rock filling zone
changes the speed of the waves. We need
to learn in future how to control these
uncertainties.

It is difficult to use reflection mode
measurement to detect erosions deep inside
a core, since most of the radar signal
energy would be reflected at the core top
interface, and little energy penetrates
into the core. For such investigations,
vertical boreholes crossing the problem
area may be needed in order to perform
tomographic tests between the boreholes
and to locate the internal erosion. NGI
has rich experience in performing
tomographic measurements in rock and soil.
Radar tomography is used for ground
investigation in tunnelling, mining and
foundation engineering.

DISTANCE   1.00m/unit

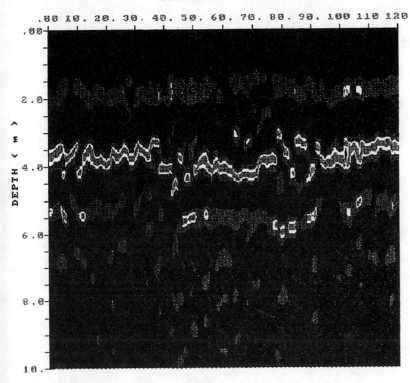

Fig. 5 Radar measurement at Nesjen Dam
(beginning section). The worst part of
the core is from 38m to 94m.

4 REFERENCES

Kong, F.N. and T.L. By, "Ground Penetrating Radar using a Frequency Sweeping
Signal", Third International Conference
on Ground Penetrating Radar, Denver,
USA, May 1990.

By, T.L., F.N. Kong and
H. Westerdahl,"Geo-radar Development at
the Norwegian Geotechnical Institute",
Fourth International Conference on
Ground Penetrating Radar, Rovaniemi,
Finland, June 1992.

Kong, F.N., H. Westerdahl, C. Lund and
T.L. By, "Geo-radar for Ground
Investigation", Nordisk Geoteknikermøde,
May, 1992.

Hydropower'92, Broch & Lysne (eds) © 1992 Balkema, Rotterdam. ISBN 90 5410 054 0

Probabilistic regional precipitation analysis

O. Cayla
SOGREAH Ingénieurs Conseils, Grenoble, France

J. P. Broch
Norwegian Institute of Technology, Trondheim, Norway

ABSTRACT: The random nature of rainfall events obliges the hydrologist to use statistical methods in order to quantify precipitation heigths and their associated recurrence intervals. But the available data represent only limited samples of the theoretically infinite parent population. The extrapolation of a statistical distribution fitted to such a sample may hence be very erroneous. The sampling error for each pluviometric post can be reduced by relating it to neighbouring stations. Supported by a mathematical theory, the Poisson Process, the relationship is established as the annual average number of events. This number should not vary throughout vast regions because the precipitation is due to large scale meteorological events.

1 INTRODUCTION

The security and economy of a hydropower project depend very much on the hydrologic study. Under-estimations of design floods may have catastrophic consequences, and over-estimations may give unecessarily high costs.

The best way of computing hydrological design values, is to divide the study into three major parts. Firstly, the rainfall data are studied to obtain the precipitation return periods. Secondly, the rainfall runoff (or discharge) relationships are established. Finally, a combination of these two studies yields return periods for the runoff and the floods.

The precipitation analysis is thus a crucial part of the study. In order to constitute good estimates of the design values, it is in an early fase important to understand the rainfall mechanisms.

2 FREQUENCY ANALYSIS

Due to the random nature of climatical events, statistical methods are introduced in order to calculate risks corresponding to design rainfall heights and flood peak discharges (and volumes). The frequency analysis is a way of summarizing observed data by a probabilistic distribution (Gumbel, for example).

The distribution would be perfectly known if the infinite series of events was recorded, that is, if the parent population was known. This has of course a theoretical sense only. In hydrology, one disposes limited samples of this parent population; the periods of observation are in the order of 10 to 100 years.

The sample may or may not be representative of the parent population. By fitting such a sample to a probabilistic distribution, the project values found by extrapolating may be very erroneous due to the sampling error.

Usually, a frequency analysis is based on one sample, and the distribution is chosen as the one giving the best fit of the data. It may however be shown that it is impossible to determine the distributon of the parent population from a hydrologic sample.

To overcome these problems, a working theory is introduced.

3 POISSON PROCESS

The Poisson Process is a well known mathematical theory. It has been developed for hydrological purposes during the two last decades by a team of French consulting engineers. The applications to hydrology was presented in it's present form by Cayla (1990). Simular methods have nevertheless been presented earlier by Duckstein et al (1976).

Given an event A that may or may not occur

(it may rain or it may not rain). To the occurence of A one assigns a measuremement x (the cumulative precipitation height, for instance). The Poisson Process is then a discrete, stationary and simple statistical process that satisfies the following basic axiom:

The probability of A occuring in the interval Δx, provided that it does not occur from 0 to x, is proportional to Δx.

With x as a time variable, this means that the probability of observing an event is the same in one interval Δt as in another interval Δt following the first. This axiom seems to be well adapted to hydrological events.

Accepting the axiom, it is mathematically shown that (Cayla (1986)):

1) The random variable of measurements, x, has an exponential distribution.

2) The number of events of A in an interval $[X_1, X_2]$ (here this interval is considered to be one year) has a Poisson distribution.

3) The annual maxima z of observed values of x have a Gumbel distribution

$$G(z) = \exp[-\exp(-az+\ln\upsilon)] \qquad (1)$$

where a is the proportionality factor of the basic axiom and υ is the annual average number of events.

4) The annual totals have a "leak" distribution (loi des fuites, see Ribstein (1983))

$$f(u) = \frac{\lambda \exp(-\lambda-u)}{\sqrt{\lambda u}} I(2\sqrt{\lambda u}) \qquad (2)$$

where u=as, λ is the annual average number of events and I is the modified Bessel function.

The "leak" distribution is not easy to treat in hydrology. It may however be demonstrated that the root-normal distribution ($\sqrt{s}$ has a normal distribution) is practically identical to the "leak" distribution if $\lambda>10$ (Cayla (1986)).

4 GUMBEL AND ROOT NORMAL DISTRIBUTIONS

When implementing the Poisson Process, the distributions of the annual maxima and totals are theoretically predetermined. The usual choice of distribution according to each specific sample is hence eliminated.

The Gumbel and Normal distributions are furthermore well known. They have both socalled probability papers which make them easy to handle.

They are finally both two parameter distributions on the form (Chow (1960))

$$x(T) = m + \sigma K(T) \qquad (3)$$

where T is the recurrence interval, m is the mean, σ is the standard deviation and K is the frequency factor. K gives the characteristics of the distribution itself, whereas m and σ are the curve fitting parameters. Graphically, K represents the probability paper and m and σ describes the fitted, straight line.

The Gumbel distribution may be expressed as

$$Z(T) = m\frac{y_0 - y(T)}{y_0 - C} \qquad (4)$$

where m is the mean (or the first parameter), y(T) is the Gumbel reduced variable, C=0.5772... is the Euler constant and $y_0 = -\ln\upsilon$ (second parameter).

The Root Normal distribution can be written on the form (Cayla and Mahboub (1977))

$$S(T) = m\frac{[\sqrt{e} + u(T)]^2}{1+e} \qquad (5)$$

where m is the mean (first parameter), u(T) is the Gaussian reduced variable, $e = \lambda + (\lambda^2 - \lambda)^{1/2} - 1$ and $\lambda = 2/Cv^2$ (second parameter). Cv is the coefficient of variation.

5 NUMBER OF EVENTS

Physically, υ and λ correspond to the average number of rainfall events. The precipitation is generated by large scale meteorological perturbations. It follows logically that the number of events should be constant throughout vast, homogenious regions. The average numbers of events are thus regional parameters.

These parameters are not, as one may expect, equal for the the maxima and totals, $\upsilon < \lambda$. This is due to the fact that the annual maxima are drawn from an under population of the set of events which, when added, constitutes the annual totals.

6 REGIONAL ANALYSIS

The regional analysis is performed by collecting data from all available raingages in a region. After a preliminary verification and control of

390

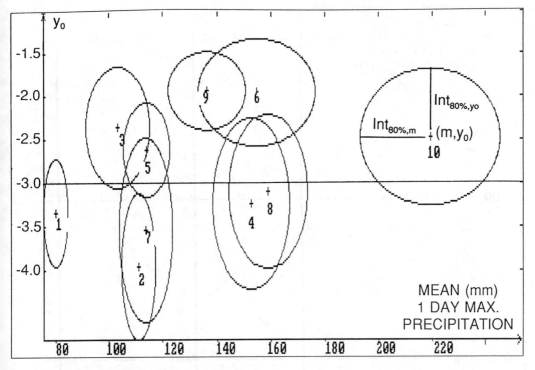

Figure 1 "Balloon" graphics. The regional parameter, y_0, versus the local parameter, m. This regional analysis yields $y_0=-3$ ($\upsilon=20$).

the data, workfiles consisting of annual totals and 1 day maxima can be constituted. One calculates then the mean and the annual average number of events and their respective 80 % confidence intervals for each of these series.

This may be presented graphically as shown in figure 1. (For the annual maxima, $y_0=-\ln\upsilon$ is used in the analysis for practical reasons.) The pairs (m,y_0) or (m,λ) are plotted, and the confidence intervals are drawn as ellipses surrounding these points. The regional average number of events is then found as the horizontal line intersecting the majority of the ellipses. If this line touches 80 % of the "balloons" with 10 % untouched on each side, y_0 (or λ) is found within the 80 % confidence interval.

The graphics represent in fact a regional parameter, y_0, on one axis, and a local parameter, m, on the other. This means that the events provoking rainfall are the same in the area, and that the mean precipitation heigth varies due to orographical effects.

y_0 is now transfered to a Gumbel probability paper, see figure 2. On this graphics, y_0 is the point where the fitted line intersects the abscissa (z=0). From figure 1, y_0 is found equal to -3,

and is, as stated, constant. The fitted lines for all stations in the region should hence pivot around this point. Series a and b in figure 2 respect directly this theory. When fitting these lines, the mean is the only parameter which has to be adjusted. On a Gumbel probability paper, the mean is situated at y=C (the reduced variable is equal to the Euler constant).

A simular analysis can be performed on the annual totals with m and λ as coordinates on the "balloon" graphics. The regional analysis for the stations in figure 3 and neighbouring raingages yields $\lambda=75$. e in equation 5 is then found equal to 148,5. The Gaussian reduced variables for wet and dry reccurence intervals of 1000 years are u(1000)=±3,1. This leaves the mean as the only undetermined variable in equation 5. Two points on the lines to be fitted on the Root Normal probability paper (figure 3) may thus be calculated by introducing m.

In practice, y_0 is calculated by

$$y_0=C-\frac{\pi}{\sqrt{6}Cv} \ , \qquad Cv=\frac{\sigma}{m} \qquad (6)$$

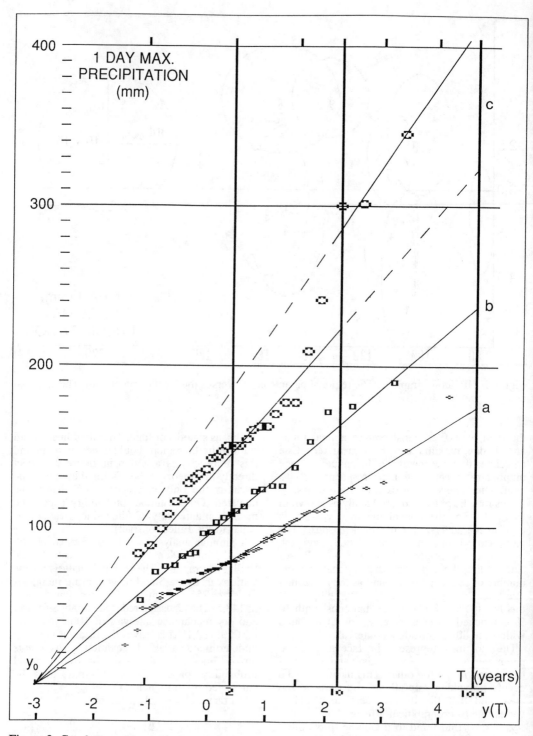

Figure 2 Gumbel graphics with three fitted samples corresponding to $y_0 = -3$. Sample c belongs to a mixed regime.

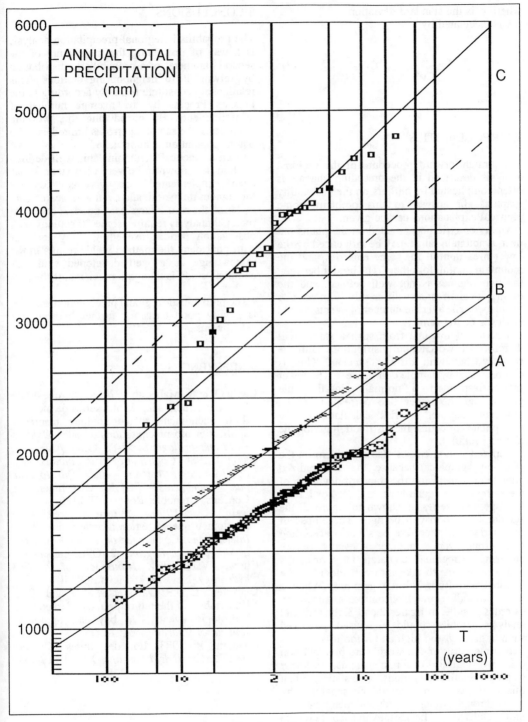

Figure 3 Root-Normal graphics with three fitted samples corresponding to λ=75. Sample C belongs to a mixed regime.

where σ is the standard deviation.
λ is calculated by

$$\lambda = \frac{2}{Cv^2} \ , \qquad Cv = \frac{\sigma}{m} \qquad (7)$$

7 IRREGULARITIES

The straightforward procedure of the regional analysis described in the preceding chapter is often complicated by outliers on the probability graphics. In answer to this problem, some physical explanations can be given.

A pluviometric post situated on the plains but near a mountain range may be influenced by the lower area rainfall for some episodes, and the mountain rainfall for others. The border between the two regimes is not well defined, and the orographical effect acts differently from episode to episode. The average number of events should however be the same.

A series of maxima from such a staion will therefore fit two Gumbel distributions with the same y_0 but with different means. This is demonstrated by sample c in figure 2. The lower line corresponds to a "normal" rainfall on the plains, and the upper line corresponds to a mountain regime. In risk calculations, one extrapolates of course the distribution giving the heaviest rainfall.

A mixed station causes also a problem on the "balloon" graphics because the regionalized parameter depends on the standard deviation of the sample (see equations 6 and 7). For a non-homogenious series as shown in figure 2, the standard deviation will be high. The ellipse of this station will therefore be situated according to a wrong average number of events. Before choosing the regionalized parameter y_0 or λ, it is hence necessary to take into concideration the Gumbel and Root Normal graphics. When a mixed rainfall regime is visualized as shown by samples c and C in figures 2 and 3, the regional analysis graphics should be revised. A mixed sample gives y_0 too high and λ too low.

In some parts of the world, the precipitation has two entirely different origins, ordinary tropical rainfall and typhons, for instance. When this is the case, they should be separated by going through typhon records and then treated as individual series. The stations in such a climate will in this way have different numbers of events for the ordinary rainfall and the typhons.

8 CONCLUSIONS

The probabilistic regional precipitation analysis is a way of restraining the uncertainty of the second parameter of a probabilistic distribution by relying it to neighbouring stations. This relationship constitute formally according to the Poisson Process by the average number of rainfall events. By introducing this working hypothesis, the sampling error is limited, and the parent population is approached.

From a practical standpoint, this well defined method has shown to be very consistent. It has been implemented in various contexts throughout the world, and even with scarce data material, the results have been good. A complete regional analysis requires some 30 stations with observation periods exceeding 20 years. But with poor data, the method yields results in the same range as a well developed study. A complete study is though a precision of the pre-study.

The probabilistic regional precipitation analysis is a very good basis for further hydrological calculations.

9 REFERENCES

Cayla O. & K. Mahboub 1977. Recherche d'une loi d'ajustement adaptée aux séries de mesures d'un phénomène météorologique discret - application aux precipitations anuelles. *Cahiers de la météorologie* 5: 57-88. Oran, Algerie.

Cayla 0. 1986. *Contribution à la connaissance de l'hydrologie yemenite - etude hydrologique par évenements.* SOGREAH Ingénieurs Conseils, Grenoble: Report NT 2182.

Cayla O. 1990. Synthèse régionale pluviométrique en région montagneuse. *Proc. International Conference on Water Resources in Mountainous Regions,* 2AO9L, Lausanne.

Chow, W.T. 1964. *Handbook of applied hydrology.* New York: McGraw-Hill

Duckstein, L.,D.R. Davis & M.M. Fogel 1976. Uncertainty in the return periods of maximum hydrologic events - a bayesian approach. *Journal of Hydrology* 31: 81-95

Ribstein, P. 1983. Loi des fuites. *Cahiers ORSTOM série hydrologique* XX 2: 119-141

Hydropower'92, Broch & Lysne (eds) © 1992 Balkema, Rotterdam. ISBN 90 5410 054 0

Floating ice as a problem to hydropower intakes

Klemet Godtland & Einar Tesaker
SINTEF NHL, Trondheim, Norway

ABSTRACT: An ideal intake structure should allow a minimum of ice to enter the forebay or the headrace tunnel. Accumulation in front of the intake should be limited to avoid the ice from being sucked into the intake or against the trash racks during normal operation. A main issue is how to divert the ice with a minimum loss of water. Several previous studies of diversion of "passive" frazil etc have been reviewed, including two recent model studies. Proper location of ice sluices combined with adjustments of structures and bottom topography can solve problems that have hampered the operation of the power plants for decades.

1 INTRODUCTION

Intake problems caused by floating ice were frequently discussed during the seventies, but are scarse in later proceedings from ice symposia. The book "River and lake ice engineering" (Ashton 1986) mentions the problem and related topics on just a few pages. The problem persists, however, as documented by recent model studies contracted in order to reduce ice problems.

One way to classify ice problems in front of intakes might follow the traditional distinction between "active" and "passive" ice:

Ice floating in supercooled water or in water of 0°C, subject to heat drain to the colder atmosphere, is classified as "active". Ice particles of this kind tend to adhere to structures or to each other, forming aggregate ice bound together by the freezing process.

Ice in water of temperature above 0°C is classified as "passive". This means that problems relate to transport and accumulation of ice particles or ice floes in a melting or neutral condition.

In this paper we shall concentrate on "floating ice" in general, as distinguished from already jammed or packed ice, or from firm surface ice attached to the river banks or structures.

An ideal hydropower intake should respond to floating ice in the following manners:

1. A minimum of ice should be allowed to enter the forebay or the headrace tunnel.

2. Accumulation in front of the intake should be avoided or easy to remove with a minimum loss of water intended for production.

Many articles have been written, describing field experience, model studies, and more or less good solutions to the floating ice problems. In this paper an overview will be given, including some not yet published examples.

2 COMMON PROBLEMS AND MEASURES

The character of problems and measures depends to some extent on whether ice floes of significant sizes can approach the intake or not, together with smaller ice particles. Ice floes can usually be looked away from only in cases where there are turbulent rapids just upstream of the intake.

2.1 Small ice particles only

Intrusion of frazil or small ice particles may be the direct result of too turbulent approaching water. In most cases, however, the intrusion will usually be preceded by an accumulation of ice in front of the intake structure, above the trash rack. Subsequent consequences may be partial clogging of the trash rack or accumulation of ice in the forebay or other parts of the headrace. The development is very sensitive to the operation of the power plant, both to the discharge as such and to sudden discharge variations in particular.

If not deposited, moderate amounts passing with the water are usually of little harm to turbines and operation.

2.2 Ice floes and mixtures

Single ice floes arriving in front of the intake structure may be caught in a revolving eddy and as such be harmless. The problems arise when the intake flow becomes strong enough to overturn the floe and suck it down against the trash rack bars, causing an immediate partial blocking of the passage. Possible consequences are: dramatic temporary head reduction and loss of production, in extreeme cases dry running turbine followed by shock return of discharge, damage to the trash rack bars, disintegration of the ice floe and passing of fragments big enough to damage the turbine.

The overturning risk increases with the turbine discharge, therfore the arrival of ice floes may require carefull or reduced operation of the power plant.

If more floes arrive, they are likely to dive and slide underneath each other, finally affecting the trash rack even without overturning.

Mixtures of floes and finer particles may form a very complex pack that is difficult to handle.

2.3 Common measures

Mitigating measures may be built in as permanent parts of the design or be applied only when needed.

Permanent arrangements include favourable design of intake geometry, special ice flushing gates, ice skimming structures and guide walls.

Temporary measures include manual or mechanical removal of ice, reduction of the power production, flushing through the flood gates, blasting etc.

Examples of measures will be given below. A governing principle for use of such measures is the minimizing of production losses i.e. minimum water used for flushing, short shut down or reduced capacity periods and minimum head loss through the intake.

3 FIELD AND LABORATORY EXPERIENCE

We shall refer examples from the well known and frequently reported Burfell powerplant in Iceland, as well as from two not previously published Norwegian cases, viz. the Rygene and Fiskumfoss plants. All three have been studied in models at the SINTEF NHL in Trondheim, and the results have been matched with field observations.

3.1 The Burfell power plant

Figure 1 shows the intake lay out of this much referred plant (Sigurdsson 1970). The plant is characterized by up to 40 m^3/s discharge of loosely packed frazil ice (about 15 m^3/s solid ice) in winter discharges as small as 100 m^3/s. There is also considerable sand transport in the river. The intake structures were designed to cope with both these problems, and later experience has confirmed that this has been achieved to a great extent.

The main principle of the design consists of a three level inlet structure, Figure 2, with sand flushing tunnels at the lower level, water passage at mid level, and an ice skimming "trough" at the top level. The ice trough is discharging ice and water into a separate ice canal with better transport conditions for removal of ice below the dam than the original river bed could provide.

Other features of the design that are important for the ice handling are the rather concentrated stilling pool in front of the inlet structure, the jetty dividing the overflow and gated sections of the dam, and the slightly increasing distance between the inlet and jetty, providing diverging flowlines in the pool.

It should also be noted that there is a 3 km long intake pond between the diversion inlet and the power plant intake, such that ice passing the

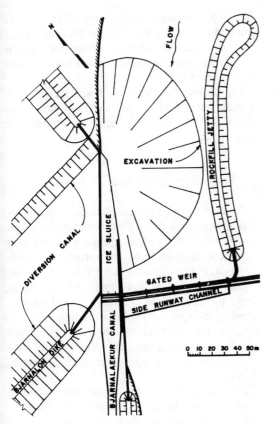

Fig 1 The intake layout for Burfell power plant

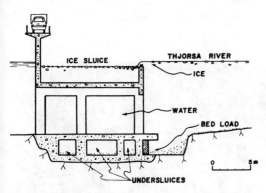

Fig 2 Three level intake structure at Burfel!

the initial design. It was the intension to use the ordinary flap gates for ice skimming. Model tests proved that the trough would require less water for the ice flushing and therefore be more economical. More details will be given below.

3.2 The Rygene power plant

Figure 3 shows some details from the corner between the intake and the old dam of Rygene power plant, including the proposed location of a new ice flushing gate.

The power plant is located in the southern part of Norway where ice problems may vary from severe to none from year to year. There is a rapid section upstream of the dam, rarely ice covered, but not turbulent enough to prevent ice floes of significant size from reaching the dam unbroken. During cold periods the rapids produce significant amounts of frazil ice, which together with possible ice floes have caused serious intake problems through many years.

The old ice flushing gate, a 1.5 m x 8 m flap gate, had proved of little value, catching only

inlet will not immediately affect the intake or turbine functions. But the pond has not enough accumulation capacity for ice arriving during the whole season, hence the need for skimming.

The ice skimming trough was not included in

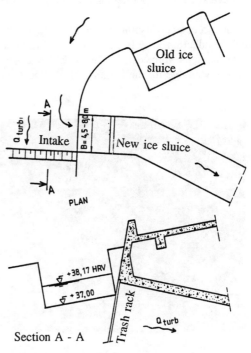

Fig 3 Detail from Rygene power plant

about 25% of the arriving ice during normal operation of the plant. Attempts to guide the ice to the gate by booms improved the situation only slightly. Arriving ice floes gathered in front of the thrash racks and were frequently sucked down against the bars. The result was added head loss, even to the extent that the plant might suddenly be without the minimum water needed for operation. Frazil and slush ice gathered both in front of the wall and inside in the forebay.

Flushing of ice floes or frazil already accumulated in front of the intake was not succesful, even during full shut down of the plant. Big floes were sometimes hooked and towed away manually from the wall and through the ice flushing gate.

In connection with reconstruction of the dam, a new location of the ice flushing gate was proposed, as indicated on Figure 3, and tested in a model (Godtland 1989). The model results were very promising. It was possible to flush both floes and frazil through the new ice gate even with full winter operation of the power plant. The gate was actually less effective in the case of a closed plant, apparently because the increased total flow during operation caused a transverse flow along the intake wall towards the gate.

The necessary flushing water represented the only production loss. Both 4.5 m and 8 m wide gate versions vere tested. The widest one was most favourable for the flushing of ice floes, but required generally more flushing water. A minimum depth across the flap for flushing of ice floes was 1.2 m. The 8 m version was chosen mainly because the old gate could be tranferred to the new place.

The winters have been mild without ice since installation of the new gate, but it is reported that the ice gate works efficiently for flushing of floating trash.

3.3 Fiskumfoss power plant

Figure 4 shows a typical situation in cold winters at the intake to the Fiskumfoss power plant. Frazil has been packed against the inlet and obstructs the trash rack. Only manual methods are available for removal of the accumulated ice.

Use of an existing flap gate in the dam, 16 m from the intake wall, Figure 5, could remove most of the approaching and accumulated ice as long as the power plant was not in use. Even small discharges through the plant caused 80% - 90% of the ice to remain in front of the intake, however.

A model study was commissioned in order to find solutions to the problems at this plant as well as to study ice flushing in more general terms (Godtland 1992). The study involved e.g. testing of a new ice flushing gate close to the intake, as indikated on Figure 5. The new gate was a definite improvement. It was now possible

Fig 4 A winter situation at Fiskumfoss

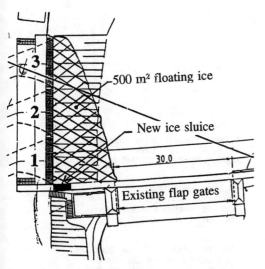

500 m² floating ice

New ice sluice

30.0

Existing flap gates

Fig 5 The intake and flap gate at Fiskumfoss

to flush out the accumulation shaded on Figure 5, but the flushing time was rather dependent on the actual combination of flushing discharge and production discharge. See also section 3.5.

3.4 Operation of stations in case of accumulation

Based on theory and experience including model studies, certain guide lines for the operation of power stations in order to minimize ice problems have emerged:

1. Steady operation is better than intermittent or varying if the purpose is to avoid ice intrusion or clogging of the intake. The obvious reason is that submergence of ice particles is the consequence of turbulent activity which will be increased during unsteady flow situations as compared to steady flow. Overturning of jammed ice floes is also activated by accelerations and turbulence.

It is possible to analyse these processes by methods similar to the sand transport theory as well as by hydrodynamic methods. See section 4.

Based on calculations or site experience, recommended safe upper limits for production during ice events may be established.

2. Intermittent plant operation may be useful to loosen and break up ice pack in connection with use of flushing gates. Usually sufficient effect is obtained by manipulation of the flap gates alone,

however. This has been well confirmed during operation of the Burfell plant. In other cases a sudden stop in production followed by careful increase again may be tried.

3. A temporary full stop may sometimes be the best solution also to handling of floating ice, mainly of the active type. This will halt the accumulation in front of the intake. If the result is the forming of a solid ice cover in the intake reservoir, it may reduce production of new ice and may also relocate accumulation patterns in a favourable manner.

3.5 Principles for flushing through flap gates

In all three power plants mentioned above, flushing through flap gates have been planned and tried with varying success.

In the Burfell plant all the flood gates were constructed as flap gates in order to function also for ice flushing. Model tests proved that a considerable part of the available winter flow might have to be spent for the flushing of ice at periods. The flap gates tended to draw ice-free water from underneath the floating ice in stead of discharging only surface water where the ice was. In this manner much water was lost without purpose. As a consequence the skimming trough shown on Figures 1-2 was installed and tested. The difference in performance is demonstrated by the Figures 6-7, quoted from the test report (Carstens and Tesaker 1966). Although presentations differ for the two cases, it is easily seen that the trough design gives noteable production savings, particularly for small discharges.

From the test results it was possible to produce an operation manual for optimizing production when discharge was limited and flushing was needed. The experience so far has confirmed the test findings with small modifications, and it has been possible to operate the plant through severe ice situations.

The quantitative findings at Burfell are of local nature, since flow pattern and water depths in the pond and over the gates will vary from plant to plant. The qualitative difference is of principal nature, however, since it has to do with the vertical flow pattern in front of the skimming devices. Water flow approaching the flap gates under the ice layer will cause an upwellig in

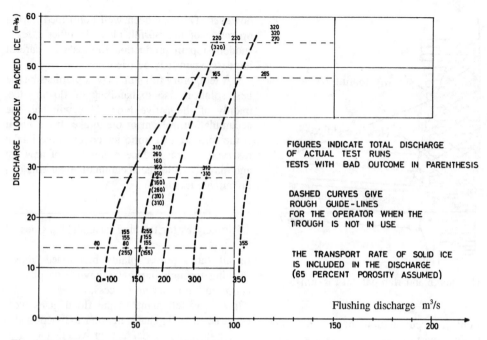

Fig 6 Flushing over flap gates at Burfell power
plant. Model results.

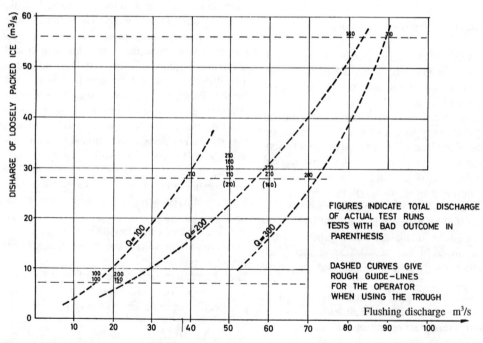

Fig 7 Flushing through skimming trough at
Burfell power plant. Model results.

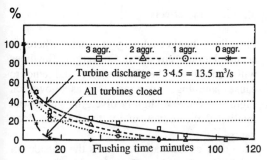

Fig 8 Remaining ice as function of time for various idling situations and 110 m³/s flushing over ordinary flap gates

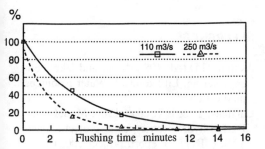

Fig 9 Remaining ice as function of time for two different flushing discharges with all turbines stopped

front of the gate sill that will retard the arriving ice. In front of the skimming trough most of the excess flow will escape underneath the trough, causing negligible upwelling, and free flow of ice into the trough.

The upwelling effect is a main reason also for the problems that have been experienced while attempting to flush accumulated ice in the two other plants, at Rygene and Fiskumfoss. Even when the whole river flow is beeing diverted over the ordinary flood flaps, the effect in drawing ice from a stagnant or backwater area is small, unless the distance is small and the up-welling motion is beeing limited in some way.

By placing the flushing gate as close as pos-sible to the intake, both those conditions are beeing approached: The distance to the accumu-lated ice is short, and the operation of the power plant provides an escape for the underflow.

Ordinary flap gates some distance away from the intake meet an opposite effect of the flow through the intake: The flow under the ice will have a component against the transport route towards the flushing gate and will therefore counteract the flushing. This is reflected in the results of the model tests:

1. It was observed that operation of the power plant in some cases could rather improve than reduce the effect of the new ice gates close to the intake, e.g. at Rygene.

2. The flushing capability of the ordinary gates was gradually reduced by increasing the produc-tion flow.

Figures 8-9 demonstrate this, showing the time needed for flushing of an area of 500 m² covered by ice, i.e. the shaded area on Figure 5.

Figure 8 shows results of a test series using 110 m³/s flushing water while the turbines are either idling at 4.5 m³/s each or completely stopped. The effect of even such small turbine discharges is striking. The effect decreases, however, when more flushing water becomes available, as shown in Figure 9 for 110 m³/s and 250 m³/s. in both cases with all turbines idling at totally 13.5 m³/s.

4 SOME HYDRAULIC PRINCIPLES

Some comments of general hydraulic nature have already been mentioned, e.g. in 3.5 about the upwelling effect and the function of ice skim-ming structures. A few more may help in further understanding of the ice problems at intakes.

4.1 The movement of ice particles

Ice transport under a continous ice cover or a stagnant accumulation of ice may be looked upon and studied as an upside down sediment pro-blem:

1. Shields' stability criterion apply for start of movement or for deposition, inserting the sub-merged buoyancy of ice for the submerged density of the particles in the formula. For a case study, see Tesaker (1975).

2. Ice floes resting under ice cover or a jam will move or overturn according to similar rules as for rock slabs on the river bed. See e.g. Uzuner (1975).

3. Concentration of ice suspended in the water

depends on and increases with the turbulent activity.

Ice particles, blocks and floes on the surface follow the mechanics of bouyant bodies and will overturn or dive under booms, ice edges or accumulated ice under conditions studied by e.g. Uzuner and Kennedy (1972) and Larsen (1975).

Since both shear stress and turbulence increase with flow velocity, and during accellerations, it is a general rule to operate the power plant as steady as possible, or at least with slow changes, if ice intrusion is to be avoided.

4.2 Eddies and whirls

Horizontal eddies have notable influence on the accumulation and flushing conditions. Small modifications of the intake area design may therfore change the situation. This was studied in some detail for the Fiskumfoss plant (Godtland 1992). Figure 10 shows the eddy situation before installation of the ice gate (Figure 5), which reduced the eddy when in operation.

Vertical whirls should be avoided, as the ice has low buoyancy and even large blocks are easily sucked down towards the trash rack.

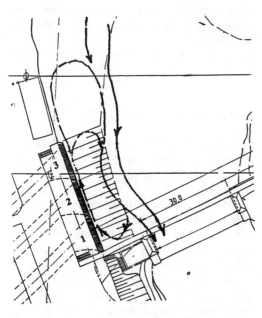

Fig 10 Eddy in front of the Fiskumfoss intake

4.3 Flow line effects

Flow lines should be divergent in areas where ice accumulation is unwanted. Converging flowlines will concentrate the floating ice and tend to cause jamming. An example was shown on Figure 1, where a slightly diverging guide wall facilitated the motion of ice towards the flood gates.

Ice drift concentration towards the outside of a curve is a flow line effect caused by the centrifugal accelleration forcing the surface water in a spiralling motion towards and down along the outer bank. Ice sluices should accordingly be placed at outward curving locations.

REFERENCES

Ashton, G.D.(ed) 1986. River and lake ice engineering. Littleton,Col, Water resources publications.

Carstens, T. & E.Tesaker 1966. Model tests on Burfell hydroelectric power project,Iceland. VHL (SINTEF) report, project 600115, Trondheim.

Godtland, K. 1989. Aust-Agder kraftverk. Modellforsøk ny dam Rygene. SINTEF report STF60 F89066, Trondheim (in Norwegian).

Larsen, P. Notes on the stability of floating ice blocks. IAHR Ice Symp., Hanover, N.H.

Michel, B. 1986. Packing in front of a forming river ice cover. IAHR Ice Symp., Iowa.

Godtland, K. 1992. Nedre Fiskumfoss kraftverk. Modellforsøk. Isproblemer. SINTEF report STF60 F92001, Trondheim (in Norvegian)

Sigurdsson, G. 1970. The Burfell project. A case study of system design for ice conditions. IAHR Ice Symp., Reykjavik.

Tesaker, E. 1975. Accumulation of frazil ice in an intake reservoir. IAHR Ice Symp., Hanover, N.H.

Uzuner, M. S. & J.F.Kennedy 1972. Stability of floating ice blocks. J. Hydr.Div, ASCE, Vol. 93, pp 2117-2133.

Uzuner, M. S. 1975. Stability of ice blocks beneath an ice cover. IAHR Ice Symp., Hanover, N.H.

Hydropower'92, Broch & Lysne (eds) © 1992 Balkema, Rotterdam. ISBN 90 5410 054 0

Displacement and stability analysis for gravity dam foundation by rigid body kinematic method

R. Kou
State Energy Investment Corporation, People's Republic of China

S. Wang
Science Academy of China, People's Republic of China

ABSTRACT: This paper introduces concepts for the analysis of movements and stability of rock foundations for gravity dams. The rock masses are separated into many rock blocks by discontinuous surfaces. The methods are derived from kinematics of rigid bodies, material mechanics and vector algebra. In this approach, dispacements of rock blocks within the calculation area, termed rigid body movements, are the variables of set of simultaneous equations.

1. INTRODUCTION

In the stability analysis of rock foundations, several approaches have been developed. Limit equilibrium method, finite element method, block theory by Richard E. Goodman and Gen–hua Shi, and general method of limit equilibrium analysis by Sarada K.Sarma and E. Hoek are some methods which can be used in practical engineering. In the limit equilibrium method and Sarma's method, the deformability of block boundary and moment equilibrium are not taken into account. Therefore, these methods are unable to give displacements of blocks and stress distribution of block boundaries. In block theory, the possibility of block to remove can be evaluated, but the magnitude of displacement can not be calculated.

In the method of displacement and stability analysis proposed in this paper, rock masses are cosidered to consist of two elements: discontinuities which are deformable and have any suitable constitutive law and rigid rock blocks which are separated by the discontinuities (see Fig. 1).

A genral plane motion of a two dimensional block has three levels of freedom or movement components: two translations and one rotation. There are totally 3Xn movement components as unknowns for a rock mass which has n blocks. The displacements at every point on the block can be determined by the

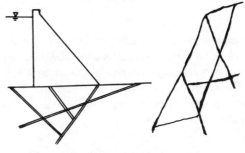

(a) Rock foundation   (b) Rock slope

Fig.1 Rock masses composed of faults and rigid blocks

rigid body movements and the position of the point relative to a fixed position about which the block rotates. There are two displacement components for any boundary of the block: normal and tangential. Both of them can be expressed by geometrical parameters of the boundary and the rigid body movements of the block.

Stress distributions on the boundary planes of a block are decided by its displacements relative to neighboring block and mechanical properties of interval materials. For every block, we have three equilibrium equations (equivilent to two vecter equilibrium equations): two for forces, one for

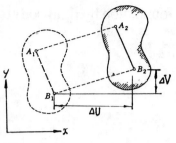

(a) Translation

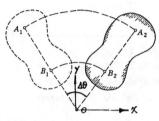

(b) Rotation

Fig.2 Block movement of rigid body

moments. There are totally 3Xn equations for the rock mass within the calculation area. So, all movement components of rigid blocks can be solved. Consequently, stress distribution, failure area, local and overall safety factors will be able to be calculated.

2. PROCESS OF ANALYSIS

To begin with considering a rock mass which is separated by discontinous structural surfaces into n blocks, every block will be considered as an element which will undergo general plane motion. Taking a block element into consideration, we will analyse its block movements and stresses caused by the block movement on its boundaries.

2.1 Block movements

We assume that $\triangle U$ and $\triangle V$ are block translation infinitesimal in x and y directions, $\triangle \theta$ is rotation infinitesimal which will be positive when it rotates clockwise (see Figure 2).

In order to simplify the following derivation, we express the total translation infinitesimal $\overline{\triangle \delta}_t$ and

rotation infinitesimal $\overline{\triangle \theta}$ of the block by two vectors:

$$\overline{\triangle \delta}_t = \triangle U \vec{i} + \triangle V \vec{j} \tag{1}$$

$$\overline{\triangle \theta} = \triangle \theta \vec{k} \tag{2}$$

2.2 Linear displacement at arbitrary point on a block

The displacement caused by block translation at point P (x, y) is equal to the translation of the block. It is uniformly distributed at all points of the block as equation (1). The linear displacement $\overline{\triangle \delta}_r$ caused by block rotation at point P (x, y) is equal to the cross product of $\overline{\triangle \theta}$ and $\vec{r}$.

$$\overline{\triangle \delta}_r = \overline{\triangle \theta} \times \vec{r} \tag{3}$$

where
$$\vec{r} = x \vec{i} + y \vec{j} \tag{4}$$
is the position vector of point P relative to O. Therefore

$$\overline{\triangle \delta}_r = \begin{vmatrix} \vec{i} & \vec{j} & \vec{k} \\ 0 & 0 & \triangle \theta \\ x & y & 0 \end{vmatrix} \tag{5}$$

or expressed as

$$\overline{\triangle \delta}_r = (-y \triangle \theta) \vec{i} + (x \triangle \theta) \vec{j} \tag{6}$$

The total linear displacement caused by translation and rotation at point P (x, y) is

$$\overline{\triangle \delta} = (\triangle U - y \triangle \theta) \vec{i} + (\triangle V + x \triangle \theta) \vec{j} \tag{7}$$

or simply expressed as:

$$\overline{\triangle \delta} = \triangle \delta_x \vec{i} + \triangle \delta_y \vec{j} \tag{8}$$

2.3 Boundary displacement components in normal and tangential directions

The boundary displacement can be decomposed into normal and tangential components shown as Fig.3.

A boundary line formed by points $P_i(x_i, y_i)$ and $P_j(x_j, y_j)$ can be described by equation

$$(y_j - y_i)x - (x_j - x_i)y + (x_j y_i - x_i y_j) = 0 \tag{9}$$

404

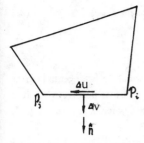

Fig.3 Decomposition of boundary displacements

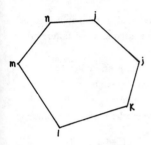

Fig.4 Clockwise arrangement of node Numbers

let

$$a = (y_j - y_i)$$
$$b = -(x_j - x_i)$$
$$d = -(x_j y_i - x_i y_j)$$
(10)

be the coefficients of boundary i which belongs to the block, (9) becomes:

$$ax + by = d \qquad (11)$$

The unit normal vector of the boundary plane will be

$$\hat{n} = \frac{\vec{n}}{|\vec{n}|} = \frac{1}{(a^2 + b^2)^{1/2}}(a\vec{i} + b\vec{j}) \qquad (12)$$

The node numbers should be clockwise arranged in describing the block, so that the positive outside unit normal vector (12) is outward(see Fig.4).

The normal displacement infinitesmal $\overrightarrow{\Delta v}$ equals the projection of vector $\overrightarrow{\Delta \delta}$ on $\hat{n}$.

$$\overrightarrow{\Delta v} = \frac{\overrightarrow{\Delta \delta} \cdot \vec{n}}{|\vec{n}|} \cdot \frac{\vec{n}}{|\vec{n}|} \qquad (13)$$

Substituting (7) and (12) into (13), we have

$$\overrightarrow{\Delta v} = \frac{a(\Delta U - y\Delta \theta) + b(\Delta V + x\Delta \theta)}{(a^2 + b^2)}(a\vec{i} + b\vec{j}) \qquad (14a)$$

or simply expressed as:

$$\overrightarrow{\Delta v} = \Delta v_x \vec{i} + \Delta v_y \vec{j} \qquad (14b)$$

in which,

$$\Delta v_x = \frac{a^2(\Delta U - y\Delta \theta) + ab(\Delta V + x\Delta \theta)}{(a^2 + b^2)}$$
$$\Delta v_y = \frac{ab(\Delta U - y\Delta \theta) + b^2(\Delta V + x\Delta \theta)}{(a^2 + b^2)}$$
(14c)

The tangential displacement $\overrightarrow{\Delta u}$ is equal to the projection of vector $\overrightarrow{\Delta \delta}$ on the boundary line, it can be calculated by

$$\overrightarrow{\Delta u} = \overrightarrow{\Delta \delta} - \overrightarrow{\Delta v} \qquad (15)$$

Substitute (7) and (14a) into (15),

$$\overrightarrow{\Delta u} = \frac{b(\Delta U - y\Delta \theta) - a(\Delta V + x\Delta \theta)}{a^2 + b^2}(b\vec{i} - a\vec{j}) \qquad (16a)$$

or simply expressed as:

$$\overrightarrow{\Delta u} = \Delta u_x \vec{i} + \Delta u_y \vec{j} \qquad (16b)$$

Where

$$\Delta u_x = \frac{b^2(\Delta U - y\Delta \theta) + ab(\Delta V + x\Delta \theta)}{a^2 + b^2}$$
$$\Delta u_y = \frac{-ab(\Delta U - y\Delta \theta) + a^2(\Delta V + x\Delta \theta)}{a^2 + b^2}$$
(16c)

The positive direction of the boundary displacement is shown as Figure 5.

$\overrightarrow{\Delta u}$ is perpendicular to $\overrightarrow{\Delta v}$, therefore

$$\overrightarrow{\Delta u} \cdot \overrightarrow{\Delta v} = 0 \qquad (17)$$

The magnitudes of normal and tangential boundary diplacements caused by block movement are given by

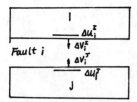

Fig. 5 Positive directions of boundary displacements

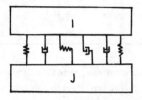

Fig. 6 Links of fault beween blocks

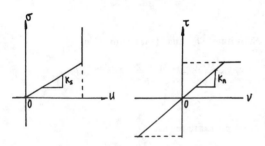

(a) σ — u curve (b) τ — v curve

Fig. 7 Relationship between stress and displacement

$$\triangle v = \frac{1}{\sqrt{a^2 + b^2}} \; [a(\triangle U - y\triangle\theta) + b(\triangle V + x\triangle\theta)] \quad (18)$$

$$\triangle u = \frac{1}{\sqrt{a^2 + b^2}} \; [b(\triangle U - y\triangle\theta) - a(\triangle V + x\triangle\theta)] \quad (19)$$

2.4 Boundary stresses expressed by relative movements between two blocks

Blocks are linked by faults which will resist relative movement between two blocks (see Fig. 6).

The normal and tangential stress infinitesmal $\overrightarrow{\triangle\sigma}$ $\overrightarrow{\triangle\tau}$ of boundary ① between block I and block J are expressed as:

$$\overrightarrow{\triangle\sigma}_i = k_n(\overrightarrow{\triangle v}_i^I + \overrightarrow{\triangle v}_i^J) \quad (20)$$

$$\overrightarrow{\triangle\tau}_i = k_s(\overrightarrow{\triangle u}_i^I + \overrightarrow{\triangle u}_i^J) \quad (21)$$

where

$\overrightarrow{\triangle\sigma}$, $\overrightarrow{\triangle\tau}$ — — Normal and tangential stress infinitesmals of fault ① . $\triangle\sigma$ is positive when it is compresive. $\triangle\tau$ is positive when it makes the fault rotate anticlockwise.

$\overrightarrow{\triangle v}_i^I$, $\overrightarrow{\triangle u}_i^I$ — — normal and tangential boundary displacement of block I.

k_n, k_s — — normal and tangential stiffness of fault ① .

The constitutive law used in the paper is shown as Fig. 7. Coulomb law is used as shear failure criterion.

when the fault is compressed: k_n = constant
when $\tau < C + \sigma tg\varphi$: k_s = constant
when the fault is pulled: $k_n = k_s = 0$

The constitutive law can be expressed by matrix:

$$\begin{bmatrix} \vec{\sigma}_i \\ \vec{\tau}_i \end{bmatrix} = \begin{bmatrix} k_n & 0 \\ 0 & k_s \end{bmatrix} \begin{bmatrix} \vec{v}_i^I + \vec{v}_i^J \\ \vec{u}_i^I + \vec{u}_i^J \end{bmatrix} \quad (2.22)$$

let:

$$\left. \begin{array}{l} \vec{\sigma}_i = \sigma_{xi}\vec{i} + \sigma_{yi}\vec{j} \\ \vec{\tau}_i = \tau_{xi}\vec{i} + \tau_{yi}\vec{j} \end{array} \right\} \quad (23)$$

then

$$\left. \begin{array}{l} \sigma_{xi} = k_n(v_{xi}^I + v_{xi}^J) \\ \sigma_{yi} = k_n(v_{yi}^I + v_{yi}^J) \\ \tau_{xi} = k_s(u_{xi}^I + u_{xi}^J) \\ \tau_{yi} = k_s(u_{yi}^I + u_{yi}^J) \end{array} \right\} \quad (24)$$

2.5 Boundary stress distribution expressed by rigid body movement

Substituting (14c) and (16c) into (24), we will obtain the stress distribution expressed by rigid body movement of block I and block J.

406

$$\sigma_x = k_s \sum_{i}^{IJ} \left[\frac{a(U-y\theta)+b(V+x\theta)}{a^2+b^2} \quad a\right] \quad b]$$

$$\sigma_y = k_s \sum_{i}^{IJ} \left[\frac{a(U-y\theta)+b(V+x\theta)}{a^2+b^2} \quad b\right]$$

$$\tau_x = k_s \sum_{i}^{IJ} \left[\frac{b(U-y\theta)-b(V+x\theta)}{a^2+b^2} \quad b\right]$$

$$\tau_y = k_s \sum_{i}^{IJ} \left[\frac{-b(U-y\theta)+a(V+x\theta)}{a^2+b^2} \quad a\right]$$

(25)

2.6 Resultants and moments of distributed stresses acting on a boundary

Resultants acting on boundary i can be obtained by integral of stresses on the boundary. These stresses are caused by both block I and J. Let $\overrightarrow{dF}$ be the resultants infinitesimal of stresses on the Resultants acting on boundary i can be obtained by integral of stresses on the boundary. These stresses are caused by both block I and J. Let $\overrightarrow{dF}$ be the resultants infinitesimal of stresses on the boundary, and $\overrightarrow{dM_0}$ be the moment of the stresses with respect to O, they can be calculated by

$$\overrightarrow{dF} = (dF)_x \vec{i} + (dF)_y \vec{j} \tag{26}$$

which is the vector of resultant force infinitesmal acting on dl. The moment of the resultant dF with respect to O is:

$$\overrightarrow{dM_0} = \vec{r} \times \overrightarrow{dF} \tag{27}$$

(26) and (27) can be expressed by stresses:

$$\left.\begin{aligned}
(dF)_x &= (\sigma_x + \tau_x)dl \\
(dF)_y &= (\sigma_y + \tau_y)dl \\
dM_0 &= [(\sigma_y + \tau_y)] x - (\sigma_x + \tau_x)y] \, dl
\end{aligned}\right\} \tag{28}$$

Substituting (25) into (28), and integral along l from i to j:

$$(F_x) = \sum_{}^{IJ} \left\{ \frac{1}{(a^2+b^2)} \int_l \left[(k_s a + k_s b)(U-y\theta) \right.\right.$$
$$\left.\left. + ab(k_s - k_s)(V+x\theta) \right] dl \right\}$$

$$(F_y) = \sum_{}^{IJ} \left\{ \frac{1}{(a^2+b^2)} \int_l \left[ab(k_s + k_s)(U-y\theta) \right.\right.$$
$$\left.\left. + (k_s a + k_s b)(V+x\theta) \right] dl \right\}$$

$$(M_0) = \sum_{}^{IJ} \left\{ \frac{1}{(a^2+b^2)} \int_l \left[ab(k_s + k_s)(U-y\theta) \right.\right.$$
$$\left. + (k_s a + k_s b)(V+x\theta) \right] x dl)$$
$$- \int_l \left[(k_s a + k_s b)(U-y\theta) \right.$$
$$\left.\left. + ab(k_s - k_s)(V+x\theta) \right] y dl \right\}$$

(29)

and can be expressed by matrix

$$\begin{bmatrix} F_x \\ F_y \\ M_0 \end{bmatrix} = \begin{bmatrix} k_{11} & k_{12} & k_{13} \\ k_{21} & k_{22} & k_{23} \\ k_{31} & k_{32} & k_{33} \end{bmatrix}^I \begin{bmatrix} U \\ V \\ \theta \end{bmatrix}^I +$$
$$\begin{bmatrix} k_{11} & k_{12} & k_{13} \\ k_{21} & k_{22} & k_{23} \\ k_{31} & k_{32} & k_{33} \end{bmatrix}^J \begin{bmatrix} U \\ V \\ \theta \end{bmatrix}^J \tag{30}$$

In (30), the first item on the right side of the equation is integral coefficient matrix of the boundary belonging to block I; the secont item, block J. When the fault is a rectangle, boundary coefficients of block I are equal to those of block J, so we have:

$$\begin{bmatrix} F_x \\ F_y \\ M_0 \end{bmatrix} = \begin{bmatrix} k_{11} & k_{12} & k_{13} \\ k_{21} & k_{22} & k_{23} \\ k_{31} & k_{32} & k_{33} \end{bmatrix}^I \begin{bmatrix} U^I + U^J \\ V^I + V^J \\ \theta^I + \theta^J \end{bmatrix} \tag{31}$$

The elements of the coefficient matrix are calculated by:

$$k_{11} = \frac{1}{L}(k_s a^2 + k_s b^2)$$

$$k_{12} = k_{21} = \frac{ab}{L}(k_s - k_s)$$

$$k_{22} = \frac{1}{L}(k_s a^2 + k_s b^2)$$

$$k_{13} = k_{31} = \frac{1}{L}[(k_s a^2 + k_s b^2)\psi_y - ab(k_s - k_s)\psi_x]$$

$$k_{23} = k_{32} = \frac{1}{L}[(k_s a^2 + k_s b^2)\psi_x - ab(k_s - k_s)\psi_y]$$

$$k_{33} = \frac{1}{L}[(k_s a^2 + k_s b^2)\psi_{xx} + (k_s a^2 + k_s b^2)\psi_{yy}$$
$$- 2ab(k_s - k_s)\psi_{xy}]$$

(32)

407

in which:

$$L = \int_l dl = (a^2 + b^2)^{1/2}$$

$$\psi_x = \int_l x\,dl = \frac{L}{2}(x_i + x_j)$$

$$\psi_y = \int_l y\,dl = \frac{L}{2}(y_i + y_j)$$

$$\psi_{xx} = \int_l x^2\,dl = \frac{L}{3}(x_i^2 + x_i y_j + x_j^2)$$

$$\psi_{yy} = \int_l y^2\,dl = \frac{L}{3}(y_i^2 + y_i y_j + y_j^2)$$

$$\psi_{xy} = \int_l xy\,dl = \frac{L}{6}(x_i(y_i + 2y_j) + x_j(y_j + 2y_i))$$

(2.33)

2.7 Equilibrium equation for single block

There are three equilibrium equations for a 2–D block, two for force, another for moment, they are

$$\Sigma F_{xi} = m\frac{\partial^2 U}{\partial^2 t}$$

$$\Sigma F_{yi} = m\frac{\partial^2 V}{\partial^2 t}$$

(34)

$$\Sigma M_{oi} = I_0\frac{\partial^2 \theta}{\partial^2 t}$$

where

F_{xi}, F_{yi}, M_{oi} -- all forces acting on the block;

m — mass of the block;

I_0 -- mass moment of inertia of the block with respect to O.

t — time.

In static conditions, the equilibrium equation expressed by matrix become

$$\begin{bmatrix} K_{11} & K_{12} & K_{13} \\ K_{21} & K_{22} & K_{23} \\ K_{31} & K_{32} & K_{33} \end{bmatrix}\begin{bmatrix} U \\ V \\ \theta \end{bmatrix} + \begin{bmatrix} X \\ Y \\ M \end{bmatrix} = 0$$

(35)

in which,

$$K_{ij} = \sum_1^m k_{ij}$$

m is the number of boundaries excluding free surfaces. $\{X, Y, M\}^T$ is matrix of external forces which are not related to boundary movements.

2.8 Stiffness matrix for multi–block rock masses

In case that there are n blocks, the equilibrium equation is as follows:

$$\begin{bmatrix} [K]^{11} & [K]^{12} & \cdots & [K]^{1J} & \cdots & [K]^{1n} \\ [K]^{21} & [K]^{22} & \cdots & [K]^{2J} & \cdots & [K]^{2n} \\ \cdots & \cdots & \cdots & \cdots & \cdots & \cdots \\ [K]^{I1} & [K]^{I2} & \cdots & [K]^{IJ} & \cdots & [K]^{In} \\ \cdots & \cdots & \cdots & \cdots & \cdots & \cdots \\ [K]^{n1} & [K]^{n2} & \cdots & [K]^{nJ} & \cdots & [K]^{nn} \end{bmatrix}\begin{bmatrix} \{\delta\}^1 \\ \{\delta\}^2 \\ \cdots \\ \{\delta\}^I \\ \cdots \\ \{\delta\}^n \end{bmatrix}$$

$$= \begin{bmatrix} \{R\}^1 \\ \{R\}^2 \\ \cdots \\ \{R\}^I \\ \cdots \\ \{R\}^n \end{bmatrix}$$

(36)

where

$$\{\delta\}^I = \{U^I, V^I, \theta^I\}^T$$

(37)

is the displacement vector of block I.

$$\{R\}^I = \{X^I, Y^I, M^I\}^T$$

(38)

is the resultant vector of external forces.

$$[K]^{IJ} =$$

$$\begin{bmatrix} K_{3I,3J} & K_{3I,3J+1} & K_{3I,3J+2} \\ K_{3I+1,3J} & K_{3I+1,3J+1} & K_{3I+1,3J+2} \\ K_{3I+2,3J} & K_{3I+2,3J+1} & K_{3I+2,3J+2} \end{bmatrix}$$

(39)

(i) When $I = J$, the elements of $[K]^{IJ}$ are:

$$K_{ij} = \sum_1^{m_I} k_{ij}$$

(40a)

where

k_{ij}—boundary coefficients of block I

m_I--Number of boundaries of block I excluding free surfaces

(ii) When I≠J, and they are neighbouring blocks

$$K_{ij} = k_{ij} \qquad (40b)$$

where

k_{ij} — — boundary coefficients of fault between block I and J.

(iii) When I≠J, and they are not neighbouring blocks:

$$K_{ij} = 0 \qquad (40c)$$

The physical meaning of $[K]^{IJ}$ is the forces caused by unit movement of block I on the block J.

2.9 Calculation of external forces acting on the blocks.

2.9.1 Dead weight

A m−gon block can be subdivided into (m−2) triagles (see Fig.8). The area of a triagle formed by three points $P_1(x_1, y_1)$, $P_2(x_2, y_2)$, $P_3(x_3, y_3)$ can be calculated by a determinant of their coordinates:

$$A_i = \frac{1}{2} \begin{vmatrix} 1 & x_1 & y_1 \\ 1 & x_2 & y_2 \\ 1 & x_3 & y_3 \end{vmatrix} \qquad (41)$$

The coordinats of weight center of the triangle are:

$$\left. \begin{aligned} x_{ci} &= \frac{1}{3}(x_1 + x_2 + x_3) \\ y_{ci} &= \frac{1}{3}(y_1 + y_2 + y_3) \end{aligned} \right\} \qquad (42)$$

The dead weight and the moment of a triangle block are given by

$$\left. \begin{aligned} \vec{F}_d &= (\gamma_r \times A_i)\vec{j} \\ \vec{M}_d &= \vec{r}_{ci} \times \vec{F}_d \end{aligned} \right\} \qquad (43)$$

in which,

γ_r is the unit weight of rock block,

$\vec{r}_{ci} = x_{ci}\vec{i} + y_{ci}\vec{j}$ is the position vector of weight center of the triangle.

Resultant forces and moment of dead weight are:

$$\left. \begin{aligned} X_d &= 0 \\ Y_d &= \gamma_r \sum_1^{m-2} A_i \\ M_d &= \gamma_r \sum_1^{m-2} (x_{ci} A_i) \end{aligned} \right\} \qquad (44)$$

2.9.2 Seismic forces

Assuming S_x, S_y are accelaration factors of seismic force in directions x, y respectively, distributed seismic forces can be replaced by resultant forces acting at the weight center:

$$\left. \begin{aligned} X_s &= \gamma_r S_x \sum_1^{m-2} A_i \\ Y_s &= \gamma_r S_y \sum_1^{m-2} A_i \\ M_s &= \gamma_r [S_y \sum_1^{m-2} x_{ci} A_i - S_x \sum_1^{m-2} Y_{ci} A_i] \end{aligned} \right\} \qquad (45)$$

2.9.3 Hydraulic forces and moments

Boundary ③ is assumed to be formed by points 1, 2, water pressure at points 1 and 2 are p_1 and p_2 respectively (see Fig. 9). The resultants of hydraulic pressure are:

$$\left. \begin{aligned} X_{hi} &= -\frac{a_i}{2}(p_1 + p_2) \\ Y_{hi} &= -\frac{b_i}{2}(p_1 + p_2) \\ M_{hi} &= Y_{hi} x_{hi} - X_{hi} y_{hi} \end{aligned} \right\} \qquad (46)$$

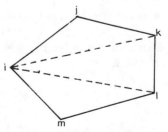

Fig. 8 Subdivision of a m−gon into (m−2) triangles

in which,

$$x_{hi} = x_1 + \frac{p_1 + 2p_2}{3(p_1 + p_2)} b_i$$

$$y_{hi} = y_2 + \frac{p_1 + 2p_2}{3(p_1 + p_2)} a_i \tag{47}$$

The resultants of all forces acting on the block are

$$X_h = \sum_1^{\cdot} X_{hi}$$

$$Y_h = \sum_1^{\cdot} Y_{hi} \tag{48}$$

$$M_h = \sum_1^{\cdot} M_{hi}$$

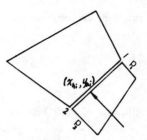

Fig. 9 Calculation of hydraulic forces acting on block boundaries

2.10 Determination of failure area and definition of safety factors

After displacement of all blocks are calculated, the normal and tangential forces on all boundaries is to be obtained. Therfore, based on Coulomb law, failure areas and safety factors can be calculated. With the assumption of $\sigma_j > \sigma_i$, there are following four different situations (shown as Fig. 10):

(a) $\sigma_i > 0$, and $|\tau| < f\sigma_i + C$, no failure area;

(b) $\sigma_i > 0$, $|\tau| > f\sigma_i + C$, and $|\tau| > f\sigma_j + C$, shear failure length will be:

$$l_{sf} = \frac{C - \tau - f\sigma_i}{f(\sigma_j - \sigma_i)} l \tag{49}$$

(c) $\sigma_i < 0$ $\sigma_j > 0$, Area of pulling open is:

$$l_{sf} = \frac{\sigma_j}{\sigma_i + \sigma_j} l \tag{50a}$$

Shear failure area is:

$$l_{sf} = \frac{\tau - C}{\tau - C + f\sigma_j} \cdot \frac{\sigma_i}{\sigma_i + \sigma_j} l \tag{50b}$$

(d) $\sigma_i < 0$ $\sigma_i < 0$, all area of the boundary is pulled open.

Safety factors include point safety factor, boundary safety factor and overall safety factor.

(1) Point safety factor

$$K_i = \frac{C + \sigma t g\varphi}{|\tau|} \tag{51}$$

(2)Boundary safety factor

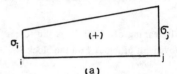

(a)

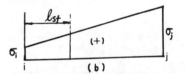

(b)

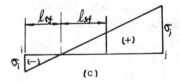

(c)

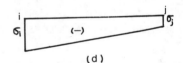

(d)

Fig. 10 Stress distribution and determination of failure areas

$$K_{bi} = \frac{C_i l_{ni} + N_i t g\psi_i}{|S_i|} \tag{52}$$

where

N_i, S_i —— normal and tangential resultant forces of boundary ①

l_{ni} —— length of non–failure area of boundary① .

(3)Overall safety factor

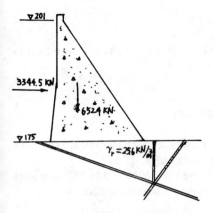

Fig. 11 Cross section of Barrett Chute, Canada

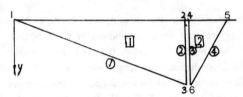

Fig. 12 Numbering of blocks, boundaries and nodes

Table 1　Coordinates and water prssure at nodes

Node No.	X(m)	Y(m)	$P(KN/m^2)$
1	0.0	0.0	258
2	20.7	0.0	0
3	20.7	7.5	180
4	21.7	0.0	0
5	25.9	0.0	0
6	21.0	7.5	180

Table 2　Mechanical properties of faults

Boun.No.	k_n	k_s	C(KPa)	$\varphi(°)$	(σ_t)
1	5000	2000	5	40	0.0
2,3	10000	4000	10	45	0.0
4	3000	1200	3	35	0.0

Table 3　External forces

Block No.	F_X(KN)	F_Y(KN)	M(KN・m)
1	3344.5	6524.8	15474.5
2	0.0	0.0	0.0

Table 4　Block movement

Block No.	U(cm)	V(cm)	$\theta(°)$
1	8.7941	0.7805	0.2382
2	4.3832	9.9902	−0.1306

Table 5　Node displacements

Node No.	X−direction(cm)	Y−direction(cm)
1	8.7941	0.0780
2	8.7941	8.6825
3	5.6765	8.6825
4	4.3832	5.2042
5	4.3832	4.0875
6	6.0925	5.2042

Table 6　Boundary safety factors

Boundary No.	Safety factors
1	0.4483
2,3	1.5073
4,	12.0781

We define

$$K(l_i) = \frac{\sum K_{bi} l_i}{\sum l_i} \tag{2.53}$$

to be the overall safety factor along a possible failure routine. l_i are boundaries of the failure routine. Definition of the overall safety factor is:

$$K_0 = Min\ [K(l_i \cdots)] \tag{53b}$$

411

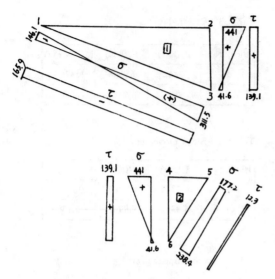

Fig. 13 Stress distribution on block boundaries

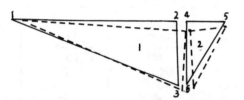

Fig. 14 Displacement situation of rock blocks

3. COMPUTER PROGRAM AND A PRACTICAL CALCULATION EXAMFLE

3.1 Computer program

Computer program for movement and stability analysis of 2−D multi−block rock masses(RBM2) is written by Fortran 77. RBM2 can be used to calculate displacements, boundary stresses and safety factors of rock masses which composed of many blocks. The program can take dead weight, water pressure and seismic forces, anchorage forces and other external forces into account.

3.2 Calculation of foundation of Barrett Chute

Barrett Chute dam is a concrete dam. the cross section, dam foundation and seepage pressure are shown as Fig.11. The dam foundation is composed of two blocks.

3.2.1 Input data

Input data are given by Fig. 12, Table 1, Table 2 and Table3.

3.2.2 Output data

Calculation results are given by Table 4, Table 5, Table 6, Fig.13 and Fig.14.

The overall safety factoe is 3.18. The most possible failure routine is boundaries 1 and 4

4. CONCLUSIONS

By establishing the relationship between block movement and stress distribution of block boundaries, we are able to calculate the block displacements and stress distribution along faults, evaluate the stability of rock foudations, and find out the most possible sliding routine. This method can also be used in the stability analysis of rock slopes.

Acknowlegement

The authors would like to express gratitude to Dr K. Y. Lo and Dr. C. F. Lee for their beneficial suggestions to the paper during my stay in UWO, Canada from 1987 to 1988.

REFERENCES

Brown, E. T. 1987. Analytical and Computational Methods in Engineering Rock Mechanics. Allen Unwin, London, England.

Cundall, P. A. 1988. Formulation of a three−dimensional distinct element model−part I: a scheme to detect and represent contact in a system composed of many polyhedral blocks. Int. J. Rock Mech. Min. Sci. Geomech. Abstr. (25)3,107−116.

Goodman, R. E. and G. Shi 1985. Block theory and its application to Rock Engineering, NJ: Prentice−Hall.

Hart, R.,J. L. Cundall and J. Lemos 1988. Formulation of a three–dimension distinct element model– Part II. Mechanical calculations for motion and interaction of a system composed of many polyhedral blocks. Int. J. Rock Mech. Min. Sci. Geomech. Abstr. 3,117–126.

Hoek E. and J. W. Bray 1977. Rock Slope Engineering, 2nd edn. IMM, London.

Sarma, S. K. And M. V. Bhave 1979. Stability analsis of embankments and slopes. J. of the Geotech. Engng. Div., ASCE, GT12.

Hydropower'92, Broch & Lysne (eds) © 1992 Balkema, Rotterdam. ISBN 90 5410 054 0

New concepts in spillway design

D. K. Lysne
Norwegian Institute of Technology, Trondheim, Norway

ABSTRACT: Spreading out the energy dissipation over the length of the spillway channel represents a different concept in spillway design. This may be achieved either trough rough surface energy dissipation or be adapting stepped spillway designs. The design implies moderate erosion/cavitation intensity, which in turn allows for unlined, rough surface spillways as a safe low-cost spillway alternative.

1 INTRODUCTION

Every dam and spillway system is tailor-mode and a very wide variety of different spillway designs exist. Nevertheless, there is a strong tendency towards similarity in design. According to (1) most spillways on the large rivers of the world consist of four main elements; radial gates, overfall, chute and ski-jump. As a consequence the main effort of recent research have been concentrated on the improvement of these elements.

The most striking development has been the improved design of bottom aerators for chutes to prevent cavitation erosion when the velocity exceeds 25 to 35 m/s. Connected to this research is also concrete quality and surface finish.

In general terms this research has consentrated on improving the detailed design of well known spillway layouts, of which the chute spillway is the most common. A characteristic feature of this concept is that some 95 percent of the potential energy is converted into kinetic energy and is maintained for energy dissipation at the outlet. This may not cause design and cost problems in case of a ski-jump outlet, providing rock scouring downstream of the ski-jump is not a problem. Nevertheless, this most common type of spillway layout does lead to a consentration of kinetic energy. New concepts are being developed with the objective to spread out the energy dissipation over the length of the spillway. Basically two concepts have been introduced:

- The unlined free surface spillway design, discussed in section 4.
- The steped spillway design, discussed in section 5.

Characteristic for these spillway designs is that erosion/cavitation intencity is significantly reduced as much more of the spillway area takes part in energy dissipation.

The lower erosion intencity does also allow for unlilned, rough rock surface spillways as a safe and low-cost spillway alternative.

2 ROUGH SURFACE ENERGY DISSIPATION

The bed friction in a mountain river or a rough unlined rock channel may have a Manning n = approx. 0,04. Artifically rough surface, figure 1, has been tested in the laboratory. The Manning coefficient is in the range, n = 0,04 to 0,045. It is known that Mannings n-value is a function of water depth (hydraulic radius) for rough surfaces, though the functional relationship is not known.

Table 1 demonstrates surface energy dissipation, assuming n = 0,04. We can read from the table that, for instance, a rough spillway shute with a slope S = 1:6,5 will maintain uniform flow for a unit discharge of 60 m^3/s, m. The velocity being approximatly 20 m/s.

This possibility for dissipating energy is seldom discussed in text books and is often overlooked as a low-cost spillway design. Most spillway channels will have a slope in the range 1:5 to 1:10. Providing good concrete quality and no sediment transport (normally the case for spillways, but not for bottom sluice ways) we know that a rough concrete surface will not be subject to cavitation erosion for velocities not exceeding 20 to 25 m/s. This being the experience and normally adopted criteria for chute block/baffle block energy dissipation basins.

Also unlined rock surfaces of reasonable

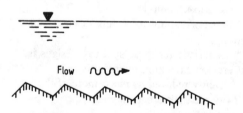

Fig. 1 Artificial surface roughness.
Manning coefficient in the range, n = 0,040 to 0,045.

rock quality is known to withstand velocities up to 20 to 25 m/s. Moderate faults may open up, but these can easily be sealed off by bolting and/or steelfiber reinforced shoterete.

3 EXPERIENCE WITH UNLINED TUNNEL SPILLWAYS

This type of spillway has been most widely used in Norway. A common layout (sections) is shown in figure 2.

The general layout shown in fig. 2 has been applied in some 35 cases. A programme for thorough inspection of the spillways and gated bottom outlets was carried out in 1987 to 1989. Table 2 summarizes the main data for 14 spillways and 6 gated outlets.

The most important findings from this field survey were the following:
- All the spillways and gated outlets were unlined with no or only a very minimum of rock support, i.e. bolting or shotcrete. No concrete lining was encountered.
- The rock quality varied over a relatively wide range, as seen in table 2.
- No critical rock erosion was observed throughout the survey. Moderate wash out was observed at two falts crossing the spillway tunnel and at one tunnel outlet where the rock was highly fractured.

One interesting observation (or lack of observation) was that no trace of cavitation/erosion was observed. One explanation may be that cavitation bubbles which form at the top of a roughness element, collapse in the mass of water downstream, away from the rock surface. Opening of cracks due to hydrodynamic pressure pulsations was

Table 1. Energy slope (S_e) as a function of velocity (v) and unit discharge.

V (m/s)	10	10	10	15	15	15	20	20	20
v (m)	1	2	3	1	2	3	1	2	3
q(m^3/s.m)	10	20	30	15	30	45	20	40	60
S_e(approx.)	1/6.2	1/15	1/27	1/2.8	1/7	1/12	1/1.5	1/4	1/6.5

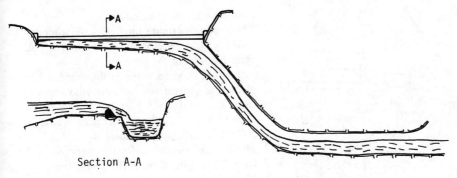

Section A-A

Fig. 2 Unlined tunnel spillway.

Table 2. Main results from field inspection survey of unlined spillways and gated bottom
outlets.(2)

Case no	Charact. geology	Head (m)	Vol. of water (mill. m³)	Q_{max} (m³/s)	V_{max}* (m/s)
	Spillway tunnel				
1	Marble	48	50	85	20
2	Quartizite	13	80	36	16
3	Greenstone/ Quartzite	90	560	178	27
4	Granittic gneis	35	160	45	25
5	Amph.gneis/ Micaschist	25	1040	183	16
6	Phyllite	30	330	46	?
7	Quartz-phyllite	23	330	50	19
8	Amph.gneis	13	78	13	?
9	Gneis	75	50	32	16
10	Anorthosite	29	110	17	12
11	Quartz-phyllite	36	340	60	20
12	Quartizitic schist	21	12	7	9
13	Monzonite	59	180	127	25
14	Hornb.-mieaschist	57	20	18	16
	Gated bottom outlet				
15	Quartzitic schist	60	1000	13	15
16	Gneis	54	6000	160	27
17	Gneis	62	8000	250	29
18	Anorthosite	28	700	(unknown)	20
19	Micaschist	50	200	336	26
20	Micaschist	45	230	346	25

* Estimated velocity

observed, however, as discussed in some
detail in section 4.2, with reference to the
spillway tunnels for the Virdnejavri Dam.

4 NEW DESIGN CRITERIA FOR TUNNEL SPILLWAYS

The tunnel spillway layout given in figure 2
may be subject to blockage, though no

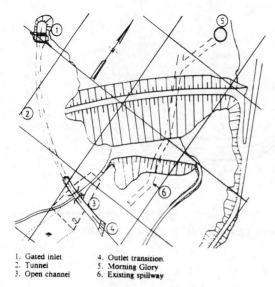

1. Gated inlet   4. Outlet transition.
2. Tunnel   5. Morning Glory
3. Open channel   6. Existing spillway

Layout of dam and spillways

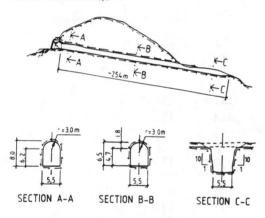

SECTION A-A   SECTION B-B   SECTION C-C

Longitudinal section through spillway

Fig. 3 Venemo Dam supplementary
        spillway.

incidents of blockage has been reported in
Norway. Possible causes may be build up of
ice and snow during the winter, drifting ice
in the spring or drifting trees due to
intensive rains and/or slides.

It is assumed that blockage may be more
of a problem in a tunnel with sharp shifts in
the alignment, vertically or horizontally.
The recommandable way to avoid blockage
hazard is to choose an open channel
spillway. This may not always be possible,
however, as dams very often are located in

a valley with steep sideslops.

New design criteria for free surface tunnel
spillways has therefore been developed. The
main features of the new consept is to
maintain a constant slope in the vertical
plane (for construction reasons, maximum
slope is approximatly 1:7) and constant
radius curve in the horizontal plan. These
criteria have been applied for the
supplenentary spillway for the Venemo
Dam (3) and the main spillway for the
Virdnejavri Dam (4).

4.1 Venemo Dam, Supplementary
spillway.

The Venemo Dam is an approximately 50
m high rockfill dam. The original spillway
is a drop-inlet spillway with an estimated
capacity of 200 m^3/s.

Updated flood analyses gave the result:
design outlet flow, Q_{1000} = 256 m^3 and the
probable maximum outflow flood of
approximately 500 m^3/s.

The dominating rock type in the area is
granitic gneiss with moderate fracturing.

A longitudinal section of the spillway is
shown in fig. 3. The tunnel alignment is a
constant radius curve, R = 250 m, and
longitudinal slope 1:7. Thus the flow is
supercritical and the curved walls cause
deflections and standing waves with
maximum wave height along the outer and
inner wall of the tunnel. These standing
waves were estimated based on theory for
wall defelctions and verified by model
studies. The maximum wave height was 50
cm while the wave length varied between
35 m and 70 m. The standing wave patern
was therefore not a critical parameter for
the design of the spillway tunnel in this
case.

The layout of dam and spillway is shown
in fig. 3.

4.2 The Virdnejavri Dam spillway.

The Virdnejavri Dam is an 140 m high
concrete arch dam. The dam is part of the

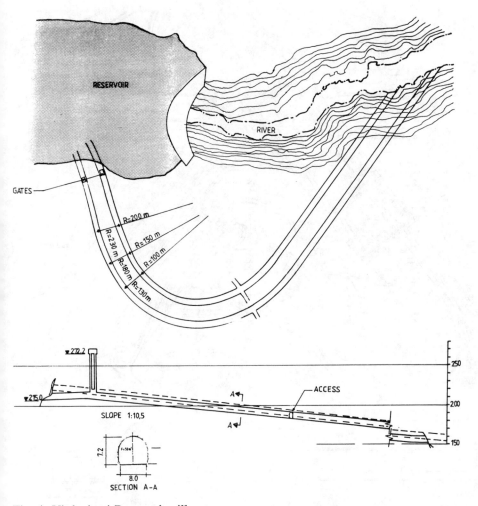

Fig. 4   Virdnejavri Dam and spillway.

Alta Hydropower Scheme, commissioned in 1987.

The spillway system in the initial plans consisted of flood gates in the dam and a fixed overflow crest with a total capacity equal the estimated probable maximum outflow flood. This layout was studied in hydraulic model tests with respect to possible rock scouring downstream of the dam. A thorough assessment of the rock stability of the side slopes in combination with the possible extent of scouring led to the conclusion that a different concept of spillway system should be considered.

The spillway concept adopted is shown in fig. 4, consisting of two spillway tunnels as the main spillway, each with a capacity of 650 m³/s. For floods exceeding 1300 m³/s, a fixed overflow crest and tree by-pass gates in the dam serves as additional spillway capacity.

The spillway tunnels have a constant slope of 1:10,5. The tunnel inlets are submerged and gated. The curvature is moderately increasing in the upper reach while the downstream streight part takes off at a tangent. The hydraulic performance of the spillway tunnels was studied in a hydraulic model study (4).

The geology in the area consists of

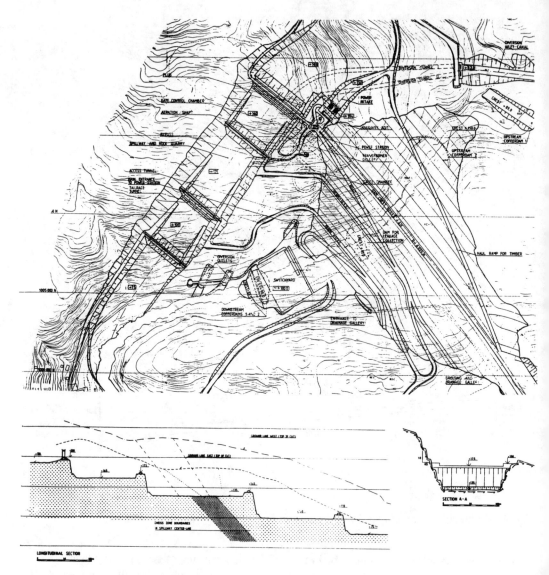

Fig. 5 Paunglaung dam spillway.

schistic rock. It was decided not to provide any rock support to be able to study the scouring effect of the high velocity flow on the unlined schistic rock surface. The tunnels have been operated a few times since commissioning of the project, varying between a few hours and a few days.

Two faults crossing the tunnels at an angle opened up immidiately during the first field test run, one fault approximately 50 m downstream from the gates and the other one further downstream. The maximum depth in the tunnel bottom was approximately 4 m. These scour wholes were filled in with ordinary concrete while the tunnel sides and top were sealed off with steel fiber reinforced shotcrete.

After this repairwork was carried out only very moderate scouring has been observed and the rock stability in the tunnels is considered perfectly safe. An interesting observation has been that some thin cracks

420

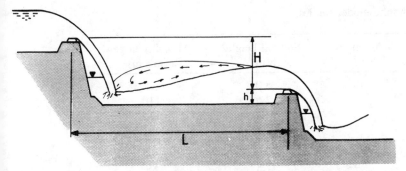

Fig. 6 Definition sketch, hydraulic performance of steped spillway.

(2-10 cm) open up, obviously due to hydrodynamic pressure pulsations. These cracks vary in depth from 0,5 m to 4 m. The cracks are spread out along the bottom and sides and do not effect the overall rock stability.

5 STEPED SPILLWAY DESIGN

A kind of steped spillway design was adopted for the La Grande 2 hydropower plant, one in the series of the Lan Grand Complex in Canada. The spillway channel is unlined and the channel invert is formed as steps. The energy dissipation along the spillway channel is mostly due to the unlined rock surface roughness.

It is reported that this principle also have been applied for some smaller spillways in other countries, but details have not been published.

This concept was developed further for the Paunglaung Hydropower Project in Myanmar (Burma). Detailed feasibility studies have been worked out, but the project is pending financing for its implementation. The project comprises a 120 m high dam and a spillway designed for, $Q_{1000} = 5000$ m³/s and $Q_{PMF} = 10000$ m³/s.

Figure 5 shows a general layout of the hydropower project and section through the spillway.

The spillway design was extensivly tested in a hydrualic model study with respect to

hydraulic performance and possible rock scouring (5).

5.1 Hydraulic performance of steped spillway.

The main objective of the hydraulic performance is to dissipate the kinetic energy in the falling jet of water for each step. That implies that the flow situation should be as indicated in the following definition sketch, figure 6.

The main parameters are: the unit discharge, q, the stepheight H, the step length, L, and the end sill height, h. These variables were studied in a series of tests with respect to maximum unit discharge q, maintaining close to 100% energy disspation within each step. The results of these tests are given in table 3.

5.2 Possible rock scouring

The spillway concept is based on stepwise energy dissipation with drop height at each step in the range of 25 to 35 m. For the design discharge of 5000 m³/s the unit discharge will be in the range of 30 to 50 m³/s. Under these conditions rock scouring downstream of each step may be expected as indicated in the following definition sketch, fig. 7.

The extent of scouring to be anticipated was studied in the hydraulic laboratory. The laboratory tests were based on the following

Table 3. Results of hydraulic model studies.

Step ehight H(m)	Step length l(m)	End sill height h(m)	Max discharge q(m³s,m)
35	150	5.0	54
		7.5	110
		10.0	115
		15.0	111
35	130	5.0	54
		7.5	89
		10.0	74
		15.0	60
35	110	5.0	60
		7.5	69
		10.0	53
		15.0	46
25	110	10.0	83
		15.0	81

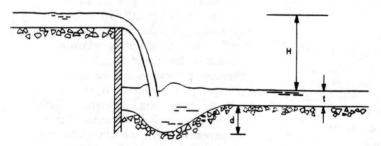

Fig. 7 Definition sketch, rock scouring tests.

assumption:

As fissures exist or develop at the rock surface a jet of water will gradually break losse blocks which then will be eroded due to the intensive turbulence in the plunge pool. This process will continue down to the depth where the pressure fluctuations in the jet no longer can move the blocks.

The results of the laboratory tests are given in fig. 8. Note that the results are given, not as single points, but as a range between two curves.

This type of tests, based on fractured rock does give conservative results. The extensive laboratory studies led to the following as a design criteria.

Maximum scour depth: d = H - t

The criteria is very conservative, however, as spill of water causing scouring will only take place a few days per year while scouring develops gradually with time. The spillway is easily inspected and if precautionary measures are needed to seal off cracks, there is ample time beteween flood seasons to have the work done.

The unlined steped spillway is a law cost solution compared with a conventional chute spillway, and the energy dissipation is spread over a much larger area.

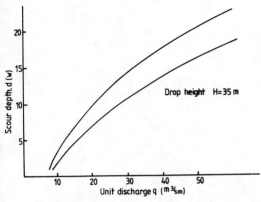

Fig. 8 Scour depth versus unit discharge
for drop height, H = 35 m.

References

Hanssen, D.H., Halsnes, P.H. and Riste, Ø.
"Spillway tunnels and gated outlets, a
field survey" (in Norwegian, "Flaumsjakter
og tappetunnelser"), Thesis 1988, The
Norwegeian Institute of Technology.

Sand, K., Berg, A., Soknes, S. and
Tvinnereim, K. "Virdnejavri Dam, Model
Studies", 1984 SINTEF report STF 60 F
84045 (in Norwegian). (Data also
available from State Power Board, Oslo,
Norway).

Sveen, A, Lysne, D.K., Petterson, L.E.
"Design of supplementary spillway".
XVI ICOLD-Congress (San Francisco), Q-
63, R-34, 537-553.

Sønsthagen, P., Tvinnereim, K., Aamodt, T.
"Paunglaung Hydropower Project, Model
Study", 1982, SINTEF report STF 60
F82086. (Information also available from
Norconsult International A.S., Oslo,
Norway).

Vischer, D. (1988). "Recent developments
in spillway design", Water Power & Dam
Construction, January 1988. (P. 8 to 9).

Hydropower'92, Broch & Lysne (eds) © 1992 Balkema, Rotterdam. ISBN 90 5410 054 0

Reservoir debris – Safety, economic and environmental considerations

N.M.Nielsen
B.C. Hydro, Vancouver, B.C., Canada

ABSTRACT: Reservoir debris can affect the safety of dams, the efficiency of power production and the attractiveness of the reservoir for other uses. Increasing awareness of dam safety, the need to maximize power production at existing facilities prior to constructing new ones, and public awareness and attitudes on the use of reservoirs has led to increased debris management programs at B.C. Hydro's reservoirs. The cost of managing debris can be high especially where the reservoir was not cleared prior to filling or where large volumes of debris are entering from the watershed. Present debris management practices at B.C. Hydro reservoirs are described, as well as an approach to coordinate the concerns of the various interested groups.

1 INTRODUCTION

B.C. Hydro is a publicly owned corporation which supplies most of the electrical energy for British Columbia, Canada. More than 90% of the annual total of about 50,000 GW.hr is generated by hydroplants. The reservoirs behind the dams vary in size from small run of river headponds to immense bodies of water and the reservoir rims are for the most part steep and covered by coniferous forest.

Reservoir debris comprises floating and submerged logs, beached timber along the shoreline and stumps and snags where the original reservoir area was not fully cleared of standing timber. The sources of floating debris include waste from past or present logging practices, debris torrents, landslides and initial reservoir filling.

Reservoir debris could make a dam unsafe if the discharge facilities were to become blocked during a major flood event. Frequent clogging of power intake trashracks causes reductions in power generation capability and efficiency. Debris may also affect the ecological, recreational and commercial use of the reservoir, with public safety, aquatic resource and wildlife implications, as well as aesthetic considerations.

B.C. Hydro is managing debris at its reservoirs to ensure dam safety, reduce head losses from trashrack blockage and provide appropriate recreational use. Methods include protection of the discharge facilities by booms, cleaning of trashracks, disposal methods for the debris on the reservoir and ways to reduce additional ingress of debris from the catchment. Consideration is being given to providing funds to investigate innovative approaches to debris disposal, especially where burning of collected debris is prohibited.

2. B.C. HYDRO'S RESERVOIRS

B.C. Hydro has 60 dams at 43 locations (Fig. 1).

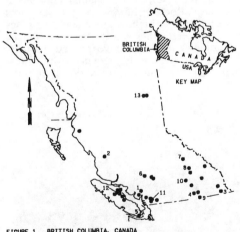

FIGURE 1   BRITISH COLUMBIA, CANADA
● B.C. HYDRO DAMS

1. Cheakamus	6. Lajoie	11. Stave Falls
2. Clayton Falls	7. Mica	12. Strathcona
3. Elko	8. Revelstoke	13. WAC Bennett
4. Keenleyside	9. Seven Mile	
5. Jordan	10. Shuswap Falls	

The prime purpose of these dams and reservoirs is the production of hydroelectric power. The safe and economic operation and maintenance of these facilities is, therefore, a predominant consideration and the effect of reservoir debris on the safety of the dams and power generation capability can be extremely important.

Since its incorporation in 1962, B.C. Hydro has cleared new reservoirs of woody debris to various standards prior to and after reservoir filling. In most cases the clearing standards specified in the dam Water Licence or elsewhere have been achieved or exceeded. Recently, in response to pressure from the local public, additional cleanup of floating debris and rooted debris in areas of high public usage has been undertaken, beyond that which was previously deemed acceptable. The cause is frequently debris that has entered the reservoir via its tributary streams and rivers or from shoreline trees that have become uprooted and have toppled into the water.

B.C. Hydro has also acquired a number of older reservoirs which had only minimal debris clearing prior to filling and has become responsible for debris management on these reservoirs as well.

3. DESCRIPTION OF DEBRIS

3.1 Types of Debris

- Rooted or fixed debris

Rooted stumps, snags and drowned trees are generally the result of the original flooding of reservoirs. Rooted stumps are generally less than 2 to 3 metres high, whereas standing snags are dead trees either along the shoreline or protruding significantly above the water surface. Drowned trees are only evident at the lowest water level. Fig. 2 shows rooted debris at Stave Lake. The examples in Section 7 describe approaches to fixed debris.

- Loose debris

Loose debris can be either floating, submerged or beached, and may be any portion of a tree which has been uprooted or logged. Floating debris is the most prevalent form of loose debris and may range from branches to full trees. It may form a mat and remain for a considerable time in one location or move up and down the reservoir dependant on wind patterns and reservoir geometry. Submerged debris is the waterlogged form of floating debris and may float just below the reservoir surface or be attracted by currents to low level outlets and power intakes. Beached debris is left along the shoreline by the action of winds, waves and ice or high reservoir levels. It may remain in the same place for many years. Fig. 3 shows loose debris on Jordan Reservoir. The examples in Section 7 illustrate how floating debris could become a dam safety concern.

3.2 Nature of Debris

The species of trees causing debris are predominantly Douglas Fir, Cedar, Hemlock and Balsam in the southern areas of B.C. and Spruce in the northern regions. Douglas Fir and Cedar tend to be larger and are more readily marketable, however for this reason they may have been selectively logged prior to reservoir filling. The butt size, length and propensity to degrade or sink are all variable factors for each species.

3.3 Sources of Debris

- Reservoir Area

The standard to which reservoir areas were

Fig. 2 - Rooted Debris at Stave Lake

Fig. 3 - Floating Debris at Jordan Diversion

cleared prior to filling is a major factor in the source of debris. While newer impoundments usually had substantial clearing undertaken. many of the older reservoirs had little or none. The immense volume of timber inundated by the largest reservoirs together with the less stringent clearing standards of the time, resulted in these reservoirs having considerable amounts of debris remaining.

- Reservoir Rim

Shoreline erosion, logging activities and landslides are the major source of debris from the reservoir rims. The action of wind generated waves on the shoreline can form beaches and undermine trees causing them to topple into the reservoir. Logging on steep reservoir slopes can leave a residue of debris which, if not carefully managed, can reach the reservoir. Slope failures can occur on natural slopes, logged areas or as a result of inadequate drainage associated with roads; this can introduce large volumes of debris into a reservoir.

- Tributary Streams

Tributary rivers and streams can be the source of considerable amounts of debris entering the reservoir. In addition to logging activities and slope failures in the catchment, other sources of debris include riverbank erosion and debris torrents. Debris torrents occur on steep hillsides and streams during peak runoff and can carry substantial amounts of debris.

4. RESERVOIR DEBRIS ISSUES

4.1 Dam Safety

B.C. Hydro has a program in place to ensure that its dams are safe and, in utilizing the natural resources of inhabited areas, do not impose an unnecessary risk to life or property.

Reservoir debris could make a dam unsafe if the discharge facilities became blocked at a time of high discharges. Peak inflows, maximum reservoir levels, high winds and minimum freeboard could combine with large debris volumes. Debris during a major flood could increase significantly due to floating of previously beached logs and ingress of debris from swollen tributaries. If the wind and current loading on the debris mat cause the debris boom(s) to break, a slug of debris could sufficiently decrease the discharge capacity of the dam to cause overtopping and possible dam breach. The management of debris (control/removal/disposal) includes consideration of the potential consequences of such an event. The approach taken is to either:

- protect the discharge facilities from blockage by debris, based on risk commensurate with the consequences of a dam breach on downstream population and property damage.
- allow the debris to pass through the discharge facilities prior to maximum discharges (debris booms to fail during the early phases of a major storm).
- design the debris booms to not fail but to allow debris to pass over or under them during extreme conditions.
- size the discharge facilities to safely pass any debris sizes and volumes that might occur (allowed for in the original design or part of dam safety improvements).

To date, debris boom replacement or upgrading to meet dam safety requirements has been completed or is planned at twelve projects. The cost of the work is in the order of CAN. $0.5 million. Table 1 gives examples of some of the debris booms and anchorages presently used to provide dam safety at B.C. Hydro's dams.

Table 1. Debris Boom and Anchorages at some B.C. Hydro projects.

Project	Debris Boom	Anchorage
Quinsam Storage	Single stick of logs chained end to end.	Cable wrapped around stumps.
Seven Mile	Single stick of logs clipped to a support cable.	Grouted rock anchors.
Revelstoke	Two log raft clipped to a support cable.	Grouted rock anchors.
Strathcona	Three log bundle clipped to a support cable.	Concrete anchors buried in dam and abutment.

4.2 Power Production

Maintaining the power intake trashracks clear from debris/trash ensures the continued capability of the hydroelectric facility to generate power. At generating plants where debris builds up against the trashracks, it is cleared on a regular basis to minimize head losses and, in the extreme, to prevent trashrack failure, possible damage to the generating facilities and closure of the plant. The choice of when and how to clear debris is based on an economic evaluation of power generation losses due to debris. Energy losses are determined by measuring head loss across the trashrack for varying levels of debris build-up. The rate of debris build-up (regular, seasonal, intermittent) is assessed as well as operational characteristics of the reservoir and powerplant. Alternative methods, frequency and costs of removing debris are investigated and the economics evaluated based on B.C. Hydro's "value of electricity". This program is part of our "Resource Smart" initiative to obtain the maximum economic output from our existing plants. Recently commenced, an initial review has identified some 50 GW.hr yr valued at CAN $1 million/yr that could be realized by optimizing trashrack cleaning methods and frequency.

Table 2 gives examples of some of the methods and frequencies of trashrack cleaning presently used at B.C. Hydro's dams.

Table 2. Trashrack Cleaning Methods and Frequency at some B.C. Hydro Projects.

Project	Cleaning Method	Frequency
Clayton Falls	Trashrakes (shallow intake).	Monthly
Strathcona	Divers (cleaning by hand or using a crane with a debris basket or grapple for removal).	Five years or as required.
La Joie	Drawdown (using rakes or prybars from small boats or ground surface).	Annually
Cheakamus	Debris Grapple (mounted on a barge).	Differential head > 3.0 m.
WAC Bennett	Debris Grapple (mounted on a crane with TV camera for guidance).	As required.
Shuswap	Load Rejection (back surge flushes trashrack; debris collected or passed over spillway), or Divers.	Few times per year.
Elko	Removal of Trashracks.	As required.

4.3 Reservoir Use

The reservoirs impounded by B.C. Hydro's dams have varying amounts of public usage, and in a few cases commercial usage. In most cases this use is growing and changing.

For public and commercial use, the occurrence of floating and rooted debris can cause hazards to navigation and public safety and interfere with recreational pursuits. For some of our major reservoirs, commercial use was the primary reason for initial and ongoing debris removal programs. More recently in response to public requests, B.C. Hydro developed a comprehensive plan in 1989 to remove debris from reservoirs. This plan identified nine reservoirs as having highest priority and allocated funds for clean-up programs. Work is continuing, with input from local communities, and in some cases partnership programs. Annual expenditures for reservoir clean-up since 1989 have been between CAN $1 and $2 million and this level of effort is expected to continue. The majority of the expenditures are at our two largest reservoirs; Williston Lake, behind WAC Bennett Dam and Kinbasket Lake, behind Mica Dam.

Reservoir debris also has environmental effects as the reservoirs support a natural biological, aquatic and wildlife ecosystem. Stumps and snags may be considered unsightly in high use recreation areas or adjacent to transportation corridors, particularly at low water level. The process of removing debris can in itself create environmental damage.

However on the positive side, debris can provide good habitat for aquatic and terrestrial wildlife.

5. METHODS TO MANAGE RESERVOIR DEBRIS

5.1 Control

Debris from the reservoir rim and perimeter streams and rivers, can be reduced by:
- Cooperation with forestry companies on ways to ensure their practices do not lead to waste from forestry operations entering streams that flow into the reservoir. This could include leaving standing timber barriers, providing adequate drainage measures, rapid replanting and minimizing strip clearing practices.
- Assessment of reservoir slopes that might fail and methods to reduce risk particularly if related to human activities (i.e. logging, road construction).
- Creation of debris traps on the streams before they enter the reservoir.
- Where debris source is an upstream hydroelectric project (perhaps owned by others), cooperate with joint approach to debris management.

Little success has been achieved in managing the continued ingress of debris to the reservoir from above the shoreline or from streams and tributaries feeding the reservoir. For some reservoirs, this source can be substantial, particularly where extensive logging operations are being undertaken.

5.2 Collection and Removal

Debris on or around the reservoirs can be managed by:
- Collecting floating debris with bag booms and removal to disposal area.
- Cutting snags and removing stumps in shallow areas or drawdown zones.
- Raising the reservoir level to float off debris around the reservoir rim.
- Placement of debris containment booms where debris bearing streams enter the reservoir.
- Construction of debris containment dykes. These can be constructed in relatively shallow water or at a suitable bend in the reservoir to guide debris into a holding area. A shear boom can assist in guiding debris toward a suitable disposal area.

5.3 Protection

The project facilities can be protected by:
- Booms to restrain debris.

- Shear booms to deflect debris.
- Skirt booms to protect structures.
- Booms in front of trashracks to keep debris from power conduits or discharge conduits.
- Passing debris over the spillway or weir.

The debris booms protecting the structures are sometimes single logs (boom sticks) connected to each other by chains and attached to fixed objects on the abutments (e.g. concrete blocks, trees). Booms recently constructed or for major projects usually comprise boom sticks attached to a continuous steel cable. Often two or more logs are used in parallel in a number of configurations such as rafts or bundles.

5.4 Disposal

Methods of debris disposal which can be used include:
- Burning, using heavy equipment to create piles or windrows.
- Piling and natural decay.
- Retrieval of marketable timber, including shake cutting.
- Cutting and sale as firewood.
- Chipping and subsequent use for pulp or cogeneration.
- Burial in depressions or disused borrow areas.

To date, most efforts to manage reservoir debris after reservoir filling have involved a collection and disposal program together with debris booms protecting the structures.

The collection and disposal programs can be quite extensive and costly. At both Williston and Kinbasket Lakes, a fleet of tugs are used to corral the debris and steer it into holding areas. As the reservoir level falls during the winter the debris is beached, piled, and eventually burnt. While selling the debris for pulp is considered, it normally has to be sorted by species for this purpose which can be difficult after it has been floating for a number of years. Also the sand in the debris can damage the commercial chippers. Collection, removal and disposal work on the reservoir is either handled directly by B.C. Hydro staff or by local contractors.

B.C. Hydro is involved in research to investigate new methods of debris disposal. Recently a proposal was initiated to "evaluate methods of debris disposal other than by burning". Mulching followed by enhanced composting is one method that will be given consideration.

6. APPROACH TO DEBRIS MANAGEMENT

Management of debris in B.C. Hydro's reservoirs is the responsibility of the local Production area. Ongoing maintenance of booms and repair/replacement is also managed by Production with technical assistance from Engineering. Production seeks the input of local communities in developing management programs in each area and liaises closely with forestry companies logging in the catchment.

Funding for debris management comes from several sources. Generally, decisions on debris clearing and the allocation of funds are based on a priority system with guidance provided by Environmental Resources. The Director of Dam Safety funds studies to assess debris issues relating to dam safety and improvements to debris booms or other works. Studies on clearing debris from trashracks to economically increase power generation are funded by Resource Management under the Resource Smart program.

B.C. Hydro is currently working towards a comprehensive assessment of reservoir debris issues. To accomplish this objective, a task force with stakeholders from each group with an interest in reservoir debris is producing a set of guidelines to be used to assess reservoir debris concerns and identify alternatives for satisfying these concerns. (Fig. 4)

Using the guidelines, a management plan will be developed to address the debris concerns on each reservoir. This plan will include a debris inventory, a summary of major debris concerns, alternative methods to manage the debris, means to assess priorities and an implementation schedule. Cost savings could be realized, priorities could be adjusted to reflect all the debris issues on each reservoir, and any control methods, clean-up, and debris boom construction could be designed to benefit overall reservoir management.

Fig 4   Reservoir Debris Issues

The local Production office will have overall responsibility for applying the guidelines and managing the debris program.

7. EXAMPLES OF DEBRIS MANAGEMENT

7.1 Upper Campbell/Buttle Lakes

In the late 1950's Strathcona Dam was constructed (Fig. 5) creating Upper Campbell and Buttle Lake Reservoirs. These reservoirs have a combined shoreline length of 113 km.

One of the preconditions to construction of the dam was that the reservoir area be cleared of timber prior to flooding. With the subsequent creation of Strathcona Park and the surrounding area being more accessible to public use and view, the remaining rooted stumps have in some instances proven dangerous and unsightly.

B.C. Hydro has worked with B.C. Parks to improve the situation. In recent years Hydro has provided funds to clear shoreline areas within Strathcona Park including the beach area at the Buttle Lake campground as well as picnic and boat launch sites along the shoreline.

Now, Hydro is developing a comprehensive stump management plan with local contractors hired to do the work. Once stumps are cleared, an unattractive margin still remains around the reservoir during drawdown. Hydro is now looking at a combination of removal and re-vegetation to solve the problem and a series of planting trials will be carried out over the next two years. The revegetation solution was prompted by the success of this method in controlling dust problems along the shores of the Arrow Lakes which are impounded by the Keenleyside Dam.

The advantages of planting vegetation are considerable. Not only does it make the area more attractive, but where stumps remain the vegetation camouflages them. In addition the vegetation provides young fish with food, as well as cover to hide from predators. The vegetation provides food for deer, elk and birds, some of whom nest in the wetland plant communities. In the long run there are also financial advantages as stump removal is very costly.

Debris also creates a dam safety concern at this project. Strathcona Dam is an earthfill structure with flood discharges routed through a spillway structure. During any large flood event, higher reservoir levels would float substantial amounts of debris which are presently beached around the rim of the reservoir. The resulting debris mat, if not restrained, could jam the spillway and reduce discharge capacity. If this led to dam overtopping, a dam breach could result with a potential for high loss of life and property damage in downstream communities.

During dam safety investigations in 1984-86, the existing debris boom was found to be inadequate to restrain any appreciable amounts

Fig 6 - Debris Boom at Strathcona Dam

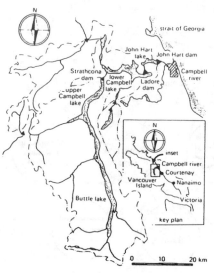

FIG. 5 LOCATION OF STRATHCONA PROJECT

of debris. A replacement boom comprising three-log bundles attached to a continuous galvanized steel cable (Fig. 6) was installed in 1987 and anchored to concrete blocks on each abutment. Douglas fir was used for the boom sticks which, combined with heavy connectors, caused the boom to float quite low in the water.

During a recent storm and ensuing spillway discharge, the boom cable separated and the entire boom passed through the spillway. Investigations are underway as to the cause of the failure.

Power generation is frequently reduced by debris at Strathcona. The power intake tower is a free standing concrete structure with intakes some 30 m below normal reservoir levels. While booms protect the tower, submerged debris regularly clogs the trashracks. Cleaning is an expensive operation involving barges, divers and a crane as well as shutting down the plant. An economic evaluation has indicated appreciable benefits of a more regular cleaning schedule.

The debris issues described for Strathcona Dam were addressed independently. Preparation of a reservoir debris management plan could have allowed a more integrated approach.

7.2 Stave Lake Reservoir

Stave Lake Reservoir is part of the Stave Falls-Alouette-Ruskin generating complex which was constructed between 1910 and 1925, and is approximately 60 km east of Vancouver (Fig. 7). Stave Lake Reservoir has an area of 60 square km and receives the drainage from 1150 square km of mountainous, forested terrain, which supports large forestry operations. The forestry practices, together with the steepness of the terrain and high precipitation (often flash floods), can result in large amounts of floating debris being carried into the reservoir on a regular basis. In addition, the reservoir area was not cleared prior to filling and substantial amounts of fixed debris are visible. These include both stumps and snags and are particularly noticeable in the many shallow areas on the lake. Stave Lake is close to Vancouver and this makes it a potentially attractive recreation area. However, the lake has not been fully developed, partly because of the debris volumes and partly because of alternative lakes where recreational use is encouraged. The cost of completely clearing the reservoir and disposing of debris would be very high (a study indicated some CAN $50 million over ten years).

During the 1970's a public waterfront park was developed as a joint project of the Forest

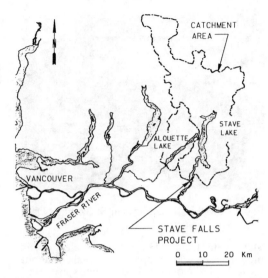

FIG. 7 LOCATION OF STAVE FALLS PROJECT

Service, the Provincial Corrections Branch and B.C. Hydro. A Hydro work boat has been used to clean debris from the lake since the 1920's. The Corrections Branch also continues to control floating debris on the lake. Most of the wood debris gathered is hauled to provincial campgrounds where it is used as firewood. Some of the larger cedar stumps are also recovered for use in the shake and shingle roofing industry.

As the Stave Falls powerplant is old, and in frequent need of maintenance and as the powerplant has only about half the hydraulic capacity of the downstream project, consideration is being given to redevelopment. The new larger powerplant would be operated with considerably less drawdown of the Stave Lake reservoir. This would reduce the amount of visible fixed debris as well as increasing average water depths and thus improving navigation.

The effect of debris on the safety of the dam is being investigated. The discharge facilities consist of six gated and four stoplogged spillway bays and four low level gated outlets. The spillway structure is in an embayment protected by a debris boom consisting of single boom logs. Prevailing wind patterns normally keep the log boom clear of debris. However, if significant amounts of debris were drawn to the spillway during a major flood and the boom was to fail, the spillway gates could become blocked. This would result in higher reservoir levels and perhaps overtopping.

Blockage of the power intake trashracks by debris is an ongoing and significant problem at

431

Fig. 8 - Blocked Trashracks at Stave Falls Dam

Debris can either be left alone, or controlled, collected, removed and disposed of. Facilities can be protected using various types of debris booms. The examples presented show the diversity of debris concerns and problems and how an integrated approach to debris management can provide a more effective way to assess different debris issues. B.C. Hydro is actively working towards this approach.

Stave Falls Dam. The racks are cleaned by mobile crane and grapple at intervals of about six to twelve months. The methods used and frequency will be assessed as part of the program to economically reduce head losses across trashracks. Debris and ice loading caused failure of four of the five power intake trashracks in 1989; this lead to plant downtime, repair costs, and lost power generation. (Fig. 8).

Debris on Stave Lake Reservoir presently has a major effect on the long term use of the reservoir. The lake supports a rich natural environment with the swampy areas and decaying trees providing good habitat. A program to remove the existing floating debris, minimize new debris from entering the reservoir (in cooperation with forestry companies), reduce drawdown of the reservoir by redeveloping the powerplant and leaving some shallow areas undisturbed could enhance the overall use of this reservoir.

8. CONCLUSIONS

Reservoir debris can affect the safety of dams, efficiency of power production, and the usefulness of reservoirs for the public and the environment. With increased awareness, more effective debris management programs are being developed for B.C. Hydro reservoirs. It has been found that the cost of debris control can be very high, especially if the reservoir was not cleared prior to filling or if large volumes are entering from the watershed.

In general, large volumes of large sized floating debris are the most significant concern from a dam safety perspective. Smaller logs and branches, particularly those that become waterlogged, can plug intake trashracks and reduce power production, whereas floating and rooted debris often affects public use of the reservoir.

Hydropower'92, Broch & Lysne (eds) © 1992 Balkema, Rotterdam. ISBN 90 5410 054 0

Overtopping of the core in rockfill dams – Internal erosion

A. Wörman
Swedish State Power Board, Vattenfall Utveckling AB, Älvkarleby, Sweden

M. Skoglund
SINTEF NHL, Trondheim, Norway

ABSTRACT: During extreme flood events, there is a risk for internal overtopping of the core in rockfill dams and resulting erosion on the top surface of the core or of the sand filter above it. If the geometric filter criterion is not satisfied in the interface between two soil layers subjected to through-flow, both a theoretical investigation and observations from experiments at prototype-scale clearly indicate that erosion starts at the downstream side of the lower, finer stratum and propagates rapidly upstream. The time-scale of the process is markedly sensitive to variation in the ratio of the grain sizes in the layers and the void ratio of the upper stratum; however does not depend significantly on the height of overtopping. If only a small fraction of the erodible grains is retained at the interface by mechanical interaction (filtration), erosion ceases rather quickly and the degradation of the interface is limited.

1 INTRODUCTION

Recent research on hydrological conditions in Scandinavia indicates that design floods for dams may have been under-estimated (af Klintberg et al., 1990). The conclusions from these studies have raised the related question of whether temporary overtopping of the central moraine-core in earthfill dams can be allowed during extreme floods. However, since overtopping of the core could lead to erosion at its interface toward the fill-material covering its top, an understanding of the possible erosion process is needed.

The purposes of the present study have been dual, i) to conduct laboratory experiments at full scale, in which the core or the sand filter above it is overtopped, and ii) to develop a theory describing the erosion process. Full-scale laboratory experiments offer a valuable opportunity to observe and gain understanding of the process as it may occur in a real dam. These insights are particularly important in light of the fact that field observations of internal erosion processes are inherently difficult to conduct and, to date, have not been done. However, because the number of parameters involved in the problem is large, interpretation and generalization of the results require both analyses and physical descriptions of the phenomena affecting the erosion process.

The analytical approach followed in the present study is based on solution of the governing equations accounting for the two-phase motion formed by sediment and water. The credibility of the mathematical formulation relies to a great extent on a constitutive equation relating sediment transport rate with state variables of the overflow. The existing transport relationship is valid for a uni-size sediment for which clogging of the transported grains does not occur at the interface between the eroded material (core or filter) and the above stratum (Wörman and Olafsdottir, 1992). One purpose has been therefore to formulate tentatively a transport relationship that accounts for the clogging mechanism observed in the experiments.

Important issues are the sensitivity of the erosion process to variations of the independent variables, like grain size, grain size distribution, porosity and height of overtopping. For instance, if separation during the construction has caused a deviation from the conventional geometrical filter criterion, would that change the erosion sensitivity significantly in case of overtopping ? Hydraulic filter design during the last decade has been recognized as an important part of the design of rockfill structures (de Wittman, 1981, de Graauw, et al. 1983, Brauns, 1991).

2 EXPERIMENTS OF INTERNAL EROSION AT PROTOTYPE-SCALE

A physical model was built at a prototype-scale of the zone most sensitive to internal erosion, the down stream side of the uppermost part of the core, including the sand filter covering the top of the core and the rockfill above (Fig. 1). The length of the test equipment was 2.5 m, the height 1.5 m, the width 1 m, and the thickness of the sand filter 0.5 m. One of the side walls was glassed, which facilitated observations of possible internal erosion. The upstream water reservoir was separated from the soil zones by a permeable wall.

Two different tests were executed; both had the same sand filter, but different rockfills placed above the sand filter. The sand filter and the rockfill prepared in the first set-up clearly did not satisfy the Norwegian filter criterion. In the second set-up, the criterion was still not satisfied, though, the difference in grain size between the layers was smaller compared to the first experiment. The grain size distribution curves for the investigated materials are given by Skoglund (1991). In both tests, the upstream water table was raised slowly to allow most of the pore space in the sand filter to be saturated. No erosion occurred at the interface between the sand filter and the core. However, as the sand filter was overtopped, a water wave propagated in a few seconds across the top surface of the filter and thereafter, the downstream edge of the filter began to erode rapidly.

In the first experiment, erosion continued until, after 2-5 minutes, the moraine core came into direct contact with the rockfill. At this time, model effects, such as problems of evacuating the eroded material from the test equipment, began to dominate and the experiment was stopped.

Similarly, in the second experiment, erosion started with the overtopping of the downstream edge of the sand filter. However, the degrada-

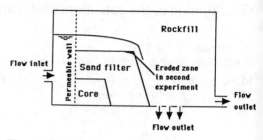

Fig. 1   Experimental set-up.

tion quickly diminished, and after about one and a half minutes the process stopped. The eroded volume was wedge-shaped, and its sides measured 13 cm in the horizontal direction and 8 in the vertical direction. The sand filter remained stable as the upstream water table was raised.

3 A TWO-PHASE PHENOMENON

Governing equations

At the downstream edge of the core crest, the curvature of flow is sharp and induces high hydraulic gradients and intense erosion. The resulting degradation of the core crest imposes a corresponding change on the overflow, and the perturbations in the two phases (water and sediment) propagate upstream. If no mechanisms act to retard the transport of sediment, the disturbances soon reach the upstream side of the core crest and the lowering of the core crest rapidly continues until it reaches the base of the core.

The temporal and spatial variation in erosion can be described by the partial differential equations that govern the system. If the elastic response of the soil is omitted, the governing equations in a one dimensional form are similar to those for alluvial streams (c.f. de Vries et al, 1990)

Momentum equation of the flow and resistance relationship

$$-\frac{d\phi}{dx} = 2\,f\,\frac{q\,|q|}{g\,D_{15}}, \qquad (1)$$

Conservation of water mass

$$\frac{d}{dx}(q\,H) = 0, \qquad (2)$$

Conservation of sediment mass

$$\frac{\partial z}{\partial t} + \frac{1}{1-n_2}\frac{\partial G}{\partial x} = 0, \qquad (3)$$

Sediment transport relationship

$$G = F(\text{Flow} ; \text{Soil parameters}), \qquad (4)$$

where ϕ is piezometric head, $\phi = p/(\rho_w\,g) + z$, q is flow per unit area, g is acceleration of gravity, D_{15} is the square sieve mesh size that passes 15 % of the grain mass, f is a friction factor defined by eq. (1) and equal to 10 for inertia-dominated flow (Bear, 1988), p is pressure, z is the elevation of the interface relative to some arbitrary datum line, ρ_w is density of water, H is water depth, G is the sediment transport rate per unit width, n is the porosity, x is the space coordinate in the direction of flow and t is time. The indices 1 and 2 denote the upper, coarser and the lower, erodible strata, respectively. Further, large D represents the grain size of the coarser stratum in distinction to small d that represents the grain size of the erodible stratum.

Eq. (2) is based on the assumption that the celerity of propagation of a water surface perturbation is much higher than the celerity of propagation of a disturbance in the interface; i.e. the flow is quasi-steady. For many applications of practical interest, this simplification has been shown to apply (Wörman, 1991). If the flow is acceptably quasi-steady and one-dimensional, the credibility of the mathematical formulation rests on the constitutive equation relating sediment transport rate with state variables of the porous medium flow, Eq. (4).

A variety of physical effects can affect the mobility of the interface grains; of these, the pressure and seepage forces and gravity are the dominant in many practical situations. The following expression has been found to apply to erosion induced by flow parallel to a horizontal interface between two non-cohesive soil strata of significantly different grain sizes (Wörman and Olafsdottir, 1992)

$$G^* = \left(0.56\,\frac{e_1}{\frac{\rho_s}{\rho_w}-1}\,\frac{\partial \phi}{\partial x}\,\frac{D_H}{d_{85}}\right)^{12} n_1. \qquad (5)$$

The transport parameter G^* is defined as $G/(V\,d_{85})$, ρ_s is the density of the erodible grains, e is void ratio and the subscript H implies harmonic mean by volume. Eq. (5) is valid only if dispersion perpendicular to the interface and adsorption of the transported grains on the coarser grain-matrix is negligible.

Transport and filtration of grains smaller than 100 µm is affected by inter-granular forces due to electrostatic interaction between charged particles and London - van der Waals forces between dipole-dipole, such as some clay minerals (Tien and Payatakes, 1979, McDowell-Boyer et al., 1986). Electrochemical forces result in cohesion within a medium and adhesion (and possible adsorption) between transported particles and the transport medium. Analysis of the migration of clay minerals could require the inclusion of the effects of these phenomena. Dispersion is also an essential phenomenon in the transport of clays.

Retardation mechanisms

The transport relationship, Eq. (5), applies strictly to uni-size sediments for which the interface properties are constant in time and the bed can be represented by a single grain size, d_{85}, and density. However, the core in rockfill dams, in Scandinavia usually consists of widely graded sediments collected from moraine deposits. For such materials, erosion leads to enrichment of the coarser grains both as a result of i) selective erosion of the different sizes and ii) retention of larger grains caused by mechanical interaction between the transported grains and the pore channels of the coarser matrix. In particular, the latter mechanism may result in clogging of grains in the vicinity of the interface and a successive reduction in the transport rate. With time, a filter cake forms that effectively prohibits further erosion. Experimental studies show that such straining filtration gradually becomes significant as the size ratio d/D grows in the range 0.05 to 0.1 (McDowell-Boyer et al., 1986, Wörman and Olafsdottir, 1992). The upper limit implies that the erodible

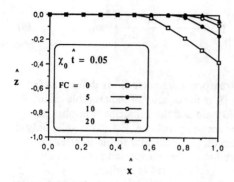

Fig. 2 Shape of the interface of erosion at the time, $\hat{\chi_0} \hat{t} = 0.05$, for different values of FC.

grain is too large to move at all in the pore channels of the coarser matrix.

For simplicity, two assumptions were made herein; i) that the change from a state of non-retardation to a filtration state is abrupt with respect to a change in the size ratio, and ii) that the transport of the erodible grains can be described by eq. (5). Further, the coarser fraction of the erodible grains, that are enriched at the interface due to clogging, causes the interface area available for transport to diminish with time. As a linear approximation the reduction in the transport rate can be accounted for by the formulation

$$G^*_{\text{eff}} = G^*_M A, \qquad (6)$$

where subscript "eff" denotes "effective", G_M is the transport rate per unit width on the area available for transport and A is the area percentage available for transport. Initially and for uni-size sediments, the area percentage available for transport is given by the porosity of the coarser stratum, n_1. Eq. (5) is based on this postulation.

If all erodible grains larger than a certain size are clogged at the interface, the clogged area, A_{cl}, can be related by continuity considerations to the mean grain size of the clogged grains, d_{cl}, the volume fraction of these grains in the parent material, η, and the cumulative erosion rate, in a manner similar to that proposed for an armoured area by Karim and Holly (1986) and Wörman (1992). The clogged

area can be written as

$$A_{cl} = \frac{\varepsilon}{d_{cl}} \frac{\eta}{1-\eta} \int_{t'=0}^{t} \frac{\partial G}{\partial x} \, dt', \qquad (7)$$

where ε is an empirical coefficient varying within the range $0 < \varepsilon < 2$ depending on i) the percentage by volume of the grains in the parent sediment that have the potential to clog the pore channels that actually do so and ii) their orientation in the clogged positions. Substitution of Eq. (3) in Eq. (7) shows that the clogged area increases linearly with the degradation of the core surface. At equilibrium, the area available for transport is zero. If the available area is given as $(n_1 - A_{cl})$, the transport relationship can be written as

$$G^* = \left(n_1 - \frac{\varepsilon}{d_{cl}} \frac{\eta}{1-\eta} \int_{t'=0}^{t} \frac{\partial G}{\partial x} \, dt' \right)$$

$$\left(0.56 \frac{e_1}{S_s - 1} \frac{\partial \phi}{\partial x} \frac{D_H}{d_{85}} \right)^{12},$$

$$(n_1 - A_{cl}) \geq 0, \quad (8a)$$

$$G^* = 0, \qquad (n_1 - A_{cl}) < 0, \quad (8b)$$

S_s is the specific density of the grains. If no grains are retained due to clogging ($\eta = 0$), the second term of the right hand member of Eq. (8) is zero and, hence Eq. (5) yields as a special case. If the second term is not zero, irrespectively of how small, the transport decays towards zero at equilibrium.

Numerical simulations

A well posed problem (a unique solution) requires two boundary conditions and, for each phase, one initial condition. The initial conditions are the phreatic surface and the interface curve at time, $t = 0$. The upstream boundary condition for the flow is the level of the water table and, on the downstream side, the hydraulic gradient is constant and close to unity (free fall). In the one-dimensional formulation, the boundary conditions were approximated as

436

$$\phi(x=0, t\geq 0) = \phi_{up} \tag{9a}$$

$$\frac{\partial \phi}{\partial x}(x=L, t\geq 0) = 1 \tag{9b}$$

The field equations for the water and sediment phases can be deduced in a similar way, as shown by Wörman and Olafsdottir (1992). However, to take account of filtration, a modification of the expression for the rate of degradation should be based on Eq. (8). The rate of degradation can be derived in a non-dimensional form by reducing Eqs. (3) and (8); after some manipulation the result is

$$\frac{1}{\chi_0}\frac{\partial \hat{z}}{\partial \hat{t}} = \left[1 + FC\int_{\hat{t}'=0}^{t} \frac{\partial \hat{z}}{\partial \hat{t}}dt'\right]$$

$$\left[11.5\left|\frac{\partial \hat{\phi}}{\partial \hat{x}}\right|^{10.5}\frac{\partial \hat{\phi}}{\partial \hat{x}}\frac{\partial|\partial\hat{\phi}/\partial\hat{x}|}{\partial \hat{x}} + \right.$$

$$\left|\frac{\partial \hat{\phi}}{\partial \hat{x}}\right|^{11.5}\frac{\partial^2 \hat{\phi}}{\partial \hat{x}^2}\Bigg] +$$

$$+ \left|\frac{\partial \hat{\phi}}{\partial \hat{x}}\right|^{11.5}\frac{\partial \hat{\phi}}{\partial \hat{x}}\frac{\partial\left(1 + FC\int_{\hat{t}'=0}^{t}\frac{\partial \hat{z}}{\partial \hat{t}}dt'\right)}{\partial \hat{x}},$$

$$\tag{10}$$

where $\hat{x} = x/L$, $\hat{z} = z/L$, $\hat{\phi} = \phi/L$, $\hat{t} = (q_0/n_1)\, t/L$, subscript 0 denotes initial value at a defined section on the boundary, L is the crest width, the initial parameter is defined as

$$\chi_0 = \frac{n_1}{1-n_2}\sqrt{\frac{g\, D_{15}}{2\, f}\frac{d}{L}\frac{1}{q_0}}$$

$$\left(0.56\frac{e_1}{S_s-1}\frac{D_H}{d_{85}}\right)^{12}, \tag{11}$$

and the filtration coefficient is defined as

$$FC = \varepsilon\frac{\eta}{1-\eta}\frac{1-n_2}{n_1}\frac{L}{d_{cl}}. \tag{12}$$

The implication of the initial parameter is that this combination of variables gives the scale for the time required to reach a certain state of the erosion process. Therefore, the parameter constituted by the product $\hat{\chi_0}\, t$, inevitably, is the time-scale parameter for the process. This result is salient for two reasons; i) by the intro-duction of $\hat{\chi_0}\, t$, a number of variables have been reduced from the original PDE's describing the process, and ii) the sensitivity of the time scale for variations in material properties can be analytically evaluated. The latter statement can be illustrated by the following fact; if the void ratio of the coarser stratum, e_1, increases from 0.4 to 0.55, the change of the value of $\hat{\chi_0}\, t$ indicates that the time to reach a certain state of degradation reduces by a factor of 45. The sensitivity in time to apparently small changes in material properties is marked.

4 DISCUSSION OF OBSERVATIONS AND THEORY

For quasi-steady flow, the solution to the flow of water is not coupled to the sediment movement during each time step of the calculation. Hence, the degradation rate is explicitly given by the first and second space-derivatives of the piezometric head according to Eq. (10) and can be calculated by the use of a standard Runge-Kutta method.

Fig. 2 shows the degradation along the interface between the sand filter and the stratum above it at an instant of dimensionless time, $\hat{\chi_0}\, t$ = 0.05, and for different values of the filtration coefficient. In Fig. 3, the evolution of degradation at the downstream edge of the sand filter is depicted for the values of the filtration coefficient used for Fig. 2.

In the first experiment, during which no significant filter development occurred (FC = 0), erosion started at the downstream edge of the sand filter where the hydraulic gradients are high and then propagated upstream. In a comparatively short period of time (about one minute), the upper surface of the sand filter attained a shape close to that represented by the curve in Fig. 2 associated with FC = 0. The evolution of the deformation of the interface in the experiment was found to be similar to that

437

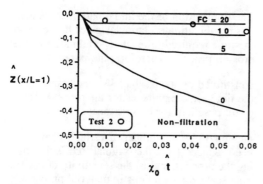

Fig. 3 Evolution of the degradation at the downstream side of the sand filter for different values of FC according to the theory.

obtained from the computations (Fig. 3 for FC = 0), though the times of occurrence deviated by a constant factor.

If the experimental values are inserted in $\hat{\chi}_0 t$, the time to reach the state of degradation depicted in Fig. 2 for the non-filtration case (FC = 0) is found to be fractions of a second, whereas the measured time was in the order of one minute. Clearly, the time scale is highly sensitive to small changes both in the empirical exponent twelve and in the variables in the base of the exponential factor; hence time predictions can only be qualitative. Nevertheless, the conclusion can be made that, for most conditions, equilibrium is reached in a period of seconds or, at most, a few minutes.

The theory shows that neither the shape of the interface nor the time scale of the process is significantly affected by a change in the height of overtopping. The result is partly due to the facts i) that not only the available energy increases with the height of overtopping but also the surface of the granular matrix on which friction occurs and ii) that the degradation is predominantly governed by the constant boundary condition on the downstream side. As a consequence, a certain permissible height of overtopping cannot readily be determined. In other words, if a small but finite height of overtopping is acceptable, a larger height is more or less equally acceptable (from the point of view of internal overtopping erosion).

In the second experiment, a limited amount of erosion occurred at the downstream edge of the sand filter. After the initial erosion of basically the finer grain sizes from the core surface, erosion diminished within a few seconds as a result of clogging of transported grains at the interface. An increase in the height of overtopping did not lead to additional erosion. A comparison of the experimental results and Fig. 2 shows that the equilibrium shape of the interface can be accurately predicted by the theory for a value of the filtration coefficient of about 15 to 20. If all erodible grains larger than $D_{15}5$ are clogged at the interface (d = 4 mm and $\eta = 10\%$), the enrichment factor, ε, is about 0.3. According to Fig. 3 (FC = 20), the degradation process ceases rather quickly, and thereafter the interface becomes static. Only for the special case of FC equals zero, does the degradation process continue monotonically. The practical implication of the latter result is that once even the slightest fraction of the erodible grains are clogged at the interface, no matter how small the fraction, the degradation process eventually ceases.

5 SENSITIVITY TO VARIATION IN PARAMETERS

Some aspects of the sensitivity of the erosion process to variation in parameters are treated in the discussion of filtration and the time scale of degradation. A high sensitivity can be given a dual interpretation; i) small changes of the governing parameters cause large variations of the erosion process (both in reality and in theory) and ii) minor faults in the empirical coefficients are allianced with large deviations between theoretical predictions and reality. For instance, the marked non-linearity of Eq. (11) implies that either a small change in the grain size ratio for the two layers between which erosion occurs or a small change of the void ratio of the coarser stratum results in a large change in times of occurrence. Furthermore, a small fault in the empirical exponent in Eq. (11) makes time predictions uncertain. However, since the solution to Eq. (10) is not sensitive to variations in this exponent, the prediction of the spatial distribution of erosion is not so uncertain in this respect.

The exponent twelve has been found to be valid close to inception of motion (the investigated ranges were: $10^{-6} < G^* < 10^{-3}$ or $0.03 < d\phi/dx < 0.3$), and in this interval the non-

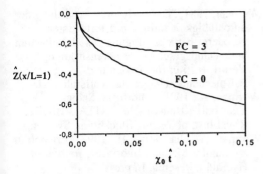

Fig. 4 Evolution of degradation at the down-stream edge of the sand filter for a case with a limited filtration (FC = 3) and for a non-filtration case (FC = 0).

linearity of the transport relationship is pro nounced. Studies of incipient motion in open-channels show that G^* varies with Shields´ parameter with an exponent of some eight to sixteen (Paintal, 1971, Pazis and Graf, 1977). For stronger transport, a lower value of the exponent was suggested by Parker (1991). Because the upper limit of the investigated interval of the hydraulic gradient (0.3) is exceeded in the simulations, a lower value of the exponent twelve might better explain the time scales measured in the experiments. Comparisons of the theoretical predictions and measured times indicate that the theory yields times that are too short. A valuable further in-vestigation would be to conduct experiments with hydraulic gradients ranging from 0.3 to unity, in particular, during filtration.

As stated previously, if the slightest fraction of the erodible stratum (the sand filter) is re-tained at the interface due to clogging, the ero-sion finally ceases. Fig. 3 shows the evolution of erosion at the downstream edge both for a non-filtration case (FC = 0) and for a filtration case for which only 1 % by volume of erodible grains are retained at the interface (FC = 3). Initially, the evolution of erosion is the same for the two cases, but the difference in filtration, causes the processes to deviate markedly at a later time. The equilibrium degradation in the filtration case occurs for z/L = 0.3, whereas the degradation continues monotonically in the non-filtration case.

6 CONCLUSIONS

Prototype-scale experiments simulating a situa-tion in which a core and the sand filter covering the top of the core are suddenly overtopped were compared with analogous numerical simulations. If the geometrical filter criterion between the sand filter covering the core and the above rockfill is not satisfied, both simulations showed that erosion starts at the downstream side of the filter and the distur-bance propagates upstream. Since the geometri-cal filter criterion between the core and the sand filter was fulfilled, no erosion was observed at this interface. The results also indi-cate that the time scale of the process is highly sensitive to variation in the ratio of the grain sizes used for the sand filter and the rockfill. Therefore, even small deviations from the geometrical filter criterion enhance signifi-cantly the risk of erosion due to internal over-topping.

In one experiment, a successive development of a filter cake at the contact surface between the erodible layer and the above stratum re-sulted in a decay in the sediment transport. Consequently, after the initial erosion of basi-cally the finer grain sizes from the sand filter surface, erosion ceased. The computations indi-cate that even if a very small volume propor-tion of the erodible grains are retained at the interface, erosion ceases rather quickly and the degradation can be conceived as being rather limited.

Furthermore, the height of overtopping does not significantly affect the erosion process, which means that if a small but finite height of overtopping is acceptable, a larger height of overtopping is almost equally so acceptable (from the point of view of internal overtopping erosion). Taking this into consideration and due to the marked retarding impact from filtration on the erosion process, some support is given to the idea of allowing temporal overtopping during extreme flood situations.

Future studies should focus on filtration effects and on an extension of the investigated range of the hydraulic gradients used as a basis for the transport relationship.

ACKNOWLEDGMENTS

The present study has been financed under a

research and developing program concerned with dam safety by Vattenfall AB (FUD) and by VAST and Norsk Konsesjonsavgiftfondet (The Norwegian Fund of Fees). The authors are indebted to Nils Johansson, John McNown and Einar Tesaker for comments on the paper.

REFERENCES

af Klintberg, L., Bergström, S, Ehlin, U., Ohlsson,P-E, Sjöborg, K-Å, 1990, "Riktlinjer för bestämning av dimensionerande flöden för dammanläggningar", slutrapport från flödeskommittén, Swedish State Power Board, Svenska Kraftverksföreningen, SMHI.

Brauns, J., 1991, "Filters and Drains", in in Advances in Rockfill Structures, edited by E. Maranha das Neves, NATO ASI Series, Serie E Vol. 200.

de Graauw, A., van der Meulen, T., van der Does de Bye, M., 1983 "design criteria for granular filters", publication No. 287, Delft Hydraulics.

de Vries, M., Klassen, G.J., Struiksma, N., 1990, "On the use of Movable-Bed Models for River Problems: A State of the Art", Int.J.Sed.Res.,Vol. 5, No. 1, pp 35 - 47.

Karim, M.F., Holly, F.M., 1986, "Armouring and Sorting Simulation in Alluvial Rivers", Journal of Hydraulic Engineering, Vol.112, No.8, pp 705-715.

McDowell-Boyer,L.M., Hunt,J.R, Sitar,N., 1986, "Particle Transport through Porous Media", Water Resour.Res., Vol. 22, No. 13, pp 1901- 1921.

Parker, G., 1991, "Surface-based transport relation for gravel river", J.Hydr.Res., Vol. 28, No. 4, pp 417-436.

Paintal, A.S, 1971, "A Stochastic Model of Bed LoadTransport", J.Hydr.Res., No. 4, pp 527-554.

Pazis, G.C., Graf, W.H., 1977, "Weak Sediment Transport", J.Hydr.Div., Vol. 103, No. HY7, pp 799-802.

Skoglund, M., (1991), "Vannstrøm over tettningskjernen og sandfilter i stenfyllings-dam",Rapport STF-60 A91089, SINTEF, NHL, Norge

Tien,C., Payatakes, A.C., 1979, "Advances in Deep Bed Filtration", J.American Inst. Chemical Eng., Vol. 25, No.5, pp 737-759.

Wittman, 1981, "Die analytische ermittlung der durchläßigkeit rolliger erdstoffe unter besoderer berücksichtigung des nichtlinearen widerstandsgesetzes der porenströmung", Veröff. Institut für Boden und Felsmechanik, Universität Karlsruhe, 87.

Wörman, A., 1991, "Interfacial Sedient Transport", Bulletin No. TRITA-VBI-152, Royal Inst. of Techn., Stockholm, Sweden.

Wörman, A., Olafsdottir, H., 1992, "Erosion in a Granular Medium Interface", Journal of Hydraulic Reserch, In press.

Wörman, A., 1992, "Incipient Motion during Static Armoring", Journal of Hydraulic Engineering, Vol. 118, No. 3.

Hydropower'92, Broch & Lysne (eds) © 1992 Balkema, Rotterdam. ISBN 90 5410 054 0

Overtopping of the morain core rockfill dam, state of flow and permeability

Øivind Solvik
SINTEF NHL, Trondheim, Norway

ABSTRACT: The paper deals with hydraulic flow problems coming up if the morain core in rockfill dams is overtopped. The state of flow, being laminar, turbulent as in a transition state is discussed. Discharge calculation methods are presented and like wise permeability tests with the intention of relating permeability coeffisients to the grading curve of the fill.

1 INTRODUCTION

In spite of the fact that Norwegian rockfill dams are not designed to take water over the top of the morain core, such situations should be considered as exceptional load. In this case the state of flow over the core will vary from being completely laminar through the sand filter, completely turbulent in crown protection and probably in a transition state between laminar and turbulent in the intermediate zone. Both the level and the thickness of the different dam zones are often available on new dams as settlement meters usually are installed. Also grain size distribution curves may excist and thus serve as a basis for the estimation of permeability coefficients. If no such informations are, it should be required to take up samples, may be in combination with the use of a georadar.

2 THE STATE FO FLOW

The rate of flow through a horisontal permeable layer on top of the morain core can be calculated with sufficient accurrancy by solving a simple differensial equation.

If the flow is laminar the equation for the spesific flow rate is:

$$q_1 = (k_1/2L) \ H^2 \qquad /1/$$

If the flow is turbulent a simular procedure would lead to the following equation

$$q_t = (k_t/3L)^{1/2} \ H^{3/2} \qquad /2/$$

$$\text{LAMINAR} \quad v = k_l \cdot \frac{dy}{dx} , \ q_l = \frac{k_l}{2L} \cdot H^2$$

$$\text{TURBULENT:} \ v^2 = k_t \cdot \frac{dy}{dx} , \ q_t = \left(\frac{k_t}{3 \cdot L}\right)^{1/2} \cdot H^{3/2}$$

Figure 1. Simplified model for discharge calculations.

Formula /1/ and /2/ are developed for a single layer. They can easily be applied to a multy-layer situation as shown on Figure 1:

$$q_2 = k_1 / (2L_1) \cdot (H_2{}^2 - (H_2 - H_1)^2) + (k_t/3L_2)^{1/2} (H_2 - H_1)^{3/2}$$
$$\text{assumed laminar} \qquad \text{assumed turbulent}$$

For fine graded sand the flow is always laminar. Even for coarse sand the flow will still be laminar in such a situation when the hydraulic gradient usually is mild. The state of flow should be expressed by:

$$i = v/k_1 \qquad\qquad /3/$$

If the gross velocity amounts to some cm/sec the flow will be partly turbulent following this equation:

$$i = v/k_1 + v^2/k_t \qquad /4/$$

Additional increased velocity and grain size will lead to complete turbulent flow:

$$i = v^2/k_t \qquad\qquad /5/$$

Figure 2 shows th different formulas used in fine grained sand and Figure 3 shows the same situation in coarser material. In this case it is safe to use the turbulent flow equation or the combined. Assuming laminar flow in such a case means overestimating the dishcarge and underestimating the dimensions of the area occupied by the water. A drainage system will in such a case consequently be designed with a too small capacity.

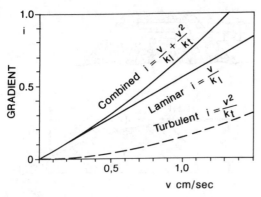

Figure 2.

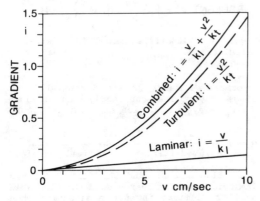

Figure 3.

If the turbulent flow formula is used in doubtful cases the Reynold's number $Re = vd/v$ should be checked.

Here: i = hydraulic gradient
v = gross velocity
k_1 = permeability if laminar flow
k_t = permeability if turbulent flow
v = kinematic viscosity

To secure a discharge error less than 10%, Re should not be less than 600 when turbulent flow is assumed. If Re is less than 600, the calculation should be carried out using the combined formula or the turbulent permeabiltiy should be corrected using this permeability:

$$k_c = v \cdot k_1 \cdot k_t/(v \cdot k_1 + k_t) \qquad /6/$$

The flow through the core, as well as through the sand filter is allways laminar and follows Darcy's law. Even for large dams the rate of flow through the core is negligible with regard to the stability of the dam toe because the permeability is low. For the sand filter, the permeability is much higher but still low enough to secure laminar flow and the rate of flow over the coretop is also now negligible with regard to the stability of the dam toe.

The flow through the transition zone is more problematic as the state of flow often is in the transition between laminar and turbulent flow. It is recommended to calculate according to (5), assuming turbulent flow and than check the Reynolds number and if necessay repeat the calculation with a corrected permeability k_c (6) which is equal to the use of equation (4).

The flow through the supporting fill, and the crown of the dam is allways turbulent and equation (5) should consequently be used.

3 THE PERMEABILITY COEFFICIENTS

For the core and sandfilter the permeability coefficients have usually been found by laboratory tests. Since the discharge through these zones is negligible with respect to the stability of the dam toe, accurate values of the permeability coefficents are therefore not required.

For the transition zone the coefficients may be calculated based on grain size distribution curves as permeability tests are rare for this zone. We have found that d_{10} can be used for a safe calculation of the permeability coefficients. This usually leads to values on the safe side especially if the sieve curve is flat between d_{10} and d_0.

The basic formulas for the permeability coefficients are:

$$k_1 = 1/\alpha_0 \cdot (n^2/(1-n)^3) \cdot g \cdot d^2_1/v \qquad /7/$$

$$k_t = 1/\beta_0 \cdot (n^3/1-n) \; g \; d_t \qquad /8/$$

α_0 = grain shape coefficient for laminar flow (= 1600 for crushed rock)
β_0 = grain shape coefficient for turbulent flow (3,6 for crushed rock)
n = porosity
d = grain size
g = acceleration of gravity
v = kinematic viscosity
μ = d_1/d_t

We have emphasized that exact knowledge of the permeability coefficient is less important for the fine grained material including the sandfilter. And the reason for that is that this discharge in any case will be small.

As a consequence of this we have paid more attention to the permeability of the transition zone and the support fill. The idea of the research has been to use the grain size distribution curve in a safe way for the calculation of the determining d_t to be used in formula (8).

We have carried out a lot of permeability tests with different sieve distribution curves, but we have not reach to a final conclusion regarding how to use the curves for the determination of the determining grain size. Three methods will be shortly described in the following:

Method 1: This is the simplest way of calculating the determining grain size to be used in the calculation. The method is based on one point d_n on the sieve curve multiplied by a factor x:

$$d_t = x \cdot d_n$$

The test results are summarized in the following table:

d_n	x-value	Relative error
10	1,76	7,02
20	1,28	5,96
30	0,95	6,75

It seems that using d_{20} is some better than using d_{10} multiplied by a factor x:. This conclusion may be wrong. The most influencing part of the sieve-curve is the lowest one, below d_{10} and d_{20}. A steep sieve-curve below d_{10} is more adequate for this method and will give more reliable results than a mild sloping curve because it is the finest grains that influence the permeability most. To be on the safe side regarding

discharge calculation one should assume a steep curve below d_{10} (or d_{20}).

Method 2: Based on what has been mentioned above we have worked out a method taking into account the entire sieve-curve. The main idea is that the method should put most weighting on the lowest part of the sieve-curve which is the most decisive for the rate of flow.
Figure 4 shows how the method works.

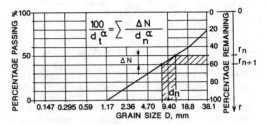

Figure 4.

$$100/d_t^\alpha = \Sigma\Delta n/d_n^\alpha$$

d_f = determining grain size
Δn = 10%, adequate intervall for calculation
d_n = corresponding grain size from the sieve curve
α = exponent to be determined by the tests.

The tests showed very little α-sensitivity and the following table shows the results from 12 sieve-curves:

α-value	Relative error
1,25	6,49
1,50	6,43
1,75	6,51
2.00	6,79

It seems that α = 1,50 is the best value, but it is obvious that the α-value is not most decisive. The lowest part of the sieve-curve where reliable informations are hard to get is always the most important.

Method 3. Finally we have introduced a third method involving not only a fixed point d_{n+1} on the sieve-curve but also the slope, here defined by $1/(d_{n+1}-d_n)$

The formula for the determining grain size d_t is:

443

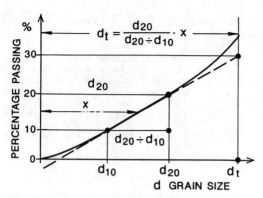

Figure 5.

used in the drainage system. That will also bring the sieve-curve more in line with the sieve-curves that the tests are based on.

$d_t = d_{n+1}/(d_{n+1}-d_n) \cdot x$ (see Figure 5)

Here we have to introduce a auxiliary factor x which has a length dimension. This factor x has been found to vary less with the fixed point d_{n+1} on the curve, but much more with the slope. Figure 6 shows how x varies with the slope of the grade curve and with the fixed point d_{d+1}. The conclusion is that using $d_{n+1} = d_{20}$ gave an average error of 4,61% while d_{30} gave 10,66% relative error.

We have so far not concluded the permeability tests. One could say that all methods offer reasonable results if used with common sence. The objection is that the lower part of the grading curve is most decisive and thats is the part of the curve which is hardest to have sufficient information about. In discharge calculation we therefore recommend to calculate on the safe side by assuming a steep slop below d_{10}. In dimensioning a drainage system this will lead to underdimensioning of the system. In such a case on should be on the safe side by removing all fine grained material to be

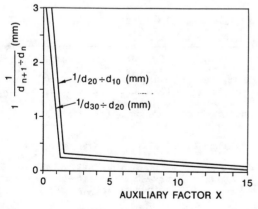

Figure 6.

444

Hydropower'92, Broch & Lysne (eds) © 1992 Balkema, Rotterdam. ISBN 90 5410 054 0

Steel fibre reinforced shotcrete and concrete in hydraulic structures

Marc Vandewalle
NV Bekaert SA, Zwevegem, Belgium

1. INTRODUCTION

Millions of cubic metres of concrete are used in the construction of dams and their appurtenant structures.

In many of these structures, mass or weight is the primary characteristic needed.

However, there are also critical areas in many hydraulic structures which do not require large masses of concrete but do require extremely high quality and strength.

The concrete used in discharge elements of dams is subjected to waterborne forces of cavitation, erosion and impact not normally experienced by structural grade concrete.

These elements include the agree cost of a spillway, the spillway itself, the spillway apron, slinceways or outlet works, guide walls. High velocity water passages are spillways and tough slinceways or outlets can cause cavitation and erosion of the concrete. In areas of lesser velocity water movement, such as the spillway aprons and stilling basins designed for energy dissipation, erosion and impact can be a problem. Impact from waterborne objects can damage exposed concrete.

Cavitation : occurs when vapour bubbles that have formed in the water flow enter an area of higher pressure. These bubbles then collapse with a great impact, and for a short interval of time, the concrete adjacent to the bubble is subjected to tremendous compressive and tensile stresses.

The concrete will then fail locally, with small pieces breaking out of the concrete and being carried off by the flowing water, the irregular surface that results aggravates the problem and more cavitation then results at an increasing rate.

Erosion : is the actual wearing of the concrete surface due to the abrasive action of the water and the particles that it is carrying. The particle size varies from sands upwards to big boulders. Other may-made debris also contribute to the problem.

Impact : loading can result from larger particles of submerged debris bounding along the bottom of a submerged element, from floating debris in fast-moving water.

Impact loading also occurs from severe wave action as such breakwater armour units as tetrapods or dolosse.

2. STEEL FIBRE REINFORCED CONCRETE (SFRC)

Concrete has always been a relatively brittle material. No serious effort has been made in the past to quantify impact resistance and toughness. However, there are often critical properties which can be primary cause of failure in concrete subjected to severe cavitation/erosion. These properties are markedly improved with the use of steel fibre reinforcement which probably is the reason why, in many field applications, SFRC has performed better than conventional concrete of higher compressive strength.

The material properties most affected by steel fibres are cavitation/erosion resistance, toughness, strain-capacity, impact resistance and fatigue strength. It is important to understand the basic reasons why fibres alter these properties, so that the material's behaviour can be better understood.

For an ideal SFRC mixture, each internal microcrack and flow is reinforced with a fibre near the crack tip.

Under stress and without the presence of fibres, the microcracks propagate with load or stress application until an ultimate failure occurs in the form of total separation of the concrete mass into two or more pieces. The energy required is low, usually only about half that required to initiate one.

With fibres present, the energy that otherwise would propagate is distributed through bond to the cement matrix around it. More energy is required to propagate the crack, and when it does, sufficient energy must be applied to pull the fibre out of the surrounding matrix.

Even after the ultimate load is reached, the fibres will continue to hold the broken fragments of concrete together and carry some load.

Deformed steel fibres with improved anchorage perform much better requiring much more energy to be pulled out. These fibres, however, must have a high tensile strength to avoid they should be stretched to ultimate load which would transform SFRC again in a brittle material.

The fibre contents of practical mixes used in dam work normally constitutes about 0.5 (40 kg/m3) to 1.2 (100 kg/m3) percent of the concrete volume.

Typical SFRC requires at least 0.5 % by volume of steel fibre content in order to gain beneficial effects, while concentrations in excess of about 1.2 % may cause workability problems.

Just as the different types have different characteristics, concrete made with these fibres will have different properties.

Steel fibres have little or no effect on compressive strength. The flexural beam or slab test, however, will clearly show how the fibre proportion of the concrete is performing.

A brittle fibre, although with good bond characteristics, will produce concrete with a few toughness, impact resistance and strain capacity.

Steel wire fibres that have no end anchorage will pull out and have a constantly decreasing load-carrying capacity with an increasing deflection soon after "first crack" is reached.

Steel wire fibres with high strength and end anchorage may show an increasing load-bearing capability even after the first crack point has been reached. After the ultimate load peak, there is only a gradual drop of the load-bearing capacity with continued deflection.

The resulting toughness and impact resistance will be completely different for these several fibre types.

The more efficient fibres made with high-strength steel wire and bent ends, develop higher impact resistance and other properties at substantially lower fibre concentrations.

When steel fibres with high strength deformed wire are used, the fibre dosage can be reduced by 25 to 50 percent compared to lower quality fibres, without loss of hardened material properties.

3. APPLICATIONS

SFRC has been used in various features of dam construction since about 1970.

SFRC has proven to be a field-adaptable and useful material in critical areas of new construction and in repairs or modifications to existing structures. The satisfactory performance of new constructions and repairs in areas where conventional concrete has previously failed has given the designers confid confidence in using SFRC in critical areas as a part of original construction.

SFRC has a highly improved impact and abrasion resistance compared with plain concrete.

The most useful research concerning the applicability of SFRC to areas in large dams which could be subjected to high-velocity flows and cavitation forces was conducted at the Detroit Dam (USA) high-head test facility.

These tests clearly showed that under true field conditions SFRC slabs resist the damage of cavitation/erosion approximately three times better than high-quality conventional concrete made with the same aggregates and mixture ingredients.

SFRC has been used to :
* Increase cavitation resistance of spillways and outlets
* Enhance corrosion resistance and impact strength of spillway aprons, plunge pools and stilling basins.

1. S p i l l w a y   a p r o n s

The spillway apron begins at the toe of the spillway and extends downstream to the stilling basin.

The water moving over the apron is generally not at velocities which would cause cavitation damage.

Large boulders and cobbles along with man-made debris can collect, however, in eddy currents over the apron, and through continual rolling and bouncing along the bottom, erode the concrete away.

2. T u n n e l s

2.1. SHOTCRETE

Shotcrete is a well suited method of placing SFRC. Practical applications have been completed on rock slopes for stabilisation and in hydroelectric power plants for tunnel and powerhouse linings.

In tunnelling the excavated underground uses to be stabilised using shotcreting and/or anchor bolts together with other reinforcing techniques. Depending on rock quality, the shotcrete should be reinforced.

A standard shotcrete reinforcement used to be a wire mesh of 4 to 8 mm diameter wires and 100 mm to 150 mm wide meshes. This reinforcing technique however, takes a lot of time and does not permit a rationalisation. Due to the constantly rising labour cost, reinforcing takes more and more a higher part of the shotcreting process cost.

Mesh installing is a highly time consuming operation and increases the total construction time, which is of vital importance in tunnelling.

Using conventional reinforcement, voids can be created behind the steel mesh, which in the worst case can act as a drain-pipe.

Filling the voids behind the mesh in draped-over areas takes extra shotcrete, more than what is required by the minimum specified thickness. It may happen even that the mesh remains uncovered.

Improvements in flexural stength, ductility and toughness, addition of steel fibres the shotcrete mix, are sufficient to enable steel fibres to be used as a replacement for steel mesh reinforcement.

Steel fibre reinforced shotcrete can be applied with conventional equipment, considering the steel fibres as one more aggregate.

Total construction time in big tunnel projects can be reduced by approximately 10-15%. Reducing shotcreting time reduces the whole operation period for all other tunnelling machines and consequently reduces the total hire cost.

A robot or spray arm is of considerable advantage, especially in tunnel constructions as it does not only replace the work of the nozzle operator but it is also capable of utilizing the full capacity of the machine.

The sequence movements for spraying can even be programmed.

This new technology makes it possible not only to drive the tunnel by machines but also to carry out the work by the machine outside the danger area.

2.2. SEGMENTAL LINING

For the construction of several hydraulic tunnels, steel fibre reinforced concrete has been used for the manufacturing of precast elements for the preliminary stabilizing tunnel lining.

Dramix® steel fibre reinforced concrete segments enable supporting and lining to be performed in our operation immediately behind the face.

The system has mainly been used in underground conditions where relatively important ground movements were expected.

The usage of steel fibre reinforced precast elements has proven following advantages as compared with other types of reinforcement:

* Crack resistance and/or crack arrestor

The elements are usually manufactured in a plant, built at the job site. The moulds are used more times per day with SFRC.

After demoulding, the concrete strength is still relatively low, although it has to resist several tensile stresses :

--> caused by its dead weight during handling;

--> caused by impact due to shocks;

--> caused by temperature changes (day-night difference, sunshine etc.) giving thermal stresses.

* Toughness :

This tunnelling technique is mostly used in ground giving big deformations. The ductile behaviour of SFRC is a very big advantage, as it allows a high bearing capacity even at important deformations.

3. Breakwater armour units

In the selection of an ideal material for breakwater armour units, the physical properties that should rank high are density, strength, toughness, resistance to impact and abrasion, and resistance to deterioration in a marine environment.

SFRS appears to possess these properties.

Eureka, California (USA), two jetties which protect the entrance of Humboldt Bay have dolosse placed on them for protection from the 12 m weaves which are prevalent during the winter season.

A dolosse is similar to the letter "H" with one leg rotated 90 degrees.

The impact forces of the wave action on an unreinforced dolosse can cause the concrete in the fluke to crack and with repeated loadings cause the crack to propagate until the fluke falls off, thus reducing the effectiveness of the dolosse. Even with conventional reinforcement, the crack can extend to the reinforcement where the seawater and air can then attach the steel. In time, the steel will be corroded to a point where it will fail under the impact loading, and the fluke can again fall off.

SFRC improves the first crack strength and impact resistance of the concrete and, thus, should prolong the life of the dolosse compared with plain and even conventionally reinforced concrete elements.

A visual inspection of the 40 tons dolosse after two winter seasons revealed no damage to the SFRC dolosse (50 kg/m3 Dramix® ZC 50/.50). Some of the unreinforced dolosse on the jetties have fai-

led after the first season.
In November 1979 surgebreakers were placed at Basin Bayon State Park in Florida
as an erosion control device to rebuild
the beach.
A surgebreaker module is a prism shaped
with a 4 ft x 6 ft base and is 4ft high
at the apex.
The two end surfaces are parallel with
the two sloping sides at different angles with respect to the base.
Each unit weighs about 4000 pounds.
The surgebreaker erosion control system
acts like an offshore reef.
The surgebreakers were precast in
Valparaiso, Florida and shipped to
the beach where they were placed
offshore by helicopter.
Steel fibres (40 kg/m3 Dramix® ZC
50/.50) were added to the concrete mix
to give the necessary reinforcement, in
addition to the rebars in the ends, and
to increase flexural strength and
resistance to impact, cracking, abrasion
and corrosion.

Hydropower'92, Broch & Lysne (eds) © 1992 Balkema, Rotterdam. ISBN 90 5410 054 0

Introduction to multi-orifice discharge conduits in Xiaolangdi Project

T.Xiang
Xiaolangdi Project, Design Institute of Yellow River Conservancy Commission, Zhengzhou, People's Republic of China

J.M.Cai
Hydraulic Engineering Department, TsingHua University, Beijing, People's Republic of China

G.C.Chai
Nanjing Hydraulic Research Institute, People's Republic of China

ABSTRACT: As an energy dissipator, the orifice plate has never been used in any large scale hydraulic project over the world . China is now proposing to adopt this measure in Xiaolangdi Project on the Yellow River . In this paper are introduced the research work accomplished in China on the hydraulics concerning the energy dissipation of orifice plate and the layout of the middle gate chamber appropriate to orifice plate dissipators.

1. General Introduction

Xiaolangdi Project , located on the main stream of the Yellow River , is the only one controlling project which can provide a large
storage at the downstream reach of Sanmenxia Dam . The main purpose of the project is expected to control floods ,prevent ice ,and to reduce sediments at the downstream, with water supply ,irrigation , power generation as well as storing clear water and releasing sediment-laden water as its secondary functions . The project is now in construction.

The dam in Xiaolangdi Project is of the rockfill type with a sloping clay core, the maximum height of which is 154m . The flood-discharge structures, which consist of nine tunnels, one chute spillway and an emergency spillway , are concentrated in or on the left abutment . The total capacity of the discharge structures amounts to 20,000CMS . The nine tunnels are: three open-flow discharge tunnels at different elevations ; three discharge tunnels with a series of closely spaced orifices as energy dissipators; three sediment sluice tunnels . Besides these ,there are six power tunnels which lead to the underground power station with a total capacity 1,800MW(6x300MW) .

The three tunnels with orifice energy dissipators are reconstructed from diversion tunnels , having a total head 130-140m and a diameter 14.5M (Fig. 1). For each tunnel , behind the intake , there is a vertical bend (elbow) followed by three orifices to dissipate energy step by step, which reduce the inner-pressure and velocity consequently. A middle gate chamber is located behind the orifice plates to control the releases, keeping the pressure in the energy dissipating chambers high enough . The tunnel behind the gate chamber is of open-flow type.

To ensure the safety and reliability of the orifice plates under the condition of releasing floods, a great deal of research work has been accomplished in many research institutes and universities. The characteristics and behaviour of energy dissipation ,cavitation and structure vibration of the orifice plates , together with the flow pattern in the middle gate chamber and its structure were studied, obtaining a large amount of result which affords scientific basis for the design of orifice dissipators . Further research work is now being carried on .

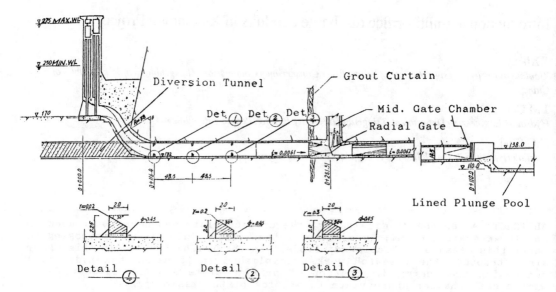

Fig.1. Xiaolangdi Project Orifice Tunnel Profile

2.Research Work on Orifice Dissipation

In the design , the three orifice plates in each tunnel are installed in the upstream pressure section at 3xD intervals (3x14.5M).The three orifice plates,each 2.0M thick,have various Beta values,namely,0.69, 0.724,0.724 from upstream to downstream respectively, where Beta is the ratio of orifice diameter d to tunnel diameter D . The orifice edge angle is 30 degree as shown in Fig. 1. The total head dissipated is found to be as high as 50 to 60 meters.

The orifice gives the flow a sudden contraction and then a sudden enlargement, resulting in a high head-loss. Flow-field data obtained by 2-D LDV have revealed the fact in detail: the high velocity gradient and intense shear between the submerged jet issued from the orifice and the quasi-stable backflow region is the main source of turbulence. Associated with the subsequent convection and diffusion, a large amount of mechanical energy is dissipated. Consequently , the values of pressure head and average velocity are lowered . This is precisely the result we expected to obtain by reconstructing the three

diversion tunnels with orifice plates , which can solve the problem arisen from the facts that the elevation of the tunnel lines is too low and the tunnels have to release silt-laden flow.

Basing upon a large amount of research work , the shape of orifice plates shown in Fig. 1 were adopted.It is found that the badly-concerned head-loss ccoefficient is dependent upon the thickness, diameter ratio , shape of the inner edge , and the combination or arrangement of the three orifice plates . The head dissipated is written as : $h=k\ V_1^2\ /2g$, where V_1 is the average velocity of orifice , and k is the head-loss coefficient.

Fig. 2 shows the relation between k and Beta when the radius of inner edge r is equal to 0.02m(r/D=0.0014).K equals to 1.0 and 0.96 when Beta=0.69 and 0.724 ,respectively. Hence , the head dissipated by each orifice amounts to about 20 meters, while the average velocity is lowered to 9-10M/S .Enlarging r will result in an obvious decrease of k , for example , when r is enlarged from 0.02M to 0.4M, k will decrease from 0.96 to 0.60 (diameter ratio:

450

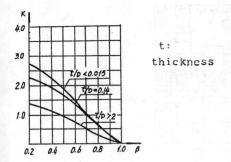

Fig.2 Head-loss coefficient vs Beta
(single Orifice plate)

0.724). This shows the great influence of inner edge shape of orifice plate on head-loss coefficient . The data obtained by 2-D LDV showed that , with the same shape and size of orifices the flow-field between neighbouring chambers are different to some degree . Generally speaking , the length of back-flow region behind the first orifice is shorter than that behind the second one . Though there is some difference in head-loss coefficient among the three orifices, it is found that there is little difference between the average coefficient of the three orifices and the coefficient of a single orifice plate (see Fig. 3).

As the Yellow River is a highly silt-laden river and the average silt content of the flow during floods is 50Kg/m³ , another important study focused on how the variety of silt content can affect the energy dissipation. Experiments have been conducted to compare the head-loss

coefficients in clear water and those in silt-laden water of different silt contents. The comparison showed that the head-loss coefficients in clear water and those in silt-laden water were nearly the same , when Reynolds number was greater than 100,000-10,000,000(see Fig. 4).

Cavitation usually occurs behind an orifice plate,even though the pressure line at that position is above zero. In all the cavitation tests conducted by native researchers, cavitation was observed . Hence ,care must be taken not to enlarge the energy dissipated arbitrarily ; and the outlet area of the middle-gate chambers should be selected appropriately to control pressure behind the last (third) orifice plate .

Surely,cavitation in a multi-orifice conduit is quite complicate . Because of the sudden change of boundary geometry, it is possible that there will occur some eddies stochastically in some corners . Since these eddies are difficult to detect in a time-mean pressure experiment , sufficient margin should be left for any measure of preventing cavitation . Further research work on improving the boundary geometry of orifice plates is also being carried on .

Besides the conflict between cavitation and dissipation , the vibration of two abutments and ridges possibly induced by the vibration of orifice plates and by the great energy of turbulent flow, is concerned by engineers. Dynamic

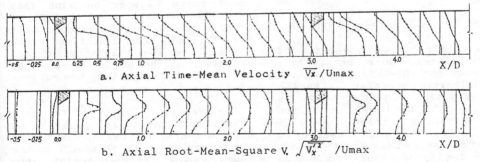

a. Axial Time-Mean Velocity $\overline{V_x}$/Umax

b. Axial Root-Mean-Square V. $\sqrt{\overline{V_x'^2}}$ /Umax

Fig.3 Velocity Distribution in 2-orifice plates chamber with LDV.
(Umax is the maximum velocity downstream the orifice)

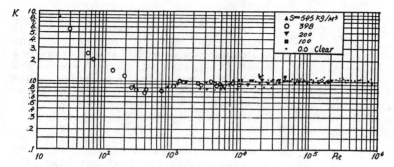

Fig.4 Head-loss Coefficient k vs Re number
(clear water and various silt-content water)

analyses and experiments revealed that the ring structure was strong enough and , under the fluctuating pressure, vibrated weakly; no resonance was detected among the orifice plates, tunnel lining and surrounding rock .

The multi-orifice conduits in Xiaolangdi Project will be the first engineering application in the world , no example can be studied and referred . To verify the laboratory experiments, sponsored by the Ministry of Water Resources , two orifice plates of the same type and at the same interval (3xD) as in Xiaolangdi were installed in the pressure section of a sediment sluice tunnel (4.4M Dia.) in Bikou Power Station.

Characteristics and behaviour of dissipation ,cavitation ,fluctuation and, static and dynamic response of structure were detected and measured. A lot of valuable information was obtained , which agreed with that of laboratory experiments mentioned above.

3.Layout of the Middle-gate Chamber

To ensure the normal operation of orifice plates under pressure , downstream the three orifice plates a middle-gate chamber with two acentric-hinge radial gates is built, followed by the open flow section . After careful selections , the gate chamber is located downstream the grout curtain of the dam foundation; the access tunnel are designed to work as

drainage galleries of the dam . Thus , the pressure section with high inner pressure will lie in the region of high external pressure ; the gate chamber will stand near the drainage system where the external pressure is lower ; and the open-flow section of each tunnel will lie in low external pressure region . Consequently, the forces exerted on the whole structure of each tunnel will be minimized, which will greatly reduce the consumption of concrete and steel . This layout scheme is believed to be economical and rational.

The outlet area of the gate chamber is an important parameter which controls the pressure in the orifice chambers . As mentioned above , the outlet will not only keep the pressure lines above zero , but also adjust the pressure lines and velocity distribution, so as to prevent the orifice chambers from cavitation. Cavitation tests recommended $52.0M^2$ as the appropriate outlet area . The double-gate scheme will reduce the total force exerted on each gate under 140.0M water-head to half of that in a single-gate scheme; as a result , this scheme will abate the difficulties in manufacturing and operation . The outlet area $52.0M^2$ will be distributed into two outlets of 4.8M x 5.4M each .

The average velocity of the jets issued from the outlets will be as high as 33M/S. To prevent cavitation and ensure good ventilation , at the downstream of the each outlet , the following details have been

452

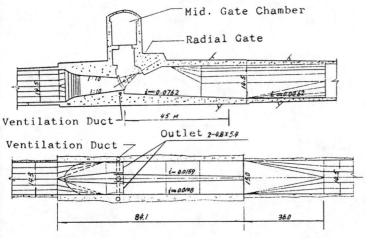

Fig.5 Middle Gate Chamber in an Oorifice Tunnel.

designed : a sudden widening of 0.5M on both sides ; a downward drop at the floor connected with a 7.62% slope of 45M longitudinal length at the downstream. As we expected, under different reservoir water levels , the jets will drop on the slope , and the ventilation ducts beneath the floor drop won't be submerged by backflow eddies . Beyond the sudden widening of the outlets , the water way will be further enlarged by flaring the sidewalls and the middle pier,so as to minimize the negative pressure on walls, and prevent them from cavitation damage. Hydraulic experiments and cavitation tests varified the rationality of the design (see Fig. 5).

We have given some information about the design and research work of the multi-orifice conduits in Xiaolangdi Project . However, because of its importance issued from dense population and developed economy of the province where the project stands , further research is being carried on , even though the construction of the project has begun .

REFERENCES

Ding,Z.Y., Cai,J.M. & Zhan ,K.X. 1988. A Study of Multi-orifice Energy Dissipators in Outlet Works.Proceedings of the 6th Congress Asia and Pacific Regional Division International Association for Hydraulic Research, Kyoto,JAPAN: July 20 - 22, 1988.

Zhang, LongRong & Wang, GuoDong, 1985. Energy Dissipation of Clear Water and Mud Flows in Multistage Orifice Dissipator.YELLOW RIVER, June,1985.

Hsu,S.T.,Ding,Z.Y., & Cai,J.M.,1988. Headloss Characteristics of Closely Spaced Orifices for Energy Dissipation. Proceedings of International Symposium on Hydraulic for High Dams (Organized by CHES and Cosponsored by IAHR), Beijing ,CHINA. Nov. 15-18 ,1988.

Shen,Xion,Yan,Y.Y. & Gao Jiansheng,1988. Turbulent Velocity Measurements in Orifice Pipe Flow with an Improved on-axis 2-D LDV System. Proceeding of the 4th International Symposium on Application of Laser Anemometry to Flow Mechanics,Lisbon.

Monitoring of the Urugua-í dam

Sara M. Yadarola & César H. Zarazaga
Inconas S.R.L., Professional Engineering Consulting Services, Córdoba, Argentina

ABSTRACT: The Urugua-í dam is located in the state of Misiones, Argentine Republic. It is a gravity dam, of roller compacted concrete with a low content of cement. Its maximum height is 80 m. and the crest length is 687 m. The seepage control through the dam is performed by a 2 mm thick continuous PVC membrane abutted to concrete precast panels. Behind the membrane there is a conventional concrete wall with a variable width. The use of a PVC lining to make the upstream of a dam of this height impervious, is a new experience, therefore it is considered very interesting to learn about its behaviour after 2 years of operation. The fundamental objective of this paper is to inform the results of monitoring performed with the instruments installed. The seepage, the uplift, the internal pressures, the deformations and the displacements are analyzed and submitted.

1. INTRODUCTION

The Urugua-í dam, built on the creek of the same name, tributary of the Higher Paraná River, is located in the province of Misiones in the Argentine Republic, approximately 1000 km N-NE from the Capital City (Buenos Aires). In the area of the site, the average daily temperatures in summer and winter are 25°C and 16°C, respectively, the characteristic isotherm of the zone being 20°C.

It is a gravity dam with a maximum height of 80 m, built with roller compacted concrete (RCC), the first dam in Argentina of its type.

The main impervious element of the dam consists in a continuous PVC membrane, 2 mm thick, a method that so far has not been used in dams of a height similar to this development.

It is for this reason that is has been considered interesting to publish the results of the monitoring performed during the first two years it has operated.

312 instruments were placed in the concrete dam, selected according to the parameter that it was wished to measure, that provide information on its behaviour in general and in especially chosen points.

2. DESCRIPTION OF THE DAM

The Urugua-í dam is a gravity structure with a straight axis, built with Roller Compacted Concrete (RCC), and is the first dam of this type in Argentina. The purpose of this dam is to produce electricity for provincial consumption. The installed capacity is of 120 MW. Its maximum height above the foundation is 80 m and the length of the crest is 687 m. The upstream face is vertical and the downstream slope is graded with an average inclination of 1V:0,8H. The spillway is built into the dam and flows freely. The total volume of concrete is 677.000 m^3 and the reservoir capacity is of 1200 hm^3. Figure N° 1 shows a transversal profile of the dam in its highest area.

The upstream face is formed by 5,05 m wide, 2,20 m high and 0,10 m thick premolded panels of reinforced concrete, that work as a lost formwork and are connected to the concrete by anchor bars. On the inside these panels have an embedded continuous PVC membrane 2 mm thick, that is connected to the foundation rock by means of a concrete block placed in a small trench opened especially for this purpose. This membrane was designed as the main impervious element of the dam.

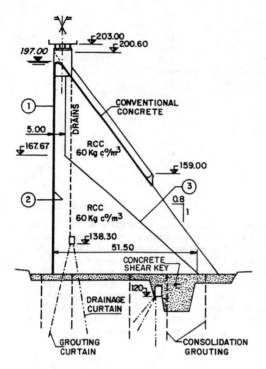

Fig.1 Main dam. Spillway section

(1) Premolded concrete panels with continuous PVC membrane
(2) Conventional concrete facing. Variable thickness
(3) Limit area with systematic bedding
mix treatment between layers

Nevertheless, bearing in mind the lack of
experience in this type of solution for
dams of this height, an area of conventional concrete was built behind the membrane with varying thicknesses between
0,90 m in the lower part and 0,50 m in
the upper part. In this area the concrete
has a content of 220 kg of cement per cubic meter, with contraction joints every
15 or 20 m, which have double water stops
and further on a drain that discharges
into the inspection gallery.

The mass of the dam is formed by 7 roller compacted concrete blocks (RCC) with
a content of 60 kg/m³ of cement, placed
in layers of 0,40 m thickness and compacted with a smooth vibrating roller.
These blocks are protected by transversal
joints in which polyethylene sheets have
been slightly inserted during construction. In the upstream area, with a minimum width of 5 m, the horizontal joints
have been treated with a bedding mix. The

width of the treated joints has been calculated in terms of the sliding stability
and at the same time to contribute to im-
perviousness.

To build the grades of the downstream
face, metallic molds were used.

The dam has an inspection and drainage
gallery 555 m long, 2,00 m wide and 3,20
m high, that was built by means of a gravel fill that was removed afterwards. Besides, the internal drains discharge in
this gallery.

With the object of lowering hydrostatic
pressures that might occur between the
PVC membrane and the conventional concrete wall, a horizontal drain was built
behind the membrane at level 143,80
throughout the length of the dam at that
level. The drain is built of crushed rock
and has a slotted pipe with two outlets
towards the gallery.

To ensure the resistance to sliding in
a certain part of the rock foundation, a
conventional concrete shear key was
built, with a gallery and drainage boreholes that lower the uplift (Figure N° 1).

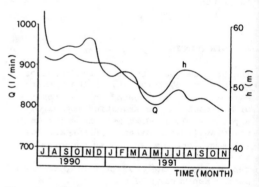

Fig. 2 Seepage through the foundation

3. SEEPAGE

3.1 Seepage through the foundation

Seepage through the foundation is measured by means of the existing drains,
distributed as follows: 78 drains in the
concrete shear key, 33 in the diversion
block, 43 at the toe of the dam and 193
in the drainage curtain the flows of
which discharge in the inspection gallery. In total there are 347 drains.

The flows discharged by these drains
are shown in Figure N° 2. The values for
the month of November 1991 are the following: at the concrete shear key 205

l/min., at the diversion block 501
l/min., at the toe of the dam 2 l/min.
and at the drainage curtain 79 l/min.,
amounting to a total seepage of 787 l/min.

The relation between the flow discharged by the drains and the corresponding hydrostatic loads produced by the reservoir has been approximately constant
during the last months, although the reservoir level decreased in around 6 m and
later started to rise again.

Considering that the seepage through
the foundation is now stable in relation
to the hydrostatic load and that the maximum values measured are relatively low,
it can be assumed that the behaviour of
the dam as regards seepage through the
foundation is normal.

3.2 Seepage through the dam

The loss of water through the concrete
dam has been classified in the following
manner:

3.2.1 Seepage through the PVC membrane

Seepage detected behind the PVC membrane
that is gathered by the horizontal drain
installed at level 143,80 with two outlets to the inspection gallery, has displayed an irregular behaviour from the
start that does not respond to the variations in the hydrostatic load and that
increased constantly even when the reservoir remained at an approximately constant level. The maximum flow registered
was 2932 l/min in April 1991, with the
reservoir at level 192. The relation between the flow and the corresponding hydrostatic load was 30 in the month of July, 1990 and increased to 59 in May 1991.
During the last months it has remained at
values of approximately 57. In November
1991 a flow of 2849 l/min was registered,
and the relation with the hydrostatic
load was 57,58 (Figure N° 3).

Presuming that the membrane was cracked
in the area corresponding to the horizontal drain, an inspection was performed by
divers in the area corresponding to the
drain, but no problems were detected. It
is possible that the inflow of water is
produced in the area where the drain is
covered by the embankment or by the joint
between the membrane and the foundation
rock or the abutment, therefore the investigations shall continue.

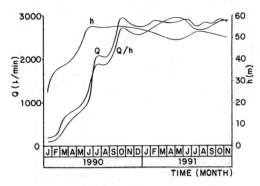

Fig. 3 Seepage through the membrane

3.2.2 Seepage through the contraction joints

All the contraction joints in the conventional concrete area have a double lock
with water stops plus a drain that discharges into the inspection gallery. Of
these joints, the one with the largest
flow is N° 15 that registered 568 l/min
in November 1991, with the reservoir at
level 192,48.

The total loss through the joints, under the abovementioned conditions in the
reservoir, amounted to 874 l/min.

To solve the problem of these joints, a
grouting treatment has been foreseen, but
cooling of the concrete has not finished
and in the meantime the temperature is
checked frequently.

3.2.3 Seepage through the concrete

The seepage that appears through the concrete in the inspection gallery or
through the internal drains amounted to
202 l/min in November 1991.

3.2.4 Total seepage throughout the dam

Losses due to seepage throughout the dam,
measured in the inspection gallery,
amounted to 3925 l/min. in November 1991,
as detailed below (Figure N° 4).

Losses collected by the membrane drain	2849 l/min.
Losses through the joints	874 l/min.
Seepage through concrete	202 l/min.
Total	3925 l/min.

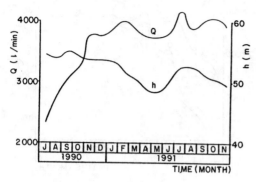

Fig. 4 Total dam seepage

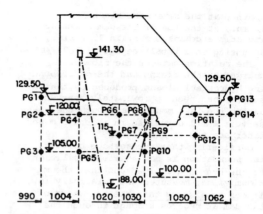

Fig. 5 Transversal profile G. Foundation piezometers

3.2.5 Total Losses

Adding up all the losses measured, the following result is obtained:

Losses through the foundation   787 l/min.
Losses through the concrete
dam   3925 l/min.
Total   4712 l/min.

From these values it is deduced that 17% of the losses are produced by the foundation and 83% by the dam. In turn, the membrane drain contributes 73% of the seepage that enters into the inspection gallery.

As a conclusion it can be stated that the losses through the foundation can be considered normal and the main problems related to seepage are located in the PVC membrane and in joint N° 15. The membrane drain problem continues under investigation to detect the places where the water enters into the drain. As regards the joints, whose water stops have defects, they shall be grouted with cement or polyethylene foam once the concrete has finished its cooling process.

On the other hand, the fact that this dam was designed and built with a double protection line against seepage: the PVC membrane and an area of conventional concrete behind the membrane, must be borne in mind.

For the time being, taking into account that although the seepage rate is high, it does not imply an important economic value, nor is there any risk regarding the safety of the dam, the PVC membrane is considered effective, therefore it continues to perform as the principal impervious barrier avoiding the transfer of the total hydrostatic load directly to the conventional concrete area.

4. UPLIFT

Uplift in the foundation is measured by means of vibrating wire piezometers installed in three transversal sections. Figure N° 5 illustrates the location of the instruments in the section corresponding to the concrete shear key.

Readings were performed for all instruments as from the date of their installation, with variable frequencies according to their location and the period considered (construction, first fill and normal operation).

The pressures registered were analyzed performing graphic representations in terms of time, compared with the levels of the reservoir and with other piezometers installed at the same level. Besides, they were compared with the uplift calculated for the dam project.

The curtain drains discharge into the inspection gallery and when the uplift does not reach this level, the drains act as piezometers, measuring the water table. The efficiency foreseen in the project of the drains has been complied with in all cases.

The piezometers located along the line of the upstream face, register approximately the level of the reservoir, the following piezometers are influenced by the drainage curtain and further back, by the drains of the concrete shear key, that maintain the pressure around level 120, while pumping is performed. If pumping is interrupted, uplift may rise up to level 134 as a maximum, this being the level of the pumping well head.

Analyzing the values of pressures registered when the reservoir was at a le-

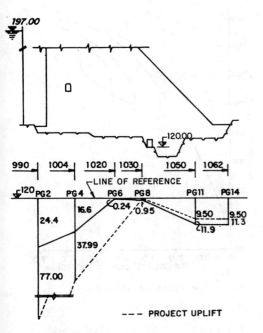

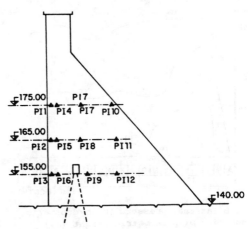

Fig. 7 Transversal profile I. Internal piezometers

Fig. 6 Uplift diagram

vel higher than that of the crest of the spillway, the following results were obtained. In Figure N° 6, between piezometers 2 and 4 located upstream and downstream of the grout curtain, the uplift decreased in 34%. In piezometers 6 and 8 near the concrete shear key, the decrease reached 99% and 96% respectively, due the effect of the drains discharging at level 120.

Later the reservoir commenced to decrease and towards end November it was at level 192,50.

In all cases the uplifts measured were lower than those foreseen in the project.

In the other two sections in the foundation checked with piezometers, the same analyses were performed with similar results.

5. INTERNAL PRESSURES

In order to analyze the behavior of the area of conventional concrete, 24 vibrating wire piezometers were placed, distributed in two transversal sections.

Figure N° 7 shows the location of the instruments in Section I where the highest values were registered.

The first line of piezometers was located 0,20 from the upstream face and the second at 0,50 m, all of them within the conventional concrete. The remainder are distributed in the RCC mass.

When the level of the reservoir was above the spillway crest, the pressures registered at the level where piezometers PI-3 and PI-6 were installed, read 20,0 m and 16,16 m respectively, and the corresponding hydrostatic load of the reservoir was 42,50 m. Consequently, the pressure at piezometer PI-3 was 48% of the load and PI-6 was 38%, signifying an important decrease of pressure between these two piezometers that are spaced by only 0,30 m. Piezometers PI-9 and PI-12 located downstream of the line of drains, registered very reduced pressures that barely represented 2% and 1% of the hydrostatic load. Consequently, in the rear area of the dam, pressure is practically nil.

The rest of the piezometers installed inside the dam, registered pressures lower than those mentioned above.

The results obtained indicate that the water of the reservoir is acting behind the membrane, exerting pressure on the conventional concrete, but in a reduced manner. Besides, these pressures disappear after the line of drains.

In Figure N° 8 the pressures measured at the piezometers of section I at level 155 are illustrated, as related to the reservoir loads.

6. TRANSVERSAL DISPLACEMENTS

Transversal displacements in the dam are measured with two direct and two inverted

459

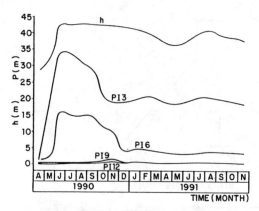

Fig. 8 Internal pressure (h: hydrostatic load (m) - p: piezometric height (m))

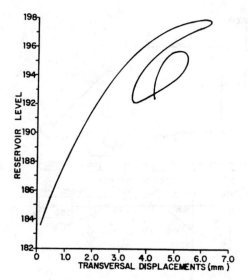

Fig. 9 Transversal displacements

pendulums installed inside the dam and also by means of a geodesic triangulation linked to the reference points located in the crest and in the downstream slope.

The maximum displacement measured at the crest was of 6,3 mm with the reservoir at the level of the spillway crest. In Figure N° 9 the displacements in terms of the hydrostatic load are illustrated. These displacements are considered normal according with deformation analysis previously performed.

7. INTERNAL DEFORMATIONS

Deformations produced inside the dam are measured with extensometers of the vibrating wire type installed in the concrete mass, especially in the contraction joints. In the joints visible from the inspection gallery, triaxial mechanical extensometers are placed.

In the area of roller compacted concrete (RCC) polyethylene sheets were slightly inserted in the joints and in these joints extensometers with long bases were installed (L = 3,00 m). In the majority of the cases the maximum deformations registered were from 2 to 4 mm, but in joint EJ-5 the deformation reached 7 mm.

Although some of the deformations measured implied important gaps in the joints, the neighboring piezometers indicated moderate pressures and the flows discharged by the internal drains were very reduced, thus it can be assumed that there are no important flows in these gaps.

In the area of the diversion block (Figure N° 10), that was built during the

first stage, 8 long base extensometers were installed (L = 3,00 m) in correspondence with the construction joints but 8 m further up, within the RCC that was built later. Of all the extensometers installed, only 2 registered important deformations (ED-2A and ED-4A) (Figure N° 11).

As these extensometers are not precisely in the joints but in the concrete mass, the specific deformations have been taken in mm/m in order to compare them with the elongation capacity obtained by means of laboratory tests carried out on RCC samples of 60 Kg of cement per m³. The maximum deformations measured with these extensometers were 0,7 and 0,8 mm/m, these values are much higher than those for admissible deformations according to laboratory tests, which leads to presume that some crack has been produced.

Careful inspections were performed on all the surface of the downstream slope, but no crack has been detected. Besides a finite element model was produced to quantify the deformations and stresses, under the hypothesis that a crack existed, but it was concluded that in the state of hydrostatic load plus self weight no tensions are produced in the area and that the relative displacements tend to close the supposedly existing crack.

Other possibilities of failures were also considered, but the conclusion is always that if the crack does exist it will not spread.

Although the stability of the dam is

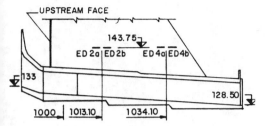

Fig. 10 Diversion block section

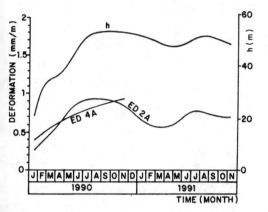

Fig. 11 Internal deformations

ensured, careful observation of deformations and uplifts continues, thus assuring correct drainage of the area and its
foundation.

8. CONCLUSIONS

During 1990, the spillway of the dam operated during approximately 6 months, thus
the dam has been submitted to maximum
normal load conditions. After this the
reservoir decreased but has been kept at
high levels. This circumstance has made
it possible to compare the main significative parameters of behaviour of the dam
(pressures, deformations, seepage, etc.)
under similar load conditions.

The information obtained from the monitoring instruments has been conveniently
processed and analyzed, allowing the following conclusions to be reached.

8.1 Seepage

Total seepage through the foundation (787
l/min in the month of November, 1991) is

considered acceptable for this type of
development and are stable in terms of
the hydrostatic load produced by the reservoir. Consequently, as regards seepage
through the foundation, the behaviour of
the dam is normal.

Seepage through the dam, on the above
mentioned date, totalized 3925 l/min, 73%
of this value corresponded to the membrane drain.

This distribution of seepages indicates
that the main problem is located in the
membrane or in its connection with the
foundation, thus investigations continue
in order to detect the places where water
enters the drain.

Although the seepage flow is high, it
does not have an important economic value
nor does it imply risk to the safety of
the dam, therefore the membrane continues
to be operated as the main impervious
barrier, without transferring the total
hydrostatic load directly to the area of
conventional concrete.

8.2 Uplifts

Uplifts measured in the foundation, with
a full reservoir, are lower than those
foreseen in the project for the stability
calculations, thus the results are satisfactory.

8.3 Internal pressures

Piezometers installed in the conventional
concrete, immediately behind the PVC membrane, registered moderate pressures that
are considerably reduced towards the inside. The remaining piezometers located
downstream of the line of drains indicate
almost nil pressures. Consequently, the
internal pressures of the dam are considered satisfactory.

8.4 Transversal displacements

Transversal displacements in the crest of
the dam, measured with pendules, are in
agreement with the geodesic measurements
and are considered normal for this dam.

8.5 Internal deformations

Of the deformations measured inside the
dam, those registered by two extensometers installed in the diversion block are
outstanding due to their magnitude. Nevertheless, although careful investiga-

461

tions have been performed, it has not been possible to detect cracks and the stability analyses do not indicate danger of failure, thus the stability of the dam is ensured. In spite of this, although the deformation has maintained a constant value during more than 6 months, close observation of all instruments placed in that area is continued.

Hydropower'92, Broch & Lysne (eds) © 1992 Balkema, Rotterdam. ISBN 90 5410 054 0

DASMOS and its application to Longyangxia Hydropower Project

Zhuang Wenzhong
Northwest China Hydropower Design Institute, Xian, People's Republic of China

ABSTRACT: DASMOS is a new comprehensive, highly user-oriented, general purpose dam-safety and environmental information processing system. It integrates a very convenient and natural man-machine communication interfaces, data generation and correction facilities, data interpretation procedures and tools, interactive dam-safety analysis and assessment techniques, graphics and tabular output and decision-making features into a single system. It is currently being set up at Longyangxia Hydropower Project and is frequently being run to perform synthetic information necessary and throuth multi-approach intuition, statistical and determinate models, as well as calling up and comparing to field-inspection records, helping the operator to follow the behavior of monitored structure, during the initial impounding stage.

The DASMOS System

DASMOS is an integrated dam safety monitoring system for managing and processing data on the structure behavior of monitored dams, obtained by means of appropriate instrumentation and field inspection, helping the dam operators and experts to evaluate the state of the structure and to follow its evolution closely. It is a new comprehensive, highly user-oriented, general purpose dam-safety information processing system. The DASMOS software system integrates a very convenient and natural man-machine communication interfaces, data generation and correction facilities, data interpretation procedures and tools, interactive dam-safety analysis and assessment techniques, graphics and tabular output and decision-making features into a single system. No prior knowledge of computers, computer operation, or computer programming is required in order to effectively use it.

The software package includes different modules and subsystems for pre-processing, storage, analysis, post-processing and display of the Results, all of which can be thus called up automatically by means of menu-driving mode to respond to all operator demands. The Menu-General, shaped as a menu-tree, is divided into eight subsystems and 52 modules. It mainly includes:

● Inquiry and consultation subsystem (2.1) for inquiry and display of dam safety volume, the design feature of project, the layout and summary of instrumentation, sensor installation parameters and their characterastic values.

● Two input and pre-processing subsystems (2.2 and 2.3) by which the user can choose options for input of observation readings and field inspection records, automatic data generation and data error-checking and correcting. The data stored are coded according to a data base which combines the need for a reduced occupation of the system's memory with the rapid retrieval of information. The modules convert raw data into values corrected of temporary files and permanent archive files for further analysis and diagnosis.

● Data base management subsystem (2.4) for interactive editer, addition, update, retrieve, reorder and backup of data in the unified data base. The DASMOS not only manages specific processings provided for by its program library, but also makes it possible to transfer data files outside the system itself for further processing of any type.

● Two analysers (2.5) which are the post-processing subsystems and guide the operater in the choice of stress calculation, statistical and determinate model predications, multi-approach analysis and assessment. The statistical reference model is useful to understand the dam's structural behavior, the effects due to variations in environmental cause, and is still used when it is not possible to have all available data required for identifying the system. The analysers are also designed to train operators so that they can master the multi-approach concepts used and therefor be able to evaluate fully the significance of abnormal reading. The system is also designed specific graphic display features for some very important monitoring measurements, such as the opening of G4 fracture zone, the uplift distribution along the dam sections, the space displaycement distribution on the dam body, as to control its evolution directly.

● Output module (2.6) for graphic and tabular output of various information in display, printing and ploting modes. This makes possible for the correlation between raw or corrected readings from different sensors, the combination of some of the parameters measured , the ploting of various kinds of graphics in selecting periods and scales and the printing-out of all the quantities data in standard formats.

● Operating and management subsystem (2.8) for entering protection by passwords at different levels of the hierarchy, operating journal, on-line help facilities for user, system and data security facilities and virus diagnosis.

Application to Longyangxia

The Longyangxia concrete gravity-arch dam, in Northwest China, the first cascade on the upper reaches of Yellow River,was built between 1978 and 1989. Its first generating unit entered operation in 1987. The dam is 178 M high, 1226 M long with a reservoir of 24.7 billion cubic metres.The associated powerplant has an installed capacity of $4\times$ 320 MW. Due to the complicated geological condition at the damsite and foundation, the huge reservoir with a series of large landslides along the lake-banks near the dam, and high earthquake intensity, so the project is operated on the way of staged impouding, observing and analyzing while rising water lever, and the initial storage level is kept at maximum 2570 M, 30 M lower than normal storage level 2600 M.

Developed and supported by Northwest China Dam Safety Centre, the DASMOS is currently being set up at one of the largest dams in

China, Longyangxia Hydropower Project and controlled by desk-top computer, completely compatible with COMPAQ, HP, SUN 286/386 etc. The project has been instrumented from outset. In order to monitor and control the perfomence of the structure, including the stabilities, deformations, phreatic levels, uplifts and seepage volumes of the dam and abutment banks, we gave the emphasis on 8 pendulum wires in dam and foundation, 36 phreatic level monitoring holes, 7 uplift monitoring sections and deformation monitoring meters both inside and outside. Plenty of instrumentation readings and various inspection records have been put into the unified data base and will be gradually collected in years to come, including environmental observation, various displacement measurements, strain and stress measurements, seepage and phreatic level measurements and field inspection records etc.

The DASMOS system, however, is frequently being run to perform synthetic information necessary to describe the scope of data value, data developing regularity, tendeny, rate cause-result relation, correlations between data, quantified distribution in the space of dam body and foundation(see Fig.1, Fig.2). The aboved-processed information is valuable for operators and authorities throuth multi-approach intuition, statistical and determinate models, as well as calling u and comparing to field-inspection records, t interactively analyze, assess and follow the behavior of project during operation, especially during the initial impounding stage and to meet dam safety operating requirement

Conclusions

The DASMOS system frees the operator from tedious and repetitive tasks, which are

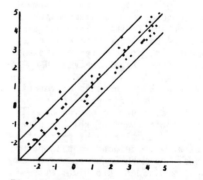

Fig. 1. Correlation between measured and computed displacements

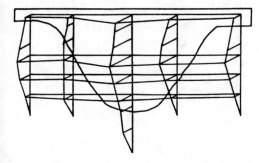

Fig. 2. Displacement quantified distribution at dam body and foundation

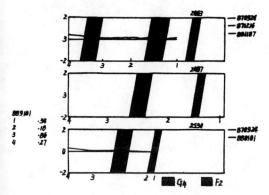

Fig. 3. The opening and closing of G4
and F2 fractures

still smaller. The extension values observed for the G4 fracture aproximated 1.0 mm
(see Fig. 3). Considering the normal condition with the dam and its foundation, the
maximum storage level in 1989 reached el.
2575.4 M two years ahead of time. Therefor,
the electricity power generation increased
3.688 TWh due to the extra amount of 7.5
billion cubic metres water storage valume
obtained.

REFERENCES

Ding, G.Q., Zhao, B.Z & Zhuang, W.Z. 1990.
    The dam safety monitoring and feedback
    Techniqus for high dams. A report to the
    national conference.
Zhuang, W.Z., Zhang, J.B. 1991. An integra-
    ted dam safety system for Longyangxia
    project. Jounal of water power of Long-
    yangxia.
Zhuang, W.Z. 1991. Studies and application
    on the criteria of dam safety monitoring.
    Hongzhou: National dam safety conference.

time-consuming and inevitably the sources
of errors. It can also manage the storage
of data automaticly and ensure more elaborate and reliable processing and provide
rapidly a synthetic interpretation and significant results, helping the dam operator
to evaluate the perfomance of the structure
and to follow its evolution closely.

During the staged impounding period, on the
basis of analysis and assessment of all the
quantities of existing data measured, it is
very important to cite the following results.
THe maximum radial displacement observed on
the arch dam crown at el. 2600 M is 16.28mm,
close to the values both from predication
model and laboratory test. The displacements
along the river flow and perpendicular to
flow at both banks were less than 2.0 mm.

The post-impounding settlements of abutment
rock have not been observed significantly
and were mostly less than 2.0 mm. The values
of uplift at the foundation sections were

4 Condition monitoring of hydropower systems

Hydropower'92, Broch & Lysne (eds) © 1992 Balkema, Rotterdam. ISBN 90 5410 054 0

Capacitive sensor technology: A key for better monitoring of hydropower generator

Jean-Marc Bourgeois
IREQ, Institut de Recherche d'Hydro-Québec, Que., Canada

Marc Bissonnette
Vibro-meter Inc., Canada

ABSTRACT: This paper describes the wide range of application of a new approach in capacitive sensor technology. Examples, based on data gathered from utilities around the world , show the benefit of this type of sensor particularly in the monitoring of hydrogenerators. The dynamic air-gap measurement system is the first commercially available product using this technology. The direct monitoring of stator bars vibration is also feasible with dedicated capacitive sensors. Many other monitoring problems could be addressed by this growing technology.

1 INTRODUCTION

Until the seventies, Hydro-Québec, was in a relatively comfortable situation. Economy was on the growing side. Many new installations promised a lot of cheap power. Sufficient power production was attainable with regular and simple maintenance activities. Occasionally, special diagnosis activities had to be undertaken to clarified obscure problems. The things are now very different. In the new economic context, every dam, every generator, every piece of equipment has to produce more efficiently with a maximum availability.

Hydro-Québec as many other utilities, is facing two different problems. On one hand, the aging of equipment and on the other hand, newly design generators; more efficient but more complex and less over designed. In spite of his wide experience in hydrogenerator operating and maintenance, the traditional ways of doing things is no longer suitable. To overcome those difficulties all utilities are now focusing on monitoring in order to reduce unexpected outages and optimize schedule downtimes. Well done monitoring relies in great part on the choice of significant parameters to be measured. Frequently this selection is upseted by the unavailability of suitable sensors for such parameters. In practice, a lot of indirect measurements followed by extensive data manipulation enable a fair evaluation of those parameters. In some special cases, sporadic measurements are required involving extraordinary equipment. Both solutions are inconsistent with continuous surveillance needs. To ovoid this dead end, Hydro-Québec like some other utilities, invested in research to improve the monitoring technics and to activate the development of new sensors.

2 NEW SENSORS DEVELOPMENT

In this way of thinking, Hydro-Québec's maintenance supervisors enjoined there Research Institute (IREQ) to do more R&D in the area of new sensors. As expert in electronic instrumentation we were rapidly involved in this new approach.

2.1 *Air-gap measurement*

Out of many items, air-gap appeared as a key parameter in the surveillance of hydrogenerators. Major stator to rotor rubs or severe vibration problems plead for this choice. However, the main difficulty had always been to deduce significant dynamic air-gap figures from manually taken static air-gap data. The first project we received target on the development of an electronic system apt to monitor the dynamic air-gap of any hydrogenerator in all operating modes. The objective was ambitious and during the progression unexpected difficulties were encountered. After 4 years of R&D, the goal was reached. We could meet all objectives of maintenance peoples with a totally new capacitive sensor technology.

The sensor itself is completely passive, immune against magnetic and electric interferences, made of non-metallic material, wafer thin, simply glued on the stator wall, easily installed in any types or sizes of hydrogenerators and linked to conditioning electronic unit through out a 10 meter long cable. The typical system is composed of 8 sensors, electronics units, a key phasor probe and an IBM PC or compatible computer to store and display data. Detail analysis could be obtained from data manipulations. Figure 1 presents a typical installation. Case study will show more in part 3.

2.2 *Stator bars vibration measurement*

This leading technology gives access to new area of application in the monitoring of hydrogenerators. The second example focuses on the crucial domain of integrity of stator winding insulation. Changes in insulation materials and new wedging technics have led to specific troubles. Looseness of bars in stator slot induces insulation wear until short. Slot discharges analysis was a first attempt to evaluate this phenomenon. Now, a more direct and permanent method can be incorporated to any monitoring system through the use of dedicated vibration capacitive sensors. A wedge-shape like sensor can be installed in selected stator slots and

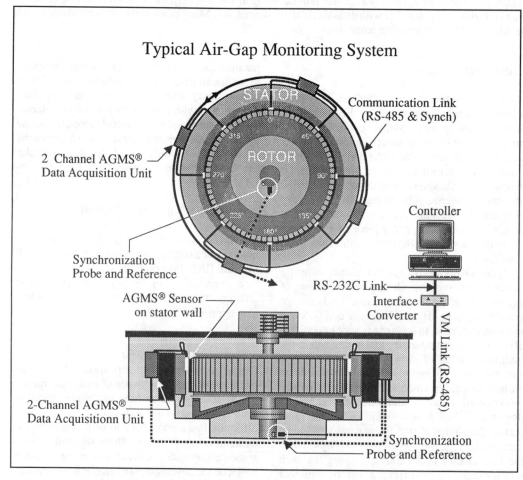

Typical Air-Gap Monitoring System

Figure 1: Typical installation, on a large hydrogenerator, of an 8 capacitive sensor Air Gap Monitoring System (AGMS®) as recommended by Vibro-Meter inc.

connected to a small electronic unit 10 meters apart. This arrangement gives direct reading of top bar radial vibrations. With this method one can track the evolution of absolute vibrations, evaluate the insulation wear and plan maintenance interventions. Section 3 will present some collected data

2.3 *Surveillance of shaft expansion*

The range of application of this capacitive technology grows beyond the domain of hydrogenerators monitoring. A third apparatus concerns the surveillance of shaft expansion during the thermal run-up of turbo-generators. In this case the capacitive sensor is use for its property of doing non-contact measurement on rotating parts. The output signal tracks the exact displacement of the coupling flange between the generator and the vapour turbine on the full expansion range. The signal is not affected by the the nature of the shaft material, the temperature, dust and oil deposits or humidity level.

2.4 *Future developments*

Continuous developments of this technology guarantee other applications. Specific works are done on sensors, electronic units or in the integration to monitoring systems. New sensor shapes, new material capable of withstanding higher temperature, operation into oil or other mediums with larger frequency response authorise new parameters surveillance. Capacitive technics and other technics can be combine to give access to new on-line parameters monitoring. As an example, one can merge data of instantaneous position of a selected pole (top of the generator) with data of instantaneous position of a target on the shaft (bottom of the generator). By calculations one has access to the dynamic torsional warp of the shaft under different operating modes. Briefly, any particular installation or custom application that requires some of the unique features of capacitive sensors could be addressed to IREQ or Vibro-Meter's peoples to get attention.

3 CASE STUDY

We are specialists in electronic systems not in generator design nor maintenance. Most of the time, explanations of specific problems shown in this section, come from specialists and experimented users working with our capacitive sensors. Here is some real data showing the benefit of capacitive sensor technology in the monitoring of hydrogenerators.

3.1 *Dynamic Air-gap*

The first broad application of our capacitive sensor are in the field of air-gap monitoring. Access to dynamic air-gap measurements permits the direct observation and exact measurement of previously only estimated phenomenon. More, it opens the way to on-line monitoring for prime surveillance and better maintenance planning. Any experimented mechanical engineer would say that the dynamic air-gap is different from static air-gap. But who can predict the actual dynamic behaviour of a particular hydrogenerator based only on static on indirect datum. With an AGMS® system, the maintenance peoples can assess manufacturers design and track for emerging problems before the guarantee goes off.

Our first example concerns La Grande 3 generating station (LG3). In the early stage of operation some of the 200 MW hydrogenerators began to show abnormal structural vibrations. In some cases the level was so high that the stator mechanical retaining system began to wear. Unfortunately, static air-gap measurements was taken showing quite uniform air-gap values. The AGMS® system gave a much better look into the problem. Stator and rotor revealed ellipse like shapes. At stop, with pole #1 upstream, the two ellipse's large axises coincided explaining these good static values. However, in rotation the rotor large axis eventually coincide with the stator short axis giving small air-gap twice per revolution and inducing cyclic deformations of stator core. Figure 2 shows the shapes and air-gap values before and after the correction made by the manufacturer.

The second example shows the effect of a particular stator design. The phenomena is known as clover shape. The stator is built by linking together large pieces sometime preassembled and delivered on the site. The stiffness is different at assembling links than elsewhere inducing a non-uniform deformation on load. Figure 3 shows this particular stator shape. Accurate data of the dynamic stator shape allows efficient corrective procedures.

A third example is related to a specific rotor design. Shortly, we refer to a method where the

471

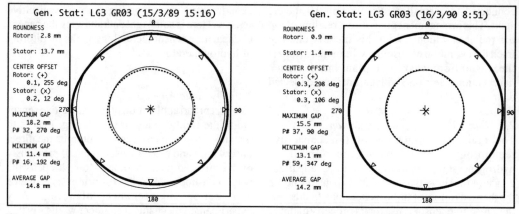

Figure 2: Rotor and stator shapes before and after the repair by the original manufacturer

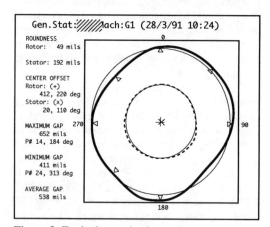

Figure 3: Typical stator's clover shape.

rotor rim is heated then shrink fit in its position and support in the radial direction by the spiders. In Rapides-des-Iles GS the rim was fit too loosely and "float" in respect to the spiders, inducing vibration in the spiders and cracking in hub and arm weld. Machine #4 was refurbished by the manufacturer "A" in 1979. The rim was tightly fit by adding shims at spider/rim interference. Figure 4 reveal the dynamic rotor shape after this corrective was applied. Machine 1, identical to machine #4, was also repaired by another manufacturer "B" with a different technic. The spider system was replaced by a solid metal plate leaving a better rotor shape as shown on figure 5. Clearly, benefit of dynamic air-gap measurements come from the possibility of assessing corrective procedures.

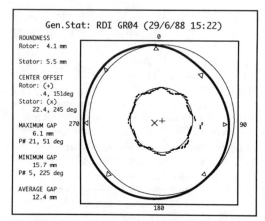

Figure 4: Rotor shape after repair by the original manufacturer "A". The position of the 8 spiders are clearly visible.

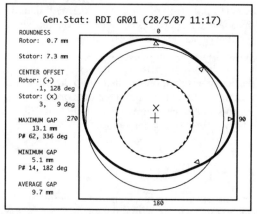

Figure 5: Rotor shape after repair by an other manufacturer "B".

472

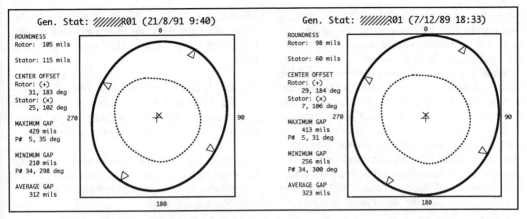

Figure 6: Stator shape displayed at extrema air-gap values. On left, smallest values near August-september. On right largest values around december-January.

The last example focuses on the benefit of continuous monitoring of dynamic air-gap as a mean to protect hydrogenerators from rotor to stator rub induced by seasonal variations of air-gap. These cycles are probably induced by concrete temperature variations. Figure 6 shows the stator shape at limit values. December exhibits an acceptable stator and August reveals an ellipse like stator shape leading too small air-gap. Any additional failures could conduct rapidly to a rotor to stator rub if the generator in not equipped with a monitoring system with alarm signals. In this case, egg shaped rotor causes stronger structural vibrations during August-september than elsewhere.

3.2 Stator bars vibrations

The second and newest development of our capacitive sensor technology applies to stator bars surveillance. Better insulation materials and evolving wedging methods authorise tighter generator design. But, many problems still exist. Now, capacitive sensors give access to direct readings of stator top bars radial vibrations. Many experienced engineers think that insulation wearing problems begin with wedging looseness. Under this condition a succession of interacting phenomena developed: vibrations, material packing, Faraday screen abrasion, slot discharges, insulation wearing and finally short to stator core. By controlling the vibration level over time, one can detect the process at early stages, track the evolution and schedule interventions before major damages occur.

Preliminary data, gather on hydrogenerators with prototype sensors and recently with commercial system, begin to reveal tendencies. In one specific case we have even established the benefit of the method. Refer to figure 7 for data summary. On the Bersimis generator #13 maintenance peoples suspected high vibration level of stator bars. Wedging system was considered poor but this generator wasn't on the maintenance schedule yet. Based on vibration values acquired on February 1989 (one of the bars was showing a steady 90 μm peak-peak displacement on load) this machine was stopped for a major inspection. Stator insulation was in so bad condition that they decided to proceed a conductive paint injection and a complete rewedging. Without those vibration

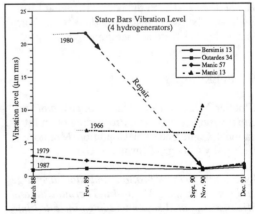

Figure 7: Summary of sporadic measurements on 4 hydrogenerators. Curves are tagged with there respective rewedging date.

473

values this generator would probably be running until a short with catastrophic results and higher repair cost. Vibration values taken on November 1990 shows a perfect behaviour.

Actually, only sporadic measurements have been taken as shown on Figure 7 but general patterns emerge. Recently rewedged generators present lower vibration levels than older ones (Outardes #34 being our reference). Too less data are now available to establish a reliable progression curve. Shortly, extended data collection and analysis will lead to a better knowledge of the phenomena. On-line monitoring of those vibrations will become a new way of tracking the stator insulation status.

4 CONCLUSION

This capacitive sensor technology is really innovative giving access to never seen parameters. It's one of the best choices in the evolving approach of on-line monitoring. With control over dynamic air-gap and stator bars vibration we cover a wide area of mechanical and electrical problems on hydrogenerators. Adding classical parameters such as temperature, power output, current, voltage, etc., gives a relatively simple and efficient monitoring system. The standard AGMS® system coupled with the ZOOM® (Zero Outage On-line Monitoring) system have already demonstrated there utility through customers' comments.

5 ACKNOWLEDGEMENT

The authors wishes to thank Mr. P. Ménard from Hydro-Québec and Mr R. Beaudoin from Vibro-Meters Inc. for there cooperation to this paper.

6 REFERENCES

Cloutier, M. et Al. 1987. *Statics and dynamic air-gap measurement system.* Proceeding of an International Conference On Condition Monitoring. University College of Swansea, U.K.

Lyles, J.F. 1975. *Vibration and slot discharge problems with Epoxy Mica-Mat, Thermalastic and Micarex insulated stator windings installed into hydraulic generators.* Technical Information Bulletin 75B-1, Ontario Hydro.

Ménard, P. et Al. 1988. *Monitoring system for AC generator stator bars vibration.* Fifty-fifth Annual International Conference of Doble clients. Boston, USA.

Ménard, P. and Bourgeois, J.M. 1989. *Using capacitive sensors for AC generator monitoring.* International Conference on Large High Voltage Electric Systems CIGRÉ, Montréal, Canada.

Harriman, F.B. 1985. *Considerations regarding rotor rim shrink on waterwheel generators.* Canadian Electrical Association, Hydraulic Power Section, Montréal, Canada.

Hüttner, H. and Al. 1986. *Quelques aspects des méthodes de diagnostic et de surveillance en exploitation pour les grands alternateurs.* international Conference on Large High Voltage Electric Systems CIGRÉ. Paris, France.

Lihach, N. 1984. *More Power from Hydro.* EPRI Journal. USA.

Ward, B.E. 1982. *Quality Evaluation Tests for Generator Stator Insulation.* B.C. Hydro Research and Development. CEA contract 76-22. Montréal, Canada.

Wayne, J.M. 1981. *Rotor rim stability.* Canadian General Electric. Canadian Electrical Association. Québec city, Canada.

Hydropower'92, Broch & Lysne (eds) © 1992 Balkema, Rotterdam. ISBN 90 5410 054 0

A review of modern turbine control systems for hydropower plants with high pressure tunnels

Hermod Brekke
Norwegian Institute of Technology, Trondheim, Norway

Li Xin Xin
SINTEF, Norway

Halvard Luraas
KVAERNER, Norway

ABSTRACT: The goal of this paper is to present modern methods for mathematical modeling and development of stabilizing systems for high head Hydroelectric power plants with long and complex high pressure tunnel systems or traditional tunnels and penstocks . The first part of the paper gives a very brief presentation of the Structure Matrix Method which is the basic tool for stability analysis in the frequency domain. The analysis includes a frictional damping term of the pressure oscillations as function of the frequency and the flow amplitudes. The influence of the turbine characteristics is also included. (Ref. 1.) The second part of the paper includes a review of different so called water column compensating systems compared with a recently developed direct pressure feedback system. For complex conduit systems the damping effect obtained by throttling of special selected shafts in the system has been proven to be a powerful tool and such system is also presented.

1 INTRODUCTION

Various methods have been developed in order to obtain an accurate analysis of the turbine governing for high head plants with long and complex conduit systems. During the last decade a mathematical model has been developed for analyses of turbine and conduit systems including nonlinear dynamic friction terms and the turbine characteristics presented in this paper. (Ref. 1). The accuracy of the presented theory opens up for detailed analyses of the influence from pressure compensation systems and local throttling of selected shafts leading down to the main tunnel in complex conduit systems. The latest development in pressure compensation system presented in this paper is different from some previously developed water column compensation systems. Comparative analyses with computer simulations of the stabilizing effects of different water column compensation systems will be given in this paper.

2 HYDROPOWER PLANT MODELING BY MEANS OF THE STRUCTURE MATRIX METHOD

The Structure Matrix Method (SMM) is related to the matrix method used for frame work and Finite Element Method (FEM) for solids. Due to the authors work (Ref. 1), the approach has been adopted and modified for mathematical modeling of oscillatory flow in conduit systems by means of a frequency domain complex plane analysis. Verification of this theory has been made by various frequency response tests in full scale at the power plants. The dynamic behavior of the elastic fluid in an elastic pipe (or tunnel) can be described by the Alleives equations with a friction term added:

$$\frac{\partial h}{\partial x} = \frac{Q_o}{gAH_o}\left(\frac{\partial q}{\partial t} + Kq\right) \tag{1}$$

$$\frac{\partial q}{\partial t} = \frac{gAH_o}{a^2 Q_o}\frac{\partial h}{\partial t} \tag{2}$$

where

$$h = \frac{\Delta H}{H_o} \quad \text{and} \quad q = \frac{\Delta Q}{Q_o}$$

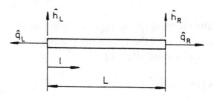

FIG. 1.

The element matrix for a tunnel or pipe derived according to the above equations can be written assuming the relative flow is positive out of the pipe in both ends and increasing pressure is positive in both ends.

$$\begin{bmatrix} \dfrac{-s}{2h_w\,z\,\tanh(Lz/a)} & \dfrac{s}{2h_w z\,\sinh(Lz/a)} \\[2ex] \dfrac{s}{2h_w\,z\,\sinh(Lz/a)} & \dfrac{-s}{2h_w z\,\tanh(Lz/a)} \end{bmatrix}$$

$$\begin{bmatrix} h_L \\ h_R \end{bmatrix} = \begin{bmatrix} q_L \\ q_R \end{bmatrix} \tag{3}$$

where

$$K = (\tau_s + \tau_d)\,\pi D/(\rho Q_o q) \tag{4}$$

and $z = \sqrt{s^2 + Ks}$

τ_s and τ_d: static and dynamic shearstresses respectively.

The dynamic shear stress for tunnels developed in (Ref. 1) has been based on a work by Jonsson on damping of sea waves on a rough sea bed (Ref. 2). The dynamic term yields;

$$\tau_d = 0.5\rho f_d C_m^2 = 0.5\rho f_d (Q_o|q|/A)^2 \tag{5}$$

a phase shift between the friction force and the shear force in an oscillatory flow has been proven both theoretically and experimentally. The friction force may then be expressed as follows for the actual range of frequencies and amplitudes in turbine governing:

$$f_d = |f_d|\cos(\pi/8) + i|f_d|\sin(\pi/8) \tag{6a}$$

then

$$K = (f_s Q_t + 0.5 f_d Q_o|q|(\cos(\pi/8) + i\sin(\pi/8)))\cdot\pi D/A^2 \tag{6b}$$

According to (Ref. 2) and converted to be valid for a tunnel cross section as shown (Ref. 1).

$$f_d = e^{-5.977 + 5.213(AK_r\omega/(Q_o|q|))^{0.194}} \tag{7}$$

This formula is valid for $Q_o|q|/(AK_r\omega) \geq 1.57$. For $Q_o|q|/(AK_r\omega) < 1.57$ the following modification has been found valid (Ref. 1)

$$f_d = 0.4725 AK_r\omega/(Q_o|q|) \tag{8}$$

Based on the frequency response tests in tunnels and pipes (Ref. 1), a fictitious roughness K_r can be determined by substituting in eq (7) for f_d the Darcy Weisbach friction factor f_s and for $Q_o|q|/\omega = 400$ i.e. for a very low frequency and a large amplitude. Then one gets:

$$K_r = (400/A)(\ln f_s + 5.977/5.213)^{5.155}$$
$$\text{(if } f_s > 0.00256) \tag{9}$$

further

$$K_r = 0 \qquad \text{(if } f_s \leq 0.00256) \tag{10}$$

Surge shafts and air accumulators:
The element matrix expression for a surge shaft or an air accumulator may be written as either a two-terminal element (Ref. 1):

$$\begin{bmatrix} 1 & 0 \\ 0 & -sH_o A_{eqv}/Q_o \end{bmatrix}\begin{bmatrix} h_R \\ h_L \end{bmatrix} = \begin{bmatrix} q_R \\ q_L \end{bmatrix} \tag{11}$$

For air accumulators the A_{eqv} may be found by following equation (see Ref. 1):

$$A_{eqv} = 1/(1/A_w + n\,H_{ao}/V_o) \tag{12}$$

where A_w denotes the water level area and n denotes the gas polytropic exponent. For an open surge shaft, $n = 0$ and $A_{eqv} = A_w$.
Local losses. The matrix for the damping caused by local variation in cross sections, for the velocity head subtraction at measuring points or for T joints of shafts may be established as follows:

$$\begin{bmatrix} -1/K_{re} & 1/K_{re} \\ 1/K_{re} & -1/K_{re} \end{bmatrix}\begin{bmatrix} h_L \\ h_R \end{bmatrix} = \begin{bmatrix} q_L \\ q_R \end{bmatrix} \tag{13}$$

For variation in cross section,

$$K_{re} = Q_o Q_t/(gH_o)\,\zeta\left(1/A_s^2 - 1/A_t^2\right)$$

For T joints

$$K_{re} = kQ_o^2 / \left(3gA_s^2 H_o\right)|q| + Q_t Q_o / \left(gA_t^2 H_o\right)$$
where k = geometry constant
A_t = cross section of the tunnel
A_s = min. or shaft cross section

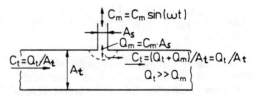

FIG. 2. A T joint.

This element gives a large contribution to the damping of surges in narrow shafts especially when the velocity in the main tunnel C_t, is large, see Fig. 2.

Representation of the turbine characteristics: The normal range of frequencies for the turbine governing stability analysis in the frequency domains will normally be between 0.002 rad/sec and 6.0 rad/sec. In this frequency range one may assume the turbine characteristic diagram for small amplitudes to be valid. Then a strict theoretical analysis as shown in (ref. 1) leads to the following general Element Matrix for all kinds of turbines.

$$
\begin{bmatrix}
-B & -C & B & -Q & 0 \\
J & L & -J & K & M \\
B & C & -B & Q & 0
\end{bmatrix}
\begin{bmatrix}
h_R \\ y \\ h_L \\ n \\ p
\end{bmatrix}
=
\begin{bmatrix}
q_R \\ 0 \\ q_L
\end{bmatrix}
\quad (14)
$$

Where:

$B = 0.5(1 - Q_n); Q = Q_n; C = Q_y + Q_\phi F_\phi$

$J = \left(3 - E_n - (1 + E_q)Q_n\right)\big/\left(2(1 + E_q)\right);$

$K = Q_n + E_n\big/(1 + E_q)$

$L = Q_y + Q_\phi F_\phi + (E_\phi F_\phi)\big/(1 + E_q)$

$M = -1\big/(1 + E_q)$

The characteristic parameters are defined as follows (referring to Fig. 3):

$Q_n = \left(*\underline{Q}\ \underline{n}_o\big/(\underline{Q}_o * \underline{n})\right)\left(d(\underline{Q}/*\underline{Q})/d(\underline{n}/*\underline{n})\right)_o$

$Q_y = \left(*\underline{Q}\ Y_o\big/(\underline{Q}_o * Y)\right)\left(d(\underline{Q}/*\underline{Q})/d(Y/*Y)\right)_o$

$Q_\phi = \left(*\underline{Q}\ \phi_o\big/(\underline{Q}_o * \phi)\right)\left(d(\underline{Q}/*\underline{Q})/d(\phi/*\phi)\right)_o$

$E_q = \left(*\eta Q_o \sqrt{*H}\big/(\eta_o \sqrt{H_o} * Q)\right)$

$\quad \left(d(\eta/*\eta)/d(\underline{Q}/*\underline{Q})\right)_o$

$E_n = \left(*\eta n_o \sqrt{*H}\big/(\eta_o \sqrt{H_o} * n)\right)$

$\quad \left(d(\eta/*\eta)/d(\underline{n}/*\underline{n})\right)_o$

$E_\phi = \left(*\eta \phi_o \sqrt{*H}\big/(\eta_o \sqrt{H_o} * \phi)\right)$

$\quad \left(d(\eta/*\eta)/d(\phi/*\phi)\right)_o$

$\underline{Q} = \dfrac{Q}{\sqrt{2gH}} \qquad \underline{n} = \dfrac{n}{\sqrt{2gH}}$

Here the asterisk denotes the best efficiency condition. For Francis turbines $Q_f = E_f = F_f = 0$ and for Pelton turbines $Q_n = Q_f = E_f = F_f = 0$.

[For Kaplan turbines Q_f, E_f and F_f will be different from zero.]

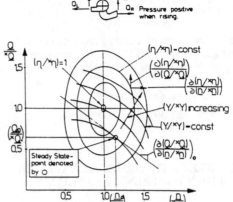

FIG. 3. Turbine characteristics diagram.

Turbine speed governors: The matrix for the most commonly used turbine governors yields:

$$
\begin{bmatrix}
D & E & F \\
0 & 1 & 0
\end{bmatrix}
\begin{bmatrix}
n \\ p_i \\ y
\end{bmatrix}
=
\begin{bmatrix}
n_{ref} \\ P_{ref}
\end{bmatrix}
\quad (15)
$$

Where

$\quad D = K_n(1 + T_n s)/(1 + 0.1 T_n s);$

$\quad E = -b - 1/G; \ F = b_p/C + 1/(GC);$

$\quad G = (1 + T_d s)/(b_t T_d s);$

$\quad C = 1/(1 + T_y s)(1 + T_x s)$

$\quad y \quad =$ guide vane position

$\quad n_{ref} =$ speed reference

$\quad P_{ref} =$ power reference setting

$\quad p_i \quad =$ actual power reference (different from p_{ref} if there is a water level controller).

Structure matrix of a turbine governor-generator unit:
A turbine-generator unit includes the turbine, the turbine speed governor and the generator. The Structure Matrix for the complete generating set may be formed as follows by 6 independent variables: h_R (1), n (2), p_i (3), y (4), p (5), h_l (6) (different orders may also be used). The Structure Matrix then may be formed accordingly (eq. 11). For turbines with a water level governor, a 7 x 7 Local Structure Matrix may be used (Ref. 1).

477

$$\begin{bmatrix} -B & -Q & 0 & -C & 0 & B \\ 0 & D & E & F & 0 & 0 \\ 0 & 0 & 1 & 0 & 0 & 0 \\ J & K & 0 & L & M-J & 0 \\ 0 & -T_as & 0 & 0 & 1 & 0 \\ B & Q & 0 & C & 0 & -B \end{bmatrix} \begin{bmatrix} h_R \\ n \\ P_i \\ y \\ P \\ h_L \end{bmatrix} = \begin{bmatrix} q_R \\ n_{ref} \\ P_{ref} \\ 0 \\ P_g \\ q_L \end{bmatrix}$$

$$(16)$$

The element matrix of a controlled valve or turbine guide vane with stroke = y, pressure and flow may be derived as in (ref. 1):

$$\begin{bmatrix} -B & -C & B \\ 0 & 1/K_q & 0 \\ B & C & -B \end{bmatrix} \begin{bmatrix} h_R \\ y \\ h_L \end{bmatrix} = \begin{bmatrix} q_R \\ y_{ref} \\ q_L \end{bmatrix}$$

3 EXAMPLE OF THEORETICAL ANALYSIS BASED ON THE PRESENTED THEORY

Comparison between the nonlinear and the traditional friction terms in an analysis of a short tunnel with a Pelton turbine connected.

By regarding a system with a single pipe from a reservoir with a valve or a Pelton turbine in the down stream end as shown in Fig. 4 the following matrix can be established. Note $Q_n = E_f = F_f = 0$. Further we assume $Q_y = K_q = 1$ in the valve matrix shown earlier

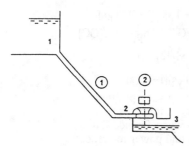

FIG. 4.

Further $h_1 = 0$ and $h_3 = 0$ because of constant pressure, then the matrix is reduced to

$$\begin{bmatrix} -0.5 - T_1 & -1 \\ 0 & 1 \end{bmatrix} \begin{bmatrix} h_2 \\ y \end{bmatrix} = \begin{bmatrix} 0 \\ y_{ref} \end{bmatrix} \qquad (17)$$

substituting for

$T_1 = s/(2h_w z \tanh(zL/a))$

$h/y_{ref} = -2h_w (z/s) \tanh(zL/a)/$

$\qquad (1 + h_w (z/s) \tanh(zL/a))$

Remembering

$z = (s^2 + Ks)^{0.5} = (-\omega^2 + iK\omega)^{0.5}$ for $s = i\omega$

$|h/y|_{min} = |-T_w K/(1 + 0.5T_w K)| \approx -T_w K$ (18)

for $\omega \to 0$

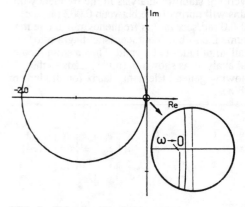

FIG. 5. Complex Plane Plot of h/q (linear damping).

Further $|h/y|_{max}$ occures for

$$\omega = ((1 + 2n)\,\pi a/(2L))\left(-K + (1 + K^2)^{0.5}\right)^{0.5}$$

where
$n = 1, 2, 3 \,........$ and then one gets if $K/\omega \ll 1.0$

$|h/y|_{max} \approx |-2.0/(1 + KL/(0.5ah_w))|$ (19)

Further min values occur for $\omega = n\pi a/L$ and $n = 1,2,3 \,...........$

$$|h/y|_{min} = \left|\left(\frac{-0.5\ T_w K}{(1 - ia/(2\pi L\ n))}\right)\Big/(1 + 0.5\ T_w K)\right|$$

$$\approx 0.5\ T_w K$$

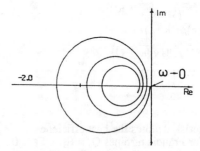

FIG. 6. Complex Plane Plot of h/q (nonlinear damping eqs (4), (5) ....(10)).

In Figs. 5 and 6 the complex plane plots of the computer simulations of the pressure response of a 1000 m long tunnel are shown. The cross section area of the tunnel is 5 m^2, the steady state flow is 10 m^3/sec and the Manning´s number = 40 i.e. a Darcy Weisbach steady state friction term f = 0.011. The roughness is equivalent to a normal relatively smooth blasted tunnel. By comparing the complex plane plot of the traditional theory with K = constant in Fig. 5 and the new theory used for the result presented in Fig. 6 one will find a significant difference especially for high values of ω. (Note $K_d = f(\omega, q, A)$).

4 FREQUENCY RESPONSE TEST AT SITE FOR VERIFICATION OF THE THEORY

Test at Skjaak power plant

The frequency response test at site was carried out at Skjaak power plant. The Skjaak power plant possesses a penstock of 2355 m with an average diameter of 900 mm. The power house is furnished with one Pelton turbine of 31.5 MW output. The nominal head is 638 m and the design flow is 5.6 m^3/sec. Two measurements were made during the test at low-load: $P_0 = 6.86$ MW, $H_0 = 679$, $Q_0 = 1.25$ m^3/s. $Y_0 = 17$ mm (the full needle stroke is 117 mm) and at high load with $P_0 = 26.13$ MW, $H_0 = 665.7$ m, $Q_0 = 4.44$ m^3/sec and $Y_0 = 76.5$ mm. This test and the computer simulations proved the consistency between the theory and the measurements as in eight other previous test (Ref. 1). In Fig. 8 and fig. 9 the results of the high load measurement and the low load measurement against the theoretical curves are shown respectively.

5 IMPROVING GOVERNING STABILITY BY WATER COLUMN COMPENSATOR SYSTEMS INCLUDING THE NEW PRESSURE FEEDBACK SYSTEM

It is well known that the anti-regulation power of the water column in the penstock has been a main problem for obtaining turbine governing stability in a hydropower plant, especially if the penstock of the plant is long. Several researchers have made efforts to develop so called water column compensators to in order obtain sufficient stability margins of the governing systems.

The compensator system can be such as schematically shown as in Fig. 10. Various transfer functions of the feedback compensator have been proposed: Araki and Kurabara compensator yields:

$$C(s) = \frac{T_w s}{(T_a s + F_n)(0.5 T_w s + 1)} \quad (20)$$

Murty and Hariharan (Ref. 4, 1984) modified the compensator:

$$C(s) = \frac{T_w s}{(T_a s + F_n)(0.33 T_r^2 s^2 + 0.5 T_w s + 1)} \quad (21)$$

Shen (Ref. 5, 1987) also proposed a compensator has the following transfer function

$$C(s) = \frac{T_w s}{(0.33 T_r^2 s^2 + 0.5 T_w s + 1)} \quad (22)$$

Lein (ref. 6) used measured pressure across the turbine as feedback signal, see Fig. 11. The transfer function chosen by Lein is

$$\Theta(s) = \frac{T_d s}{(T_H s + 1)} \quad (23)$$

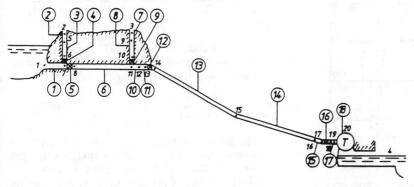

FIG. 7. Layout of Skjaak power plant.

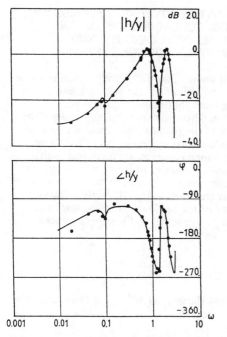

FIG. 8. Pressure response high load (Skjaak).

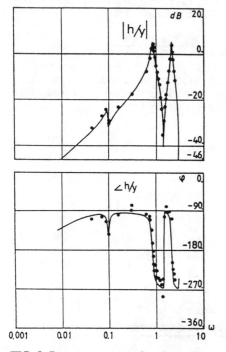

FIG. 9. Pressure response low load (Skjaak).

which is referred to as a DT1 (derivative one time constant) element. However, the time constant T_H is not clearly defined in his paper.

The authors of this paper have made an analysis of these compensator systems and some comments can be given.

In the approaches of Araki and/or Murty, a problem rises in case where Fn = 0, (this value was used in Murty´s analysis for deriving the stability limited curves). In such a case, the compensator will cause an unacceptable high permanent speed droop. The resulting permanent speed droop will be the ratio of T_w/T_a, which may be as big as 30%.

The approach of Shen or Lein does not have the problems mentioned above. However, according to the authors study, these approaches don't always improve the stability of the complete system. The principle of pressure feedback systems which in the authors opinion is the most advanced system, may be illustrated in the following simplified block diagram.

In the diagram, the notation f = 2 h_w tanh (Ls/a) is used. By assuming Qn = 0, the following n - n_{ref} open loop transfer function can be obtained.

$$\frac{n}{n_{ref}} = \frac{(T_d s + 1)(1 - \Theta)}{b_t T_d s\left(1 + 0.5\Theta\left(1 - 2\left((T_d s + 1)/T_d s\right)\Theta\right)\right)T_a s}$$

(24)

By introducing
$$0.5(1 - 2((T_d s + 1)/T_d s)\Theta) = b$$

we obtain

$$\frac{n}{n_{ref}} = \frac{(T_d s + 1)(1 - \Theta)}{b_t T_d s(1 + b\Theta)T_a s}$$

(25)

$$\frac{p}{y} = -\frac{1 - b}{2b} + \frac{(1 + b)}{2b}\frac{\left(1 - 2bh_w \tanh(Ls/a)\right)}{\left(1 + 2bh_w \tanh(Ls/a)\right)}$$

(26)

Lets now consider that if a DT1 element is used for Q as suggested by Lein and let $T_H = T_d$ and with an amplification factor k:

$$\Theta = \frac{kT_d s}{2(1 + T_d s)}$$

(27)

The factor b is then reduced to b = 0.5 (1-k) which becomes a constant. By studying eq. (26), one finds that if (27) is used, the frequency characteristics of p/y is a circle depending on the values of k, see Fig. 12.

In cases where the pressure shaft is not too

480

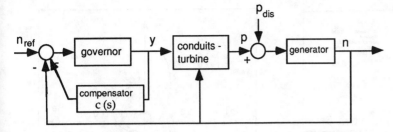

FIG. 10. Schematic diagram of a water column compensation system.

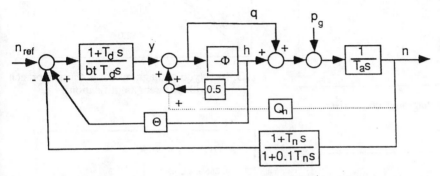

FIG. 11. Schematic diagram of a pressure feedback compensation system.

long, such a decreasing of the diameter of the circle often increase to some extent the stability margins. However, this approach may decrease stability margins if the pressure shaft is long. A typical long penstock power plant, Skjaak Power Plant, described earlier in this paper, is chosen to illustrate the problems.

The single Skjaak Pelton turbine (where $Q_n = 0$) is governed by a traditional PI regulator. By

means of computer simulation of the open loop frequency responses of the governing system with a DT1 (proposed by Lein) element as pressure feedback compensator, the stabilizing or

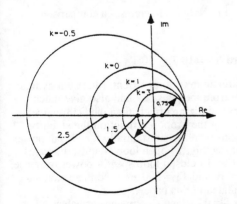

FIG 12. The DT1 element reduces the p/y circle as function of k (see eq. 27).

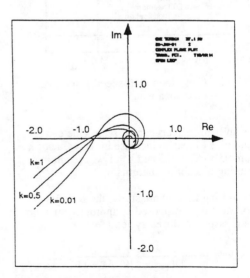

FIG 13. Nyquist diagrams for pressure with a DT1 element.

481

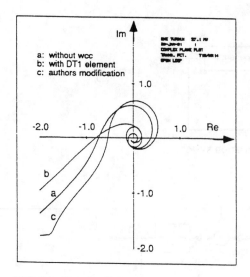

FIG. 14. Nyquist diagrams, comparison.

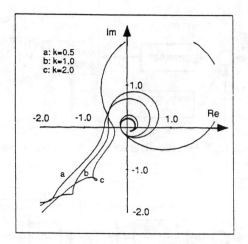

FIG. 16. Nyquist diagrams for authors approach
with different amplifications.

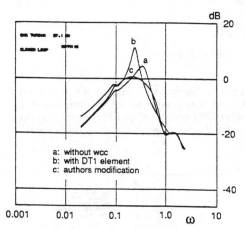

FIG. 15. Close loop load-speed responses,
- a comparison.

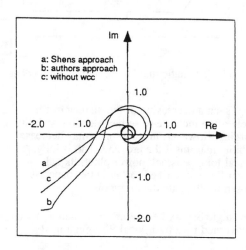

FIG. 17. Nyquist diagrams, - a comparison.

destabilizing effects can be seen. One can see
from Fig. 13 that an increase in the value of k
from 0.01 to k = 0.5 and k = 1 will result in decreasing of stability margins.

According the authors study, the following
modification is proposed as an improvement of
the pressure feedback system:

$$\Theta(s) = \frac{K\,T_1 s}{(T_1 s + 1)(T_d s + 1)(0.5 s + 1)} \qquad (28)$$

(K = 1 except for Fig. 16)

where $T_1 = 10\,T_d - 50\,T_d$

This function to some extent, filters out hydraulic
noise from the penstock and also give much
better stabilizing effects than the DT1 element.
By using the proposed function the stability
margins increase substantially for Skjaak power
plant, see Fig. 14 open loop Nyquist diagrams
and Fig. 15 closed loop speed - power response.
($T_d = 10$ and $T_1 = 500$ are used in all examples
calculated in this paper).

For Skjaak, Shen's compensator behaves in a
similar way as the pressure feedback with a DT1
element, see Fig. 17 and Fig. 18.

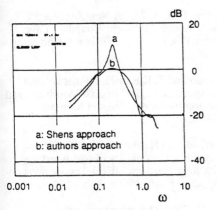

a: Shens approach
b: authors approach

FIG. 18.Close-loop load speed response

6 STABILITY IMPROVED BY "DAMPING SHAFTS"

The "Jostedal Stability" system is described in (Ref. 7) The tunnel system, includes a big reservoir, a 30 km main tunnel system and 19 shafts draining a glacier and small rivers from the catchment area. The outstanding feature of this stabilizing system is that instead of a separate surge shaft the last inlet shaft in the complex tunnel system is designed as a "damping shaft". The damping shaft is made in such way that the damping of the water hammer waves passing through the shaft will be large. The damping is, however, at the same time small enough to cause strong decay of the water hammer oscillations that pass the shaft on the way up in the conduit system. Therefore, the energy also in these oscillations are drained out by the damping shaft. The damping is achieved by a 10 m long concrete plug with a small opening located in the downstream end of the shaft.

The opening through the plug is only $.3 \, m^2$. The cross section of the widened junction to the tunnel is 50 m^2, the shaft has only 2.54 m^2, while the main tunnel has 34 m^2. The distance from the turbine to the damping shaft is 5200 m and a traditional open shaft would give a complete unstable system. For the stability analysis of the complex system of Jostedal the "singu-

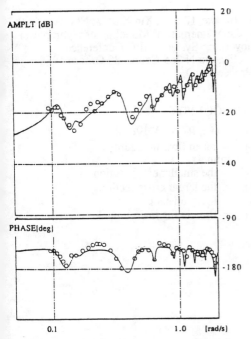

FIG. 19.Frequency response test at Jostedal with throttled shaft.

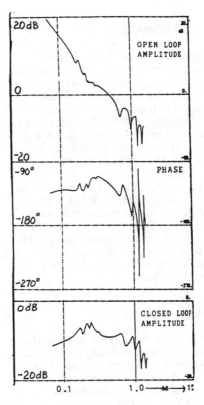

FIG. 20. Stability analysis of Jostedal with throttled shaft.

483

larity margin" has been introduced which is described in Ref. 7.

Frequency response tests of penstock pressure as a function of needle stroke at the power plants show good agreement with frequency response simulations proving that the mathematical modeling of the waterways has been successful. Therefore the open and closed loop simulations should also be correct. In Fig. 19 is shown the simulation compared with results of the frequency response test of the pressure response.

In Fig.20 the results of the computed stability analysis of the system with the throttled shaft is shown.

The experiences from the Jostedal Hydro Power Plant demonstrate thatstability can be obtained by means of damping shafts.

The requirements for a damping shaft is not very fancy except for the design tool. For such systems the analysis of the stability must be carried out more accurate than with traditional simulations. The theoretical analysis presented in this paper has proven its ability for this task.

CONCLUDING REMARKS

By means of the presented mathematical modeling tool based on the Structure Matrix Method, very good agreement with frequency response tests has been obtained. This tool has made possible a through study of the water column compensating system. This paper shows that the authors approach to the pressure feedback system and the throttling shaft system have an improved stabilizing effect for power plants with a long penstock.

The first plant in operation with pressure feedback system to the governor where a large compressed air accumulator is substituted by the new system will be Svartisen Hydro Power Plant. In this plant two 350 MW turbines designed for 580 m net head will be installed and approximately 20 mill NOK is saved by the new governor from Kværner, Norway.

REFERENCES

1. Brekke, H. 1984, "A stability study on Hydro Power Plant Governing including the influence from a quasi nonlinear Damping of oscillatory Flow and from the Turbine characteristics" Dr. Dissertation: The Norwegian Institute of Technology, NTH may, 1984.

2. Jonnson, I. G., 1978, "A New Approach To Oscillatory Rough Turbulent Boundary Layers", series paper 179, I. S. V. A. Technical University of Denmark.

3. M. Araki and T. Kurabara, "Water column effect on speed control of hydraulic turbines and governor improvement", Hitachi Review, Vol.22,1973.

4. Murty, M. and Hariharan, M. "Analysis and improvement of the stability of hydro turbine units with long penstock" IEEE Transactions on power apparatus and system, Feb. 1984.

5. Shen, Z. "Water hammer pressure compensation in turbine regulation systems" Water Power, Aug. 1973.

6. Lein, G and Maurer, W. "Advance in control of hydropower plants", (Will be published in book series for Hydraulic Machinery, Vol.9, Gover)

7. Lurås, H. 1988, "Damping of Oscillations in Narrow Shafts and its Stabilizing effect on Hydro Power Plants". Proceedings of the IAHR Symposium on Progress within Large and High-Specific Energy Units, Trondheim, Norway, June 1988.

8. Brekke, H. & Li Xin Xin: "A New Approach to the Mathematical Modeling of Hydropower Governing Systems" IEC Conference Publication N 285 Conrol 88 Oxford 1988 (Proceedings).

SYMBOL LIST

$\tau_s = \left(f_s \, \rho \, Q_t / A^2 \right) Q_o |q|$

Q_t = mean flow in regarded tunnel

A_s = the smaller cross section

A_b = the larger cross section

ζ = loss constant

Q_t = flow in the main tunnel

Hydropower'92, Broch & Lysne (eds) © 1992 Balkema, Rotterdam. ISBN 90 5410 054 0

Research on surge-control facilities for hydropower plants in Sweden

Xiao L. Yang & K. Cederwall
Hydraulics Laboratory, Royal Institute of Technology, Stockholm, Sweden

C. S. Kung
Department of Water Resources, Royal Institute of Technology, Stockholm, Sweden

ABSTRACT: This paper reports part of the research programs aimed at the evaluation of surge-control facilities for hydropower plants. The facilities investigated include simple, throttled, closed and differential types of surge chamber. The stability of surge oscillations is analyzed by the direct method of Liapunov and the phase-plane technique. The two procedures lead to the same results for small surge perturbations about the steady state. Certain dynamical features of the hydraulic system such as Hopf bifurcation can only be achieved through nonlinear analysis. Water-hammer pressure wave transmission is also examined for some of these surge tanks. The purpose of it is to find out which type of tank transmits less water-hammer pressure under the same flow condition.

1 INTRODUCTION

Surge-control facilities in hydroelectric power stations refer mainly to different types of surge tank or their combination. Traditionally, the analysis of a surge chamber falls into two categories: analysis of the stability of surge oscillations in the chamber, and analysis of its ability to minimise the transmitted portion of the pressure fluctuation into the tunnel upstream of the chamber.

The rigid-column approach with certain assumptions is generally used in the analysis of surge stability (Chaudhry 1987). The lumped dynamical system is governed by momentum and continuity equations. For a closed surge chamber with an air-cushion, the equation of state of gas can be used. The flow condition at the turbine is the principal boundary condition, and its effect on the stability is critical. A turbine unit operating in an isolated power grid is usually assumed to maintain a constant power output following any load change.

The pressure transmitted into the tunnel upstream is usually referred to as water-hammer pressure wave transmission. Its magnitude is an important measure of the hydraulic response of a surge tank (Mosonyi & Seth 1975). For a given condition of flow in the system, a superior surge tank allows little of pressure wave to be transmitted.

It should be pointed out that relatively slow surge oscillations are a result of the introduction of the surge tank itself and a secondary effect of the pressure release. Compared with the studies on such oscillations, the importance of the basic function of surge tanks to reduce pressure wave transmission is sometimes overlooked.

Research activities devoted to the evaluation of the surge-control facilities in Sweden are conducted mainly at the KTH Hydraulics Laboratory and Vattenfall Utveckling AB. This paper reports a part of that research.

2 LIAPUNOV STABILITY TECHNIQUE

In some cases, the governing equations of oscillations can be reduced to a second-order differential equation of the form

$$\frac{d^2x}{dt^2} + B(x, \frac{dx}{dt})\frac{dx}{dt} + G(x)x = 0, \qquad (1)$$

where x denotes the departure of the water level from its steady-state position in the tank. A characteristic of Equation (1) is the nonlinear damping term, which is a function of both x and dx/dt. The equation can be written as $d^2x/dt^2 = f(x, dx/dt)$, which is equivalent to the following pair of equations

$$\frac{dx}{dt} = y \qquad (2a)$$

$$\frac{dy}{dt} = f(x, y) \qquad (2b)$$

Eqs. (2) comprise an autonomous system since time t does not appear explicitly in f(x, y). The number of its equilibrium states (solutions of dx/dt = 0 and dy/dt = 0) depends on the boundary condition at the turbine. The case of constant power leads to three solutions. Two of them are real and the third is virtual. The solution for which both values are zero (0, 0) corresponds to the steady-state grade line and is of practical account.

2.1 The direct method of Liapunov

The concept and the direct method of stability developed by Liapunov have found applications in many fields (Thompson & Stewart 1989). Its concept refers to the stability definition of Liapunov; it pertains to perturbations of the initial conditions of a physical system, rather than of its parameters or its differential equations. The method employs phase or test surfaces, and they are usually called Liapunov functions. These functions sometimes make possible the determination of the course of solution curves in the x-y phase plane or in the x-y-E(x,y) phase space, and the stabilities of the equilibrium states can thereby be determined. With this method, the domain of initial conditions generating stable solutions can be found. However, to produce a form of E(x,y) that will determine the stability for a particular case is the difficult part of the analysis.

Based on the theory, it can be proved that the function E(x, y) defined by

$$E(x, y) = \frac{y^2}{2} - \int_0^x f(\phi, 0) \, d\phi \qquad (3)$$

is a weak Liapunov function for the system represented by Eq. (2) if the following three conditions are valid in the region close to the equilibrium state (0, 0):

(a) E(x,y) and its partial derivatives are continuous;

(b) dE/dt = [f(x,y) - f(x,0)]y =< 0; and

(c) $\int_0^x f(\phi,0) \, d\phi < 0.$

The function E(x,y) represents the total specific energy of the system: the first term on the right-hand side of Eq.(3) is the kinetic energy, and the second the potential energy. dE/dt is the temporal rate of change of the energy, and it must be negative for damped oscillations.

2.2 Asymptotic stability criterion

Conditions (a) and (c) can be satisfied. The condition (b) can be reduced to a criterion of the form (Yang et al. 1991):

$$Y = < \frac{2}{(1-X)^2} [(1-X) - \frac{F_0}{F}], \qquad (4)$$

where $X = (M_0/H_0)x$, $Y = (F/Q_0)y$, H_0 = net head on turbine, $M_0 = 1 + n(P_0 + P_a)F/V_0$, Q_0 tunnel discharge, F = water-surface area in the chamber, V = air-cushion volume, P = air-cushion gauge pressure, P_a = atmospheric pressure, n = air constant (1 < n < 1.4). The quantity F_0 is the water-surface area defined by Svee (1972) for asymptotic stability of small perturbations about the steady state.

For open surge chambers, the parameter M_0 is unity and F_0 is equal to the Thoma area (Thoma 1910).

Expression (4) is an inequality that identifies the condition for oscillations about the steady state that are asymptotically stable. A family of curves can be produced if condition (4) becomes an equality, and it is plotted in the (X, Y) Cartesian plane in Figure 1.

For a given value of F/F_0, the corresponding curve divides the plane into two regions. The region below the curve, where condition (4) is valid, is a stable area, whereas the region above it is an unstable area. The stability of the steady state can be determined therefore by finding in which of the two regions the steady state is located and also by its distance from the curve. With an increase of F/F_0, the curve moves upwards relative to the origin. For the case $F = F_0$, the curve passes through the origin. This case is the limit of stability that corresponds to the Svee or Thoma criterion. If the surge-tank area F is greater than F_0, the origin falls in the stable region. Large perturbations around the steady state can be stable. The value of F/F_0 determines the maximum amplitude of damped surge oscillations.

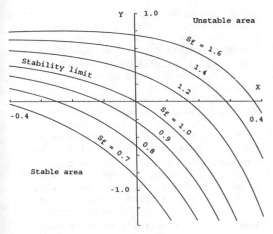

Fig.1 Stability diagram resulting from the direct stability method of Liapunov

2.3 Periodical oscillations

In non-conservative systems, any periodical motion is independent of initial conditions. The requirement for it should satisfy (Szemplinska-Stupnicka 1990):

$$\int_{mT}^{(m+1)T} [\frac{dE(x,\ y)}{dt}] dt = 0 \qquad (5)$$

in which T = surge period and m = integer (m >= 0). Eq.(5) states that the exchange of energy during one cycle is zero.

If the surge motion is a sinusoidal wave so that x = Asin(2πt/T), equation (5) can be reduced to

$$\frac{F_C}{F_0} = \frac{1}{\sqrt{1 - (M_0A/H_0)^2}} \qquad (6)$$

where A = surge amplitude and F_C = critical chamber area for periodic motions. F_C is a function of A for given H_0 and M_0. Equation (6) relates the condition for periodic motions with the Svee (Thoma) criterion for infinitesimal small surge perturbations. The stabilization of larger amplitude obviously requires larger chamber cross-section.

Similar analyses can be made for the second real solution. Stability conditions similar to (4) and (6) can also be obtained.

3 PHASE-PLANE TECHNIQUE

The use of the phase-plane technique in surge stability analysis dates back to the late fifties. This method is a geometrical device and is used extensively to depict on phase planes such features as singularity, periodicity, damping, unlimited growth of physical systems (Jordan & Smith 1987, Chaudhry 1987, Li & Brekke 1989).

For both throttled and unthrottled surge chambers, either closed or open, the governing equations can be transformed into two equations of the form (Yang 1991)

$$\frac{dZ}{dt} = f_1(Z,\ Q_s) \qquad (7a)$$

$$\frac{dQ_s}{dt} = f_2(Z,\ Q_s) \qquad (7b)$$

where Z is the water level in the chamber with the upstream reservoir level as reference and Q_s is the flow into the chamber. Equation (7) is characterized by nonlinearities, which arise from the head loss in the headrace tunnel, from the throttling effect due to the orifice if any, and from the boundary condition at the turbine. Both linear and nonlinear analyses are needed to reveal the stability feature of the system. Z and the tunnel discharge Q is chosen to define the phase plane Q-Z.

3.1 Linear stability

In the Q-Z phase plane, three singularities occur, and their locations are physically equivalent to those in the Liapunov method. Corresponding to these singularities, the flow into the chamber is always equal to zero. Therefore, the presence of the orifice does not affect the locations of the singularities. Both throttled and unthrottled chambers have the same singularities.

The linearized procedure specifies the type of singularity in the Q-Z plane and identifies the stability criteria near them. The analysis shows that the Svee (Thoma) stability criterion must be satisfied for asymptotic stability in the closed (open) surge chamber, with the precondition of $h_{w0} < H_0/2$. The quantity h_{w0} is the head loss in the tunnel. In this case, the singularity corresponding to the steady-state grade line is a stable spiral, while the second singularity is a saddle point. The second singularity can only be stable if the head loss exceeds one third of the gross head of the plant in question. Such a high loss is obviously uneconomical.

The results of stability criteria from the linear analysis coincide with those from the Liapunov method.

3.2 Nonlinear stability

The linearization of nonlinear equations (7) around the equilibrium states gives an indication of stability features only in a local sense. If the surge amplitude is large, the conclusions are not applicable. Besides, the nonlinear damping effect due to the orifice can not be included in the procedure. Nevertheless, the linearized approach forms the basis of any further nonlinear investigation.

The nonlinear analysis is made by means of numerical integration of equation (7). The fourth-order Runge-Kutta procedure is used. The result indicates that, in the presence of the orifice, the system manifests itself as a supercritical Hopf bifurcation. This is shown in the Q-Z-Λ phase space in Figure 2. Λ is the chamber area (and also the air-cushion volume for the closed tank). The bifurcation point Λ_0 corresponds to the Svee (Thoma) condition for the closed (open) chamber and does not depend on the orifice (Yang & Kung 1991a).

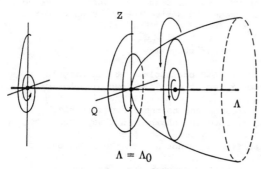

Fig.2 Occurrence of a supercritical Hopf bifurcation in the presence of a throttled type of surge chamber

If the chamber size is larger than the Svee (Thoma) condition ($\Lambda > \Lambda_0$), the phase portrait consists of an asymptotically stable spiral around the origin (the steady state). Even at the bifurcation point ($\Lambda = \Lambda_0$), the phase path is still asymptotically stable. In the case of $\Lambda < \Lambda_0$, a stable limit cycle surrounds an unstable spiral at the origin. The change in the chamber size

gives rise to an exchange of stability from a stable focus to a stable limit cycle without passing through the stage of being a neutral centre at Λ_0. The analysis shows also that an unstable limit cycle exists outside both before and after the bifurcation point.

By comparison, the unthrottled chamber exhibits itself as a subcritical Hopf bifurcation, which is illustrated in Fig.3. It has the same bifurcation point as the throttled case. For the case of $\Lambda < \Lambda_0$, an unstable limit cycle occurs, which encircles the stable steady state. At the bifurcation point, the limit cycle shrinks into a cycle in a point sense. For the case of $\Lambda > \Lambda_0$, the phase trajectory originating from close to the steady state becomes an unstable focus going towards infinity.

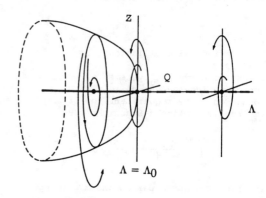

Fig.3 Occurrence of a subcritical Hopf bifurcation in the presence of an unthrottled type of surge chamber

The distinction between the supercritical and the subcritical Hopf bifurcation rests with the throttling orifice.

3.3 Differential surge chamber

The differential surge chamber is typified by three first-order ordinary differential equations of the form

$$\frac{dQ}{dt} = f_1(Q, Z_1, Z_2) \tag{8a}$$

$$\frac{dZ_1}{dt} = f_2(Q, Z_1, Z_2) \tag{8b}$$

$$\frac{dZ_2}{dt} = f_3(Q, Z_1, Z_2) \tag{8c}$$

where Z_1 and Z_2 are water levels in outer tank and in the riser, respectively. These equations are also nonlinear. Besides all the nonlinear terms that occur for the throttled chamber, an additional nonlinear term is caused by the flow exchange over the top of the riser.

Similar analyses are also made on the stability of the differential surge chamber (Yang & Kung 1991b). Two phase planes Q-Z_1 and Q-Z_2 are used, in which the phase portraits are displayed. The linearized approximation to equation (8) shows that the Thoma stability criterion applies to the differential surge chamber. A super-critical Hopf bifurcation is also demonstrated in both the Q-Z_1-Λ and Q-Z_2-Λ phase planes in the nonlinear analysis. The controlling parameter Λ is the total surge-chamber area. The bifurcation point corresponds to the Thoma area.

4 WATER-HAMMER PRESSURE WAVE TRANSMISSION

If a surge chamber can release water-hammer pressure more efficiently, there will be less pressure wave transmission into the upstream tunnel. Moreover, the resulting pressure along the penstock will be reduced and the governing stability of turbine will be improved. A surge chamber resulting in low transmission leads to an economical design of the system.

As shown in Figure 4, surge tanks are evaluated in terms of the water-hammer pressure they transmit into the tunnel upstream: (a) is a throttled type, (b) a is Johnson differential type, and (c) and (d) are variants of the Johnson type. An important feature of (c) and (d) is that, in contrast to (b), the riser is separated from the outer tank and the orifice connects the outer tank directly with the waterway.

Analysis (Yang & Cederwall 1991) shows that the Johnson type (b) transmits less pressure than does the throttled type (a). The riser plays an important role in the reduction of the transmission. For the same surge-tank dimensions, the two modified types (c) and (d) transmit even less water-hammer pressure than does the Johnson type. The difference is obviously caused by the change of location of the orifice. The superiority of type (c) over the Johnson type in transmission has been verified experimentally (Wang & Yang 1989).

For the Johnson type, the throttling orifice is located in the wall of the riser near its bottom. This is unfavourable to both the construction and the structural

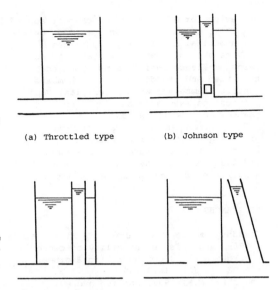

(a) Throttled type   (b) Johnson type

(c) Variant of (b)   (d) Variant of (b)

Fig.4 Comparison of surge tanks based upon water-hammer pressure wave transmission

stability of the thin riser. For the modified types (c) and (d), the orifice is in the bottom of the tank so that they avoid these weaknesses. Besides, the riser of (c) can be rectangular and built adjacent to the wall of the surge tank. In this way, it can also serve as a gate shaft for the gate that is normally installed at the upstream end of the penstock.

5 CONCLUDING REMARKS

The Liapunov method can be regarded as an extension and generalization of the energy method in mechanics. Its physical interpretation is that, if the total energy of a system is always decreasing with respect to time in the neighbourhood of a given equilibrium state, then this equilibrium is asymptotically stable. However, one difficulty is often encountered: the class of systems for which the total energy can serve as a Liapunov function is limited, and generally it is difficult to find a Liapunov function. As a matter of fact, the main drawback of this method is that no established procedure exists for constructing a Liapunov function for a given dynamical system.

The linearized approach in the phase plane method gives the same results as the Liapunov method. They both lead to the Svee criterion for a closed surge chamber and to the Thoma criterion for an open one. Only the numerical nonlinear analysis can provide the information of stability as to large amplitude oscillations. Both the throttled and the differential chambers exhibit themselves as supercritical Hopf bifurcation, whereas the simple unthrottled chamber as subcritical one.

Water-hammer pressure wave transmission gives a good indication of the characteristic of a surge chamber. The two modified Johnson types (c) and (d) are both superior to the original Johnson type.

REFERENCES

Chaudhry, M.H., M.A. Sabbah & J.E. Fowler 1985. Analysis and stability of closed surge tanks. J. of Hydraulic Engineering, ASCE, 111(7): 1079-1096.

Chaudhry, M.H. 1987. Applied hydraulic transients. New York: Nostrand Reinhold.

El Naschie, M.S. 1990. Stress, stability and chaos in structural engineering: an energy approach. London: McGraw-Hill.

Fox, J.A. 1977. Hydraulic analysis of unsteady flow in pipe networks. London: The Macmillan Press Ltd.

Jordan, D.W. & P. Smith 1987. Nonlinear ordinary differential equations. Oxford: Clarendon Press.

Kapitaniak, T. 1991. Chaotic oscillations mechanical systems. Manchester and New York: Manchester University Press.

Leipholz, H. 1987. Stability theory. Chichester: John Wiley & Sons.

Li, X.X. & H. Brekke 1989. Large amplitude water level oscillations in throttled surge tanks. J. of Hydraulic Research, IAHR. 27(4): 537-551.

Mosonyi, E. & H.B.S. Seth 1975. The surge tank - a device for controlling water hammer. Water Power & Dam Construction 27(2): 69-74 and 27(3): 119-123.

Novak, P., A.I.B. Moffat, C. Nalluri & R. Narayanan 1990. Hydraulic structures. London: Unwin Hyman.

Szemplinska-Stupnicka, W. 1990. The behaviour of nonlinear vibrating systems. Dordrect: Kluwer Academic Publisher.

Svee, R. 1972. Surge chamber with an enclosed compressed air cushion. Proceedings of the International Conference on Pressure Surges: G2/15-24.

Thoma, D. 1910. Zur theorie des wasserchlosses bei selbsttaetig geregelten turbinenanlagen. Munchen: Oldenburg.

Thompson, J.M.T. & H.B. Stewart 1989. Nonlinear dynamics and chaos. Chichester: John Wiley & Sons.

Thorley, A.R.D. 1991. Fluid transients in pipeline systems. England: D & L George.

Wang S.R. & X.L. Yang 1989. Experimental investigations of new type of surge chamber. Proc., the 6th International Conf. on Pressure Surges: 373-382.

Yang, X.L. & K. Cederwall 1991. Surge tanks evaluated from the point of view of pressure wave transmission. Proceedings, the XXIV Congress of IAHR: D115-122.

Yang, X.L. 1991. Stability of oscillations in air-cushion surge chambers. Licentiate Thesis, Bulletin No. TRITA-VBI-155. Stockholm: KTH-Hydraulics Laboratory.

Yang, X.L., C.S. Kung & K. Cederwall 1991. Large-amplitude oscillations in closed surge chamber. J. of Hydraulic Research, IAHR (accepted for publication March 1991)

Yang, X.L. & C.S. Kung 1991a. Stability of air-cushion surge tanks with throttling. J. of Hydraulic Research, IAHR (accepted for publication September 1991).

Yang, X.L. & C.S. Kung 1991b. Nonlinear stability of differential surge chambers. Journal of Hydraulic Engineering, ASCE (submitted for possible publication December 1991).

Hydropower'92, Broch & Lysne (eds) © 1992 Balkema, Rotterdam. ISBN 90 5410 054 0

Installation of surge damping devices in existing water power plants

N. Dahlbäck
Vattenfall Utveckling AB, Sweden

ABSTRACT:
Water power stations have a long life span, usually, 60-100 years. During this time period conditions can be changed in such a way that the original criteria for dimensioning no longer provide satisfactory safety. The need for improving the surge safety in existing underground water power stations and a method of achieving this with low building costs and short shutdown periods is presented in this paper.

1. NEW DEMANDS ON DESIGN OF SURGE DEVICE

Water power stations have a long life span, usually, 60-100 years. Changes in dimensioning values in a system immediately give rise to the question as to which parts ought to be reconstructed in the system. For instance, it is obvious that some parts in the electrical system have to be changed when the power output is increased.

When dealing with transient phenomena the situation can be much more complicated. Without any change in the performance of the system there can still be changes in its use. New operational demands and a new generation of control systems and operating personnel can suddenly result in an operational situation that was highly unlikely at the time of construction to occur every other year.

Slow transients with small damping, such as surge in underground water power stations with long tunnels are especially dangerous, it is rare that a surge gallery is dimensioned for a multiple event (e.g more than two load changes). In several water power systems the time before a surge is damped out can be of the order of an hour. If the power station is not capable of a full load acceptance after a load rejection this will cause unacceptably long shut down after every stop. The present situation, with water power stations more frequently used for peak load demands, both increases the probability of multiple events and the economic benefit of high availability.

The new remote control systems with control and increased automation decrease the possibility of a human operator sensing something irregular that would have made him extra careful in regulating the load.

The consequence after a multiple event not dimensioned for can be seen in picture 1.

A 250 MW power station was flooded from the downstream surge gallery after several load acceptances and load rejections at 50% load . The adjustments were due to tests performed on different parts in the electrical system after a shutdown period for overhaul. This example and the discussion above demonstrate the need for devices that can damp slow surge in underground power stations.

2. SURGE DAMPING POSSIBILITIES

The power loss of the surge can be said to be proportional to

$$Q \times H_L$$

where Q is the periodic flow in the tunnel or surge gallery and H_L is the head loss caused by friction in the tunnel or by singular loss in an orifice in the surge gallery inlet.

The energy loss is a time integration of the above expression.

Installing an orifice in the surge gallery without emptying the tunnels implies that it has to be placed above the stationary water level. This means that only during the upper half of the surge will there be a loss through the orifice. Theoretically two orifices above the water level can give the same result in energy loss as one at the bottom of the surge gallery.

When placing an orifice with a horizontal cross section the impact of the vertically moving water surface must be taken into consideration, otherwise high pressure will occur when the water hits the orifice.

There is always the possibility of dividing a surge basin into cells, compare with Johnson's differential surge tank. (Reference 1). For instance,

FIG 1. Seitevare water power station was flooded in June 1987. 800 m^3 of water violently passed through the machine hall to the bottom of the power station.

cylinders with several orifice openings can be put down in the surge basin. The force on such cylinders will depend on the water levels inside and outside which in turn depend on the size of orifices and the speed of the surge.

There is also the possibility of building into the control system devices that try to make small changes in the load, thus getting an additional surge out of phase of the surge that should be damped. Such a system always gives rise to new discussions of safety because of the probability of causing unwanted load regulations.

3. A METHOD FOR DAMPING SURGE IN INCLINED SURGE GALLERIES

The target of the work presented here is water power stations that use inclined transport tunnels from the construction period as surge galleries. This is a common situation in Sweden as it minimizes the volume of total blasted rock. Furthermore, it is common that these surge galleries are connected to

the access tunnel to the machine hall. This type of construction is often necessary since the air flow caused by the surge escapes through the access tunnel. Flooding in this type of water power station can cause severe accidents.

In surge galleries made from transport tunnels it is possible to damp the surge in another way, utilizing the fact that the water is not only moving in a vertical direction. Obstacles placed in the inclined tunnel create large energy losses when the water level passes through them. The energy loss is not only a friction loss, but is also the result of other mechanisms. When there is an inflow to the gallery the obstacle forces the water to rise, thus transforming the kinetic energy to potential energy, which in its turn is lost behind the obstacle when the water falls. When the cross section is nearly filled the obstacle causes losses almost like an orifice. When there is an outflow from the gallery the obstacle causes an increased loss in the same way as a stepped flume (reference 2). Obstacles such as walls also have the effect of keeping a certain amount of the water and releasing it very slowly. In this case there

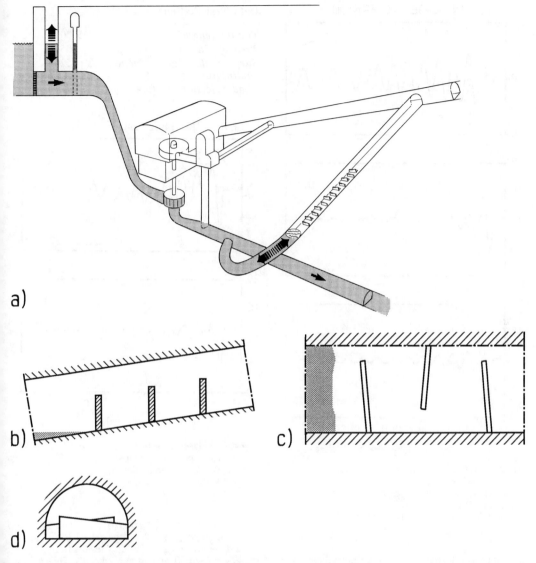

Fig 2. Damping device in form of Walls in inclined surge gallery
- a) View of underground water power station
- b) Length section through walls
- c) Plane secton
- d) Cross section

will be an energy loss when the water is moving down the slope. Furthermore, some of the surge will be put out of phase and in this way cause a damping. All these mechanisms combine together and cause a considerable damping of the surge.

An effective and easily constructed obstacle is a wall with a height of up to 60% of the tunnel height and a width of up to 95% of the tunnel width (see figure 2). Several walls can be placed one after the

other in the tunnel. In opposition to an orifice such obstacles have their best effect when they are placed near the stationary water level (i.e. there is no need to empty the tunnel to get a better effect).

This is due to the fact that the velocity of the water surface has its maximum when it passes the stationary level.

Another advantage with the construction of walls is the fact that the inertia and the possibility to fast

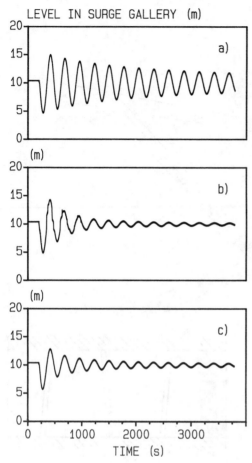

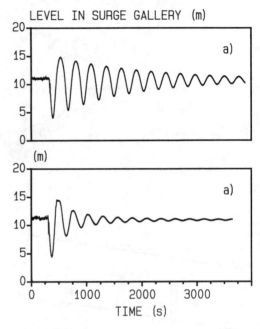

Data Test Station

Tunnel length	5000 m
Area of Tunnel	105 m^2
Surge gallery cross section	52m^2
Inclination	1:7
Load rejection of flow	50 m^3/s

Fig 3. Surge calculations for test station.
    (Data see fig 4).
    a) Without damping device
    b) With damping walls
    c) With an orifice (14 % opening)

Fig 4. Model tests for test station
    (with presented main data).
    a) Without damping device
    b) With damping walls

movements of the mass in the surge gallery are not affected in the same way as by an orifice. This could be an essential factor when there are problems with too high or too low pressures after load rejections.

The wall has several parameters that can be varied, such as height, width, angles and shape of the openings for return flow. We have performed model tests to guide the dimensioning of walls and measure the damping effect in different situations.

4. TEST RESULTS

The diagrams in figs 3a-c show results from surge calculations. The used main data are presented in fig 4. The three different calculations presented are made with only friction loss (a), with a crude

simulation of 2 repeated walls (b), and with an orifice in the bottom of the surge gallery (c).

The diagrams in figs 4a-b show results from model tests with 0 and 2 repeated walls. Thus a comparison can be made between figs 3a-4a and between figs 3b-4b. It is seen from the model tests that it is possible to reduce the time for damping considerably. With only friction loss the time to reach a surge amplitude that is 10% of the maximum is 3-4 times longer than in the case with two walls.

The calculations demonstrate the difference between the orifice and the walls. The walls do not affect the first minima as they are dry until the return of the surge. Therefore the orifice is more efficient at the beginning, but already after two surge periods the total energy loss caused by the walls is larger than that caused by the orifice with only 14 % opening area. Furthermore such a small orfice will cause too high pressure surges while the walls do not affect the pressure surge at all.The energy loss is roughly proportional to the number of walls, but the other parameters such as height, width angles etc are dependent on the slope of the surge tunnel and of its bending.

5. DISCUSSION

The power station in figure 1, was reconstructed a few years ago. An extra surge gallery was added to the original one to avoid the new operational restrictions that caused a reduced availability. The total cost of this new gallery was nearly 2 million dollars (or ECU). By installing a few walls in the old surge gallery the same purpose would have been achieved at a cost of less than 15% of this.

There will always be many possibilities of varying the design of the walls . The shape of the surge gallery is the main factor but the available construction technique at remote places can also affect the design. The main idea of having vertical walls covering about half the cross sectional area will probably still be valid.

The short assembly time and the low cost for construction will make it possible to introduce this device in several water power stations and thus increase safety. Discussion as to whether to invest in a reconstruction of the surge gallery in the likelihood of a multiple event taking place, can now be avoided.

The method discribed in chapter 3 is pat.pending in Sweden, no 9200423-3.

REFERENCES

Johnson, R.D. The Differential Surge Tank. Trans. Am. Soc. Civ. Eng., 1324, Vol. 78, 1915.
Stephenson, D. Energy dissipation down stepped spillways. Water Power Dam Construction, Sep. 1991.

Vibration monitoring of hydropower machinery

R. Husebø
Nybro-Bjerck/Norconsult, Asker, Norway

ABSTRACT: Vibration analysis is used both in trouble shooting and as general condition control on hydro turbines and generators. In Norway, a large number of vibration monitoring systems are installed, mainly based on measurements of the shaft vibrations. Systems have been developed for performing condition control with a suitable combination of on-line monitoring and periodic measurements.

Vibration measurements has improved the quality of condition control, by giving information on the condition in a machine while running and without excessive dismantling. Maintenance planning is supported by this form of condition control.

In this article, examples are given on use of systems measuring and monitoring vibrations in hydro units.

MONITORING SYSTEMS

A vibration monitoring system is the only surveillance of the mechanical condition in a hydro power unit. Electric and hydraulic properties are most often closely monitored, while temperature monitoring is the only parameter except vibration giving an indication of change in bearing forces.

Shaft vibration and shaft position in the guide bearings have proved to be useful parameters when locating the forces acting on turbine, generator, shaft or bearings. In a simple way the amplitude and direction of the forces in different running conditions may be found, hence allowing a detection of changes in forces. Although large and heavy, hydro units are sensitive to changes in forces. Knowing the bearing forces is of great help in locating the defects and suggesting action.

Since 1985 most new hydro plants in Norway are equipped with vibration monitoring systems. Vibration monitoring systems are also installed on a large number of existing hydro machines.

The systems used are mainly shaft vibration monitoring systems, measuring the shaft vibration in two directions at each guide bearing. They are often supplied with additional accelerometers or velocity sensors at the turbine, the axial thrust bearing and the generator stator. The monitoring unit is placed in the control room, to simplify periodic readings of vibration amplitudes.

Shaft vibration sensors are placed near the guide bearings. Since shaft vibrations may have nodal points located in the bearings, it is an advantage to place the sensors apart from the bearing center, even at the cost of some accuracy when evaluating bearing forces.

Shaft vibration monitoring will normally detect changes at low frequencies, from shaft speed or lower up to a few times the rotational speed. Hydro turbines and generators have large masses, and the displacement amplitude at

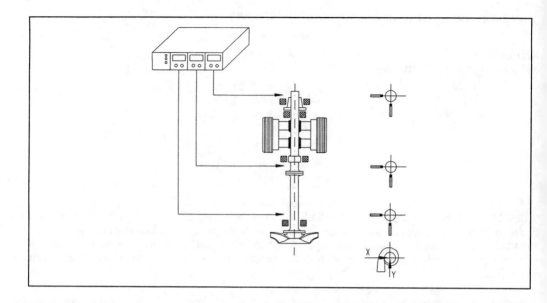

higher frequencies will be small compared to the normally predominant speed frequency. Therefore, accelerometers or velocity sensors should be used to monitor frequencies above 5 - 10 times the speed of the unit, to detect changes at an early stage. (When faults at higher frequency get severe, however, they may be detected at lower frequencies as well.)

On vertical hydro units accelerometers or velocity sensors are used on the turbine, on the generator stator and on the axial thrust bearing. Often the axial thrust bearing is supported on the generator stator, and vibrations from both parts can be measured with one sensor at the bearing.

At stationary conditions in the normal operating range, there is normally small changes in shaft vibration for hydro turbines. Small changes in load, in headwater level and in temperatur have only minor effect on shaft vibrations. The shaft vibrations therefore give reliable longterm readings, and changes in condition may easily be detected, even if running conditions differ between each reading. Changes in tailwater level, however, may have considerable influence on the turbine vibrations.

Monitoring the frequencies below 1000 Hz is sufficient for most hydro turbines with sleeve bearings. For the generator, except for the

speed frequency, the frequency of most importance to monitor is 100 Hz (passing frequency of the rotor poles). For the turbine, a wider range of frequencies is of importance, mainly the blade passing frequency and several multiples of the speed, the number of runner vanes, guide vanes and stay vanes, and frequencies connected to the oscillating periods in the penstock, draft tube etc. The thrust bearing need monitoring of all the frequencies necessary for the generator and the turbine, as well as the speed times the number of pads.

Proper alarm limit setting is far more important for shaft vibration monitoring systems than for systems monitoring the bearing housing vibration. Most monitoring systems have two alarm relays for each monitored parameter. As a general rule, it is recommended that one limit is set to give warning when vibrations change from the normal values, and that one is set to give a shut-down when the amplitude exceeds 80 % of the bearing clearance. Also, signals should be given when shaft position gets too close to the bearing segments, indicating large pulling forces.

The international standards VDI 2059-5 and ISO-7919-5(draft proposal) cover shaft vibrations in hydraulic machinery. Limits are based on statistical values, and these values may also be a guide when setting proper limits.

Hydro turbines often have an operating range at part load where the turbine shaft vibrations are higher than in the normal operating range. The alarm limits have to cover the total operating range, and even if the part load vibrations may be supressed by use of alarm delay or load dependent limit values, the limits must often be set considerably higher than the vibrations in normal operation. Due consideration must be paid to such effects when setting the alarm limits, to avoid false trips, and to achieve a highly reliable monitoring system.

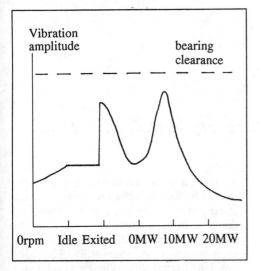

Load dependence of turbine shaft vibrations for a Kaplan turbine.

Vibration measurements in condition control can be divided into four different parts which may be performed separately or as a supplement to each other:

- "Fingerprint" measurements
- Periodic analysis
- On-line monitoring
- Trouble-shooting

Considering the long life-time of a hydro generating unit, monitoring a few vibrational parameters will be sufficient, combined with periodic analysis and a fingerprint at the first measurement. Several systems are available for on-line monitoring and analysis of vibrations in rotating machines. The systems are often expensive, with large extents of instrumentation required at each unit, and with a large amount of data to be dealt with.

A measuring system, "IMPULS", has been developed to perform measurements of shaft vibrations in hydro units. The system uses a PC for measurements, and gives an easy-to-read presentation of the shaft vibration and shaft position. This enables the user to draw conclutions on the condition of the machine in a simple way. The system is based on use together with vibration monitoring systems, and is a natural extension of the shaft vibration monitoring systems already installed in many plants.

EXAMPLES OF USE
MISALIGNMENT

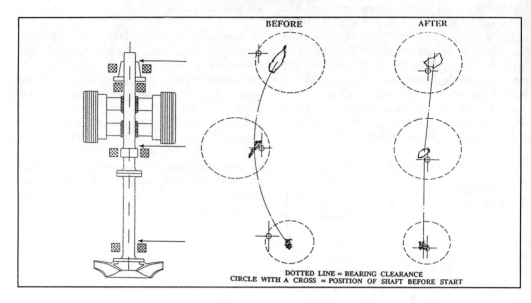

BEFORE   AFTER

DOTTED LINE = BEARING CLEARANCE
CIRCLE WITH A CROSS = POSITION OF SHAFT BEFORE START

Description:
A vertical 200 MW Francis turbine is equipped with a shaft vibration monitoring system. Two sensors are installed at each guide bearing, perpendicular to each other and in the same vertical plane at all bearings.

The unit has three hydro-dynamic guide bearings, the upper bearing is a combined thrust and guide bearing.

The figure shows the shaft motion in the guide bearings, at idle speed without energized generator.

Problem:
Elevated temperature in the guide bearings.

Analysis:
To find the clearance in each guide bearing the shaft was forced from side to side at stand-still while its position was read from the monitoring equipment. In the figure the clearance is drawn with an indication of the shaft position at stand-still. The shaft position at stand-still is chosen as reference.

Afterwards the machine was started and the position of the shaft was recorded during start-up and normal operation.

Solution:
By comparing the clearance of the guide bearings to the change in shaft position from standstill to idle speed, it could be seen that the shaft position canged in different directions in the lower generator bearing compared with the turbine- and the upper generator bearing. This indicated a misalignment of the guide bearings.

The lower generator bearing had to be moved approximately 0.5 mm.

When looking for changes in foundations, this method is in most cases well applicable.

DYNAMIC BALANCING

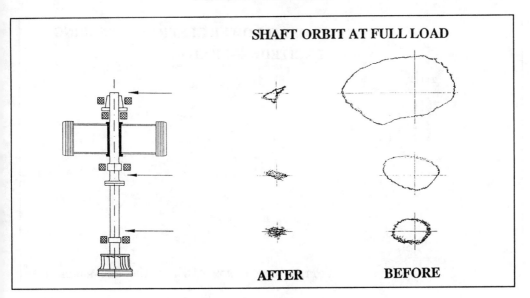

SHAFT ORBIT AT FULL LOAD

AFTER   BEFORE

Description:

A vertical low-head Francis turbine is equipped with shaft vibration monitoring system, measuring shaft vibrations in two directions at each bearing. The unit has three guide bearings, with a thrust bearing at the top of the generator.

Problem:

Large shaft vibrations at all three guide bearings, at the rotational speed of the unit. The vibration amplitude was close to the generator guide bearing clearance, indicating radial forces of considerable size acting on the bearings.

Also, a vibration amplitude close to the bearing clearance complicates the settings of alarm limits in the monitoring system, since an alarm must be triggered at amplitudes smaller than the bearing clearance.

Analysis:

Shaft vibrations increasing with speed indicated a mechanical unbalance. A decrease in vibration when exciting the generator indicated a magnetic imbalance in a different direction from the mechanical imbalance.

Solution:

Balancing of the rotor was recommended. Based on one measurement with the non-contacting sensors in the monitoring system, necessary balancing weights were calculated and location stated. 80 kg of weights were placed at a given position, as a compromise between mechanical and magnetic balancing.

At runaway speed the magnetic forces are not present. When balancing a generator to compensate for magnetic forces, the vibrations at runaway speed must therefore be considered.

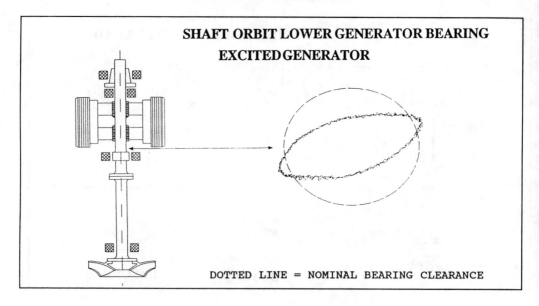

SHAFT ORBIT LOWER GENERATOR BEARING EXCITED GENERATOR

DOTTED LINE = NOMINAL BEARING CLEARANCE

Description:

A 20 years old vertical high-head Francis turbine is equipped with shaft vibration monitoring system, measuring shaft vibrations in two directions at each bearing. The unit has three guide bearings, with a thrust bearing at the top of the generator. The generator guide bearing is of the tilting pad type.

Problem:

Large shaft vibrations at lower generator guide bearing.

Analysis:

The shaft vibration amplitude was modest at start-up and idle speed, increasing when exciting the generator rotor. Shaft position did not change when exciting rotor, the bearing oil temperature did not rise during a period with increase in shaft vibrations. The reason was found to be wear of the tilting pad supports in the lower generator bearing.

Solution:

The unit was kept in operation, spesial attention was paid to the development of the shaft vibrations. Over a period of 6 months the shaft vibration increased with more than 25 %, also exceeding the nominal bearing clearance. The bearing was dismantled and the tilt pad construction was modified to increase the strenght of the pad supports.

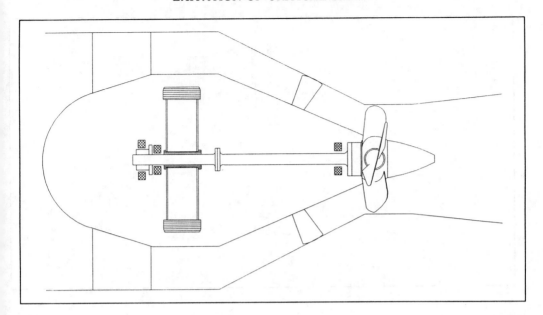

Description:

Two identical 20 MW Bulb-turbines are installed in one power station. The units have 4 turbine runner blades, and a speed of 100 rpm.

Problem:

Excessive vibrations in one of the turbines when operating within the normal operating range.

The vibrations are caused by the 1 order natural frequency excited from the blade passing frequency during normal operation.

40-50 % of the 4-blade Bulb and Kaplan turbines may have a blade passing frequency close to the 1 order natural frequency, dependent of design and spesific speed. This may lead to severe damage if hydraulic disturbance occure, even if the unit operates smoothly in normal operation.

Analysis:

The shaft 1st. order natural bending frequency is 25% above run-away speed. Run-away speed is approximately 3 times the operating speed, giving 1st order natural frequency close to 4 times unit speed. The unit has 4 runner blades, hence the blade passing frequency excite the 1.order natural frequency of the unit during normal operation.

Solution:

The stay-ring of the unit was stiffened, to increase the turbine bearing stiffness and increase the 1.order natural frequency. The vibrations were reduced after this modification. A vibration monitoring system was installed to detect possible development of foundation/stiffnes changes.

MAGNETIC PULL IN GENERATOR

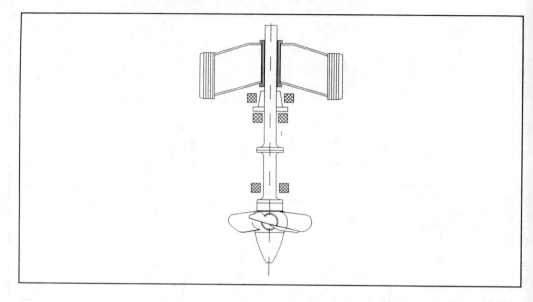

Description:

A vertical 100 rpm Kaplan turbine with two guide bearings and umbrella type generator has been in operation for 30 years.

The unit is equipped with a shaft vibration monitoring system, monitoring the shaft vibrations at each guide bearing, and the axial bearing housing vibrations.

Problem:

Break-down of the generator guide bearing. Installation of a spare bearing similar to the original bearing also resulted in bearing break-down.

Analysis:

At idle speed the shaft vibrations were low, with a 0-peak amplitude of 30 microns. When exciting the generator rotor, the shaft vibration remained unchanged. The shaft position changed severely, indicating a magnetic pull from the generator stator.

Solution:

The guide bearing lubrication was improved, and the stator housing ajusted to ensure an even air gap between rotor and stator.

Monitoring the shaft position will give an indication of further changes in the stator shape and air gap, enabling adjustments of the air gap before bearing damage occur. As long as the shaft position is close to the bearing center, the monitoring of shaft position will be very sensitive to changes in magnetic pull from the generator stator.

Hydropower'92, Broch & Lysne (eds) © 1992 Balkema, Rotterdam. ISBN 90 5410 054 0

Magnetic field measurement

François Lalonde
IREQ, Institut de Recherche d'Hydro-Québec, Que., Canada

ABSTRACT: This paper describes a system to measure and analyse magnetic flux in the gap of an electric machine. Our research group has contributed to new monitoring systems in the past. With these new instruments, we have been able to determine that some machines still show strong vibrations without any rotor/stator deformations. Magnetic unbalance is suspected, but can not be measured while the generator is operating. The following paper describes a sure way to do so, the preliminary results and the future goals.

1 INTRODUCTION

Most of Hydro-Québec's power come out of hydraulic generators. For many years, Hydro-Québec has acquired a wide experience in hydropower generator operating, maintenance and dynamic behaviour. Like many utility companies, Hydro-Québec is concerned with equipment aging. In order to maintain high plant availability, we need better monitoring. Choosing the parameters that must be monitored is not a simple task. We could tried to monitor everything, taking the risk of being submerge with data. Many new instruments are tested on rotating machines for better monitoring, but not all of those are beneficiary. On the other hand, some instruments give us useful information about the machine behaviour. The dynamic air-gap monitoring system, which provides information on rotor and stator shape under various operating conditions, is a good example. This system was developed by our research group specially for large generators at Hydro-Québec. We are not a rotating machinery specialist team, our expertise is on electronic instrumentation.

With time we have gained some experience on large generator equipment. The research on air-gap monitoring lead us to the development of other instruments. One of latest instruments tried at Hydro-Québec is the magnetic field monitoring system. We suspect that magnetic unbalance may be the cause of many rotating machine vibrations. When we suspect this kind of defect there is no simple method available to check the integrity of the rotor.

2 METHODS

Many different procedures are used to detect winding shorts circuits. Our intention is to find the simplest method specifically designed for hydraulic generators!.

2.1 Pole Impedance Measurement (PIM)

During a normal maintenance outage, pole impedance measurements are a complicated task that require a lot of manpower and time. This method gives indirect results which only allow one to guess the operational uniformity of the magnetic field. Nevertheless, shorted poles can be detected, but is this enough? External influences that modify the magnetic field are temperature, dynamic air-gap changes and mechanical stress. These influences are present during the operation of a generator and must be measured on-line, without stopping the generator, especially during peak periods.

2.2 Time Domain Reflectometry (TDR)

According to many authors, the TDR technique may be used to detect a short on a winding. If a pulse is applied to the rotor winding circuit, a short winding causes reflections that can be analysed. This technique is well described in J.W. Wood (1986) and is called recurrent surge oscillograph (RSO). Up to now, this testing technique may be used on site. Future development may lead research to on-line TDR testing.

2.3 Rotor Shaft Current (RSC)

According to Z. Posedel (1991), a short in a pole winding will cause a current in the shaft. Using suitable technique, one can analyse the harmonic content of the shaft current to detect the presence of a short. This technique may be used on-line on some machines where only one side of the shaft is grounded. In that case, this may be very difficult to apply to hydrogenerators.

2.4 Split Phase Current (SPC)

Unbalance split phase current may be induced by magnetic unbalance. Also, other problems on the generator may cause unbalance split phase current. In this case, the method gives us indications of the state of the machine, but we can't be sure what is the source of the problem.

2.5 Search Coils (SC)

The magnetic flux variation in the air-gap will induce a voltage in a coil placed in this air-gap. It is obvious that this search coil will detect a short winding. Many are reluctant to

install this type of device in the air-gap because it could hazard the machine or personnel according to P.J. Tavner and al. (1986). Also, many have found this sensor difficult to design and install. Usually the search coil is installed on the stator wedging system in turbine-generators. The coils are made of many turns of wire on a small form. One of the benefits of this technique is that it is used on-line.

2.6 Resume of the techniques

Some other techniques are described in many papers but the above are the only one selected to be useful in hydrogenerator applications.

Table 1. Application of the selected techniques.

Method	Site testing	On-line
PIM	Yes	No
TDR	Yes	Maybe
RSC	No	Yes
SPC	No	Yes
SC	No	Yes

3 TECHNIQUE SELECTION

Our goals are to get the simplest technique to detect magnetic unbalance on hydrogenerators in normal operating condition. So, the first two methods are rejected. We have now to look at Rotor Shaft Current and Search Coil.

The Rotor shaft current method could be very attractive but we rejected it because it is not always applicable on hydrogenerators and also it can't tell us where the defective pole is. It's the same with the split phase current method.

The coil in the air-gap may be difficult to apply to every large generator since the air-gap can be very small (<10mm).

None of those techniques are exactly what we are looking for. Consequently, we decided to look what else could be done combining techniques to get a simple method, compatible with computer means for the development of a positive analyses package.

4 THE NEW SYSTEM

In order to avoid hazard in the generator we used our experience in air-gap measurement with flat probes. We know how to build flat sensors that can be installed on the stator surface. For many years we had no problem since the probes are made of special conductive material over a tin material support.

With a different arrangement coupled to a specially design electronic conditioning circuit we have formed a new system to measure magnetic flux in the gap of a rotating machine. The probe is easily installed on the stator. Usually it could be cemented on the stator without removing the rotor. The probe is installed on the stator wall by inserting it between two poles. We have selected a fast setting cement (1min) for installation.

The probe is linked to electronic conditioning circuits through 30 meter cables. It is now possible to monitor, on-line, the magnetic field of a large rotating machine. With the addition of a key phasor probe, it is possible to identify the magnetic field value of each passing pole.

Modern computer technique will allow the user to gain access to magnetic fields at different time intervals, operating conditions and to compare the field of each pole. Access to the data is always available, therefore discarding the need for outages during pole impedance measurements. In the near future, we would like to incorporate the newly developed Magnetic-Field-Measurement-System to the Air Gap Monitoring System (AGMS ®), which would display, on a polar plot, the rotor/stator roundness and centres and also the magnetic field shape.

5 PRACTICAL RESULTS

The first 4 probes were fixed on a 60MW hydrogenerator equipped with the Air Gap Monitoring System. The method has proven good results since April 1991. The system showed that it may not give absolute measurement value, yet it provides very good relative value. In other words, it can't tell the exact magnetic flux value in Wb unit but it can be used to compare all the poles in order to get the difference between them. A well balanced machine should have poles with the same magnetic flux value. If an interturn short-circuit occur on a pole, the magnetic flux of this pole will be reduced and should be easily detected.

All the following data are from generator #1 in Manicouagan river. The magnetic values are relative, and the air-gap values are in mm. All the measurements are synchronised to pole 49 so correlation are possible. This machine have 72 poles.

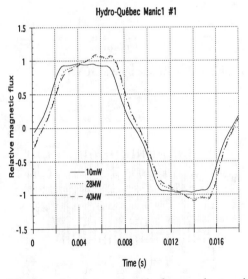

Graphic 1. Raw signal from electronic conditioning circuit. (10, 28 and 40MW from a 60MW machine)

From graphic 1, we can't say very much about the machine, since we see only two consecutive poles. If we look at the zero crossing of the signals, we can see that when the power output increases, the phase angle changes. Also, notice the slight slant on the top of the pole at high power, witch seemed to be normal. Signals like this one give us confidence in the method since it appeared to be close to reality. If we zoom to the top of the poles, we see some peaks that could be the damping bars as we can see in graphic 2. This information may be useful in the future to detect faulty damping bars. More sophisticated data analyses should be done for this kind of fault detection.

It is difficult to analyse the magnetic condition of the machine from the information in graphic 3. The next step may be to zoom the full size value and to see only the peaks from the flux signal as shown in graphic 4.

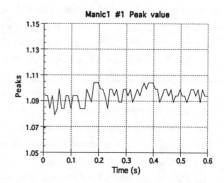

Graphic 4. Peak value of relative magnetic flux for 1 turn.

Graphic 4 is a much better view of the magnetic uniformity. Some may like it better in a polar representation. With the use of proper data acquisition and computer treatment, both representations are easily done as we can see on graphic 5.

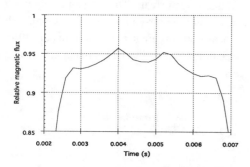

Graphic 2. Top of one pole at 10MW

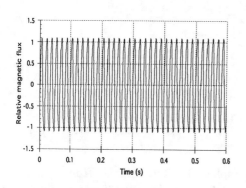

Graphic 3. Signal from electronic conditioning circuit for one complete turn. (72 poles at 40MW)

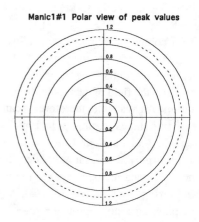

Graphic 5. Peak value on a polar plot to observe the roundness of the magnetic field.

508

The slight flux variations on this particular machine visible on graphic 4 and 5 indicate that there is no problem on the poles. Those waves may be caused by air-gap variation. Air-gap is easily measured with the AGMS. In graphic 6, we can look at air-gap and relative magnetic flux at the same time. This machine has no problem. It is mechanically and magnetically round. In the future, we would like to install a temporary short in a pole to identify the particular pattern for this kind of problem.

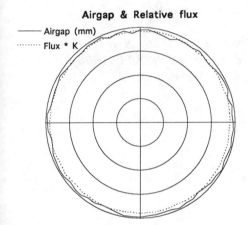

Graphic 6. Absolute air-gap and relative magnetic flux.

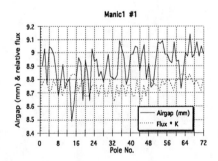

Graphic 7. Zoomed absolute air-gap and relative magnetic flux for 1 turn.

Note that full integration of air-gap and magnetic field data analyses are not done yet. So graphic 6 and 7 are real data plotted on the same view but taken at different time on the same machine. The information were manually treated to fit on those graphics.

From all the information that we get from our first set-up in Manicouagan 1, we think that there is no single instrument able to get positive diagnostic of generator vibration. Magnetic field measurement may be very useful especially when it is coupled with air-gap and power measurement. In this case, only a single magnetic sensor may be enough.

Magnetic flux measurements will detect interturn short on most machines. According to computer simulation, on a 72 poles machine with 20 turns per pole, a single winding short on one pole will reduce the magnetic flux by ≈5% on this pole as shown on graphic 8.

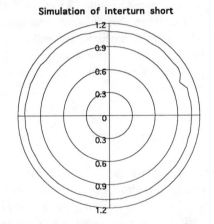

Graphic 8. Simulation of a interturn short causing a 5% magnetic flux diminution in relative unit.

6   FUTURE GOALS

In the future, we would like to understand more about the measuring system and the influence of magnetic field unbalance on the machine. More sensors will be permanently installed on machines and more specific tests will be tried to study this system. The first tests with the equipment was done on a machine in good condition. In the future, the method will be applied where we suspect rotor problems. Also, we will focus on data acquisition, software analyses and integration of other parameters like air-gap, power etc... For this task, we will cooperate with our subsidiary which commercializes the Air Gap Monitoring System. (Vibro-meter)

7 CONCLUSION

There is a lack of magnetic field measurement devices in the generator instrument market. The development of the described system will be very useful even if it does not give calibrated data. Relative data is enough to compare magnetic flux on each pole and detect interturn shorts. More development will lead us to more sophisticated analyses that could help us. One day, maybe the method will be applied on most of the generators as a standard instrument.

8 ACKNOWLEDGEMENT

The author wishes to thank Mr. J.M. Bourgeois from Hydro-Québec, the power plant personnel in Manicouagan region and Vibro-meter staff for there cooperation in the whole project.

9 REFERENCES

Albright D.R. 1971. *Intern short-circuit detector for turbine-generator rotor windings*. IEEE Transaction on power apparatus and systems, Vol. Pas-90, No. 2, March/April 1971.

Conolly H.M. & Lodge I. & Jackson R.J. & Roberts I. 1971. *Detection of interturn faults in generator rotor windings using airgap search coils*. IEE Conf. Publ. 254, 1985.

Posedel, Z. 1991. *Arrangement for detecting winding shorts in the rotor winding of electrical machines*. United States Patent, 5,006,769, Apr. 9, 1991

Tavner, P.J. & Gaydon, B.G. & Ward, B.A.1986. *Monitoring generators and large motors*. IEE PROCEEDINGS, Vol. 133,Pt. B, No. 3, May 1986.

Wood, J.W. 1986. *Rotor winding short detection*. IEE PROCEEDINGS, Vol. 133,Pt. B, No. 3, May 1986.

Hydropower'92, Broch & Lysne (eds) © 1992 Balkema, Rotterdam. ISBN 90 5410 054 0

Monitoring of hydraulic performance of power plants

Leif Vinnogg & Martin Browne
Nybro-Bjerck AS/Norconsult International AS, Nesbru, Norway

Johan Hernes
Nord-Trøndelag Elektrisitetsverk, Steinkjer, Norway

ABSTRACT: Continuous effort is made to utilize hydropower systems so they give maximum benefit to their owners. With this objective, hydraulic condition monitoring (HCM) has recently gathered interest, partly due to recent developments in measuring equipment and computer software.

This paper comments on measuring methods that can be used for calibration of a HCM system, areas of application for HCM and describes a system equivalent to HCM that has been in operation since 1960 at Tunnsjø Powerplant in Norway.

1 INTRODUCTION

In Norway electricity production is nearly a 100% hydropower. Of our total domestic energy consumption hydropower accounts for approximately 50%, fossil fuels 37% and solid fuels 13%.

Obviously it is of great economical importance to utilize our hydropower resources as efficiently as possible. This is important also from an environmental point of view. We would prefer more of our energy production transferred from the burning of fossil fuels to hydropower, but we want to keep as many as possible of our rivers untouched.

We must therefore put a lot of effort into perfecting our hydropower energy production system (including the way we operate it). This involves tasks such as refurbishing/uprating equipment or even changing the total layout of the powerplant(s) and using advanced computer programs for better production planning. Such computer programs already deal with problems like:
- weather forecasts
- runoff models
- energy consumption models
- transfer lines
- detail models of powerplants and systems of plants
- price prediction models

The level of sophistication is already high. However there are still improvements to be made, especially regarding condition monitoring (HCM).

Condition monitoring will be of great value for the following tasks:
- continuous flow measurements
- reservoir management
- trend determination of headlosses
- optimization of load distribution
- determination of a turbines "hillchart"

In Norway we have had a system equivalent to HCM in operation since 1960 at Tunnsjø powerstation. There however the purpose of the measuring system is to monitor the flow through the powerstation accurately.

This paper gives a description of what we consider to be necessary in order to establish a HCM system, areas of application and possible benefits. A brief description of the arrangement and experience with HCM at Tunnsjø powerstation will also be given.

2 DEFINITIONS

In this context condition monitoring (HCM) is:
An activity where <u>measured data</u> are compared with <u>numerical values</u> at regular intervals.

Measured data are process data from a powerplant, such as: pressures, waterlevels and guidevane positions.

The numerical values originate from a mathematical description of the powerplant. We assume that the powerplant consists of a hydraulic system, turbine(s) and generator(s).

Regular intervals can be any chosen time period ranging from continuous measurements to annual registration.

Principal symbols:

$f[x_1, x_2]$	A function of x_1, x_2
g	Gravitational constant (9.81 m/s^2)
H	Head (m)
P	Output (MW)
Q	Flow (m^3/s)
η	Efficiency
ρ	Density (kg/m^3)

Indices:

g	Generator
G	Gross
n	Net
N	Nominal
p	Plant (total)
t	Turbine
w	Waterways, hydraulic system

3 NUMERICAL DESCRIPTION OF A POWERPLANT

We need an accurate description of the powerplant for comparison with measured data. In this section we will make some comments on the necessary relations that should be established in order to incorporate a HCM system.

3.1 The turbine

The efficiency of a turbine is the ratio between the power output on the turbine shaft and the natural power given by the hydraulic head and flow through the turbine. A typical example for a Francis turbine is shown in Figure 1. The example shows the turbine efficiency at nominal head. In the case of a large head variation, the

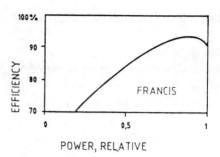

FIGURE 1: Example of turbine efficiency for a Francis turbine.

whole "hill chart" of the turbine must be known.

$$\eta_t = f_1[Q, H_n] \text{ or } \eta_t = f_2[P_t, H_n]$$

3.2 The generator

The generator efficiency gives the ratio between the mechanical energy of the turbine shaft and the electrical energy delivered from the generator. Figure 2 shows a typical generator efficiency curve. (For simplicity we have neglected the influence of reactive load.)

$$\eta_g = f[P_g]$$

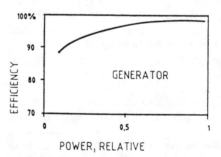

FIGURE 2 : Example of generator efficiency

3.3 The hydraulic system

For simplicity and practical purposes we will assume steadystate conditions in the hydraulic system. The hydraulic system affects the net head (H_n) which must be known. For a plant with several turbines the net head may be

512

different for each turbine depending on the flow through the turbine and usually also the flow through the neighbouring turbine.

Assuming one turbine, the efficiency of the hydraulic system can be expressed as:

$$\eta_w = f[Q] = (H_G - \Delta H)/H_G = H_n/H_G$$

Here ΔH is the sum of all the hydraulic losses from the intake to the outlet (example shown in figure 3. The losses for separate parts of the waterways may of course be computed individually (thrash racks, tunnel, penstock etc).

An example of an efficiency curve for the hydraulic system is shown in Figure 4.

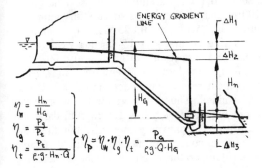

FIGURE 3 : Headlosses/efficiencys for a typical power plant

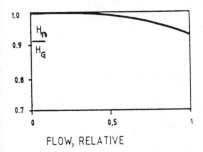

FLOW, RELATIVE

FIGURE 4 : Example of efficiency for a hydraulic system

3.4 Plant efficiency

A powerplant's total efficiency is the product of the mentioned efficiencies when they are a function of corresponding values of flow or output.

$$\eta_G = \eta_w \, \eta_t \, \eta_g = P_g/(\rho \, g \, H_G \, Q)$$

Figure 5 shows an example of a plant's total efficiency curve. In the same figure the turbine efficiency curve is also drawn. Note the difference in location of the best efficiency point! In other words it is necessary to know all the "efficiency curves" to determine the optimum point of operation.

The above may be well known, but the knowledge has not fully been turned to account. However, today the need for optimum operation of plants has increased. Especially the large electricity boards put a lot of effort into getting reliable data on head losses and efficiency data. New plants are usually measured for verification of guarantee terms, but most older plants will probably find it profitable to make repeated efficiency measurements.

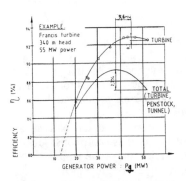

FIGURE 5 : Turbine Efficiency and Total Efficiency

4 CALIBRATION MEASUREMENTS (TURBINE EFFICIENCY)

It is of interest to make a registration of the status when establishing a HCM system. In other words it is an advantage to have a set of values for comparison with measurements.

Measurement of headlosses are usually performed together with the turbine measurements. Good measurements can be difficult to obtain for several reasons: difficulty in keeping steady conditions, accuracy in flow, access to measuring sites, lack of good pressure tappings etc. Headloss measurements will not be discussed regarding establishing status. Instead we will give some comments on measuring turbine efficiency.

There are several methods for measuring the efficiency of hydro turbines. They are described in the IEC-code no. 41 and 198 /1/. Several primary methods giving the absolute values of efficiency have been used:

- thermodynamic
- current meters
- pressure-time (Gibson)
- tracers
- acoustic
- orifice plate
- weirs
- volumetric

In Norway the thermodynamic method is totally dominating. The company Norconsult has carried out about 70 thermodynamic measurements and 5 currentmeter measurements since 1980.

4.1 The thermodynamic method

The theory of the method is old, but in the late 1950'ies the method was developed for practical use as suitable measuring equipment became available. Norwegian research, by professor Alming and his colleagues at the Norwegian Institute of Technology, has contributed greatly to the practical application of the method.

The basic principle is to measure the temperature difference between the inlet and the outlet of the turbine, ie. the temperature rise due to heat generation caused by the hydraulic losses in the turbine. This energy loss in relation to the available hydraulic energy gives the loss percentage and consequently the hydraulic efficiency, see Figure 6. Thus the flow measurement is avoided.

Thermodynamic measurements according to the IEC code are directly applicable as commissioning tests for hydraulic heads above 100 m. The reason for the head limitation is the limitation in the accuracy of temperature measurement. This can best be explained with an example:

If all the potential energy in water 427 m above a certain level is converted to heat for the same amount of water, the temperature will rise 1°C. If the same water is led through a turbine with 90% efficiency then 10% of the energy

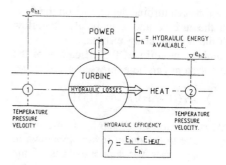

FIGURE 6 : Basic principle og the Thermodynamic method

will be converted to heat, giving a temperature rise of 0,1°C. Assuming a gross head of a 100 m and the same loss of 10% the temperature rise will be 0,023°C. To obtain a measuring accuracy of 1% means we have to measure a temperature difference of 0,0023°C.

With modern temperature measuring equipment resolutions better than 0,001°C can be obtained. Thus it is possible to measure the efficiency at heads below 100 m with the same accuracy as competing methods, but then the measuring conditions must be favourable (stable water temperature). However the uncertainty will increase with reduced head. Figure 7 shows the relation between estimated uncertainties and head for some thermodynamic measurements Norconsult have performed. We have carried out successful measurements for heads as low as 60 m.

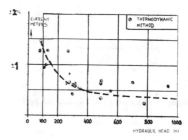

FIGURE 7 : Calculated uncertainty of measured turbine efficiency (Thermodynamic Method)

Note however that we are not talking about absolute temperature accuracies, only accuracy in temperature differences. It is also important to point out that for thermodynamic

514

measurements of turbine efficiency we do not need to know turbine output or flow, thereby reducing one source of error.

Fig. 7 indicates the relationship between the uncertainty and the head. Some calculated uncertainties from actual measurements are plotted.

It appears from the diagram that for heads somewhere between 50 m and 100 m the uncertainty will be ± 1,5%, which is the same uncertainty as can be obtained with currentmeter measurements of efficiency. For higher heads the thermodynamic method will usually be superior.

Thermodynamic measurements have the advantage of being cheaper than most other methods. Rigging time is shorter and performing measurements usually takes only a few hours.

4.2 The acoustic method

This method is relatively new with regard to practical application. It is not yet fully accepted as an absolute method for flow measurements, as the experience of use is limited. The velocity profile in the measuring sections are subject to about the same requirements as currentmeter measurements. The uncertainty under normal conditions is estimated to ± 1,5% for the flow measurements, ie. about ± 1,7% for the efficiency in closed conduits.

An interesting aspect with the method is that it is well suited for permanent installation because:
-the equipment does not block or disturb the flow
-the repeatability is very good.

The equipment is expensive today, but this will probably change as more manufacturers turn up and more experience is gained.

4.3 Relative flow and efficiency measurements

Absolute values of headlosses and efficiencies are not vital when establishing a HCM system. Some times only changes with time are of interest. For such purposes relative flow and efficiency measurements can be used.

The relative discharge can be a chosen

function of one of the following:
-differential pressure originating from Winter-Kennedy pressure tappings, a converging part of penstock, a bend or friction loss for a part of the hydraulic system.
-needle or wicket gate opening (or more available, the servo motor stroke)
-the acoustic method can also be put in this category.

The Winter-Kennedy and the servomotor stroke method has been used for permanent monitoring for 30 years the Norwegian plant; Tunnsjø, described in the next chapter.

5 FLOW MONITORING AT TUNNSJØ POWERPLANT

5.1 Plant description

The layout of Tunnsjø is shown schematically in Figure 8. The plant has the following characteristic data:
- Underground powerplant in operation since 1960
- 1 Francis turbine
- Maximum output 30 MW
- Nominal head 58 m
- Arrangements for accurate computation of flow

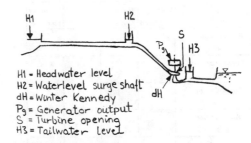

H1 = Headwater level
H2 = Waterlevel surge shaft
dH = Winter Kennedy
Pg = Generator output
S = Turbine opening
H3 = Tailwater level

FIGURE 8 : Tunnsjø Powerplant, location of measurements

Tunnsjø powerplant gets its water from a reservoir which has a natural runoff to Sweden. This water is compensated from a reservoir in another Norwegian watercourse. A condition for building the plant was that the Swedish watercourse should not be influenced by the transfer of Norwegian water.

515

This calls for arrangements for accurate flow measurements. Both the Swedish and Norwegian authorities are supervising the measurements. The flow measuring systems have been calibrated with currentmeters 3 times until now.

5.2 Measuring systems

Three different measuring systems are used for calculation of discharge:
1) Servostroke - head
   ($Q = f[$turbine opening, $H_n]$)
2) Winter-Kennedy
   ($Q = f[$differential pressure in spiral])
3) Efficiency method
   ($Q = f[P_g, \eta_g H_n, \eta_t]$)

There is actually a fourth, based on the headloss in the headrace tunnel, but it is normally not used.

All methods have to be calibrated through accurate measurements. Having done this, we then get independent comparable measuring systems. Table 1 shows which measurements that are used in each measuring system. Figure 8 shows the location of measurements in the powerplant.

Table 1: Measuring system and measurement dependancy

	Stroke and head	Winter Kennedy	Efficiency	Headrace tunnel
Reservoir level H1				X
Surge chamber H2	X			X
Turbine spiral dH		X		
Servo stroke S	X			
Generator output Pg			X	
Tailwater level H3	X		X	

Absolute calibration (with currentmeters) has been performed every 10 year.

Until last year recording of measured values (waterlevels, pressure, servostroke and power) was done by registration on hardcopy devices (pen recorder). The process was time-consuming, the recordings had to be collected, read and digitalized before the calculations could be done. Last year an automatic system was installed. The new system has digital transducers, data transmission and computer analysis. Thus it is very efficient and irregularities can be discovered quickly.

5.3 Measuring equipment

The new automatic measuring equipment consists of the following:

-**Waterlevels** are measured using digital encoders (optic) connected to floats in separate measuring wells. They are calibrated regularly against manual waterlevel probes lowered into the wells.
-**Winter-Kennedy** pressure difference is measured with an analogue precision transmitter(4-20 mA). Signals are converted in a computer. Measurements are calibrated regularly with an U-tube water manometer.
-**Servomotor stroke** is measured using a digital encoder wheel where linear movement is transferred to rotation by wire. Values are calibrated using the fixed mm-scale on the regulating ring.
- **Generator output** is measured using values from the power stations control system.

5.4 Data processing

The measurements go directly into the computer in the power station where they are stored. The flow is then calculated with all three methods. Computed flows have the following priority:

1) The servostroke-head method
2) Winter-Kennedy method
3) The efficiency method
(The headloss method is not fully automatic.)

The measured and calculated mean values are normally printed every hour, or on demand. Results appear in 11 columns:

- Time
- Water level in surgechamber

- Tailwater level
- Servostroke
- **Discharge** calculated from servostroke and head
- Differential pressure
- **Discharge** calculated from Winter- Kennedy
- Energy production in one hours
- **Discharge** calculated from efficiency
- Chosen value for DISCHARGE
- Volume of water discharged per hour and day

Every hour the three discharge values are compared to see if the deviation is reasonable and within defined limits. The chosen discharge will then be according to a set of rules, using the priorities mentioned above.

As the powerstation is unmanned and remote controlled, data are also transferred to a computer in the control centre of the electricity board by modem. Accumulated values per week and month are computed and the water resource account is completed.

If the telephone connection should fail, data for 10 days can be stored locally at Tunnsjø in the computer. Output to the printer in the powerstation will continue every hour.

5.5 Flow measurement uncertainty

The uncertainty in the different measuring methods has so far been investigated by comparing the methods with each other. Between half and full load the differences varies between these limits:

    Servostroke-head / Winter-Kennedy $\pm 1\%$
    Servostroke-head / Efficiency $\pm 2\%$

Close to the best operating conditions for the unit differences are significantly smaller. However, the differences are systematic and an adjustment of the formulas can reduce deviations considerably so that only random errors will dominate. This will be done in connection with the absolute calibrations planned for next year, after a main service on the unit.

6 HYDRAULIC CONDITION MONITORING (HCM), AREAS OF APPLICATION

Hydraulic condition monitoring has lately been discussed a lot among engineers in Norwegian powerplants and in the related research institutes. The motivation for this has been mentioned in the introduction of this paper, but recently also the means of achieving the object are becoming more available, such as:
- The development of affordable high precision pressure gauges with outstanding long term stability
- Low priced sophisticated software for statistical analysis and error detection of measurements
- Affordable communication equipment (which is becoming all the more common as powerstations are gradually becoming remote controlled anyway)
- Low computer costs

The question is what can we expect to achieve with HCM?

Monitoring the flow accurately as a part of the water resource control, can definitely be done to a high degree of accuracy as proved at Tunnsjø.

Monitoring of changes in turbine efficiency is more difficult and even doubtful, because of limited measuring accuracy. In a typical case the reduction of turbine efficiency may be 0.5% in 10 years. This is probably close to the life expectancy of the measuring equipment. Only in cases where there is an severe wear could we expect to use HCM to detect trends for change in turbine efficiency.

Using Tunnsjø as an example we can suggest what kind of information that HCM can give. The three measuring methods behave differently when a change in one part of the system occurs:
- The Winter-Kennedy method depends only on the quality of the pressure tappings in the spiral. It is not influenced by change of headloss in the penstock or reduced efficiency of the turbine (worn wicket gate or damaged runner).
- The servostroke-head method is influenced mainly by worn wicket gates and servomotor connection, and to a certain extent by increased headloss in the penstock.

- The efficiency-method is influenced by changes in losses in penstock, turbine and generator, in other words by all changes.
- The friction in the headrace tunnel depends only on occurring losses between the intake basin and the surge shaft.

Consequently we can detect irregularities such as:
- clogged thrash rack (ice, debris etc)
- increased tunnel friction (rockfall, air pockets etc)
- obstacles stuck in the turbine
- sudden damages to the turbine
- increased friction in the penstock
- instrumentation failure
or other changes such as:
- the effect of a turbine rehabilitation
- change of headloss due to sandblasting and painting a penstock

HCM will also provide new information not previously available, such as the complete hill chart for the prototype turbines.

The benefit of the monitoring must be considered in each case and will affect the arrangement and procedures for the HCM system.

For the time being we are of the opinion that normal wear in the turbine should rather be detected by regular inspections or special efficiency measurements. Headlosses can easily be measured separately.

7 AUTOMATIC OPERATION OF POWERPLANTS

In the future modern process control systems will surely include what we have called HCM, software for computing optimum setpoints for each unit and the ability to automatically change the units setpoints.

Modules for such systems are already available today. Norconsult has had software for calculating optimal setpoints commercially available since 1985. We have also, together with ABB, made a special version of this that directly communicates with the powerplants control system. Such a system has been in operation for approximately a half year. Finally

we are preparing the development of a system for HCM.

The benefits of such a system should be obvious. Maximum production would be achieved through:

- optimum setpoints at any time
- updated optimization input
- early warning of errors can prevent serious damage
- warnings of excessive headlosses
- deciding optimum intervals for maintenance

The disadvantage could be that fully automizing usually means less staff at site, which in turn means less detailed knowledge about each plant. This can sometimes turn out to be a safety risk.

8 CONCLUSIONS

Developments in measuring equipment, software and computer technology have made hydraulic condition monitoring more feasible.

Practical experience shows that systems can be made to monitor flow with an accuracy close to $\pm 1 \%$ and that this can probably be improved.

Changes in hydraulic losses is one obvious area of application for hydraulic condition monitoring, whereas detection of changes in turbine efficiency is less likely.

Intelligent systems will be developed incorporating modules for condition monitoring and optimization (of setpoints).

Reference
1. International Electrotechnical Commission, Code no. 41 and 198 "International Code for the field acceptance tests to determine the hydraulic performance of hydraulic turbines, storage pumps and pump-turbines".

Hydropower'92, Broch & Lysne (eds) © 1992 Balkema, Rotterdam. ISBN 90 5410 054 0

Decision support for condition-based maintenance of hydropower plants

Thomas C.Wiborg
ABB Energi AS, Drammen, Norway

ABSTRACT: A system for condition-based maintenance planning is being developed by ABB in cooperation with, Nybro-Bjerck, EFI, Kværner Eureka and some of the utility companies in Norway. The project started in 1990 and it will be finished in 1992. The project has been supported by the Norwegian Government and the utility companies' branch organization, VR. The goal has been to establish the most salient tools and methods necessary for practical implementation of "condition-based maintenance" (CBM) in hydropower plants. For this purpose, it is established a comprehensive computer-based archive and a knowledge-based system for interpretation of measurements and observations.

1 A QUESTION OF PHILOSOPHY:

ABB Energi has noted an increasing interest among utility companies for better decision support routines related to optimized operation and maintenance of the hydropower plants.

20-30 years ago, maintenance planning systems were introduced in many utility companies. A card-based system called "Helledal systemet" found wide acceptance. To day we see related systems in computerized form being implemented.

While these administrative systems serve their purpose very well, they are to a large extent based on a traditional maintenance-philosophy: "Time-based preventive maintenance".

Industries of to day have often realized that a more flexible approach is more economic. On the economic balance sheet there is always a question of weighing inspection and repair costs against the gains in safety, efficiency and availability. Another important consideration is to determine the optimal time for investing in a major refurbishment, often including uprating of the plant to higher output rating and improved efficiency.

The gospel of "Condition-based Maintenance" (CBM) has often been given only lip-service by those responsible for plant operation. One may wonder why our branch of the industry has been slower than aviation, shipping offshore and process industries in this respect. There are probably several reasons, one of them being that we are concerned with equipment with overall outstanding reliability and "economical life-span" compared to the other branches mentioned.

2 NOT AS EASY AS IT SEEMS...

Common sense tells us that we shall not spend efforts and money on actions which are not called for. Sticking strictly to pre-determined maintenance intervals regardless of the actual need for overhauls etc. is therefore nonsense.

The problem of CBM is that it is meaningless without a rather comprehensive system for information management and thorough knowledge with respect to how measurements and observations are to be interpreted.

This information management requires standardization and a lot of discipline. Knowledge concerning the condition of components requires "hands-on" practical experience. The latter is not easy with remotely controlled power stations. In the "good old days", experienced engineers listened to the "melody of the plant" and could notice subtle changes in the "tune". To day, the ears, eyes and smelling ability of the maintenance engineers need to be replaced by sensors and computers and indeed this is what has happened. The problem is, however, that the experienced human mind is not (as yet) to be replaced by a computer: When you experience a health problem, you prefer to consult a specialist in medicine. You may have to go through a number of tests, and as a layman you are often incapable of interpreting the results and decide the best remedy to the problem. If you are in luck, there are data available concerning your past "performance". If it is, it will help the physician in evaluating your most recent test data. Would you consult the vendor of the testing equipment instad of the doctor? Probably not! Using this as a parable to machinery diagnosis, we must realize that those who sell advanced instruments for CBM may have limited knowledge about how to take actions based on the measured data.

3 THE NORWEGIAN PROJECT

The conception of the project was linked to the merger between ASEA and BBC and the ABB's acquisition of EB in Norway. People with experience in the field of "Condition Monitoring and Diagnostics" know that the "bottleneck" within CBM is the detailed diagnostic knowledge related to the various inspection and test methods. As a result of the merger, there was an opportunity to pool together and consolidate a vast amount of experience from several countries and this is indeed what we have done. To get the optimal result, we established a project and cooperation not only between the ABB companies, but also between ourselves and the utility companies in Norway, EFI, Kværner and Nybro-Bjerck. In course of the 3 year project, we have seen that this was a good strategy for bringing the CBM-philosophy into practical results for the industry.

There were other influencing factors as well:

1: An increased focus the later years on how to obtain better utilization of existing hydroelectric plants, -as an alternative to new development projects.

2: Many plants have come to an age where it is natural to consider repair and refurbishment to maintain reliable operation.

3: The strong decline in construction of new hydropower plants the later years has led to a "manpower drain". This will also be a

"know-how drain" unless the experience from development and operation is better systemized and documented for the future.

4: The general proliferation of low cost computer resources and the recent advances in development of "Knowledge -Based Systems" (KBS) have made implementation of computer based archives and decision support systems more attractive.

3.1 Goals

The goal has been to enable personnel in the utility companies to implement CBM.

A basic tool in this respect is a set of CBM handbooks which are guidelines into how various components may be monitored, how they may fail etc. The books describe the various components with respect to function, wear and ageing as well as the salient measuring and inspection methods available for assessing their condition. Most important; the books describe standardized report sheets to be used during measurements and inspections. Also described are the principles in translating historical data, the "anamnesis" of the components, and the current observations, -into advice for maintenance actions.

Said in other words: The books explain how and where it is feasible to implement CBM and they specify the input to the administrative system.

Also to be developed is a database archive and "interpretation modules" based on KBS technology which serve as the man-machine interface.

Fig. 1 illustrates the principles of such a database/decision-support system.

To be further developed are also the criteria for interpretation of measured data. This involves

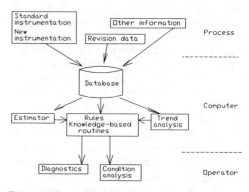

Fig. 1 KBS-based decision support.

development of observation and measurement techniques and systems for data reduction. Data programs which extract information from a database containing diagnostic data from a vast number of tests and inspections will be developed.

For the generator part of the system, this enables ABB to supply the individual KBS systems supplied to customers with the appropriate decision criteria. As already mentioned, the cooperation between ABB companies allows pooling of vast amount of experience from the diagnostic work in the companies.

It shall be possible to transfer data from the utilities data archives to ABB so that expert assistance in interpretation and decision making may be supplied.

A prototype of the system will be tried out in a hydropower plant.

4 RESULTS AS OF MARCH 1992

4.1 Mapping the needs for CBM

The damage statistics from The Norwegian Power Pool, ("Samkjøringen") were studied to establish which were the components causing most loss of production-hours and most frequent outages. As it is seen from table 1, the generator has the highest fault-hours and the control systems had the highest fault frequency.

Table 1

Fault-hours and faults per plant-year. From 995 plant-years 1984-88

	F.-hours	F.-number
Contr. syst.	3.2	0.6
Turbine etc.	2.8	0.1
Generator	14.7	0.2

These values should be used with caution as it is known that faults occur that are not reported. We also know that a small number of dramatic events tend to raise the value of the fault-hours data.

For the generator, it is the stator that accounts for most of the faults and this is hardly surprising. The stator windings and the stator core are prone to ageing from thermal-, electrical- and mechanical stresses (The "TEAM stresses").

4.2 Handbooks

We can not expect CBM to be implemented in hydropower plants before instructions about measuring programs and basic rules for interpretation of measured data are made available. As of to-day, draft versions of handbooks for

* The stator winding and core system
* The stator frame
* The rotor
* Shaft vibration measurements

-have been made available for the advisory group of the project.

Handbooks for bearings, the turbine and the control system will be ready within a few months. The photo in fig.2 shows the 4 handbook drafts.

Fig. 2 The handbooks.

These books are "textbooks" for understanding the "hows and whys" of diagnostics. The main chapters are:

1 Component descriptions: The typical designs and their different attributes.

2 Damage mechanisms and their interactions. Factors influencing the ageing/damages and consequences of a progressing damage development.

3 Measuring and inspection methods. This also includes advice as to the various methods' usefulness for detection of the various failure types, risk of interferences and misinterpretation etc.

4 Measurement programs. This is a description of measurements and observations to be carried out and guidance as to how to quantify the outcome so that it may be compared to previous and later inspections/measurements. It also includes standardized report sheets.

5 Principles for decision making based on analysis of recorded data and trend development

As for the last item, the handbooks can not give exact criteria for evaluating the individual machine but they demonstrate how the data archive may be used for decision support. When implementing CBM according to the guidelines in the books, the user will need expert support to determine the decision criteria parameters for the individual technical installation.

4.3 Decision support criteria

There has been 2 main activities for establishment of improved criteria for decision support:

- Further development of Partial Discharge Analysis (PDA) techniques

- Establishment of a common diagnostic database within ABB

Within PDA, several consecutive measurements have been made on generators before and after various maintenance actions to study the methods' sensitivity and repeatability. Similar tests have been done in laboratories where defined faults and changes in environmental conditions can be simulated under controlled conditions. The equipment used have been the Canadian system from Ferguson and the PRPDA (Phase Resolved PDA) developed by ABB. While the former has been used extensively by many utility companies for many years, the latter has special merits in giving more detailed information about the state of the insulation system. This, however, poses a certain problem when it comes to reducing the amount of parameters to be included in trend-analysis of the machines.

Another ABB PD-system, "PAMOS", has also been tested and compared with the two others. PAMOS has better possibility for continuous monitoring of PD activity.

The efforts with establishing a common ABB diagnostic database have good progress. The type of database chosen is ORACLE-SQL and the format of all the data is agreed upon. The data pooled together this way does not include proprietary data such as customer names etc. but only generic information regarding the generator design, its condition and measured values. This common database will be a valuable tool for ABB when giving advice from diagnostic measurements and when setting up CBM systems for the individual customers.

4.4 EDP tools

Efficient use of CBM is hardly possible without using state-of-the art EDP tools. The system outlined on fig. 1 is near to complete when it comes to the data archive part. The data archive consist of all the data specified in report sheets in the "handbooks".

The decision support rules are contained in "Interpretation modules" designed by the use of KBS technology. The rules follow the logical model illustrated in fig. 3:

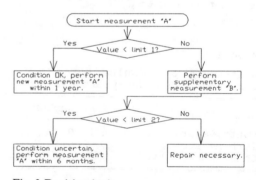

Fig. 3 Decision logics

The criteria needed for the various decisions depend on a number of factors pertaining to the generator design and dimensions, age, type of operation etc. A matrix of such parameters has been established. It is the experience from generators with similar configuration of these parameters, combined with the "as new fingerprint" of the individual machine, that establish the decision criteria.

It must be kept in mind, however, that a KBS system for decision support as described above will not be capable of substituting the expert consultant. The measured and inspected data and the documented history ("anamnesis") of the machine will be the facts needed by the specialist in his, or her, judgements.

4.5 A KBS system simulating generator cooling

This system is a stand-alone unit for generator cooling diagnosis. The system, which development is nealy finished, contains a complete electrical and thermal model for a generator. Based on the design data and the main electrical parameters (voltage, current, KVA, Cos φ etc.), the theoretical thermal balance of the machine is calculated. This way, the interdependence of the cooling system parameters such as input and output cooler temperatures, stator core temperatures etc. may be determined. The plant personnel may now compare every recorded temperature value to those of the mathematical model. In case of discrepancies, the KBS system suggest possible reasons according to 150 pre-programmed rules. When several hypotheses are possible, the one with greatest probability is presented first.

The mathematical model makes it possible to follow up trends in temperature development with far better accuracy than when relying on the "non-normalized" temperature readings. Detection of a gradual abnormal development such as increased losses from hot-spots in the stator core lammellations, or less dramatic things such as unclean coolers, -has now become easy.

The system carries out consistency checks and it will give a warning if the data entered are mutually inconsistent. If one of the many temperature sensors in the stator core displays abnormal values, while the others are as calculated, the system will suggest "sensor error" as a possible cause.

A special application of the system is to determine the effect of operating the generator above the rated load limit. This may be a desired option in some instances. In many cases the turbine has higher rated power than the generator. The KBS system calculates the temperature on the field windings, -a temperature which often is a dimensioning limit for the generator.

4.6 The ROtor SEnsor (ROSE)

Most of the project activities have been directed toward systematic use of information for decision support in CBM. This subproject, however has developed a low cost measuring system for transmission of signals from the rotor to the non-rotating environment.

Telemetric equipment for transmission of signals from rotating units without the use of brushes and slip-rings have been available for many years. The major disadvantage with the commercially available equipment has been bulkyness and high cost.

Fig. 4 The ROSE transmitter

ROSE is based on transmitting digital signals from the rotor to the stator with infrared light, - the same principle used when remote controlling a TV or audio equipment. The unit is very compact and it has acceptable production cost.

The transmitter unit (fig. 4), which also contains the multiplexer and the sensor system electronics (except for the miniaturized sensors), is contained in a unit sized like a 10 mm thick credit card. The first version developed contains 8 channels of which 7 are temperature sensors and No. 8 measures field coil voltage. Several (up to 32) transmitter units may be fitted on the rotor. It is specified that it shall withstand a centrifugal field of at least 1000 G and have a measuring accuracy of $\pm 1^0$ K

Power supply to the transmitter is obtained from the voltage drop across one or two field coils. Fig. 5 shows the principles of the transmission system.

The detector unit has a standard RS 232 interface and display of the measured tempera-

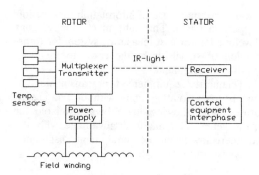

Fig. 5 ROSE -signal transmission principle

tures can be done using a PC or some other computer system.

The prototype system has so far been tested on a laboratory generator and installation in a hydropower station is planned in the near future.

5 FINAL COMMENTS

The project "Condition Monitoring of Hydropower plants" will be finished by the end of 1992. We have seen that a national cooperation, with the utility companies and others, -paired with the cooperation between regional units of ABB, seem to be very successful.

Most Norwegian utility companies seem determined to introduce more efficient maintenance systems. They favor the idea of shifting from "time-based" to "condition-based" preventive maintenance. In our opinion, this change is not merely a question of adopting a new philosophy, - it is just as much a question of training, standardization and having the most salient tools at hand.

In this project, we have taken a practical approach to provide these tools by using the combined skills of the cooperating parties.

6 ACKNOWLEDGEMENTS:

Apart from ABB, the project has obtained economical support from:

- The Royal Norwegian Council for Scientific Research (NTNF)
- The Water System Management Association (VR)
- Statkraft

- Drammen Energiverk
- Sira-Kvina Kraftselskap
- Bergenshalvøens Kommunale Kraftselskap
- Hydro Energi

The technical cooperation has involved

- Nybro-Bjerck AS.
- Norwegian Electric Power Research (EFI)
- Kværner Eureka AS
- ABB Corporate Research, Norway
- ABB Drives, dept. UME, Switzerland
- ABB Generation, Sweden
- ABB Energi AS

Special thanks to mr. Knut Vik who was project manager until he left ABB earlier this year.

5 Optimization of electricity production on regional, national
and international levels – Considering analytic, operational
and environmental issues

Hydropower'92, Broch & Lysne (eds) © 1992 Balkema, Rotterdam. ISBN 90 5410 054 0

Optimizing operational levels in a major hydropower plant

A. Afshar
University of Science and Technology and Mahab Ghodss Co., Iran

A. Salvaitabar
Mahab Ghodss Co., Iran

ABSTRACT: Utilized water power accounts for only 11 percent of the known hydro-power potential in Iran. Implimentation of Karoun IV reservoir will result in 10 percent increase in national peak-power generation. This article discusses the methodology developed and applied to this reservoir site in order to optimize the developed Energy. This optimization is achived by developing an optimum, minimum operating rule curve and associated operating policy. The methodology employs a simulation model to simulate power generation for a given installed capacity. By parametric investigation (Step increament), the best combinations of maximum & minimum rule curve is determined. Determination of the optimum rule curve for a given installed capacity facilities the determination of the contract level of the energy.

1 INTRODUCTION

Karun IV reservoir is one of the major dams in Karun development master plant with main objective of hydropower development. The reservoir will be located at about 670 km. Upstream of river mouth with a drainage area of more than 30,000 km^2. The average annual discharge of the river at the dam-site is in the order of 170 m^3/s.
Although power generation is the main objective of the development, it is believed that the reservoir will regulate the surface runoff and bring about the regulating benefits to the downstream Power Plants and irrigation system. According to meteorological data, the annual evaporation at the site is about 1756 mm per year. To conclude the input data, the total sediment inflow to the reservoir is estimated as 200 MCM for a 50 years useful life.

The main objective of this study is to come up with an operational level (i.e., normal water level as well as dead storage) which is close enough to the optimum. To do so, it was needed to establish a relationship between the power generation indices (i.e., guaranteed output, average annual energy,...) with reservoir operational parameters (i.e., normal and minimum water level).

2 METHODOLOGY

To develop the above mentioned relationships

between the hydroenergy and operational indices, a simulation model was used. In the simulation model, the equal power generation principle is utilized in which the power generated in dry periods should be equal or axceeds a minimum predefined value. Guaranteed output (GP), firm energy (FE), and mean annual energy (MAE) are the hydroenergy indices considered in this study. It is clear that the defined hydro-energy indices are functions of the discharge as well as the net head on the turbines. Therefore, to conduct the study, the following procedure in simulating the system was followed:
-selection of a feasible range for NWL
-preliminary analysis and screening of the relevant dead water level
-simulation study to establish the relationship between NWL and hydroenergy indices
-decision making based on the indices

3 DEAD WATER LEVEL

The minimum permitted water level in the normal operation of a reservoir is defined as the dead water level (DWL). In a hydro-power reservoir, the DWL selection has a direct and prononced effect on the power that is generated. Low DWL will reduce the total amount of spill, hence, will increase the regulated discharge. On the other hand, low DWL will reduce the average head acting on the turbines, resulting in a partial reduction of power. Analysis of the effects of DWL was carried out on the maximum guaranteed output. This criteria was chosen based on the fact that, in the peak regulating power plants, the guaranteed output plays an important role in the operation of the plant in the grid system.

Results of the simulation runs for different NWL's are presented in Fig.1. It can be seen that as the DWL approaches the NWL, the power plant acts like a run-off river plant. In this case the difference between the guaranteed outputs of the selected NWL's are only due to different heads on the turbines. By lowering the MWL, the guaranteed output increase. The rate of increase becomes very slow after the first 30-40 meters in high NWL's.

4 RESERVOIR SIMULATION

As mentioned earlier, the principle and mode of operation is the equal power generation. Therefore, in any month, it is supposed to produce the guaranteed output unless it is impossible. To manage the reservoir operation, the actual power output is determined based on the actual water level at the end of each month. If the water level is higher than the NWL, the discharge is increased to maintain the normal water level. In this case, a comparison is done between the installed capacity and the power output. The smaller valve should be chosen as the actual power output. Conversly, if the water level is lower than DWL. If the water level is between DWL and NWL, the guaranteed output will be the actual output. To do the simulation, a computer program was developed and used in this study.

Through the reservoir simulation study, the relationships between the normal water levels and the hydroenergy indices were developed. Results are graphically depicted in Fig.2. From the simulation results, it may be concluded that:

The relationships between the normal levels and the average annual energy potentials are nearly linear. The average energy potential increase 245 G.w.h for each 20 meters increase in the normal water level.

The relationships between the guaranteed outputs and the normal water level, above the elevation of 980 m.a.s.l is also nearly linear. For each 20 meters rising of the normal water level, the guaranteed output increases 24 MW. But below the elevation of 980 m.a.s.l, the increment of guaranteed outputs is only 20 MW for each 20 meters rising of the normal water level.

The firm energy, increase fastly above the elevation of 1020 m.a.s.l, i.e. about 207 G.w.h for each 20 meters increase in normal water level. Below the elevation of 1020 m.a.s.l, the average increment value is about 135 G.w.h for each 20 meters rise in normal water level.

5 INSTALLED CAPACITY VS.ABSORBED ENERGY

Relationship between the installed capacity and the average annual absorbed energy is given in Fig.3. It is interesting to note that the average annual absorbed energy is a linear function of the installed capacity below the value of the guaranteed output. For installed capacity exceeding 400 M.W., the rate of increase in annual absorbed energy is very small. In fact, an installed capacity of 1000 M.W. (i.e., Plant factor of 21.32%) will absorb 99.3 percent of total annual potential energy.

Fig.4 is a graphical representation of the power duration curve of the Karun IV reservoir. It shows that about 46.7 percent of the time the proposed power plant will operate at guaranteed output and 53.3 percent of the time will produce more output. With an installed capacity of 1050 M.W., the average annual absorbed energy has been extimated as 2135 G.w.h., out of total potential of 2147 G.w.h.

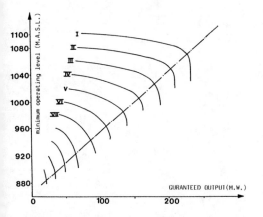

FIG.1 MINIMUM OPERATING LEVEL V.S. GURANTEED
OUTPUT FOR DIFF. N.W.L.

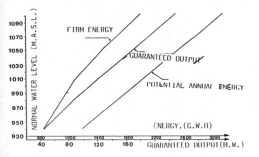

FIG.2 NORMAL WATER LEVEL V.S. SELECTED
ENERGIE INDICES

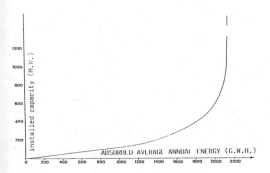

FIG.3 INSTALL CAPACITIES V.S. ANNUAL
ABSORBED ENERGIE

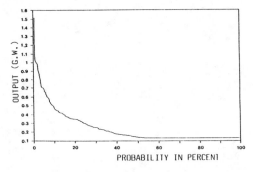

FIG.4 POWER OUTPUT DURATION CURVE

REFERENCES

WARNICK, C.C.1984. HYDROPOWER ENGINEERING

CREAGER, W.P. 1950 HYDROELECTRIC HAND-
BOOK

BARROWS, H.K. 1955 WATER POWER ENGINEER-
ING

CHOW, V.T. 1964 HANDBOOK OF APPLIED
HYDROLOGY

Hydropower'92, Broch & Lysne (eds) © 1992 Balkema, Rotterdam. ISBN 90 5410 054 0

Coordination of hydropower production with thermal generation for enhanced multinational resource utilization

O.J. Botnen, A. Johannesen & A. Haugstad
Norwegian Electric Power Research Institute (EFI), Trondheim, Norway

I. Haga & S. Kroken
State Power Board/Norpower, Oslo, Norway

ABSTRACT: Hydro-dominated power systems and thermal power systems often have widely different characteristics in terms of investments as well as operational aspects. Coordinated utilization of such systems is inherently beneficial, and provides potentially for savings for all parties involved.

The paper illustrates the benefits of such coordination by way of overview discussion and evaluation of two cases;

- the NORDEL interplay which is described in terms of respectively 'limited' (i.e. today's) ambition level and an 'extended' ambition level as to coordinated effort, and

- hydro systems as peak power suppliers, exemplified by analyzing a 'pump storage' type cooperative interplay between Norway and Continental Europe.

1. INTRODUCTION

Thermal and hydro-dominated electric power systems have characteristics which make coordination between such systems inherently desirable. Main potential benefits of coordination:

- large variations in annual hydro-power generation can be balanced by importing power from thermal-based systems during dry years and exporting surplus hydro power to thermal-based systems during wet years,

- capacity limited thermal power systems can utilize surplus power capacity in energy limited hydropower systems during peak load,

- the generation capability in cost-efficient base load thermal power can be utilized better thanks to the regulating capability of hydro power,

- normal operating conditions can be restored more quickly aided by imports from neighbouring countries after major operational disturbances in own system.

The mutual operational benefit of coordinating the utilization of power systems varying from purely hydro to purely fossil-fuel based, has been the main incentive for cooperation between the Nordic countries of Norway, Sweden, Finland and Denmark through NORDEL. The NORDEL collaboration is based on utilizing the interconnecting links between the involved countries for coordinated operation of their generation systems.

2. COOPERATION BETWEEN NORDEL COUNTRIES

The first Nordic electric power network was established in 1912 when agreement was reached on the building of an interconnection across the sea between Sweden and Denmark (Zealand). NORDEL was founded in 1963 to expand and organize Nordic collaboration on electric power. This collaboration has been a success, and today the interconnections between the Nordic countries have sufficient capacity to let the NORDEL-system potentially be regarded as a well balanced single system in operation [1]. **Figure 1** shows the capacity picture of the inter-Nordic transmission system as of 1990.

Figure 1. NORDEL-system transmission capacity picture as of 1990 [1].

2.1 System Description

Each of the four interconnected Nordic countries have initially dimensioned their electric power generation systems to meet respective domestic demand. Hydropower has been exploited where possible, mainly in Norway and Sweden. Nuclear power plays a significant role in Swedish and Finnish electric power supply, while the Danish system is wholly fossil-fuel based.

The Norwegian power production system is almost entirely based on hydropower (over 99 %). Annual hydropower generation may vary from 93 TWh to 128 TWh, depending on inflow, with an average of 108 TWh produced annually. Total installed capacity is 27 000 MW. This capacity is divided between 650 plants situated throughout the country, with few plants having a capacity of over 500 MW. Total reservoir capacity is about 80 TWh, about 25 TWh (10 units) of this being large pluriannual storage reservoirs. Annual domestic consumption is about 105 TWh [2].

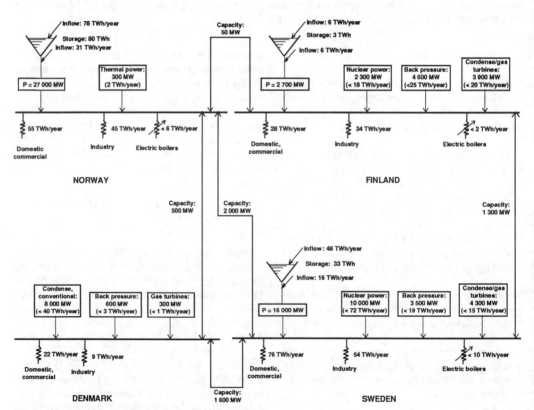

Figure 2. Overview description of the NORDEL system (1990).

The Swedish electric power system contains a mix of hydro, nuclear and fossil-fuel based generation units. The average annual hydropower production is about 64 TWh, with variations in the range of 51 to 72 TWh/year due to inflow variations. Total installed hydro power capacity is 16 000 MW, and the total reservoir capacity is about 33 TWh. Nuclear plants in Sweden have an installed capacity of 10 000 MW, and can at the most produce about 72 TWh/ year. Fossil-fuel based generation in Sweden is normally only 5-10 TWh/year, even though installed capacity in these plants is 7800 MW. Annual domestic consumption is about 140 TWh [2].

Electric power generation in Denmark is based entirely on fossil-fuel fired units, with imported coal being the major fuel. Installed capacity in Danish coal-fired units is 8000 MW. Domestic electric power consumption is about 31 TWh annually [2].

Finland's power generation system consists, as the Swedish, of a mix of hydro, nuclear and fossil-fuel based units. Average annual hydro production is about 12 TWh (installed capacity 2700 MW). Total installed capacity in nuclear units is 2300 MW, allowing an annual production of up to 18 TWh. Fossil-fuel based generation averages 23 TWh/ year. The installed capacity in these units is 8500 MW. Annual domestic consumption is about 62 TWh. Finland's annual import of electric energy is about 11 TWh/year [2].

Figure 2 gives an overview of the electric power systems in these four countries as of 1990.

2.2 NORDEL Rules for Cooperation

Covering domestic demand for electrical power is considered the primary task for the respective national power systems. Foreign trade with electrical power has not been deemed a major activity, - only a marginal short-term activity based on surplus/incremental power exchange, or assistance in cases of component/subsystem failure. This operation-oriented exchange has been motivated by the reductions in operational costs thus attained in the process of meeting domestic demand.

Nordic power collaboration implies economic gains for all participants. The trade with electrical power across borders is based on agreements on exchange of both spot and firm power. The NORDEL organization does not involve itself in expansion planning for the participating nations. Only recommendations are issued as guidelines. Among NORDEL recommendations on Nordic power exchanges are that the participants should:

- utilize the incremental cost principle when scheduling, i.e. let the running costs for the various production units determine their utilization,

- apply comparable methods of calculating the incremental costs,

- use the calculated incremental cost as a basis for agreements on power exchanges,

- aim for equal profit-sharing from bilateral power exchanges,

- limit the price for power exchanges to a certain maximum amount (at present 75 SEK/MWh) above the seller's incremental cost [1].

A prerequisite for participation in the NORDEL spot power market is that each participating nation must have the capacity, both in power (MW) and energy (TWh), to meet its own demand for electric power. Each nation must in other words in principle be self-sufficient in electric power, when agreements on bilateral firm power exchange are taken into account. The NORDEL spot power market is thus one of restricted access. This reflects the mainly operational nature of NORDEL co-operation up to now.

Although NORDEL rules for spot power exchange to a large extent are based on a series of 'gentlemen's agreements' (not written and signed treaties), they may for a purely hydro-based system like the Norwegian, be interpreted to mean that average spot energy import should not exceed average spot export in the long run. Up to now this has not presented any problems, as Norway has been a net exporter of electric energy. This situation could, however, change in the future. Studies presented later in this paper show the consequences of adhering to this interpretation as opposed to 'free access', i.e. full coordination.

3. SYSTEM MODELLING AND SIMULATION

The studies presented in the following parts of this paper are based on system performance simulations using a multireservoir model referred to as the **Power Pool Model** [3]. The simulations were performed using 50 years of historical inflow, with the time resolution being 1 week split into 4 periods (peak load, off-peak day, night and weekend).

The Power Pool Model consists of two parts:

- A **strategy evaluation** part which computes regional decision tables in the form of expected incremental water costs for each aggregate regional subsystem. These calculations are based on use of a Stochastic Dynamic Programming (SDP) algorithm for each subsystem, with an overlaying hierarchical logic applied to treat the multireservoir aspect of the problem.

- A **simulation** part which evaluates optimal operational decisions for a sequence of hydrological years. Weekly hydro and thermal-based generation is in principle determined via a market clearance process based on the incremental water cost tables calculated for each aggregate regional hydro model. Each region's aggregate hydro production is (for each time step) distributed among available plants using a rule-based reservoir draw-down model with a detailed model of each region's hydro system.

Figure 3 shows the regional layout of the system model used for performance simulation. Each region may contain hydro as well as thermal generation capacity in addition to local consumption.

In the system model used, the Norwegian hydro system was modelled fairly detailed, with a total of about 550 plant/reservoir modules distributed between the 12 regional subsystems in figure 3. The Swedish and Finnish hydropower systems, however, were modelled only as aggregate regional models.

4. BENEFIT OF 'LIMITED' (I.E. OPERATIONAL) COORDINATION

4.1 Operation of present system

The Power Pool Model is in this particular study used to quantify the economic benefit of present NORDEL-type operational coordination. Two cases are simulated using the Power Pool Model:

I. Interconnected/coordinated operation: Power exchange according to present NORDEL rules.

II. Disconnected operation: No spot power exchange allowed between countries.

In both cases the present (1991) NORDEL power system is modelled.

Simulation results are shown in (1). All countries

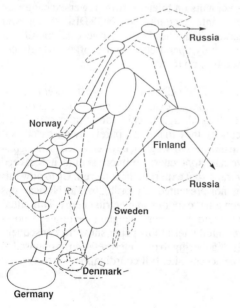

Figure 3. The system model's regional layout.

Economic benefit of limited (i.e. operational) coordination in the NORDEL power system:		
Variable costs, coordinated operation	:	8666 MNOK*)/year
Variable costs, disconnected system	: 9974	-"- (1)
Economic benefit of operational coordination	:	1308 -"-
Economic benefit related to total power delivery	:	0.0043 NOK/kWh

*) MNOK = mill. Norwegian KRONER.
1 Norwegian KRONE (NOK) ~ 0.15 US$.

benefit from coordinated operation. For Norway, the benefit is related to less spillage of water in wet years by way of the export option, and to reduction of curtailment in dry years afforded by the import option. For Denmark and Finland, the benefit is related to the import of surplus hydro power (from Norway and Sweden) at lower cost than own thermal variable costs, and to profitable exports to hydro systems in dry years. For Sweden, the benefit is a combination of similar effects related to Norway and Denmark/ Finland.

Economic benefits in (1) do not precicely include the effect of some minor bilateral firm power contracts within the NORDEL system.

4.2 The Effect of Local Load Increase

Two cases are simulated to illustrate the effect of a load increase in Norway of 2.9 %:

I. The reference system; the present (1991) NOR-DEL power system, as case I in section 4.1.

III. 2.9 % load increase in Norway; otherwise the system is unchanged from case I.

In both cases power is exchanged according to present NORDEL rules, requiring national self-sufficiency in power supply as a prerequisite for access to the NORDEL spot market (see 2.2). In this case we interpret these rules to mean that average spot-type imports to a purely hydro-based system like the Norwegian must not exceed corresponding exports.

In the performed simulations, Norway adheres strictly to this requirement. In case I the stated balance requirement is non-binding, since Norway is a net exporter. In case III however, the restrictions becomes binding.

Simulation results are shown in (2).

The specific costs of this load increase are about 0.20 NOK/kWh. The cost of introducing new generation capacity, on the other hand, would be about 0.25-0.30 NOK/ kWh. So even limited coordination, with Norway strictly adhering to our interpretation of NORDEL rules as stated above, presents a profitable option to increasing generation capacity.

Cost to the NORDEL system of 2.9 % load increase in Norway (3 TWh/year) with Norway strictly adhering to the required national balance in spot exchange:

Variable costs, reference system	:	8666 MNOK/year
Variable costs, 3 TWh load increase in Norway:	9264	-"-
Cost of 3 TWh load increase in Norway:	598	-"-
Specific cost of of 3 TWh load increase in Norway:		0.20 NOK/kWh

(2)

5. BENEFIT OF 'EXTENDED' COORDINATION

In this section the Power Pool Model is used to illustrate the benefits of full operational coordination, giving participating nations unrestricted access to the NORDEL spot market regardless of their national power supply balance. Load increases - wherever they occur, - are covered by optimally scheduling the entire NORDEL system as a single system. Relative to prevailing rules for operation, this strategy will provide for reduced generation expansion in the NORDEL region.

5.1 The Effect of Local Load Increase

Two cases are investigated to yield the main effect of a load increase in Norway of 2.9 %, now with full operational coordination:

I. Reference system, as in section 4.2.

IV. 2.9 % load increase in Norway (3 TWh/ year); otherwise the system is unchanged from case I.

Simulation results are shown in (3).

In case IV Norway has now become a net importer of electric energy. As (3) shows, the specific costs to the NORDEL system of covering a local load increase of 3 TWh/year in Norway, have now been reduced to 0.13 NOK/kWh, compared to 0.20 NOK/kWh in the case of limited coordination in section 4.2.

The Power Pool Model has limited precision in describing system peak load conditions. For checking/optimization of MW-capacities, dedicated power adequacy analysis models have to be applied. With reference to (2) and (3) it is expected that, at the most, small and local MW-reinforcements will be required to sustain the specified load increase of 2.9 % in Norway.

5.2 The Effect of a General Load Increase

In this section the following two cases are investigated:

I. Reference system, as in section 4.2.

V. 2.9 % general load increase; the load is increased by 2.9 % in all four participating countries.Otherwise, the system is unchanged from case I.

The load increase amounts to 9 TWh/year in the NORDEL system. Simulation results are shown in (4).

The specific cost of covering a total load increase of 9 TWh/year in the NORDEL system with existing generation capacity, is evaluated to ~0.14 NOK/kWh, assuming unrestricted (i.e. 'single-owner') operational scheduling of all NORDEL production- and transmission facilities, and the present level of main world market fuel prices.

Increases in peak load will per se eventually require system extensions in terms of MW-capacities, but this is expected to only moderately in-fluence the overall cost picture of the considered scenario.

The cost implied by individually having to introduce new production capacity to sustain a national power balance, will be in the range 0.25-0.30 NOK/kWh.

Thus it has been illustrated - by way of considering the present NORDEL-system -that potentially great savings can be attained by wisely utilizing the combined potential of an interconnected hydro-thermal system.

6. HYDRO SYSTEM AS PEAK POWER SUPPLIER

6.1 General

In thermal power systems MW capacity - and hence also a significant part of total cost of power - is determined by the requirement of power sufficiency during (daytime) peak hours. Electrical energy in terms of off-peak kWhs, is not likely to become a scarce commodity.

In a hydro-dominated power production system (like e.g. the Norwegian), there may at times inherently be an abundancy of peaking capability, since economic system design to some degree provides for MW-capacity to cope with excess inflow situations. Additional peaking capacity may often be established at moderate cost by upgrading and/ or extending existing facilities.

These characteristic differences between thermal and hydro-based production systems give incentive to a 'pump storage' type cooperative interplay between the two types of systems: The hydro system provides for (daytime) peak power, while the thermal system in principle returns the same amount of energy in the course of off peak hours.

6.2 Illustration of Peak Power Supply from Norway to Continental Europe.

Studies have been performed to illustrate utilization of the Norwegian hydro production system as a pump storage facility to typical thermal systems like those found in Continental Europe. Preliminary cost-benefit figures have been established based on the following two studies, using the Power Pool Model:

- Study of present 1991 operation of NORDEL and a Continental European system.

- As above, but now introducing a 1000 MW sea cable between Norway and the Continent, with an associated peak power delivery agreement.

The peak power of 1000 MW is presumed delivered from Norway to the Continent 80 hours/week during daytime periods. Yearly energy supplied is about 4.2 TWh. The specified Continental European system is in this study supposed to return the energy during night and weekend periods with duration 88 hours/week, adding up to about 4.6 TWh/year. Thus the thermal system is responsible for covering all transmission losses (about 5 % each way) related to the deal.

Peak power delivery is modelled as a firm power obligation in Norway and a corresponding peak power injection in the Continental European system. The redelivery during night/weekend periods is modelled similarly, but in the opposite direction.

Operationally, the Continental European system will experience a reduction of its start/stop-costs. Fuel consumption could rise marginally because of the stated covering of transmission losses. The Norwegian/NORDEL system will see only a marginal difference in variable costs.

The need for new power capacity in Continental Europe may be reduced or delayed as a result of an agreement on peak power delivery from Norway. The Norwegian hydro production system may on the other hand have to expand its MW-capacity and its transmission facilities. The net benefit of such a transaction scheme will be influenced by the following main cost elements:

- Investments related to new and reinforced transmission facilities and hydro capacity expansion.

- Reduction or postponement of investments in new peak power capacity in Continental Europe.

- Changes in operational costs in Norway and Continental Europe related to fuel consumption and start/ stop of units.

The results of this study indicate that a Norwegian peak power delivery scheme could prove economically feasible.

7. CONCLUSIONS

Considerable savings can be attained by wisely utilizing the combined potential of an interconnected hydro-thermal system.

Models have been developed that can evaluate the energy-, economy- and emission-related behaviour of such an interconnected system, - under varying assumptions as to e.g. rules of trade between participants in the power market.

Applying these models, the paper evaluates and discusses two cases of interest; the NORDEL cooperation, and 'pump storage' type cooperative interplay between Norway and Continental Europe.

REFERENCES:

[1] NORDEL 1989. Annual Report.

[2] NORDEL 1990. Annual Report.

[3] O.J.Botnen, A. Johannesen, A. Haugstad, S. Kroken, O. Frøystein:
"Modelling of Hydropower Scheduling in a National/International Context."
Hydropower '92, Lillehammer, Norway. June 16-18, 1992.

Hydropower'92, Broch & Lysne (eds) © 1992 Balkema, Rotterdam. ISBN 90 5410 054 0

Long-term scheduling of hydro-thermal power systems

Anders Gjelsvik, Tom A. Røtting & Jarand Røynstrand
Norwegian Electric Power Research Institute (EFI), Trondheim, Norway

ABSTRACT: Methods for solving the long-term scheduling problem in a hydro-dominated power system are reviewed. Especially the paper aims at a mathematically oriented exposition of the single-reservoir model frequently used in Norway. Two other techniques that allow more detailed models are more briefly described, and modeling problems related to the new structure of the Norwegian power market are discussed.

1. INTRODUCTION

The objective of the long-term generation scheduling in a hydro-thermal power system is to find a hydro release policy that minimizes average operating costs. These costs include thermal generation costs, social costs in case of load curtailment, and the net expence of buying and selling of power.

In operation scheduling, the results of the long-term scheduling are used to set targets for the seasonal and short-term scheduling tasks that take care of the detailed operations of the hydro system. The tools of long-term scheduling are also used in generation and/or transmission expansion planning, to carry out simulation studies. Since the water storage capacity of the reservoirs is limited, the scheduling problem essentially is to strike a balance between present and future use of water for hydropower generation. If too much water is used at present, shortage may occur later, leading to high costs. If one is too conservative, spillage may occur.

The long-term scheduling problem is stochastic, the main reason being stochastic inflows. Mathematically, it can be formulated as a stochastic optimal control problem. A formal solution can be obtained by a stochastic dynamic programming (SDP). This approach was suggested about 30 years ago (Stage and Larson, 1961; Lindqvist, 1962) and known as the incremental water value method, it has been in use in Norway for many years (EFI, 1974; Killingtveit and Reitan, 1981). The method is described in some detail in the present paper; one reason for this is that it is difficult to find detailed accounts of the method in the English literature. It offers a good starting point for discussion of other methods.

Due to the well-known dimensional problem of dynamic programming, this approach can only deal with one or two reservoirs.

For the multireservoir case, several approximate methods have been reported in the literature. Most of them make some decomposition so that the individual parts can be treated by DP; examples are (Egeland et al. 1982, Turgeon, 1980 and Sherkat et al., 1985).

Recently, Pereira (1989) has published an approach, called stochastic dual dynamic programming, (SDDP) that seems promising for stochastic scheduling of multireservoir hydroelectric systems. This method is briefly described in section 4 of the paper.

2. MATHEMATICAL SYSTEM MODEL

The planning horizon in long-term scheduling is usually set 3-5 years ahead. We divide this time in N stages, usually of duration one week each.

2.1 *Reservoir and power plant*

The system we look at, may have several reservoirs, say m. We define the following quantities, for i = 1, ... ,m and k=1,2,..., N

$x_i(k)$ volume of water stored in the i-th reservoir at the beginning of the k-th stage.

$q_i(k)$ release from the i-th reservoir during stage k.

$s_i(k)$ spillage from the i-th reservoir during stage k.

$v_{1i}(k)$ controlled inflow to the i-th reservoir during k.

$v_{2i}(k)$ uncontrolled inflow to the power station associated with the i-th reservoir

The uncontrolled inflow v_1 goes directly to a power station and cannot be stored. v_1 and v_2 are stochastic variables. There are also other uncertain quantities in the problem, such as demand and market prices, but here we consider only the inflow as stochastic.

Stored water in the reservoirs obey

$$x_i(k+1)=x_i(k)-q_i(k)+\Sigma'q_j(k)-s_i(k)+\Sigma's_j(k)+v_{1i}(k) \quad (1)$$

$$\underline{x}_i \leq x_i(k+1) \leq \bar{x}_i, \ i=1, \ldots, n \quad (2)$$

Σ' runs over the reservoirs immediately upstream to the i-th power plant. $\underline{x}$ and $\bar{x}$ are reservoirs limits. In the following, we shall often denote limits by bars, as in (2), without further explanations.

For the i-th power plant, the total water available during stage k is $q_i(k)+v_{2i}(k)$, and we write the corresponding energy generated as:

$$P_i(k)= \begin{cases} \eta_i(q_i(k)+v_{2i}(k)) & \text{if } \ q_i(k)+v_{2i}(k) \leq Q_i^{max} \\ \eta_i Q_i^{max} & \text{otherwise} \end{cases} \quad (3)$$

η_i is the energy conversion factor of the i-th plant, and Q_i^{max} is the maximum discharge. In long-term planning it is considered adequate to have η independent of the release.

2.2 Power market and electrical network

In long-term sheduling the electrical network is lumped into a small number (A) of areas with one bus in each (e.g. Johannesen and Flatabø, 1989). Power may be transferred between the buses, subject to upper limits on transfer.

In each area there is a firm power demand D_j and a power market where power can be sold and/or

bought, according to daily or weekly decisions. Among the options in the market are power for electrically heated boilers, transactions with the power pool, thermal generation and rationing. We define

$L(k)$ cost of market transactions at stage k

M total number of possible market transactions

M_j set of possible transactions in area j

$c(k)$ price of the i-th transaction, negative for selling

$y_i(k)$ amount traded in the i-th transaction

$t_{ij}(k)$ amount of energy transmitted from area i to area j at stage k.

One has

$$L(k) = \sum_{i=1}^{M} c_i \ y_i(k) \quad (4)$$

$$0 \leq y_i(k) \leq \bar{y}_i(k) \quad i=1, \ldots, m \quad (5)$$

Then the mean power balance for area j can be written:

$$\Sigma'P_i(k) + \sum_{i \in Mj} \sigma_i y_i(k) + \Sigma'' \alpha_{ij} t_{ij}(k) - \Sigma'' t_{ji}(k)$$
$$=D_j(k) \quad j=1, \ldots, A \quad (6)$$

Σ' runs over all power stations in the j-th area and Σ'' runs over all areas connected to area j. α_{ij} is a factor slightly less than 1.0, to correct for losses. $\sigma_i=1$ if the i-th transaction is buying, and -1 if the i-th transaction is selling.

2.3 Objective function

Since the inflow is stochastic, the operating cost of the system is also stochastic, and we decide to minimize

$$J=E\{\sum_{k=1}^{N} L(k) - S(x_1(N+1), \ldots x_n(N+1))\} \quad (7)$$

where $E\{\bullet\}$ denotes expectation over all possible future inflow sequences. S (....) represents the value of stored water at the final stage k=N. It must be included since the horizon is not set at infinity.

540

3. THE ONE-RESERVOIR MODEL

3.1 *Model description*

In section we describe the one-reservoir optimization model based on incremental water values that has been widely used in Norway the last 20 years or so (EFI 1974, Killingtveit and Reitan 1981). This model is a specialization of the model in the preceeding section to the case with one reservoir, one power station and one area.

We try to describe the basic elements in the model but we do not present the details of how the computations are organized . The mathematical derivation to be presented here is along the lines in (Gjelsvik 1980); it is believed to be more compact than earlier analyses.

The incremental water value method for the one-reservoir model arises when dynamic programming is applied to the stochastic optimal control problem of minimizing (7) for one reservoir.

We assume that the inflow sequence is a white noise sequence, that is, the inflows during various stages are independent of each other (see also section 3.6). The storage x can then be regarded as the state variable of the system. We can now apply dynamic programming (Bellman 1957) to the problem.

Further, we assume that for a given stage the release can be adjusted according to the inflow observed.Then we can write the recursive Bellman functional equation of dynamic programming as:

$$\alpha_k(x(k)) = E\{\min_{v_1(k),\, v_2(k)} [L(y_1(k),...y_M(k)) + \alpha_{k+1}(x(k+1))]\}$$

$$k = N, N-1, \ldots, 1 \tag{8}$$

the minimization subject to

$$x(k+1) + q(k) + s(k) = x(k) + v_1(k) \tag{9}$$

$$\underline{x}(k) \le x(k) \le \bar{x}(k) \tag{10}$$

$$\underline{q}(k) \le q(k) \le \bar{q}(k) \tag{11}$$

$$0 \le y_i(k) \le \bar{y}_i(k) \quad i = 1, \ldots, M \tag{12}$$

$$P(k) + \sum_{i=1}^{M} \sigma_i y_i(k) = D(k) \tag{13}$$

$$P(k) = \begin{cases} q(k) + v_2(k) \text{ for } q(k) + v_2(k) \le Q^{max} \\ Q^{max} \qquad\qquad\quad \text{otherwise} \end{cases} \tag{14}$$

$x(k)$, $v_1(k)$, $v_2(k)$ are given quantities and L is given by (4). $\alpha_k(x(k))$ is the expected cost in going from state $x(k)$ at the beginning of the k-th stage to the final stage in an optimal manner. One has

$$\alpha_{N+1}(x) = S(x) \tag{14b}$$

The constraints (9)-(14) are obtained by specializing the corresponding equations of Section 2 to one reservoir and omitting the reservoir and area indices. Reservoir storage is measured in electric energy units so that $\eta=1$ in (3).

In the standard DP approach one uses (14b) and constructs a table of α_N (7) using (8) with k=N. Then α_{N-1} is constructed from α_N in the same manner, and so on.

In the water value method one proceeds similarly, but instead tables of $-\partial\alpha/\partial x$ are constructed. $-\partial\alpha/\partial x$ is the incremental value of stored water. For convenience we introduce:

$$\kappa_k(x(k)) = -\partial\alpha_k(x(k))\,/\,\partial x(k) \tag{15}$$

3.2 *Inflow data used*

It is customary to use 30 to 50 years of observed data for the inflow. For each week, this gives 30 to 50 discrete outcomes for v_1, and v_2. In practical computing one extracts from these observations a few (e.g.7) values that are used to represent the inflow distribution.

Thus, for a given x(k) the minimum in (8) is computed for each discrete inflow value.

3.3 *Finding the optimal release*

This subproblem arises in two contexts: 1) when solving (8) and (2) when the water value tables are used in actual system operation.

Given a reservoir volume x(k) and inflow forecasts v_1 and v_2, and a table of $\kappa_{k+1}(x(k+1))$, finding the release q(k) corresponds to solving the minimization problem inside the brackets in (8). We name the minimum $\Phi_k(x(k), v_1(k), v_2(k))$:

$$\Phi_k(x(k), v_1(k), v_2(k)) =$$

$$\min[L(y_1(k), \ldots, y_M(k)) + \alpha_{k+1}(x(k+1))] \tag{16}$$

subject to (4) and (9)-(14).

We form the Lagrangian

$$L = L(y_1(k), \ldots y_M(k)) + \alpha_{k+1}(x(k+1))$$

$$- \mu \, [x(k) + v_1(k) - x(k+1) - q(k) - s(k)]$$

$$+ \lambda \, [D(k) - P(k) - \sum_{i=1}^{M} y_i(k)] \tag{17}$$

μ and λ are multipliers for the water balance (9) and the power balance equation (13), respectively. The signs are chosen so that the multipliers will be non-negative. The simple bounds (10)-(12) are retained.

From the Lagrangian we obtain the Kuhn-Tucker conditions for optimality (Luenberger, 1973). We shall not state all possible combinations of binding constraints, but will show a few important conditions.

We have, using (4):

$$\frac{\partial L}{\partial y_i(k)} = \frac{\partial L}{\partial y_i(k)} - \lambda = c_i(k) - \lambda$$

Thus, at optimum

$$c_i(k) - \lambda \begin{cases} < 0 & \text{if } y_i(k) = \bar{y}_i(k) \\ = 0 & \text{if } 0 \le y_i(k) \le \bar{y}_i(k) \\ > 0 & \text{if } y_i(k) = 0 \end{cases} \tag{18}$$

that is, λ is equal to the price prevailing in the market if the market exchange is not at a step. When the release $q(k)$ does not hit a boundary:

$$\frac{\partial L}{\partial q(k)} = \mu - \lambda \frac{\partial P}{\partial q} = 0 \;): \mu = \lambda \frac{\partial P}{\partial q} \tag{19}$$

(19) says that the marginal value of an increase in $x(k)$ (or $v_1(k)$) is equal to the marginal value of power, the marginal conversion factor taken into account. With (14), $\partial P/\partial q = 1$.

When $x(k+1)$ does not hit a boundary, one obtains

$$\frac{\partial L}{\partial x(k+1)} = \frac{\partial \alpha_{k+1}(x(k+1))}{\partial x(k+1)} + \mu = 0$$

$$): \mu = - \frac{\partial \alpha_{k+1}(x(k+1))}{\partial x(k+1)} = \kappa_{k+1}(x(k+1)) \tag{20}$$

If $x(k+1)$ hits the lower or the upper boundary, then (20) does not hold. In the case where overflow occur, $s(k)$ is nonzero, giving

$$\frac{\partial L}{\partial s(k)} = \mu = 0 \tag{21}$$

If the lower limit is hit, or if the upper limit is hit without overflow, then $q(k)$ is adjusted to reach the limit in question, and consequently $q(k)$ is in its interior and (19) holds.

In practice, the minimization problem is solved by a search in $q(k)$. Essentially, it is a question about satisfying the single firm power demand (13), using the resources available.

Given a tentative $q(k)$; $x(k+1)$ is computed from (9), and using (20) μ is found from the table of κ_{k+1}, using linear interpolation, or from (21) in the case of overflow. $P(k)$ is given by (14), and the $y(k)$ are then found from (13), loading according to increasing incremental cost. The optimality conditions are checked, and if necessary, $q(k)$ is further adjusted. We note that $q(k)$ can be found without knowing α_{k+1} itself.

μ and λ are found from the incremental cost of the resources in the final solution. Because of (20), together with (24) below, the incremental water value method has often been phrased as "operating along constant water value". (19) and (20) shows that this is not completely true.

3.4 Computing the water value tables

Given a table of κ_{k+1}, we have outlined how to find $q(k)$, given $x(k)$, $v_1(k)$ and $v_2(k)$. In order to step recursively backwards, we must be able to build tables of κ_k from those of κ_{k+1}. Equations (8), (15) and (16) yields

$$\kappa_k(x(k)) = - \frac{\partial \alpha_k}{\partial x(k)} \bigg|_{v_1(k),v_2(k)} = E \left\{ - \frac{\partial \Phi_k}{\partial x(k)} \right\} \tag{22}$$

To find $\partial \Phi_k / \partial x(k)$, we note that $x(k)$ is a fixed parameter in the minimization, appearing in the constraint (9), with the multiplier μ. Sensitivity theory (e.g. Luenberger, 1973) therefore gives

$$\frac{\partial \Phi_k}{\partial x(k)} = \frac{\partial L}{\partial x(k)} = - \mu \tag{23}$$

Inserted in (22), this gives

$$\kappa_k(x(k)) = \underset{v_1(k),v_2(k)}{E} \{\mu(x(k),v_1(k),v_2(k))\} \tag{24}$$

where we have explicitly shown the dependence of μ on $x(k), v_1(k), v_2(k)$.

Equations (23) and (24) are valid whatever $x(k+1)$ is at a boundary or not. The different cases are reflected in the value of μ, the Lagrange multiplier for the water balance equation. Equations (19), (20) and (21) give the possible values.

S in (7) is not known. This problem is avoided by instead requiring that

$$\kappa_{N-52}(x) = \kappa_N(x) \qquad (25)$$

In the computations, the last 52 weeks are repeated until (25) is fulfilled to a satisfactory degree. Proceeding as described here, one obtains tables of κ_k for $k=N, N-1, \ldots, 1$. They are usually computed for 51 equidistant x- values at each stage. Figure 1 shows curves for equal water values for an example.

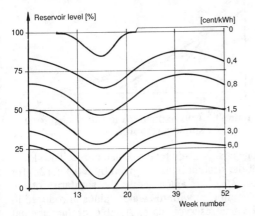

Figure 1. Example of incremental water values for a single reservoir.

3.5 Extensions to the one-reservoir model

A system with two reservoirs can be dealt with in a way similar to the one-reservoir model, but the computation time is much longer. To handle multireservoir systems one may aggregate the reservoirs and plants into a one-reservoir model. However, this may not always be satisfactory. Therefore the one-reservoir model discussed above has been extended into the so-called Power Pool Model, which is an approximate multireservoir and multiarea model (Egeland et al. 1982; Johannesen and Flatabø, 1989). It consists of several one-reservoir systems that are coupled electrically and where the subsystems are optimized iteratively, by incremental water values.

In the one-reservoir model the inflow may

show significant sequential correlation. This can be dealt with by adding a state variable for the inflow. This increases the computational effort.

4. MULTIRESERVOIR SYSTEMS. THE STOACHASTIC DUAL DYNAMIC PROGRAMMING APPROACH

Stochastic dual dynamic programming is a recent approach to stochastic scheduling of hydropower systems (Pereira, 1989, Gorenstin et al. 1990, Røtting and Gjelsvik 1991). As we shall see, it has much in common with the dynamic programming approach to the one-reservoir model, but it can handle multireservoir systems. For a detailed description of the method the reader is referred to (Pereira 1989).

We return to the multireservoir formulation of section 2. The objective to be minimized is given by (7). The inflow model used in the SDDP algorithm is somewhat different from that used in section 3. It is assumed that the inflow distribution in each stage can be approximated by a few (I) discrete values. (This was also done in the one-reservoir model on the level of practical implementation.)

Storage in a reservoir will therefore in principle develop as shown in Figure 3. Separate values of $q(k)$, $s(k)$ and $y(k)$ are assigned to each one-stage transition, shown as "branches" in Figure 3.

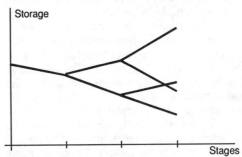

Figure 2. Reservoir development assumed in SSDP.

The solution of the optimal scheduling problem can in principle still be obtained from the Bellman equation (8), which now reads

$$\alpha_k(x_1(k), \ldots, x_n(k))$$

$$= E\{\min[L(y_1, \ldots, y_M) + \alpha_{k+1}(x_1(k+1), \ldots, x_n(k+1))]\} \qquad (26)$$

in the multireservoir case.

543

A direct solution like that of the one-reservoir case is not feasible with several reservoirs. The main idea of SDDP is that with the present linear model each cost-to-go function α_{k+1} can be represented by a set of hyperplanes, as shown in (Pereira 1989). In this way many-dimensional water value tables are avoided. The hyperplane approximations of SDDP are built iteratively, to give increasing accuracy.

Introducing the hyperplane representation, the one-stage subproblem for the minimization in the Bellman equation (26) can be written:

$$\min [L(y_1(k),y_2(k) \dots ,y_M(k),k) + \alpha_{k+1}] \qquad (27)$$

subject to (1)-(6) and

$$\alpha_{k+1} + \sum_{i=1}^{n} \mu_i^r(k)x_i(k+1) \geq \beta_{k+1}^r \qquad (28)$$

$$r = 1,2, \dots , R$$

Here $x_1(k),\dots,x_n(k)$ and inflows $v_{11}(k)$, ..., $v_{1n}(k),v_{21}(k), \dots , v_{2n}(k)$ are given quantities. μ_i^r, $r = 1,2, \dots , R$ are coefficients representing the hyperplanes for α_{k+1} and β_{k+1}^r are corresponding constants. R is the number of hyperplanes generated so far, and increases for each iteration. In this way, the problem is broken down to one-stage subproblems.

The algorithm starts without any hyperplanes, or with a lower bound on the α_k if available. Each iteration cycle consists of a simulation run forward in time and a backward recursion.

One the forward run, the system is simulated for a random sample of inflow realizations. At each stage computing the releases by solving (27) with the constraints (1), (16) and (28). The simulation is carried out for only a modest number of the possible realizations of the inflow. This keeps the computational work acceptable.

The forward run gives the x-values that are to be used when updating the α_k approximations on the backward run. On the backward run, one starts with $\alpha_{N+1}(\dots) = S(\dots)$ and computes improved representations of α_N, α_{N-1} and so on, using (26) with the subproblem as in (27). At a stage x(k) on a simulated trajectory, (27) is solved for all the I possible discrete inflows (not only those used in the forward run). The dual variables of (1) are recorded in each subproblem, and their average computed. This gives new hyperplane coefficients $\mu_i^{R+1}(k)$.

This corresponds to (25) for the single reservoir model, so we see that there is a strong similarily.

The backward run gives a lower bound $\underline{J}$ of the expected operating cost. Averaged over all inflow realizations, the forward simulation gives a corresponding upper limit $\bar{J}$. The algorithm is said to have converged when the cost bounds $\underline{J}$ and $\bar{J}$ are sufficiently close. Because of the sampling variations they cannot be expected to meet exactly. In experiments (Røtting and Gjelsvik 1991) a tolerance of the order of the sample standard deviation of the cost was used. The sample of realizations must not be too small.

Returning to the cost representations, Figure 3 shows a one-reservoir case of SDDP. The straight lines are the "hyperplanes" that represent the cost.

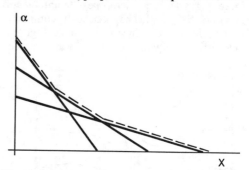

Figure 3. Approximations to future cost for a single reservoir.

The cost is given by the dotted curve. The slopes of the straight lines are given by $\mu^r(k)$ in (28) for r=1,2,3. This again shows the similarity of the $\mu^r(k)$ in (28) with the water values introduced in the one-reservoir case. A table of incremental water values corresponding to Figure 3 would have given slope of the dotted line as a function of x. One may note that in Figure 3, the first cost representation would be a single straight line; the other lines being added as the iterations proceeds, making the cost function more detailed.

In SDDP, the subproblems (27) are solved by linear programming, possibly with specialized techniques (Røtting and Gjelsvik 1991).

We have run demonstrations of the SDDP on stochastic seasonal planning, and the technique seems promising. For application in long-term planning, convergence behaviour with a large enough number of stages must be closer investigated. Another problem is that of finding a good inflow approximation in multireservoir systems.

A difficulty to be further investigated lies in the final stage value function, since the periodicity requirement (25) of the one-reservoir model cannot be implemented here.

5. A TWO-STAGE STOCHASTIC PROGRAMMING APPROACH

In the two previous methods, the stochastic variable has been modeled as able to change at each stage.

An approach that might be computationally less demanding, is to regard the inflow as known up to time N_1 (one to three weeks ahead, say). After that, a number of separate inflow scenarios are possible, say three, each with a given probability.

With linear submodels as in earlier sections, this problem may be formulated as a linear program. It can be decomposed using Benders' algorithm (Benders 1962). The master problem, is taken to be the stages with given inflow, up to $k=N_1$. The remaining stages belong to the subproblem.

In the subproblem, the reservoir contents at stage N_1 are to be regarded as fixed. The subproblem thus decomposes into three separate problems.

If there are no cascaded plants in the system, the subproblems become minimum cost network flow problems. They can be solved very efficiently.

Although the master problem and the network flow subproblems have to be solved several times before Benders' algorithm converges, limited experiments indicate that the decomposition approach is considerably faster than solving the same problem directly by linear programming.

The two-stage stochastic programming approach has the drawback that each scenario is considered too optimistically since the future inflows are not really known. On the other hand, the approach is a simple improvement over deterministic models.

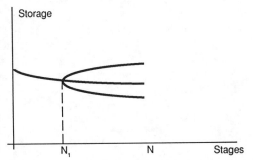

Figure 4. Reservoir storage scenarios.

6. ON THE FORM OF THE SOLUTIONS

In the case of the one-reservoir model, we have a feedback policy. At a particular stage, given the reservoir content, we can find from the water value tables what the release ought to be. This means that during a simulation the water values need not be recalculated unless other parameters change, such as the market description.

With SDDP, the whole computation is carried out with given initial reservoir volumes. The hyperplane approximations obtained for the future cost can be used over some stages, but they will become less accurate if the actual trajectory departs much from the states encountered during optimization. Thus, rerunning of the scheduling algorithm will be needed more frequently than in the case of the one-reservoir model.

With the scenario model, results can at most be used for the first N_1 stages so that frequent recomputations will be necessary.

7. NEW MARKET STRUCTURES

The Norwegian power system has recently become more market-oriented.

The changes that have been made so far, does not change much the market model and scheduling philosophy used here. Better time resolution is needed, but this must be taken care of in the seasonal and short-term scheduling.

It has, however, been considered to introduce a futures market for power. This means that contracts are made for delivery of relatively small amounts of power at some future time. These contracts can be traded by both parts.

In SDP, the process is assumed to have white noise, and the decision on q in (8) is for the first time stage. With futures contracts, one can at each stage make decisions about those contracts, decisions that affect later stages. This means that the state vector should be enlarged to carry information about the contracts, to obtain a Markov process again. However, this would be very difficult due to the dimensionality of the problem. Also, there is the problem of forecasting price changes. This is an issue that will have to be further investigated.

CONCLUSION

We have reviewed methode for long-term scheduling.

For the one-reservoir incremental water value method we have shown that the computation of the water value tables can be interpreted and unified in terms of the Lagrange multiplier for the reservoir balance equation in the one-stage subproblem.

Of the two methods for multireservoir systems discussed, the SDDP approach has a great deal of similarity with the incremental water value method. We regard SDDP as very interesting, although more computational experience is needed with the high number of stages occurring in long-term scheduling.

Considering the new power market in Norway, the most difficult point on the mathematical side, will probably be to deal with futures contracts.

REFERENCES

Bellmann, R. 1957. *Dynamic Programming,* Princeton University Press.

Benders, J.F. 1962. *Partitioning procedures for solving mixed-variables programming problems,* Numerische Mathematik 4: p. 238-252.

EFI 1974. *Economical operation of power systems.* (Committee report, in Norwegian). Trondheim,: Electric Power Research Institute (EFI). Report 1609.

Egeland, O., J. Hegge, E. Kylling, J. Nes 1982. The extended Power Pool Model. Operation Planning of Multiriver and Multireservoirs. Hydrodominated Power Production System A Hierarchical Approach. CIGRE Report 32.14.

Gjelsvik, A. 1980. Stochastic long- term optimization in hydroelectric power systems. Dr.ing. Thesis, University of Trondheim, The Norwegian Institute of Technology, Div. of Eng. Cybernetics, Trondheim.

Gorenstin, B.G., N.M. Campodonico, J.P. Costa, M.V.F. Pereira 1990. A framework for marginal cost evaluation in hydroelectric systems with transmission network constraints. *Proc. Tenth Power Systems Computation Conference.* Graz., Austria, 19-24 August.

Johannesen, A., N. Flatabø 1989. Scheduling methods in operation planning of a hydro-dominated power production system, *International J. of Electrical Power & Energy Systems* 11: 189-199.

Killingtveit, A., R. Reitan 1981. A mathematical model for planning of short and long term operation of hydro power production system. *Proc. Int. Symposium on Real-Time Operation of Hydrosystems:* 1-19. Waterloo, Ontario: University of Waterloo.

Lindqvist, J. 1962. Operation of a hydrothermal electric system: A multistage decision process, *A.I.E.E. Journal,* April: 1-7.

Luenbenger, D.G. 1973. *Introduction to linear and nonlinear programming,* Addison-Wesley.

Pereira, M.V.F. 1989. Optimal stochastic operations scheduling of large hydroelectric systems, *International J. of Electrical Power & Energy Systems* 11: 161-169.

Pereira, M.V.F., G.C. Oliveira, C.C.G. Costa, Kelman, J. 1984. Stochastic streamflow models for hydroelectric systems, *Water Resources Research* 20: 379-390.

Røtting, T.A., A. Gjelsvik 1991. Stochastic dual dynamic programming for seasonal scheduling in the Norwegian power system. Paper 91 SM 487-9 PWRS, IEEE PES Summer Meeting, San Diego.

Sherkat, V.R., R. Campo, K. Moslehi, E.O. Lo 1985. Stochastic long- term hydrothermal optimization for a multireservoir system. *IEEE Trans. Power Apparatus and Systems.* PAS-104.

Stage, S., Y. Larsson 1961, Incremental cost of water power. *AIEE Transactions (Power Apparatus and Systems),* August: 361.

Turgeon, A. 1980. Optimal Operation of multi-reservoir power systems with stochastic inflows. *Water Resources Research* 16: 275-283.

Hydropower'92, Broch & Lysne (eds) © 1992 Balkema, Rotterdam. ISBN 90 5410 054 0

Implementation of an integrated system for short term scheduling in Alta Power Plant

A. Hjertenæs
Siemens A/S, Power Division, Trondheim, Norway

A. Gjelsvik
Norwegian Electric Power Research Institute (EFI), Trondheim, Norway

Ø. Skarstein
Norwegian Institute of Technology, Trondheim, Norway

ABSTRACT: A complete system for optimal utilization of available water resources for hydro electric power plants is described. The system is collecting automatically or manually, information from the Norwegian Meteorological Institute (DNMI), field measurement stations and the general power control system. These informations give the operator the following decision support:

- optimal total discharge for turbines,
- optimal total discharge for escape valves,
- optimal setpoint distribution on turbines,
- optimal discharge distribution on escape valves,
- trend forecasts of forebay level.

The decision support system described in the report has been tested in Alta Power Plant since January 1992.

1 INTRODUCTION

The present paper describes a decision support tool that has been developed for use in Alta Power Plant. The main objective of the system is to give a more efficient use of water and reduce spillage. The system is based on modern equipment available for collecting meteorological and hydrological data in real time. These data are input to a mathematical runoff model named the HBV-model (Nesse et. al., 1982) that predicts water inflows to reservoirs. Based on this prediction decision support to the use of the available water resources in an optimal manner is given.

Alta Power Plant is difficult to operate without water spillage during snow melting season because the reservoir is small. On the average, about 20 % of the water flows past the turbines. Simulation studies, using 70 years of measured flow in the Alta river, indicate that it is possible to increase the energy generated by 2-3 % by improved operation planning (Statkraft, 1991). This shows that scheduling tools are useful.

The system to be described has been developed for Alta Power Plant. However, the system has a modular design and can be adjusted to other power plants. In Section 2 we will give a short description of Alta Power Plant. Section 3 describes the mathematical system model and Section 4 the implementation. Finally, some results are briefly described in Section 5.

2 DESCRIPTION OF ALTA POWER PLANT

Figure 2.1 shows the Alta Power Plant system. The power plant is located in the Alta-Kautokeino river system (Gåserud et al., 1992). The catchment area is 6011 km^2. The average precipitation is 360 mm/year, which is much less than annual average per km^2 in other parts of Norway.

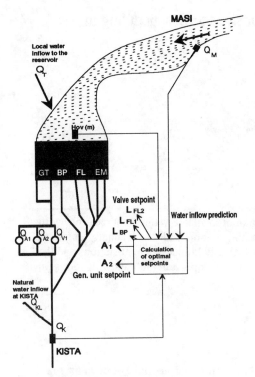

Figure 2.1 The Alta Power Plant system. The dam has several gates and valves. GT are gates for the tunnel system to the turbines, BP are by-pass valves, FL are flood gates and EM are emergency gates.

In summer it will take one or two days before the inflow to the reservoir responds to rainfall in remote parts of the catchment area. In dry periods, the time constant of inflow decline is of the order of one to two weeks.

Alta Power Plant was put into operation during the spring 1987. The average production is 625 GWh over the year. The production is 425 GWh during the summer and 200 GWh winter power. There are two Francis turbines with 110 MW and 55 MW rated power, at 170 m average water head. The maximum total discharge through the turbines is 100 m^3/s.

The manoeuvering regulations for the power plant are rather detailed, and vary through the year. The most important constraint is that downstream flow is required to deviate no more than +/-10% from a "natural" flow.

Figure 2.2 shows a typical water inflow time series over one year. There is a large peak corresponding to the snow melting season. Since the reservoir is small compared to the volume of the peak, some spillage is inevitable. In late spring, the short-term inflow prediction uncertainty increases substantially. This leads to an increased probability of spillage.

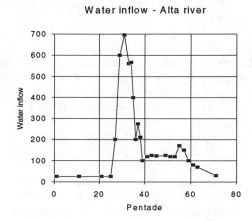

Figure 2.2. A typical year of water inflow in (m^3/s) for the Alta river, given as pentade values (5-day averages).

3 MATHEMATICAL SYSTEM MODEL

The hydro scheduling (Gjelsvik, 1992) is carried out in three steps, where each step gives input data to the next:

1. Seasonal analysis a few months ahead. This gives reservoir target one week ahead for step 2 of the planning.

2. Optimal scheduling for the first week, with target from 1. The results are hourly releases.

3. Optimal allocation of the hourly releases to the two generating units.

Here we shall focus on the scheduling for one week in step 2 above. The mathematical model to be presented, is the one applicable in summer and autumn.

We look N=168 hours ahead, and wish to find a release schedule that maximizes

$$J = \sum_{k=1}^{N} q_1(k)P(k) + q_2\, E(N) \qquad (1)$$

Here:

$P(k)$ = energy generated (MWh) in hour k

$q_1(k)$ = incremental value of energy in hour k in NOK/MWh.

q_2 = marginal value of potential energy in stored water at the end of the week, in NOK/MWh.

$E(N)$ = potential energy in stored water at the end of the week, in MWh.

Formula (1) expresses that one wants to maximize the value of the energy generated, while taking into account that some of the energy should be stored for later use. The sequence $\{q_1(k)\}$ can describe a power spot price that varies with time, as in a power market. In Alta, utilization of the resources has been given priority, so that at present all $q_1(1)$ .... $q_1(N)$ are put equal to 1.

q_2 reflects the value of stored water at the end of the present week. The ratio between the q_1-values and q_2 determines whether one should generate maximally or store water for future use, and is crucial for the nature of the schedule that results. q_2 should be chosen from seasonal considerations, as in step 1) in the planning hierarchy mentioned above.

3.1 Power station model

To simplify the problem, we ignore costs related to start-up and stop of units (mainly loss of water). The power station is modelled as a segmented piece-wise linear relation between the release Q_A and P, the energy generated:

$$P(k) = \sum_{j=1}^{N_{seg}} k_j\, \Delta Q_j(k) \qquad (2)$$

Here is

N_{seg} = number of segments used

k_j = energy conversion coefficient for segment j

$\Delta Q_j(k)$ = release corresponding to segment j in time k.

This representation is for the power station as a whole, and it implies that one uses the best combination of generation units for each value of the release. After the total release has been found, the load on each unit must be determined. This is carried out in step 3) in the scheduling hierarchy.

The energy conversion factors k_j, $j=1,...,N_{seg}$ are proportional to the water head, which in principle is unknown before the schedule has been found. In order to obtain a linear model we ignore the head dependence during optimization. This is corrected for by rerunning the linear model with head updating after a tentative schedule has been found.

3.2 Reservoir model and streamflow constraints

We define the following quantities, all in million m^3/h:

$Q_A(k)$ = total release through the turbines hour k

$Q_L(k)$ = total release through gates hour k

$Q_M(k)$ = estimated inflow to the reservoir hour k

$Q_K(k)$ = reference for downstream river flow at location Kista

$V(k)$ = volume of stored water at the end of the k-th hour.

We have:

$$Q_A(k) = \sum_{j=1}^{N_{seg}} \Delta Q_j(k) \qquad (3)$$

$$E(N) = \sigma\, V(N) + constant \qquad (4)$$

σ in (4) is a fixed coefficient. The allowed range for the final reservoir is given as

$$V_{fmin} < V(N) < V_{fmax} \qquad (5)$$

We assume that the allowed range in (5) is

small enough for (4) being a good approximation without including head variation.

The water balance for the reservoir becomes:

$$V(k+1)= V(k) - \sum_{j=1}^{N_{seg}} \Delta Q_j(k) - Q_L(k) + Q_M(k) \quad (6)$$

$$0 < V(k) < V_{max} \quad (7)$$

The downstream flow constraint mentioned in Section 2 becomes:

$$0.9Q_K(k) < Q_A(k) + Q_L(k) < 1.1Q_K(k) \quad (8)$$

Estimates for Q_M and Q_K are obtained from the runoff model (see Section 4) for the respective catchment areas (cf. Figure 2.1).

3.3 Objective function

With (2) and (4) inserted in (1) the problem is to find

$$maxJ = \sum_{k=1}^{N} q_1(k) [\sum_{j=1}^{N_{seg}} k_j \Delta Q_j(k)] + q_2 \sigma V(N) \quad (9)$$

when the initial reservoir $V(0)$ is given and (5), (6), (7) and (8) are to be satisfied.

3.4 Finding the optimal schedule

With fixed k_j in (9), the optimization problem turns out to be a linear optimal network flow problem (Kennington and Helgason, 1980) that can be solved with very efficient software. In the graph corresponding to the problem, there are $N+1$ nodes corresponding to reservoir balance equations (6) and N nodes corresponding to the streamflow condition (8). The variables $V(1)$, $V(2)$,..., $\Delta Q_1(1)$, ..., $\Delta Q_{N_{seg}}(N)$ etc. are represented as arcs in the graph.

The linear model for Alta Power Plant has 337 constraints and about 1700 variables and is optimized by a general network flow code. Building the model for the network flow code is carried out by a model generator written

specifically for the project.

The schedule obtained has releases for N=168 hours. However, one may desire to perform a rescheduling long before that time has passed, for instance when the inflow deviates from the forecast. Thus we expect that a new schedule is computed almost every hour.

3.5 Finding the endpoint conditions

Some input parameters to the model may be difficult to set. These are the parameters describing the desired condition one week ahead, i.e. q_2 and the target range (V_{fmin}, V_{fmax}). To aid in this, a model with time horizon several months has been designed to cover the seasonal planning, step 1) of the planning hierarchy. It has the same structure as the one presented here, but each time step is one week. For this model the end conditions have to be set in October or November. From this model targets for the first week are found and submitted to the one-week model.

4 IMPLEMENTATION

The decision support system for optimal utilization of water resources is based on mathematical algorithms described in Section 3. The complete system is divided into 2 main parts, the HYDRO PRED system and the HYDRO OPT system (Hjertenæs and Furu, 1992).

Figure 4.1 gives an overview of the HYDRO PRED system. HYDMET is a system for automatic collection of data, control and correction of data. The system includes an application library and report generator.

The box marked HBV contains the socalled HBV runoff model. It is a dynamic hydrological model on state space form. It's states are updated by hourly hydrological and meteorological measurements through a Kalman filter. The parameters are estimated based on historical data from the actual catchment. The model predicts runoff from

550

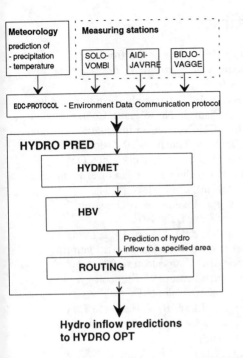

Figure 4.1 Hydro prediction system (HYDRO PRED).

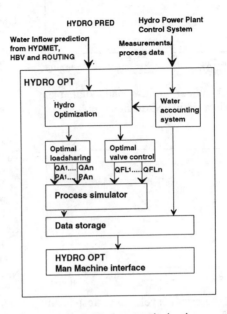

Figure 4.2 Hydro optimization system (HYDRO OPT).

power generated), a detailed simulation can be run. The system also includes a water accounting system, which contributes to the input to the optimization.

the catchment hour by hour one week ahead (Nesse et al., 1982).

The ROUTING module is included to calculate the water inflow to a specific geographical point on the basis of known/calculated inflow to another point. The output of the HYDRO PRED system is predicted water inflow to the reservoir at the dam. This is the main input to the HYDRO OPT system.

Figure 4.2 shows the HYDRO OPT system. The mathematical algorithms described in Section 3 are implemented in the module HYDRO OPTIMIZATION giving as output optimal total water discharge through turbines and escape valves. The optimal loadsharing calculates the optimal distribution of power on available units (turbine + generator). The optimal valve control finds the optimal discharge distribution on the available escape valves. The process simulator includes detailed models for reservoir, pipe, river, turbines and generators. For instance, based on the calculated optimal discharges (and

5 RESULTS

The system described in this paper was set on test and evaluation at Alta Power Plant in January '92, so we have few results from actual operation in the power plant. However, extensive simulations have been carried out. A typical scedule is shown in Figures 5.1 and 5.2 below. The effect of the +/-10% constraint is clearly seen in Figure 5.1. Also, by looking at Figure 5.2, one can see that in the time interval from about hour 75 until hour 165, the system is trying to save water in order to increase the head, thereby increasing the total output over the planning period.

6 CONCLUSION

We have described a system for operator's decision support in Alta Power Plant. We believe that the most important feature of the

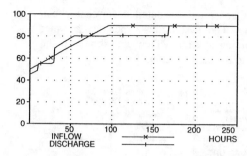

Figure 5.1 Water inflow and total plant discharge in a simulated example (m^3/s).

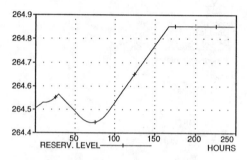

Figure 5.2 Reservoir level (m) corresponding to Figure 5.1.

system is the integration of the measurements and mathematical models.

The HBV runoff model is believed to give significantly improved inflow forecasts, compared to older manual procedures. Based on the inflow forecasts, the short-term hydro optimization part finds schedules so as to reduce spillage and operate the units at the best efficiency possible within the operating constraints. The most difficult part in the hydro optimization is probably the setting of the future reservoir targets. A special model has been devised to deal with that.

Prior to the project, simulation studies estimated a possible increase in generated power of 2-3%. The system is about to be put in operaion, and is expected to be in active use from May '92, which is when snow melting is likely to commence.

REFERENCES

Statkraft 1991. More power from water resources (in Norwegian). *Fossekallen* 4/5:8-9.

Gjelsvik, A. 1992. Optimal scheduling in Alta Power Plant (in Norwegian). Course lecture, The Norwegian Institute of Technology, January 8-10. Oslo: NIF.

Gåserud, Ø. & L. Mannsverk 1992. On optimal scheduling in Alta Power Plant (in Norwegian). Course lecture, The Norwegian Institute of Technology, January 8-10. Oslo: NIF.

Hjertenæs, A. & O. T. Furu 1992. Decision support system for Alta Power Plant (in Norwegian) Course lecture, The Norwegian Institute of Technology, January 8-10. Oslo: NIF.

Kennington, J.L. & R. V. Helgason 1980. *Algorithms for network flow programming* (John Wiley).

Nesse, L., Olaussen, E., Aam, S. 1982. A hydrological model for a catchment in the Sira-Kvina Power Company system. Practical experiences and further developments. *Decision Making for Hydrosystems:Forecasting and Operation.*Edited by T. E. Unny and E. A. McBean. 433:450 Colorado: Water Resources Publications.

Hydropower'92, Broch & Lysne (eds) © 1992 Balkema, Rotterdam. ISBN 90 5410 054 0

Economic-environmental scheduling of hydro thermal power systems

Terje Gjengedal & Oddbjørn Hansen
Norwegian Institute of Technology, Trondheim, Norway

Tom A. Røtting
Norwegian Electric Power Research Institute (EFI), Trondheim, Norway

ABSTRACT: The present paper describes a framework for expanding the hydrothermal operation task from traditional cost minimization to also include environmental objectives. The basic idea is to maintain the physical value of the impact through the decision process, rather than assigning a specific monetary value. A trade-off curve, representing all alternative operation policies, is derived by assigning relative weights to the pollutants. The suggested method, which is based on multiobjective Dual Dynamic Programming, identifies both minimum cost and minimum emission operation of hydrothermal systems. The main contribution of the approach is to indicate how the operation of a hydrothermal system changes as function of the total environmental weight. The approach is illustrated and discussed in a case study.

1.0 INTRODUCTION

In recent years, great attention has been focused on the threat of global warming. Based on models of the atmosphere, many scientists have predicted that the buildup of certain anthropogenic air emissions will cause a "greenhouse effect" by which the earths's temperature will increase. At this time, anthropogenic emission of CO_2 appear to be the principal contributor to this effect. Increasing attention has also been given to the problem of acid deposition caused by emission of SO_2 and NO_x. These gases are primarily emitted during the combustion of fossil fuels.

Environmental externalities will be further more important part in the future resource planning. This favors the newer and more efficient resource options (Cohen 1990, Putta 1990). While much work remains to be done on this important issue, we believe the time is ripe to extend such efforts beyond planning to power system operation. Planning can only go so far, since it is the use of existing facilities that determines the level of environmental pollution from electric generation. Therefore, we want to address the externality issue also in operational decisions, with focus on hydro thermal operation.

Several analysts have studied methods for incorporating environmental effects when scheduling thermal power systems (Delson 1974, Kermanshahi 1990, Bernow 1991). Much of this work is based on monetary valuation of environmental impacts from electricity generation, using a method for monetization, such as damage cost, cost of abatement, etc. The scheduling is performed based on the sum of operation cost and environmental cost during normal operation.

The optimal operation of a hydrothermal system is usually formulated to determine a generation schedule for each plant that minimizes the expected operation cost along the planning period, while meeting, operational and physical constraints. Due to the complexity of the problem, it is usually divided into an hierarchy of subproblems ranging from long-term scheduling to medium-term and short-term scheduling. The results from one subproblem is used as input to the consecutive problem.

The hydrothermal operation task is a single objective problem in that the only objective explicitly included in the criterium function is to minimize the expected operation cost. Throughout the last decade, many authors have contributed to this topic and different methods and approaches have been used for solving the long-term/medium-term operation problem.

The present paper describes a framework where the traditional hydrothermal problem formulation is extended by calculating emissions from electricity production of fossil fuel power plants. The emissions are explicitly incorporated into the criterion function.

The hydrothermal scheduling task is formulated as a multiobjective problem of minimizing operation cost and emission of multiple pollutants. We show how the technical, economic and environmental behavior of the hydro thermal system is modelled. Finally, we describe a framework to solve the multiobjective problem, and illustrate this through a test case.

2.0 PROBLEM FORMULATION

Here we briefly describe the mathematical formulation of the hydro scheduling problem in deterministic form. A more extensive description of models for scheduling is given in (Johanessen and Flatabø, 1989).

The test problem come from a seasonal planning problem with time horizon several months ahead. The initial contents of all reservoirs are known, and ranges and value functions for the final reservoir contents are assumed given (from the long-term planning procedures). We regard the future inflow as known.

2.1 Reservoir and hydro power plant model

Let T be the number of stages in the scheduling problem and N the number of reservoirs. We define for stage t, t=1,...,T, and reservoir i, i=1,...,N:

s_{it} = spillage from the i-th reservoir

q_{it} = inflow for the i-th reservoir

v_{it} = content of the i-th reservoir at the beginning of stage t

For given head, it is assumed that a piecewise linear function is adequate for describing the output from a power station. If P_{it} is the power output then

$$P_{it} = \sum_1 \eta_{ij} u_{ijt} \qquad (1)$$

where for the i-th plant, η_{ij} are the energy conversion factor and u_{ijt} is the release corresponding to the j-th segment.

Let v_t, u_t, s_t, and q_t be vectors whose components are v_{it}, u_{ijt}, s_{it} and q_{it} respectively for all i (power stations) and j (segments). The water balance equations for the reservoirs can then be written

$$v_{t+1} = v_t + Bu_t + B's_t + q_t \qquad (2)$$

where B and B' are related to the incidence matrices for the hydro system (release and spill can go different ways). In addition, there are upper and lower reservoir limits

$$\bar{v}_t \geq v_t \geq \underline{v}_t \qquad (3)$$

In particular note that (3) include special limits imposed on the final volumes v_{T+1}.

2.2 Power balance equations

In a typical model for seasonal planning, the power grid is somewhat simplified (Johanessen and Flatabø). The system is divided into regions, and the network within each region is represented by an equivalent bus. The region firm power demand is allocated to this bus, as well as possible thermal resources and options for selling and buying power to/from the other regions and locally. Power can be transmitted from one region to another subject to transfer limits. All costs, including cost of thermal resources and rationing, are taken as piecewise linear. The energy balance equation for stage t can then be written

$$Hu_t + Gy_t = p_t \qquad (4)$$

where

y_t = vector of market transactions and exchange with other subsystem, including thermal power plants

p_t = vector of the bus firm power demands

G = matrix that describes connections in the power grid and allocation of resources to buses

H = matrix that contains energy conversion factors for hydro power stations from (1)

In (4) there is one equation per region in the system function

2.3 Thermal plant model

Generally, there is a nonlinear relation between fuel input and power output. However, in the medium and long term scheduling, it is convenient to use a piecewise linear approximation. Thus the production cost is given as

$$C_{T,it} = \sum_k \varphi_{T,ik} P_{T,ikt} \qquad (5)$$

where $C_{T,it}$ is the production cost for the i-th thermal plant in stage t, and $\varphi_{T,ik}$ and $P_{T,ikt}$ are the incremental cost and power output corresponding the k-th segment.

The emission to air is given as a piecewise linear relation found by multiplying the fuel input with the emission coefficient. For pollutant number j we then have

554

$$E_{j,it} = \sum_k \beta_{j,ik} P_{T,ikt} \qquad (6)$$

where $E_{j,it}$ is the emission of pollutant number j from the i-th plant in stage t, and $\beta_{j,ik}$ is the incremental emission in (kg/kWh) for the k-th segment.

2.4 Criterion function

The criterion function for the multiobjective scheduling task can be formulated as

$$\text{Min } [C_{tot}, E_j] \qquad (7)$$

where j=1,2...,J denotes the number of pollutants included in the criterion function and E_j represent emission for pollutant j.

The operation cost, C_{tot}, include net expenses for buying power and for thermal generation, minus the value of the water remaining in the reservoir after stage T. The operation cost can thus be written

$$\sum_{t=1}^{T} d_t^T y_t - \sum_{i=1}^{N} c_i(v_{iT+1}) \qquad (8)$$

where

d_t = Cost vector for market transactions
N = Number of reservoirs
$c_i(v_{iT+1})$ = Function giving value of stored water at the final time, i=1,...,N

The end point water values are assumed to be available from the long-term planning. In the following we assume that all $c_i(v_{iT+1})$ are piecewise linear functions, which means that (8) can be transformed into a linear objective function.

Emission to air is given by the emission from individual thermal plants. Total emission of pollutant j is

$$E_j = \sum_{i=1}^{M} \sum_{t=1}^{T} E_{ijt} \qquad (9)$$

where j=1,2,..,J and M is the number of thermal plant.

3.0 SOLUTION METHOD

One major complication introduced into the hydro-thermal problem is that the cost and emissions to air are heterogenous and non-commen-surable, that is, they cannot be measured in a common unit. In addition the objectives of minimizing operating cost and emissions are usually incompatible, that is, they are often in conflict with each other in the sense that lowest emission operation often gives highest cost and vice versa.

If monetary values reflecting pollution damages were available, the problem is to minimize a single objective with respect to given constraints. However, there are numerous difficulties and obstacles in quantifying air pollution damages. Thus, because of the conflict between cost and pollution, any solution to the hydro thermal problem requires a compromise between the conflicting interest of minimizing cost and minimizing pollution.

In this paper we seek to generate all alternative operation schedules, that is, the set of schedules from which a best compromise schedule should be selected.

This is accomplished through a trade-off analysis between cost and emissions, yielding a trade-off curve containing the efficient set of schedules.

3.1 The weighting method

One commonly used approach for generating the trade-off curve is the weighting method which converts the multiobjective criterion function (7) into a weighted single objective problem (8). The weighted problem is then solved, yielding one point on the trade-off curve. By permuting the weights, the trade-off relations between the objectives are calculated.

The weighting method benefits from the fact that existing optimization algorithms can be applied. However, when the number of objectives grows, it becomes difficult to extract information from the growing number of trade-off curves. E.g. limiting our study to three pollutants (SO_2, NO_x and CO_2) gives a four dimensional solution space spanned of the cost parameter and the pollutants.

A major simplification of the problem is obtained if some of the objectives are combined into a single criterion, thus reducing the dimension of the problem. In this paper, we are combining the emission objectives into a single criterion; total weighted emission E_{tot}:

$$E_{tot} = \sum_{j=1}^{J} \alpha_j \cdot E_j \qquad (10)$$

were α_j are the relative value of impact from pollutant j with j=1,2,....,J. The α_j's must satisfy the relation:

$$\sum_{j=1}^{J} \alpha_j = 1 \qquad (11)$$

Thus, the multiobjective problem is reduced to a bicriterium problem where the trade-off between C_{tot} and E_{tot} is determined by solving:

$$MinF = wC_{tot} + (1-w)E_{tot} \qquad (12)$$

$$0 \le w \le 1 \qquad (13)$$

The weight w represents the p.u. weight assigned to costs, and (1-w) represents the p.u. weight assigned to total environmental impacts.

It should be emphasized that the weight w is no value judgement of the relative importance of the two objectives C_{tot} and E_{tot}.

There is no inherent significance to w. The trade of curve is calculated by successively increasing w from 0 to 1. However, we may obtain a monetary interpretation of the weight w in (NOK/kg) by rewriting (12):

$$MinF = w\left[C_{tot} + \frac{1-w}{w} E_{tot} \right] \qquad (14)$$

where

$$F' = C_{tot} + \frac{1-w}{w} \sum_{j=1}^{J} \alpha_j E_j \qquad (15)$$

The eq. (12), (14) and (15) have identical solutions. The term (1-w)/w may be interpreted as total environmental cost, while terms $(1-w)\alpha_j/w$ may be interpreted as individual environmental cost for pollutant "j".

3.2 Solution method

For a given w, the single objective problem F' is a linear program with a staircase structure corresponding to the time stages. The problem is also a discrete linear control problem that can be solved by Dynamic Programming. However, to maintain the detailed representation of the hydro system an other approach is required due to the well known dimensionality problem in traditional Dynamic Programming (Bellman 1975). In this paper, the single objective problem is solved by Dual Dynamic Programming where the conventional tables in DP is avoided (Pereira 1989, Røtting and Gjelsvik, 1991).

The flow-chart in Figure 1 illustrates a solution procedure using the above described formulation. The procedure is able to solve the stochastic hydrothermal scheduling problem. However, for illustrative purposes we have

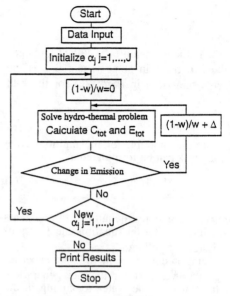

Figure 1   Solution procedure

applied it to a deterministic case. The major results will be valid even to the stochastic case. A set of relative weights is initially specified. The single objective problem F' is solved for the entire range of environmental cost (1-w)/w from zero to a very large value. For practical reasons the calculation is terminated when a change in emission no longer is observed.

4.0   TEST CASE

4.1   Test system

To illustrate the multiobjective approach to hydrothermal operation, we apply the procedure in Figure 1 to a test system.

The multireservoir hydro system consists of 9 reservoirs with a total capacity of 1700 Mm^3, and 7 hydro power plants with a total generation capacity of 1000 MW. Average inflow to the system is 3100 Mm^3.

The end point water values are given from the long term planning. Here, the same values are used for the entire range of the environmental cost (1-w)/w. In practice, the water values should be recalculated for each set of environmental cost.

The thermal system comprises of 5 plants with a total capacity of 100   MW. The characteristics of the thermal system, fuel prices, - heat rate constants, and emission coefficients for each plant, are given in Table 1. The units 1,2 and 3 represents sales options and units 4 and 5

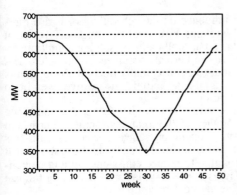

Figure 2   Firm power

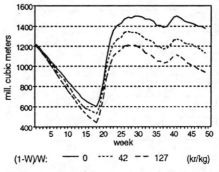

(1-W)/W: —— 0   ···· 42   – – 127   (kr/kg)

Figure 3   Development of reservoir content

Table 1 Characteristics of the thermal system

UNIT	Capacity (MW)	Prod.cost (øre/kWh)	Emission coefficients (kg/kWh)		
			SO_2	NO_x	CO_2
1.Coal 2.375%S	20	14.0	0.0163	0.002	1.09
2.Coal 0.87%S	20	18.0	0.0085	0.000975	1.43
3.Oil 1.5%S	20	20.0	0.008	0.003	1.09
4.Oil 0.75%S steam	20	22.0	0.0041	0.001	0.862
5.Gas steam	20	21.0	0	0.0014	0.6084

represents purchase options. The firm power demand is given in Figure 2.

As indicated in Table 1, we limit our study to include emission of 3 pollutants (J=3). Thus a fully written version of (15) is

$$\text{MinF'} = C_{tot} + (\frac{1-w}{w})\left[\alpha_1 E_1 + \alpha_2 E_2 + \alpha_3 E_3\right] (16)$$

where α_1=0.71, α_2=0.287 and α_3=0.003 are relative weights assigned to SO_2, NO_x and CO_2 respectively.

4.2   Test results

We want to illustrate how the operation schedules change as function of the total environmental weight (w) and also the relative weight (α_j) assigned to individual pollutants. The set of relative weights (α_j) is based on environmental cost studies (Ottinger 1990). Observe that only the set of relative weights is obtained from the environmental cost study. The total environmental cost (1-w)/w is varied continuously from zero to a very large value; i.e. until a change in emission is no longer observed.

Figure 3 shows how the total reservoir content changes as function of total environmental cost (1-w)/w for the given set of relative weights. Figure 4 shows the similar effect on the hydro system power generation.

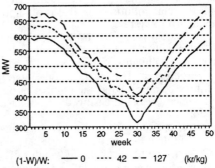

(1-W)/W: —— 0   ···· 42   – – 127   (kr/kg)

Figure 4   Total hydro power generation

—— Sale   ···· Purchase

Figure 5   Average sale and purchase

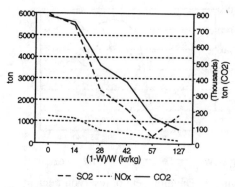

Figure 6   Total emission

Figure 5 shows how the market transactions changes as function of total environmental cost (1-w)/w for the given set of relative weights. Figure 6 shows the similar effect on the total emission of SO_2, NO_x and CO_2 respectively.

The relative weights of environmental effects is of great importance to the operation of the thermal system. A different set of relative weights will change the dispatch order of the thermal plants (Gjengedal et. al. 1992). However, it is the total pollution cost that is of importance to the hydro system.

Our test system is limited to three pollutants only. Conceptually the model allows for including other air pollutants - such as CO, CH_4, VOCs, particulates. However, a presentation problem may be observed as the number of pollutants increase, also increasing the volume of information.

The set of α's is based on an environmental cost study in the test case. A more general approach would be to permute the weights. The model allows for such a permutation. However, the same information problem may be observed.

One natural extension of the hydrothermal problem is to allow for investments in flue gas control equipment, and/or less polluting units.

In practice, an extension of the test case is necessary. This will include a stochastic inflow model. In addition, the planning horizon should be extended to several years. Then, the seasonal operation of the hydro power system will be independent of the initial end point water values.

5.0   CONCLUSION

We have described a method for including environmental externalities in hydrothermal operation. The simple test case gives only an indication on how environmental externalities

will influence on the scheduling of hydro thermal systems. We evaluate trade-off between costs and air pollutants by solving a multiobjective hydrothermal optimization problem.The method, can conceptually treat all air-pollutants by assigning an individual, relative, weight to each pollutant.

An extract from our test results show the importance of the total environmental cost and how the operation schedule changes with this cost. In particular, the analysis illustrates that it is possible to reduce emissions without switching fuels at thermal plants or increasing the use of flue gas control equipment.

The test results are limited to the simplifications of the test case. However, we consider the proposed method to be an appropriate contribution to solving the problem of how to deal with environmental impacts from electricity production.

From a society perspective, it is very expensive operating electric facilities according to traditional economic practice. While implementation of environmental externalities will not be a simple matter, the large potential benefits to society make the effort very worthwhile.

REFERENCES

Bellmann R.,Dynamic programming,Princeton University Press,Princeton,N.J.,1957.

Benders, J.F., Partitioning procedures for solving mixed-variables programming problems, Numerische Mathematic 4, 1962, pp. 238-252.

Bernow S., Biewald B., Full-cost dispatch: incorporating environmental externalities in electric system operation. The Energy Journal, March 1991.

Cohen S.D, Eto J.H Goldman C.A. , Survey of state PUC activities to incorporate environmental externalities into electric utility planning and regulation. Lawrence Berkeley Laboratory, LBL-28616, May 1990.

Delson J.K, Controlled emission dispatch, IEEE Transactions on Power Apparatus and Systems, 1974, V 93, no 5.

Gjengedal T., Johansen S., Hansen O., A qualitiative approach to economic-environmental dispatch: treatment of muliple pollutants. IEEE 92 WM 038-0 EC, New York January 1992.

Johannesen A., Flatabø N., Scheduling methods in operation planning of a hydro-dominated power production system, International Journal of Electric Power & Energy Systems, Vol. 11, no. 3, pp. 189-199, July 1989.

Kermanshahi B.S. Environmental marginal cost evaluation by non-inferiority surface. IEEE Transactions on Power Systems, Vol.5, No. 4, November 1990.

Ottinger R.L Environmental cost of electricity production, Pace University Center for environmental Legal Studies, 1990.

Pereira, M.V.F., Optimal stochastic operations scheduling of large hydroelectric systems, International Journal of Electric Power & Energy Systems, Vol. 11, no. 3, July 1989.

Putta S.N, Weighting externalities in New York State, The Electrisity journal July 1990.

Røtting T.A., A. Gjelsvik, 1991. Stochastic dual dynamic programming for seasonal scheduling in the Norwegian power system. Paper 91 SM 487-9 PWRS , IEEE PES Summer Meeting, San Diego.

Hydropower'92, Broch & Lysne (eds) © 1992 Balkema, Rotterdam. ISBN 90 5410 054 0

The trans-national supply of hydropower

K.Goldsmith
UK

ABSTRACT: Economy requires hydro schemes to be developed to their optimum size regardless of the power demand in their vicinity. The exploitation of power markets further afield can lead to progressive interconnection of national electricity networks and to substantial trans-national power flows. Realisation of the benefits that can be achieved in this way tends to encourage closer operational integration of neighbouring power systems with national boundaries becoming increasingly less relevant. The nature, significance and future prospects of this development are discussed in this paper.

1 THE CASE FOR INTERCONNECTION

The merit of hydropower over other methods of large-scale electricity production is becoming increasingly recognized. Its renewable nature and total reliance on indigenous energy sources, its stable costs once a scheme is built and its comparatively benign characteristics all contribute to its attractiveness. The major part of the still unexploited hydro resources lies in the developing world where only about 10% of the net exploitable potential has been developed. Current estimates therefore foresee a much more rapid expansion in the hydropower capacity in the developing countries than elsewhere:
- by 5.0% per annum, or 150GW in total, over the 15-year period 1985-2000 (compared with 1.8% per annum, or 115GW, for the industrialised world)
- by 8.6% per annum, or 600GW in total, over the 35-year period 1985-2020 (compared with 1.6% per annum, or 235GW in total, for the developed countries).

These scenarios do of course assume that the very large funding requirements for such programmes can be met and that growth of the installed hydro capacity is limited only by demand and physical resources. They assume also that no constraint is placed on the scale of development at any particular site.

All power production benefits from economies of scale but the advantage of exploiting a given site to its full potential is particularly marked in the hydro case where the site-dependent costs dominate the investment requirements. It is clear that the demand which a new hydro scheme has to satisfy is not necessarily related to the available resource potential. This is particularly so in the developing world where the resource potential at a number of promising sites is large and dwarfs the required supply capacity in the respective country, sometimes by a factor of 10 or more even if the incremental demand is aggregated over a number of years to come. The question to be resolved is then how to reconcile the scope for building relatively large hydro plants, and the developmental and economic attractiveness of doing so, with the limited market for the output from such plants that may be locally available. This brings into focus the need to look for additional power markets further afield and outside national boundaries. Particularly receptive markets will be those relying to a large extent on thermal power generation; these are subject to the vagaries of uncertain and potentially escalating fuel prices and also to a highly polluting resource.

Exploitation of hydro-thermal complementarities and the consequential uni-directional power flows provided the original incentive for trans-national electricity exchanges in Europe and North America. Once cross-boundary transmission lines were built and power exchanges realised, other advantages soon became apparent. They included:

- the possibility of rendering mutual assistance in emergencies and thus reducing the amount of standby capacity each of the interconnected power systems had to carry,
- the scope for phasing the installation of additional power plants in the interconnected networks and thereby reducing the overall rate of new investment,
- the opportunity of meeting particular load requirements - peak loads for example - more economically than local power sources would permit,
- centralised system control facilities leading to a more reliable and often cheaper supply.

The extent of the advantage derived from a particular interconnection arrangement must depend on the characteristics of the power plants and networks involved but it can be considerable. Integrated operation with neighbouring networks does mean abandoning some independence in planning, financing and operating the national power system and can therefore - and frequently does - meet considerable opposition. The progression from the export of hydropower surpluses to full integration of neighbouring networks brings into play new concepts of international cooperation which have been much studied but as yet rarely achieved.

2 ENERGY EXCHANGES

The relative amounts of electricity transferred across national boundaries are not yet very significant as is brought out in Table 1. Reluctance to abandon control over essential national electricity supplies, difficulties of financing multi-national schemes and inadequate exploitation of hydro resources have combined to limit the importance of trans-national power flows up to the present time.

The continent-wide figures conceal some important cases where the energy production or supply pattern is dominated by international exchanges. Countries where exchanges have become significant can be divided into two groups.

A Countries of the alpine or pre-alpine type possessing a large hydro storage capacity which can provide winter, peaking and emergency energy to predominantly thermal power systems on a virtually tidal flow pattern. Energy is imported during the spring

Table 1. International Electricity Exchanges (1989)

	Imports in % of consumption	Exports in % of production
World	2.5	2.3
Europe (excl.USSR)	7.6	6.4
USSR	0	2.3
North America	1.1	1.1
South America	5.0	0.6
Africa	1.0	1.7
Asia	0.2	0.2
Austria	12.3	16.4
Norway	0.3	12.8
Switzerland	34.5	37.6
Benin/Togo	88.9	0
Bhutan	38.1	88.0
Laos	4.7	62.6

and early summer when the reservoirs are filling up and also often during the night when import of thermal generation surpluses permits stored hydro energy to be saved. The tidal flow of energy is particularly marked in countries having also some thermal or nuclear generation, e.g in Austria and Switzerland where energy imports at certain times of the day or season can lead to appreciable savings in marginal generation costs as well as in stored hydro energy.

In an all-hydro system such as Norway energy exports rely on excess capacity but are time regulated for maximum benefit to both supplier and user so that full advantage can be taken of the available hydro storage; the basic hydro capacity is adequate to cope with most of the national demand throughout the year.

Supplies of regulated energy from storage originally formed the basis for trans-national power exchanges throughout Europe and they are still the most important element in the exchanges although now overlaid to some extent by the need to balance thermal and nuclear generation in the interconnected networks.

B Developing countries, both mountainous and tropical, which have hydro capacities well in excess of national requirements and can provide relatively large uni-directional bulk power deliveries throughout the year. In some cases, in Bhutan and Laos for example, power exports from dedicated hydro schemes become effectively tied to the neighbouring network. The geographical disposition of demand centres across the country can make it attractive to supply some of them from across the border rather than from inland hydro plants; this can result in a comparatively large power import even where the country, taken as a whole, has a large power surplus available for export.

Tied schemes generally supply large receptive power markets in which the power imports form only a small fraction of the total consumption but where marginal cost savings arising from reduced thermal generation are economically attractive. This is the case with the exports from Bhutan to India (0.5% of consumption) and from Laos to Thailand (1.6% of consumption).

The situation of Benin and Togo is exceptional. Both countries rely almost entirely on electricity imports from the 880MW Akosombo scheme in Ghana even though they have their own hitherto unexploited hydro resources. The arrangement maximises the utilisation of a central hydro source and overcomes the unattractiveness of developing small local schemes for a dispersed demand even though, with the anticipated load growth, local plants will ultimately have to be built. Cooperation was fostered in this case by a common development and financing agency which could ensure optimum utilisation of physical and financial means.

As shown in Table 1, the importance of energy exchanges is greatest in Europe. The interconnection capacity between the 12 countries in the west and south of the Continent, currently about 65GVA, is some 56.5% of their hydro output capacity or 15.2% of the total installed generating capacity, including thermal and nuclear. However, the quantity of energy exchanged between these countries in 1990, import and export, was only 4% of consumption. The large excess transmission capacity is accounted for by the diversity of the power flows between the participating countries. This situation is very different to that experienced elsewhere in the world where power exchanges are generally restricted by the transmission capacity. It greatly encourages flexibility in the operation of the hydro plants to the benefit of all the interconnected systems.

3 SUPPLY CONDITIONS

Preconditions for trans-national power transfers are that:
- power markets of the right size and characteristics are available within feasible transmission distances,
- complementarity between generation and demand patterns can be achieved,
- funding for the often considerable investment requirements can be secured,
- managerially and financially equitable arrangements spanning a sufficiently long period of time can be made between all the parties involved (bearing in mind that considerable investments in generation and transmission facilities may be needed).

These preconditions have been largely met in interconnection scenarios in the industrialised world but do cause problems with developing countries.

It is clear that network inter-connections become feasible only if they bring advantages to both supplier and purchaser. Proper sharing of these advantages can be difficult in situations where a free choice of alternatives is impeded by scarcity of investment capital. Open power markets of the kind seen in Europe, with free choice of power sources and the possibility to exploit momentary supply-demand comple-mentarities, have not yet developed elsewhere. Energy trade still relies essentially on unidirectional power flows arising from excess capacity of particular hydro sources.

The supply arrangements will therefore remain more basic, at least for the foreseeable future. Costs of imported power must remain competitive with local production costs, with due allowance made for the quality of the energy supplied in terms of time and seasonal values.

The receiving utility needs to have sufficient

resilience in its production pattern to absorb variations in incoming supplies without curtailing the service to its consumers. Such variations may be caused by uncontrolled seasonal run-off conditions of hydro plants or restricted withdrawal from hydro storage schemes (in order to conserve water), or indeed by force majeure. Unfortunately, the very circumstances that have led a country to rely on bulk power imports in the first place usually discourage the provision of an adequate amount of standby capacity, with the result that any curtailment of power imports is likely to lead to load rejection. The current situation in Mozambique is a case in point; this country imports 67% of its electricity consumption and suffers from almost daily loss of load since, because of transmission limitations, the 2400MW Cahora Bassa scheme in the north of the country is at present not in operation.

3.1 *Tied Schemes*

With the exception of Europe and North America, tied schemes are still the primary source for the international energy traffic.

Where bulk power transfers between adjacent countries have involved some communality in the design and financing of the production facilities, the power sources will in some way remain tied to the foreign power market either physically or contractually. Joint development of frontier rivers provides typical examples of physically tied schemes (e.g Salto Grande (1620MW, Argentina/ Uruguay), Itaipu (11,900MW, Brazil/Paraguay), Kariba (1270MW, Zambia/Zimbabwe)) or the Iron Gates plant (2100MW, Romania/Yugoslavia)). Similar schemes are under construction or in the planning stages. Some of the main river basins offer a vast hydro potential which could satisfy the electricity demand in the riparian countries for many years to come. The Mekong River, for example, which borders Thailand, Laos, Cambodia and Vietnam, is thought to have an exploitable potential of more than 15,000MW. The production capacity of joint schemes can greatly exceed the demand of one of the partners and his surplus can then be exported to the other; this is the case with Uruguay, Paraguay and Zambia in the examples mentioned above.

Contractually tied power plants are sited within the territory of one country and are owned and operated by it but a large proportion of the output is delivered on a long-term basis to the neighbouring country and in effect forms an integral part of the latter's own national supply. The rationale for this arrangement rests on the competitiveness of the energy subscribed for. A distinction need be made in the appraisal of competitiveness between:
-    the substitution value of the energy from local facilities that might have been built instead,
-    the marginal value of the saving in local thermal generation achieved through the import of hydro energy.

When tied schemes are funded by loans in convertible currency, the foreign purchaser of bulk power from such schemes has to contribute to the loan redemption payments in convertible currency also. This means that imports of power from tied schemes may have to be paid for, at least partially, in the same way as imports of oil or foreign goods and cause the same drain on foreign currency reserves.

This issue requires careful handling by financing institutions who can often act for both supplier and purchaser and ensure that acceptable terms are negotiated. The deal will have the greatest chance of success if the economy of the purchaser is reasonably strong and his power imports are no more than marginal to his total requirements. Cases in point are the 2400MW Cahora Bassa scheme on the Zambesi River in Northern Mozambique which is to supply some 90% of its output to Zimbabwe and South Africa when circumstances permit or, on a considerably smaller scale, the 150MW Owen Falls station on the Nile in Uganda which supplies 25% of its output to Kenya. In both cases, the supplier countries are among the less developed and derive considerable economic benefit from the export of electricity. A tied scheme inevitably implies abandoning some of the supplier's independence. Some countries fear that, in the absence of competition, this could result in total domination of their electricity supply, or even their economy, by a foreign purchaser.

This has led to the deferment of some promising developments, for example on some of the large rivers flowing south from Nepal into India, such as the Karnali or Gandak.

3.2 *Utility Cooperation*

Untied power transfers occur in several places where hydropower surpluses, even if only seasonal, can reduce fossil fuel based generation and where transmission costs between network infeed points do not impair the economic merit of the arrangement. Power is transferred from the network as a whole rather than from specific plants. Power flows remain mainly uni-directional; reverse flows - under emergency conditions for example - are infrequent except where capacity reserves are shared. Supplies are charged for under negotiated tariffs and there is generally no involvement by the purchaser in the supplier's financing arrangements.

Several international development agencies have initiated studies to determine whether closer cooperation between regional groups of countries could bring more enhanced economic benefits. The objective is to achieve not only operational coordination on the European model but also to ease the financing burden for future investment and hence lower the long-run marginal cost component in the consumers' electricity tariffs. Perhaps the most thorough study undertaken hitherto was sponsored by the United Nations and dealt with the six Central American countries - Guatemala, Honduras, El Salvador, Nicaragua, Costa Rica and Panama. Their aggregate demand in the year 2000 is estimated to be 11,000MW and their consumption 60TWh/a. Existing and potential generating capacity in the region comprises 12,000MW of hydro plant (average generation 57TWh/a) and 1100MW of geothermal plant (tentative output 8TWh/a). Because of considerable seasonal variations in hydrology, some 2000 to 3000MW of steam plant will also be needed to compensate for shortfalls in hydro production but the main weight of power generation will be carried by the hydro plants.

Three scenarios were set out and compared with the costs of continued independence of each national network:

A Independent planning of each system but full transfer of surpluses between systems and operational coordination, including sharing of reserve capacity

B Fully integrated regional system expansion planning with centralised and optimised system operation without regard to national boundaries

C An intermediate case with independent planning but common and optimised plant installation programmes; restricted operational coordination with power transfers limited to 20% of each national demand.

The capacity and energy savings computed were considerable for all cases. Even with Scenario A, the oil consumption for electricity generation would be halved throughout the region. With Scenario B, the overall benefits achieved with A would be more than doubled. Scenario A is now to be implemented step-wise but Scenario B is unlikely to materialise for some time to come; it will probably be preceded by Scenario C. There is still some reluctance to abandon national control over an essential service especially as no precedent has been set to date, except on a small scale. National autonomy over electricity supply has not been set aside anywhere in the industrialised world.

A similar though less ambitious exercise is now in hand under World Bank sponsorship for the SADCC region of Southern Africa which extends from Tanzania to the South African border. Ten countries are involved but the work will concentrate initially on the five core countries of Botswana, Malawi, Mozambique, Zambia and Zimbabwe. As in the case of Central America, the first objective is to establish in what way trans-national interconnections could improve the utilisation of existing hydro facilities in the North of the region (eg of Cahora Bassa (2400MW) and of Inga (1400MW) in Zaire). Power would be wheeled from there through the core countries to replace thermal generation in the South of the region, including ultimately South Africa. Attention is also to be paid to savings in reserve capacity which interconnected operation may bring about. The work is likely to be extended to larger areas of Africa if the results of this study are sufficiently promising.

The aim of studies of this nature is to set the framework for closer cooperation between national utilities in the development and operation of their power systems and thereby to foster socio-economic progress in the developing world. As in Latin America and East Asia also, there is a vision of regional electricity grids within which national boundaries are no longer relevant for electricity supply. Hydropower will remain the primary source around which regional electricity grids can be fashioned. The inter-action of an economically favourable production potential and of effective means for absorbing its output is likely to greatly encourage the development of relatively large central hydro plants where physical conditions are favourable and institutional and financial constraints can be overcome. It is to be hoped that all agencies concerned with economic and social development will devote attention to the issues that need to be resolved if more extensive electrification, especially of the developing world, is to be achieved.

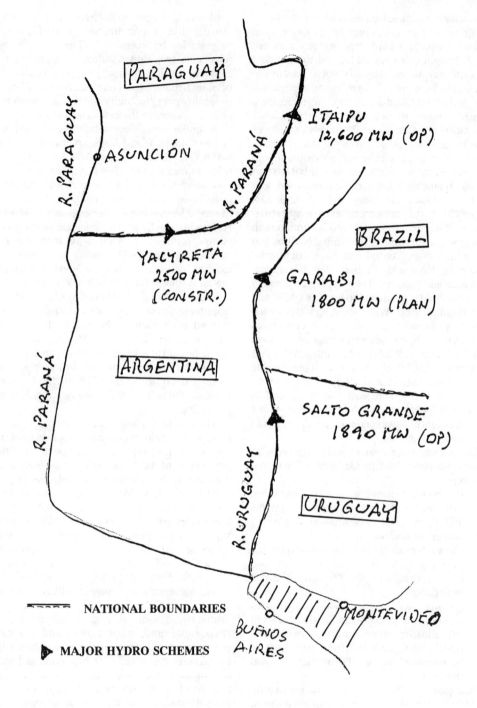

PARAGUAY

R. PARAGUAY

ASUNCIÓN

R. PARANÁ

ITAIPU
12,600 MW (OP)

BRAZIL

YACYRETÁ
2500 MW
(CONSTR.)

GARABI
1800 MW (PLAN)

R. PARANÁ

ARGENTINA

SALTO GRANDE
1890 MW (OP)

R. URUGUAY

URUGUAY

NATIONAL BOUNDARIES

MAJOR HYDRO SCHEMES

BUENOS
AIRES

MONTEVIDEO

FIG. 1 TIED HYDRO SCHEMES

IN LATIN AMERICA

$\longleftrightarrow$ **EXISTING**

$\longleftarrow - - \rightarrow$ **PROPOSED**

   MAJOR HYDRO SCHEMES

FIG. 2   POWER SYSTEM INTERCONNECTIONS

IN THE SADCC REGION

Hydropower'92, Broch & Lysne (eds) © 1992 Balkema, Rotterdam. ISBN 90 5410 054 0

Peak power – The optimum operation of a hydropower plant

Torodd Jensen & Anders Korvald
NVE, Norwegian Water Resources and Energy Administration, Oslo, Norway

ABSTRACT: The paper describes peak power operation as an alternative for energy operation of hydropower plants. Changing political situations create new possibilities for international cooperation within distribution of electric energy. The possible connection to markets dominated by thermal power plants can result in new hydropower plants meant for peak power only. The paper gives information on two hydropower schemes in Norway and Lesotho and discuss the environmental benefit from hydroelectric versus thermoelectric peak power.

1 INTRODUCTION

Development of hydropower resources for peak power operation has up to now been considered uneconomical in a hydropower country like Norway because the peak power demand could easily be covered by the installed capacity constructed for energy production during winter time. In addition limited connections to other countries where electricity is produced by thermal power plants has reduced the possibilities for other markets.

The need for uprating/refurbishing of Norwegian hydropower plants has focused on price for peak power because peak power in addition to increased and improved energy production will be important for the economical feasibility of these projects.

There are several hydropower plants which can be enlarged in Norway. A study on environmental impacts and construction cost indicates projects of several 1000 MW. The development of these projects, however, depends on markets outside Norway.

Hydropower plants are easily regulated to meet the electricity variation in a marked during daytime. Thermal powerplants, operation for peak power, are run with lower efficiency than thermal power plants covering the basic load. Hence, power from hydropower plants will have positive influence on the environment referred to air pollution per. kWh. The conclusion is that new markets and international cooperation can give a total energy production with less pollution.

Similar possibilities can be found in other countries, where a hydropower plant

designed for energy production can be a marginal economical project. Developed for peak power production, however, it can be a good investment. The Muela powerscheme in Lesotho, designed for interconnection to the grid in the Republic of South Africa is an example.

Our conclusion is therefore that development of hydropower plants to day should include the possibilities for peak power operation. The design for hydropower plants should include several stages of implementation in order to allow for peak power operation in the future.

2 THE ADVANTAGE OF HYDRO-ELECTRIC VERSUS THERMO-ELECTRIC PEAK POWER

Due to adnormal weather conditions there has been a large energy surplus in Norway in 1989-90 and many have argued in favour of exporting energy to Denmark and the continent. However, political signals indicate difficulties in implementing longterm export agreements. "Pump-power" agreements are more politically feasible. The idea is to swap electricity by selling hydropower during the day in exchange for energy from thermal power plants at night. The Norwegian power plants can be turned off at night in order to refill the reservoirs.

Hydroelectric power plants are possible to regulate from zero up to maximum load and back again in a few seconds, while the regulation of thermal powerplants take hours for conventional powerplants and days for nuclear powerplants. The power supply has to follow the daily demand which varies a lot (figure 1).

In a system dominated by thermal powerplants the peak power demand is met by gas-, oil- and coal-fired thermal plants. Power plants with the ability of regulating the production have a low efficiency which results in a highly polluting and expensive power production. This is why the price of peak power supply is up to four times higher than the price of basic load.

From 1990 to -91 both the capacity demand and energy demand in Norway have increased by 2.4 % . Current capacity is about 3000 MW higher then maximum demand, but with an annual growth of 2 %, balance would be reached by the years 2000-2003. Energy production and demand is, however, expected to be in balance already by 1996.

Implementation of the large capacity potential indicated on figure 2, will affect the capacity balance and the export possibilities. The recent agreement of 440 MW export through a new interconnection to

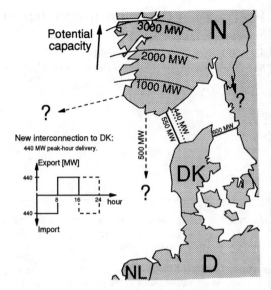

Fig.2 Capacity potential and possible interconnections

Denmark (completed in 1994) will not affect the prospected balance, pursuant to the contract Norway can stop exporting peak-power for ten days each year in order to supply domestic consumers.

Focusing on possibilities of capacity exports and increasing domestic demand, NVE has looked into the cost of expanding power plants in South-West Norway. The increased capacity can be achieved by extending existing water tunnels and installing new turbines in addition to the old ones. Enlargement of power plants with long tunnels and a large capacity potential is usual solved by building a completely new power plant parallel to the existing one. The result of the study is presented in figure 3 where the value of energy gained is assumed to be 0 NOK/kWh and 0.15 NOK/kWh. If one goes further north along the west coast of Norway, there is a much larger capacity potential with a cost less than 2500 NOK/kW.

Figure 3 shows the total costs for each kW referred to the power plant. In order to compare the costs of production with peak power prices in Germany and the Netherlands one has to add the cost of new submarine power cables and converters. Costs for possible extensions of onshore transmission lines have to be added. Total investment in the new interconnection to Denmark is estimated to be NOK 1200 million where the cable itself costs 400 million and the converters is calculated to about NOK 2x400 million. Norway and the Netherlands are considering

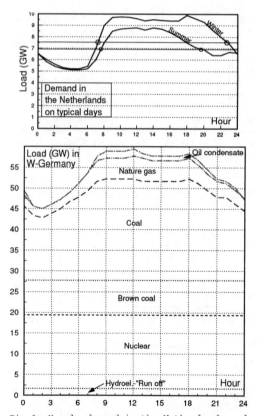

Fig.1 Hourly demand in the Netherlands and supply in W-Germany on day of maximum demand [UCPTE].

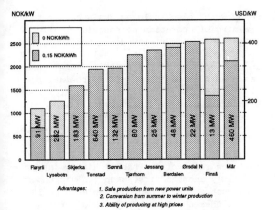

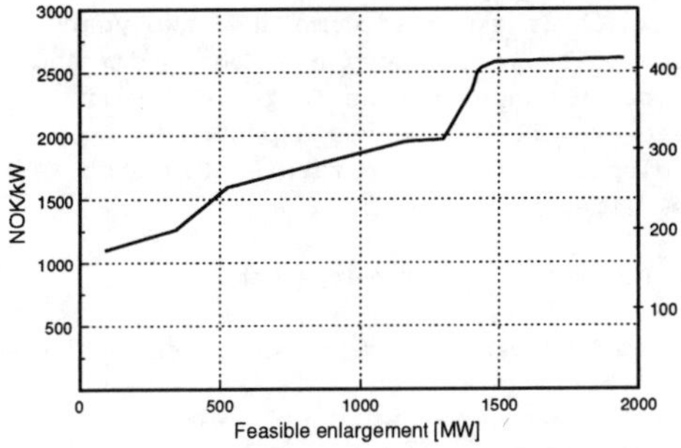

Fig.3 Costs versus feasible capacity increase in 12 power stations in south western
part of Norway.

laying a 550 MW's cable between the to countries. This power cable is estimated to cost
about NOK 2400 million. One of the reasons
behind this agreement is the problem the
Netherlands have in supplying peak power and
the expected taxes on polluting energy production. Peak power exports to the Netherlands exchanged with cheap energy imports
during low load periods would perhaps be
more interesting for Norway than exports/im-
ports with Danish power plants where the
ability of peak power production is higher.

Increasing the capacity will usually enhance the value of a power plant due to:

✿ less spill of water and gaining of head
  gives a higher production.
✿ increased safety when the production is
  partly moved to the new installation.
✿ increased ability of selling power at
  high prices.
✿ positive environmental effects by increasing the efficiency of power plants on
  the continent, caused by the evening out
  of the varying curve of demand. The result is less emission of the greenhouse
  gas CO_2.

✿ small local environmental impact thanks
  to optimal exploitation of existing power plants.

Submarine power cables to the continent will
make this possible, but in addition to the
positive effects of capacity exchange the
interconnection will make it possible to
export energy in wet years and import it in
dry years.

3 PEAK POWER IN NORWAY WITH EMPHASISE ON
TONSTAD POWER PLANT

Physical installation of more capacity is
limited by the amount of water available and
impacts caused by changing discharge flows.
Larger capacity for the purpose of peak
power supply would therefore increase the
daily variation of waterlevel both upstream
and downstream. At Tonstad the impact is
small due to the big storage reservoir and
outlet in a lake.

The Norwegian market for peak power production and the long headrace tunnel is a
limitation for increased capacity at Tonstad. New power cables to both Denmark and
the Netherlands will considerably increase
the regional demand. Because Tonstad is
located close to the continent, necessary
extensions of the transmission lines could
be kept to a minimum if Tonstad were able to
deliver the peak power. Tonstad is very well
suited for peak power delivery.

The existing power plant was inaugurated
in 1970, with the installation of four 160
MW units. The mean annual production is
about 3700 GWh. In 1989 the capacity was
increased by the installation of a new unit
of 320 MW, which makes a total of 960 MW.
When this last unit was installed there was
a discussion about expanding the cross-section of the headrace tunnel, but the length
of the tunnel (32 km) and the expected value
of lost production during construction time
made it unprofitable. Because the tunnel is
constructed for 4x160 MW, the head loss at
maximum load today is 2.5 times higher then
before installation of the 320 MW unit. As
one can see from the efficiency curve below,
peak power supply from Tonstad would increase the total losses. A new power plant and
tunnel would lift the curve of efficiency
and increase maximum output capacity (see
figure 4).

The existing tunnels are already too small
so an expansion of Tonstad implies the construction of a new power plant and tunnel,
see figure 5. The total cost of such a project is estimated to 2000 NOK/kW (300
USD/kW).

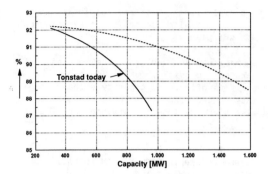

Fig.4 Total efficiency at Tonstad.

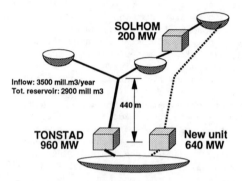

SOLHOM
200 MW

Inflow: 3500 mill.m3/year
Tot. reservoir: 2900 mill m3

440 m

TONSTAD
960 MW

New unit
640 MW

Fig.5 Existing powerplant and possible
enlargement.

Possible extension at Tonstad:
  • Installation: 2x320 MW
  • Length of tunnel: 22 km
  • Cross-section of tunnel: 115 m²
  • Total cost: 1300 mill.NOK

This is an environmentally acceptable scheme
both in relation to the impacts in the construction area and the expected reduction of
air pollution from thermal power plants.

4   PEAK POWER IN SOUTHERN AFRICA WITH EMPHA-
    SISE ON MUELA HYDROPOWER SCHEME, LESOTHO.

4.1 Introduction

Lesotho's technically exploitable hydroelectric potential is estimated at about 2000
GWh (450 MW). The cost of harnessing this
potential, however, is high because of the
lack of natural heads and erratic seasonal
flow patterns. Today only four small hydroelectric power plants are in operation in
the country. Two of them, Mantsonyane (2500
KW) and Semonkong (200 KW) are financed by

NORAD which also included assistance for
training of operation personnel.
    Lesotho's electricity distribution is
today totally dependent on sales from ESCOM
in the republic of South Africa (RSA). RSA
has overinvested in coal fired thermal power
plants, and the electricity price from these
plants are heavily subsidized.
    Lesotho Highland Water Project (LHWP)
which is a project for large export of water
to RSA has been planned since early 1950's.
The project has three phases and everyone
includes reservoirs. Phase I has two steps
(IA, IB), and the first step (IA) is under
construction and expected completed in 1996.
It includes the Katze dam with diversion
tunnels for export of water to RSA. Phase
1B, which is expected completed two years
after phase IA includes the Mohale dam and
diversion tunnel to the Katze reservoir.
These are the two phases expected to be
developed in forseeable future. Fig 6 shows
the different projects.

4.2 The Muela hydropower scheme

The head between the dams in Lesotho and the

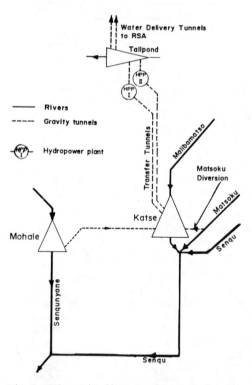

Fig.6 Schematic diagram of reservoir system
in the Lesotho Highland Water project, phase
IA and IB. Muela hydropower pl. is indicated

area in RSA is several hundreds m and originally the water supply scheme was planned with narrow tunnels in order to kill the energy and reduce the construction cost by smaller tunnels. It was soon clear that the increasing of the cross section of the tunnels could be done with acceptable additional cost and thus make it possible to utilize LHWP also for hydroelectric energy production. A suitable place for a powerstation is in Muela.

The Muela hydropower project has been identified as the most economic opportunity for the development of an electric generating capability in Lesotho. It will virtually eliminate Lesotho's present almost total dependence on imported electricity from the RSA.

The Muela underground hydropower project is designed with a capacity of 72 MW for phase IA and another 40 MW in phase IB. The headrace tunnel from Katze dam is 45 km long (4.4m cross section). There is not planned an intake pond at the end of the tunnel before the pressure shaft for the powerstation in phase I A. There are, however, planned a reservoir in Muela which is the tailwater for the powerstation. This reservoir will allow Muela to be operated for a variable load. The need for such an operation is shown in fig 7.

Average load factor in Lesotho is 0.45. Not surprising is the need for power at its highest during winter months June, July and August.

Compared to RSA/ESCOM prices the Muela project will have an economic rate of return of approximately 7 %. This situation can be changed in a positive way for the hydropower scheme if taxes for airpollution from coal

fired thermal powerplants are introduced, if subsidizing of thermal power plants is stopped or if the design of Muela is changed so that peak power operation also serving the RSA is possible.

4.3 Peak power from Muela

To be able to operate the Muela powerplant for peak power deliverance, a reservoir between the transfer tunnel from the Katze dam and the pressure shaft has to be constructed. A headpond at Sentelina has been investigated in several alternatives. It has been shown that a headpond has a lower net benefit than the scheme without a headpond. The reason is that by a headpond it is not possible to utilize the total head from the Katse reservoir, hence reduced energy output must be faced with the headpond solution.

The benefits of the hydropower plant is defined as import savings from the resulting reduction in power imports from ESCOM (RSA). Import savings comprise capacity benefit and energy benefit.

Capacity benefit is defined as the cost savings from reducing the peak power demand from ESCOM. A reduction of 1 kWh in any month would have saved Lesotho US$ 5,5. The problem is, however, that Lesotho need additional security from ESCOM, gas turbine generators or other generation options if Muela should fail. This security means that the possible reduction of peak power charge from ESCOM is not much, even if Muela is constructed for peak power operation.

The energy benefit is defined as the cost savings from reducing energy demand from ESCOM. A reduction of 1 kWh would have saved Lesotho US$ 0.01. A possible additional benefit would be future energy and peak power sales to RSA/ESCOM. However, to date no agreement exists on such sales.

The result is that Muela has been designed as a load-following plant. This alternative allows fully utilization of the head from Katse reservoir by using the transfer tunnel as a pressure tunnel for the power plant. It will minimize the need for supplementary power imported from ESCOM.

Changing political situations and better cooperation between Lesotho and RSA can, however, result in possibilities for peak power from Muela. There exist peak power problems in RSA because of the domination of coal fired thermal powerplants for electricity generating, and it is possible to include a headpond in the further development of Muela as part of phase 1B in LHWP.

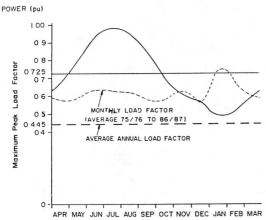

Fig.7 The variation of load factor during a year.

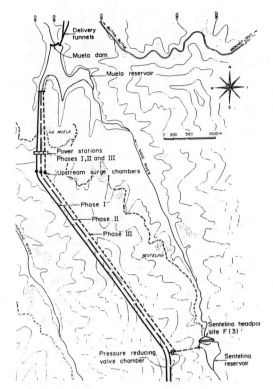

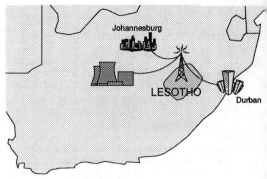

Fig.8 Muela hydroelectric power scheme is designed for Phase IA and IB in the LHWP. For Peak power operation it is possible to include a headpond when Muela is enlarged as a consequence of the implementation of phase IB.

4.4 Conclusion

For the moment it does not seem that a peak power plant will be developed for serving the marked in RSA. One reason for this is that Lesotho has little experience with operation of hydroelectric power plants. Hence it is expected that the forced outage rate will be high during the first year of operation.

Peak power from Muela and other high head power projects in Lesotho could give Lesotho the opportunity to play an important role in the future cooperation with RSA. Water for watersupply, irrigation and hydropower is resources for Lesotho like oil and gas is sources for other countries. By combining hydropower with thermal power to encrease the efficiency in thermal power plants Lesotho can earn money and contribute to reduced airpollution from RSA.

REFERENCES

Norge som energinasjon. EFI - TR 3848, 1991.

O/U-prosjektet. Kostnad ved effektutbygging i Sørvest-Norge. NVE-publ. nr 23/1991.

Operation of the West European Interconnection system, UCPTE, 31.10.91.

Hydroelectric power in Lesotho, NVE-V report No 3 1987

Transmission network system development, Lesotho by Steinar Grongstad NVE, august 1991.

Lesotho Highlands Water Project, hydropower design contract, stage 1 studies, Gibb/ Sogreah Joint Venture January 1989

Staff Appraisal Report, Kingdom of Lesotho, Lesotho Highlands Water Project (Phase IA), June 2, 1991

Muela Hydropower Project and Related Activities, Summary, Donors Conference - Maseru, October 1991

Hydropower'92, Broch & Lysne (eds) © 1992 Balkema, Rotterdam. ISBN 90 5410 054 0

Modelling of hydropower scheduling in a national/international context

O.J. Botnen, A. Johannesen & A. Haugstad
Norwegian Electric Power Research Institute (EFI), Trondheim, Norway

S. Kroken & O. Frøystein
State Power Board/Norpower, Oslo, Norway

ABSTRACT: The Norwegian power production system has an installed capacity of about 27 000 MW, 99,2 % of which is in hydroelectric plants. Depending on political concensus and the cost of production alternatives, present capacity may conceivably be doubled. Offshore gas resources amount to about $3 \cdot 10^{12}$ standard m^3, 20% of which is developed or decided developed. Potentially, a strong increase can be foreseen in the electricity-related use of gas from the continental shelf, both for export and domestic purposes.

Under the auspices of the State Program for Utilization of Natural Gas ('SPUNG') a system modelling scheme has been developed to enhance the utilization of national gas- and hydropower resources. The model is process-oriented and allows for simulating power system performance in terms of economy, quality of supply and environmental stresses, in a Nordic or wider European context.

Following a brief overview of problem formulation, the paper focuses on the developed model and solution methods involved in system performance modelling and evaluation. Furthermore, the paper presents illustrative results from sample studies.

1. PROBLEM OVERVIEW

The exploitation of national energy resources involves modelling of system performance on three main levels of planning:

- **Energy-policy level**. The utilization of national resources is by and large considered an agent for fullfilling agreed-upon socio-political objectives. Such objectives would e.g. be full employment, diversity of industrial base and national trade balance. System models applied on this level are normally aggregate, econometric and non-physical of concept. Main output from this planning level is politically feasible resource utilization scenarios.

- **Overview system level**. Based on scenario premises from the planning level above, the task is to evaluate the cost- and robustwise best system expansion strategy in view of uncertainty adhering to many premises. Models applied are often process-oriented, aiming at describing the electrical and economical performance of each main design under study. Principal elements of a system expansion strategy; type, tentative capacity and location of power production plants, layout of grid network, supply option(s) for covering heating and cooling demand.

- **Detailed subsystem level**. With premises produced on previous levels, subsystem and component design is finally conducted. This is often afforded by applying specific models that describe the local design problem in depth while representing the external system in some simplified way.

The paper addresses itself to level 'overview system planning.' Following a brief description of system characteristics, the paper focuses on developed models and solution methods for system performance evaluation. Illustrative results from sample studies are included.

2. SYSTEM CHARACTERISTICS

2.1 The Norwegian Power System

The power production system is strongly hydro-dominated with an installed total power capacity of about 27 000 MW in close to 650 plants. Average yearly hydro production is about 108 TWh. New hydro power may be developed to a limited extent. Total thermal capacity is approx. 200 MW. Wind- and wave-power is being developed in some areas, but the contribution from these renewable resources will remain marginal.

The main transmission system consists of ~12 000 km of interconnecting lines including a number of ties to neighbouring countries. On the demand side contractual power is supplied to industry and the general public. In addition to contractual demand, there is a considerable spot market type electrical demand, defined by available options to switch between electricity and fuels for heating, and between electricity and hydrocarbons in chemical processes.

The power system is made up of a number of main interconnected regional systems as illustrated in **figure 1**. Within each there is an extensive hydro production subsystem and thermal generation is available in some of them. The regional power market will normally include firm power demand as well as multiple spot market sales options.

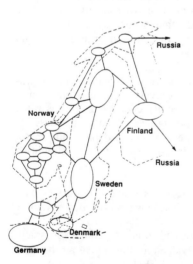

Figure 1. Regional subsystems in a model of the Nordic power system.

2.2 Power Exchange within NORDEL

Power exchange between participants in the NORDEL[*] system is operation-oriented. Agreed-upon recommendations urge the participants to:
- utilize the incremental cost principle when scheduling,
- apply comparable methods of calculating incremental costs,
- use the calculated incremental cost as a basis for agreements on power exchanges,
- aim for equal profit-sharing from bilateral power exchanges.

Each nation must in principle be self-sufficient in electric power, when agreements on firm power exchange are taken into account. The NORDEL spot power market is thus one of restricted access. This reflects the mainly operational nature of NORDEL cooperation up to now.

2. 3 Norwegian Gas Resources

Offshore gas resources amount to about $3 \cdot 10^{12}$ standard m^3, 20 % of which is developed or decided developed. In the future a strong increase is foreseen in the electricity-related use of gas from the continental shelf, both for domestic and export purposes.

Electricity-related use of gas may involve scheduling processes of variable complexity:

If the gas available for power production has no (immediate) alternative utilization, the scheduling of a gas-fired plant is trivial; it corresponds in principle to the handling of run-of-river hydro plants. The challenge in optimization is to utilize all available hydro storage facilities so as to maximize the benefit of the available combined aggregate of hydro- and gas resources.

If gas available for power production faces alternative options for use (via e.g. a sub-sea pipeline infrastructure), the scheduling of a gas-fired plant may become non-trivial: contracts will normally specify available gas volumes on a (multi)seasonal or (multi)annual basis, and this will provide for

---

[*] **NORDEL-Framework for cooperation between the power systems of Norway, Sweden, Denmark and Finland [1].**

degrees of freedom to optimally adapt gas power production to current electrical market conditions.

3. SYSTEM PERFORMANCE MODELLING

Overview system planning studies are based on comprehensive system performance simulations, presuming a 'normal' state of system operation. These simulations are performed using a multireservoir model referred to as the Power Pool Model [2]. This is a stochastic model for optimization and simulation of system operation in cases where hydro power plays an important role. The optimal scheduling of hydro-resources is sought in relation to uncertain future inflows, thermal generation, power demand or supply obligations, and options for doing transactions in domestic as well as international electrical spot markets. Because it allows simulation of large (i.e. consisting of many rivers, plants, reservoirs) hydro systems with a relatively high degree of detail, the Power Pool Model is deemed well suited for comprehensive studies on a national or international scale.

The Power Pool Model consists of two parts:

- A **strategy evaluation part** computes regional decision tables in the form of expected incremental water costs for each of a defined number of aggregate regional subsystems. These calculations are based on use of a stochastic dynamic programming-related algorithm for each subsystem, with an overlaying hierarchical logic applied to treat the multireservoir aspect of the problem.

- A **simulation part** evaluates optimal operational decisions for a sequence of hydrological years. Weekly hydro and thermal-based generation is in principle determined via a market clearance process based on the incremental water cost tables calculated for each aggregate regional subsystem. Each region's aggregate hydro production is for each time step distributed among available plants using a rule-based reservoir drawdown model containing a detailed description of each region's hydro system.

Time resolution in the model is 1 week, or optionally fractions of a week (e.g. peak load day, off-peak day, night, weekend).

3.1 The System Model

In the Power Pool Model the modelled interconnected power system is divided into regional

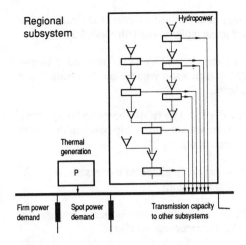

Figure 2. Subsystem description.

subsystems, as shown in the sample system in **figure 1**. System subdivision may be based on hydrological or other characteristics having to do with the local hydro systems, or it may be based on bottlenecks in the transmission system.

Within each subsystem hydro power, thermal power and consumption (firm power or spot power demand) may be modelled, as illustrated in **figure 2.** In addition the transmission system between subsystems is modelled with defined capacities and linear losses. Certain transmission fees may be modelled.

The **hydro power** system within each region/ subsystem may be modelled in detail. Based on standard plant/reservoir modules as shown in **figure 3**, even large/complicated river systems may be modelled. A model of the Norwegian hydro system may for example involve from 500 to 800 plant/reservoir modules, depending on the degree of detail. Figure 2 shows an example of a small regional hydro system modelled using standard

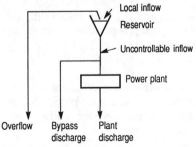

Figure 3. Standard plant/reservoir module.

modules. The following properties may be attached to each plant/reservoir module:

- a reservoir, defined by its volume and relationship between water volume and elevation above sea level,

- a plant, defined by its discharge capacity and a piecewise linear relationship between discharge and generation,

- inflow (weekly) either to the reservoir or directly to the plant,

- different routes for plant discharge, bypass discharge and reservoir overflow,

- variable constraints on reservoir contents and waterflow (plant or bypass discharge),

- pumping capability, either reversible turbines or dedicated pumping turbines.

Thermal generation units are usually defined by their variable costs (defined by fuel costs etc.) and capacity, and are modelled as such. Both costs and capacities are modelled as functions of time (maintenance work cycles may be included). This type of modelling assumes fuel can be purchased and used as needed. This is often the case with coal- and oil-fired plants, nuclear plants, and possibly some gas-fired plants.

Typical of some fossil-fuelled plants, however, is that they are contractually or otherwise bound to receiving a specified 'inflow' of fuel. This is particularly the case with gas-fired plants. The fuel inflow may be specified continually, or for example annual or pluriannual volumes may be specified. Thermal units bound by this type of constraint on fuel inflow are either modelled by fixed energy series injected directly into the power system (specified volume per week or fraction of a week, no local fuel storage) or by equivalent hydro plants. The latter may be used both in the case where local fuel storage is possible, and in the case where fuel volumes are specified only for longer periods of time, for example annually.

Two types of **power consumption** are modelled; firm power demand and spot power demand, where consumption per time step is a function of market price:

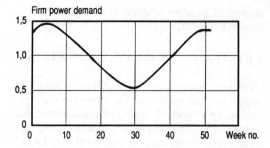

Figure 4. Typical yearly profile for firm power demand from Norwegian households.

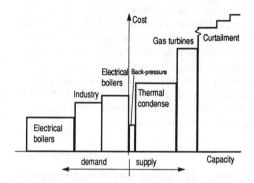

Figure 5. Sample spot power market, thermal system and curtailment costs for a specified week.

Firm power demand is modelled as specified power consumption week by week (or for fractions of a week) as illustrated in **figure 4**. Inability to deliver firm power entails buying curtailment power at high costs.

Spot power demand within each subsystem is modelled as a stepwise price-quantity relationship for each week (or fraction of a week). This market consists mainly of electric boilers and some industrial consumption. **Figure 5** illustrates a model of this market for a specified week. As the figure shows, curtailment power and thermal generation capacity (assuming fuel can be purchased and used as needed) are modelled principally in the same way as spot power demand.

Power exchange between countries, or between any interconnected subsystems for that matter, may be spot exchange or contractually fixed exchange:

578

Optimal **spot exchange** between subsystems is one of the results of the market clearance process in the Power Pool Model, given by incremental power costs, transmission capacity, transmission losses, and any transmission fees which might be incurred. 'Transmission fees' in the model might in fact not be fees for transmission of power, but instead be the profit required by a country or subsystem before being willing to sell power to another subsystem. Or it could be a totally fictitious 'fee' used simply to limit spot exchange, for example to limit annual power import.

This type of exchange 'fee' could actually be of interest for studies of future scenarios in the Nordic power system, since national sufficiency in power supply in principle is a prerequisite for participation in the Nordic spot market. For a purely hydro-based system such as the Norwegian, this could be interpreted to mean that average spot import should not exceed average spot export in the long run. For some scenarios spot import might have to be restricted in order to fulfill this requirement, and the best way to do this would be to use an exchange 'fee' in terms of a Lagrangian multiplier.

Contractually fixed exchange between subsystems is modelled as a firm power obligation for the exporting subsystem, and as a fixed energy inflow injected into the importing subsystem. Transmission capacities for spot exchange would have to be modified to take into account the transmission of firm power.

One interesting case of firm power exchange is the aspect of using a hydro system as a supplier of peak power to a thermal system. At peak load periods each week firm power would be exported from the hydro to the thermal system. At off-peak hours the same energy could be returned as firm power, or exchange at off-peak hours could be based on spot exchange. Studies have been conducted using the Power Pool Model to study the profitability of such arrangements.

3.2 The Power Pool Model: Strategy part

To limit the computational burden, the strategy part of the Power Pool Model is forced to utilize an aggregate model representation of the hydro system within each regional subsystem, i.e. an aggregate energy reservoir with an equivalent power plant and energy time series for controllable and non-controllable inflow. Otherwise the subsystem models are as indicated earlier (**figure 2**).

Given the stated multireservoir model description, the objective of the long-term optimization process is to establish an operation strategy that for each stage in time, produces the 'best' decision vector, given the system state at the beginning of the stage. By 'best' decisions is understood the sequence of turbined and spilled water volumes that contribute to minimizing the expected operational costs during the period of analysis. By system states is understood regional reservoir storage levels and per stage hydrological inflows.

The stated problem from optimal control can in principle be solved by stochastic dynamic programming (SDP) described by the recursive equation

$$\alpha_t^*(X_t) = \mathop{E}_{A_t|X_t}\{\mathop{Min}_{U_t}(C_t(U_t) + \alpha_{t+1}^*(X_{t+1}))\} \qquad (1)$$

subject to the constraints that water balance equations and bounds on states and decision variables must be fullfilled at each stage [3]. The interpretation of terms in (1) is as follows:

t : index of stage

X_t : state vector at the beginning of stage t

$\alpha_t^*(X_t)$: expected value of the operation cost from stage t to the end of the planning period under the optimal operation policy

$A_t|X_t$: the distribution of inflow volumes A_t conditioned by state X_t

$E\{\}$: represents 'expected value'

U_t : decision vector for stage t

$C_t(U_t)$: immediate cost associated with decision U_t

The solution of (1) requires the definition of discretized states. The number of such states increases exponentially with the number of state variables in the problem. Thus formal SDP-solution becomes infeasible when the number of reservoirs exceeds 2-3.

For practical solution of the multireservoir decision problem an approximate methodology has been developed [2]. A stochastic DP-related algorithm is used as the 'nucleus' for solving each regional subproblem, and an overlaying hierarchical logic is applied iteratively to treat the multireservoir aspect. The process is illustrated in **figure 6.**

• A regional decision table in terms of incremental water cost is first calculated for each subsystem decoupled from the others. A version of back-

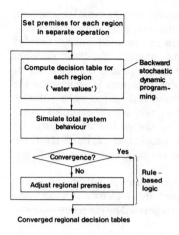

Set premises for each region in separate operation

Compute decision table for each region ('water values')

Backward stochastic dynamic programming

Simulate total system behaviour

Convergence? — Yes

No

Adjust regional premises

Rule - based logic

Converged regional decision tables

Figure 6. Main logic to handle multireservoir problem.

ward SDP called the 'water value method', is used to this end [4].

• Simulation of total system behaviour is next performed using the computed decision tables to determine energy production in each subsystem, energy exchange between subsystems and transactions with neighbouring countries.

• Feedback is then executed conditionally: If a stable and satisfactory solution is found, the process is finished. If not, the result from the simulation is used to adjust regional premises, and return then made to regional decision table computation.

3.3 The Power Pool Model: Simulation

In the simulation part of the Power Pool Model system performance is simulated for a chosen sequence of hydrological years. Based on the incremental cost tables calculated previously for each aggregate regional hydro system, weekly operational decisions on power generation (hydro, thermal) and consumption (spot consumption, curtailment of firm power consumption) are made in what can be termed a market clearance process. A detailed rule-based reservoir drawdown model affords the distribution of each subsystem's aggregate hydro production among available plants for each time step. Historical inflow series covering a period of typically 50 years are the basis for simulation.

Figure 7 illustrates the weekly operational decision process, summarized in the following points:

For each week:
- Based on current reservoir levels and incremental cost tables for stored hydro energy, optimal generation, spot consumption and exchange are calculated on an aggregate subsystem level for all periods within the week (e.g. peak load, off-peak day, night, weekend). This is afforded by a network flow algorithm.

For each period within the week, the following is repeated for each subsystem with a local hydro system:
- A rule-based reservoir drawdown model seeks to distribute the desired hydro production among available plants. Constraints in the hydro system may cause the reservoir drawdown model to deviate from the generation found to be optimal at aggregate subsystem level. In a case where increasing hydro generation will cause loss of water (e.g. bypass past plants placed in cascade) the cost of increasing local hydro production is weighed against the cost of deviation from desired production. The cost of deviation is calculated on the basis of a stepwise cost-quantity function showing power supply and demand as a function of price from neighbouring subsystems

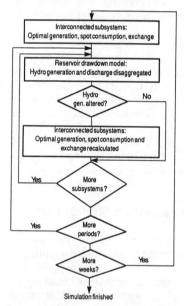

Interconnected subsystems:
Optimal generation, spot consumption, exchange

Reservoir drawdown model:
Hydro generation and discharge disaggregated

Hydro gen. altered? — No

Interconnected subsystems:
Optimal generation, spot consumption and exchange recalculated

More subsystems? — Yes

More periods? — Yes

More weeks? — Yes

Simulation finished

Figure 7. The weekly decision process in the Power Pool Model's simulation part.

as well as local thermal capacity and spot power demand. This function has to be constructed specifically for each subsystem and for each step in time, as it is a function of reservoir state in all other subsystems.

- If resulting hydro production deviates from 'desired' subsystem production, then optimal generation, spot consumption and exchange are recalculated at the subsystem level using the network flow algorithm. This time, however, hydro generation is fixed for those subsystems that have already been scheduled by the reservoir drawdown model.

At the end of each week, the aggregate reservoir level is updated with results from the reservoir drawdown model, and hence premises are set for next week's operational decisions.

As stated earlier, the disaggregation of regional subsystem storages into individual reservoir storages and subsystem hydro production into individual plant production is afforded by a detailed reservoir drawdown model which utilizes a rule-based logic for reservoir depletion [5]. The model operates with 2 types of reservoirs:

- Buffer reservoirs, whose operation is defined by guidecurves. These are mainly reservoirs with low storage capacity in relation to inflow (e.g. run-of-the-river type).

- Regulation reservoirs, which are operated according to a general reservoir drawdown strategy (rule-based).

The basic goal of the reservoir drawdown strategy is to produce a specified amount of energy in such a way as to minimize expected future operational costs. This goal is sought fulfilled by:

a) seeking to minimize risk of overflow during that part of the year when inflow is greater than discharge,

b) seeking to avoid loss of power capacity caused by empty reservoirs during that part of the year when discharge is greater than inflow.

In the Nordic countries this implies dividing the year into a 'filling' season (late spring, summer and early fall with high inflow, low power consumption) and a 'depletion' season (late fall, winter and early spring with low inflow, high power consumption).

Results which may be extracted from the Power Pool Model's simulation part include:

- hydro system operation (reservoirs, generation, water flows),
- thermal generation,
- power consumption, curtailment,
- exchange between subsystems,
- economic results,
- energy consumption-related emission figures (SO_2, NO_x, CO_2, dust/soot, ashes)
- incremental benefit figures of increasing the capacity of various facilities.

4. SAMPLE STUDIES

4.1 Hydro-based Peak Power Supply to Thermal Systems

At present, power exchange capacity between Norway and Denmark is about 550 MW. The transmission grid consists of two HVDC undersea cables. The grid was established on the basis of an agreement between Norway and Denmark on supplying peak power to the Danish thermal-based system using Norwegian hydropower. While allowing the Danish system a respite in the construction of new thermal capacity, as well as providing reduced fuel consumption in years of high inflow in Norway, the agreement gives Norway the option of relatively inexpensive imports in low-inflow periods.

Studies have lately been performed related to an expansion of the transmission capacity by 440 MW between Norway and Denmark. The studies were performed using the Power Pool Model, and involved simulation of two cases:

- Assumed 1995 NORDEL system operation with today's transmission capacity of 550 MW between Norway and Denmark.

- As above, but now introducing an upgraded capacity of 990 MW by means of a new sea cable.

Figure 8 illustrates the simulated utilization of the undersea cables with and without increased capacity.

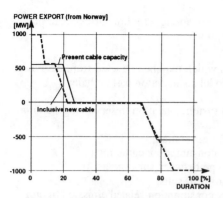

Figure 8. Simulated utilization of undersea cable between Norway and Denmark, before and after increasing capacity.

The studies outlined here indicated that an investment in increased transmission capacity would be profitable. Based in part on these studies, the decision has now been made to realize plans for a new undersea cable, which will be completed by late 1993.

4.2. Prospective Interplay between Norwegian Hydro- and Gaspower Resources

Complex scheduling processes are implied when gas-fired plants are integrated into a purely hydro-based production system like the Norwegian. Depending on the agreed-upon premises for delivery and use of gas, such plants may in principle be scheduled according to either price or volume signals. If the gas price is the agreed-upon principle premise for resource scheduling, the planning problem is similar to the one met with when dealing with conventional thermal generation. If available gas volume per defined unit of time is specified for a given project, the design and scheduling problem may get more involved. If the specified unit of time is small, implying constant gas flow, an optimally sized buffer storage for gas may enhance system economy by affording increased flexibility of operation.

Case studies have been performed relating to the introduction of gas-fired power plants in Norway, scheduled either according to price or according to volume signals. **Figure 9** shows simulated annual output from a 1000 MW gas-fired plant, which in this study was scheduled according to a specified gas price at some future scenario. As the figure shows, optimal gas-based power production (and

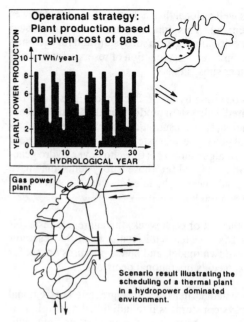

Scenario result illustrating the scheduling of a thermal plant in a hydropower dominated environment.

Figure 9.

thus gas consumption) depends to a great extent on inflow variations in the hydro system. This type of operation, though, assumes delivery of gas can be taken as needed.

Similar studies have also been performed on gas-fired plants scheduled according to volume signals, where contractual gas volumes were specified for periods varying from 5 to 50 years.

5. CONCLUSIONS

For overview planning of the development and use of national resources in the form of hydro power and gas, a system modelling and simulation scheme has been developed based on stochastic dynamic programming, network flow linear programming and heuristics. Simulation results, including sensitivity signals from performance models, have been found efficient for use in guiding the setting of design variables.

REFERENCES:

[1] Botnen, O.J., Haugstad, A., Johannesen, A., Haga, I., Kroken, S.: *'Coordination of Hydro-power and Thermal Generation Systems for*

Enhanced Multinational Resource Utilization.'
Hydropower '92, Lillehammer, Norway, June
16-18, 1992.

[2] Egeland,O., Hegge,J., Kylling,E., Nes,E.:
*'The Extended Power Pool Model. Operation
Planning of a Multiriver and Multireservoir,
Hydrodominated Power Production System.'*
CIGRE 1982, Paper 32-14.

[3] Pereira, M.V.P.:*'Optimal Stochastic Operations Scheduling of Large Hydroelectric Systems.'* CEPEL-Report, May/1987.
Rio de Janeiro, Brazil.

[4] Killingtveit, Å. and Reitan, R.:
*'A Mathematical Model for Planning of Short
and Long Term Operations of a Hydro Power
Production System.'* Proceedings of Interna-
tional Symposium in Real Time Operation of
Hydrosystems. June 24-26, 1981.
Univ. of Waterloo, Ontario, Canada.

[5] EFI Technical Report No. 3483: *'EFI's Models
for Hydro Scheduling',* 1988.

Hydropower'92, Broch & Lysne (eds) © 1992 Balkema, Rotterdam. ISBN 90 5410 054 0

The modular interactive river system simulator – A tool for multi purpose planning and operation of hydropower systems

Marit Lundteigen Fossdal
Water System Management Association, Norway

Ånund Killingtveit
Norwegian Institute of Technology, Trondheim, Norway

ABSTRACT: A computer system called the *Modular Interactive River System Simulator* is now under development in Norway. The simulator is aimed for two main purposes: 1) Multi purpose water resources planning and 2) Optimal operation of hydro systems. A plan for the program system was prepared in 1989. Detailed specification started in April 1990 and the whole project is planned to be finished by 1993. The simulator is planned to include more than 15 different water related models within a common user interface and with one common database to facilitate the data communication between different models and simplify data storage and data ingegrity. The models included in the simulator can be grouped into four main classes: 1) input from the watershed to the river system, 2) Processes within rivers and lakes, 3) Hydropower simulator models, and 4) Consequences for humans and ecosystems. The user interface will be based on modern principles, with emphasis on graphics and on-line help for the user.

1 INTRODUCTION

A general simulation system for water resources planning will need to include a large number og different models, for example rainfall-runoff models, hydraulic models, water temperature models, water quality models and so on. The list could easily have been made much longer and still some users would find one or more important models missing. The situation today is that a large number of models which is capable of solving one special problem exists, for example to compute a runoff forecast, to compute water level or to compute water temperature. A real planning situation, however, require a number of different computations performed with different models, where output data from one model can be input to one or several others. This is not always easy to perform, due to different data storage and user interface for each model. A complete program system, with all types of different models integrated under one common user interface, is termed a *"Modular Interactive River System Simulator"*.

The idea of such a computerized "River System Simulator" was first introduced in 1986, and then planned in detail in 1989 by a working group established by Water System Management Association (Water System Management Association 1990). This plan constitute the basis for the future work on the simulator, and this paper summarizes the main ideas and concepts in the plan.

A total of at least 15 different models used in water resources analysis will be integrated under one common user interface an with one common data base system. Mostly existing models will be used, with some adaption of the user dialogue and new routines for database access, but with most of the computational algorithmes unchanged. The system concept is illustrated in Figure 1.

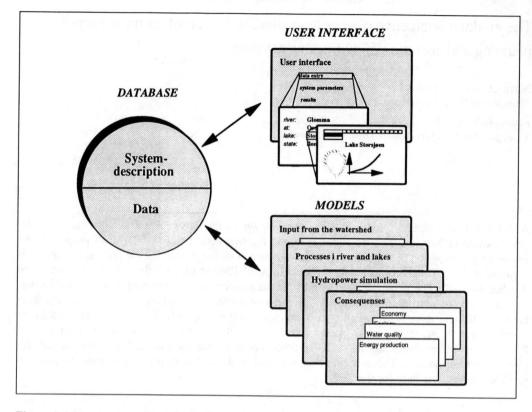

USER INTERFACE

DATABASE

System-description

Data

User interface
Data entry
system parameters
results

river: Glomma
at:
lake: Sto
state:

Lake Storsjøen

MODELS

Input from the watershed

Processes i river and lakes

Hydropower simulation

Consequenses

Economy

Water quality

Energy production

Figure 1. Main structure. The simulator consists of three parts: the database, the user interface and the computational routines.

A common logical *data model* for the simulator has been established. This work was considered to be essential for the future development of the simulator.

2 PROJECT PURPOSE

The main objective of the project has been *".. to develop a computerized tool for the multi purpose planning and operation of river systems"* (Water System Management Association 1990). The program system is designed to be used by many different user groups with varying background and responsibilities, for example:

- Hydro power companies
- Research institutions
- Educational institutions
- Consultants
- Public administration and Water Authorities

The main goals for the use of the simulator will be:

"To improve decisions concerning different user groups' needs for discharge, water level, water covered area etc. during the planning of new hydropower developments, or when old regulations are reviewed.

To specify environmental requirements and contribute to cost-efficient measures.

To improve operational decisions during special (difficult) situations, for example during floods or droughts, when ice-problems occur etc."

The results from the simulations can also be used for public information and as a method to visualize effects of different strategies in discussions of how to operate multi-purpose water projects. It is also expected that the simulator will become important for educational purposes where it can be used in the same way as a ship or flight simulator to train the operational staff to cope with critical situations which will not be encountered during normal operation.

3 MAIN STRUCTURE OF THE SIMULATOR

3.1 A modular and expandable system

The basic idea behind the planning of the River system simulator is to utilize existing and well documented models as far as possible, and to create one common user interface and one common database system for all different types of models. It is important to design a *modular* system so that each individual model can be included or removed from the simulator without affecting the other models. This principle is illustrated in Figure 2.

3.2 Main types of models in the simulator

The different models which is being incorporated into the simulator can be grouped into four main categories or model classes as indicated in Figure 3:

1) Models for the input *from* the watershed *into* the river system. Rainfall-runoff and sediment yield models are examples of this type.

2) Models of physical, chemical and biological processes *within* rivers and lakes in the watercource. Water-level and water temperature models are examples of this type.

3) Hydropower simulation models, simulating reservoir operation, hydropower

production and economical value of power production.

4) Models describing *results* or *consequences* for human population or an ecosystem. Fish habitat models are examples of this type.

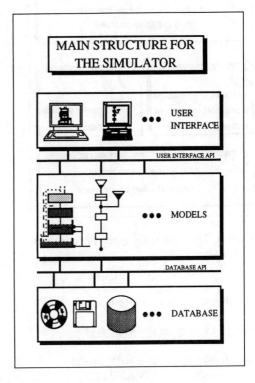

Figure 2. The modular structure in the simulator.

Several different models within each class were evaluated by the planning group. The final decision of models to include in the simulator was made in December 1991, after detailed planning and studies of available models. An international expert group was invited to a workshop for final discussions together with a panel of users and Norwegian experts. The discussions and conclusions from this workshop (Wathne 1992) was an important basis for the final decisions, where 15 different models were chosen.

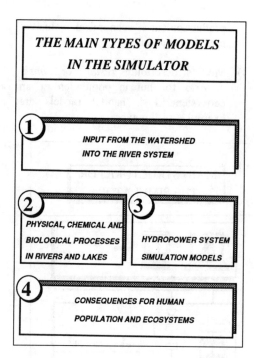

THE MAIN TYPES OF MODELS IN THE SIMULATOR

① INPUT FROM THE WATERSHED INTO THE RIVER SYSTEM

② PHYSICAL, CHEMICAL AND BIOLOGICAL PROCESSES IN RIVERS AND LAKES

③ HYDROPOWER SYSTEM SIMULATION MODELS

④ CONSEQUENCES FOR HUMAN POPULATION AND ECOSYSTEMS

Figure 3. The four main types of models in the simulator.

3.3 Models chosen for use in the simulator

Type 1: Input from the watershed

Rainfall-runoff: The HBV-model
Urban runoff: NIVANETT

Type 2: Physical, chemical and biological processes in rivers and lakes

Hydraulic models: DAMBRK
                    (Unsteady flow)
                    HEC-2
                    (Stationary flow)
Water temperature: RIVICE
Ice: RIVICE
Water quality: FINNECO, QUAL-II
Erosion: Not decided yet

Type 3: Hydropower systems simulation models

Multi-reservoir systems: VANSIMTAP
Single reservoir systems: ENMAG

Type 4: Consequences for human population or ecosystems

Fish habitat: River modelling system
                (FBV-model)
                Two new models:
                BIORIV, BIOLAKE
Wildlife: Futher development of the
                FBV-model
Tourism: Further development of the
                FBV-model

4   HARDWARE AND SOFTWARE PLATFORMS

4.1 Standardization

The simulator is intended for use on workstations and powerful personal computers. The following standards is used:

- Operating system: UNIX
- Programming languages: FORTRAN 77 and
                      90 and C++
- Database access: SQL
- Graphics: PHIGS-based systems
- User interface: Motif

4.2 User interface

Since the simulator is intended for use by many different user categories, the user interface design will be important, but also difficult. An important part of the project will therefore be to plan, design, implement and test the user interface for a number of different user groups. The possibility of using different user interface "shells" is being investigated at present. The use of GIS-systems as an interface to the simulator will be tested in 1992.

4.3 Data model and database design

The use of one common fysical model and one physical database for all the different models is considered to be very important. A commercial relational database will be used and will be accessed with embedded SQL statements in the models to export and import data to and from the database. A number of relational database systems have been investigated, and SYBASE has been chosen for the development work and first implementations. The use of SQL database access language should, however, ensure easy porting to other relational databases as ORACLE or INGRES.

5 PROJECT ORGANIZATION

The development of the simulator is a complex task and will involve many different institutions. The project is initiated and financed by the Water System Management Association. Norwegian Hydrotechnical Laboratory (SINTEF NHL) in the SINTEF Group has the coordinating responsibility and project leader. A reference group with 6 members has been established.

6 TIME SCHEDULE

The model is now under development, and the main structure will be implemented before the end of 1992. The 15 different models will be implemented in 1992 and early in 1993. The testing of the model system will be done in at least 3 different rivers in 1993. The final release of the system is planned 1/1-1994. By 1993 a training program will also be developed, both in Norwegian and English. All program documentation and user dialogue will be available in English.

7 REFERENCES

Water System Management Association, 1990. *Plan for the development of a River System Simulator for Multi Purpose planning and operation in river systems.* (In Norwegian). Asker, January 1990.

Wathne, M. (ed.) 1992. *River system simulator. The development of a decision support system.* SINTEF NHL report STF60 A92010, ISBN No. 82-595-7023-8. Trondheim.

Hydropower'92, Broch & Lysne (eds) © 1992 Balkema, Rotterdam. ISBN 90 5410 054 0

Instream flow requirements and hydropower development – A case study from the Gjengedal development in Norway

Ånund Killingtveit
Norwegian Institute of Technology, Trondheim, Norway

Ola Lingaas
Sogn og Fjordane Energy Company, Norway

Kjetil Arne Vaskinn
SINTEF NHL, Norway

ABSTRACT: In the planning of hydropower developments the instream flow requirements are often decisive of both the project economy and environmental effects. This is especially important in many salmon rivers, where the preservation of salmon stock is of great importance. The instream flow is also of vital importance for many other users, including agriculture, water supply and recreation. In this process both economical, technical and environmental effects have to be considered and optimal balance between competing goals has to be found. The use of simulation models has proved to be helpful in this process. By combining simulations of hydropower production, river flow, physical parameters in the river and fish habitat, the effect of varying instream flow requirements can easily and accurately be investigated. A direct comparison between project economy and usable fish habitat has been possible in the case study Gjengedal. The different types of simulation models used are described, and also how these models have been used in the planning process.

1 INTRODUCTION

The potential energy resources which could be developed as hydropower in Norway is about 170 TWH. About 110 TWH is developed today. Considering the advantages with hydropower compared to other energy resources such as oil and coal, it is easy to conclude that Norway should consider to develop more.

Computer models have been used in Norway since before 1970 for simulation of the hydropower production. These models are fairly good. Good tools such as simulation models have on the other hand not been available for simulation of the environmental effects of possible changes in the flow regimes until fairly recently. This paper describes the combined use of both such types of models, in the Gjengedalen river in western Norway where the Sogn og Fjordane Energy Company wanted to use the models as a practical tool during the planning process of a hydropower development.

2 GJENGEDAL RIVER: THE STUDY AREA

The Gjengedalen River is situated on the western part of Norway. see Figure 1. The catchment area is about 170 km² with a mean annual runoff of 12 m³/s. The variability in the runoff is rather high. The normal discharge during the spring flood period caused by snow melting is 30-50 m³/s. During low flow periods in wintertime the runoff is about 1-2 m³/s. The highest elevation in the catchment is 1300-1400 m.a.s.l. The upper part is an open mountainous area, while the lower part is characterized by a deep narrow valley. The area is an important recreational area. Along the river in the valley there is farmland. Salmon fishing is an import recreational activity in the lower 10 km. of the river. The upper part off the river is important as habitat for raising fish stock.

The planned hydropower project in the river consists of two reservoirs at 500 m.a.s.l. Some smaller rivers will be diverted to the reservoirs. The outlet from the power plant will be at 40 m.a.s.l.

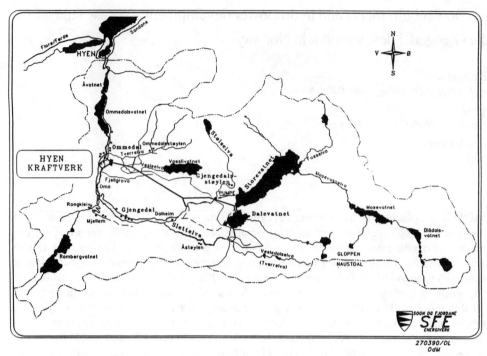

Fig. 1   Gjengedalen.

3   MODELS

3.1 *The hydropower simulation model*

The ENMAG program system is a simulation model developed for design and operation of hydropower systems (Killingtveit, 1987). It can also be applied to simulation of flood control schemes, irrigation, water supply or such projects in cooperation with hydropower. The model is available on IBM-compatible PC's.

The ENMAG model is modular in the sense that the system being studied can be modelled by a small number of building blocks, called modules. The 4 types of modules used are:
  * Reservoir
  * Diversion structure
  * Hydropower station
  * Control point

Each module can be connected to any number of other modules by an adressing method defined in the data set. With these four types of modules and the connectivity method, even very complex systems can be modelled. The total system can, however, only have one storage reservoir. A number of other types of data can be entered to set up a simulation for a given system:
  * Hydrological data
  * Energy market data
  * Restrictions
  * Operational strategy

Hydrological data are entered into separate preprocessing programs, which prepare special files for use during the simulation. The results can be presented in a number of different reports or written to files for later use by post-processing programs. The most important results from a hydropower system simulation will normally be the average production, the economical value of the production and the firm power reliability given in % or as number of years with firm power deficit in the simulation period.

592

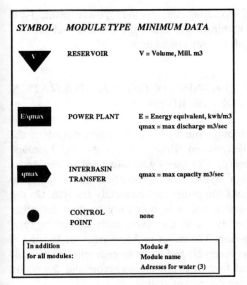

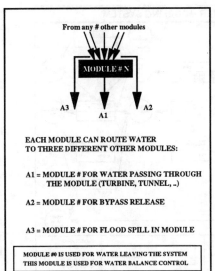

Fig. 2 The ENMAG System.

The different types of modules are shown in Figure 2. Here also the minimum amount of data for each module is shown. More detailed data can be added when needed. The principle for connecting modules together is also shown in Figure 2. The ENMAG model for the Gjengedal system is shown in Figure 3.

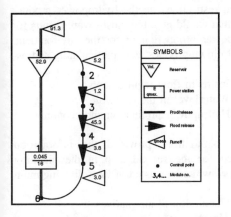

Fig. 3 The ENMAG system for Gjengedalen.

3.2 The River Modelling System

The main purpose of the River Modeling System (RIMOS) (Vaskinn, 1988) is to analyze the hydraulic conditions in a river and to show how these conditions are related to the discharge. The most important hydraulic parameters to analyze are depth, velocity and surface area. These parameters are, together with substrate and cover, important for classifying the habitat for fish.

The RIMOS system is made up of several simulation models. The different models are shown in Figure 4. The circles represent different databases and data storage systems, while the squares represent different simulation models.

Through the integration of hydraulic, hydrological, hydropower production and habitat simulation models in the system, it is possible to simulate in great detail how a planned hydropower development will affect river discharge, hydraulic parameters and finally the fish habitat in the river.

3.3 Physical habitat simulation

The hydrologic and hydraulic conditions in the Gjengedalen river was investigated in detail for the years 1986 and 1987. The data served as the basis for calibration of the RIMOS model. Using the calibrated model it was possible to

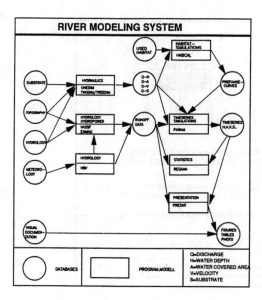

Fig. 4 The River Modelling System.

perform detailed simulation of the instream flow conditions. The most important fish species in the river are Atlantic Salmon and Brown Trout. The habitat criteria for these species can be presented as a curve which shows what habitat the fish prefer and what kind of habitat the fish try to avoid, Figure 5 (Heggenes et al. 1988).

By combining the results from the physical habitat simulations with the habitat preference curves, is it possible by means of RIMOS to simulate the distribution of total area of the river with habitat that the fish prefers, and habitat that the fish tries to avoid.

Further, by combining this information with data for river flow for different alternative ways

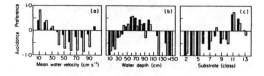

Fig. 5 Habitat preferences by Atlantic salmon (solid bars) and brown trout (open bars) parr in the River Gjengedalselva, 1987-88. (a) Mean water velocity. (b) Total water depth. (c) Substratum.

of operating the planned power plant, is it possible to get a linkage between habitat and hydropower production.

4 PLANNING OF OPTIMAL UTILIZATION OF THE RIVER

Through the hydropower development in the Gjengedalen River, the Sogn og Fjordane Energy Company was interested in getting as high hydropower production as possible, to make the project economically feasible. On the other hand, it is necessary to consider other user types in the river, such as farming and fishing. The major problem areas are located at the reservoir and at the river reaches with redu-ced flow between the reservoir and the outlet from the power station.

By using the two model types, ENMAG and RIMOS, it was possible to provide good infor-mation to all the different groups. The models were used in two steps in this work. During the first step, simulation of energy production and the effects for a large number of alternatives on the environment was done. The output from these simulation, supplemented by other types of information made it possible to study the main environmental effects of different alterna-tives. Doing this, it was possible to find out which restrictions on the hydropower produc-tion that would give the most severe results for the power production or for the environment. The following issues were particularly addres-sed:

- Instream flow requirements
- Waterlevel in the reservoir.
- Hydropower production and economy

The simulations provided the information needed for the second step in the planning process, which consisted of the following main steps:

* Generation of alternatives
* Simulations
* Analysis and trade offs.

Several different alternatives for minimum instream flow and reservoir filling were formu-lated. Some of these alternatives were formed in such a way that the condition for the fish

594

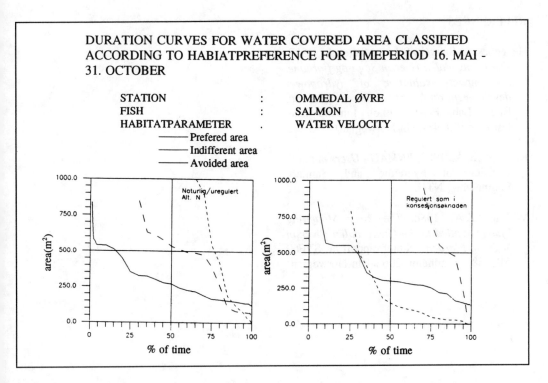

Fig. 6 Duration Curves.

and recreation should be as good as possible, other alternatives were formulated as compromise between the different users.

All the alternatives were then simulated in the models. In this way we were able to quantify and compare both energy production and value and environmental effects.

The final stage in the planning process was the analysis and comparison between all the simulated alternatives. During this stage of analysis the results from the computer simulations were extensively used. All the combinations of the flow restrictions and reservoir restrictions gave different habitat and economical values. These values were presented in a matrix. For each combination of restrictions we also produced graphical presentation of the environmental effects. Figure 6 gives an example of how this was done. This figure shows the duration curve for what is preferable habitat with regard to water velocities for atlantic salmon on one location given one set of alternatives. The figure also show the natural conditions.

Sogn og Fjordane Energy Company has evaluated several alternatives and has put in an application for permission to start the development of the hydro power plant. The results from the simulations are made available to the public authorities and other organizations. By doing this we feel that we have created a common platform for the evaluation of the project. The consequences for the hydropower production and the environment of choosing different alternatives are clearly illustrated. We feel that the RIMOS system has proved to be a valuable tool for planning of hydropower plants.

ACKNOWLEDGEMENTS

This project was made possible due to a comprehensive cooperation between Freshwater Ecology and Inland Fisheries Laboratory at the University of Oslo (LFI), NVE, SINTEF NHL and Sogn og Fjordane Energy Company. The Fund for Licensing fees (Konsesjonsavgifts-fondet) has also been funding the project.

REFERENCES

Heggenes, J., Saltveit, S.J., og Sæter, L., 1988. *The Gjengedalen river in Sogn og fjordane. An impact evaluation of hydropower development on Atlantic Salmon and Traut.* Rapp. Lab. Ferskv. økol. Innlandsfiske, Oslo, nr 100, 48s. (In Norwegian)

Killingtveit, Å., 1987. *ENMAG - Users manual.* Division of hydraulic and Sanitary Engineering, NTH.

Vaskinn, K.A., 1988. *The River Modelling System applied to the Gjengedalen River in Sogn og Fjordane.* Sintef rapport nr. STF60 A88068, Trondheim, 206 s. (In Norwegian)

Hydropower'92, Broch & Lysne (eds) © 1992 Balkema, Rotterdam. ISBN 90 5410 054 0

State-of-the-art plant control systems link turbines and dispatch centers

C.J. Meier
Water Resources Management Staff Group, Sulzer Escher Wyss Ltd, Hydraulics Division, Zürich, Switzerland

ABSTRACT: Despite operation of large-scale hydropower schemes by computerized plant management systems, fast and effective secondary regulation on regional or national levels is limited by insufficient operational information on the participating power plants and the weak optimizing effect of dispatch centers. Optimal dispatching as a constant decision-making process requires more precise operational data than is possible with conventional hard-wired control systems. State-of- the-art plant control systems can handle practically unlimited operational complexities but need extensive engineering and a functional rather than technological approach. Computer simulations considering hydrological and economic aspects lead to energy management concepts demonstrating the overall profitability of resultant operational procedures. Benefits such as fast response under economic constraints are complemented by an overall efficiency improvement based on continuous measurement, filing and exploitation of detailed process information in the automation system.

1 OPTIMIZING SECONDARY REGULATION

1.1 Introduction

Operation of today's large-scale hydropower schemes depends on computerized plant management. However, hardly any of the computer systems currently used transmit their data directly to the generating units. Most of them advise the operating engineers by supplying them with data presented in tables, diagrams and charts. Until now, fast and effective secondary regulation was limited by insufficient operating mode information on the participating power plants and the weak optimizing effects of dispatch centers.

The forecasted power demand is distributed by the dispatcher among the power plants according to a daily schedule. Since the actual demand cannot be exactly foreseen, a regulation system must compensate for all deviations from the prediction so that combined frequency and power regulation is carried out within the grid system. Only those power plants equipped with power or speed governors are involved in grid regulation. This primary regulation ensures the output needed by the dispatcher as well as an appropriate frequency

stabilization contribution, but cannot fully compensate for frequency deviations when load fluctuations are too strong. The dispatch centers correct the residual error by new settings for nominal output or a corrected frequency power droop for the power stations.

Generally, this secondary regulation is carried out in accordance with the overall criteria for grid safety. Nevertheless there is now a growing trend towards considering the financial aspects of this procedure as well, i.e. power stations are optimized so as to ensure the necessary output at minimum power consumption. The criteria for this are as follows: minimum line losses, adequate spinning reserve and minimum primary energy consumption. The optimization project described in the following succeeded thanks to the high information level of the dispatch centers allowed by digital data communication. Such optimization processes are of maximum benefit in countries with regionally integrated production.

1.2 A practical example of production optimization

The Austrian grid system is fed by various regional

power companies, and two so-called "Sonder-gesellschaften" (Oesterreichische Draukraftwerke OeDK und Tauernkraftwerke TKW) with the explicit order to maintain network regulation. The TKW are equipped with a regulation capacity of 900 MW installed in storage plants. In order to minimize regulation costs, the following procedure has to be adopted:

•   The current energy demand must be distributed among the different generating units participating in grid regulation according to the overall optimum efficiency (Seidl, 1991).
•   For optimal adjustment of the generating units to the current situation, the grid master computer at the regional dispatch center must receive continuously updated information on grid regulation quality and energy production.

This includes not only current efficiency data on all power plants in operation, but also deviations of the output from the sum of current optimum efficiency points, size of regulating ranges at different levels of efficiency and incremental flow rates of the various units. Since the strong dependence of plant efficiency on actual discharge is difficult to formulate, this relationship has to be simulated numerically if high losses are to be avoided. This filed information and current operating data on all power plants forms the inputs and constraints of an optimization procedure based on a Lagrangian algorithm. The grid master computer then calculates the optimal load distribution and regulates the units in operation through the remote regional control systems.
The filed unit data is checked against measurements carried out every six months, and updated as necessary.

1.3 *Restrictions*

The main advantage of this centrally optimized secondary regulation is the possibility of cost-effective participation in the energy market, for example energy can be exported at the highest possible profit. However, this kind of optimization is only possible for:
•   power companies with a large seasonal storage capacity, since this process ignores all hydro-meteorological boundary conditions;
•   countries with centralized grid regulation.

2.   DELEGATED OPTIMIZATION TASKS

2.1 *New optimization tasks*

Since centralized network regulation is not always possible (for instance if a large number of suppliers within one hydrological system have different kinds of operating targets), it is better to adopt a more decentralized approach that will also avoid excessive operation costs:

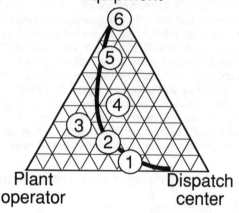

Supplier of Hydromechanical Equipment

Plant operator   **Dispatch center**

1   Regional economic dispatching
2   Secondary regulation with Kaplan units
3   Water resource management
4   Optimal multi-unit operation
5   Improved plant instrumentation
6   Optimal runner blade control for Kaplan units

Delegated optimization tasks
Fig. 1

Delegating the various optimization tasks saves time and minimizes the amount of information to be exchanged. They must be delegated at a hierarchical level which is high enough for optimization procedures to result in substantial cost savings:
•   Load distribution in the different power plants according to overall incremental flow rates
•   Load distribution among the different units of large power plants by the plant management system according to accurate individual incremental flow rates.

Another advantage of this kind of decentralized optimization is that it allows extended exploitation of optimization potential according to the criterion of maximum resource economy, through detailed integration of regional hydrometeorologic influences. Both the centralized and the decentralized forms of optimization system, each with its own characteristics, require modern control system technology. For optimization by the dispatcher, the plant management system must exchange large amounts of data with the dispatch center (high communication capability), and for decentralized optimization the control system must be capable of handling large amounts of data and performing extensive computation.

2.2 Requirements for plant management systems

This leads to new requirements for a state-of-the-art power plant management system:
• Aims and priorities for optimizing functions as a basis for plant control system performance setting (analyses and simulation)
• High data handling capacity at the various control levels (background computing units)
• Remote control interface for plant management systems (AC telegraphic system, ISDN/WAN)
• High requirements on the data logging concept, which can be complemented with simulation programs for determining operating data difficult to measure or predict.

The data accuracy necessary to achieve an energy gain in the order of 1 % must remain unchanged when transferred to higher control levels. Since multiple transformations of analog signals cannot meet this demand, signals must be transmitted digitally by high speed serial data bus.

3. NEW ENGINEERING TASKS

3.1 Engineering Steps

For all parties involved, the layout of computer assisted automation systems for plant control poses new engineering problems.

First of all it is necessary to make an analysis in order to see which parts of the plant or which aspects of the operational management concept have not been exploited up to their maximum potential. For example an assessment of all energy transformation processes for sufficient measuring intrumentation enabling the dispatcher to optimize energy production:
• Flow measurement systems for each individual unit
• Determination of actual efficiency
• Inflow and water level measurement devices

or other improvement possibilities by automated optimization of hydraulic relationships as follows:

• Minimization of water consumption in the turbine (optimal runner blade control).
• Distribution of a given load according to the best combination of units and corresponding power or discharge setpoints (unit commitment and economic dispatch).

Secondly, an adequate operational concept for the targeted efficiency improvement must be designed and evaluated according to economic criteria. Simulation of the various extreme operating situations plays a very important role in the plant automation and operational management concept. The results must be optimized to achieve a maximum annual energy output.

The third step is implementation of the results by specifying the extent and capability of the automation systems and instrumentation required.

3.2 Methodology

In view of these problems, we come to the conclusion that the extensive engineering required should adopt a functional rather than technological approach. Clearly, increased complexity of the functions to be implemented always results in decreased transparency of the technological structure. Hence we require greater specialization on one hand and a new methodology on the other (Glattfelder, 1984). The methodology of systems engineering based on functional groups has produced very good results in practice. This method basically consists of definition, classification and progressive refinement of the main functions, starting at the highest functional level according to the top-down principle. Selection of the first, highest detail level and resultant structural arrangement of the main functions has a decisive influence on the possibilities for optimal operation. After completing this functionally oriented structuralization, the plant

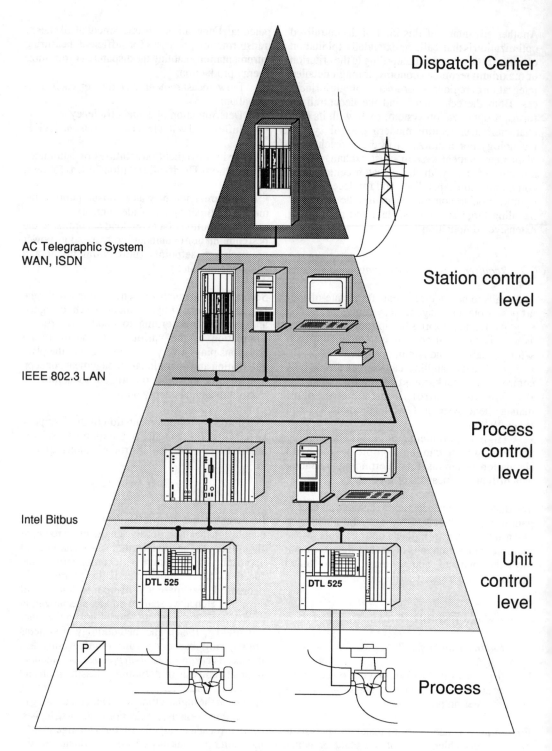

Dispatch Center

AC Telegraphic System
WAN, ISDN

Station control
level

IEEE 802.3 LAN

Process
control
level

Intel Bitbus

Unit
control
level

DTL 525

DTL 525

P / I

Process

Station management system
Fig. 2

is stratified according to the different levels of automation (see figure 2). When selecting the power control system the following points must be taken into consideration:

- Is the chosen system open and accessible?
- Is it supported by current standards?
- Does the system allow for long-term flexible extension?
- Can it be implemented on a step-by-step basis according to the bottom-up principle?

In order to provide each customer with the decisive fundamentals most suitable for his needs, Escher Wyss offers the above mentioned steps as separate engineering services. We also help customers solve complex operating problems such as level control and flow regulation, storage scheme operation and optimized unit dispatching as well as the setting of operational priorities, quality criteria and optimizing algorithms. In this way we contribute our specific hydraulic know-how for implementation in power plant optimization and preventive maintenance schemes. Modern data-oriented process control technology enables the full potential of hydraulic power plants to be exploited on a cost-effective overall basis. An indispensable prerequisite for effective implementation, however, is well-organized project engineering.

4 ESCHER WYSS CONTRIBUTION

4.1 *Engineering*

- Simulation of run-of-river power plant operation with "Floris" simulation software for the following applications: level regulation, primary grid regulation support, operation of cascade power plants.

The FLORIS program computes the dynamic response of waterways by solving De Saint Venant's equations for arbitrary systems of one-dimensional open channels (Fäh, 1991). It also incorporates turbine, grid and governor characteristics in order to run simulations (HNT Hydraulic network transients, Német, 1979). This facilitates the design of control systems and supplies advanced data on the correct operating settings, especially for configurations of extreme discharge conditions which cannot be tested in practice.

- Plant operation concept "Augst"
An annual energy calculation for this kind of plant

with completely different kinds of units is used as basis for presetting the plant control system. By filing and evaluating all central operating parameters during this phase, the plant master computer is supplied with highly detailed turbine characteristics for optimal operational control.

4.2 *Software*

- Optimizing monitoring software: "Egglfing"
This patented optimum runner blade control software works as a separate monitoring program on the unit-related background computers, thus relieving the turbine governor from data processing tasks with such high measuring and computing requirements.

- Turbine characteristics:
Bispline hill charts allowing partial differentiation for use in Lagrangian algorithms, as a basis for all efficiency optimization programs. These turbine characteristics are updated by means of direct or indirect efficiency measurements.

4.3 *Hardware*

- Advanced flow measurement systems: "Tarbela Plant Monitoring System"
By installing ultrasonic flow metering devices, on-line efficiency determination can be incorporated in the plant monitoring system. This allows optimal operation of various units differing considerably from each another even under extreme seasonal head variations between 49 and 135 m, while still observing strict safety regulations for operation at the capacity limit of the world's largest pressure tunnels.

- Network compatible hardware components: "Weser Project"
The Governor & Control Department of Escher Wyss was awarded a contract in 1991 for supplying 18 turbine governing systems in 6 power stations on the middle stretch of the river Weser between Minden and Bremen. Along this stretch, the river Weser has been canalized and fitted with large capacity locks placing high demands on level control systems.
The level and discharge control system design for the cascade plants, with a new remote control and monitoring centre at Robert Frank station, required

a completely new data communications layout for the AEG plant control system.

The turbine governing systems are integrated in the Intel Bitbus station communication network. This is designed to interconnect different types of open system for transmission of short information within a hierarchical system. The integrated Bitbus is considered a current industrial standard for linking personal computers, process stations and logic control units.

5   CONCLUSION

The approach described in this article brings more effective automation of hydro power stations than is possible using conventional control systems with high hardware investment costs but relatively weak data handling capacities:

*   Improved primary energy exploitation of approx. 1 to 2.5 % leads to a short pay-back period on additional investment in optimization systems (Rux, 1991, Hässig, 1990)
*   Higher load contribution at peak-load hours through fast delivery capability
*   Cheaper external energy purchase thanks to coverage of extreme load peaks by own production facilities (Meier, 1992)
*   Avoids continuous part-load operation of generating units, thus reducing cavitation damage (Seidl, 1991).
*   Provides advanced criteria for maintenance scheduling.

6   REFERENCES

Fäh, A., Beffa, C., Kühne, A. 1991: "Floris" User Handbook, Version 2
Laboratory of Hydraulics, Hydrology and Glaciology, ETH Zürich, Switzerland
Glattfelder, A., Steinbach, J. 1984: Decentralized hydro power control - a systems engineering approach.
Water power & Dam contruction July 1984:16-20
Hässig, P. 1990: Umbau des Kraftwerks Niederried-Radelfingen.
"Wasser, Energie, Luft" 1990, H9: 233 - 237
Meier, C.J., Steinbach, J. et al. 1992: Erneuerbare Energie optimal genutzt.
Escher Wyss brochure, 1992

Német, A. 1979: Water hammer calculation and security of hydraulic installations
Escher Wyss brochure, 1979
Rux, L.M. 1991. An improved incremental economic dispatch method for hydroelectric generators with non-convex energy production cost curves.
Waterpower proc: 1218 - 1227
Seidl, A., Ottendörfer, W. 1991: Die Kraftwerks-einsatzoptimierung bei den Tauernkraftwerken.
e&i 1991/H6: 262 - 274

Hydropower'92, Broch & Lysne (eds) © 1992 Balkema, Rotterdam. ISBN 90 5410 054 0

PERESE: The new Hydro-Québec optimization model for the long-term scheduling of reservoirs

André Turgeon
Hydro-Québec (IREQ), Varennes, Que., Canada

ABSTRACT: A description of the PERESE model, that is presently developed at Hydro-Québec to replace the EGERIE model built in the seventies, is given. The hierarchical model has four levels: At level 1, the optimal operating policy of the system, represented by a model in which all hydroelectric installations are aggregated together, is determined by stochastic dynamic programming on a semi-annual time basis. A level 2, the semi-annual results, for the period designated by the user, are splitted up between the months. Level 3 dispatches the monthly hydroelectric production to the rivers, and level 4 distributes the production assigned to a river between its installations.

1 THE SYSTEM

Hydro-Québec is a public utility that has for mandate to produce, transport and distribute electricity throughout the province of Québec. The company presently operates 54 hydroelectric powerplants, 1 nuclear plant, 1 thermal plant and several gas turbines and diesel generating units. The combined installed capacity of these plants is equal to 31100 MW. The production in 1990 amounted to 141 TWh (135 TWh from the hydro plants and 6 from the thermal).

In addition to the powerplants, seasonal reservoirs have been built on most developed rivers in order to store the excess water of the spring flood for the winters where demand is high and natural water inflow is low. There are presently 16 such reservoirs in the system. Moreover, there are 5 inter-annual reservoirs in the system in which the surplus water of wet years is stored for dry years.

In the next 15 years, 9 new hydroelectric powerplants and 4 seasonal reservoirs are planned to be built to meet the increasing demand for electricity, which means that by the year 2006 the hydraulic sub-system might consist of 63 powerplants and 25 seasonal and inter-annual reservoirs.

The operating policy of this large system is determined at Hydro-Quebec for a long-term, middle-term, short-term and very-short-term horizons. The long-term horizon may vary from 10 to 20 years, depending on the study which is done. That study may be, for example, to test different generation expansion scenarios in order to find out whether demand will be met with the desired reliability. The study may also be to test different load scenarios or, else, to determine whether additional firm energy can be sold to neighboring utilities. The middle-term horizon may vary from 1 to 2 years. The operating policy determined for that horizon is used, for instance, to establish which equipments will be available for maintenance during that period. It is also used to prepare schedules for importation and exportation of energy to neighboring producers.

The short-term planning is done on a daily time basis over an horizon of 1 month. It serves to coordinate outages of major transmission and generation equipments and to prepare detailed production and exchange programs. These programs are transmitted to the dispatching center where they are used as guidelines for the very-short-term planning, that is the real-time operation of the system. The objectives of the very-short-term planning are to manage the generation and transmission systems as well as the power interchanges with neighboring utilities, and to meet the frequency, voltage, generation and transmission requirements of the system in order to ensure a safe and reliable operation.

Many factors, such as the water flow delay between powerplants in series, must be taken into account in the short-term management of the installations but not in the long-term. Indeed, the closer we get to the real-time management, the greater is the number of factors, constraints and limitations that have to be taken into account in the models of the system. On the other hand, forecasting the river flows and demand for electricity is much more difficult and risky when the horizon is 15 years than when it is 1 day. In fact, because an accurate forecast of the river flows cannot be done for a long time in advance, river flows are represented by stochastic variables in our long-term generation planning model. We would like to do the same for the demand for electricity, but the available data do not permit to determine its probability distribution. In the middle-term, short-term and very-short-term models, there are no stochastic variables. However, this is not important here since the long-term generation planning is the subject of the paper.

2 THE APPROACH

The long-term generation planning model is used at Hydro-Québec for three main activities:

a) To determine guidelines for the middle-term planning studies;

b) To construct the generation expansion plan;

c) To carry out special studies for the top management and different departments of the company.

Most of the work done with the model consist in studying different scenarios for generation expansion and load growth. For each scenario, the model determines whether or not demand will be satisfied with the desired reliability, and what will be the expected benefits and costs.

Since the number of scenarios to study can be large and since the people who request the studies usually want the results rapidly, often in the same day, it is extremely important that a run of the program on the computer takes a few minutes only. That was the first constraint that the users imposed upon us when they decided to replace the old EGERIE model built in the seventies by a new model to be named PERESE.

Of course, solving a stochastic optimization problem having more than 20 state variables in a few minutes of computer time is simply not possible yet. On the other hand, the people who request the study of different scenarios for generation expansion and load growth are usually not interested in knowing the production of every powerplant but solely the total production of the system. Consequently, the reservoir and hydro plants can be aggregated together, as Arvanitidis and Rosing [1970] and Turgeon [1980] did, for this type of study. Having done that, the optimal operating policy of the aggregate system can rapidly be determined by stochastic dynamic programming.

At the first level of the PERESE model, the optimal operating policy of the aggregate system is determined on a semi-annual time basis. This time basis was selected for three reasons:

1. Solving the problem on a semi-annual time basis over a period of 15 to 20 years is much faster than solving it on a monthly time basis.

2. In order to determine the spillages and shortages, and hence the probability of not meeting demand, it is necessary to know the level of the aggregate reservoir on November 1st, just before the winter, and on May 1st, just before the spring flood. This means that the year must be broken down into at least two periods of six months: November 1 st to April 30 and May 1 st to October 31.

3. The third and most important reason is to preserve the long term variance and covariance of the hydrologic process [Valencia and Shaake, 1973]. More precisely, the serial correlation of the yearly energy inflow to the aggregate reservoir is about 0.4 at Hydro-Québec. The variance of the total energy inflow over a period of 2, 3 or 20 years is consequently larger when this correlation is taken into account than when it is not, and if the variance is larger it is likely that the probabilities of having spillages and deficits will be larger too. If the problem at the first level would be solved on a monthly time basis, this inter-annual correlation would not be preserved.

The semi-annual operating policy found at the first level of PERESE can be broken down between the months of the half-year if the user wishes to. This is done at the second level of PERESE. The reasons for going to this second level may be that:

1. The user wishes to have the results for the calendar years and not for years beginning on November 1st and ending on October 31.

2. For the winters where the reservoir is low and the probability of a deficit is high, the user may want to break down the semi-annual operating policy between the months so as to obtain more accurate results.

3. The user may need to determine the production of some of the real installations. This is done on a monthly time basis in the model because, for small installations, determining a semi-annual operating policy may not give relevant results.

In brief, PERESE is a hierarchical model having four levels:

<u>LEVEL 1</u>: Determine the optimal operating policy of the aggregate system on a semi-annual time basis.

<u>LEVEL 2</u>: Decompose the results of the first level between the months.

<u>LEVEL 3</u>: Dispatch the monthly hydroelectric production found at level 2 to the rivers.

<u>LEVEL 4</u>: Distribute the production assigned to a river at level 3 between the installations of that river.

Levels 1 and 2 are already completed and in operation. Level 3 will be added in 1992 and level 4 in 1993.

3 THE AGGREGATION

The purpose of this section is to show how the aggregate model of the hydroelectric installations is built.

First, the initial content of each reservoir is converted into potential energy. This is done by replacing the water by its downstream generating capability. For reservoir i, this amounts to calculating $\sum_{j \in I_i} h_j^* x_{i,o}$, where:

$x_{i,0}$ = initial content of the reservoir (hm^3)

h_j^* = mean conversion factor GWh/hm^3 for plant j,

I_i = set of plants downstream of reservoir i.

If the water leaving reservoir i can take more than one path, the path that maximizes the potential energy is assumed to be taken.

Next, S_1, the initial content of the aggregate reservoir, is determined by summing up the content in potential energy of all reservoirs in the system.

In the step after, the planning horizon, which may be 10 to 20 years long, is broken down into periods of 6 months. Periods begin on November 1st and May 1st and are called respectively winter and summer. Following that, the program estimates the inflows of potential energy that would occur in each of the periods, or seasons, if the river flows would be equal to those registered in the 45 past years. The results are stored in the variables $Y_{k,n}$, where the index k represents the season and n the year the river flows were registered. Note that the index k is absolutely necessary here because the number of plants and reservoirs is growing with time, so that the same natural inflows to the system will produce more energy in year 2000 than in the current year.

With the 45 inflow data for season k, the program determines $P_k(y/z)$, the probability that the inflow be equal to y GWh given that the inflow in the preceding summer was equal to y GWh. A conditional probability is determined in order to take into account the very important year-to-year correlation mentioned above. The probability is not assumed to depend upon the inflows of the preceding winter because inflows in winters are small.

Not all inflows to the aggregate reservoirs can be stored if there are run-of-river plants in the system of if minimal flow constraints exist on some rivers. The program will determine the portion of $Y_{k,n}$ that must immediately be released and denote it by UMIN$_{k,n}$.

If the inflow to a run-of-river plant happens to be greater than the capacity of the plant, spillage will occur. Therefore, a part of UMIN$_{k,n}$, called DMIN$_{k,n}$, will be spilled. DMIN$_{k,n}$ may of course be equal to zero.

The program determines afterward up to which point the release from the aggregate reservoir can be increased without increasing the spillage. This point is denoted by UINT$_{k,n}$. Hence, if the release U_k is kept between UMIN$_{k,n}$ and UINT$_{k,n}$ the spillage will be equal to DMIN$_{k,n}$. Past UINT$_{k,n}$, the spillage is assumed to increase linearly with the release, and past UMAX$_{k,n}$ the release is entirely spilled, as depicted in Figure 1 below.

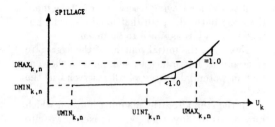

Figure 1: The spillage curve.

Since there are 45 years of inflow data, 45 curves similar to that in Figure 1 will be drawn for each season. Regression analysis is used to establish relations between UMIN, UINT, UMAX, DMIN, DMAX and the inflow. These relations are designated by $UMIN'_k(y_k)$, $UINT'_k(y_k)$, $UMAX'_k(y_k)$, $DMIN'_k(y_k)$ and $DMAX'_k(y_k)$, where y_k represents the inflow in season k.

Knowing the inflow and release from the aggregate reservoir, it is now possible to estimate the portion of the release that will be spilled and the portion that will produce electricity. Let $\hat{U}_k$ designate the last portion. The electrical energy generated will be equal to $\hat{U}_k$ GWh only if the conversion factors at the time the release is made are equal to the mean factors used above to convert the water into potential energy. Since conversion factors vary continuously with the level of the reservoirs behind the plants, it is thus unlikely that the electrical energy produced will be exactly equal to $\hat{U}_k$ GWh. The program takes that fact into account in the following way. First, the average content of the aggregate reservoir for the season is estimated to be equal to:

$$S^*_k = c_k S_k + (1 - c_k) S_{k+1} \qquad (1)$$

where S_k represents the content (GWh) of the reservoir at the beginning of the season, S_{k+1} the content at the end of the season, and where c_k is set to 0.5 for k = winter and to 0.25 otherwise. Next, S^*_k is distributed between the seasonal and inter-annual reservoirs according to a predefined rule. Knowing the content of reservoir i the program then determines h_i, the corresponding conversion factor for plant i. Finally, the production G_k is estimated from the following equation

$$G_k = \left[1 + \sum_i \left(\frac{h_i - h^*_i}{h^*_i} \right) \frac{YR^*_{i,k}}{Y^*_k} \right] \hat{U}_k \qquad (2)$$

where $YR^*_{i,k}$ stands for the mean inflow (GWh) to reservoir i in season k, and Y^*_k for the mean inflow to the whole system.

4   THE OPERATING POLICY

An operating policy for the aggregate hydroelectric system is paid feasible if it satisfies the following equations:

$$S_{k+1} = b_k S_k + SNEW_k + Y_k - U_k, \forall k \qquad (3)$$

$$\underline{S}_{k+1} \leq S_{k+1} \leq \overline{S}_{k+1} \qquad\qquad , \forall k \quad (4)$$

$$U_k \geq UMIN'_k(y_k) \qquad\qquad , \forall k \quad (5)$$

$$S_2 \text{ given} \qquad\qquad\qquad (6)$$

where b_k is the revalorization factor for period k, $SNEW_k$ the volume of potential energy stored in the new reservoirs added to the system in season k, and where $\underline{S}_{k+1}$ and $\overline{S}_{k+1}$ are the lower and upper bounds for S_{k+1}, the content of the aggregate reservoir at the beginning of season $k+1$. b_k will be greater than 1.0 only if new hydroelectric powerplants, located downstream existing reservous, are added in season k. The energy value of the water stored in those reservoirs will jump by a certain amount when the new plants will come into operation, which is taken into account by the factor b_k.

As was shown in section 3, the U_k GWh of potential energy released in period k will produce G_k Gwh of electrical energy. To determine the value at which U_k must be set, the program does the following. First, it builds the load-duration curve for the season from the monthly load-duration curves that are supplied by the user. Next, it finds the expected maximum power, L_k, the hydroelectric system will generate in season k, and places it under the load-duration curve, as shown in Figure 2.

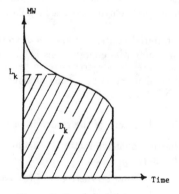

Figure 2: Load-duration curve

The area under the load-duration curve that is not hatched is then filled with energy produced by peaking generating units, which is easy to do. The problem is to decide whether the hydroelectric generation G_k should be set equal to, greater than, or smaller than D_k, the hatched area under the load-duration curve in Figure 2. If G_k is set to a value smaller than D_k, the difference $D_k - G_k$ will have to be satisfied by some other means. These can be the thermal and nuclear plants, importing energy from neighboring producers, or shedding load. Whatever means are selected there is a cost attached to it. That cost will be assumed to be equal to $C_k (D_k - G_k)$. On the other hand, if G_k is set to a value greater than D_k, the excess generation $G_k - D_k$ will be sold to neighboring producers for a revenue of $R_R (D_k - G_k)$.

The function $C_k(D_k - G_k)$ is made piecewise-linear and convex. The first segment corresponds to the mean with the smallest marginal production cost, the second to the mean with the second smallest marginal production cost, etc. The function $R_k(D_k - G_k)$ is also made piecewise-linear but concave. The first segment corresponds to the market having the largest marginal revenue, the second to the market having the second largest marginal revenue, etc.

The functions $C_k(D_k - G_k)$ and $R_k(G_k - D_k)$ can easily be combined together to give the benefit function $B_k(D_k - G_k)$. If the objective of the company is to maximize its expected benefits, then G_k, and hence U_k, should be selected to

$$\text{maximize} \sum_{k=1}^{K} E \{B_k(D_k - G_k) - A_k(L_k)\} \quad (7)$$

subject to constraints (3) - (6). In (7), the symbol E stands for the mathematical expectation and $A_k (L_k)$ for the cost of the energy produced by the peaking units to fill the area under the load-duration curve above the level L_k. Here, L_k is a function of S_k and S_{k+1}, D_k a function of L_k and hence of S_k and S_{k+1}, and G_k an function of S_k, S_{k+1}, Y_k and U_k.

Problem (7) is solved in PERESE by dynamic programming. Briefly, this consists in solving equations (8) or (9) below, depending on whether k corresponds to a winter or a summer, starting in season K with $F_{K+1} (s, z) = 0 \; \forall s, z$ and going backward in time:

$$F_k(S_k, Y_{k-1}) = \begin{array}{c} E \\ y_k/y_{k-1} \end{array} \begin{array}{c} \text{maximum} \{B_k(D_k - G_k)\} \\ U_k \end{array}$$
$$- A_k(L_k) + F_{k+1}(S_{k+1}, Y_{k-1})\} \quad (8)$$

$$F_k(S_k, Y_{k-2}) = \begin{array}{c} E \\ y_k/y_{k-2} \end{array} \begin{array}{c} \text{maximum} \{B_k(D_k - G_k)\} \\ U_k \end{array}$$
$$- A_k(L_k) + F_{k+1}(S_{k+1}, Y_k)\} \quad (9)$$

In practice, we get rid of the effect of setting arbitrarily $F_{K+1} (s, z) = 0 \; \forall s, z$ upon the results for seasons 1 to k by adding a few years, identical to the last, at the end of the planning horizon.

Let $U_k^o(S_k, Y_{k-1}, Y_k)$ and $U_k^o(S_k, Y_{k-2}, Y_k)$ be the solutions to problems (8) and (9) respectively. If U_k in equation (3) is replaced by these solutions, the equation becomes a function of S_k, Y_k and Y_{k-1} or Y_{k-2}. Since $P_k(Y_k/z)$, the probability distribution of Y_k conditioned on the inflows of the preceding summer, is known, the probability distribution of S_{k+1}, can then be determined. If suffices to solve recursively equation (10) or (11) below, depending on whether k represents a winter or a summer, starting in season 1 with the known values of S_1, Y_o and Y_{-1}, and with PR_1 $(S_1, Y_o) = 1$ or $PR_1 (S_1, Y_{-1}) = 1$, and going forward in time:

$$PR_{k+1}(S_{k+1}, Y_{k-1}) = \sum_{S_k} \sum_{Y_k} PR_k(S_k, Y_{k-1}) \cdot$$
$$P_k(Y_k \mid Y_{k-1}) \quad (10)$$

$$PR_{k+1}(S_{k+1}, Y_k) = \sum_{S_k} \sum_{Y_{k-2}} PR_k(S_k, Y_{k-2}) \cdot$$
$$P_k(Y_k \mid Y_{k-2}) \quad (11)$$

Here, $PR_k (S_k, Y_{k-1})$ designates the joint probability distribution of S_k and Y_{k-1}. The probability distribution of S_k alone is obtained by summing up $PR_k (S_k, Y_{k-1})$ over all possible values of Y_{k-1}.

Once the probability distributions of Y_k and S_{k+1}, $k = 1, \ldots, K$, have been determined, the program calculates the probability distributions of the hydroelectric and thermal generations, of the amounts of energy imported and exported, and of spilling water and not meeting demand. The program also verifies whether the reliability criteria are satisfied or not. If they are not, the user receives a message to modify the cost function for not meeting demand, and to re-submit the program.

5 THE MONTHLY GENERATION

At the second level of PERESE, the total generation for a season, determined at the first level, is splitted up between the months. This is done in the following way.

Let $G_k^o(S_k, Y_{k-1}, Y_k)$ be the hydroelectric generation corresponding to the optimal release U_k^o (S_k, Y_{k-1}, Y_k), and let $T_k^o(S_k, Y_{k-1}, Y_k) = D_k - G_k^o(S_k, Y_{k-1}, Y_k)$. If $T_k^o(S_k, Y_{k-1}, Y_k) > 0$, meaning that the hydroelectric generation alone will not satisfy demand, then means of supplying energy other than hydro should be used, for otherwise load will be shed. Suppose that n such means exist and that those have been ranked by increasing cost of supplying energy. Let $\overline{W}_{i,k,m}$ denote the maximum energy that mean i can supply in month m of season k at a cost of $c'_{i,k}$ \$/MWh.

Provided that no constraints are violated, the program will distribute the $T_k^o(S_k, Y_{k-1}, Y_k)$ GWh between the six months in the following way. Starting in month 1, the energy to be supplied by mean 1 will be set equal to

$$W_{1,k,1} = \min\left\{\overline{W}_{1,k,1}, T_k^o(S_k, Y_{k-1}, Y_k)\right\} \quad (12)$$

If $W_{1,k,1} < T_k^o(S_k, Y_{k-1}, Y_k)$, the program will advance to month 2 and fix the energy to be supplied by mean 1 in that period equal to

$$W_{1,k,2} = \min\left\{\overline{W}_{1,k,2}, T_k^o(S_k, Y_{k-1}, Y_k) - W_{1,k,1}\right\} \quad (13)$$

If $W_{1,k,1} + W_{1,k,2} < T_k^o(S_k, Y_{k-1}, Y_k)$, the program will advance to month 3 and set

$$W_{1,k,2} = \min\left\{\overline{W}_{1,k,3}, T_k^o(S_k Y_{k-1}, Y_k) - \sum_{j=1}^{2} W_{1,k,j}\right\} \quad (14)$$

If necessary, mean 1 will also be used in months 4,5 and 6. If $\sum_{m=1}^{6} \overline{W}_{1,k,m} < T_k^o(S_k, Y_{k-1}, Y_k)$, the program will repeat the same procedure for mean 2. If demand is still not met, mean 3 will next be used, and then means 4,5 6, etc.

Provided that all constraints are respected, the procedure is optimal for two reasons. First, since $c'_{1,k} < c'_{2,k} < c'_{3,k} \ldots < c'_{n,k}$, it minimizes the cost of supplying $T_k(S_k, Y_{k-1}, Y_k)$ GWh. Second, it maximizes the water head because the contri-

bution of a mean is placed at the beginning of the season rather than at the end when the choice exists. This reduces the hydroelectric generation in the first months of the season, which increases the water head for later months.

If $T_k^o(S_k, Y_{k-1}, Y_k) < 0$, the excess hydroelectric generation will be sold to neighboring producers. Suppose that there are L such producers, or markets, and that those have been ordered so that $r_{1,k} > r_{2,k} > \ldots > r_{i,k} > \ldots > r_{L,k}$, where r_{ik} is the revenue for a GWh sold to producer i. Provided that no constraints are violated, the sales will be splitted up between the months of the season in the following way. Starting in month 6 and going backward in time, sales will first be made to producer 1. If that producer cannot buy all excess energy, sales will next be made to producer 2, starting also in month 6 and going backward in time. If this is not sufficient, sales will then be made to producers 3,4,5 and so forth.

This procedure is optimal for also two reasons. First, it maximizes the revenues since $r_{1k}, > r_{2,k} > \ldots > r_{Lk}$. Second, it maximizes the water heads because the sales are made at the end of the season rather than at the beginning when the choice exists.

As it was clearly stated, the procedures just described for splitting up $T_k^o(S_k, Y_{k-1}, Y_k)$ between the months are optimal provided that all constraints are respected. Those constraints, which are similar to constraints (3)-(6), can be mathematically written as follows:

$$S'_{k,m+1} = b'_{k,m} S'_{k,m} + SNEW'_{k,m} + Y'_{k,m} - U'_{k,m} \quad (15)$$

$$\underline{S}'_{k,m+1} \leq S'_{k,m+1} \leq \overline{S}'_{k,m+1} \quad (16)$$

$$U'_{k,m} \geq UMIN''_{k,m}(Y_{k,m}) \quad (17)$$

$$\sum_{m=1}^{6} Y'_{k,m} = Y_k \quad (18)$$

$$S'_{k,1} = S_k \quad (19)$$

where $S'_{k,m}$ designates the energy content of the aggregate reservoir at the beginning of the m th month of season k, $b'_{k,m}$, the revalorization factor for month m, $SNEW'_{k,m}$, the potential energy stored in the new reservoirs commissioned in month m, $Y'_{k,m}$, the inflow of potential energy in month m, and $U'_{k,m}$, the release.

Once $T_k^o(S_k, Y_{k-1}, Y_k)$ has been divided between the months, by the procedure described above, and the hydroelectric generation in each month becomes therefore known, the program can compute $U'_{k,m}, m = 1, \ldots, 6$. Whether these monthly releases satisfy or not constraints (15)-(19) depends to a large extent on the inflows $Y'_{k,m}, m = 1, \ldots, 6$. These monthly inflow, like the semi-annual inflows, are stochastic, and their probability distributions can be determined from the 45 years of historical flow data. However, the monthly inflows cannot take all possible values here, but only those values that satisfy constraint (18). Valencia and Shaake [1973] have developed a model for generating monthly inflows that satisfy (18) while maintaining all relevant statistics of the variables. Their model take the sample form:

$$Y'_{k,m} = a_{1,k,m} + a_{2,k,m}Y_k + \sum_{j=1}^{5} d_{j,k,m}Q_{j,k}, \quad (20)$$

where $Q_{j,k}, j = 1, \ldots, 5$, are independently distributed standard normal deviates, and where the coefficients $a_{1,k,m}$ $a_{2,k,m}$ and $d_{j,k,m}, j = 1, \ldots, 5$, $m = 1, \ldots, 6$, are selected so that the expected means, variances and convariances of the generated data are equal to the historical means, variances and covariances, and so that constraint (18) is satisfied.

For each value of Y_k, PERESE generates 25 sequences of monthly inflows with equation (20). Then, for each sequence, it verifies whether constraints (15)-(19) are satisfied or not. If they are not, it modifies the way $T_k^o(S_k, Y_{k-1}, Y_k)$ was divided between the months. For instance, suppose that k stands for a winter and that the content of the reservoir decreases below the level $\underline{S}'_{k,4}$ at the end of month 3. In this case, a greater portion of $T_k^o(S_k, Y_{k-1}, Y_k)$ will be assigned to months 1 to 3 and a smaller portion to th rest of the season. The program may ever increase the contribution of the non-hydraulic sources to a value larger than $T_k^o(S_k, Y_{k-1}, Y_k)$ in order to prevent shedding load unnecessarily.

Since T_k^o is a function of S_k, Y_{k-1} and Y_k, the procedures described in this section will have to be repeated for all possible values of S_k, Y_{k-1} and Y_k. In PERESE, these variables all allowed to take 100, 11 and 11 different values respectively, which means that the task of splitting up $T_k^o(S_k, Y_{k-1}, Y_k)$ between the months may have to be repeated up to 121000 times for each of the 25 sequences of inflows in season k. This neverthe-

less takes only a few seconds of computer time in PERESE.

Knowing the probability distribution of S_k, which is obtained from solving (10) or (11) and assuming that the 25 sequences of monthly inflows have an equal probability of occurrence, the program finally calculates the probability distributions of $S'_{k,m}, m = 2, \ldots, 7$. Note than the probability distribution of $S'_{k,7}$ may be different from the probability distribution of S_{k+1} calculated at the first level of PERESE for the following reasons. Firstly, for some sequences of monthly inflows the total release from the reservoir in season k may have to be modified to satisfy constraint (16). Secondly, the spillages may be different. Thirdly, the correction for the water heads used each month to calculate the generation may be different from the correction for the whole season used in (2).

6 THE GENERATION OF THE RIVERS

The third level of PERESE will distribute the total monthly hydroelectric generation between the rivers. Since this level is not constructed yet, we will, in this section, only briefly outline the approach that will likely be retained.

In section 3, we have shown how to split up $T^o(S_k, Y_{k-1}, Y_k)$, between the months of season k. The logical next step would be to dispatch the resulting monthly generations to the rivers. However, considering that this exercise would have to be repeated for all possible values of S_k, Y_k and Y_{k-1}, and all possible ways of dividing S_k between the rivers and Y_k between the months and rivers, we think that the computer time would be excessively long. The only alternative approach that we can think of is simulation.

Suppose that we are in season k and that Y_{k-1}, Y_k and X_{ik}, the energy content of river i at the beginning of season k, are known. Summing X_{ik} over all i gives S_k, and, for a given value of S_k, Y_{k-1} and Y_k, the solution of the first level gives the hydroelectric generation for the season and hence T_k^o. T_k^o can be splitted up between the months by the method presented in section 3. This gives, among other things, $G'_{k,m}$ the hydroelectric energy to produce in month $m, m = 1, \ldots 6$. The step after, which has not been resolved yet, is to dispatch $G'_{k,m}$ to the rivers.

The procedure described in the previous paragraph can be repeated season after season, starting in season 1 with the known value of $X_{i,1}, i =$

609

$1, \ldots n$, and $Y_k, k = 1, \ldots, K$, and going forward in time. The value of Y_k can be randomly generated with equations (21) or (22) below, depending on whether k is a winter or summer.

$$Y_k = \alpha_{1,k} + \alpha_{2,k} Y_{k-1} + \alpha_{3,k} \xi_k \quad (21)$$

$$Y_k = \alpha_{1,k} + \alpha_{2,k} Y_{k-2} + \alpha_{3,k} \xi_k \quad (22)$$

Here, $\alpha_{1,k}, \alpha_{2,k}$ and $\alpha_{3,k}$ are constants and ξ_k represents and independent standard normal random variable.

Thousand of sequences of inflows Y_k, $k = 1, \ldots, K$, can be generated with equations (21) and (22), and hence thousands of simulations can be done. The results of these simulations can be used afterward to estimate the probability distributions of $X_{i,k}$, of the generation of each river, spillage, thermal production, etc.

If thousands of simulations need to be carried out to obtain good estimates of the probability distributions, then it is extremely important that the task of dispatching $G'_{k,m}$ to the rivers be done rapidly. Let:

X'_{ikm} = content in energy of river i at the beginning of month m of season k;

q_{ikm} = inflow of energy to river i in month m.

V_{ikm} = the release from reservoir i in month m,

b''_{ikm} = the revalorization factor for the energy stored in river i at the beginning of month m.

$XNEW_{ikm}$ = the potential energy stored in the new reservoirs on river i commissioned in month m,

H_{ikm} = the electrical energy generated by river i in month m. It is a function of $X'_{ikm}, X'_{i,k+1,m}, q_{ikm}$ and V_{ikm}

$f_{ikm} (X'_{ikm}, XC_{ikm})$

= the expected marginal value of the energy stored in river i at the beginning of month m given that the content of the river is equal to X'_{ikm} GWh and the content of all the other rivers combined is equal to XC_{ikm} GWh.

The dispatching policy, that is the choice of

values for $V_{ikm}, i = 1, \ldots, n$, will be feasible if it satisfies the following constraints:

$$X'_{i,k+1,m} = b''_{ikm} X'_{ikm} + XNEW_{ikm} + q_{ikm} - V_{ikm}$$
$$(23)$$

$$\underline{X}'_{i,k+1,m} \leq X'_{i,k+1,m} \leq \overline{X}_{i,k+1,m} \quad (24)$$

$$V_{ikm} \geq VMIN_{ikm}(q_{ikm}) \quad (25)$$

If there exist several dispatching policies that satisfy constraints (23)-(25), then the policy found by the following algorithm should be implemented.

STEP 1: Fix the value of γ.

STEP 2: For every river i, determine the value of V_{ikm} that satisfies the following equation.

$$f_{i,k+1,m}(X'_{i,k+1,m} XC_{i,k+1,m}) - \gamma \frac{\partial H_{ikm}}{\partial V_{ikm}} = 0 \quad (26)$$

If that value violates constraint (24) or (25), adjust V_{ikm} to the nearest feasible value. Find the corresponding value of H_{ikm}.

STEP 3: Verify whether the following equation is satisfied

$$\sum_{i=1}^{n} H_{ikm} = G'_{km} \quad (27)$$

If it is not, change the value of γ and return to step 2.

In brief, if thousands of simulation need to be carried out, the task of dispatching G'_{km} to the rivers can be done rapidly with the above algorithm.

The dispatching policy found with the above algorithm would be optimal if the function $f_{i,k+1,m}$ $(X'_{i,k+1,m}, XC_{i,k+1,m})$ in (26) would be replaced by the function $f'_{i,k+1,m}$ $(X'_{1,k+1,m}, X'_{2,k+1,m}, \ldots, X'_{n,k+1,m}) \triangleq \partial F_{k+1,m} (X'_{1,k+1,m}, \ldots, X'_{n,k+1,m}) / \partial X_{i,k+1,m}$, where $F_{k+1,m} (X'_{1,k+1,m}, \ldots, X'_{n,k+1,m})$ represents the expected future benefits that the company would realize if an optimal operating policy would be followed. But since determining $F_{k+1,m} (X'_{1,k+1,m,\ldots}, X'_{n,k+1,m})$ is not possible when n is large, we use the function of two variable proposed by Turgeon [1980]. Lederer, Torrion and Bouttes [1983] from Electricité de France have used these functions of two variables to determine the production of their rivers, and not merely to dispatch a given production G'_{km} to the rivers as we do here. They write, in the conclusion of their paper, that the optimality gap has never exceeded 0,8% in the tests they made.

The function $f_{i,k+1,m}(X'_{i,k+1,m}, XC_{i,k+1,m})$ can be determined by solving a stochastic dynamic programming problem of two states variables on a monthly time basis [Turgeon, 1980]. The objective function should be similar to those used in (8) and (9), but designed for the month. The serial correlation of the river flows must unfortunately be neglected, for otherwise there would be four state variables in the problem and not two. The problem can of course be solved beforehand and the results stored until needed by the simulations. If there are 10 rivers in the system, solving 10 dynamic programming problems will certainly consume a fair amount of computer time. However, if the user wishes to determine the production of, for example, two particular rivers only, then only three dynamic programming problems should be solved: one for each river and one for all other rivers combined. Note that it may also be possible to aggregate nearby rivers with highly cross-correlated inflows, solve one dynamic programming problem for the group, and afterward, disaggregate the result with simple rules, like those proposed by Johnson, Stedinger and Staschus [1991]. This is the type of possibility that we intend to study when we will construct that level of PERESE.

A last point about the inflow. The semi-annual inflows generated by (21) and (22) must be disaggregated here not only between the months, but between the rivers also. The results obtained must satisfy the following equation:

$$\sum_{i=1}^{n} \sum_{m=1}^{6} q_{ikm} = Y_k \qquad (28)$$

This disaggregation can be done with a model similar to (20), but designed for many sites. It can also be done with one of the models proposed by Grygier and Stedinger [1988], namely the LAST model, the Generalized Stedinger - Pei-Cohn model or the SPIGOT model.

7 THE PRODUCTION OF EACH PLANT

The fourth and last level of PERESE will distribute the production H_{ikm} assigned the river i in month m of season k between the installations of the river. Since this level will be added to the model in 1993 only, we have not chosen yet the approach that will be followed. One possibility, however, is to use the principal component analysis method of Saad and Turgeon [1988].

Suppose that river i has five reservoirs, and let Z_{mj} be the content of reservoir j at the beginning of month m. The idea of principal component analysis is to search for five linear combinations of the type

$$\xi_{mp} = b_{mp1} Z_{m1} + b_{mp2} Z_{m2} + \ldots + b_{mp5} Z_{m5}$$

$$= \sum_{j=1}^{5} b_{mpj} Z_{mj} = b_{mp}^T Z_m, \qquad (29)$$

where T denotes the transpose, that have the following characteristics: (1) ξ_{m1} has the largest possible variance, (2) ξ_{m2} is orthogonal to ξ_{m1} and has the largest variance after ξ_{m1}, (3) ξ_{m3} is orthogonal to ξ_{m1} and ξ_{m2} and has the largest variance after these variables, and (4) so forth for ξ_{m4} and ξ_{m5}. The linear combinations, in other words, allow a set of variables $(Z_{m1}, Z_{m2}, \ldots, Z_{m5})$ to be transformed into an equivalent set of variables $(\xi_{m1}, \xi_{m2}, \ldots, \xi_{m5})$ that has two interesting attributes: The variables are (1) uncorrelated and (2) in decreasing order of variance.

An interesting property of matrix $B_m = [b_{m1}, b_{m2}, \ldots, b_{m5}]$ is that it is orthonormal, which means that if

$$\xi_m = B_m^T Z_m$$

then

$$Z_m = B_m \xi_m \qquad (30)$$

Suppose that ξ_{m1}, the first component of ξ_m, accounts for 95% of the variance of the five variables, and the remaining four for only 5%. Since each of the last four components has a very small variance, replacing them by their mean should not have significant effect on the solution. (30) can therefore be approximated by

$$Z_{mj} = b_{m1j} \xi_{m1} + \sum_{p=2}^{5} b_{mpj} \mu_{mp}$$

$$= b_{m1j} \xi_{m1} + b_{moj} \quad ; j = 1 \ldots 5, \quad (31)$$

where μ_{mp} is the mean of ξ_{mp} and b_{moj} a constant equal to $\sum_{p=2}^{5} b_{mpj} \mu_{mp}$.

When approximation (31) is found acceptable, the releases from the reservoirs in month m can easily be determined as follows:

STEP 1: Fix the value of $\xi_{m+1,1}$

STEP 2: For every reservoir j, set

$$Z_{m+1,j} = b_{m+1,1,j} \xi_{m+1,1} + b_{m+1,0,j}$$

If $Z_{m+1,j}$ violates some constraints, adjust it

to the nearest feasible value, and then find the corresponding release from reservoir j. If that release is not feasible, change it for the nearest feasible value and modify $Z_{m+1,j}$ accordingly.

STEP 3: Determine the production of the river. If that production is not equal to H_{ikm}, modify the value fo $\xi_{m+1,1}$ and return to step 2.

8 REFERENCE

Arvanitidis, N.V. & J. Rosing. 1970. Optimal Operation of multireservoir systems using a composite representation. IEEE Trans. Power App. Syst. PAS-89 (2): 327-335.

Grygier, J.C. & J.R. Stedinger. 1988. Condensed disaggregation procedures and conservation corrections for stochastic hydrology. Water Resources Research, 24 (10): 1574-1584.

Johnson, S.A. J.R. Stedinger & Staschus 1991. Heuristic operating policies for reservoir system simulation. Water Resources Research, 27 (5): 673-685.

Lederer, P., P. Torrion & J.P. Bouttes. (1983). A global feedback for the French system generation planning. 11th IFIP Conference on System Modeling and Optimization, Copenhagen.

Saad, M. & A. Turgeon. 1988. Application of principal component analysis to long-term reservoir management. Water Resources Research, 24 (7): 907-912.

Turgeon, A. 1980. Optimal operation of multireservoir power systems with stochastic inflows. Water Resources Research. 16 (2): 275-283.

Valencia, D. & J. C. Shaake. 1973. Disaggregation processes in stochastic hydrology. Water Resources Research 9 (3): 580-585.

Hydropower'92, Broch & Lysne (eds) © 1992 Balkema, Rotterdam. ISBN 90 5410 054 0

Integrated control centres – A new concept for integration of hydro scheduling programs into energy management systems

J. E. Værnes
South Trøndelag Power Company, Norway

N. Flatabø
Norwegian Electric Power Research Institute (EFI), Norway

ABSTRACT

The paper presents an open system solution for the integration of application programs into existing or new concepts of control centre systems. The standards and solutions chosen for establishing the integrated concept are described. Further a brief presentation is given of the application programs for hydro production scheduling and inflow forecasting, that will be available through the integrated control centre concept.

The development work is a cooperation between the power industry, manufacturers, and research institutes. A pilot installation will be accomplished during the year of 1993.

1 INTRODUCTION

Control centres are normally equipped with system control and data acquisition system (SCADA system) for handling the remote control for daily operation of the hydro power system. A limited choice of programs for scheduling of the power production are also available. Network analyses programs are more available on the market. Scheduling methods for hydro power systems have for years been developed at EFI in cooperation with the power utilities.

These programs have been used off-line for production planning. The planning programs need measurement data as input from the SCADA system. The transfer of data from the on-line SCADA system to the off-line planning programs is cumbersome, and thus the planning process is difficult to finish in due time.

The project called Integrated Control Centres (ICC) has the superior goal to develop a concept for control centres where application programs for hydro production planning, inflow forecasting, load forecasting, and network analysis are integrated. The measurement data from the SCADA system will be available to the planning programs through a data communication procedure.

The paper describes the integrated concept, and specially focus on standards and solutions chosen for system architecture, data handling, and man machine interface system. Further, the application programs for hydro scheduling will be described, and steps in the development work will be outlined.

The first Integrated Control Centre is estimated to be installed at a pilot utility within the year of 1993.

2 INTEGRATED CONCEPT

2.1 Open system demands

In a prestudy report [1] an open system solution was described. In an open system solution specific requirements are given to
- system architecture
- man machine interface
- data handling
- data communication.

The specified solution shall support the following requirements:
  i) the system solution shall easily be adapted to different system architectures
  ii) the application programs can be transferred to different hardware platforms dependent of changing demand.
  iii) the application programs can be adapted to different users need and amount of data
  iv) the established solution shall be based on standard components and standardized architecture thus supporting maintainability of the system.

To ensure an open system solution, UNIX is chosen as operating system.

A relational database system is chosen, and SQL is specified as interface to the database.

Communication between SCADA system and the ICC-system is supported by the ELCOM-90 communication protocol.

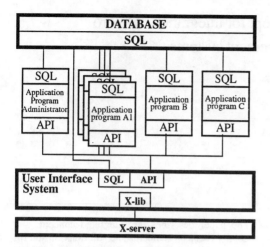

Fig. 1 The architecture of the ICC system

Man machine interface shall be separated from the application programs.

A system architecture that complies with the above mention demands, is shown in Figure 1.

The ICC system architecture consists of three main layers: the database system, the application programs, and the user interface system. The application programs have one application programming interface (API) against the database system and one against the user interfaces system.

Data is stored in a relational database system. An entity-relationship model visualizes the entities and attributes in the database and the logical connections between the entities.

The application programs, shown in Figure 1, are just examples of possible programs. Some of these programs will be described in later chapters of this paper.

The user interface system handles the user interaction and the interaction to the application programs. The interface gives the user different views to data manipulation and task specification.

The architecture of the ICC system differs in many ways from existing software systems. In present software solutions the application programs include user interface, algorithms, and data handling in the same program.

The algorithms which were the kernels of the old programs, will in the new solution be the application programs, and be separated from user interface and data handling.The system architecture of the ICC system enforces the separation between user interface, application programs, and data handling.

The chosen system simplifies the configuration of an actual system. The maintenance can be performed on separate software modules. Application programs can be maintained and changed independently of the user interface system and data base system.

2.2 Man machine interface

The ICC system has a distributed architecture. At runtime it consists of a set of computer processes (applications) with interconnections.

When the ICC system is started, at least one window appears on the screen. This window provides the controls necessary to enable a working session with the ICC system.

Which processes to run and the network node each of them are running on, can be interactively defined by the user.

As a consequence of the distributed ICC architecture, the applications are running as separate computer processes. This is normally transparent to the user, but gives him the ability to physically start and stop applications. The word physically is used to focus that a stopped application does not occupy any computer resources. A computer system which logically consists of several applications might be linked together forming one executable image.

The dialog texts in the user interface of the ICC system is stored in several ways.
- Text explaining the different messages is stored in the database.
- Textual information used in Help tool is stored on separate files.
- Text strings to appear in the menus are stored together with the menu description on separate files.
- Text strings to appear in the UIS models are stored together with the model description on compiled files.

The dialog text is stored in such a way that the transformation into another language should be as easy as possible.

The ICC system will follow the Motif style guide and the design principles described there.

The Motif widgets will be used to implement the ICC menu system and some dialogue boxes.

The ICC system will use multiple windows.

Each application will normally introduce a new set of windows. All the windows will have a consistent layout and interface.

The windows used in the ICC system are controlled by the current window manager. This means they can be resized, moved, iconized and put in front of or behind other windows. The window manager function can be activated with the mouse on the window decoration frame.

The ICC system will use two types of menus.
- Pulldown menus connected to a menu-bar on top of the windows
- Pop-up menus that can be activated on window backgrounds or on graphical objects.

The ICC user interface will provide multiple views towards the same data.

This means that the user is given a flexible choice on how to present and edit data.

Copy and Paste will be supported to enable efficient exchange of information, controlled by the user, both between different applications within ICC, and internally in an application.

Selecting information from data organized in tree structures will be widely used. The ICC interface will therefore support mechanisms that contribute to an efficient solution of such tasks.

The major part of the data in the database are related to power system components that forms a network structure. A topological model gives a representation of the "real world" and will therefore be convenient for a user in all kind of data administration.

This module will be used to:
- Visualize the system model currently loaded
- Modify system model layout
- Show system status by means of dynamic graphical object presentation
- Do operations on selected objects.

The view which contains the system model is controlled by zoom and pan sliders. The graphical objects used in the model will have a dynamic description to reflect some kind of status information.

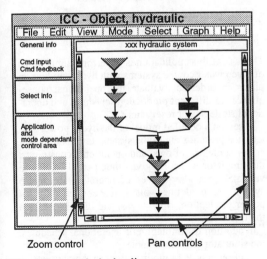

Zoom control   Pan controls

Fig. 2 Layout of a hydraulic system

tion. Percentages of reservoir contents, and text boxes showing current value are examples of such dynamic information.

An example of a topological model of a watercourse (reservoir, power plant and waterroute) is shown in Figure 2. The tool is not limited to operate just on hydraulic systems, but can also be used for presentation of electric networks, field station layout for the HYDMET application, etc.

2.3 Communication interface

All the software components of the ICC system (database, user interface, application programs) are computer processes. They will interact through an event-orientated message system.

The communication with processes outside the ICC system shall be handled by the ELCOM-90 communication protocol [2].

Some basic communication interfaces are listed below:
- ELCBAS-90
- Ethernet (TCP / IP)
- EDC - protocol
- TS-ENVOY

3 APPLICATION PROGRAMS

3.1 Overview of the scheduling problem

In the ICC concept programs for solving the hydro scheduling problem will be implemented. A brief description of the hydro scheduling problem will be given as an introduction to the solution chosen for the ICC concept [3].

The optimal scheduling of hydro resources poses severe computational problems in system operations planning. This necessitates problem decomposition. In the EFI approach, production scheduling is split into three main levels:
1) Long-term scheduling
2) Mid-term scheduling
3) Short-term scheduling.

The long-term scheduling aims at evaluating the seasonal and pluri-annual handling of regional storages in view of firm power supply obligations, inflow statistics, main hydraulic and electric transit limitations, and the available options of substituting other forms of energy in the power markets by hydroelectric power. Main input to the long-term scheduling process is inflow statistics, firm power forecasts, and spot-market-related forecasts. The long-term scheduling provides the boundary conditions for the mid-term scheduling level.

The mid-term scheduling assumes a deterministic near-future period of k (1-3) weeks and aims to match supply and demand optimally within this period, taking into account the stochastic nature of the problem over the remaining part of the mid-term planning period. Depending on the time of the year and on system conditions, the mid-term scheduling period may vary from a few weeks up to a year. The mid-term scheduling provides the boundary conditions for the ensuing short-term scheduling level.

The objective of the short term scheduling is to utilize the available resources given by the mid-term scheduling and maximizing the profit within the period in consideration by exploiting the options for buying and selling in the spotmarket. The constraints related to this process are:
- Coupling to mid-term scheduling
- Hydraulical constraints
- Electrical constraints
- Firm load obligations
- Contractual demand.

Short-term inflow forecasting is confined to the nearest future (say 2-5 days) depending on how reliable meteorological forecasts of precipitation and temperature are expected to be. The objective of a short-term inflow forecast is to have a prediction of catchment run-off as a time series with an appropriate time resolution, given a weather forecast.

A long-term inflow forecast reflects the uncertainty of expected inflow. Normally this is achieved by using historical inflow series or a probabilistic model reflecting the statistical properties of the historical observations. When there is snow in the catchment, the amount of snow will certainly influence and overrule the stochastic parameters governing the future long-term inflow.

The main operational objective is to have that sequence of operational decisions implemented, that contributes to the minimization of the expected total variable cost, taking into account proper quality of energy supply constraints, responsibilities relating to the sharing of network operational duties, and the rules of electrical energy spot-price market clearance.

Main operational decisions are:
- scheduling of power output from own generator units
- selling/buying of power on the electrical spot price power markets, and possibly
- curtailment of firm power delivery during periods of critical inflow shortage.

The scheduling process is split into three main levels, as described in Figure 3.

In the ICC project application programs will be developed for most of the scheduling and forecasting functions shown in Figure 3. Overview of the appli-

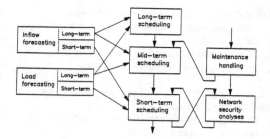

Fig. 3 Structural overview of production scheduling and related planning activities

cations to be included, and the priority in development are given in the next chapter.

3.2 Priority in implementation of application programs

The application programs to be implemented in the ICC concept are given a priority by an enduser advisory group. The chosen priority of the application programs is as shown below:
- ICC Basic UIS software
- Hydro Power Plant Operation Simulator
- Short Term Inflow Forecasting
- Mid Term Price Plant Ranking
- Mid Term Inflow Forecasting
- Short Term Scheduling, Optimization
- Long Term Scheduling, Optimization, Simulation
- Mid Term Scheduling, Optimization

3.3 Description of application programs

3.3.1 Simulator for operation of the Hydro Power Plants

The task of this application is to perform simulations of given water course systems with hydro production plants in order to evaluate the operational consequences of different production schedules and operational strategies of reservoirs.

Based on a detailed model of the hydro power system, the present state of the system described by reservoir contents, inflows, and operational constraints, the simulator will for a given time period be able to validate the consequences of operating the hydro system on predefined plans for combinations of power production, rule curves, and discharges from the reservoirs.

In addition to simulation of operational schedules, the simulator will also handle
- consequences of modifications of the physical hydro system

- validation of the consequences of different operational constraints
- improved inflow calculation
- calculation of efficiency factors
- state validation of some of the measurements used in the simulation.

The simulator is run as one application program in the ICC system, as shown in Figure 1.

The simulator might be running together with other application programs. In addition two or more duplicates of an application may be running at the same time, e.g. two simulators are running at the same time on two different problems.

The simulator may use inflow forecast data from the HBV-model described in chapter 3.3.3.

The simulator can handle the following objects in a water course system:
- reservoir
- water gate
- water route
- water route junction
- hydro generating unit
- hydro pumping unit
- hydro power plant.

The simulator can be used under four different modes of operation, or in combination of them.

i) Tight plan

Power production plans or reservoir releases are specified for all power plants or reservoirs

ii) Rule curve simulation

Rule curves for all reservoirs are specified. The simulator will either turn off the power production, or run at best efficiency point for the combination of generating units that gives the best approximation to the rule curves.

iii) Slack reservoirs

The sum of production for the system is specified for each time interval. Rule curves are defined as reservoir trajectories and slack reservoir contents. The simulator is governed by the condition that the reservoir deviation from the trajectory curve compared to the slack reservoir content should be relatively equal for all reservoirs. The operation of power plants connected to the reservoirs will be in accordance with ii)

iv) Slack bus (power plant)

All the reservoirs and power stations, except for one reservoir with an underlying power plant (slack bus), are specified either by power production, discharge, or rule curve. The total production is specified for each time interval and is fulfilled by the slack bus production if possible.

v) Combinations of simulation modes

The modes of operation may be combined in the same simulation in the following manner:
- tight plan + rule curve simulation

- tight plan + rule curve simulation + slack reservoirs
- tight plan + rule curve simulation + slack bus

Without regards to the length of the simulation period, the time resolution during the simulation is one hour. Results from a simulation are stored at given intervals depending on the length of the simulation period, as described below:
- 1 hour interval when length of period <= 14 days (336 hours)
- 4 hour interval when length of period is between 15 days and 60 days
- 24 hour interval when length of period > 60 days.

The simulator shall be able to calculate the actual inflow to all the reservoirs based on known reservoir trajectories, productions, and water gate positions for every time interval.

Based on the efficiency description of the turbines and generators, and head loss description of the watercourse, efficiency curves, and effective head for a specified state shall be calculated on request.

Output from a simulation is for every time interval
- reservoir content for all the reservoirs (Mm3)
- waterflow for all the water routes   (m3/s)
- production for all the units   (MWh/h)
  (the production may differ from the schedule)
- flagging for overflow and constraint violations
- relative efficiency factor
  (actual efficiency / best efficiency for the plant).

3.3.2 Short term scheduling

The task of this application is to establish a power production schedule for each time interval and for all power plants. The production schedule is established by use of optimization techniques. The production schedule obtained can be used as starting condition to the simulator when used in tight plan simulation mode.

The objective of the short-term scheduling is to utilize the available hydro resources and to maximize the profit within the period in consideration by exploiting the options for buying and selling in the spotmarket.

The constraints related to this process are:
- Coupling to mid-term scheduling
- Hydraulical constraints
- Electrical constraints
- Firm load obligations
- Contractual demand

Coupling to mid-term scheduling

The main results from the mid-term scheduling are desired reservoir endpoints and the incremental value of water. These results form the interface between the short-term and the mid-term scheduling.

Hydraulic constraints

The hydraulic constraints make the problem coupled between the different time intervals.

Examples of hydraulic constraints are:
- Hydraulic coupling between plants situated along the same water course.
- Minimum and maximum reservoirs.
- Limits on discharge from reservoirs.
- Rates of change in flow between time intervals.
- Rates of change on reservoir levels between time intervals.

Electric constraints

In many cases, hydro plants are located far from the consumers. This causes transport of power over long distances and may cause problems for system operation such as:
- Stability problems (Voltage collapse, Angle stability)
- Overload / Cascading in outage conditions.

In the long-term and the mid-term planning, simplified network representations have been used combined with mean values for the power transfer capacity. In the short-term planning, more details of the transmission system have to be modelled since the daily load variation has to be accounted for. Normally, only few hours in operation are critical, and it is important not to constrain the system too much. Therefore, an optimal operation requires an adequate modelling where the relevant variables can be monitored and controlled to be within an acceptable range of values.

Problem solution

In a hydro system, the use of water resources is time constrained, and the calculations have to be coupled over the time period in consideration. The size of the problem will be enormous if it has to be solved as one problem.

However, the problem has some properties which makes it suitable for decomposition into several smaller problems and to use specialized algorithms to solve each problem.

The solution is separated into a hydro subproblem and an electrical subproblem. The hydro subproblem takes care of all constraints related to the flow in the water courses, and to the limited available resources for the planning period in consideration. The electrical subproblem takes care of the constraints in the electrical network and the balance between load and generation.

The decomposition approach adopted between the hydro subproblem and the electrical subproblem is a price decomposition scheme.

The hydro subproblem takes care of all hydraulic constraints and examples of these are
- Coupling between plants
  - Release to another reservoir
  - Spillage
- Coupling to the mid-term scheduling (desired reservoir endpoints)
- Constraints on the flow in the water courses
  - Minimum and maximum releases
  - Rate of change between different time intervals
  - Rate of change on reservoir levels.

The hydro subproblem has no equation for the balance between load and generation for each hour. This implies that there is no coupling between the different water courses, and gives the opportunity to decompose the problem further. Due to this decoupling, the hydro subproblem is solved separately for each water course. There are several advantages of this. The building of the model is simpler and more flexible and, in principle, all water courses can be solved in parallel. It is also easier to handle water courses with special constraints since more general algorithms can be used if necessary.

Since there is no equation for the balance between the load and generation, the only information used to improve the hydro schedule from one iteration to the next, is the incremental value of power calculated in the electrical subproblem. In each iteration, the hydro subproblem operates as the available resources can be sold at a fixed price in a market. Therefore, the available resources will be utilized in the periods with highest incremental worth of power.

In the electric subproblem, a detailed model of the electrical network can be used. This implies that all buses and main transmission lines can be included in the optimization process. However, it is also possible to just exchange power on one busbar.

Main results

The main results from the short term scheduling are:
- A detailed hourly production plan for the period in consideration.
- A plan that is hydraulic feasible with respect to:
  - Coupling between plants.
  - Coupling to mid-term scheduling.
  - Flow constraints in the water course.
- Incremental cost of power at all the buses in each hour.
- A plan that is security constrained due to defined security criteria in the network.

3.3.3 Short term inflow forecasting

Short-term inflow forecasting is confined to the nearest future (say 2 - 5 days) depending on how reliable meteorological forecasts of precipitation and temperature are expected to be. This may of course be dependent on the general climatic conditions.

The objective of a short-term inflow forecast is to have a prediction of catchment run-off as a time series with an appropriate time resolution, given a weather forecast.

For practical use as a tool in an operative short-term scheduling procedure, an inflow forecast model should be relatively simple with a minimum of data input requirements.

In Norway the use of the HBV-model has shown very good results, and this concept is developed to a commercial product.

The run-off is described as a function of aerial precipitation, air temperature, and the actual state of the catchment. At present, the time resolution is one day.

The HBV-model consists of four reservoirs arranged in a cascade as shown in Figure 4.

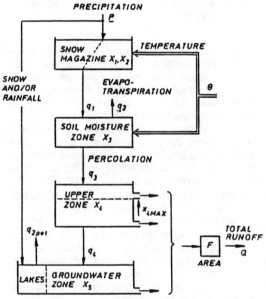

Fig. 4 The cascade arrangement of reservoirs in the HBV-model

3.3.4 Mid term scheduling

The mid-term scheduling assumes a deterministic near-future period of k (1-3) weeks and aims to match supply and demand optimally within this period, taking into account the stochastic nature of the problem over the remaining part of the mid-term planning period.

Figure 5 summarizes the scope of the Mid-Term Scheduling and relates it to the Long-Term Scheduling. Depending on the time of the year and on system conditions, the mid-term planning period may vary from a few weeks up to a year. Time resolution is a fraction of a week. Each time step represents a given load level within the week.

The Mid-Term Scheduling Model can be used as:

1. An optimal discharging model together with a long-term scheduling model such as the Power Pool Model.
2. A separate hydro-thermal scheduling model.
3. A link between the long-term models and the short-term model.
4. A model for analysing the spot price market.
5. A decision model for maintenance planning.

The mid-term scheduling provides the boundary conditions for the ensuing short-term planning level. These conditions are given as target reservoir contents at the end of week k (typically k=1) together with incremental water costs associated with the end-point storages, and a tentative thermal production schedule for the coming k weeks.

In the mid-term scheduling model the planned production for the next time increment is determined on the following bases
 - the whole planning period is taken into account

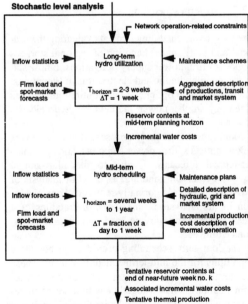

Fig. 5 Long-Term and Mid-Term Scheduling Activities

- the hydro production system has a detailed description
- hydraulic and electric constraints and limitations are dealt with
- the thermal production system has a production cost description
- interchange of energy between regions is dealt with
- the expected values of the end point reservoir contents are taken into account
- the hydro production efficiency curves are utilized
- the transmission system is described as a simplified network with given capacities where losses are taken into account
- the inflow has either a simplified stochastic or a deterministic description.

The mid-term scheduling model gives as its main results:
- hydro plant generation
- incremental cost of water
- discharge from reservoirs
- thermal production
- interchange of power between areas
- incremental cost of power within each area
- buying and selling of energy on the spot price market.

3.3.5 Long term scheduling

The long-term scheduling aims at evaluating the seasonal and pluri-annual handling of regional storages in view of firm power supply obligations, inflow statistics, main hydraulic and electric transit limitations, and the available options of substituting other forms of energy in the power markets by hydroelectric power.

Main input to the long-term scheduling process is inflow statistics, firm power forecasts, and spot-market-related forecasts. Owing to the pluri-annual aspect of reservoir handling, a period of analysis of two to three years is normally required. Time resolution in the analysis is one week or fractions of a week representing varying load and market conditions within each week.

The long-term scheduling provides the boundary conditions for the mid-term scheduling level. These are the feasible range of reservoir contents at mid-term planning horizon together with a set of incremental water costs associated with the end-point reservoir storage range, as indicated in Figure 5.

The production, transit, and demand system is

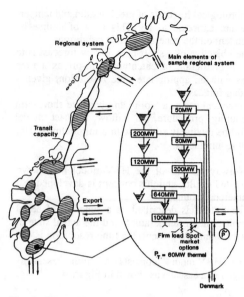

Fig. 6 System overview and sample subsystem description

modelled in a power pool model. Each region in Figure 6 is represented by an aggregate regional model at the strategic level, and an associated reservoir-drawdown model with a detailed description of the hydro system. The interconnecting network is modelled with limited transit capacity and linear losses.

For each region with local hydro production there is an associated reservoir drawdown model, which gives a fairly detailed representation of reservoirs and power plants in cascade or parallel, and which takes into account constant or time-variable constraints on reservoir contents and volume discharges. The effect of head variations is also to a certain extent taken into account.

Import and export markets are normally modelled by aggregate regional models of neighbouring countries, where each country's demand, spot-market options, and hydro- and thermal power are modelled. Exchange of power on a firm power and/or a spot-market basis can be modelled.

Given the stated multireservoir model description, the objective of the long-term optimization process is to establish the operation strategy that for each stage in time produces the "best" decision vector given the system state at the beginning of the stage. By "best" decision is understood the sequence of turbined and spilled water volumes that contribute to minimizing the expected operation cost along the period of analysis.

4 REFERENCES

[1] "Integrert driftssentral. Forstudie" (Integrated Control Centres. Prestudy), EFI TR 3722, September 1990. (In Norwegian)

[2] J. Hegge, A. Larsen: "The ELCOM Utility Communication Concept", IEEE Transactions on Power Systems, Vol. 6, November 1991.

[3] "Software for Production Scheduling in Hydro Systems", EFI TR 3798, February 1991.

Hydropower'92, Broch & Lysne (eds) © 1992 Balkema, Rotterdam. ISBN 90 5410 054 0

The dynamic production model DYNPRO

A.Wolf
Vattenfall Utveckling AB, Älvkarleby Laboratory, Sweden

ABSTRACT: The dynamic production model DYNPRO is presented. DYNPRO is a new type of model for calculating the generation of electricity and its economic value for a cascade of hydropower stations. The program simulates the processes in the waterways between stations in order to calculate the production at each station. DYNPRO can show its capabilities, especially during short term regulation and in river like reservoirs.

DYNPRO has two modules; a hydraulic module and a production module. In the hydraulic module a dynamic routing based on the St.-Venant equations is executed. In the production module the generated energy is calculated from the simulated water levels, the turbine flow and the efficiency curve of the turbine. The economic value of the generated energy, the volume of the turbine flow and the volume passing the spillways are also calculated at each station.

Typical applications of DYNPRO are discussed. The use of DYNPRO to calculate the benefit from dredging between two stations with short term regulation is presented in detail.

1 INTRODUCTION

One-dimensional mathematical models for simulating flow and water levels in open channels have become a reliable engineering tool. They are used to solve a wide variety of problems. These models are based on the St. Venant equations that describe conservation of mass and momentum. Water levels and flows are simulated with good precision in space and time. If, however, the flow pattern is typically 2- or 3-dimensional these models can not be applied.

The final goal of hydropower industry is to design and operate hydropower stations and the connecting waterways in such a way that the profit is maximized. Dynamic routing models are used as part of many investigations design to improve the income from hydropower stations. There is a need for a model that can simulate the whole system of both waterways and hydropower stations. A model that can simulate both the hydraulics in the waterways, the production of energy, and the income with respect to the actual energy price is a powerful tool that can be used to increase the income from a cascade of stations.

2 THE STRUCTURE OF DYNPRO

DYNPRO is a general system that can be applied to practically any configuration of channels and power stations. The waterways can be connected in any way. The flow in the channels can be diverting, converting or looped.

In DYNPRO a power station consists of turbines, spillways and legal restrictions of reservoir levels. A power station can have any number of turbines. Each turbine is simulated separately. A turbine can be placed at any point within the modelled area. Turbines belonging to the same power station can be placed in different channels. What has been said for turbines is also valid for spillways. If reservoir water levels are outside the allowed band width DYNPRO issues a warning.

The geometry of the channels is described by sections across the channel. At a section the width of the channel is given at several levels. Between two consecutive sections the width is interpolated. The number of sections that must be specified depends on how detailed the channels are to be described. The roughness of the channel is specified by the Manning coefficient.

For each turbine the efficiency is specified as a function of flow. The efficiency curve will normally be related to the sections adjacent to the turbine. However, during set-up of the model for a specific area, any two sections can be chosen that are appropriate in the actual case. The operation of each turbine is given as a hydrograph.

The price for the produced energy can vary with time. The energy price is the same for all turbines.

At the boundaries of the modelled area, time series of flow or water level, or a relation between flow and water level, must be given. The initial values of water level and flow are explictly given.

At each time step of the simulation the water level and flow is calculated for each section, turbine and spillway. The production is then calculated at each turbine from the simulated water levels, the flow and the specified efficiency curve.

The output from DYNPRO consists of diagrams and tables. The diagrams can be simultaneously displayed on the screen during the simulation. They are updated after each time step. The tables can be printed or they can be used for graphical display or statistical analysis.

DYNPRO can be used on a PC, a work station or a main frame. The code is written in FORTRAN. The graphical features are so far restricted to an IBM-PC.

3 DEVELOPMENT OF DYNPRO

The hydrodynamic module of DYNPRO was developed on the basis of DAMBRK (Fread, 1988). The following features have been added:

1. The dynamic routing can be carried out for a net of channels. In DAMBRK routing is restricted to one channel.

2. Boundary conditions may be water level as a function of time, or flow as a function of time, or flow as a function of water level. These conditions can be used at any boundary.

3. In a reach where the St. Venant equations are not valid a relation between flow, water level and time can be specified. Weir operation, for example, can be specified as a time series of flow. The head loss in a reach with a 3- dimensional flow pattern can be measured in a physical model and can then be specified as a relation between flow, upstream water level and downstream water level.

4. The change of water level at each

iteration within a time step is restricted to a percentage of the actual water depth. With this improvement the algorithm becomes more robust. A break down due to a "negative area" that occurs sometimes in DAMBRK is avoided.

5. Initial flow and water levels can be explicitly specified.

The production module was developed fro scratch. The hydrodynamic module and the production module are independent of each other but they are connected by a short interface routine.

The first stage in the development of DYNPRO has been reached and DYNPRO is now in operational use at Älvkarleby Laboratory. Work is going on to improve the abilities of DYNPRO.

4 RANGE OF APPLICATIONS

DYNPRO can be used to improve the coordination of several hydropower stations. The coordination can be extended to all stations within a river basin. There are two tasks that must be solved.

To begin with, only the sum of the production at all power stations within the river basin is considered irrespective of ownership. The first task is to optimize the operation within the planning horizon in respect to the income for all stations together for a given amount of water. The second task is to share the extra income among the owners of the power stations. This second task is as important as the first. The best operation on an overall basis must by no means be the best for each station. There is no company that will accept a reduction of energy sales if there is no compensation from those who benefit from the changed operation scheme.

DYNPRO offers a solid basis for solving these tasks. The economic result of different operation schemes can be evaluated with satisfying accuracy. This will eliminate problems that are caused by different subjective judgements. Negotiations between power companies can be restricted to agreement on what situation is normal before improving overall operation.

With DYNPRO the hydraulic design of waterways and turbines can be evaluated. It is important to point out that the design cannot be calculated but it can be evaluated. The hydrodynamic simulation in DYNPRO is restricted to typically one-dimensional flow. It is therefore not possible to calculate the efficiency curve of a turbine from its geometry. The efficiency must be measured in a physical model. The measured efficiency curve is then used in DYNPRO to calculate production under sitespecific operation conditions. In this way the most suitable turbine can be chosen for a hydropower station.

The design of the waterways can be evaluated by DYNPRO and the benefit from dredging in the waterways can be calculated. Several alternatives can be investigated to find the solution with the best relation between cost and benefit. If the flow is one dimensional the head loss after dredging can be calculated by DYNPRO. Otherwise the head loss must be calibrated. An example is shown in detail further down.

Another application is the training of operation staff. Operation conditions during extremely high flow can be simulated. Extremely high flow may occur just once or twice in a lifetime but the actions taken during such an extreme event may be crucial to avoid damage. Using DYNPRO as a simulator for extremely high flow gives one of the few possibilities of training for such an extreme event.

5 CASE STUDY: BENEFIT OF DREDGING IN A DROWNED RAPID

DYNPRO has been used to investigate whether dredging in the drowned rapid Bergsforsen will be profitable. Bergsforsen is situated in the river Umeälv/Sweden between the power stations Pengfors and Stornorrfors (see fig. 1).

The daily regulation is high in both power stations. In Pengfors the flow is normally 0 for some hours during the night. In Stornorrfors the flow may not be less than 105 m3/s. Pengfors is dimensioned for 450 m3/s and Stornorrfors for 1045 m3/s.

When production is increased in Stornorrfors the water level downstream of Bergsforsen falls much faster than the water level upstream of Bergsforsen. After the initial phase the head difference over Bergsforsen increases and flow through Bergsforsen increases. For low water levels at Stornorrfors flow in Bergsforsen becomes critical and depends only on the water level upstream of Bergsforsen.

Bergsforsen divides the reservoir of Stornorrfors into two parts. The downstream part of the reservoir is smaller and responds faster to a flow change than the upstream part.

Dredging in the Bergsforsen will result in a faster response of the upper part of the reservoir, higher upstream water levels at Stornorrfors and an increase in the production at Stornorfors. Another consequence may be that the downstream water level in Pengfors will be lower after dredging.

Figs. 2 to 4 show how the reservoir Stornorrfors reacts to flow changes. At time 0 the flow in Pengfors was 10 m3/s, in Stornorrfors 150 m3/s and in Vindelälven 140 m3/s. After 1 hour the turbine flow at both stations is increased by 340 m3/s within a quarter of an hour. Then the flow is kept constant. The inflow from Vindelälven is unchanged. Two simulations were done; one before dredging in Bergsforsen and one after dredging.

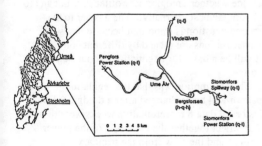

Fig.1 Map of Sweden and the area of the case study. Abbreviations: h=water level, q=flow, t=time.

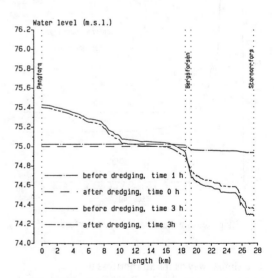

Fig.2 Water level profile between Pengfors and Stornorrfors

625

In fig. 2 water level profiles from Pengfors downstream to Stornorrfors upstream are shown. At time 0h only Pengfors benefits from dredging. The water level downstream of Bergsforsen is independent of the dredging because the water level at about 25 km is not allowed to rise above +75.00 m.

The length profile of the water level is also shown at time 3h, that is 2h after flow increase. Dredging causes lower water levels upstream of Bergsforsen and higher water levels downstream of Bergsforsen. The head gain due to dredging is higher in Stornorrfors than in Pengfors. This corresponds to the fact that the smaller part of Stornorrfors' reservoir is downstream of Bergsforsen.

The development of the upstream water level at Stornorrfors is shown in fig. 3. Dredging does not affect the water level during the first half hour after flow increase. Then the head gain due to dredging continuously increases until it becomes steady after about 7h. In reality great changes in flow occur within 7 hours due to short term regulation and the changes at Stornorrfors and Pengfors are different, so the system will practically never be in a steady state. A dynamic production model can predict benefits from dredging with a significantly higher accuracy than a sequence of steady simulations.

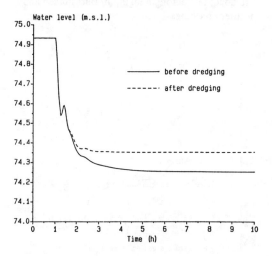

Fig. 3 Upstream water level at Stornorrfors

In a similar way as for Stornorrfors the downstream water level at Pengfors is shown (see fig. 4). The steady state is reached in a shortertime than at Stornorrfors.

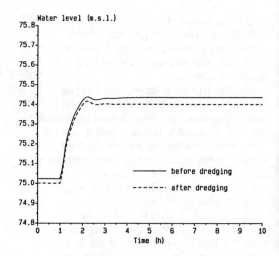

Fig.4 Downstream water level at Pengfors

The bottom of Bergsforsen is quite irregular, the water is shallow and the reach is situated in a bend. It is not wise to use a 1-dimensional mathematical model to predict how much the head loss will decrease due to dredging. Therefore a physical model of Bergsforsen was built. Four dredging alternatives were tested. Each alternative gave a table with the upstream water level as a function of the flow and the downstream water level (h-q-h - relation).

These tables were used in DYNPRO as a part of the hydrodynamic simulation when calculating the benefit of the dredging. Two questions were investigated for each dredging alternative:

1) How great will the extra income due to dredging in Bergsforsen be?

2) How much will the income increase in Pengfors and how much in Stornorrfors?

The first question is important in order to decide whether dredging is profitable or not and to choose the best alternative. The answer to the second question is important because the two power stations are owned by different companies. It will be a basis for negotiations about sharing the costs.

The assumption was that the regulation pattern will be unchanged after a dredging. From historical data 8 representative weeks were derived that reflect together the operation of the power stations and the flow from the tributary Vindelälven.

Table 1: Energy price used when planning investments with long term pay off time.

Price during period	Week number		Price ($/MWh) day	night/ weekend
1	45-12	66	48	
2	13-16, 37-44		57	46
3	17-36	37	33	

For planning reasons 3 periods with different energy prices within the year (see table 1) were chosen. The prices depend on the length of pay off time, the interest rate and inflation.

Roughly speaking the price period 1 is winter, 2 is spring and autumn and 3 is summer. Within each price period the price is higher during the daytime from Monday to Friday and lower during the night time and weekends. The price variation within a day is highest in winter and lowest in summer. In winter the price is about 40% higher during the daytime than during the night time/weekend. In summer the price difference is about 10%.

The representative weeks were derived from the weekly mean flow in the years 1958 - 1986. The weekly mean flows were classified after price period and size. The flow varies most during price period 3. In this period the flow can be greater than the design flow at Stornorrfors. Therefore the flows in period 3 were divided into 4 classes, whereas the flows in period 1 and 2 were divided into two classes each. These representative mean flows were then distributed within the week according to the experience of a person who has worked for a long time with regulation at Stornorrfors.

DYNPRO is executed for all representative weeks. In the first series of runs no dredging has been carried out in Bergsforsen. This is the present situation. Then the series of runs is repeated for each dredging alternative, that is for each h-q-h - relation in Bergsforsen. The total income during a year is calculated as a weighted mean of the representative weeks. The weight of each representative week is proportional to the number of weeks that were used to derive the respective representative week. Finally the benefit due to each dredging alternative in Bergsforsen is calculated.

Large areas of the reservoir are covered by ice during the winter. Neglecting this fact will lead to an underestimation of the benefit of a dredging. The simulation of flow in ice covered rivers has top priority in the future development of DYNPRO. We expect that this feature will be implemented before the end of 1992.

ACKNOWLEDGEMENT

DYNPRO is developed by Vattenfall Utveckling AB, Älvkarleby Laboratory. The development of DYNPRO is financed by Vattenfall AB.

REFERENCES

Fread, D.L. 1988. The NWS DAMBRK model - theoretical background / user documentation. National Weather Service, Silver Spring, Maryland.

6 Electricity supply – Choosing the energy source for electric power production

Hydropower'92, Broch & Lysne (eds) © 1992 Balkema, Rotterdam. ISBN 90 5410 054 0

An input-output analysis of a transition from hydropower to thermopower

M. P. N. Águas & J. J. Delgado Domingos
Instituto Superior Técnico, Lisboa, Portugal

ABSTRACT: Input-output analysis was applied to characterize the changes in the usage of different energy sources in the electricity generation in Portugal between 1971 and 1990, a period when this country has undergone a transition from a hydropower based system to a thermal one. The analysis was based on the Portuguese government data-base on energy and divides the energy system in four sectors: coal, liquid fuels, gaseous fuels and electricity. This paper describes the evolution of the national electric power production system and the methodology used to create special input-output matrices for energy purposes.

1. INTRODUCTION

In spite of the low electric consumption per capita, Portugal depends in external sources in more than 50% to satisfy the actual demand. Till 1960 electric system was based on hydroelectric power stations.

The growth of national electricity generation in the 1970's and 1980's, mainly based on fuel-oil and coal fired power stations, has changed deeply the national energy system.

This change has been widely discussed in a general context by lot of authors, just as by Delgado Domingos (1975, 1980) or Ribeiro da Silva (1990).

The present analysis characterizes the change on electricity generation by the evaluation of the energy efficiency of the total system. The calculations are based on input-output matrices for energy purposes with a treatment similar to the Leontief (1951) methodology.

In this study it is also included an analysis on the CO_2 emissions changes between 1970 and 1990 due to the enhancement of thermal electricity generation.

2. ELECTRIC POWER SYSTEM EVOLUTION

2.1 Changes in power generation

In last three decades the electricity production in Portugal has suffered a great change as it is shown in Figure 1.

The first thermopower plant started to work in 1960. Its power was 3x50 MW and based on the use of national anthracite and fuel-oil. At that time, all electric power was supported by hydropower plants with a total power of 990 MW.

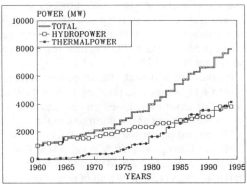

FIGURE 1: Power generation distribution
(Forecast: 1992/94)

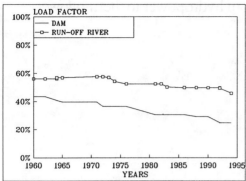

FIGURE 2 : Load factor of hydropower stations
(Forecast: 1992/1994)

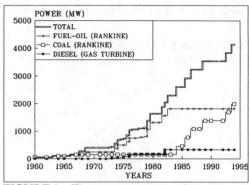

FIGURE 3 : Thermopower generation

The annual rate of growth of the electricity power system has been around 6.3%. Hydropower increased in a linear way, reaching 3070 MW in 1990. and in a similar way for dam and run-off river power stations. However, as represented in Figure 2, run-off river power stations has a bigger energy productivity.

The load factor, defined as:

$$\text{Load factor} = \frac{\text{Annual average energy production}}{\text{Annual maximal capacity}}$$

presents, for run-off river power stations, values between 50% and 60%, while in dam power stations this factor is lower due to its special use in peak demand hours, and varied between 25% to 45%. The load factor trend for both types of power stations decreased continuously with time, i.e., the latest built are the ones that presents the lowest energy productivity. An example of this is the biggest hydropower station of Portugal recently constructed, the Alto Lindoso Power Station (expected to start up in 1992), a run-off river station with 634 MW and a predicted annual energy production of 877 GWh, that represents a load factor of just 16%.

In the same period, thermopower stations showed a higher growth, reaching to 3550 MW in 1990. This growth was based on coal and fuel-oil fired power stations as it is illustrated in Figure 3.

The biggest power stations are: Fuel-oil Power Station of Carregado (750 MW), Fuel-oil Power Station of Setubal (1000 MW) and Coal Power Station of Sines (1240 MW), all of them operating in a Rankine cycle.

Gas turbine power stations, representing 330 MW, built to overcome shortages in local electricity supply in peak hours demand, are now closed due to high fuel costs.

Until 1995, it is predicted that new power stations, with a total power of 1320 MW, will start operating, corresponding to 720 MW in hydropower stations (mostly dam power stations) and 600 MW in a coal fired power station (Pego Power Station, groups 1 and 2).

2.2 Changes in energy production

During the last two decades, electricity gross consumption raised from 8100 GWh to 30200 GWh, as shown in Figure 4, representing an average growth of 6.8% per year. The figure divides the electricity sources by hydroelectricity, thermoelectricity and imported electricity.

In the 1970's, hydroelectricity was the main kind of energy generation and it represented around 70% of the total production (Figure 5).

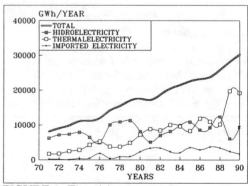

FIGURE 4 : Electricity consumption

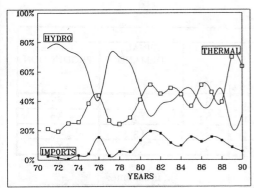

FIGURE 5 : Electricity distribution

This period includes the effect of a dry year (1976) on the thermoelectricity production and on the electricity imports.

As shown in Figure 4, the annual hydro-electricity production as been very irregular in the last two decades, varying from 8600 GWh/year to 11000 GWh/year. However, the annual average hydroelectric production has grown from 8000 GWh in 1970's to 8600 GWh in 1980's.

The thermoelectricity production, started in 1960, was in 1971 20% of total electricity consumption, reaching to 70% by 1990, with a gross production of 20000 GWh.

Up to 1985, the thermal electricity was based on fuel-oil fired power stations shown in Figure 6. After 1985, the consumption of coal increased significantly due to the Sines Coal Fired Power Station. In 1990 energy consumption in thermopower stations represented 0.4 of the national final demand (power stations consumption not included).

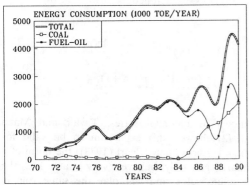

FIGURE 6 : Energy consumption in Thermo Power Stations

The electricity imported, was not very significant during in 1970's (except in the dry year of 1976). In contrast, it represented around 15% in 1980's. This growth in electricity imported was a consequence of the difference between the annual growth rate of the power generation in operation and the annual growth rate of electricity consumption .

3. THE INPUT-OUTPUT METHOD

The input-output analysis relates the interdependency between the production and consumption sectors in order to satisfy a final demand. In this application of the input-output method for energy purposes, we intend to identify the energy consumption inside the energy sectors and to relate that to total energy losses. The question that we intend to answer is the follow:

"For a given system, what is the energy efficiency related to the final demand consumption of a special kind of energy ?"

The input-output analysis starts by the calculation of the energy flows between energy transformer sectors. In this analysis the total energy structure is divided in 4 major sectors. The form of the general input-output matrix is illustrated in Table 1.

The input-output model is based on the intersectorial energy flows matrix, [S], the final demand vector $\bar{Y}$ and the total production, $\bar{X}$ and is mathematically translated as:

$$\sum_{j=1}^{j=4} S_{ij} + Y_i = X_i \qquad (1)$$

that is, for sector i, the sum of supplies from i to all energy sectors (1st factor) plus the final demand of energy i (2nd factor) represents the total production of sector i. The energy losses are supported by energy sectors.

In what refers to matrix [S], the value in the element S_{ij} means the part of energy produced by sector i that is consumed by sector j, so the values in line i represents supplies from the sector i, while values in column j represents consumptions of sector j. Matrix [F], named for economic purposes the primary factors matrix, represents the income (imports or national energy production) and the sector energy losses.

TABLE 1 : The input-output model

OUTPUTS	INPUTS					FINAL DEMAND	TOTAL PRODUCTION
	INTERSECTORIAL CONSUMPTION						
	Sec. 1	Sec. 2	Sec.3	Sec.4	TOTAL		
Sec. 1	S_{11}	S_{12}	S_{13}	S_{14}	S_1	Y_1	X_1
Sec. 2	S_{21}	S_{22}	S_{23}	S_{24}	S_2	Y_2	X_2
Sec. 3	S_{31}	S_{32}	S_{33}	S_{34}	S_3	Y_3	X_3
Sec. 4	S_{41}	S_{42}	S_{43}	S_{44}	S_4	Y_4	X_4
Intersectorial Consumption							
Imports	F_{11}	F_{12}	F_{13}	F_{14}			
Other Fuels	F_{21}	F_{22}	F_{23}	F_{24}			
Auto Production	F_{31}	F_{32}	F_{33}	F_{34}			
Energy Losses	F_{41}	F_{42}	F_{43}	F_{44}			
Primary Factors							
TOTAL CONSUMPTION	X_1	X_2	X_3	X_4			

or, in scheme:

[S]	$\vec{Y}$	$\vec{X}$
[F]		
$\vec{X}$		

It is to be noted that energy appears twice in the global input-output formulation but with different signals, positive in intersectorial flows and negative in [F].

Based on the intersectorial flows matrix and the total production vector, the technical coefficients matrix [A] is calculated as follows:

$$A_{ij} = \frac{S_{ij}}{X_j} \qquad (2)$$

which represents the required energy consumption of energy i from sector j to produce one unit of energy j.

A technical coefficients matrix of primary factors, [G], is also defined as:

$$G_{ij} = \frac{F_{ij}}{X_j} \qquad (3)$$

Substituting (2) in (1):

$$\sum_{j=1}^{j=4} (A_{ij} \times X_j) + Y_i = X_i \qquad (4)$$

or in vectorial notation:

$$[A] \times \vec{X} + \vec{Y} = \vec{X} \qquad (5)$$

When the final demand $\vec{Y}$ is known the required production of each sector is calculated by:

$$\vec{X} = ([I] - [A])^{-1} \times \vec{Y} = [Z] \times \vec{Y} \qquad (6)$$

where [I] stays for the identity matrix and [Z] relates final demand with total production.

The matrix [Z] can be calculated by:

$$[Z] = [I] + [A] + [A]^2 + [A]^3 + ... \qquad (7)$$

resulting from the Taylor's series expansion in powers of x of the function:

$$f(x) = (1+x)^m = 1 + mx + \frac{m(n-1)}{2!} x^2 + \frac{m(m-1)...(m-n+1)}{n!} x^n + ... \qquad (8)$$

known as binomial series. Taylor and Mann (1972) showed this expansion to be valid for $|x| < 1$. Accordingly (Pearl (1973)) the norm of a matrix generally behaves very much like the absolute value and will play the same role in matrix situations as the absolute value does when treating real-values functions. Defining the

634

Euclidian norm of [A], results $||A||<1$, considering that the final demand is always positive.

The series expansion represented in (7) has a physical meaning. Lets think about a 2x2 intersectorial matrix [S], referred to two energy sectors named Electricity and Coal. The total production of electricity is calculated as follows.

The first term of (7) represents an intersectorial energy flow of order 0. In this case the production of electricity will be the final demand of electricity:

$$X_{Elec}^0 = Y_{Elec} \qquad (9)$$

The other terms of (7) means the intersectorial energy consumption. The second term, representing an intersectorial energy flow of order 1, is calculated as follows:

$$X_{Elec}^1 = A_{Elec,Elec} \times Y_{Elec} + A_{Elec,Coal} \times Y_{Coal} \quad (10)$$

that is, the total electricity production of order 1 is the electricity consumption inside the electric sector plus the electricity consumption by the coal sector when this sector needs to satisfy a coal final demand Y_{coal}. High order terms, representing more complex intersectorial flow, have lower values.

Once calculated the total production for a given energy demand, the total energy losses is calculated by:

$$E.L. = \sum_{j=1}^{j=2} G_{EnergyDissipation,j} \times X_j \qquad (11)$$

and the energy efficiency will be:

$$\eta = \frac{\sum_{i=1}^{i=2} Y_i}{E.I. + \sum_{i=1}^{i=2} Y_i} \qquad (12)$$

4. ENERGY DATA BASE ANALYSIS

The intersectorial energy flows matrices are based on the official Portuguese data base on energy, comprising annual energy balances between 1971 and 1990, Direcção-Geral de Energia (1986,1991). The unit is the TOE (tonnage of oil equivalent) defined as $1TOE=10^7$ kcal. For fuels TOE/Ton is calculated in a NHV (Net Heat Value) basis, while in electricity TOE/GWh is calculated by direct units conversion, 1GWh=86 TOE.

In this study the energy system is divided in the following 4 sectors:

- Coal Sector (national anthracite, imported brown coal and coke).
- Liquid Fuels Sector (as crude, petrol, fuel-oil, JP1, JP2, naphtha,...).
- Gaseous Fuels Sector (as lpg, refinery gas, town gas, coke gas,...)
- Electricity Sector (hydro, thermo and imported electricity).

It must be taken into account in this division that:

a) it considers a gaseous fuels sector (including hydrocarbon and non-hydrocarbon fuels) due to the easy substitution between gaseous fuels and in order to allow a future study comprising natural gas (expected availability in 1995).

b) do not considers a solid fuels sector because firewood national statistics are very insufficient (it only considers comercial acountancy). This simplification will not affect the results because firewood mainly satisfies a final demand market.

Table 2 illustrates an input-output matrix.

The input-output matrices of the period 1971 to 1990 have been calculated according to the described methodology , followed by evaluation of matrices [Z] in order to relate total production with final demand by the Input Output method.

The analysis has been carried out in order to answer the question formulated in Point 3, that was, for a final demand of 1 unit of electricity, for instance, how much energy is lost.

Considering that electricity sector as the fourth sector, the result will be:

$$\vec{X} = [Z] \times \vec{Y}_{Elect} \text{ with: } \vec{Y} = \{0,0,0,1\} \qquad (13)$$

The energy efficiency will be calculated by (12).

TABLE 2: The input-output matrix for 1986.

	INPUTS						
OUTPUTS	Coal	INTERSECTORIAL CONSUMPTIONS				FINAL DEMAND	TOTAL PRODUC-TION
		Liquid Fuels	Gaseous Fuels	Electri-city	TOTAL		
Coal	0.0	0.0	58.1	764.4	822.5	618.9	1,441.4
Liquid Fuels	0.8	461.7	446.7	1,767.2	2,676.4	7,889.1	10,565.5
Gaseous Fuels	18.9	66.7	61.0	16.1	162.7	748.8	911.5
Electricity	0.8	27.6	2.7	294.9	325.9	1,671.8	1,997.7
Total Intersectorial Consumption	20.5	556.0	568.5	2,842.5			
Imports	1,355.1	10,568.0	329.8	247.2			
Other Fuels	0.0	0.0	0.0	122.4			
Auto Production	96.7	0.0	48.9	734.7			
Energy Dissipation	-30.9	-558.4	-35.8	-1,949.0			
Total External Energy Flows	1,420.9	10,009.5	342.9	-844.8			
TOTAL CONSUMPTION	1,441.4	10,565.5	911.4	1,997.7			

A similar analysis is performed for a final demand of Coal (Y={1,0,0,0}), Liquid Fuels (Y={0,1,0,0}) and Gaseous Fuels (Y={0,0,1,0}).

As the present study pretend to correlate the trend of the efficiency with the last two decades hydropower/thermopower transition, annual values of the efficiency are represented graphically as function of the percentage of annual thermoelectricity production in the total national annual electricity production, β, for the four energy demand vectors:

$$\beta = \frac{Thermoelectricity}{Thermoelectricity + Hydroelectricity} \quad (14)$$

5. RESULTS AND DISCUSSION

5.1 *Coal*

The Coal Sector relation between β and efficiency is represented in Figure 7. In this figure a significant change is identified on efficiency with the start up of the Coal Fired Power Station of Sines (1985).

Before 1985 the coal consumption was mainly national coal, forcing mining energy consumption. In this period efficiency was around 90% when satisfying a coal final demand.

In contrast, after 1985, imported coal has been dominant, mining energy consumption was done abroad and efficiency increased to 98%.

In general terms, it is not clear a dependency of efficiency on β.

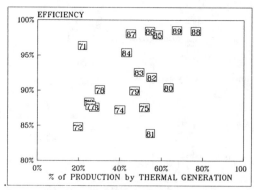

FIGURE 7 : Efficiency change for coal final demand consumption

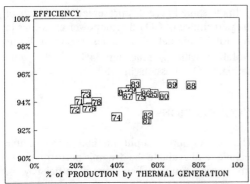

FIGURE 8 : Efficiency change for liquid fuels
final demand consumption.

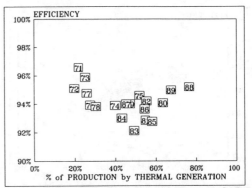

FIGURE 10 : Efficiency change in gaseous fuels
final demand consumption.

5.2 Liquid Fuels

The β .vs. efficiency relation is shown in Figure 8.
The Figure shows a slight increase of the
efficiency as thermal electricity production
increasesThis behaviour is due to the fact that the
main refinery losses, fuel-oil losses, reduces as β
increases, as shown in Figure 9. This trend is
explained by the enhancement of the residual
liquid fuel consumption in electricity generation.

5.3 *Gaseous Fuels*

As illustrated in Figure 10, system efficiency shows
an increase with β in the 1980', denoting a
variation of 93% to 97% when satisfying a
gaseous fuels final demand. However, this relation
should be analysed with caution because major
energy losses in this sector are due to distribution
and transport town gas losses.

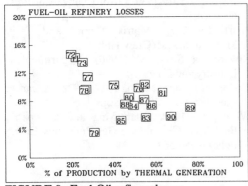

FIGURE 9 : Fuel-Oil refinery losses

Another factor, also not related with β, refers
to a change in statistical criteria change for blast
furnaces exhaust gas, Direcção-Geral de Energia
(1986,1991).

5.4 *Electricity*

The system efficiency in electricity final demand is
very sensitive to the hydro/thermopower
generation change due of the low efficiency of the
thermopower stations.

Figure 11 shows an almost linear growth of the
energy losses with the enhancement of
thermoelectricity. Linear extrapolation gives the
following values of efficiency:

Total hydropower system : $\eta=72\%$
Total thermopower system: $\eta=35\%$

5.5 *Untapped hydroelectricity*

There are still considerable untapped resources of
hydroelectric energy below unit powers of 10 MW
which were not taken into account on the official
estimates before 1980, Delgado Domingos (1980).

The most recent evaluation of around 1015
MW (forecast annual production of 4025 GWh)
and a special program was launched by the
Government to attract the private sector to
explore this sector of production, Ribeiro da Silva
(1990)

637

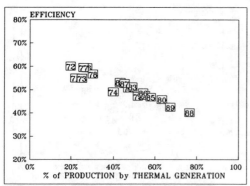

FIGURE 11 : Efficiency change in electricity
final demand consumption.

6. CO_2 EMISSIONS

Besides the energy losses analysis performed, it is interesting to note the evolution of CO_2 thermopower stations emissions in the period considered. The calculation is based on CO_2 emissions per heat release, Wilson (1990), represented in Table 3 and in annual fuel-oil and coal consumption in thermopower stations.

Fossil Fuel	$g_{CO_2} / MJ_{(LHV)}$
Oil	73
Natural Gas	49
Coal	92

TABLE 3: Specific CO_2 emissions

The values are represented in Figure 12. The evolution of emissions shows a significant growth and a dangerous non-linear trend in relation to β due to the increase of electricity generation based on CO_2 coal power stations. emissions.

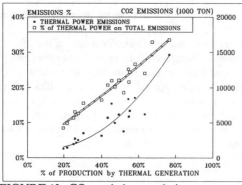

FIGURE 12 : CO_2 emissions analysis

In the same figure is represented the change of the percentage of CO_2 thermopower emissions in the total national The values shows a linear evolution with β, reaching 34% of the total national CO_2 emissions in 1990.

7. CONCLUSIONS

The use of input-output method is a suitable technique to evaluate the dynamic performance of an energy system. The energy efficiency of Portugal energy system have been analysed in the period between 1971 and 1990, a period when the electric generation system has undergone a transition from hydropower to thermopower.

The static analysis allows the evaluation of the relation primary/final energy for four kinds of energy sectors while the dynamic analysis identifies direct impact of the electric generation transition in coal, liquid fuels and electricity sectors.

The greenhouse effect due to the emissions of CO_2 present also a significant change in the period in analysis.

REFERENCES

Delgado Domingos, J.J. 1975. O Problema Energético National. Revista Técnica, 437:33-41

Delgado Domingos, J.J. 1985. Uma Politica de Consevação de Energia em Portugal. Defence Institute Conference. Portugal

Direcção-Geral de Energia 1986,1991. Balanços Energéticos.

Leontief, W. 1951. The Structure of the American Economy 1919-1939. Oxford University Press, New York.

Pearl, M. 1973. Matrix Theory and Finite Mathematics. McGraw-Hill.

Ribeiro da Silva, N. (1990). Department of Industry and Energy.

Taylor, A.E. and Mann, R.W. 1972. Advanced Calculus. John Wiley & Sons, Inc.

Wilson, D. 1990, Quantifying and Comparing Fuel-Cycle Greenhouse-Gas Emissions. Energy Policy 6:550-562.

Hydropower'92, Broch & Lysne (eds) © 1992 Balkema, Rotterdam. ISBN 90 5410 054 0

Marginal value of investments in hydro power plants

S. Alerić
Institut za Elektroprivredu, Zagreb, Croatia

ABSTRACT: Hydro power plants which are under construction or are going to be constructed in the future have more and more a multipurpose character. The construction of hydro power plants has, up to now, been financed by Hrvatska elektroprivreda, since these were traditional facilities which generated cheep and clean electricity. However, future construction of hydro power plants would certainly require more financiers of future users.
This is why I am trying in this paper to develop a procedure for determining maximum value of investment in hydro power plants which should be provided by Hrvatska elektroprivreda. This procedure will, after theoretical elaboration on mathematical basis, be tested on one concrete hydro power plant in Croatia.

1 ESTIMATE OF POSSIBLE PRODUCTION IN HYDRO POWER PLANTS (short presentation)

The basic energy feature of a hydro power plant is its possible production. It includes practically all specific elements and characteristics of the respective hydro power plant, from its power unit (turbine-generator-transformer) to topographical and hydrological data.

The estimate of possible production in hydro power plants is based on the methodology of constant and variable power and capacity (Ref. 1). Main goal of this method is to achieve the largest quantity of electricity in the shortest time possible and with highest capacity, but in a way which doesn't allow overflows (water losses).

The estimate of possible production in hydro power plants has a significant importance, specially if this production serves for further energy and economic valorisation of this plant (for example: selection of the scope of construction and number of power units, construction schedule, etc.)

Interdependence of possible electricity production in a hydro power plant and its installed flow (scope of construction) is shown on Figure 1.1.

2 METHODOLOGY OF ENERGY BALANCE IN THE ELECTRIC POWER SYSTEM (short presentation)

The methodology of energy balance

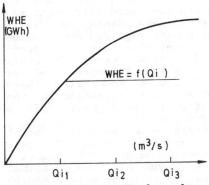

Figure 1.1 Interdependence of possible production and the installed flow in a hydro power plant

in a mixed (hydro-thermal) electric
power system is based on the same
method (Ref. 1) which is applied
for the estimate of the possible
production in hydro power plants.
The main objective of this method
is the supply of the satisfactory
amount of electricity to the customers with minimum costs. Respect-
ing the above mentioned principle
and simulating the conditions in
the electric power system, by the
construction of a particular hydro
power plant we achieve certain
effects. These are: reduction in
necessary construction of thermal
capacities, out down of fuel costs
in thermal power plants, changes
(increase or decrease) of the costs
related to electricity reductions
and increase of fixed costs in the
electric power system.

These effects are determined by
thorough analysis of circumstances
in the electric power system,
considering the same reliability of
electricity supply to the customers
(the same LOLP loss of load probability).

Figure 2.1 shows a graphic representation of the way in which the
efficiency of the new hydro power
plant is determined, i.e. smaller
need for the construction of thermal power plants. This method
compares new facilities, i.e. new
power plants, with an referential
plant or an alternative one. Since
the electro-energy balance is made
for the period of at least 40
hydrological years (for the same
period we estimate possible production in hydro power plants), and
for each month in this period,
Figure 2.1, curve k1 consists of
480 points (40 years x 12 months).
Each of the 480 points on Figure
2.1 represents required power from
thermal power plants in the respective hydrological period. In other
words, Figure 2.1 represents the
curve of required power from thermal power plants depending on the
water flow in hydro power plants
(hydrology data).

The ordinate in the neutral point
of the coordinate system - 0 point
- represents the greatest quantity
of power required from thermal
power plants i.e. the driest hydrological period (lowest electricity
generation in hydro power plants).

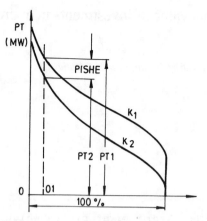

Figure 2.1 Probability curve of
required power from thermal power
plants

The curve k1, Figure 2.1, represents required power from thermal
power plants without the construction of a new hydro power plant,
and the curve k2 with the new hydro
power plant constructed. The point
01 shows the reliability of consumers' supply or LOLP (previously
determined by energy and economical
analysis). The difference between
curves k1 and k2 represents savings
in the system development (PISHE -
thermal capacities which needn't be
constructed if a new hydro power
plant is constructed. This estimate
also shows differences in fuel
costs and damages losses by reductions. When all these effects
(decrease in required power from
thermal power plants, fuel costs
and reduction losses) are converted
in cash value, we get the energy
value of the observed hydro power
plant in the electric power system.
On the other hand, the construction
of a new hydro power plant results
in increased costs of the electric
power system. The ratio of energy
value (savings and benefits in the
electric power system) and fixed
costs of the observed hydro power
plant is called the relative ener-
gy-economic value (its determination will be explained below).

3 METHODOLOGY FOR DETERMINING MARGINAL VALUE OF HYDRO POWER PLANTS

3.1 General

When we consider each and every kind of electric power facility and want to determine its real position, role and solvency the electric power system, we have to observe it within the system in which it operates or will operate after its realisation.

The observation of facilities within the electric power system is a comprehensive and thorough task which should be performed correctly and professionally with high level of quality by the contractor. The approach to each facility should be made with great care and consideration since these facilities are very expensive and, considering lifetime and exploation, the mistakes and ommisions which could occure can hardly be corrected.

For this purpose different models and program packages have been developed in the world and in our country, which enable high quality and correct analysis of electric power facilities.

3.2 Energy and Economic Value of Power Plants

The valorisation of energy contribution made by each power plant can be performed only by a complex analysis of the compliance with estimated consumption, taking into account all specific features of the system, from consumption characteristics to plants' characteristics, their availability, availability of power resources, etc. In the electric power system which has greater participation of hydro power plants' generation in the total satisfaction of requirements, the availability of hydro power, uncertainty of inflows in the required quantities and periods of time, constitutes a complex problem in system development planning and, accordingly, the valorisation of energy contribution of these facilities to the electric power system becomes an extremely complex task. In the model for the analysis of the conditions in the electric power system the operation of each facility should be simulated in a high quality manner (for example: pumping-acumulation plant with proper water inflow in the upper reservoir), and this task demands a detailed elaboration of the models belonging to each type of facility considering all specific features which are derived from the realization of the plant or the characteristics of the inflow.

The method which is going to be utilised for the valorisation of energy contribution of hydro power plants will be used in all analysis of the electric power system development. It is based on a thorough analysis of the conditions in the electric power system, and the uncertainty of hydrological conditions is avoided by the simulation of a great number of hydrological conditions (up to 40 hydrological years, item 2).

Basically, the valorisation of energy contributions of plant consists of the valorisation of the contribution in power and capacity (item 2) which can be generated in a particular plant.

Energy and economic value (EEV) of a plant includes
- savings in fixed costs for the construction of TPP capacities (alternative solution)
- savings in fuel costs
- decrease (increase) of losses caused by reductions.

The above mentioned elements could be expressed analytically in the following way:

$$EEV = BT \times PT \times QTE + TG + TR \quad (106 \text{ CRD/year}) \quad (3.1)$$

where:

EEV = energy and economic value of a plant (106 CRD/year)

BT = specific investment of the referential, supplemental, substitutional, standardised or alternative TPP (106 CRD/MW)

PT = TPP capacity at generator which needn't be constructed if the observed hydro power plant is constructed (MW)

QTE = yearly quota of expenditures in TPP (106 CRD/year)

TG = savings if fuel costs (106 CRD/year)

TR = differences in the losses caused by reductions (10^6 CRD/year)

3.3 Relative Energy and Economic Value

According to the definition, relative energy and economic value is a ratio between the energy and economic value (contribution of the observed facilities to the electric power system) and fixed yearly costs of the observed facilities.

It can be expressed in the following way:

$$REEV = \frac{EEV}{STHE} \qquad (3.2)$$

where:

REEV = relative energy and economic value

EEV = energy and economic value of the observed power plant (10^6 CRD)

STHE = fixed yearly costs of the observed power plant (10^6 CRD).

Fixed yearly costs are the product of investment value and the quota of yearly costs, i.e.

$$STHE = IHE \times QHE \qquad (3.3)$$

where:

STHE = fixed yearly costs of the observed power plant (10^6 CRD)

IHE = investment value of the observed power plant (10^6 CRD)

QHE = yearly quota of expenditures of the observed power plant (%/100)

Relative energy and economic value is a parametre which unifies practically all energy and economic features of a particular facility. Therefore, this parametre, i.e. relative energy and economic value, is used for different purposes in electro-economic analyses.

It represents the basis for the determination of main energy characteristics of hydro power plants, such as: scope of construction (Qi), installed capacity (Pi), reservoir size or the slow-down elevation, length of intake and outlet structures, etc.

Relative energy and economic value is also used for the comparison of different power resources and can be utilised for the determination of the structure of power resources in the electric power system. Finally, relative energy and economic value serves for the determination of priorities in the construction of generating facilities, and it can represent the bases for making decisions on the construction or rehabilitation of a certain facility.

From the above mentioned it can be seen that relative energy and economic value plays an important role and should be treated with particular attention in the estimates. Recently, it is estimated with average and current costs emerging in the electric power system, and is frequently accompanied by the sensibility analysis.

3.4 Marginal Value of Investments

Marginal value of investments, in this case, means the maximum amount of investments which is profitable for Hrvatska Elektroprivreda to invest in the observed facility instead in an alternative solution. This is, therefore, a maximum amount invested in the observed facility which is solvent from the energy point of view.

The amount of the marginal (maximum) value of investments will be estimated using relative energy and economic value. The equations 3.2 and 3.3 consist of fixed yearly costs or investments. In order to clearly explain the standard and procedure of the determination of marginal value of investments in hydro power plants, the equation 3.2 will be a little bit modified and expanded by the equation 3.3. Relative energy and economic value can be estimated in the following way:

$$REEV = \frac{EEV}{IHE \times QHE} \qquad (3.4)$$

Based on the equation 3.4, the marginal value of investments in a hydro power plant can be defined as:

$$GIHE = \frac{EEV}{1 \times QHE} \qquad (3.5)$$

where:

GIHE = marginal value of investments in hydro power plants (10^6 CRD)

EEV = energy and economic value of the observed hydro power plant (10^6 CRD)

"1" = relative energy and economic value of the observed hydro power plant (10^6 CRD)

QHE = quota of fixed yearly expenditures for an alternative solution (referential plant)

The equation 3.5 practically equalises the observed hydro power plant with an alternative solution. The alternative solution, a supplemental or referential thermal power plant, has always energy and economic value which equals one "1". If we know EEV (energy and economic value) and QHE (quota of expenditures) it is easy to estimate by the equation 3.5 a marginal (maximum) value of the observed electric power facility. EEV (energy and economic value) of the observed power plant is estimated by the equation 3.1 and elements (PT, TG and TR) of that equation represent the result of electro-economic analysis (energy balance statements). Quotas of expenditures for thermal power plants and hydro power plants are determined by an economic procedure (which isn't discussed in this paper) and are calculated as ratios of fixed yearly costs of operation and investments.

If the marginal value of investments (as well as relative energy and economic value) is calculated with the costs which occur in average years (years for which electro-energy balance statements are made) than we get average marginal investments in the observed facility, in this case HPP Podsused. If it is calculated with current costs and investments which occur during the lifetime of the power plant in a dynamic way, and are reduced to one referential year, we get current marginal values of investments in the observed facility.

4 AN EXAMPLE OF THE DETERMINATION OF MARGINAL VALUE OF INVESTMENTS IN HYDRO POWER PLANTS

According to the standards briefly explained above, we estimated the marginal value of investments in one hydro power plant within the electric power system of Croatia. This hydro power plant is going to be constructed at the waterflow which has an average multiannual medium flow of 310 m3/s. It's installed capacity will be 48 MW, with the average possible generation of approximately 200 GWh.

All calculations are made for the electric power system in which the planned yearly electricity consumption is 17376 GWh, and the maximum load 2878 MW. Possible generation in the existing (already constructed) hydro power plants is approximately 6620 GWh/year (38 % of total requirements), and their installed capacity is 1840 MW. In the existing (already constructed) thermal power plants and nuclear power plants 2300 MW would be available. The referential thermal power plant has the installed capacity of 300 MW, is fired on stone coal and its specific investment value is 18,925 x 106 CRD/MW (ca 1400 US$/MW).

4.1 The Results of Energy Balance Statements

Based on the above data: observed new hydro power plant, electric power system, and using program packages, we determined the effects which are achieved by the construction of a new hydro power plant.
These effects are as follows:
- savings in TPP capacities (MW)   28,4
- savings in fuel costs (106 CRD)   74,4
- decrease in reduction losses (106 CRD)   - 5,2

If we estimate the reduction in the scope of TPP construction using the

specific investment value and the quota of expenditures (11,35 %), we will get the following cash amounts:
- savings in the construction
  of TPPs (10^6 CRD)   61,0
- savings in fuel costs
  (10^6 CRD)   76,4
- decrease in reduction
  losses (10^6 CRD)   - 5,2
  TOTAL   (10^6 CRD)   132,2

The value achieved by the equation 3.1 (which defines energy and economic value of hydro power plants) amounts 132,2 x 10^6 CRD.

Using the quota of expenditures for hydro power plants and equation 3.5 we can easily calculate the marginal value of investments in a particular hydro power plant (calculated with average costs). It amounts 1467,3 x 10^6 CRD (ca 110 x 10^6 US$).

If we calculate marginal value of investments in a new hydro power plant with current costs, we get the amount of 1115,0 x 10^6 CRD (ca 83 x 10^6 US$).

5 CONCLUSION

In this paper I presented a procedure for the determination of the marginal (maximum) value of investments in hydro power plants. This procedure is based on a complex observation of the circumstances in the electric power system itself and in the system to which the observed hydro power plant belongs. Energy and economic analysis determines the value of investments which could be made by Hrvatska elektroprivreda and up to which this facility is considered to be solvent. Due to the limited number of pages in this paper, I couldn't present the accomplished sensibility analysis based on the proportion of variable power of the observed hydro power plant, the participation of hydro power in the system, the type of referential (alternative) thermal power plant, rate of actualisation, etc.

However, taking into account the remaining hydro power potential in Croatia (mainly of multipurpose character), it is expected that this procedure would be frequently used in the future.

REFERENCES

Požar, H. 1983 and 1985. Snaga i energija u elektroenerget- skim sistemima, Volume I and II, Informator, Zagreb.

Hydropower'92, Broch & Lysne (eds) © 1992 Balkema, Rotterdam. ISBN 90 5410 054 0

Developing hydropower – The only way for China's electric power optimization

Cheng Dong
China Hua-Neng Group, Beijing, People's Republic of China

Gao Xiang
Hydraulic Department of Tsinghua University, Beijing, People's Republic of China

ABSTRACT: This paper introduces the gigantic superiority of hydropower in the electric power in China. It shows with practical data and technical economic index that hydropower is the only way to optimizing electric power production in China. China is rich in resource of hydropower, and the investing environment is nice. And now, China's economic deepening reforms are going on and more foreign advanced technology is being imported to speed up the development of hydropower. This paper discusses that nice economic benefit of hydropower exploitation is needed to promote the rapid development of hydropower, and increase the proportion of hydropower in the electric power system as a whole. And the optimization of the Chinese electric power production can be realized.

1 THE CHARM OF HYDRO-ELECTRICITY GENERATION IN CHINESE ELECTRIC POWER INDUSTRY

The hydro-power resources which can be exploited in China is 370 GW. Till 1990, the exploited hydro-power was only 35.3 GW, and the exploitation rate of hydropower resources was 9.5 percent, therefor there are still great potentialities of resource. Among China's present electric power composition, hydropower accounts for 26.3 percent of the total electric power capacity, and plays a decisive role and, it is much high concerned in electrical power system. Hydro-power has great charm in China's electric power industry.

1.1 Hydropower's advantages in economy

Hydropower production is the simultaneous accomplishment of hydroenergy's being produced and electric energy's being transformed from hydro-energy. Thereby, its construction investment is more saved than the coal-burning power production and the nuclear power production as well, it has great charm economywise. For example, according to the investment statistics of coal burning power plant which Hua-Neng Group built, adding up the investment of coal-burning power plant construction, the coal consumption and transtportation, the average unit investment is 2079 RMB per KW, however, the investment statistics of ten large and medium sized hydro-power stations built in the same period, shows that the average unit investment is 1382 RMB per KW, is 34 percent lower than that of the coal-burning thermal power construction, and the unit investment of nuclear-power is several times

higher than that of the coal-burning thermal plant, it is more incomparable with hydro-power in economic benefits.

1.2 The enormous advantages in technical economy index of hydropower production and operation

The "fuel" of hydro-power is hydraulic energy, which is of inborn resource, is inexhaustible in supply. But, among the coal-burning thermal power production cost, the fuelcost accounts for 25 percent of the production cost. In addition, the comparisons of several production and operation technical economy indexes are listed in Table 1.

1.3 Hydro-power brings about fairly high additional value for electric power system

In the large electric power system of certain hydropower proportion, because the hydro-power units start flexibly, have strong abilities of peak modulation and 'climbing' with load, and its scope of adaption to load variation is wide, it plays the role of peak modulation, frequency modulation, phase modulation and rotation reserve, accident reserve for the electric power system, it produces a lot of dynamic profits for electric power production, promotes the steadiness of electric power system, guarantees the security and full-out put of electric power production, reduces the coal consumptions of thermal power units, brings about fairly high additional value for electric power system, and optimizes the electric power system. Hydro-power does not produce

Table 1.

Item	Ratio of hydropower to thermal power	Remarks
Production cost	1:3.60	sampling average value
Plant electricity consumption rate	1:33.00	the thermal-power plant's electricity consumption rate is counted as 7-8%
Average equipment accident rate	1:5.20	sampling survey of plants capacity of more than 100 MW
Adjustable hours of units	1:0.89	sampling survey of plants with a capacity of more than 100 MW
Occupation rate of production floating capacity	1:6.00	average value of statistic data

carbon dioxcide which thermal power plants will discharge, does not contaminate the atmosphere, has not the harm of acid rain. Therefor, hydro-power stations have un-replaceable advantage in improving environment.

2 A FAIRLY GOOD INVESTMENT ENVIRONMENT FOR HYDRO-POWER IN CHINA

The average annual growth rate of China's gross output value of industry and agriculture between 1985 and 1990, was 10.01 percent, however, just the same period, the average annual electric power growth rate was 8.5 percent. The capacity of facilities which consume electric power is two and half times as big as that of facilities which produce electric power in our country. The whole country is short of electricity geneating capacity of 19,000 MW. Thereby, the situation of inadequate electric power supply will last for a relatively long term and that provides a steady and ready market for hydropower developing.

At present, coal-burning power plants account for more than 70 percent of the present electric power system, but it faces double restriction of inadequate coal resources and transport capacity. At present, the situation of coal resources supplying for the built coal burning power plants is very severe. It is imperative to transfer the whole contry's electric power investment to hydroelectric power production.

China has very rich hydro-power resources, however the exploitation level is low, there are still more than 90 percent of resources needing to be exploited, and there have been a group of hydro-power plant reserves which earlier stage works are done deepgoingly and solidly, and which have been investigated and designed for

many years, therefor, the optimizing choices of hydro-power exploited places are very abundant.

The economic reforms of China are stepping into a deepgoing developing stage. For the sake of a large number of inborned advantages of hydro-power, under China's planned market economy developing, according to the law of value, seeking and enhancing the self-value of hydro-power under fair competition, setting up the system of self-circle and self-developing of hydro-power industry, strengthening the vitality of hydro-power business, it is no doubt that the investmental environment of hydro-power should be improved, and that the hydroelectric power business's developing should be promotted under the economic reform.

3 ACCELERATING THE SEVELOPING OF HYDROELECTRIC POWER PRODUCTION AND OPTIMIZING CHINA'S ELEC-TRIC POWER PRODUCTION SYSTEM DEPEND ON DEEPEN-ING ECONOMIC REFORM.

In the course of electric power production, hydro-power generation has momentous significance in the following aspects, such as stabilizing the electric power production system, guaranteeing the quality of electricity supply of electric network, boosting the dynamic economic profits of electric power production and optimizing electric power production. Nevertheless, it is necessary to change old views and deepen economic reform in order to accelerate the developing of hydro-power generation. The key points of reform are listed as following.

3.1 Change old views, spur exploitation by improving benefits

Change the hydro-electric power exploiting pattern from 'speed pattern' to 'benefits pattern'. The so-called 'speed pattern' is what, in the past we subjectively wished to promote the developing of hydropower so that we strongly asked the government for more hydraulic projects being set, strived for more constructions being started, vied for construction speed, but ignored developing benefits, the financial and material resources which the state could bear as well, some projects' investments had no guarantee after the project constructions being started, because of the lack of funds and inadequated material supply, the time limit for some hydroelectric projects had to be delayed repeatedly, there was not only less speed but also lower benefits. In the strategic thoughts of hydroelectric power developing, it should be made clear that spurring exploitation through economic benefits, striving for high speed developing on the basis of large benefits, taking the economic benefits as the main axle in every link, such as plan of choice of dam, investigation and designing, construction and installation, production and operation, etc; evaluating the project's exploitation value according to social objective economic values, that is to say, regarding the immediate benefits' high or low as the principle of making a strategic decision, breaking through all of the old fetters of "leader's consciousness" and subjective determination.

3.2 Change the industrial structure model, set up the economic order of stock and indenture system.

Hydroelectric power resources are owned by the state, this method of many kinds of stock company may be used when developing and utilizing the hydroelectric power resources. The stock mechanism can not only assemble production factors, enhance marketable competitive power, balance the relationship between production and property, but also will strengthen the budget restriction to enterprises, realize the enterprises' independent running and financial affairs, improve business management and benefits level. Hydroelectric power projects may be graded according to their sizes. For large and medium sized, high grade projects, it mainly depends on the large sized business blocs to raise funds or introduce foreign funds for exploitation or conduct Chinese-foreign joint developing. For small sized projects, it is permitted for individual or collective to develop. The stock companies can independently conduct business counting, manage their own business independently, be financially independent according to the government's business-law. It is necessary to abide by economic institu-

tions, constrcts should be signed among each related side of the joint venture to make sure the ownership of project properties, marketable occupation rate, right of using electricity, charges for electricity merged in the electric net-work, standard of interest rate, tax rate, the principle of dividing the profits by each related side and the cooperating term of validity, etc. To sign contracts according to items respectively and acquire authoritative legal protection, so as to promote hydroelectric industry to head for steady, high, efficient and orderly contract economy.

3.3 Reasonably adjust the price of electricity and accumulate funds, in order to promote hydroelectric power developing

China's economic reforms are trying to break through the pricing barriers, adjust pricing structure, form a reasonable price system according to the value-laws of commodities. For many years, electric power price has been tortured, and especially the hydroelectric power price has been more tortured, on the one hand the hydropower price is constrained by the whole electric power price system, on the other hand, high quality hydroelectric power can not have a relevant price in electric power system, and what produced dynamic additional benefits cannot be compensated. If we say the reason of hydroelectric power's slow developing is that its price has been dual-tortured, that is not exaggerated. We should fully utilize the "invisible hands" of commodity circulation, we should not try to obtain high profit through trade monopoly, but it is needed to ensure hydroelectric power's obtaining social average profit rate, so that there is possibility for hydroelectric power to accumulate funds and develop by itself. In the electric power pricing structure, it is needed to reflect the high quality and high price of hydroelectric power, to fairly compensate various kinds of additional dynamic value, which the hydropower produces for electric power system. Hydraulic work members should change their tradtional view that gentleman never talks about money, under planned commodity economic competition mechanism, profits are the direct economic base of competition, and the necessary method to develop productive forces. Various level hydroelectric power enterprise's economic legal persons, should be open and aboveboard, bold and assured to strive for profits and to seek hydropower fair price and reasonable profit rate.

3.4 Multi-levelly, omnibearingly, wide-rangingly raise funds for hydropower exploitation and developing

In terms of economic law, the key point of rais-

ing funds is to increase industrial profit rate.
In the course of China's being deepening its
economic reforms, marketable mechanism must be
prefected, after one reasonable pricing structure
has been formed gradually, the law of value will
spontaneously balance the relationship between
the supply and demand, optimize the disposition
of resources, and guide the flowing direction of
social funds. Hence if only various reforms stated
above are well carried out, and if only we attach
importance to that boosting effectiveness through
profits and facing social to carry out stock
system multi-levelly and omnibearingly, it is not
difficulty to raise hydropower developing funds.
It is also another important method of raising
hydropower developing funds to plan asking for
soft loans from abroad to purchase hydraulic
facilities with excellent behavior. At present,
some large sized hydroelectric power and thermal
power stock electric enterprises with the
generating capacity of 2000-3000 MW, which funds
were raised, are appearing in China, that can
prove that the potentiality of developing electric power through raising funds is very
tremendous. Among the living example of existing
electric power enterprises through raising funds,
the thermal power enterprises make up the
majority, but if only hydroelectric figure which
has been tortured severely is corrected and lots
of advantages of hydropower are given full play,
the tremendous force of hydro-power's raising
funds, asking for loan, and absorbing domestic
and foreign funds, can be expected.

3.5 Learn and absorb the advanced techniques and
    manage experiences of international hydro-
    electric project, readjust the inner structure
    of China's hydroelectric power industry,
    strengthen hydropower construction troops

There has been accumulated a lot of useful experiences in China through more than forty years
hydraulic engineering designing, construction,
production and operation. There is relatively high
technical level in those facts, such as , dam
construction technology, silt control when
reservoirs built on silt-carring river, energy
dissipation and erosion control of flood
discharge of high dam, etc. Nvertheless, each
country in the world has its own distinguished
industry developing level, in the fact of hydraulic construction technology, there are a lot of
advanced experiences and techniques worthy to be
learned by us. For example, increasingly developing high embankment dam, hyperbolic thin arch dam,
rolled compacted concrete dam, reinforced-concrete
face rock-fill dam, and the unique experiences of
Norweign Kindom, such as long diversion tunnel
and using underground cavern groups to exploit
hydro-power resources and improve the valid
utilization ratio of water. At the same time of
introducing advanced techniques, we must readjust

the inner structure of hydropower industry, optimize labor combination, improve management and
administration, increase labor productivity, spare
no efforts to shorten engineering construction
cycle, so that the economic benefits of hydropower
business can be more fully brought into play.

4. Concluding remarks

Hydroelectric power has great charm in electric
power system, is the only way to optimize electric power production. China has abundant hydropower reources and good investment environment
as well. However, it depends on China's deepening
economic reforms, absorbing international advanced
techniques and improving hydroelectric power
exploitation's economic benefits that we want
to boost hydropower's developing, optimize China's
electric power system. By high benefit to push
hydroelectric power's high speed developing,
thus to realize the optimization of electric
power production.

648

Hydropower'92, Broch & Lysne (eds) © 1992 Balkema, Rotterdam. ISBN 90 5410 054 0

Rapport énergie thermique – Énergie hydroélectrique

V.D. Vieira Filho
PRINCÍPIO, Tecnologia de Projetos, São Paulo, Brazil

ABSTRACT: The energy sector in Brazil has become, troughout the last twenty years, and in particular during the period of intense industrialization and development of the agricultural sector in the country, a dynamic pole, having ever assured the energy basis, essential to this progress. One of the characteristics of the Brazilian power supply is that the energy spectrum is divided to three equal parts, i.e. water power, petroleum and all other forms of energy. Another characteristic is the dominating State presence. This sector, having lately gone through a crises, face today some challenges. One of them is the problem, particularly serious not only because of the large investments, but also because the foreign market cannot be reckoned with any possible adjustments to restore the balance between supply and demand. Some particular enterprises had started to solve its own problem, updating the power supply with the gouvernment authorisation to explore power sources. The general conclusions of one analisys donne for an important Brazilian group is indicated, to show that is expected that a greater flow of private savings is essential to irrigate this sector, supplementing the endeavour of the Gouvernment to provide the country with the necessary infrastructure, enabling it to proceed on its trail without major discontinuities.

1 INTRODUCTION

Le Brésil a observé, dans les derniers vingt ans, une augmentation significative de sa production agricole, avec une expansion de la frontière des cultures vers les régions ouest et centre-ouest du pays. Cette croissance de l'agriculture a provoqué un vecteur de développement régionale, avec l'installation d'industries de transformation des produits agricoles de base, des nouveaux centres d'emmagasinage et de distribution des produits, avec une amélioration sensible de la qualité de vie locale, et une augmentation de la population. En plus, l'accroissement de l'agriculture a provoqué aussi une réévaluation sur les méthodes d'utilisation du sol en exigent ainsi des équipements d'irrigation plus modernes et aussi plus puissants (en demandant plus énergie), parce que les hauts plateaux exploitables sont réellement très larges. Le territoire brésilien couvre une surface d'environ 8,5 millions de km² (1,7% de la superficie totale du monde) avec une population estimée de 150 millions d'habitants, soit une densité de l'ordre de 17 habitants par kilomètre carré. Jusqu'aujourd'hui cette distribution rest encore hétérogène, avec une concentration urbaine d'environ 74% de la population totale. Le taux d'accroissement de la population dans les derniers dix ans fut de l'ordre de 2,2%, sauf pour la région ouest et centre-ouest où on a observé un taux de 3,1%. En considérant la consommation globale d'énergie, dans la période de 1973-1988 on a eu un taux d'augmentation de 4,3% par an, similaire au taux de croissance de l'économie nationale (5,5% par an entre 1970 et 1989). Dans le cas de l'énergie électrique, cette période a eu un taux de 9,4% par an, phenoméne remarquable où la dynamique des caractéristiques des divers centres consommateurs (maisons, industries, commerce, etc) a permis une augmentation de la consommation même dans une période de crise. Il faut observer aussi que l'économie nationale a atteint une croissance de 3,6% dans l'année de 1989 et la consommation d'énergie, dans l'époque, a depassé les 4,8%, même avec le rabaissement forcé de la tarif de US$56/MWh en 1988 vers US$39/MWh en 1989. Au niveau regionale il a été possible d'observer les variations suivantes:

Tableau 1. Demande d'énergie

	1970 GWh	1970 %	1988 GWh	1988 %
Brésil	35730	100	197639	100
Nord	348	1	7455	4
Nord-est	3037	8	30114	15
Sud-est	28273	79	124776	63
Sud	3471	10	27202	14
Ouest	601	2	8092	4

Tableau 2. Demande d'énergie - taux d'augmentation (% par an), par région

	80/70	89/80
Brésil	12,3	6,3
Nord	18,5	16,4
Nord-est	16,6	8,8
Sud-est	11,1	5,0
Sud	15,0	7,6
Ouest	19,0	10,0

Au niveau des différentes classes de consommateurs on a eu une distribution comme suit:

Tableau 3. Demande d'énergie par classe

	1970 GWh	1970 %	1989 GWh	1989 %
Total	35370	100	197639	100
Industrie	16165	45	103621	53
Residence	8288	23	43571	22
Commerce	5147	15	22345	11
Agricole	317	1	6254	3
Autres	5813	16	21848	11

Il est possible de remarquer que même dans la période de crise le pays a observé une augmentation de la demande industriel, fait que peut s'expliquer par:
- le process de modernisation, avec incorporation de technologies plus efficaces, avec substitution des equipements et des appareils de contrôle;
- le développement des industries "électro-intensives" de transformation (utilisation massive de l'énergie électrique dans la production d'aluminium, par exemple); et
- la substitution des combustibles dérivés du pétrole par sources alternatives.

Dans le cas des consommateurs liés aux activités agricoles, y inclus l'irrigation, on observe que même avec un taux de croissance de 3% en 1989, il fut le secteur qui a présenté le plus haut taux d'augmentation, fonction aussi de la modernisation et mécanisation de l'agriculture.

2 DÉVELOPPEMENT DE LA MATRICE ENERGÉTIQUE BRÉSILIÈNNE

D'accord avec les statistiques historiques disponibles, quand l'économie marche dans des conditions adéquates avec un progrés normal, le secteur d'énergie absorbe une partie significative, de l'ordre de 12%, du capital produit par l'économie.
Le tableau ci-dessous montre l'évolution de la production d'énergie primaire, par source:

Tableau 4. Évolution de la production d'énergie (en %)

	1973	1988
Pétrole	13	19
Gaz Naturel	2	4
Charbon	2	2
Nucléaire	0	0
Hydroélectr.	28	39
Bois	48	22
Cane-à-sucre	7	13
Autres	0	1

Tableau 5. Variation du taux d'augmentation (% par an)

	80/73	88/73
Pétrole	1,3	8,4
Gaz Naturel	9,3	11,3
Charbon	23,3	12,6
Nucléaire	0	0
Hydroélectr.	12,3	8,6
Bois	-0,8	0,1
Cane-à-sucre	11,8	9,8
Autres	19,7	12,5
Total	5,3	5,7

Les estimatives pour l'année 2000 sont présentés pour donner une orientation sur les expectatives de participation de chaque source d'énergie, dans le cadre des paramètres disponibles, dans le marché future:

Tableau 6. Année 2000 - participation de chaque source sur la demande globale

Source	%
Pétrole	25,5
Gaz Naturel	4,0
Charbon	5,0
Hydroélectr.	41,5
Bois	7,5
Cane-à-sucre	5,0
Alcool	2,5
Autres	9,0

3 SOURCES PRINCIPALES D'ÉNERGIE

3.1 Électrique

Études techniques et économiques ont indiqué la nécessité de continuer l'évolution du système de production d'énergie sur une base hydrique, en maintenant comme réserve l'énergie d'origine thérmique pour garantir les régions isolées et un renforcement local, où nécessaire pour la garantie du système.
Actuellement, le potentiel théorique brésilien peut être estimé en 255.000 MW, duquel 22% se trouve déjà en conditions d'exploitation, 37% encore disponibles se trouvent en étude et 41% sont tellement estimés. Des 41% cités, 42.300 MW se

trouvent dans les régions sud-est, moyen-ouest et sud, 7.000 MW dans la région nord-est et 45.950 MW dans la région nord. Seul le bassin amazonique est responsable par 74.400 MW de la puissance encore disponible.

La distribution du potentiel théorique peut être divisée de la façon suivante:

Tableau 7. Potentiel hydrique brésilien

	Nord/ Moyen ouest	Nord-est	Sud-est/ Moyen ouest	Sud
(a)	4620	10515	28010	13130
(b)	45950	6995	17306	25017
(c)	77780	1925	14154	9598

où (a) sources hydriques en exploitation ou en construction;
(b) sources encore disponibles, mais déjà étudiées; et
(c) sources encore disponibles, seulement estimées.

3.2 Pétrole et gaz

En considérant le pétrole et le gaz naturel, les réserves brésiliènnes sont de l'ordre de 440.000.000 m³ de pétrole et 116.000.000.000 m³ de gaz naturel. Presque 69% du volume de pétrole mentionné se trouve dans la mer, dans des zones profondes.

4 OUVERTURE DU MARCHÉ DE GÉNÉRATION ET DISTRIBUTION D'ÉNERGIE - AUTO-GÉNÉRATION

La politique de développement qui a été mise en route par le nouveau président, à partir de 1990, liée au marché de fourniture d'énergie, a ouverte une porte pour des entreprises privées dans un domaine qu'appartenait à l'état dès les années 40.

Le programme gouvernementale d'appui à l'auto-génération d'énergie pour la consommation interne aux aménagements industriels ou agricoles et même la création d'un processus appelé "privatisation" des petites usines thérmiques et hydroélectriques pour la gestion privée, a permis aux principales groupes économiques du pays de créer "l'industrie de l'énergie", parce que tout l'excedente de production peut être aisement vendu à l'état étant donnée la crise que le secteur de production d'énergie a sofrit pendant ces derniers dix ans, avec un retard constant dans le programme national d'implantation des grandes usines.

Cette ouverture du marché de génération et distribution d'énergie a été aussi le résultat d'un appel des consommateurs à l'état, pour garantir que les contraintes liées aux possibilités de manque d'énergie puisse être équationnée de façon a éviter de mettre en risque de paralisation la production industrielle et l'agriculture (avec tous ses coûts associés).

Dans ce nouveau scénario, des entreprises privées ont cherché des alternatives pour réduire ses coûts, avec l'investissement dans l'implantation des unités propres pour l'auto-génération.

Comme exemple, on peut citer que dans les derniers 10 ans on a eu la construction de 110 petites centrales hydroélectriques (PCH) au Brésil, dont 70% par des entreprises privées, la plupart dans la fin de cette période.

5 RAPPORT ENTRE LES SOURCES DISPONIBLES

L'entreprise que dans le passé recent était appuyée sur l'utilisation intensive des moteurs diesel, pour la production d'énergie, pour sa consommation interne, obligée à accepter cette condition en fonction du manque d'un réseau de distribution qui n'était pas d'intérêt de la concessionaire locale, et qui n'a pas pu exploiter un ressource disponible à côté de son parc industriel ou aire agricole parce que l'exploitation du même appartenait seulement à l'état, a réevalué les coûts d'implantation des unités propres, en envisagent de travailler avec des sources plus fiables et avec des prix competitifs, principalement pour les producteurs électro-intensifs.

Sans rentrer dans une discussion détaillée sur chaque source d'énergie liée à chaque environnement locale des sites, et en considérant un cas réel où les seules alternatives locales étaient les sources thermiques (pétrole) et hydriques, on a développé des analyses, entre les deux sources pour obtenir un rapport entre les mêmes.

Un groupe privée dedie à l'agriculture intensive avec irrigation, lié aussi à l'industrialisation des produits agricoles et à l'exploitation des ressources minéraux dans un état de la région du Brésil Centrale, a se proposé à étudier des alternatives pour l'alimentation de ses bésoins d'énergie à partir de l'utilisation d'un ressource hydrique disponible auprès de l'aire où se trouvent ces centres consommateurs. Ce groupe, qui pendant plusieurs années a utilisé du diesel, avec un haut coût d'opération et d'entretien des moteurs (parce que le pays encore import une grande partie du pétrole consommé) a pris cette décision basée aussi sur les expectatives de croissance de sa production et sur l'impossibilité d'obtention d'un supplément d'énergie à partir du réseau de l'état (à une distance supérieure à une centaine de kilomètres). Ce groupe a pris en compte aussi l'augmentation des prix de l'énergie dans le pays, lesquels ont été maintenus stables plusieurs années avec subside du gouvernement.

Les études de faisabilité développées à l'époque, en considérant des estimatives économiques optimistes et non optimistes, ont indiquée que l'exploitation du ressource hydrique existent dans la région d'intérêt permet d'y installer 2 (deux) PCH, qui lors de l'implantation du système pourront offrir, en plus du support pour les activités du groupe, une disponibilité d'énergie supérieur à 20 MW. Cet excédent d'énergie pourra être absorbé par le système aujourd'hui exploité par l'état, avec une distribution locale, ce qui rendre le projet encore plus intéressant.

Dans le cas de l'énergie thermique utilisée, basée sur le diesel, la composition du coût pour la comparaison entre les alternatives a considéré, entre autres, le prix d'achat des nouveaux moteurs, d'installation des unités, d'achat mensuel de combustible, de transport mensuel du même, d'emmagasinage, d'entretien des équipements, de substituion des piéces détachées, de contrôle de la pollution de l'environnement, la vente de l'excédent (dans les heures creuses) au réseau de l'état, etc.

A la fin, même dans un bilan optimiste, le rapport énergie thermique-énergie hydroélectrique a maintenu un indice supérieur à l'unité, c'est-à-dire dans le cas étudié l'énergie d'origine hydrique a été toujours moins chère que l'énergie thermique.

6 CONCLUSIONS

On peut obtenir directement des données indiquées dans les observations présentées que le Brésil se trouve encore dans une phase de développement intense, même avec les conditions d'inflaction élevée par lesquelles le pays a traversé.

Il est possible de voir aussi qu'il existe un effort très large du système il-même (soit les entreprises privées et le gouvernement) de dépasser les problèmes qui devront arriver avec la reprise de l'économie aux niveaux antérieurs, fonction de l'arrêt qui a été mis dans les grands ouvrages de génération hydroélectrique, lequel sera responsable pour un probable arrêt économique, et les risques associés.

Comme cité ci-dessus, il existe un souci général de surpasser cette phase avec une mise en service de plusieurs petites centrales hydroélectriques, ou thermo-électriques (où les conditions soient plus favorables), de façon à éviter des gros investissements, et même de favoriser un partenariat entreprise privée-état.

Dans le cas particulier de l'entreprise qu'a demandé les études, en fonction de la taille de l'organisation et des conditions exceptionnelles de développement de la région (comme indiqué dans les tableaux des indices de développement du pays), il fut possible aussi de obtenir des avantages réels dans une situation qu'au début été très défavorable (consommation direct du diesel, sans aucune autre alternative disponible, sauf la construction d'une ligne particulière, avec plus d'une centaine de kilomètres, pour obtenir énergie de la concessionaire plus proche).

L'analyse du bilan (ou rapport) énergie thermique-énergie hydroélectrique, dans le cas étudié, a montré que pour les conditions particulières et locales, même dans un écart de variation des hypothèses de calcul très large, toujours l'aménagement le plus intéressant été la source hydrique.

Dans un pays avec un vaste réseau hydrique encore non-exploité, il nous semble que dans l'analyse des sources plus économiques d'énergie électrique, au Brésil, on aura probablement des avantages pour la génération basée sur les eaux des fleuves.

Hydropower'92, Broch & Lysne (eds) © 1992 Balkema, Rotterdam. ISBN 90 5410 054 0

Québec's choice for long-term electricity supply: Hydropower

L. Dumouchel & M. Baril
Hydro-Québec, Que., Canada

ABSTRACT: This paper describes the various avenues being studied by Hydro-Québec to satisfy electricity demand beyond the year 2000. It notes their principal technical, environmental and economic characteristics and compares them on this basis. It also proposes certain types of installations suitable for each avenue. These are proposals that Hydro-Québec will be submitting in the spring of 1992 as part of a consultation process. According to Hydro-Québec's analysis, hydroelectricity is by far the most promising avenue in technical, economic and environmental terms. However, nuclear power plants and, to a lesser extent, wind power and thermal plants using biomass could make a contribution in the future. Consultations will enable the public to express opinions on the possible choices.

1. INTRODUCTION

Since the start of this century, hydroelectricity has been Québec's primary energy resource. Over the years it has become the driving force of the province's industrial development. In the 1960s, 70s and 80s, Hydro-Québec built large-scale hydroelectric projects that have provided impressive experience and know-how. In the process, the utility has earned an international reputation in the fields of engineering, construction, project management and applied electrical research.

Nevertheless, the construction of hydroelectric generating stations has not yet exhausted all of Québec's hydroelectric resources. The rivers of Québec offer a still very large theoretical potential in the order of 50,000 MW.

In 1992, Hydro-Québec will set out its major orientations for this decade and beyond the year 2000 in order to submit in 1993 its Development Plan to the government of Québec. A major segment of this plan will be devoted to the expansion plan for facilities to satisfy growth in needs for electricity.

Over the last few years, the energy context in Québec has changed significantly. The economic, technical and environmental aspects have evolved. The market for exporting electricity to the United States has changed, mainly due to the contribution of American non-utility generators of electricity and energy conservation programs. Moreover, non-utility generators in Québec can make a large contribution to the generation of electricity. The utility is also seeking to take greater account of popular social and environmental concerns. Finally, energy issues are increasingly a topic of public debate. This new context has led the government of Québec to ask the utility to hold public consultations on its long term energy choices in the preparation of its Development Plan.

These consultations will make it possible to reflect public opinion in a more structured way when drafting the expansion plan. Generating facilities will be a particularly important aspect of these discussions. In order to provide more background material for discussion of this topic, Hydro-Québec has prepared a review of current knowledge about various means of generation that might be considered after the year 2000.

2. SPECIAL CHARACTERISTICS OF SUPPLY AND DEMAND IN QUEBEC

2.1 Supply

Until now, Hydro-Québec has preferred hydroelectricity because of Québec's enormous hydroelectric potential and the economic advantages of this method of generation over others. Thus, the total installed capacity of its hydroelectric facilities was 23,140 MW in 1990. Moreover, Hydro-Québec benefits from most of the electricity generated at the Churchill Falls hydro station (5,260 MW), which it owns jointly with the Newfoundland electric utility.

The utility also depends on the electricity generated by a CANDU-type nuclear plant, which has an installed capacity of 670 MW, as well as the output of one oil-fired thermal generating station and two gas turbine generating stations, for a total installed capacity of 1,060 MW.

Expressed in percentages, 94% of Hydro-Québec's generating capacity is hydroelectric, while thermal and nuclear facilities account for 4 and 2% respectively.

2.2 Demand

In view of the large proportion of hydroelectric facilities, Hydro-Québec is heavily dependent on water inflows. These are highest in spring, when the snow melts, and substantial in fall, although to a lesser extent. As Figure 1 indicates, the needs for electricity do not correspond to the natural pattern of water inflows. Indeed, they are the reverse of each other. Actually, demand from November to April represents about 60% of annual consumption when natural inflows provide only 28% of the year's total.

In order to cope with this imbalance, Hydro-Québec has reservoirs where in spring and summer it can store the water that will be used in fall and winter to satisfy demand while inflows are reduced. In addition to these seasonal reservoirs, Hydro-Québec has multi-annual reservoirs, built to store surplus generated by heavy precipitation from one year to the next. This enables the utility to face periods of low inflows lasting several years.

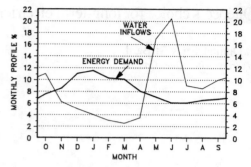

Figure 1: *Profile of water inflows and energy demand*

3. TECHNICAL AND ECONOMIC ASPECTS

This section describes the various avenues being studied by Hydro-Québec to satisfy demand beyond the year 2000. It notes their principal technical and economic characteristics. The economic method used to estimate the cost of energy is based on a present value technique. The cost of energy is expressed in terms of levelized unit energy cost (LUEC) in U.S. dollars. In this paper, 1 U.S. $ equals 1.19 CAN $. See section 6 for details on the method.

3.1 Hydroelectric Generating Stations

The undeveloped hydroelectric potential of Québec amounts to more than 50,000 MW, divided among large and small rivers. Large rivers are defined as those whose sites would be suitable for hydroelectric development of 100 MW and more, while small rivers have sites that would allow development of up to 100 MW.

3.1.1 Large Generating Stations

The potential of undeveloped large rivers represents an annual energy output estimated at 200 TWh. If this energy were to be generated at a capacity factor of 60%, the corresponding power would be 40,000 MW. Some 18,600 MW of this potential is economically viable. The following table shows the total capacity of projects under construction and for the future.

Table 1. Economically viable hydroelectric potential

	Peak capacity (MW)	Annual production (TWh)	LUEC U.S. $ ¢/kWh
Projects under construction	2,510	14.1	2.7
Future projects	16,125	84.7	3 to 3.9
TOTAL	18,635	98.8	

Hydro-Québec has already begun to develop the viable hydroelectric potential by building three generating stations. They will be commissioned between 1993 and 1995 and will have peak power of 2,510 MW. They are among the generating stations that make up Phase II of the James Bay development. The average LUEC of these 3 stations is estimated at 2.7¢/kWh. Future generating stations provide 16,125 MW. Their LUEC would range from 3.0 to 3.9¢/kWh.

Among future developments, two are on a very large scale. These are the Grande Baleine and Nottaway - Broadback - Rupert complexes. The total capacity of these two projects represents most of the economically viable potential and about 50% of installed capacity to date. The first project consists of three generating stations and has a capacity of about 3,200 MW. The second contains 11 generating stations and a capacity of about 8,440 MW. These two complexes can be built in stages or only partially so that the utility has greater flexibility in its decision-making. By proceeding in this way, Hydro-Québec can react even more efficiently to uncertainties in demand.

Some of the other projects that are part of the theoretical potential of 40,000 MW have been the subject of preliminary studies. These projects are generally located on less promising sites or are more remote from consumption centres. Preliminary studies have been used to select sites that deserve more detailed study for determination of their cost-effectiveness.

3.1.2 Small Generating Stations

The potential of small undeveloped rivers in Québec represents an installed capacity estimated at some 9,700 MW, divided among more than 400 rivers. Hydro-Québec has made evaluations or preliminary studies of several small river sites representing a technically viable potential of 7,000 MW.

Available estimates indicate that about 200 MW could be developed at a LUEC below 3.9¢/kWh. Hydro-Québec could also count on an additional 6,800 MW to the extent that the cost price is below the cost of nuclear or thermal plants, that is 5.0¢/kWh. There is a theoretical potential of 2,700 MW that has not been studied.

Under its purchasing policy, Hydro-Québec gives priority to non-utility generators (NUGs) who wish to develop hydroelectric generating stations of 25 MW or less. It ascribes them its avoided cost, that is 3.7¢/kWh. The avoided cost is based on the energy cost of its own future hydro-projects.

3.1.3 Purchases from Neighboring Systems

The long-term potential for purchasing electricity from neighboring systems consists in large part of the hydroelectric potential of Newfoundland. Between now and 2040, Hydro-Québec will continue to import about 31 TWh of energy annually (4,100 MW) under the contract which now links it with Newfoundland utility.

Hydro-Québec is now negotiating an additional agreement with Newfoundland to purchase a large share of the hydroelectric output that Newfoundland intends to commission at the start of the next decade. The project would involve the construction of three generating stations in Labrador with annual output of about 18 TWh and total installed capacity of about 3,200 MW. If an agreement is reached, this project would provide Hydro-Québec with a substantial margin for maneuver after the year 2000, especially if demand were to increase more quickly than anticipated or the utility wished to increase exports.

Long-term energy purchases from power systems other than Newfoundland's are not under consideration for strictly economic reasons.

3.2 Nuclear

In the Québec context, the most promising types of nuclear plants for satisfying needs beyond the year 2000 are the CANDU or the pressurized water reactor (PWR).

3.2.1 CANDU Reactors

CANDU reactors are designed by Atomic Energy of Canada Limited (AECL) and manufactured in Canada. Various models of CANDU-type plants are available. Their reactors have power that can vary from about 450 to 900 MW.

There are 23 CANDU reactors in operation throughout the world. Moreover, ten reactors are presently under construction. Most CANDU reactors in operation have achieved load factors that rank them among the best. They have an enviable level of technological maturity.

For Québec, CANDU reactors offer a number of advantages over other types of reactors. First is the fact that it is a Canadian design. Moreover, Canada has acquired extensive experience in building and operating this type of reactor. It therefore has the manpower and suppliers for developing a nuclear program. Finally, Canada has excess heavy water production at this time and sufficiently large uranium reserves for supplies not to be a problem.

The nuclear plant used as a reference for establishing the LUEC is a CANDU generating plant with four integrated 880-MW units. These have the lowest generating costs among CANDU plants, that is 5.0¢/kWh. One plant of this type is now under construction in Ontario, Canada.

3.2.2 Pressurized Water Reactors

The pressurized water reactor (PWR) is the most common nuclear plant used in the world. Some 53% of all nuclear installed capacity is of this type, compared to 22% for boiling water reactors (BWRs) and 5% for heavy water reactors (HWRs), which include the CANDU reactors. Moreover, PWRs represent 66% of the capacity of the reactors under construction, compared to 11% for BWRs and 16% for HWRs.

Hydro-Québec considers the most promising PWRs to be those made in France by Framatome under license from Westinghouse. Because they are highly standardized, these models can be commissioned in less time than the American models. Moreover, they can cost less. Models are available in 900, 1,300, and 1,400 MW. EDF has 34 reactors of the 900 MW series and 18 of the 1,300 MW series.

3.2.3 New Generations of Reactors

The nuclear industry is facing a number of problems that it must overcome if the public is to accept nuclear generating plants. It has therefore adopted a strategy consisting of improving the safety of nuclear installations to an even greater extent. This means working on the design of passive safety and intrinsic safety reactors.

Hydro-Québec is not relying on these new designs in the horizon considered, given the level of progress made by this technology internationally.

3.3 Base-load Thermal Generating Stations

Hydro-Québec could operate combined-cycle combustion turbines. It may also consider coal-fired stations, either conventional or fluidized-bed. It may also call upon thermal plant operated by non-utility generators.

3.3.1 Combined Cycle Combustion Turbine Plants

These plants use both combustion turbines and steam turbines. Both turbines generate electricity. By adding steam turbines it becomes possible to obtain a substantial improvement in the thermal yield of the plant.

Several fuels can be used, including natural gas, liquefied natural gas, no. 2 oil, liquefied petroleum gas and gasified coal. Natural gas is the cheapest of these fuels in Québec. A pipeline network allows access to North American reserves. However, Québec has no major reserves of this type of fuel.

Hydro-Québec does not operate any plants of this type at this time. However, a preliminary study has been conducted on a combined-cycle plant with an installed capacity of 966 MW, to evaluate its cost-effectiveness.

The LUEC cost price depends substantially on what happens to the price of fuel. In Québec, of all the fuels possible for this type of generating station, natural gas is the most economical.

If it is assumed that the price of natural gas increases according to a consumer price index increase of 3.7% per year, the average cost price of the energy would then be 4.3¢/kWh. If it is assumed that the price of this fuel increases according to its own inflation rate as estimated by Hydro-Québec, the LUEC would then be 5.8¢/kWh.

3.3.2 Conventional Coal Plants

Hydro-Québec has conducted a preliminary study of a 3,000 MW coal-fired plant containing four 750 MW generating units. This plant would be equipped with scrubbers and catalytic reducers. Such systems are required to meet the environmental standards for sulphur dioxide (SO_2) and nitrogen oxide (NO_x) emissions.

The LUEC of such a plant is extensively affected by forecasts for the cost of coal. If the cost of coal were to rise according to the consumer price index increase of 3.7 % per year, the LUEC would be 5.5¢/kWh. If the fuel were to move according to its own inflation rate as estimated by Hydro-Québec, the LUEC would then be 6.1¢/kWh.

Coal would be obtained from the world market, which would make Hydro-Québec dependent on outside sources.

3.3.3 Fluidized Bed Plants

Although there are several fluidized bed plants in operation in the world, they are mainly demonstration plants. Hydro-Québec believes that they have not reached sufficient maturity.

3.3.4 Non-Utility Thermal Generation

Non-utility thermal generation is one of the ways of filling generation needs. Hydro-Québec acts as buyer of electricity from non-utility generators (NUGs).

At this time, Hydro-Québec is negotiating long-term purchases from NUGs. It plans to buy some 750 MW by 1996, according to the average demand forecast. Some 600 MW of this total would come from thermal plants.

Decisions relative to the purchases of electricity from NUGs for the period beyond the year 2000 will depend on several factors, including the purchase price offered by Hydro-Québec, the price of fuels and their availability.

Hydro-Québec has developed a purchase policy for non-utility thermal generation based on an avoided cost of 3.7¢/kWh, which is the estimated cost of its own future hydro-projects.

Thermal plants use mainly fossil fuels, although the use of biomass is also possible. Biomass includes domestic waste, wood and peat. In Québec, the theoretical potential of domestic waste amounts to some 250 MW, while wood offers 150 MW and peat 1,000 MW. For non-utility generation of a thermal nature, Hydro-Québec prefers the use of biomass rather than fossil fuels.

3.4 Solar Energy

The study on solar energy has been limited to a review of photovoltaic (PV) systems. Tests conducted in California indicate that research on PV systems must be continued, in order to increase their reliability and reduce their production costs.

The cost of electricity generated by PV systems is closely related to insolation. At the latitude of Montreal, for example, it would take an area of 115 km^2 to generate energy equivalent to that of the Grande Baleine complex, which is 16.2 TWh. On the other hand, at the latitude of Sacramento, California, it would take only 65 km^2.

The experiment conducted in California by Pacific Gas and Electric is used as the basis for establishing the cost of production. The installation cost of their system is evaluated at \$3,210/kW and the cost of production at \$0.15/kWh. Since insolation in Québec is well below that of the test bench used by PG & E, production costs at Montreal's latitude would be about 30 to 50 % higher.

Although a reduction in PV systems costs is expected in the future due to technological progress, they are still too high to consider this method of generation in the year 2000 horizon.

3.5 Wind Energy

Wind energy is a technology that has reached a degree of maturity, in particular with respect to reliability and cost. This technology remains in development and is expected to continue to make advances.

Hydro-Québec has conducted a wind energy study program since 1975. As part of this program, with the National Research Council of Canada, it developed an experimental, vertical axis 4 MW wind generator. It is considering the use of commercially tested machines for its isolated systems and has installed a Bonus 65 kW wind generator.

There is substantial potential for wind energy in Québec, at least in theory. However, the geographic distribution of this potential would in most cases entail substantial costs for infrastructure and electricity transmission.

In the light of the California results, it is possible to envisage wind farms whose generation cost in the horizon being considered would reach 5.0¢/kWh. This cost of production is the gross cost at the site. To obtain equivalent service, this cost would have to be increased substantially, to include the cost of the transmission lines and the equipment theoretically required to compensate for the intermittent nature of this form of energy.

If included with essentially thermal generating facilities, the energy from wind generators can be directly translated into savings in fuel. This makes it very simple to determine whether wind generators are cost-effective. And it is in fact as part of essentially thermal systems that wind generators are now most successful.

On the other hand, in a mostly hydroelectric power system, like Hydro-Québec's, energy from wind generators can be expressed as the accumulation of water in reservoirs. At this time, such an operation would not be profitable for the utility. The cost of production of hydroelectricity in Québec is below that of wind energy. Moreover, in periods of high inflows, the output of wind generators could result in the intentional spillage of water. This also tends to reduce their economic benefits. That is why with respect to wind energy, Hydro-Québec is concentrating its development efforts on supplying isolated systems served by diesel generators.

3.6 Fuel Cells

The phosphoric acid fuel cell is the closest to being marketed. International Fuel Cell (IFC), for example, already offers a range of models from 200 kW to 11 MW. However, although this type of cell is at an advanced level of technology, only demonstration units have been produced to date.

The demonstration fuel cell units developed in the United States and Japan are used as reference for costs. Installation costs are still very high, in the order of $14,300/kW. When the technology is mature and in high demand, they could fall to $1,430/kW. Like thermal plants, the LUEC will depend substantially on the cost of fuel at the time such plants begin operation. For this type of generation, Hydro-Québec would be dependent on outside sources of supply.

Hydro-Québec does not think that this technology can be implemented extensively in the year 2000 horizon.

4. TECHNICAL AND ECONOMIC COMPARISON

As we have just seen, the projects belonging to the hydroelectric potential, nuclear plants and thermal plants are valid option in both technical and economic terms. All of these alternatives could therefore be called upon in a construction plan for after the year 2000. Table 2 shows the economic ranking of the reference generating facilities, based on the energy costs.

Table 2. Levelized Unit Energy Cost estimates

	Peak Capacity (MW)	LUEC (¢/kWh)
Large hydro	18,635	2,7 to 3.9
NUGs	N/A	3,7
Small hydro	200	≈ 3.9
Nuclear plant	4 X 880	5.0
CCCT	966	5.8
Coal plant	4 X 750	6.1

Nevertheless, hydroelectricity remains the most beneficial solution for Québec in the long term. In fact, the cost of energy for the hydroelectric projects in the 18,835 MW potential is from 22 to 46% less than for its closest competitor, the CANDU nuclear plants with four 880 MW reactors. These in turn have an energy cost significantly lower than that of other thermal power plants, i.e. coal-fired and combined cycle generating stations.

To a lesser extent, NUGs operating small hydroelectric generating stations (25 MW or less) or thermal generating plants add significant electricity generating capacity at a cost of 3.7¢/kWh, which is Hydro-Québec's avoided cost.

Wind power has reached an acceptable technological level. However, since the cost of energy for this option is higher than either for hydroelectric or nuclear, Hydro-Québec is not presently considering integrating such facilities into the main grid. Nevertheless, wind farms could be considered since they might be cost-effective for the isolated systems.

As far as solar energy and fuel cells are concerned, their costs are still too high to merit serious consideration. Nevertheless, expected technological progress could reduce their costs significantly.

5. ENVIRONMENTAL ISSUES

All methods for generating electricity, like other uses of natural resources, have an impact on the environment. It would not be possible as part of this report to present all the environmental data relative to each of these methods. However, to allow a comparison of various generating methods on an environmental basis, an exhaustive analysis was made, based on six general themes. These are: water, air and soil quality, natural ecosystems, global environmental conditions, land occupancy and use, perceptions and social changes and, finally, safety. These themes were selected to reflect the environmental issues most commonly discussed by both the general public and the scientific community.

The results of this analysis clearly indicated what consequences should be assumed when a given method of generation is selected. Table 3 lists the principal environmental issues for each of the methods that has shown sufficient technical maturity.

In the horizon of the year 2000, the major environmental issues can no longer be limited to merely local pollution problems. Instead they are defined in terms of sustainable development and planetary atmospheric conditions.

The Brundtland Commission on the environment and development recommended

Table 3. Principal Environmental Issues

Hydro	- water quality
	- uses of large areas
	- population disruptions
Nuclear	- waste heat
	- waste and spent-fuel management
	- risk of accident
Thermal	- waste heat
	- atmospheric pollution
	- risks associated with fuel spills or mining
Wind	- land use
	- noise
	- visual impact

that preference be given to renewable sources of energy, to avoid compromising the ability of future generations to respond to their own needs. Moreover, the destruction of the ozone layer and the accumulation of greenhouse gases in the upper atmosphere have become major concerns for the whole planet. These have resulted in international agreements intended to slow down and, wherever possible, to reverse such global pollution phenomena. With this two-fold concern in mind, the international scientific community agreees that the generation of electricity from fossil fuels must be kept to a minimum whenever another solution is available as an alternative.

In Québec, hydroelectricity and nuclear energy are the only energy options that respond adequately to these two concerns. As technical progress is made in the years to come, generating methods that are now considered marginal, like wind and solar power, will gradually be adopted. However, thermal production from biomass offers advantages when it is based on making use of residue that would otherwise be either lost or dangerous for the environment.

6. METHOD OF CALCULATION FOR ENERGY

The generation projects that could be considered to satisfy growth in demand beyond the year 2000 have different technical and economic characteristics. The size, construction cost, operating cost and fuel cost, when fuel is required, differ greatly from project to project.

As a result, to compare them, it is necessary to adopt a common base and ensure that the projects render an equivalent service adapted to the characteristics of demand in Québec. For this purpose, the main elements considered are: power and energy supplied at consumption centres; expenses related to studies, construction, maintenance and operation of the generation and transmission facilities; costs of fuel for thermal and nuclear generating stations; the useful life of the equipment, overhead and taxes.

All expenditure to be considered for establishing equivalent service is taken into account using a present value technique. In order to rank the various hydroelectric, nuclear and thermal projects, the levelized unit energy cost (LUEC) was used as the economic criterion. This LUEC is obtained by dividing the present value of all expenditure of an option over its service life by the present value of energy. In the case of other methods of generation, the economic data comes from experiments or demonstrations conducted elsewhere in the world. As a result, they do not reflect the equivalent service rendered to the Hydro-Québec power system.

7. HYDROELECTRICITY: A PROMISING FUTURE

The technical, economic and environmental review of various generating methods leads to the conclusion that hydroelectricity, nuclear power, wind power and thermal generation from biomass are the most attractive options for Québec.

Given the large hydroelectric potential of Québec and the possibility of developments of all sizes at a very low cost, Hydro-Québec has an exceptional range of choices available for the future. This allows tremendous flexibility in its actions. Moreover, the possibility of building complexes in stages or only partially adds to the flexibility and makes it possible to adjust construction plans to the changes of context the future undoubtedly holds in store.

Hydroelectric generating stations offer other important advantages. Given their very low operating costs, they virtually avoid the effects of inflation. Moreover, the useful life of hydroelectric structures is almost twice as long as those in thermal or nuclear installations.

Finally, they offer security in face of uncertainties about fuel supplies.

In Québec, hydroelectricity serves as one of the driving forces of the economy. Much has been done in the past to develop hydroelectric resources. The potential that remains to be developed is nevertheless substantial and offers a promising future.

REFERENCE

Dumouchel, L., Baltar, J., Coupal, M., Lebeuf, M., Primeau, R., Vaccaro, M., Dubreuil, L. 1992. *Moyens de production*. Hydro-Québec, Montréal.

Hydropower'92, Broch & Lysne (eds) © 1992 Balkema, Rotterdam. ISBN 90 5410 054 0

The Brazilian electric system and the insertion of solar power plants

L.C.Giarolla de Morais & Arthur J.F.Braz
CEMIG, Companhia Energetica de Minas Gerais, Brazil

ABSTRACT: Solar energy is nowadays a very important source of renewable and clean energy for the tropical countries. This work reports the experience of CEMIG in the feasibility studies of a solar power plant for future expansion of its generating system. Emphasis is given to the modelling approach and economic aspects.

1 THE BRAZILIAN GENERATING SYSTEM

One of the most important characteristics of the Brazilian generating system is the predominance of hydroelectric power plants, which correspond to more than 90% of the installed capacity. Besides this, the hydroelectric potential is one of the largest of the world, equivalent to 255 GW (corresponding to a yearly production of 1,200 TWh or about 7 million oil barrels/day) and only 30% is in operation. This way, the development of hydroelectric generation was the natural strategy to meet the demand growth.

The Brazilian system, also presents different regional characteristics due mainly to the large extension of its territory (8,5 million km²). There are two interconnected subsystems the North/Northeastern and the South/Southeastern being the latter the most developed one, representing more than 70% of the country's installed capacity. The South/Southeastern system is composed by many large reservoirs and plants, arranged in a complex topology in several river basins. The bulk power is produced by several utilities, which are multi-owned and electrically interconnected. CEMIG (Companhia Energetica de Minas Gerais), the utility of minas Gerais state, was created in 1952 and has presently an installed capacity of about 4,400 MW and 18,000 employees. CEMIG's generating system is mostly hydroelectric, having only one significative thermal power plant of 125 MW of installed capacity.

The Brazilian thermal system is composed by coal, oil and one nuclear power plant and represents only 10% of total capacity. Although this percentage is not so representative, the thermal units play a very important role in the system, because they act like a reservoir, leading to a lower operating cost and a greater offer of guaranteed energy.

During the low inflows (dry spells or critical periods), when the available stored water in the reservoirs is used, the thermal units operate at their maximum to meet the demand. On the contrary, when the natural inflows are higher, the hydraulic system is capable of meeting the demand, with the filling of the reservoirs and reduction of fuel consumption in the thermal plants. Therefore, a thermal plant operating in a predominantly hydro system consumes fuel only in a certain period of time (the dry spells), although it can guarantee an energy superior to its fuel consumption.

Thermal generation would be necessary, theoretically, only in a dry spell. But as the availability of hydroelectric energy is limited (stored water) and depends on the variability of future inflows, the operation of a hydrothermal system is a difficult task and corresponds to use the "free" stored water in the reservoirs or burning fuel in the thermal units. Consequently, there is a link between a decision in a given state (a month, for instance) and the future consequences of such a decision, as shown in Figure 1. In the particular case of the Brazilian system, since it is interconnected and the thermal units are owned by just a few of the utilities, there is still an institutional problem. This problem was solved by the creation of GCOI (Coordinating Group for the Interconnected Operation) in 1973 and thermal generation costs are shared by all utilities proportionally to their loads, without taking into account the ownership of the plants.

The technical solution to the problem of determining the most economical proportion of hydro and thermal generation is based on stochastic dynamic programming and will be

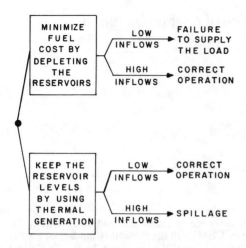

Fig. 1 Operating decision problem

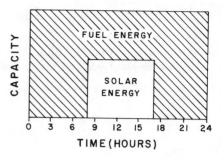

Fig. 2 Base mode operation for a typical day

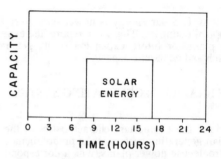

Fig. 3 Exclusively solar operation mode

discussed later with the evaluation of the energy credit of a solar power plant.

This work describes the attractiveness of solar power plants to the Brazilian system, the methods to evaluate the guaranteed energy of a typical plant and some economic aspects to compare it with other projects.

2 ADEQUACY OF A SOLAR POWER PLANT TO BRAZILIAN SYSTEM

CEMIG started analyzing the possibility of inserting a solar power plant in its system based on the specific characteristics of this kind of power plant compared with other conventional thermal power plants. The main difference between this technological conception and the conventional ones is that part of the energy output comes from a free source, the solar field, instead of burning fuel all the time.

Although CEMIG's investigations were done for a specific solar technology, this work presents a methodology that could be applied to the Brazilian system independently of the specific features of the power plant. Therefore, for different technologies, different energy conversion processes and different design approaches, electric output would be different.

Considering the operation policy for hydro and thermal power plants, some operating modes were proposed in order to make the solar power plant (SPP) suitable to the Brazilian system. These modes are the following:

1. Base mode operation: In this mode the power plant is supposed to operate in maximum load 24 hours a day. The energy produced by the solar field is complemented by a back-up system

burning fuel (oil). Figure 2 shows, for a typical day of insolation, the capacity x time curve.

2. Exclusively solar operation mode: In this mode the energy production comes entirely from the sun, i.e., the back-up system is never used, as shown in Figure 3.

Intermediate operating modes were analyzed but, knowing that these modes might permit the burn of fuel in a non-critical period, they were discarded because this is not justified when water is stored in the system reservoirs.

Considering the two proposed modes and taking into account the system's operating rules, the chosen mode for the feasibility studies was a combination of base mode and exclusively solar mode. Hence, in critical periods, SPP is called to operate in base mode. On the other hand, if there is enough stored water in the reservoirs, the adequate operating mode is exclusively solar.

3 ENERGETIC AND ECONOMIC ANALYSIS OF THE INSERTION

3.1 Solar-electric energy conversion

Survey studies for the Minas Gerais state showed the places with better insolation conditions and other important characteristics like water availability, soil conditions, topography etc. Among 20 analyzed places, 4 were studied in detail (Flachglas 1989). Finally the place chosen was Januaria (approximately 15 km from the Januaria township and 582 km from Belo Horizonte, the state capital) for getting the best score in all considered aspects.

The main problem found in estimating the solar energy production was the lack of adequate radiation data. CEMIG has been measuring solar radiation since 1984, with many installed measuring equipments. The problem was that those equipments measured only global radiation, whereas direct solar radiation was necessary for the calculations. CEMIG decided to investigate the methodologies to derive direct radiation from global radiation data. This work and its results are shown in CETEC/CEMIG (1991).

After obtaining direct radiation, an hourly series was elaborated for a whole year combining data in order to build what was called a typical year. Other relevant meteorological data like wind velocity, sun angles, precipitation etc were put together to calculate the energy output from the solar field. The calculations resulted in an annual energy production of about 123,516 MWh.

3.2 Evaluation of the energy credit

As stated before, the operation of hydrothermal generating systems is a difficult task because of the particular characteristics of these systems:

. There is a coupling in time (a given operating decision has consequences in the future);

. The problem is essentially stochastic, due to the nature of the inflows;

. The cost of generation in a hydro plant is evaluated by the amount of fuel cost savings (indirect cost of operation);

. There is a tradeoff between supply reliability and operating costs.

The technical solution to the problem corresponded to the development of computational models to determine the most economical proportion of hydro and thermal generation, on a monthly basis (strategic model) and a simulation model to evaluate the energy production of each plant.

The strategic model was developed in 1977 by a joint effort of ELETROBRAS (the Brazilian federal power agency) and CEPEL (the federal electric energy research institute). This model is based on stochastic dynamic programming and gives the optimal amount of thermal generation as a function of the reservoirs' levels. Because of the well known "curse of dimensionality" in dynamic programming, all system reservoirs were aggregated into one equivalent reservoir. To find the optimal thermal generation for each state (month) the model uses two state variables: 1) the stored energy in the system equivalent reservoir (water levels); and, 2) the conditioned hydrologic trend (represented by a stochastic streamflow model).

The output of this model is given through decision tables with the optimal thermal generation for each system state.

Given the thermal decisions for every month, it is necessary to evaluate the energy produced by each plant. This is done through a simulation model ("SIMULADIN") which was developed by CEPEL and CEMIG experts. The model has an individualized representation of each plant and is fed by a series of synthetic streamflows to represent the stochastic process.

The evaluation of the energy credit is done according to the Brazilian energy supply criterion, 5% annual risk of energy shortage. To obtain the system's guaranteed energy, simulations are done for a given target load. The number of years with any shortage divided by the total simulated years gives the annual risk of shortage for that load. The process is iterative and finishes when the risk is equal to 5%. The final load will be the system's guaranteed energy.

The guaranteed energy production of each plant is obtained through the identification of the several droughts in the simulation to evaluate the effective contribution of each plant in many periods (sceneries). These dry spells correspond to the period of time when all the reservoirs of the system are depleted to meet the load and the thermal units operate at their maximum. Figure 4 shows a typical dry spell.

Note that, for both the strategic model and the simulation model, the energy production of the

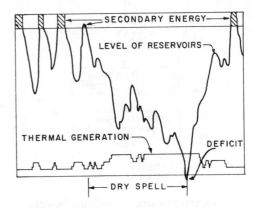

Fig. 4 A typical dry spell

exclusively solar operation mode was represented as a continuous mean production which came from solar to electric conversion calculations, and the oil back-up system was dispatched according to the system needs. This was done to simplify the modelling process.

As the result of the simulations, the energy credited to the back-up system was 333,756 MWh/year.

Figure 5 shows the entire process for determining the guaranteed energy.

3.3 Economic analysis

The economic analysis was performed considering two main indicators of projects attractiveness: the internal rate of return (IRR) and the energy cost (US$/MWh).

3.3.1 Internal rate of return

The annual cash flow was built taking into account the following:

. project investment cost;
. operations and maintenance costs;
. revenue considering estimated tariffs according to official forecasts;
. energy production calculated as described in the previous section.

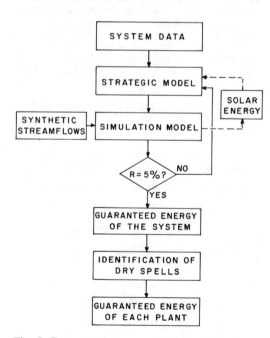

Fig. 5 Guaranteed energy calculation flowchart

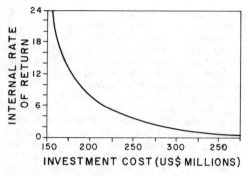

Fig. 6 IRR x Investment costs

The internal rate of return was calculated for many different investment costs in an adequate range for IRR's. Figure 6 shows the curve of IRR as a function of the investment costs.

The investment costs for some noticeable points are shown in Table 1.

Table 1 Internal rate of return

IRR (%)	Investment cost (US$ million)
6	225
10	187
12	178

3.3.2 Cost of energy

The Brazilian electric sector uses the cost of energy as a comparison parameter between alternative generating sources. The lower the cost of energy, the greater the profit possibilities. This index is calculated through the following expression:

$$CE = \frac{INV.CRF + OMC + AFE}{GEC} \qquad (1)$$

where:

CE = Cost of energy in US$/MWh;
INV = Investment cost of project in US$, with interest during construction included;
CRF = Capital recovery factor considering the project's life cycle of 25 years and a discount rate of 10% a year;

OMC = Yearly operations and maintenance costs, in US$;
AFE = Average yearly fuel expenditure in US$;
GEC = Guaranteed energy credit in MWh.

Considering the investment costs obtained in Table 1 and data contained in Table 2, the cost of energy was calculated for the same IRR levels. The results are shown in Table 3.

Table 2 Data for energy cost calculation

Item	Unit	Value
CRF	-	0.110168
OMC	US$ million	4.8
AFE	US$ million	3.0
GEC	MWh	457,272

Table 3 Cost of energy for some investment costs

Investment cost (US$ million)	Cost of energy (US$/MWh)
225	71
187	62
178	60

The cost of energy should be compared with other indexes as for example the marginal cost for generation expansion.

4 CONCLUSION

CEMIG's decision criterion for the specific studies of the proposed solar power plant at that time was the internal rate of return. The Company established the range of 10 to 12% of IRR for declaring the interest in the project.

This paper presented some characteristics of solar power plants that make them attractive for the Brazilian generating system.

The modelling approach for the evaluation of the guaranteed energy of the plant was presented as well as the guidelines for the economic evaluation of a solar power plant's competitiveness.

Although the solar power plant studied in detail was shown not competitive, the Company and its technicians believe in the importance of solar energy and expect for the near future the appearing of competitive solar power plants capable of being included in the Brazilian generating system.

REFERENCES

Braz, A.J.F. 1988. Hydroelectric power plants dimensioning with probabilistic energy supply criteria, (portuguese). MSc thesis, UFMG. Belo Horizonte, Brazil.

CEMIG - 16.101-GM/GR1 -001/90. 1990. Januaria solar thermal power plant pre-feasibility studies, (portuguese). Belo Horizonte, Brazil.

CEMIG - 12.101-GM/GR1-001/91. 1991. Januaria solar thermal power plant - feasibility studies, (portuguese). Belo Horizonte, Brazil.

CEPEL. 1988. SIMULADIN Program system manual, (portuguese). Rio de Janeiro, Brazil.

CEPEL/ELETROBRAS. 1977. Equivalent reservoir system model manual, (portuguese). Rio de Janeiro, Brazil.

CETEC/CEMIG. 1991. Survey on direct solar radiation in Januaria-MG Region, (portuguese). Belo Horizonte, Brazil.

Flachglas Solartechnik/CEMIG. 1989. Site survey report for large scale solar power plants in Minas Gerais. Cologne, Germany.

Flachglas Solartechnik. 1990. Solar electric generating system for Brazil - A feasibility Study. Cologne, Germany.

Pereira, M.V.F. 1984. Hydroelectric system planning. IAEA Technical report series 241:303-336. Vienna, Austria.

Pereira, M.V.F. 1984. Stochastic streamflow models for hydroelectric systems. Water resources research Vol 20,3:379-390.

Terry, L.A. et alii. 1986. Coordinating the energy generation of the Brazilian national hydrothermal electrical generating system. Interfaces 16:36-38.

Ventura Filho, A. et alii. 1991. Primary sources competitiveness for electric power production, (portuguese). SPSE/BRACIER: 1-46. Rio de Janeiro, Brazil.

Hydropower'92, Broch & Lysne (eds) © 1992 Balkema, Rotterdam. ISBN 90 5410 054 0

Hydropower generation development in Tanzania

S.J.Kimaryo
Tanzania Electric Supply Company Limited, Tanzania

ABSTRACT:

Tanesco (Tanzania Electricity Supply Company) is a State owned power utility entrusted with the tusk of generating, distributing and selling electricity in Mainland Tanzania. The utility dates as far back as 1908 when it was initially owned by the Germans to supply power to the railway workshops and part of Dar es Salaam town. In 1931 two separate power utilities were established. One known as Tanganyika Electricity Supply Company (TANESCO) and the second one Dar es Salaam Electricity Supply Company (DARESCO). The former was established around Pangani Falls near Tanga and was mainly supplying electricity to the Sisal estates; the later was initially supplying power to the towns along the Central railway line namely Dar es Salaam, Dodoma, Tabora and Kigoma.In 1964, three years after independence the government purchased all the shares of TANESCO. The two companies were then merged to form the state owned power utility known as Tanganyika Electricity Supply Company (TANESCO). In 1968, after the formation of the Union of Tanganyika and Zanzibar to form Tanzania the name of the Company was changed into Tanzania Electricity Supply Company (TANESCO). Since then the major task of the company has been to ensure that the country is supplied with cheap, abundant and reliable electricity. This has, to a large extent, been achieved by developing a number of hydro potentials and construction of transmission lines to get the power to the load centres.

1. DISCUSSION

The hydro power potential of the country is estimated at 4,000 MW. To date only about 350 MW has been developed to meet over 80% of the power demand of the country. There are four hydro plants namely Kidatu, Mtera, Hale, Pangani falls and Nyumba ya Mungu which form the National Grid Network (Ref. table 1). The stations are interconnected and connected to the load centres by transmission lines ranging from 66kV up to 220kV. (Ref. Fig.1).

Table 1. Existing Hydro Power Plants in Tanzania

Plant	Capacity(MW)	Commissioned	Head(m)
1.Pangani Falls	17.5	1934	94.5
2.Hale	21.0	1964	63.4
3.Nyumba Ya Mungu	8.0	1969	29.3
4.Kidatu	200.0	1975	175
5.Mtera	80.0	1985	92
Total	326.5		

The oldest hydro power plant (Pangani Falls) dates as far back as 1934.

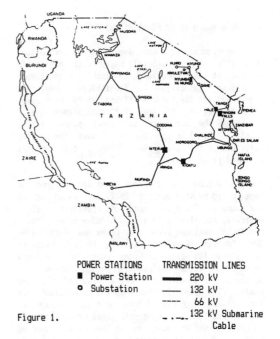

POWER STATIONS	TRANSMISSION LINES
■ Power Station	▬▬ 220 kV
○ Substation	──── 132 kV
	---- 66 kV
	_.._.._ 132 kV Submarine Cable

Figure 1.

NOTE: Information is at mid 1989

1.1 Pangani Falls Hydro Plant (17.5 MW)

As mentioned earlier on this is one of the oldest
hydro power plant with an installed capacity of 17.5
MW. The Plant was initially developed to assist in
the development of the sisal industry in Tanga.
However, it is now interconnected to the National
Grid Network and is currently capable of injecting a
maximum of 15MW into the National Grid System after
an intensive rehabilitation which was completed in
1991.

The plant has 2 x 5MW and 3 x 2.5MW horizontal
Francis Boving turbines priming English Electric
generators. The rehabilitation, was carried out by
Tanesco engineers and technicians after acquiring
skills from a Norwegian contractor M/S National
Industrie who trained them through rehabilitation of
another plant (Hale).

Through the rehabilitation the 2 x 5 MW and 2 x 2.5
MW units have been revived and are capable of
generating up to 15 MW. The plant with a head of
94.5 meters is a surface type of plant and the
water is lead from the storage intake to the
turbines through two steel pipes. Access to the
power plant is by a cable trolley rolling up and
down on rails. The plant is more or less a run-off
the river type with a small pond allowing for
limited daily regulation.

The plant is interconnected to the grid system
through two 33kV lines running from the plant to a
33/132kV s/s located about 15km up stream.

1.2 Hale Hydro Power Plant(21 MW)

With the increasing power demand in Dar es Salaam,
especially the industrial sector in the early
sixties, it was necessary to look for additional
generation capacity. A feasibility study, was
conducted and development of the 21 MW Hale power
plant proved to be the list cost alternative of
providing the necessary additional power to the
country. The load in Dar es Salaam (the Capital
City) was by then being met by running diesel plants
situated in the city.

The 21 MW under-ground hydro plant was designed and
constructed by M/S Balfour Betty who owned part of
the company. The plant came into operation in 1964.
Basically the plant with a head of 63.4 meters has
2 x 10.5MW vertical Francis turbines manufactured by
M/S English Electric driving English Electric
generators. The heavy machines parts were and are
moved in and out of the plant through an access shaft
by the use of an overhead crane situated over the
shaft. The water is lead into the power plant from
the intake which is 2 km upstream through an
underground tunnel. Along with this development and
in order to get the power to the load centre (Dar es
Salaam) a 283 km 132kV transmission was constructed.

In 1974, with the assistance of Swedish expertise,
Tanesco stopped the plant, primarily to investigate
into the cause of heavy vibrations of the units
which had been in existence since commissioning of
the plant. In the process of carrying out this
investigation it was necessary to strip down the
units. Upon examination of the runners it was found
out that the lower shroud had been machined
eccentric; presumably to establish static balance of
the runners(Ref.Fig. 2a & 2b) With this
eccentricity on the runner a pumping action was
being experienced when the runner was in motion
hence causing the vibrations. The lower shroud of
the runners were therefore machined back to
concentricity and static balance done by casting
appropriate lead weights at appropriate location on
top of the runners.

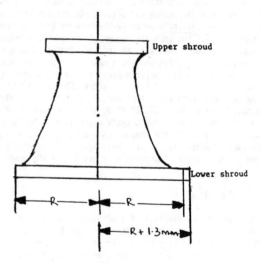

Fig 2a Runner before Machining

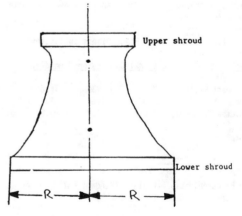

Fig 2b Runner after Machining

The machining exercise of the two runners was quite challenging and interesting to Tanesco for the very reason that there was no lathe machine in the country or neighbouring countries for that matter, to handle such large diameters. A methodology was therefore improvised whereby a tractor was used to achieve the task of machining the runners. One of the hind tyres of the tractor was removed and through an attachment jig the runner was fixed to the rear axle of the tractor. Another jig was fabricated to facilitate positioning of a lathe tool head next to the runner to be machined. As the tractor rotated the runner, the machinist worked the tool along the surface to be machined. After machining the shrouds of the runners to concentricity the runners were statically re-balanced by casting lead weights at appropriate positions on top of the runners.

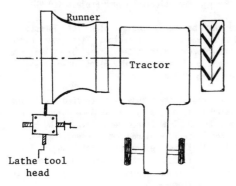

Figure 3. Machining runner using tractor

In 1988 through NORAD financing the plant was rehabilitated whereby the worn out parts were refabrished. The plant is now capable of generating up to its maximum installed capacity but due to lack of water storage (the storage is completely silted) the plant generates up to a maximum of only 15MW.

1.3 Nyumba ya Mungu hydro plant (8 MW)

In 1969 a large rock fill dam was constructed at Nyumba ya Mungu upstream of the Hale power station. The dam has a capacity of 470,000 acre ft.

The main purpose of the dam was to facilitate irrigation downstream of the river, fishery, water storage for annual regulation of hydro plants down stream namely Hale and Pangani. In order to maximize the use of the water released from the dam an 8MW Hydro plant was constructed at the foot of the dam. The plant has 2 x 4MW vertical Francis turbines. Through the same NORAD financing the two machines were rehabilitated in 1991 under the supervision of a Norwegian Company - M/S National Industrie. The governor regulating mechanism which was giving a lot of problems was replaced with a new one.

1.4 Kidatu Hydro Plant (200 MW)

With the increasing power demand in the country in the seventies and the rising cost of foisil fuels it was decided to construct a 200MW Hydro plant on Great Ruaha river which originates from the South West Mountain ranges of the country. The Plant was planned and executed in two phase:

1. The first phase consisted of construction of the 200 MW power house (civil works) and installation of 2 x 50MW hydro generating sets. This was started in 1970 and completed in 1975. The first phase included also the construction of a daily regulation earthfill dam with a capacity of 125 million cubic meters.

2. The second phase consisted of installation of the remaining 2 x 50 MW together with construction of a concrete dam upstream to facilitate for annual regulation of the Kidatu hydro plant. The dam has a maximum storage capacity of 3,200 million cubic meters. This phase was started in 1977 and completed in 1980

Along with the development of the Kidatu hydro plant, transmission lines were constructed to get the power from the plant to the load centres mainly Dar es Salaam (the capital), Moshi, Arusha and Tanga in the North East of the country and Zanzibar. Extensions have since then been made to several other towns as shown in Table 2 below through 220kV and 132kV transmission lines.

Table 2. Transmission lines of the Grid Network

LINE	VOLTS (KV)	LENGTH (km)	COMMISSION	FINANCIER
Kidatu - DSM	220	300	1975	CIDA
Kidatu-Iringa	220	298	1985	ADB,OPEC CDC
Iringa-Mbeya	220	216	1985	Govt.Italy
Iringa-Dodoma	220	235	1985	Govt.Italy
Dodoma-Mwanza	220	550		Govt.Yugo.
Mwanza-Musoma	132	205		Govt.Italy
Shy-Tabora	132	200		Govt.Italy
Hale-DSM.	132	283	1963	
Chal-Moro	132	82	1967	CIDA
Hale-Moshi	132	275	1975	CIDA
Moshi-Arusha	132	70	1983	Govt.France
DSM-Zanzibar	132		1980	Govt.Zanzibar

With the system expansion, it was felt necessary to introduce a grid control centre. In Phase II of Kidatu project a package of establishing a grid control centre was included and this was financed through a 20 million DM grant from the then government of West Germany. The grid control centre was installed in Dar es Salaam and commissioned in 1988. For all the two phases the World Bank played a leading roll in the financing of the projects. The other participants were SIDA, NORAD, KFW, KUWAIT and Italian Government.

The particulars of each financier's participation is as shown in Table 3. below:-

Table 3. Financing of the Great Ruaha Power Project.

Project	Financier	Amount(Million)
Kidatu Phase I	IBRD	USD 35.0 (softloan)
	SIDA	USD 17.5 (grant)
	CIDA	USD 15.5 (grant)
Kidatu Phase II	IDA	USD 30.0 (softloan)
	KFW	DM 60.0 (grant)
	SIDA	SKR 80.0 (grant)
Mtera Plant	IDA	SDR 32.5 (softloan)
	Govt.Italy	USD 29.5 (softloan)
	KFW	DM 58.5 (grant)
	Kuwait	KD 4.45 (softloan)
	NORAD	USD 13.5 (grant)
	SIDA	SKR 94.5 (grant)

The Kidatu Hydro Plant with 4 x 50MW generating Sets is the largest power plant which supplies power to most parts of the country. It has a head of 165 metres and located under ground. The water is lead from the dam to the power house through an 11km tunnel and four penstocks. The two vertical Francis turbines installed in 1975 are of Yugoslavia make coupled to Yugoslavia generators. The other two turbines installed in 1980 are Voith make coupled to Yugoslavia generators.

The power is transmitted to the switch yard through twelve 11.5/220kV single phase power transformers located in the power house and lead from the transformer bay to the pothead yard through oil field cables located in a vertical cable shaft. From the pot-head yard the power is then transmitted to the switchyard through four short transmission lines (1.0 km route length).

In 1986 the top and bottom covers for the first two turbines had to be refabrished due to heavy pitting.

In 1988 one of the first two generators had a serious flash-over whereby the whole excitor was damaged and two poles of the main rotor were also damaged. The main cause of the accident was due to accumulation of carbon powder on the generator windings and the rotor which weakened the insulation.

Through a SIDA grant the excitation system for all four units is now being replaced with new ones of the static type. Along with the replacement of the excitation system rehabilitation of all the turbines and generators is being carried out. The exercise is to be completed in the 1st quarter of 1993.

1.5 Mtera Hydro Plant (80 MW)

As mentioned earlier on Mtera dam was commissioned in 1980 to store water for the 200 MW Kidatu hydro plant. In order to maximize the release of the water from the dam an 80MW hydro plant was constructed and commissioned in 1988.

This is also an under-ground hydro power plant with a head of 92 metres. The water ways consist of a short head race tunnel and a long tail race tunnel (10.0 km). The two ABB (Italy) 40MW generators are primed by two Kvaerner Brug turbines. The power is lead from the power house up to the switchyard, situated on top through 11.5kV cables located in a vertical cable shaft. In the switchyard 2 x 45 MVA 11.5/220kV transformers are located to facilitate connection of the plant to the grid net work.

1.6 Pangani Redevelopment Hydro Project (60 MW)

While doing an energy master plan for Tanesco in 1985 it was found out that the next hydro development candidate after Mtera plant was the construction of a 60MW hydro plant down stream of the existing 17.5 MW plant.

A full feasibility study was carried out and completed in 1990 by a joint venture team of M/S IVO and NORPLAN Consultants and financed by Nordic Countries (Norway, Finland and Sweden). The study confirmed that the redevelopment of Pangani falls is the least cost option for providing the necessary additional power to the grid network in the mid nineties.

The feasibility study came up with a total cost estimates of the project as being to be USD 100 million, comprising of construction of a 66 MW underground power plant, 68km of 132kV transmission lines to connect the new plant to the grid network, training of Tanesco personnel and establishment of Pangani River Water Management set up. The financing is through grant from NORAD (42%), FINNIDA (33%) and SIDA (25%). Construction work started at beginning of this year 1991 and the plant is due for commissioning in April, 1995. The project components are:-

1.   Civil works which include construction of the power house and tunneling of the water ways, establishment of camps and construction of a 10km access road. This work has been awarded to M/S NOREMCO through competitive bidding at a contract price of USD 35 million.

2.   Electrical works which include generators, switchgears, control cubicles, and substations has been awarded to M/S ABB Generation of Sweden, EB Power Generation of Norway and ABB Stromberg Distribution Ltd of Finland through direct negotiation at a total contract price of USD 26 million.

3. The turbines, governors, valves, cooling water, dewatering and drainage contract has been awarded to M/S Kvaerner Eureka A/S of Norway through direct negotiation at a total contract price of USD 10 million.

4. The transmission lines will be constructed by Tanesco personnel under the supervision of the Consultant. Procurement of the material, equipment and tools estimated at USD 7.5 million will be done in Finland through competitive bidding under the supervision of the Consultant.

5. Project supervision which has been awarded to the joint venture of M/S IVO and NORPLAN Consultants is estimated to cost about USD 16 million.

Other interesting features of the project are the establishment of a water management unit to be located within the project area. This is one of the conditions agreed between the government of Tanzania and the financiers. Within the framework of the project a unit will be established and be furnished with the necessary resources, training of the personnel will also be done through the period of the project implementation.

Another feature encooperated in the project is the training of Tanesco technical personnel by the Consultant and the various contractors. Tanesco Engineers Technicians and Surveyors have been involved in the feasibility study and detail design stages by working with the consultant at the project site and thier home offices. Similarly engineers and technicians will be seconded to the Consultant to work with him during project implementation. Some of the Tanesco technical personnel will work with the various contractors during erection period. During the of the guarantee period the consultant will continue to train Tanesco operation and maintenance personnel.

Finally a community development component has been included in the project whereby health, education and recreation facilities will be introduced to cater for the population which will grow within the project area.

1.7 Kihansi hydro plant (165 MW)

The Tanesco Power Master Plan done by M/S Acres in 1985 and up-dated in 1989 indicated that the next hydro plant to be constructed after Pangani redevelopment to meet the power demand in the late ninenties was Kihansi hydro Plant.

With the assistance of a grant from the government of Japan, M/S EPDC of Japan was commissioned to do the feasibility study of the Kihansi project. The study was completed in 1990. The feasibility study recommended development of the Kihansi project in two phases.

1. Phase one (Lower Kihansi) to consist of construction of a 153 MW hydro plant with a head of about 900 meters. This was estimated to cost about USD 206 million.

2. Phase two (Upper Kihansi) to consist of construction of a dam upstream of lower Kihansi with a storage capacity of 75 million cubic metres and establishment of a 47MW hydro plant at the dam site. This was estimated to cost about USD 261million.

In 1990 the World Bank was approached by the government of Tanzania for assistance to finance implementation of the project.
The World Bank conducted a pre-appraisal of the project in the same year and indicated that it would co-finance the Kihansi project together with other associated essential components which would then form a project known as Power VI. The other components of Power VI are the rehabilitation of the distribution network, training of Tanesco personnel and institution strengthening.

M/S Norplan have now been appointed, through international bidding, to carry out the detail design and preparation of tender documents. The Consultant started work in December, 1991 and is to submit the final documents towards the end of 1992. However, the consultant is to submit to Tanesco an inception report by May 1992 to enable the government to call for a donors meeting in May, 1992.

The construction work for phase 1 is expected to start by 1st quarter of 1993 and the plant is planned to be commissioned in 1997.

1.8 Future hydro power development

Pre-feasibility studies have been done for other hydro sites with capacities ranging from 160MW up to 1,400 MW. In the Power VI project the next hydro power generation candidate will be elevated to full feasibility level. However with the discovery of abundant natural gas reserves in the South-East Coast of Tanzania the tempo for hydro power generation may be affected to a certain extent. Introduction of electricity generation using natural gas will now be considered along with the development of the next hydro power plants - this includes even the development of the second phase of Kihansi (Upper Kihansi). Though the natural gas alternative looks very and to some extent environmentally friendly lucrative there is however the element of time needed to put the necessary institution in place to manage the production and distribution of the gas. This is a very new technology to the country and could take a while to have it adopted.

CONCLUSION:-

Hydro power generation has played a key role in the development and industrialization of the country. This has been achieved through government securing funds from various donors and financing agencies in the form of grants and soft loans. With the foreign financing it has been possible to acquire the necessary equipment and technology. Abundancy of hydro potentials coupled with the high demand for electricity in the country, will place hydro power development as the most favorable alternative of meeting the country's electricity needs in the future. However, with the discovery of natural gas and the governments commitment and decision to develop it there might be a decline on the hydro power development in the future.

A number of problems have been encountered in developing the hydro power potentials in Tanzania. The primary ones being:-

1. Despite efforts put in both by the government and financing agencies to ensure that environmental aspects are fully looked at while developing the various sites, there has been some sites where the environment has still been affected. One of the problems Tanesco is facing in some of the hydro plants already in operation is the depletion of the surrounding forests by the population living in the plant area and villages that have developed in the vicinity of the plant.

As long as firewood remains the paramount source of energy to most of the people in the country this problem will continue to exist. To a certain extent the Tanesco has reduced the problem by providing free electricity and two plate electric cookers to its employees. However, the effect is minimum because the majority of the people cutting down the forest are not employees of the utility.

In considering our next hydro power plant however we have seriously considered the option of having the plant operated by remote control in order to cut down on the number of employees to a bare minimum. This will automatically discourage villages mushrooming around the plant.

2. The mode of financing has to a certain extent affected the operation and maintenance of the various plants.

Due to conditionalities of certain donor agencies, some of their money is tied up with supply of equipment from the donor countries.

This has made it very difficult, if not impossible, for Tanesco to standardize on the equipment acquired over the years, there by presenting a very difficult and challenging situation for the operation and maintenance personnel. Furthermore with this kind of situation whereby Tanesco has ended up with a vast number of various suppliers for the same equipment, it has been denied the advantage of interchangibility of spare parts. As long as part of the funding is obtained from grants, Tanesco is bound to continue experiencing this problem.

3. Another problem, which I presume is also experienced by other utilities in the tropics, is siltation of storage dams. We already have a dam located at Hale power station that has been made to disappear completely because of siltation. The dam was commissioned in 1964 and was already completely silted by nineteen eighties - a span of about 20 years. The rest of the dams which have been constructed in recent years are also threatened by similar problem. This is an area where technology could be deployed to look for effective methods of minimizing siltation of dams in the tropical areas.

REFERENCES

Feasibility Study Report on Kihansi hydro power development project report October 1990 - by EPDC Japan.

Pangani falls redevelopment feasibility study- October 1990 - by IVO/NORPLAN

Feasibility study report on Development of the Great Ruaha power project - December 1964 - SWECO

Power Sector Development Plant 1985 to 2010- December 1985 - by M/S ACRES

Hydropower'92, Broch & Lysne (eds) © 1992 Balkema, Rotterdam. ISBN 90 5410 054 0

Accelerated hydropower development – The best option for India

H.R.Sharma
Central Electricity Authority, Government of India, New Delhi, India

ABSTRACT: India's economic hydro power potential has been assessed as 84,000 MW at 60% load factor. So far only 14% of this potential has been developed. The present and future power scenario in India has been presented. Various alternatives to meet the ever growing demand of power have been discussed. Accelerated hydro power development is considered to be the best option to bridge the gap between the demand and supply of power.

0 INTRODUCTION

0.1 Energy is the vital key input for the agricultural, industrial and overall economic development of a country. The annual per capita consumption of electricity is regarded as the measure of the overall development and living standard of a nation. The annual per capita consumption in most of the developed west European countries and Japan is of the order of 10,000 to 12,000 kWh and in Norway it is more than 25,000 kWh. In India the average annual per capita consumption of electricity is only of the order of 250 kWh.

0.2 During the last four decades of planned development, after independence, there has been a spectacular growth in the installed generation capacity in India. Ther installed capacity which stood barely at 1362 MW in 1947, has since increased to 67,168 MW (as on 31.1.92), the share of hydro power being about 28%. Table-1 shows the planned growth of installed capacity in utilities since the First Five Year Plan (1951-56).

Table-1. Growth of installed capacity (MW) (Utilities)

	Thermal	Nuclear	Hydro	Total
Dec. 1947	854		508	1362
Dec. 1950	1153		559	1712
1st 5-year Plan 1951-56	1755		940	2695
2nd 5-year Plan 1956-61	2736		1917	4653
3rd 5-year Plan 1961-66	4903		4124	9027
Annual Plan 1967	5335		4757	10092
1968	6396		5487	11888
1969	7050		5907	12957
4th 5-year Plan 1969-74	9058	640	6965	16664
5th 5-year Plan 1974-79	15207		10833	26680
Annual Plan 1980	16424		11384	28448
6th 5-year Plan 1980-85	27030	1095	14460	42585
7th 5-year Plan 1985-90	44598	1565	18566	64720
As on 31.1.1992	46521	1800	18847 (28% of total)	67168

Despite this phenomenal growth in the installed capacity most regions in the country presently face severe shortages in the availability of power and grid collapses occur regularly. The Central Electricity Authority has estimated that there is presently a peaking shortage of 17.7% and an energy shortage of 8.5%. The energy shortage has an adverse impact on all sectors of the economy as a result of which national growth rate suffers. From power planning point of view the country has been divided into five regions based on geographic considerations –

Northern region, Western region, Southern region, Eastern region and North-Eastern region. The regionwise shortages are as follows:

(All figures in percentage)

	North	West	South	East	N.East	All India
Energy	8	0.2	12.5	21.6	8.3	8.5
Peak	19.6	7.6	21.8	32.4	17.2	17.7

These estimates of shortage, however, do not take into account various measures taken by the utilities to manage the peak demand by resorting to statutory power cuts and load rostering, and the prevailing low frequency' and voltage conditions in the grid. The actual shortage based on unrestricted demand would be much higher. The shortage of power is providing to be a significant detrimental factor to economic growth. Accordingly, to accelerate the pace of development in the country, it is necessary to accord a very high priority to power development, being the most important infrastructural sector. In this context hydro power development plays a very important role.

0.3 Hydro Power is a clean renewable source of energy. It has several intrinsic advantages. It does not contribute to either air or water pollution or green house effect. The cost of generation of hydro energy is relatively low being of the order of 50 paise/kWh to 100 paise/kWh against more than 150 paise/kWh for most of the thermal power stations. Similarly, the cost of the operation and maintenance of hydro power stations is also very low. Further, because of their ability for quick start/stop operations, hydro power stations are ideally suited for providing peaking power while the thermal power stations (including nuclear power plants) meet the base load requirements. Besides providing high value peak power, hydro power stations significantly improve the performance of thermal power plants and the system load carrying capacity resulting in the optimal utilisation of the power system.

1. POWER GENERATION RESOURCES

Almost 98% of the country's power is presently generated through coal based/natural gas based thermal power stations and hydro power plants and only about 2% by nuclear power plants.

Coal: The Geological Survey of India has estimated that India has a coal resource base of approximately 192 billion tonnes (as on 1.1.91). The details are as follows:

Coal Reserves of India (Million Tonnes)

Type of Coal	Proved	Indicated	Inferred	Total
Prime Coking	4621.71	1864.55	613.45	7099.71
Medium coking	9486.53	11154.80	1418.19	22059.52
Non-coking	44875.30	71002.52	47322.10	163199.92
Total	58983.54	84021.87	49353.74	192359.15

The regional distribution of these coal resources is as follows:

Region	Proved	Indicated	Inferred	Total
Northern	-	-	-	-
Western	12120.28	20550.22	11516.07	44186.57
Southern	5278.30	1650.50	3842.55	10771.35
Eastern	41327.93	61671.86	33536.66	136536.45
N.E. Region	257.03	149.29	458.46	864.78
Total	58983.54	84021.87	49353.74	192359.15

The non-coking variety of coal is the basic resource for power generation and this is mainly available in the Eastern and Western regions of the country.

Oil and Natural Gas: The proved and recoverable reserves of oil and natural gas (as on 1.1.91) are estimated to be 757 million tonnes of oil and 686 billion cubic metres of gas. The bulk of the gas reserve is free gas off the West coast. The regionwise distribution of crude oil and natural gas is as follows:

A. Crude Oil (million tonnes)

Region	On-shore	Off-shore	Total	% of total
Western	161.71	450.74	612.45	80.86
North Eastern	144.95	-	144.95	19.14
Total	306.66	450.74	757.40	100.00

674

B. Natural Gas (Billion cubic metres)

Region	On-shore	Off-shore	Total	% of total
Northern	1.04	-	1.04	0.15
Western	92.58	457.36	549.94	80.11
North-Eastern	135.47	-	135.47	19.74
Total	229.09	457.36	686.45	100.00

Nuclear Energy Sources

Nuclear fuel, namely the Uranium reserves in India are placed at about 70,000 Te of U 308 as per the present assessment. These reserves can support an installed capacity of 10,000 MWe of nuclear power reactors in the first stage comprising pressurised heavy water reactors (PHWRs). India is endowed with significant reserves of Thorium placed at about 3,60,000 Te. The ultimate aim is to convert Thorium into U-233 in the breeder reactors for using them to generate electricity through fast breeder or thermal reactors.

Non-conventional Sources of Energy

The non-conventional sources of energy are defused in nature and as such are attractive in rural and isolated areas where grid supply is a costly proposition. The available non-conventional energy sources are; biomass, solar energy, wind power, ocean energy such as tidal, wave, ocean thermal energy conversion (OTEC) etc. According to present estimates, it may be possible to generate 6000 MW of power through biomass and 2000 MW from sugar based agricultural waste. Wind power potential has been assessed as 20,000 MW. About 3 & 3 MW of wind power is already under operation and about 8 MW is under construction.

Various forms of ocean energy are receiving attention at present. However, with the present technology, only tides can be harnessed on commercial scale. The economic tidal power potential identified so far is of the order of 8000-9000 MW. Plans have already been prepared for a 900 MW tidal power plant in the Gulf of Kachchh on the Western coast in Gujarat.

The non-conventional energy, in India, is being produced basically through demonstration projects. These sources are still very costly and considerable research and development efforts are needed to make them compete with conventional sources. As such, non-conventional energy sources are not likely to play any significant role in the Indian power scene in the next ten to fifteen years except for tidal energy.

Hydro Power Potential

Nature has endowed India with high mountains and gigantic rivers. Himalayas, spreading across a length of about 3200kms and a width of 640 kms, with falls ranging from a couple of hundred metres to a couple of thousand metres, abound in hydro energy potential, the snow clad mountains giving rise to perennial rivers. India's economic hydro power potential comprising major and medium schemes has been assessed as 84,000 MW at 60% load factor which would correspond to more than 160,000 MW of installed capacity. The annual hydro energy potential has been assessed as 600 billion units. The Great Indus, Ganga and Brahmputra river systems with their tributaries emanating from Himalayas constitute more than 75% of the country's assessed economic hydro power potential. The basin-wise distribution of the hydro power potential is given below:

Basin	Assessed Economic Hydro Power potential at 60% Load Factor
1. Indus River Basin	19,988 MW
2. Brahmputra River Basin	34,920 "
3. Ganga River Basin	10,715 "
4. Central Indian Rivers	2,740 "
5. West flowing river system	6,149 "
6. East flowing river system	9,532 "
	84,044 MW

At present (as on 31.1.92) there are 279 hydro power plants in operation in India having an installed capacity of 18,845 MW which corresponds to about 14% of the total hydro power potential of the country. The total utilised hydro power potential would increase to about 21% on completion of the on-going schemes. Thus, bulk of the hydro power potential of the country still remains to be harnessed. The status of region-wise hydro potential development is given in Table - 2.

2. ENERGY DEMAND AND OPTIONS

2.1 The latest power demand projections based on the 14th Electric Power Survey

NAME OF THE STATE	Potential Assesed @ 60% L F (MW)	Pot.Developed @ 60% L F (MW)	% Devel-oped	Pot.Under Development @60% LF (MW)	% Under Develop-ment	% of total Pot. Develop-ment & under development	CEA cle Schemes Pot. @ LF (MW)	% of CEA cleared schemes	Total % of Potential developed+ under development + CEA Cleared
I NORTHERN REGION									
1. Jammu & Kashmir	7487.00	308.33	4.12	358.17	4.78	8.90	516.33	6.90	15.80
2. Himachal Pradesh	11647.00	1797.47	15.43	733.27	6.30	21.73	771.33	6.62	28.35
3. Haryana	64.00	51.67	80.73	5.00	7.81	88.54	6.67	10.42	98.96
4. Punjab	922.00	481.33	52.21	205.00	22.23	74.44	212.67	23.07	97.51
5. Rajasthan	291.00	188.67	64.83	12.00	4.12	68.96	0.00	0.00	68.96
6. Uttar Pradesh	9744.00	1027.17	10.54	991.83	10.18	20.72	481.33	4.94	25.66
7. Sub. Total (NR)	30155.00 35.88%	3854.63	12.78	2305.27	7.64	20.43	1988.33	6.59	27.02
II WESTERN REGION									
1. Gujrat	409.00	136.67	33.41	112.67	27.55	60.96	0.00	0.00	60.96
2. Maharshtra	2460.00	1081.00	43.94	224.17	9.11	53.06	0.00	0.00	53.06
3. Madhya Pradesh	2774.00	546.00	19.68	1014.83	36.58	56.27	238.05	8.58	64.85
4. Goa	36.00	0.00	0.00	0.00	0.00	0.00	0.00	0.00	0.00
5. Sub Total (WR)	5679.00 6.76%	1763.67	31.06	1351.67	23.80	54.86	238.05	4.19	59.05
III SOUTHERN REGION									
1. Andhra Pradesh	2909.00	1381.92	47.50	34.37	1.18	48.69	23.33	0.80	49.49
2. Karnataka	4347.00	1955.33	44.98	674.50	15.52	60.50	4.83	0.11	60.61
3. Kerala	2301.00	962.33	41.82	369.13	16.04	57.86	51.93	2.26	60.12
4. Tamil Nadu	1206.00	942.17	78.12	51.83	4.30	82.42	11.33	0.94	83.36
5. Sub Total (SR)	10763.00 12.81%	5241.75	48.70	1129.83	10.50	59.20	91.43	0.85	60.05
IV. EASTERN REGION									
1. Bihar	538.00	92.50	17.19	238.45	44.32	61.51	0.00	0.00	61.51
2. Orissa	1983.00	722.17	36.42	386.62	19.50	55.91	0.67	0.03	55.95
3. West Bengal	1786.00	21.67	1.21	65.50	3.67	4.88	111.50	6.24	11.12
4. Sikkim	1283.00	13.33	1.04	44.17	3.44	4.48	281.45	21.94	26.42
5. Sub Total	5590.00 6.65%	849.67	15.20	734.73	13.14	28.34	393.62	7.04	35.38
V. NORTH EASTERN REGION									
1. Assam	351.00	105.00	29.91	90.83	25.88	55.79	6.67	1.90	57.69
2. Meghalaya	1070.00	60.00	5.61	61.67	5.76	11.37	0.00	0.00	11.37
3. Tripura	9.00	8.50	94.44	0.00	0.00	94.44	0.00	0.00	94.44
4. Ar.Pradesh	26756.00	6.17	0.02	108.33	0.40	0.43	262.00	0.98	1.41
5. Manipur	1176.00	73.17	6.22	2.00	0.17	6.39	3.33	0.28	6.68
6. Mizoram	1455.00	0.00	0.00	1.00	0.07	0.07	71.67	4.93	4.99
7. Nagaland	1040.00	0.00	0.00	81.88	7.87	7.87	0.00	0.00	7.87
8. Sub Total (NER)	31857.00 37.91%	252.83	0.79	345.72	1.09	1.88	343.67	1.08	2.96
ALL INDIA	84044.00 100.00%	11962.55	14.23	5867.22	6.98	21.21	3055.10	3.64	24.85

completed by Central Electricity Authority in March, 1991 for the terminal years of the 8th, 9th and 10th five year plans, i.e. 1996-97, 2001-02 and 2006-07, place the country's system peak load demands as 73,656 MW, 104,541 MW and 144,295 MW respectively and the corresponding system energy requirements as 416, 595 and 824 billion units. Tentatively, a capacity addition of 36,646 MW comprising 9397 MW hydro, 26,074 MW thermal (including gas based projects) and 1175 MW nuclear is envisaged during the 8th five year plan period (1992-1997). Similarly, a capacity addition of 46,733 MW comprising 17,705 MW hydro, 24088 MW thermal and 4940 MW nuclear is proposed during the 9th five year plan period (1997-2002). However, depending upon the availability of funds, this programme would be suitably modified. Even with the above envisaged additions, the peaking shortage would be of the order of 15.3% and 11.9% respectively at the end of 1996-97 and 2001-02.

2.2 As discussed earlier in para 1, the non-conventional energy sources are not likely to play any significant role in meeting the power and energy requirements of the country in the next 10 to 15 years. In view of the fact that nuclear reserves in India are very large, nuclear energy will have to play an increasingly larger role. However, its contribution during the next 10-15 years may not be very significant. Accordingly, the country has the following main options for power development in the next 10-15 years:

- Coal based thermal
- Gas based thermal
- Hydro power

2.3 It would be seen from Table-1 on page-1 that the hydro power development remained a preferred option till the beginning of the Fourth Five Year Plan (1969-1974). The proportion of hydro capacity in the overall installed capacity was 46% at the end of 1969 and thereafter it started declining and is now only 28% against the desired optimum figure of 40% for the Indian power system. From the beginning of the Fourth Five Year Plan, interest in power development tended to tilt in favour of thermal generation owing to relatively the shorter gestation period. The pre-dominance of thermal generation which at present is of the order of 70% has posed several technical and operational problems. High share of thermal in the system has resulted in sub-optimal operation of thermal plants because of backing down of these stations during off-peaking hours. The thermal generation has many other problems. The coal reserves are not evenly distributed and the bulk of the coal fields in India are located in the eastern region. As a result coal based thermal power projects have either to be location specific, resulting in major transmission problems or the coal has to be moved to different load centred power stations primarily by rail putting heavy strain on the rail system in the country. Problems are being experienced primarily in the southern and western regions due to short-falls in coal production. The coal requirements for the year 1991-92 have been estimated to be 146 million tonnes. However, according to present indications only about 130 million tonnes of coal may be supplied for power generation. At times the thermal power stations are threatened for closure due to lack of coal supplies. The operational costs of thermal power stations are also consistently increasing due to increase in fuel cost.

2.4 The disposal of fly-ash is a serious problem for the thermal generating units. At present the thermal generating plants in India are producing more than 50 million tonnes of fly-ash annually. Besides requiring large areas of land for disposal, the fly-ash poses a severe threat to the environment.

2.5 Gas based thermal power stations like storage based hydro power plants are ideally suited for meeting the peak load requirements. These power stations have a short gestation period of about 2-3 years and use relatively a non-polluting fuel source. However, a constraint in the setting up of new gas generating capacity is the fact that great difficulty is being experienced in getting gas linkages for these power plants. In a few cases even after sanction of gas linkages, adequate quantity of gas is not being supplied resulting in sub-optimal utilisation of the power plants. The cost of the energy from gas based thermal power plants is also relatively very high as compared to hydro energy.

2.6 In view of the several disadvantages (higher cost of generation, sub-optimal use, pollution etc.) and difficulties (trans-port of coal etc.) associated with thermal power stations and the fact that about 80% of the hydro power potential still remains to be harnessed, the best option for the country to meet its ever-growing requirement of power and energy, would be to resort to accelerated hydro power development. The hydro power plants not only help in meeting the peaking load requirement of the system, but also improve the overall performance of the thermal power stations as well as of the power system as would be evident from the case study presented later.

2.7 In the thermal pre-dominant regions with deficit peaking power but surplus off-peak energy, development of pump storage schemes needs to be undertaken. Out of the total identified economic pumped storage potential of more than 98,000 MW, only 10 pumped storage schemes with an aggregate installation of 4844 MW have been completed or are under execution. The pump storage schemes have several advantages although their efficiency is of the order of 70%. They help to meet the peak demand, improve the performance of thermal and nuclear power plants, control net work frequency, etc. Therefore, emphasis also needs to be placed on pump storage development. Possibility of using sea as the lower reservoir also needs to be explored to economise the overall development.

2.8 Further, to improve the peaking capability of the system, installation of additional units at the existing reservoir schemes and thereby operating them at reduced load factor would also be desirable.

2.9 The uprating, renovation and modernisation of the existing hydro-power plants can also contribute significantly towards meeting the peaking capacity and energy demands of the system.

3. IMPACT OF HYDRO CAPACITY ADDITION IN A SYSTEM - A CASE STUDY

3.1 A case study has been carried out to examine the impact of addition of 1000 MW hydro capacity in an existing storage reservoir based hydro power station with an installed capacity of 560 MW in one of the regional systems of India having a total installed capacity of 28562 MW with hydro-thermal ratio as 16.7: 83.3. The average load factor of the existing power station is 42.5%. With the proposed addition of 1000 MW (without any additional energy), the load factor of the station would be 15.25%. The detailed integrated system operation studies have been carried out for peak days of January month (Load Factor = 83.8%), both with and without the proposed addition of 1000 MW installation. To meet the load demand the run-of-river type and irrigation controlled hydro power stations and thermal power stations (including nuclear power plants) have been considered to meet the base load and storage based hydro power plants to meet the peak demand (Fig.1). The possible saving of hydro energy on holidays has also been taken into account. The results of the study are summarised below:

	Case A: Without addition of 1000 MW Hydro Station	Case B: With addition of 1000 MW Hydro Station
Thermal installed capacity in the system (MW)	23790	23790
Hydro installed capacity (MW)	4772	5772
Total installed capacity (MW)	28562	29562
Demand met (MW)	15810 (55.35%)	16813 (56.87%)
a) Thermal component(MW)	11857	11995
b) Hydro component(MW)	3953	4818
Energy met (MW hrs)	317470	338143
a) Thermal component (MW hr)	272366 (P.L.F=47.7%)	284379 (P.L.F.=49.8%)
b) Hydro component (MW hr)	45604	45604 +8160*

*Hydro energy saved from holidays

Note: Hydro saving from holidays was not considered in Case A. Had it been considered thermal utilisation in case A would have been still lower.

The study indicates that with the addition of 1000 MW hydro power capacity which is only 3.5% of the total installed capacity of the system and improves the hydro-thermal mix ratio from 16.7 : 83.3 to 19.5 : 81.5 only, the load carrying capacity of the system increases by 1003 MW (865 MW hydro and 138 MW thermal, after accounting for the outages etc.) and the thermal utilisation increases by 12013 MWhrs resulting in an average increase of 2.1% in the plant load factor

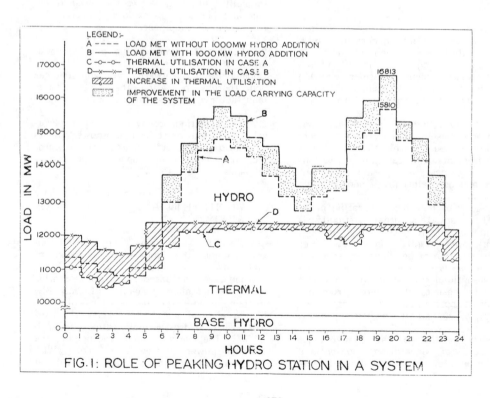

FIG. 1: ROLE OF PEAKING HYDRO STATION IN A SYSTEM

of thermal power plants. For Indian power system, a hydro thermal ratio of 40:60 has been found to be the optimum one against the present all India figure of 28:72. The case study clearly establishes the immediate need of accelerated hydro development to derive maximum advantage from the installed capacity in the power system besides several other intrinsic advantages of hydro power as stated earlier.

4. SOME MAJOR ISSUES

4.1 Notwithstanding the several inherent and intrinsic advantages over other conventional energy sources, the hydro power development tends to have a stunted growth, its share declining from 46% in 1969 to 28% at present. Some of the major issues responsible for the retarded growth of hydro power in the country are:

- inadequate funding
- delays in according environment and forest clearances
- land acquisition and law and order problems
- contractual complications and failures
- geological surprises in underground works

These are briefly discussed below:

4.2 Problems of Funding

Lack of adequate funding and proper cash flow has been one of the major constraints in hydro power development in the country. A number of hydro projects which should have been completed in the Seventh Five Year Plan (by March, 1990) are still languishing on account of fund and cash flow problems. Similarly a number of sanctioned hydro power projects are not able to take-off due to lack of funds. Externally aided hydro power projects are also experiencing shortage of matching domestic funds. Even investigations and preparation of feasibility reports of hydro power projects are suffering because of funds shortage.

4.3 Environmental Issues

Many hydroelectric projects have experienced long delays and repeated hold-ups on account of environment and forest clearances. in a few cases the hydro power projects, already in advanced stage of execution, have been suspended indefinitely. Major environmental issues relating to the hydro power projects are:

- Rehabilitation plan for project affected people
- Compensatory Afforestation Plan
- Catchment Area Treatment Plan
- Rehabilitation and relocation of archaeological, religious and historical importance

4.4 Inter-State Aspects

Inter-State aspects involving common river systems have also been one of the major factors inhibiting the hydro power development. A large number of hydro power projects are pending clearances and sanction on account of Inter-state disputes.

4.5 Local Issues

Problems of land acquisition for project works, law & order and labour problems etc. have also been responsible for delays in the execution of hydro power projects.

4.6 Geological Surprises

As stated earlier, about 3/4th of the hydro power potential in India lies in the Himalays. The latter being comparatively young, the geological surprises, especially in underground works, have also been one of the important factors causing time and cost over-runs in the execution of hydro power projects.

4.7 Contractual & Organizational

The execution of hydro power projects has also suffered due to contractual complications and due to lack of competent personnel and organisation and modern construction equipment and machinery.

4.8 Steps to accelerate Hydro Power Development

In order to accelerate the pace of hydro power development in the country, following steps are considered necessary

- All the sanctioned and on-going hydro power projects to be fully funded upto the completion stage
- Matching domestic funds to be made available on priority to all externally aided hydro power projects
- Investigations and feasibility studies for hydro power projects to be adequately funded to ensure sufficient number of hydro power projects in the shelf
- Private participation in power deve-

lopment to be encouraged

- Delays of about 2 to 3 years in creating infra-structural facilities to be avoided by taking up these facilities immediately after the projects is techno-economically cleared by the Central Electricity Authority
- All extension schemes involving addition of generating units in the existing power houses and schemes proposed at the toe of existing dams need to be exempted from environmental and forest clearance requirements
- Run-of-the river schemes which normally do not involve any significant submergence need to be cleared from environmental and forest angle without insisting on detailed rehabilitation and catchment area treatment plans
- The catchment area treatment (CAT) needs to be taken up on national level in an integrated manner. A 'national fund' needs to be created for this purpose. Cost of the CAT of areas affected by and in the vicinity of the project only needs to be loaded to the cost of the project
- To avoid delays in identification of land for compensatory afforestation, 'land bank' of non-forest land needs to be created
- Inter-state aspects need to be resolved expeditiously
- Local issues such as land acquisition, law and order problems, labour problems etc. need to be resolved on top priority by State authorities, and
- Latest techniques of design and construction, modern construction equipment and machinery and management system also need to be inducted to cut cost and time over-runs

5. CONCLUDING REMARKS

5.1 The country's economic hydro power potential has been assessed as 84,000 MW at 60% load factor. Of this, only 14% has been developed and another 7% is under development at present. As such, about 79% hydro power potential still remains to be harnessed. Of the various alternative sources of electricity generation in India, hydro power is found to be the cheapest one. Besides, it is the cleanest renewable source of energy. It does not only provide the much needed peak power but also improves the per-

formance of thermal power stations and of the entire power system. Accordingly, accelerated hydro power development is the answer to India's ever growing energy demand. Steps suggested in para 4.8 would go a long way in accelerating the hydro power development in the country.

Hydropower'92, Broch & Lysne (eds) © 1992 Balkema, Rotterdam. ISBN 90 5410 054 0

Research on the economic evaluation of hydro and thermal electric power and the optimal proportion between them in the power system of Guizhou province

Zhou Xiawen
Electric Power Bureau, Guiyang, Guizhou Province, People's Republic of China

ABSTRACT: According to abundant resources of hydropower and coal in Guizhou province, calculation and evaluation for developing hydro and thermal power in the near future is discussed in this paper. The conclusion is that all the economic evaluation indicators of hydropower are higher than those of thermal power; additionally, multipurpose utilization of hydropower is significant. Priority to develop water power is obvious.

By using different annual costs of the electric system between hydropower and thermal power, the annual cost curve gives the optimal proportion of hydropower and thermal power in the Guizhou electric system in the year 2000.

There are abundant resources of hydropower and coal in the Guizhou province, developing conditions are optimal. But, hou to evaluate economic character of hydropower and thermal electric power in developing Guizhou electric system in the near future and the year 2000, realize the optimal proportion between them, understand developing direction for the electric system, and get extreme economic effecte of the electric system ? They are very important subject to be investegaty on technology and economy.

RESOURCES OF HYDROPOWER AND COAL

In Guizhou province, there are theoretical hydropower potential of 18.7 GW, exploitable capacity is 16 GW, corresponding to 66500 GWh/year. These hydropower resources are concentrated on the rivers of the Wujiang , the Napanjiang and the Beipanjiang. For example , the Wujiang river crosses the central area of Guizhou province from west to northeast. Eleven hydropower cascades on the main stream of it had been planed. Among them nine are within Guizhou province, the total installed capacity is 8560 MW. These planed hydropower stations are near load center, distances to Guiyang city capital of Guizhou province are about 100—300 km. And project volume of these planed cascades is small, submerge for reservoir is small, too; because Wujiang river locates in the Yunnum—Guizhou plateau, all dam sites locate in gorges. Average cost for these projects is about RMB¥ 1420/kw. Therefore, developing costs for hydropower potential in Guizhou province are

economic. But the resources of hydropower are only developed 8 per cent in exploitable capacity of Guizhou province.

There are abundant coal resources in the west of Guizhou province. Total coal potentiol known are 49 Gt. Annual production is about 35 Mt. According to the electrical developing plan, there are good conditions for constructing large thermal electric plants in mine area. These planed thermal electric plants near load center, distances to Guiyang city are about 150—300 km.

Therefore, there are abundant resources of hydropower and coal in Guizhou province, and developmental condition is optimal. It is an energy base for the Southwest and the South of China.

ECONOMIC EVALUATION FOR THERMAL POWER

According to the electrical development planning of Guizhou province, by using 4×200 MW thermal electric plant as representative project for economic evaluation is suitable. In Guizhou province, transportation capacity of railroad is insufficient, production of coal is limeted by transportation capacity; therefore, cost of construction coal mine isn't considered on economic calculation for thermal power. Additionally, distances to transmission line for hydro and thermal electric power in the electric system is about equality; so, cost of transmission line of them is dosn't calculated. Only economic calculation and benefit evaluation for hydro and thermal electric projects are made in this paper.

According to price level in the year 1986, the

Table Ⅰ —The construction period , cost and unit production period for 4×200MW of thermal power project

Costruction period	year	1	2	3	4	5	6	7
Annual cost	RMB¥ ×10^6	46	73.6	92	230	202.4	230	46
Unit production	unit/year					1	1	1	1

Table Ⅱ —Total period, annual cost and unit production for the Goupitan(2000MW) hydropower project.

Total period	year	1	2	3	4	5	6	7	8	9	10	11
Annual cost	RMB¥ ×10^6	70	100	200	250	160	300	420	500	540	320	120
Unit production	unit/year									1	2	1

total cost for the 4×200 MW of thermal power plant is RMB¥ 920 million. The cost, construction period and unit production period for the 4×200 MW of thermal power project are shown in the Table Ⅰ.

Economic calculation and financial analysis for the thermal power project accoding to following parameters:
· Economic using period is 25 year;
· Depreciation rate per year is 4 per cent of total investment;
· Cost of operation and maintenance per year is 3.35 per cent of total investment;
· Expending coal is 380g/kwh;
· Cost of standard coal into the plant is RMB¥ 54/t;
· Average annual generation has 4400 GWh;
· Using electricity rate of the power plant is 8 per cent;
· Average price for the power plant sell electricity is RMB¥ 77.60/10^3 KWh; and
· The tax rate is 15 per cent.

According to economic calculation for investment, cost and gains of the thermal power project, the financial internal rate of return(FIRR) for developing thermal power is 12.1 per cent.

In addition, according to shadow price for coal, equipment and electric energy, national economic evaluation for thermal power project is discussed and calculated. The result of calculation is that economic internal rate of return(EIRR)is 10.4 per cent.

ECONOMIC EVALUATION FOR HYDROPOWER

The Wujiang is a largest river in Guizhou province. There are abundant resources of hydropower on the main river, and developing conditions are optimal. It is planed firstly developing river in the near future. At present, only the Wujiangdu(630 MW) completed, the Dongfeng(510 MW) is undur construction on the main stream. But the Hongjiadu

(540 MW) and the Goupitan(2000 MW) will be develop in the near future.

The Goupitan is a good representative project for developing conditions of hydropower. It is fifth cascade on the main stream of the Wujiang river. The project will be with a concrete gravity arch dam of 225 m high, the storage capacity of the reservoir 5690 Mm^3, the effective capacity 3660 Mm^3, the installed capacity 4×500 MW, the average annual electric capacity 9190 GWh. The distance to Guiyang city is about 140 km. The total cost of the project will be about RMB¥ 2980 million(1986). And the total construction period will be eleven year. See the Table Ⅱ.

Economic calculation and financial analysis for the hydropower project are also based on follwing the parameters, and the another parameters are the same with the thermal power.
· Economic using period of hydropower station is 50 year;
· Annual depreciation rate is 2 per cent of total investment;
· Annual cost of operation and maintenance is 1 per cent of total investment ; and
· Using electricity rate of hydropower plant is 0.1 per cent.

According to economic calculation for the investment, cost and gain of the Goupitan hydropower project, financial internal rate of return for developing the Goupitan project is 12.8 per cent.

In addition, national economic evaluation for hydropower is discussed. The result of calculation is that economic internal rate of return for the Goupitan project is 15.2 per cent.

ECONOMIC COMPARISON AND ANALYSIS EVALUATION

By evaluation of finance and national economy for the representative projects of hydro and thermal electric power, all economic index of hydropower is higher than thermal electric power, see table Ⅲ.

Table III — the comparison of property and economic effect for hydro and thermal electric power

		Hydro power of Goupitan	Thermal power in mine area
Installed capacity	MW	2000	800
Guaranteed output	MW	838	736
Average year production	GWh	9192	4400
Total period	year	11	7
Total investment	RMB¥ $\times 10^6$	2980	920
Using electricity rate for plant	%	0.1	8
Fuel cost per year	RMB¥ $\times 10^6$	0	90.29
Annual cost of production	RMB¥ $\times 10^6$	77.48	157.91
Energy cost 1986	RMB¥ /10^3 kwh	8.43	39.01
Economic using period	year	50	25
FIRR	%	12.8	12.1
Return on investment	%	16.3	10.2
Return and tax on investment	%	20.0	15.3
EIRR	%	15.2	10.4
Return period of loan	year	15.8	15.2

It is shown from the Table III, for construction period of hydropower is longer than thermal electric power, total investment is higher; but production cost of hydropower is lower than thermal power, using period is longer, energy cost is lower. Therefore, all economic evaluation indicators of hydropower are higher than those of thermal power.

In addition, multipurpose utilizations of hydropower are significant; for example, flood control, navigation, irrigation and water supply etc. After Goupitan hydropower station complated, the ship of 500t will navigate to the middle stream of the Wujiang river that could not navigate before. And Goupitan will controls flood 10 per cent of the Changjiang(Yangtze) river.

Therefore, the economic effects of hydropower are higher than those thermal power. priority to develop water power is obvious.

OPTIMAL PROPORTION

Because guaranteed output of hydropower is droped lower during the dry season, the electric system needs supplement from thermal electric power. If proportion between hydro and thermal electric power in the eletric system is suitable; then annual cost of hydropower and thermal power in the electric system should be minimum, and economic effect should be maximum.

According to the forecast of the electric developing plan, peak load of the Guizhou electric system in the year 2000 will be reached 4 GW, and annual production will be 26200 GWh, averege load will be 2.99 GW.

To meet the requirement of calculating necessary capacity of the Guizhou electric system of different proportion between hydropower and thermal power in the year 2000, installed capacity of the electric system must satisfy three conditions as follows.

· Service capacity of the electric system must be more than peak load of 4 GW;

· Production capacity of the electric system in the year must be more than 26200 GWh; and

· Guaranteed output of hydropower and thermal power in the system must be more than average load of 2.99 GW.

According to these conditions and the present conditions of the Guizhou electric system, the necessary capacity of the system of various proportion between hydro and thermal electric power in the year 2000 is calculated. Then annual cost for hydropower and thermal power in the system can be calculeted. See Table IV.

By using numbers in the Table IV, the annual cost curve for necessary capacity of various proportion between hydro and thermal electric power in the Guizhou electric system is plotted. See Fig. 1.

It is shown from the Fig. 1, the curves have a lowest point, this is a minimum point of annual cost of the system. When the proportion between hydro and thermal electric power in the system reaches optimal point, annual cost of the eletric system is lowest, and economic effect of the electric system is maximum.

It is also shown from Fig. 1, when the proportion between hydro and thermal electric power is optimal; the hydropower is 73—74 per cend of total capacity of the electric system; and thermal power capacity is only 26—27 per cent.

But proportion between hydro and thermal electric power of the electric system is usually change, because a electric system is

683

Table IV—Necessary capacity and annual cost of various proportion between hydro and thermal electric power in the Guizhou electric system in the year 2000

Proportion between hydro and thermal electric power (Per cent of total)		Necessary capacity of the system (MW)			Annual cost for necessary capacity (RMB¥ ×10⁶)		
Hydro	Thermal	Total	Hydro	Thermal	Total	Hydro	Thermal
50	50	5360	2680	2680	1599/2065	657/705	942/1360
60	40	5390	3240	2150	1542/1930	794/856	748/1078
70	30	5420	3800	1600	1489/1801	931/1000	558/801
72	28	5430	3920	1510	1481/1778	961/1031	520/747
74	26	5450	4030	1420	1477/1762	988/1060	489/702
76	24	5550	4220	1330	1492/1768	1034/1110	458/658
78	22	5650	4400	1250	1509/1776	1078/1158	431/618
80	20	5760	4600	1160	1527/1784	1127/1210	400/574
85	15	6050	5140	910	1573/1802	1260/1352	313/450

Note: Number on slant line is calculated by the result of prevailing price, number under slant line is calculated by the result of shadow price

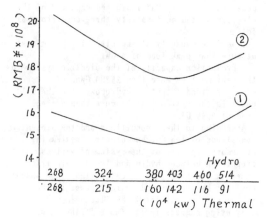

Fig. 1. The annual cost curve for necessary capacity when various proportion between hydro and thermal electric power in the Guizhou electric system, where: 1=the annual cost curve for prevailing price; 2=the annual cost curve for shadow price.

always develop (while large generation unit take part in the system). According to the shape of the annual cost curves, the hydropower proportion of 70—75 per cent in the electric system is optimal limit, and the proportion of thermal electric power of 25—30 per cent is suitable.

Therefore, when margin capacity in the electric system is considered, suitable installed capacity of hydropower in the Guizhou electric system in the year 2000 will approximately be 4100 MW, and installed capacity of thermal electric power will approximately be 1500 MW; then, total capacity of the Guizhou electric system will approximately be 5600 MW.

7 Hydropower schemes

Potential development of the Uruguay River basin

R.H.Andrzejewski & H.H.Dijkstra
ELETROSUL, Centrais Eletricas do Sul do Brasil S.A., Brazil

ABSTRACT: Electricity supply has grown at very substantial rates in Brazil during the decades of 1960 and 1970. The majority of the electricity supplied within the country derives from hydropower and future development is still very much based on this source.

Withim the more industrialized South and Southeast there is the Uruguai River basin, with a significant capacity for generation of more than 8000 MW of electric power at still very attractive costs.

Besides the low costs, other favorable aspects are present, such as the proximity to main load centers, existing transmission facilities and comparatively low social and environmental impacts.

The requirement by the Brazilian constitution that new concessions for power development are to be established through a tendering process, open an oportunity among the best in the world for investment in hydroelectricity at low production costs, within an area of substantial economic growth.

1. THE BRAZILIAN ELECTRICITY SUPPLY

During the past three decades Brazil has spent a great effort in industrialization and overall development. Hydropower has been the basis for energy supply, supported on large availability of water resources.

Installed hydropower capacity increased from 3,600 MW in 1960 to 50,000 MW at present (table 1) and trends are of increasing participation of electric power in the national energy matrix. Demand for electricity is still at low levels in KWh/inh and has a relevant potential to increase with improved living standards (table 2).

Table 1. Electric energy in Brazil

year	Electric power instaled capacity			Energy consumption			Gross national product	KWh/inh
	thermal (MW)	hydro (MW)	total (MW)	electric (TWh	total (TWh)	electric (%)	10^6 US$(85)	
1950	347	1,536	1,883					
1955	667	2,481	3,148					
1960	1,158	3,642	4,800					
1965	2,020	5,391	7,411					
1970	2,372	9,088	11,460	38.2	233	19.4	103.2	580
1975	4,801	16,323	21,124	69.9	322	22.9	166.8	1380
1980	5,768	25,584	31,352	122.8	421	29.7	236.8	1670
1985	4,359	37,437	41,796	172.6	485	37,0	249.7	1820
1990	4,669	50,537	55,203	216.6	547	39,6	276.2	1820

Table 2. Energy consumption data (1982)

Country	KWh/inh	KWh/10^3 US$ of GNP
Brazil (1985)	1,820	657
West Germany	5,952	630
Italy	3,277	600
France	4,906	560
Japan	4,906	640
U.S.A.	9,929	850
Canada	15,731	1,530
Norway	22,642	2,020

The national plan for electric energy supply for the period 1987-2000 "Plano 2010", issued by the Federal Government in 1987, projects an increase of the GNP per inhabitant from US$ 1,670 in 1985 to US$ 4,430 for the year 2010, for a population growth of 17%/year. To meet the resulting demand of electricity there will be a need to increase electric power supply at a rate of 5,65%/year, far below the recorded rates from 1960 to 1990.

The long term planning is yearly revised for a ten year construction program and it has been adjusted to the situation of the last decade of economical difficulties.

In spite of the present reduced economical growth, Brazil has a large potential for an early rehabilitation and for a fast increase in the electric power demand as occurred in the past. To prevent electric energy supply deficit, most projects which have been systematically postponed in the ten year construction plans should have their technical studies, funding schemes and construction programes retaken.

2. ELECTRIC POWER SYSTEMS

Two main electric grids account for the majority of large scale electric power transmission in Brazil. The South-Southeast interconnected system handles about 80% of the total electric power supply and the North-Northeast interconnected system takes some 20% of the electric power. Other minor grids account for the supply of isolated areas.

Almost 2/3 of the national demand for electricity is centered in the more industrialized Southeast. The region has a strongly connected power grid with the South that consumes about 15% of the total energy within the country (Figure 1).

The Southern region has a subtropical climate and a well balanced industrial/ agricultural economy supported by good infrastructure of energy, roadways and communications. Gross National Product per inhabitant and economical growth rate are above national average and social living standards are less uneven.

Within this region and including the State of Mato Grosso do Sul most of the transmission system owned by ELETROSUL, a electric power is supplied through the federal electric power utility dealing with gross power production and transmission. The region is of great

Figure 1. Brazil - Regional electric data

importance to the country's electrical
system for its large natural hydraulic and
coal resources. Its geographical situation
favors the electric connection with the
more industrialized Southeast and with the
neighboring countries Argentina, Uruguay,
Paraguay and Bolivia.

3. ELECTRIC POWER RESOURCES

About 90% of the country's electric power
production is hydroelectric, the remaining
10% being thermal (coal, oil, nuclear and
biomass). Hydropower is still the most
important resource to be developed in
Brazil. Out of 72,700 MW inventoried firm
power only 30% is commercially exploited
to date.

The known coal resources are in the
South and stand for an estimated power
potential of about 46,800 MW capacity.

Nuclear, biomass, natural gas and fuel
oil are other alternatives for power
production. Figure 2 presents energy cost
data as estimated for the "Plano 2010".

For the South-Southeast interconnected
system some 10.000 MW of hydropower is
still to be developed at costs under US$
25/MWh. Half of this potential rests in
the Uruguay River basin.

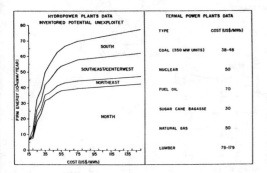

Figure 2. Energy cost (July-1986)
Source: ELETROBRÁS (1987)

4. THE URUGUAY RIVER BASIN

Part of the great Da Prata River basin,
the Uruguay River basin comprises an
extensive area of Brazil, Argentina and
Uruguay. The upper basin is totally
located in Brazil's southernmost States of
Rio Grande do Sul and Santa Catarina.
Running west till the border of Argentina,
the medium stretch of the river turns
south bordering Brazil and Argentina and
at the lower stretch bordering Argentina
and Uruguay.

4.1 The water power resources

The only existing power development along
the main stem of the Uruguay River is the
joint Argentine-Uruguayan Salto Grande
project in the lower portion of the river.

Along the medium international stretch
three major projects were indentified on
the borderline between Brazil and
Argentina: São Pedro (745 MW), Garabi
(1800 MW) and Roncador (2700 MW).

The upper Uruguay basin with an area of
75,300 square kilometers was inventoried
in 1979 by ELETROSUL when 23 schemes with
over 25 MW capacity each were selected for
hydropower development projects. Later
studies undertaken on basis of more
detailed data confirmed a potential
capacity of over 8,000 MW available along
the national stretch of the river.

With a strategical location for
international electric integration, the
Uruguay River basin is the last one with
such important hydropower resources still
to be developed nearby the important load
centers of South and Southeast Brazil.

High voltage 500 kV transmission lines
already cross the river near sites for
power development, linking South and
Southeast without further significant need
of investment on power transmission for
the first major plants to be built.

Along the border with Argentina a 50 MW
connection is being completed between
Uruguaiana (BR) and Passo de Los Libres
(AR) and a 900 MW connection is planned at
the site of the future Garabi Powerplant.
Not far northwest the Itaipu Hydroelectric
Powerplant (12,600 MW) is an important
connection with Paraguay.

4.2 The upper Uruguay River basin

All over the upper river basin the
landscape is hilly and rivers run through
well defined deep V shaped valleys. The
prevailing geology comprises overlaying
basaltic lava flows featuring competent
foundation and construction materials.

Such local physical characteristics
provide very suitable conditions for high
dam construction at low costs and reduced
land flooding.

Subtropical forests that covered great
part of the region were mostly replaced by
grazing and croplands, notwithstanding
the unfavorable topography.

Rainfall and floods are much related to
the movement of polar air masses.
Rainstorms result in fast response in
river flows due to steep slopes, low
permeability and little forest cover.

High peak discharges with fast rising river levels are frequent. The sediment transport which is mostly suspended load has litlle significance because of the basin's highly cohesive clayey topsoil. The siltation potential is very small and the reservoirs life time is computed in hundreds of years.

The prevailing regional economy is sustained by rural activities and related industries. Most land owners detain small properties with well balanced production of corn and poultry or swine. Social and political consciousness is strongly present and supported by local authorities, clergy, unions and cooperatives. A special Popular Regional Commission CRAB was organized to discuss and set actions to mitigate social impacts caused by dam construction.

4.3 Main projects data

From the upper Uruguay inventory studies 23 projects were selected and ordered in a sequence economically optimized for implementation (table 3).

During feasibility and basic design studies some changes were introduced on basis of new data and criteria for energy valuation and specially on more extensive environmental studies. The revised data of the 10 main projects that account for 90% of the inventoried capacity are shown on table 4, including the Garabi powerplant that has priority among the international schemes to be developed. Figure 3 shows the general scheme for the Uruguay basin development within Brazilian borders.

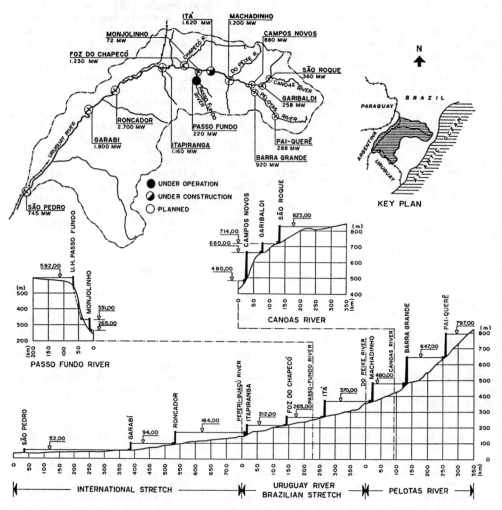

Figure 3. Uruguay River Basin - Planned hydropower plants scheme

Table 3. Upper Uruguay River basin inventory data (1979)

River	Project name	Drainage area (km^2)	Maximun normal W.L (m)	Net volume ($10^6 m^3$)	Gross head (m)	Mean power (MW)	Installed capacity (MW)	Number of units
Canoas	Barra Pessegueiro	2,200	853,0	1190	24	-	-	-
Canoas	São Roque	9,760	823,0	5165	91	127	256	2
Canoas	Garibaldi	13,200	732,0	1945	108	205	398	2
Canoas	Campos Novos	14,200	624,0	527	144	283	748	3
Pelotas	Passo da Cadeia	2,190	940,0	1310	143	48	104	2
Pelotas	Pai-Querê	6,250	797,0	1742	150	136	288	2
Pelotas	Barra Grande	12,860	647,0	3865	167	302	609	3
Pelotas	Machadinho	35,800	480,0	4510	112	550	1590	4
Uruguay	Itá	44,500	368,0	3590	103	673	1841	5
Passo Fundo	Passo Fundo	2,300	592,0	1388	261	98	220	2
Passo Fundo	Monjolinho	4,150	331,0	9	66	41	72	2
Chapecó	Aparecida	1,660	858,0	1200	91	29	64	2
Chapecó	Abelardo Luz	1,860	767,0	-	117	46	84	2
Chapecó	São Domingos	2,290	650,0	216	61	27	55	1
Chapecó	Quebra Queixo	2,670	589,0	643	159	83	162	2
Chapecozinho	Gabiroba	876	814,0	317	41	-	-	-
Chapecozinho	Bom Jesus	1,370	694,0	530	52	-	-	-
Chapecozinho	Xanxerê	1,470	617,0	-	75	14	25	1
Chapecozinho	Voltão Novo	1,570	542,0	-	112	24	45	1
Chapecó	Foz Chapecozinho	5,260	430,0	870	95	93	184	2
Chapecó	Nova Erexim	7,440	335,0	855	70	102	198	2
Uruguay	Iraí	62,200	265,0	2610	57	576	1488	6
Uruguay	Itapiranga	71,400	208,0	113	43	510	1248	6

Remark: Passo Fundo hydropower plant was already in operation.

Table 4. Main power projects data (1988)

Project	Drainage area (km^2)	Flooded area (km^2)	Gross head (m)	Mean power (MW)	5% risk power (MW) * **		Installed capacity (Mw)	Number of units	Total cost (10^6 US$)	Decenal plan 93/2002	Present status ***
Itá	44,500	104	105	802	682	795	1,620	6	1,678	1998	U.C
Campos Novos	14,200	24	180	410	339	381	880	4	852	2000	B.D.
Machadinho	35,800	234	110	626	538	577	1,200	4	1,279	2001	B.D.
Barra Grande	12,800	79	167	321	312	330	920	4	1,069	2001	F.S.
São Roque	9,650	320	109	169	146	146	360	2	1,134	-	R.I.S.
Garibaldi	12,700	28	54	110	87	101	228	2	529	-	R.I.S.
Pai Querê	6,250	48	150	145	124	128	288	2	629	2002	I.S.
Foz do Chapecó	61,340	83	53	583	505	552	1,228	6	2,041	-	R.I.S.
Itapiranga	72,700	118	47	587	514	560	1,160	8	2,488	-	R.I.S.
Garabi	115,820	810	40	660	612	630	1,800	6	2,121	2000	B.D.

*   first addition expected 5% risk garanteed power
**   long term expected 5% risk garanteed power
*** U.C.=under construction, B.D.=basic design completed, F.S.=feasibilities studies,
I.S.=inventoried, R.I.S.=inventory studies revised.

4.4 Overall planning

The national plan for electric energy supply "Plano 2010" was issued in 1987 for the period 1987-2000. This issue included many of the plants from the Uruguay River basin for their favorable characteristics and low cost. Several of these projects should be under construction in 1991.

Due to a strong decrease in the projected growth rates for electricity demand, the ten year plan was revised and most of the new projects were postponed. The energy supplied by the 18 generators, 715 MW each, from the Itaipu hydroelectric power plant which were put in commercial operation between 1984 and 1991, has been sufficient to support the demand in South-Southeast Brazil during this period of low economical growth.

For the coming years there will be need of supply from new electric power sources. The yearly revision of the 10 year plan is adjusted to forecasted demand. The present plan 1991-2000 includes construction of three of the plants from the Uruguay River basin: Itá, to be in operation in 1998 Garabi and Campos Novos scheduled for the year 2000. Most of the remaining projects are to be scheduled for the decade 2000/2010.

4.5 Present status of engineering and construction

Four projects have been completed by ELETROSUL to the stage of basic design with sufficient detailing for preparation of construction bids.

The 1620 MW Itá plant, earlier planned to operate in 1995, has been started with camp and infrastructure facilities as well as preliminary excavation. Works have been temporarily interrupted due do lack of financing, and construction bidding now depends on funding arrangements to be completed.

Negotiations with the affected local population have come to agreement in good terms. Thirty percent of the reservoir area has already been bought and relocation of the Itá town (1,100 inhabitants) is well under way. Two projects have been successfully completed for settlement of 110 families of non-owners working on rented land. These programs have come to a slow-down but will not be discontinued while construction is interrupted.

5. CLOSING REMARKS

The Uruguai River basin presents a very attractive potential for the development of hydropower projects that can supply great part of the future needs of energy in South-Southeast Brazil.

Its favorable location requires little initial investment in power transmission systems and it is the nearest hydropower potential to the major load centers, still to be developed.

Environmental impacts have been carefully evaluated and procedures for mitigation are feasible to achieve at compatible costs. Land is settled by small farmers that replaced the subtropical forest by grazing and croplands. No indigenous population is to be directly affected. The major impact on affected population has led to negotiations that resulted in well established agreements and the first relocation and resettlement programs have been successfully carried out.

The present Brazilian constitution establishes the requirement of tendering for new power development concessions, seeking competition, efficiency and minimal final cost for energy users. The Brazilian Government, is also studying alternatives to enhance external partnership for new projects, with negotiable rights for guaranteed power marketability. Hence, the Uruguay River basin is to be seen as one of the worldwide most attractive alternatives for investment in hydroelectricity.

6. REFERENCES

ELETROBRÁS, Centrais Elétricas Brasileiras S.A. 1978 Plano 2010, Plano Nacional de Energia Elétrica. 1987/2010, Relatório Geral.

Feliciano Dias, Renato et alii, 1988. Panorama do Setor de Energia Elétrica no Brasil, Memória da Eletricidade.

ELETROSUL, Centrais Elétricas do Sul do Brasil S.A. DO/DPL, 1989. Alternativas Energéticas para o Atendimento ao Mercado de Energia Elétrica.

ELETROSUL, Centrais Elétricas do Sul do Brasil S.A. and CNEC Consórcio Nacional de Engenheiros Consultores S.A., 1979. Bacia Hidrográfica do Rio Uruguai, Estudo de Inventário Hidroenergético.

Hydropower'92, Broch & Lysne (eds) © 1992 Balkema, Rotterdam. ISBN 90 5410 054 0

Atatürk Dam and Powerplant

Erdoğan Basmaci
State Hydraulic Works 16th Regional Directorate, Şanlıurfa, Turkey

ABSTRACT: Construction of Ataturk Dam and Powerplant is nearly complete. It is the biggest scheme in Turkey and located downstream of Keban and Karakaya Dams on the Euphrates river. The 184 m high dam consists of 84,5 hm³ embankment volume. The Powerplant has an installed capacity of 2400 MW with annual power generation of 8,9 TWh. Furthermore the grout curtain is one of the largest in the world, with 800 km total drilling length. Also the existing machinery and equipment used for earth moving works is the biggest fleet in recent years in the world.

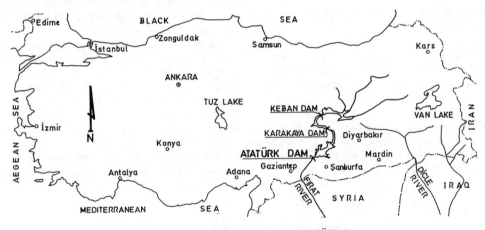

Fig 1 Location map of the ATATÜRK Dam

1 HYDROLOGY AND RESERVOIR

The catchment area of Euphrates river at Atatürk Dam site is 92 338 km² with an average annual inflow of 26,5 km³. Talweg elevation of the river is 380 m a.s.l, and the maximum reservoir level is, 542 m a.s.l, corresponding to a total reservoir volume of 48,5 km³. Normal operational minimum level is 526 m a.s.l with a wolume of 37,7 km³.

2 GEOLOGY

The base rock at the dam site consists chiefly of upper cretaceous formations, starting with an upper lithological unit of plicated limestone, underlain by bituminous limestone with chert bands and finally followed by a very thick lithological unit of dolomitic limestone.

A large fault zone (Bozova fault) passes through the left abutment of the dam, cutting the end of the spillway chute. This large fault zone is followed by

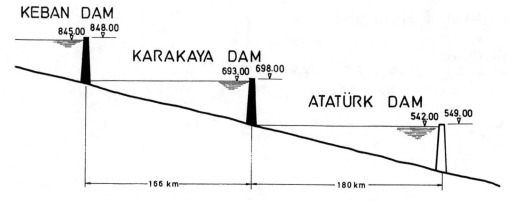

Fig 2 Development of the Lower Euphrates catchment

sub-vertical faults of smaller magnitude. Numerous sub-vertical and inclined minor faults intersect the river bed at both abutments.

3 THE OTHER DAMS AND POWERPLANTS ON THE EUPHRATES RIVER IN TURKEY

Keban Dam and Powerplant: 207 m high rockfill dam and 1 360 MW capacity power-plant with an annual power generation of 6,5 Twh, completed in 1974.

Karakaya Dam and Powerplant: 175 m high arch gravity concrete dam and 1 800 MW capacity powerplant with an annual power generation of 7,3 TWh, completed in 1987.

4 DAM EMBANKMENT

The Atatürk rockfill dam, with a central clay core as shown in Figure 3, is 184 m high above its foundation. The curved crest is 1 664 m long. The total embankment volume is 84,5 hm³ including the upstream cofferdams.

The material 12 hm³ for the impervious clay core was hauled from a borrow area located 5 km south of the dam. The alluvium material for the 3 hm³ transition and filter zones and 10 hm³ shell was extracted from the river. For the dam shells, 51 hm³ of basalt was obtained from a quarry 6 km southwest of the dam site. The random fill for the embankment (8 hm³) was taken from the excavation of the main structures.

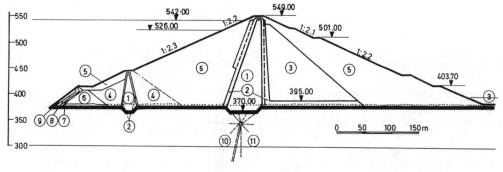

1 Core
2 Transition zone
3 Random
4 River alluvium
5 Rockfill (basalt)
6 Rockfill
7 Sand and gravel
8 Impervious material
9 Riprap
10 Grouting curtain
11 Drainage

Fig 3 Typical cross section through the ATATÜRK Dam embankment

The upstream main cofferdam is a 55 m high dam with a central clay core. The total embankmant volume is nearly 8 hm³. A closure dyke and an upstream pre-cofferdam have been constructed before commencement of the upstream main cofferdam.

5 DIVERSION TUNNELS AND BOTTON OUTLETS

There are 3 horseshoe shaped diversion tunnels with an inner diameter of 8 m. Diversion tunnels are located on the left abutment and each tunnel is about 1 300 m long.

Construction of the tunnels was performed between 1981 and 1986. The river was diverted to the tunnels in June 1986.

The design capacity of the 3 diversion tunnels is 3 x 700 m³/s at 416 m a.s.l. According to the records about 3 000 m³/s has passed through the tunnels during the 1988 flood season.

The 3 diversion tunnels have been transformed into bottom outlets being equipped with double vertical gates and one horizontal gate in each gate chamber. The maximum capacity of each bottom outlet is 500 m³/s.

Intake structures of the 2 diversion tunnels are at elevation 380 m a.s.l, whilst the intermediate structure is located at elevation 463 m.a.s.l because of constructional and hydrological reasons.

6 SPILLWAY

The radial gated spillway is located on left bank of the dam. Each of the 6 gates is 16 m wide and 17 m high. A cross section of the spillway crest is shown in Figure 4. The elevation is 525 m.a.s.l and the total discharge capacity is 16 800 m³/s. In addition to 6 main radial gates, there are 2 auxiliary gates at elevation 505 m.a.s.l. The auxiliary gates are also of radial type with the size of 5 m x 8 m and each having a capacity of 500 m³/s.

Following a total excavation of 12 hm³, 1,4 hm³ concrete was cast for the spillway. Downstream of crest structure is an open channel chute (800 m long) connected to a stilling basin (400 m by 200 m)where 6 000

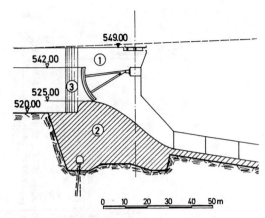

1 Tainter gate   2 Concrete sill   3 Pier

Fig 4 Cross section of the spillway

prestressed anchors are being installed to ensure the stabilty of the structure during severe operational conditions.

7 INTAKE STRUCTURE

The intake structure is located at the right abutment, adjacent to the dam body. This is a concrete gravity dam, 71 m high and 200 m long. Total excavation is 6 hm³ and total concrete volume is 585 000 m³. A typical cross section is given in Figure 5.

The intake structure is equipped with trashracks and roller gates. Also, there are stoplog gates for each entrance to provide possibility of maintenance for the roller gates. The roller gates are 4,8 m wide and 7,20 m high and rectangular shaped, a steel transition zone leads to circular shaped penstocks.

The roller gates are operated by a hydraulic hoisting mechanism whilst a gantry crane is used for the stoplog gates.

8 PENSTOCKS

A typical longitudinal section through a penstock, conveying water from the intake to the powerhouse, is shown in Figure 6. The lengths of 8 penstocks varies between 530 m for the inner one close to the dam,

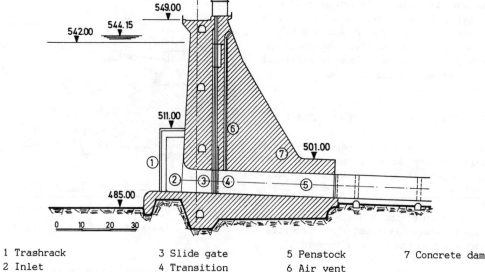

| 1 Trashrack | 3 Slide gate | 5 Penstock | 7 Concrete dam |
| 2 Inlet | 4 Transition | 6 Air vent | |

Fig 5 Cross section of the Intake Structure

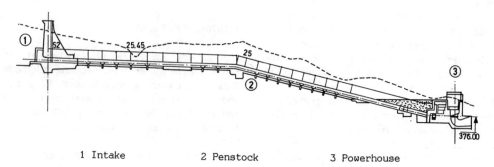

| 1 Intake | 2 Penstock | 3 Powerhouse |

Fig 6 Longitudinal section through the penstocks

and 650 m for the outer one close to the abutment.

Because of this difference in pipe lengths, the diameters vary according to position of the penstocks, so that the net head is constant for each turbine. Thus, the diameters are 6,6 m for the two innermost penstocks, 6,8 m for the next two, 7 m for the next two and 7,25 m for the two outermost. These dimensions are kept constant along the entire length of the penstocks.

The steel thickness varies from 22 mm to 43 mm. The penstocks have been manufactured at the job site. The total quantity of steel used for the penstocks is about 26 000 t.

9 POWERHOUSE

The powerhouse is located on the right abutment at the foot of the dam and has eight units, each with an installed capacity of 300 MW.

A cross section of the powerhouse is shown in Figure 7, the height is 55 m above the foundation, the width is 49 m and the length is 258 m. The total concrete cast is 340 000 m³.

Butterfly valves 5,7 m diameter control the flow from the penstocks to the turbines. Two 400 t cranes are provided for the construction, maintenance and repair of the equipment.

The vertical axis Francis turbines,

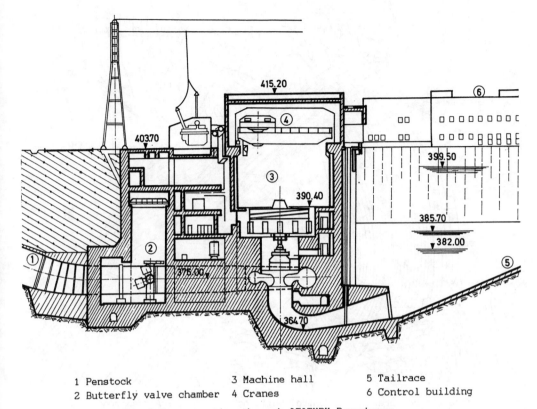

1 Penstock   3 Machine hall   5 Tailrace
2 Butterfly valve chamber   4 Cranes   6 Control building

Fig 7 Cross section through ATATURK Powerhouse

which operate at 150 rev/min, have runner diameters of 5,15 m. The output voltage of the generators is 15,75 kV; 24 transformers of 105 MVA capacity, three for each unit, step the voltage up to 380 kV.

10 TRANSMISSION LINES AND SWITCHYARD

There are four 380 kV transmission lines between the powerhouse and switchyard. Each line carries electricity from two turbine generator units. The length of each line is about 2,7 km. The switchyard is on the south of the layout and covers an area of 105 000 m². The switchyard is connected to the National Grid by eleven 380 kV and seven 154 kV lines. Four of the 380 kV lines and two of the 154 kV lines are standby. 380 kV lines will be used for long distance transmission and the 154 kV lines are for regional distribution.

11 GROUTING WORKS

The grout curtain of Atatürk Dam is one of the biggest in the world. Drillings are carried out from galleries having a total length of about 12,5 km. At the talweg the depth of curtain is 190 m and 125 m at the abutments. There are two galleries under the talweg, one of them is for grouting and the other one is for the control and drainage. At the abutments there are 4 galleries from elevation 400 to 550 m. The length of the grout curtain is 5 000 m and the surface area is 1 200 000 m².

40 drilling rigs have been used for the total drilling length of 1 800 000 m. Four grouting plants, 50 grout pumps (with 50-100 bars and total capacity of 300 m³ per hour), 70 mixer agitator sets have been used.

The two rows of grouting holes were added with a third row along the talweg and lowest galleries in the abutments.

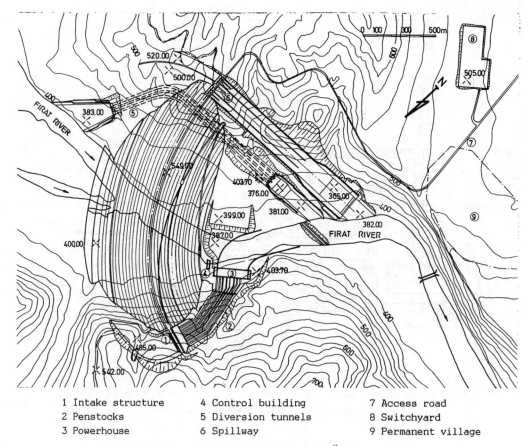

1 Intake structure   4 Control building   7 Access road
2 Penstocks   5 Diversion tunnels   8 Switchyard
3 Powerhouse   6 Spillway   9 Permanent village

Fig 8 General plan view of the ATATÜRK Dam

12 CONSTRUCTION EQUIPMENT

The total earth moving volume for the Atatürk Dam and Powerplant is more than 135 hm³, including 84,5 hm³ dam embankment. Open excavation works are about 50 hm³ and underground excavation works are about 800 000 m³ including diversion tunnels.

Among the earth moving equipment are 70 bulldozers (D8-D9-D10), 30 different kinds and excavators (ten of them have 12 m³ bucket capacity), 20 graders, 70 loaders (40 of them have 10 m³ buckets), 200 dump trucks of 85 tons capacity, 40 dump trucks of 35 tons capacity, 40 hyraulic drillers and several trailers and service vehicles.

The total capacity of five concrete batching plants is about 430 m³/h. There are 3 screening plants with a capacity of 500 ton/hour. The concrete fleet has 20 silobuses for carrying the cement from the cement factories, 20 concrete transmixers and carriers, 40 concrete pumps, 10 tower cranes, 3 conveyor belts and several types and different sizes of the auxiliary small machinery.

13 CONCLUSION

The dam embankment and most of the appurtenant structures have already been completed, whilst cabling, installation of auxiliary systems and erection of the powerhouse equipment are still in progress

Impounding of the reservoir has continued since January 1990, reaching already to about 70 % of the total height. The commercial operation of the powerplant shall commence as soon as the lake level exceeds 513 m a.s.l.

Various technical matters arose during construction of the embankment, concrete structures and also the grout curtain. Furthermore, amendments were made in the design considering the availability of new techniques, such as aeration of the spillway chute channel.

Following this general description about the Ataturk Dam and Powerplant, it is hoped that the engineers involved in the construction will analyse specific matters dam and foundation behaviour with very valuable contributions to dam engineering.

The existing construction and laboratory records, continuous data from the available large dam monitoring scheme and systematical hydrogeological investigation have been a challenge to the technical personnel so that beneficial studies are under way.

Hydropower'92, Broch & Lysne (eds) © 1992 Balkema, Rotterdam. ISBN 90 5410 054 0

Lam Ta Khong Pumped Storage Power Project in Thailand

T. Mahasandana
Electricity Generating Authority of Thailand, Thailand

T. Nishigori
Electric Power Development Co., Ltd, Japan

ABSTRACT: Thailand is developing electric power sources to cope with the rapid increase of peak power demand, and in line with the policy the Lam Ta Khong Pumped Storage Project is scheduled to commence the operation of 1,000 MW in the year 1997 and 2000.

1 INTRODUCTION

Thailand's economic growth rate in recent years have recorded more than 10%, being prominent among Asian countries. This domestic economic growth boom resulted in a sharp increase of electric power consumption. In 1990, the peak power generation rose up to a record 7,094 MW, marking a growth rate of 14% as compared with the average 10 per cent rate over the past decade. To cope with this remarkable increasing demand, implementation of various power projects have been urgently undertaken.

The load forecast shows that the peak

Fig.1 Lam Ta Khong pumped storage project

demand will grow up to 13,000-14,000 MW in 1997, and it will be necessary to commence new power facilities of about 900 MW every year. The Thai government has a policy of utilizing energy resources such as natural gas, lignite, hydropower, etc., suppressing the amount of oil import as less as possible. In line with this policy, he accomplished the self-sufficiency of energy generation of 87% in 1989, from 20% in 1980. In view of the energy policy, hydropower is one of the indispensable domestic energy resources and its development has been carried out through large-scale projects, resulting in that the remaining possible large-scale hydropower resources to cope with the peak demand are limited only to develop international rivers such as the Mekong river and the Salawin river.

Under the above situation, keen attention has been recently given to the Lam Ta Khong pumped storage project which is very close to Bangkok metropolitan area, the biggest energy consumption area in Thailand. The project is situated about 200 km north-east of Bangkok and scheduled to commence its operation in 1997.

To clarify the soundness of the project in terms of engineering, economy, finance and environment, and in response to a request from the Government of the Kingdom of Thailand, the Government of Japan decided to conduct a feasibility study on the project and entrusted the study to the Japan International Cooperation Agency (JICA). the feasibility study was carried out from 1990 to 1991.

2 LOAD FORECAST

2.1 Trend and present situation of power demand

The economy of Thailand has developed steadily with plenty of arable land, natural resources and a large labour force, achieving a high economic growth rate. Till the mid-1950s its per-capita GDP was only 80 U.S.$, however, the today's GDP per capita has reached 1,400 U.S.$ or so.

The economic performance of Thailand in recent years has been much favorable thanks mainly to the expansion of exports from mid-1986. The real GDP (gross domestic products) for 1987, 1988 and 1989 grew by 9.5%, 13.2% and 12.2% respectively.

Table 1 shows the transition of power and energy generation in Thailand for 1981-1991. Especially, sharp increases are recognized in the years of 1987-1991 due to the favorable economic performance.

Table 1. Power and energy generation in Thailand

Unit: MW, GW

Year	Power	Energy
1981	2,589 (7.09)*	15,960 (8.18)
1982	2,838 (9.63)	16,882 (5.78)
1983	3,204 (12.91)	19,066 (12.94)
1984	3,547 (10.70)	21,066 (10.49)
1985	3,878 (9.33)	23,357 (10.87)
1986	4,181 (7.80)	24,780 (6.09)
1987	4,734 (13.23)	28,193 (13.78)
1988	5,444 (15.00)	31,997 (13.49)
1989	6,233 (14.49)	36,457 (13.94)
1990	7,094 (13.81)	43,189 (18.47)
1991	8,045 (13.41)	49,225 (13.98)

* Figures in parenthesis show increases in %.

2.2 Load forecast

Load forecasts in Thailand have been based on estimates issued by the Load Forecast Working Group which comprises representatives or some experts from various authorities and institutes. In January 1990, the load forecast was prepared using annual GDP growth rates of 8.0% for the base case and 8.8% for the high case.

However, on 23 April 1990, the peak power generation reached 7,094 MW. The new peak of 1990 was higher than the 'High Case' load forecast. With consideration on the possible increasing demands due to the rapid industrialization and the urgently needed power supply, the revised load forecast was formulated with the slight adjustment on peak and energy generation of the 'High Case' load forecast.

For the benefit of demand management by trying to reduce the system peak, the time-of-day (TOD) rate has been applied for large industries which consume the electric power above 2,000 kW, effective from 1 January 1990. With the present TOD rate, the peak demand during 18:30-21:30 is estimated to decrease by about 70-100 MW or 0.9-1.2% of the total system peak. As for the revised load forecast, it is premature to involve the impact of TOD rate since experiences in metering, billing and analyses are still required to clearly identify its effectiveness and magnitude of peak power reduction.

3 ENERGY OPTION

Electricity generation by energy sources in 1980 and 1989 is shown in Table 2. The proportion of energy generation by crude oil and condensate was about 80% in 1980, however, it declined sharply to 11% in 1989. On the other hand, energy by lignite increased from 9% to 20%, and energy by natural gas accounts for 53% in 1989 by installing new facilities. Hydropower, natural gas and lignite are domestic energy resources, and the self-sufficiency rate of energy generation made a remarkable improvement from 20% (hydro + lignite) in 1980 to 87% (hydro + lignite + natural gas) in 1989.

Table 2. Comparison of electricity generation by energy sources in 1980 and 1989.

Unit: GWh

Sources	1980	1989
import	753 (5.1)*	509 (1.4)
diesel oil	349 (2.4)	1 (0.0)
hydro	1,653 (11.2)	5,249 (14.4)
fuel oil	10,672 (72.3)	4,029 (11.0)
lignite	1,327 (9.0)	7,359 (20.2)
natural gas	0 (0.0)	19,310 (53.0)
total	14,754 (100)	36,457 (100)

* Figures in parenthesis show percentage to total.

For power development planning in Thailand, the five-fuel policy has been adopted, four of them i.e., hydro, gas, oil and lignite are considered for short and medium term planning, and the other one is imported energy such as coal to be considered for medium and long term profile. Brief summaries of each energy option are described below:

1. The hydroelectric potential in Thailand is estimated about 10,626 MW. Up to 1990, only 2,566 MW of hydropower capacity or 24 percent of the total potential have been exploited. Of this amount, 2,250 MW is in operation, and 316 MW is under construction and/or committed. The remaining potentials of about 8,060 MW (76 percent of the total potential) are not easy to be developed due to environmental and forestry problems.

2. The reserve of natural gas in Thailand is estimated about 19.25 TCF, comprising 23 fields off-shore and 3 fields on-shore. It is expected to have a recoverable reserve of 16.1 TCF, of this amount, 8.6 TCF is proven reserve.

3. The reserve of crude oil in Thailand are of small quantity. the known reserves of crude oil and condensate have been estimated about 1,150 million barrels. It is expected to have a recoverable reserve of 498 million barrels, of this amount, 236 million barrels is proven reserve.

4. There are various lignite resources scattered in Thailand. The total geological reserve is estimated about 2,069 million tons. The significant reserves located in the north are about 1,598 million tons or equivalent to 77.2% of the total reserves, and the other 471 million tons (22.8%) is located in the south.

5. Natural resources in Thailand are limited. Present estimate shows that natural gas and lignite reserves are not sufficient for the increase in generating capacity in the long term. Therefore, imported coal is an alternate source of fuel besides heavy oil, which has to be included in generation expansion program.

4 POWER DEVELOPMENT PLAN

The total installed capacity of generating facilities of Thailand is 8,299 MW as of 1989. Of this capacity, 27% is for hydro and 73% for thermal as shown in Table 3.

Consumer expectation and economic growth requires the reliable supply of electricity. To meet this need, the process of power development planning considers projected demand over a planning period, available energy resources and alternative means of meeting the demand growth. The objective of power development planning is to meet the power demand with generation and transmission development at a specified level of reliability and in the least-cost manner.

The main part of future power development in Thailand is to be filled up by thermal power developments such as combined cycle, lignite-fired and coal-fired power plants.

Table 3. Total installed capacity of
generating facilities in Thailand

Unit: MW

	EGAT	Others	Total
hydro	2,238	33	2,271
steam turbine	4,006	593	4,599
combined cycle	760	-	760
gas turbine	238	-	238
diesel	29	402	431
total	7,271	1,028	8,299

Figure 2 shows the power development plan
up to the year of 2006. the figure involves
installed capacity and peak generation
requirement. The target of reserve margin
is 25% to ensure that generating capacity
will be sufficiently provided and thus the
LOLP (Loss of Load Probability) will not be
more than 1 day/year. However, the actual
reserve margin ranges 15-20%.

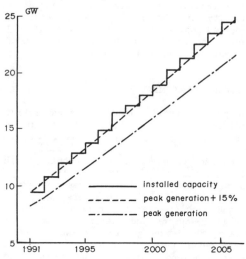

Fig.2 Power development plan in Thailand up
to 2006

5 LAM TA KHONG PUMPED STORAGE DEVELOPMENT PLAN

The development scale of the project is
1,000 MW having the capability of 8 hours'
continuous operation. It is desirable to
commence its operation of 500 MW in the year
of 1997 and the remaining 500 MW in 2000.
To finalize the optimum development plan,
keen attentions were paid to power system
stability and environmental restriction in
the project area.

5.1 Power system stability

The power system of Thailand is divided into
the following 4 regions; Region 1
(Metropolitan area and its surrounding
area), Region 2 (North-East area), Region 3
(South area) and Region 4 (North and middle
area). Each region is connected by
transmission lines of 500 kV, 230 kV or 115
kV. The trunk power lines connecting among
these regions have recently been reinforced.
The 500 kV line connecting the Mae Moh
thermal power station in Region 4 with
Greater Bangkok Area has started its
operation at 500 kV in the entire line
length. The 230 kV second Central-Southern
Tie line has been newly completed to connect
Region 1 with Region 3. Furthermore, the
500 kV line of 333 km long between Region 1
and Region 4 and the 230 kV line of 300 km
long between Region 1 and Region 2 are being
constructed.

When a power system includes pumped
storage power plans, pumping operation at
off-peak hours involves severer condition
than generating operation at peak hours from
a viewpoint of power system stability. In
pumping a great difference of voltage phase
angles between supplying generator to
pumping and pumping motor causes the power
system stability to considerably
deteriorate. If the system is not strong
enough, even a small disturbance such as a
rapid load change will cause vibrations to
both generators and motors, and will put
some of them out of step.

Judging from demand-supply balance and
preliminary study, the project scale has
been proposed 600 - 1,000 MW. However, it
is feared that this plant scale may be
limited by the system stability which
depends upon the strength of power system at
a time when the plant is commissioned,
because pumping operation causes a very
severe state of the system stability.

In order to examine possible pumping
capacity of the plant, the stability of
power system was analysed under such
conditions as power demand, plant capacity,
method of power transmission and fault
location of transmission line. The results
of the power system analysis revealed that
the capacity of pumping operation of the
project should be 500 MW or below. As for

generation of 1,000 MW of the project, there is no problem of power system stability. Hence, the project is scheduled to commence its operation of 500 MW at the first stage in 1997, and the remaining 500 MW at the second stage in 2000.

5.2 Environmental restriction

Forest areas in Thailand has been greatly reduced in the past two decades. In 1961, 57 percent of the whole country was covered by forests. Forest area declined to 37 percent in 1974 and 32 percent in 1982. Largest declines in forest area have occurred in northern portions of Thailand.

Forests have been converted to cultivated lands to provide food for a rapidly increasing population. Also export of agricultural products is economically important, providing slightly less than 70 percent of thailand's total export income per year. Currently, about 43 percent of the land area of Thailand is dedicated to permanent agricultural production.

The rapid agronomic expansion and population growth, which are causing changes in land use patterns, are of national concerns in Thailand. The Thai government has had a policy of managing forests for the sake of the benefit and welfare of the public.

The Fourth National Economic Development Plan in 1977 had the following goals for conserving the Thailand's forests:

1. Maintain at least 37 percent of the total land area in forests, which ratio is the same as in 1974.

2. Reduce the deforestation rate to the extent not exceeding 800 square kilometers per year.

3. Increase the reforestration rate to 80,000 hectares annually.

4. Increase the number of wildlife conservation areas and national parks.

5. Implement reforestation programs in all significant watershed areas.

The above cited goals highlighted the critical need for a watershed classification which would properly allocate the land areas for various uses. The need for such classifications has been recognized for over 20 years, however, only in recent years a vigorous project, the Watershed Classification Project, has been pursued.

The Watershed Classification process requires a system for establishing potential land uses concerning physical and/or environmental characteristics of landscape units. Physical characteristics of landscape units are stable features such as long-term average climate, elevation, slope, landform, geology and soils. Environmental

features of landscape units are less stable and interact with short-term climatic trends, human uses and some physical features which influence plant and animal populations.

A decision has been made to establish five major watershed classes, ranging from the class 1 to 5. Moreover, the watershed class 1 is divided into two; 1A as protective forest and 1B as commercial forest or certain restricted agricultural uses. The classes 2 to 5 are generally lower elevation lands with a broad range of agricultural uses.

The class 1A are areas of protective forests and headwater source areas usually at higher elevation with very steep slopes. In any case, the land use in the class 1A area is strictly prohibited, and these areas should remain in permanent forest cover.

The class 1B are areas of similar physical and environmental features to the class 1A, but some portions of the area have been cleared for agricultural use or occupied by villages. In case of public land use in the class 1B by governmental organizations, they should propose and report environmental effects of the project to the National Environmental Board, and get its approval for the land use.

Considering the environmental restrictions of watershed classifications, the optimum layout and the preliminary design of power facilities including upper reservoir, waterway, powerhouse, transmission line and temporary facilities were done.

5.3 Preliminary design

The project area is located in an environmental area where development is restricted by the Thai government. Construction of structures on the surface are prohibited in a certain part of the area, the class 1A. Hence, in order to evade the restricted area, underground structures are designed in principle. The plan and profile of the project are shown in Figure 3.

The upper reservoir is a rockfill dam with asphaltic concrete facing, being maximum height of 60 m, crest length of 2.2 km and total embankment volume of 6 million cubic meters.

The lower reservoir utilizes the existing Lam Ta Khong reservoir, which was constructed for the purpose of irrigation water supply by the Royal Irrigation Department.

The waterway route is designed to keep its length between two reservoirs as short as possible. The penstock route and the powerhouse passes across the area 1A, and it

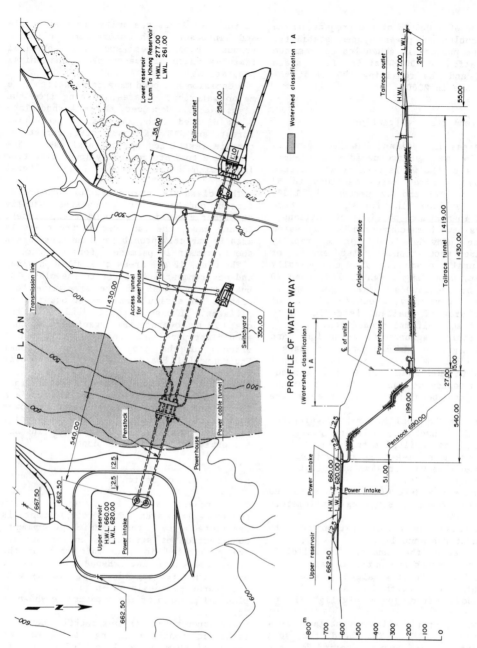

Fig.3 Plan and profile of the project

706

is installed not on the surface but in the underground so as to fulfill the environmental restriction.

The tailrace of 1,419 m long is equipped with a surge chamber to ensure full safety against the fluctuation in power generation and also to enable the plant to operate an automatic frequency control.

6 CONCLUSION

Thailand has pursued a policy of utilizing domestic energy sources to cope with the rapidly increasing demand due to the prominent economic growth. As a result of large scaled hydropower development, possible hydropower sources to be developed remain only in the international rivers. Also other medium-small hydro projects are currently difficult to realize in economical and environmental point of view. Under such circumstances, the Lam Ta Khong Pumped Storage Project has been focussed and recognized a fruitful development by the feasibility study of Electric Generating Authority of Thailand and Japan International Cooperation Agency. The project is scheduled to commence its operation in the year 1997, and the authors hope to take the next opportunity to report and/or present construction activities of the project.

Hydropower'92, Broch & Lysne (eds) © 1992 Balkema, Rotterdam. ISBN 90 5410 054 0

History of development of hydropower in India

C.V.J.Varma & A.R.G.Rao
Central Board of Irrigation and Power, Malcha Marg, Chanakyapuri, New Delhi, India

1 INTRODUCTION

From an installation of a tiny 200 KW hydro unit in tea estates of Darjeeling in 1897 to commissioning of units of 105 MW in Bansagar tons in Madhya Pradesh in March 1991 spans nearly hundred years saga of development of Hydro Power in India. The rate of growth which was modest prior to independence, accelerated during post-independence period especially when planned growth of economy was envisaged under successive five year plans realising that availability of power was the key input for economic development leading to better standard of living for millions of people. As on today, the country boasts of every conceivable type of hydro generation including run of the river schemes, storage schemes, pumped storage plants besides Hydro generation due to transbasin diversion of water either by gravity of pumping with units capacity ranging from 6 MW to 165 MW each.

In recent times, emphasis has also been given for generation of power from mini/micro and small hydro electric schemes. Hydro units have been installed in the existing irrigation reservoirs for generation of electricity which was not contemplated earlier when the dam was built.

Renovation and uprating of units which have been in service for considerable time, are being carried out. Most interesting development in recent times has been setting up of parallel station of higher capacity on basis of low load factor to suit present grid condition for meeting peak demand in lieu of old stations built some 40 to 50 years back which operated almost as base load station.

Though the need for speedy harnessing of balance of hydro potential, is realised, resource crunch besides objection of environmentalists to implementation of projects have slowed down the pace of development considerably leading to imbalance in hydro-thermal mix thus disrupting the stable operation of the Power System.

2 PHYSIOGRAPHY & RIVER SYSTEMS

India is endowed with towering mountain ranges, rolling hills, lofty plateaus and extensive plains criss-crossed by rivers affording scope thus for Hydro generation. India can be classified into seven well defined regions physiographically, as described below.

1. The Northern Mountains comprising the mighty Himalayan ranges.
2. The great plains, traversed by the Indus and Ganga-Brahmaputra river systems. As much as one third of this lies in the arid zone of western Rajasthan-The remaining area is mostly fertile plains.
3. The Central Highlands, consisting of a wide belt of hills running east-west starting from Aravalli ranges in the west and terminating in a steep escarpment in the east. The area lies between the great plains and the Deccan plateau.
4. The peninsular plateaues comprising the western ghats, eastern ghats, north deccan plateau, south deccan plateau and eastern plateau.
5. The east coast, a belt of land about 100-130 KM wide, bordering the Bay of Bengal and lying to the east of the eastern ghats.

6. The west coast, a narrow belt of land of about 10-25 KM wide, bordering the Arabian Sea and lying to the West of the western ghats and

7. The islands comprising the coral islands of Lakshadeep in Arabian Sea and the Andaman and Nicobar islands of the Bay of Bengal.

India experiences very great diversity and variety of climate and an even greater variety of weather conditions ranging from continental to oceanic, from extremes of heat to extremes of cold, from extreme aridity and negligible rainfall to excessive humidity and torrential rainfall.

Most of the Rainfall in India occurs under the influence of South West monsoon between June to September in most parts of India except for some area which receives rainfall under the influence of north-east monsoon during October and November. The rainfall in India shows great variations, unequal distribution, still most unequal geographical distribution and the frequent departures from the normal. It generally exceeds 1000 mm in areas to the east of longitude 78oE. It extends to 2500 mm along almost the entire coast and western ghats and over most of Assam and Sub-Himalayan West Bengal. On the west of the line joining Porbandhar to Delhi and thence to Ferozpur, the rainfall diminishes rapidly from 500 mm to less than 150 mm in the extreme west. The peninsulai has large areas of rainfall less than 600 mm with pockets of even 500 mm.

3 RIVER BASINS

Rivers in India fall into four categories viz. (1) Himalayan rivers (2) Deccan rivers (3) Coastal rivers and (4) Rivers of the inland drainage.

Himalayan rivers are generally snow fed and perennial besides get copious supply during south west monsoon.

The rivers in Deccan are mostly rainfed especially during south west monsoon carrying huge volume of water during rainy season and dwindling thereafter till the next monsoon. Many of these rivers are not perennial.

The coastal streams, especially on the west coast receive copious rainfall, command huge inflows from limited catchment areas, loses great heights within short length affording scope for developing of Hydro power.

The streams of inland drainage basin are mostly of ephemeral character.

Depending on the size of the catchment area, River basins are categorised as major, medium and minor basins.

Major river basins of India are

1 Indus
2 (a) Ganga
   (b) Brahmaputra
   (c)Barak
3 Sabarmati
4 Mahi
5 Narmada
6 Tapi
7 Brahmani
8 Mahanadi
9 Godavari
10 Krishna
11 Pennar
12 Cauvery

Besides, there are 22 west flowing and 24 east flowing medium river basins which along with major rivers make bulk contribution to Hydro Power development.

4 TOTAL HYDRO POTENTIAL

The total hydro potential of the country has been assessed by the Central Electricity Authority as 84000 MW at 60% L.F. which is said to be 2.9% of world's potential.

4.1 *Scenario Hydro Power development since independence*

Hydro power development till independence was rather slow. Construction of water resources projects with attendant benefit of hydro power generation received a big boost after independence under successive five year plans. The performance by any standards has been very impressive.

The installed capacity of Hydro Stations was only 508 MW against a figure of 18442.9 MW achieved till the end of March 1991 recording an increase of about 36 times. The total energy generation which stood at about 2194 million units in 1947 rose to 64817.4 million unit in 1991, an increase of about 28 times.

The following table gives progress achieved during different periods since 1947 both capacity wise as well as energy generation wise.

Bulk of hydro potential is yet to be harnessed till end of March 1991 in different areas of the country as can be seen from the statement given below.

5 INDIGENOUS MANUFACTURE OF EQUIPMENT

Rightly recognising the need for manufacture of turbines and generating units indigenously to facilitate speedy development, the Govt. of India had set up Bharat Heavy Electricals Ltd. which has been supplying the most of the equipment required for a hydro electric station. The largest hydro unit in operation at present is 165 MW at Dehar Power Station under Beas Project.

Some of the projects under construction like Sardar Sarovar, Tehri Hydro Electric Project and Nathpa Jhakri Hydro Electric Project will have units of capacity varying from 200 MW to 250 MW each which BHEL will be supplying.

6 CENTRAL ORGANISATION FOR IMPLEMENTATION OF H.E. PROJECTS

National Hydro Electric Power Corporation was set up in 1976 to establish major hydro-electric projects on regional and national considerations like Salal hydro electric project etc.

The country has developed necessary skill and expertise in design, standardisation, construction and consultancy in the Hydro Electric Power Development, including major, medium pumped storage plants besides mini, micro and small hydel schemes.

7 PRESENT SCENARIO OF HYDRO ELECTRIC DEVELOPMENT

13th electric power survey conducted by the Central Electricity Authority has indicated shortage of peak power in all the regions of the country except North Eastern region during the eighth and ninth five year plans.

The pace of development of Hydro Electric Power has slowed down considerably during recent years due to obvious reasons like environmental considerations as well as problems in acquisition of forest lands so much so an imbalance has been created in the hydro-thermal mix to permit flexibility in operation.

Only fraction of Hydro Electric Potential has so far been harnessed inspite of some major and unique projects having been commissioned.

The present situation warrants speedy

Sl. No.	Region	Total Potential at 60% L.F.	Percentage of Total Potential of the Country	Potential Developed in MW	Percentage Potential Developed	Energy Generation in MU
1.	Northern Region	30156	36%	6072	20	48116
2.	Western Region	5679	7%	2275	40	73769
3.	Southern Region	10763	13	8246	77	34021
4.	Eastern Region	5590	6	1455	26	21385
5.	North Eastern Region	31857	38	395	1.2	1187
6.	All India	84044	100%	18443	22	178478

harnessing of balance of hydro potential after ensuring that the impact on environment as well as acquisition of forest lands are kept minimum by adopting suitable measures.

Resource crunch is also one of the factors which has slowed down the pace of Hydro electric power development.

It is hoped that the present policy of Govt. of India to allow private sector to execute the power projects, will to certain extent, ease the position.

7.1 Need for accelerating Hydro Projects

Implementation of Hydro Electric Projects to harness the balance potential at the earliest is called for considering their inherent advantages as detailed below.

1 They are sources of renewable energy with minimal impact on environment.

2 They are inflation free. The cost of generation of Hydro Power will remain practically static compared to cost of generation from other sources which escalates once in three to five years. Overall tariff rate is kept at reasonable level mainly due to low cost hydro energy.

3. Considering more and more thermal station/nuclear stations that are being set up to meet base load requirements, matching hydel capacity are required to meet the peak demand.

Hydropower'92, Broch & Lysne (eds) © 1992 Balkema, Rotterdam. ISBN 90 5410 054 0

Proposed Uma Oya multipurpose hydropower project in Sri Lanka and its technical and environmental aspects

Nihal Rupasinghe
Central Engineering Consultancy Bureau, Sri Lanka

ABSTRACT: The bulk of energy generation in Sri Lanka is from Hydropower. To meet the country's future power and energy demands there are several proposals. The proposed Uma Oya multipurpose project is one proposal from the hydropower sector. This project will also help to irrigate 10,000 hectares of new land in the most undeveloped South East Dry Zone in Sri Lanka in addition to the peaking power of 150 MW and 450 GWh of energy, generated from the proposed under ground power house with 2 x 75 MW Pelton turbines, utilizing a head of 770 m in a single stage.

1 THE EXISTING HYDRO ELECTRIC SYSTEM IN SRI LANKA

The bulk of electricity generation (99.8% in 1990) in the national grid comes from hydro electric plants. Fourteen hydro power stations are operated by the Ceylon Electricity Board with an installed capacity of 1015 MW with the capacity of providing about 3512 GWh/year under average hydrological conditions. Out of this, the power stations recently constructed within Sri Lanka's longest river Mahaweli Ganga between 1978 - 1990 under the accelerated Mahaweli Programme supplies 580 MW of power and 1800 GWh of energy annually. These stations are Victoria (210 MW, 726 GWh), Kotmale (201 MW, 482 GWh), Randenigala (122 MW, 378 GWh) and Rantembe (49 MW, 216 GWh). A 120 MW Power station which supplies 317 GWh of energy on the Walawe river, which is another major river in Sri Lanka is under construction and scheduled to be commissioned this year.

2 NEED FOR FUTURE DEVELOPMENT

Also in addition to hydropower, Sri Lanka has thermal power plants having installed capacity of 230 MW comprising of Diesel and Gas Turbines. At the present growth rate of 10%, Sri Lanka needs about 8000 GWh of energy per annum by year 2000 and 21,000 GWh of energy by year 2010. To meet this demand there are several proposals for Hydro power projects, Coal power projects, and Thermal power projects.

The country has already developed most of the hydro electric potential of the main rivers Mahaweli and Kelani basins, but yet untapped potential remains on both these rivers. Several studies on the remaining hydro electric potential of the country have been completed during the recent past. Upper Kotmale Project (Mahaweli river), Broadlands Project (Kelani river), Kukule Project (Kalu Ganga) and Uma Oya Multipurpose Project (On Mahaweli Tributory) are the main future candidates from the hydro power sector. This paper outlines the findings of the Pre-Feasibility study carried out on Uma Oya Multipurpose project with respect to technical and environmental aspects.

3 BACKGROUND OF UMA OYA MULTIPURPOSE PROJECT

Uma Oya, one of the major tributaries of Mahaweli Ganga, flows Northwards and joins the Mahaweli Ganga on its right bank at Rantembe Reservoir just upstream of Minipe. Originating from the central hills at an elevation of about 2500 m.a.s.l., Uma Oya drains a catchment of 723 sq. km and has a mean annual run-off of 600 MCM. In its passage of 75 km upto the confluence with the Mahaweli Ganga, it drops by 2365 m.

Uma Oya has a gross hydro-energy potential of about 600 to 700 GWh. It has therefore attracted attention for generation of hydropower from early planning stages

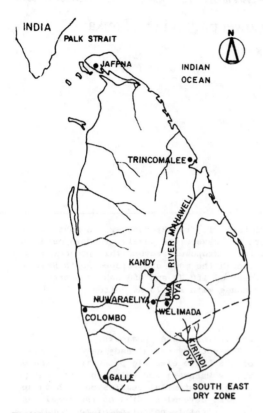

FIGURE 1 - PROJECT LOCATION

land in the Kirindi Oya and Menik Ganga basins.

Beyond this proposed trans-basin diversion, the residual drop of 828 metres and residual yield of 330 MCM will be left in the middle and lower catchment of Uma Oya for generation of Hydropower either indipendently or even more profitably by linkage with the already constructed Randenigala reservoir on the Mahaweli River.

4 PROPOSED HYDRO POWER DEVELOPMENT LAY OUT

4.1 General

Several alternative layouts were studied and the most favoured development consists of the following.

A concrete arch dam across Uma Oya 92 m high (FSL 365 m.a.s.l.) with an active reservoir capacity of 17 MCM and a side intake which will divert waters of Uma Oya through a 19 km long 3.5 m diametre trans-mountain power tunnel to the adjacent basin (Kirindi-Oya). In its passage, the supplies to the tunnel will be supplemented by.

a. From Norwegian type Brook intake at Mahatotilla Oya (Tributor of Uma Oya) which contributes about 33% of the total volume of diverted water. This will be achieved by

since 1959. At different periods, United States Operation Mission USOM, UNDP/FAO, Government of Netherlands (NEDECO), GTZ of Germany and KfW also of Germany have sponsored studies for ascertaining the feasibility of the development of hydro-power potential of Uma Oya.

All these studies in the past have invariably emphasised in basin hydropower development only. However, in Nov. 1989 motivated by the keen desire of the Government of Sri Lanka for the development of South East Dry Zone (SEDZ), the Central Engineering Consultancy Bureau conceived, a Multipurpose scheme as an alternative development encompassing a major proportion of the power potential of Uma Oya basin. Essentially, this scheme envisaged to divert the waters of Uma Oya and its tributary Mahatotila Oya from Welimada Plateau through a trans-basin diversion. A drop of 750 m could be utilized for a single stage development of about 450 GWh of hydro-energy annually (at an underground power station of 155 MW) and augment the Kirindi Oya flow by about 250 MCM which could be used to develop about 12,500 ha of

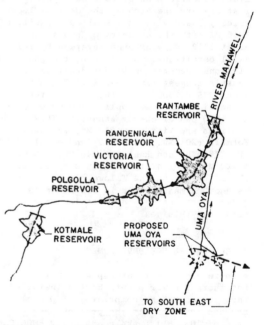

FIGURE 2 - RESERVOIRS ON RIVER MAHAWELI WITH RESPECT TO PROPOSED UMA OYA RESERVOIRS

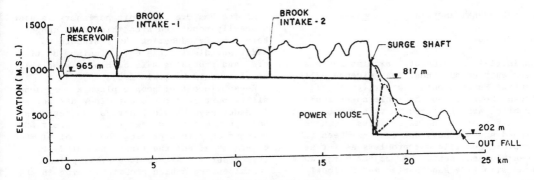

FIGURE 3 - LONGITUDINAL SECTION OF WATER WAY

a 20 m high, pick up dam of roller compacted concrete across Mahatotilla Oya nearly 5 kms upstream of the confluence with Uma Oya and a drop shaft to the tunnel passing underneath.

b. From another Brook intake from Upper Kirindi Oya (it self) at 12.1 km of the tunnel where it passes underneath Kirindi Oya. This would contribute about 12 MCM for generation of power.

The low pressure tunnel will terminate at the surge shaft of 10 m diametre with an orifice at the bottom from which steel lined high pressure shaft of 2.5 m diametre will convey the water to two Pelton turbines of 75 MW each (Utilising a head of 770 m in a single stage) installed in an underground power station with a free-flowing tailrace tunnel. The underground power station will be located about 650 m below ground with a 56 m(L) x 14 m(W) x 35 m(H) main cavern to house two 75 MW Pelton turbines and main inlet valves. The entry from the penstock to the Pelton turbine will be oblige (Norwegian Cost reduction method) so as to accomodate the inlet valves within the cavern with little increase in span of the cavern. Tailrace tunnel is horse shoe shape with the area of 25 m^2 and length of 4.25 kms.

The cost benefit ratio at 10% discount rate is 1.57 and Economic internal rate of return is 14.7%. Financial cost estimated at Rs. 11 billion (245 US $ million) with 35% is the local cost component. Also the unit generation cost is estimated to cost about 2.15 Rs/kwh or 0.048 US $/kwh.

4.2 Water Conductor System

4.2.1 Low Pressure Tunnel

There would be more than sufficient rock cover above the 19 kms long tunnel trace except at some isolated locations, such as crossing below first brook intake at Matotilla and Rawana Ella. It is intended to provide reinforced concreted lining in such isolated reaches.

The tunnel is expected to pass through an alternating sequence of gneisses, granulities, quartzites and lime stone with charnokite, with Charnokites and gneisses predominating. However, the geological evaluation in its totality leaves little doubt about the technical feasibility of the proposed tunnel alignment. Also it could be expected that about 90% of the tunnel will need no supports. The diametre of the tunnel has been kept as 3.5 m from hydrolic and practical consideration although the economic diametre worked out on basis of weighted average discharge through the tunnel has been varying between 3 to 3.3 m depending upon the unit rate at which the power generated from the project will be sold. Considering the long length of the tunnel and due to favourable Geology, use of Tunnel Boaring Machine (TBM) is the obvious and expedient choice. The alternative of equivalent modified horse shoe section for ease of construction but, this alternative was given up for reasons of economy and expediency.

4.2.2 High Pressure Tunnel and Power Station Complex

The rock type and the sequence in the zone where the pressure shafts and the underground power station will be located are generally competent and favour the technical feasibility of an underground pressure shaft which has been adopted in reference to a surface layout and surface power station after an economic study. The vertical and steel lined pressure shaft has been suggested at this stage.

715

5 ENVIRONMENT ASPECTS

5.1 General

An initial environmental examination has been carried out under the guidence of the Central Environmental Authority of Sri Lanka. However, on the whole no serious problems are envisaged in this respect, as firstly most of the major works are underground and secondly due care has been taken to avoid resettlement problems as far as possible and locate the project works away from wild life sanctuaries and national parks.

The proposed Main reservoir on the Uma Oya will inundate a land area of about 52 ha while the Dyraaba reservoir (Boork intake) would cover 3 ha. The reservoir bed area consists mainly of paddy, vegetable and potato cultivations with narrow valleys and steep slopes. The natural vegetation around the area is mainly Pathana and as such the intended project may not have a significant effect on large animals. However the disturbances caused on such habitant would have effects on other smaller animals and disturb the breeding grounds of the avifauna.

Further it would necessitate resettlement of about 500 persons (100 families) due to impounding of the reservoir and other construction activities. The resettlement figure is very much less when compared with other proposed Hydropower projects in Sri Lanka. As an example the proposed Upper Kotmale project needs 13,500 persons to be resettled and proposed Kukule project needs 9500 persons to be resettled.

But, the damming of flowing water and impounding of reservoirs should no doubt have impacts on the natural and human environment. The intended project is no exception to this. Such impact could be catergorized as impact to human ecology, impact to animal ecology, impact to aquatic ecology and reservoir downstream changes.

5.2 Human Ecology Changes

This could be divided mainly to two factors i.e. effects to people who continue to live in the vicinity of the Dam and effects to people who will be re-settled. The people who continue to live within the Dam area have little knowledge of the presence of the danger of the large body of water specially in the initial stage of reservoir filling. As a result, the lake becomes a death trap to the locals, specially when fishing, bathing and washing vehicles. This was clearly seen at Victoria reservoir which was constructed in 1984.

As the Uma Oya reservoir periphery is also a heavily populated area this could happen here too. By educating the residents on different disciplines eg. swimming, boating etc. and providing safe bathing places this problem could be minimised.

Re-settlement of people plays a very sensitive part in the process. They are normaly under psychological stress and fear. As Uma Oya project requires resettle about 500 persons only this problem too is not of much consequence. But the lands inundated from this reservoir are potatoe growing lands (potatoes are a high profitable crop in the island) an area which is very famous for good quality potatoes.

5.3 Aquatic Ecology Changes

Observations on recently built reservoirs in the hill country in Sri Lanka reveals that rate of growth of visible Aquatic plants is fairly low. The deep reservoirs where the water close to the reservoir bed is not sufficiently stirred, deoxygeneration may take place. As a result this becomes an unfavourable habitat for deep water organisms. This phenominon is true for proposed Uma Oya reservoir too.

Also the observations made on recently built lakes in Sri Lanka reveal that the rapid fluctuation of water level in the reservoir, promotes weed growth within the reservoir bed. The intermittent inundation of this vegetation can create an environment for it to decay and pollute the water. This problem could not be much at Uma Oya, as reservoir is small.

Evidence so far collected indicates that the migration of fish upstream along the rivers is not significant. As such this is not a major consideration in implementing reservoir projects in Sri Lanka. However the river habitant is completely changed when the reservoir is impounded. The indigenous species of fish living in fast flowing water is replaced by lake dwelling fish. The introduction of commercial fisheries in lakes has not been very succesful due to religious and other social factors. Therefore in Uma Oya too it will be difficult to develop the commercial fishing industry.

The slopes of the upper reaches of Uma Oya and to a certain extent that of Mahatotilla Oya are extensively cultivated. Use of pesticides and fertilizers on these cultivations are inevitable. These chemicals no doubt find their way to the water of the two Oya (Streams) along with eroded soil. These factors would effect water quality specially it will spark the blue green algae which is present frequently in Sri Lankan lakes.

This problem has been experienced in the Kotmale reservoir, very recently. The presence of a large quantity of blue green algae will drastically reduce the oxygen content in the reservoir.

The result of these habitat changes will be the elimination of certain aquatic species of fish and plants and enhancement of the others. For this reason specially the impact on the project on endemic and amphibian species need to be investigated in the next phase of the project.

5.4 Stability of Reservoir Area and Downstream

The rapid fluctuation of water levels in reservoirs give rise to instability of reservoir banks which has become a major concern in the recently built large reservoirs close to urban areas. This problem also has to be anticipated in the Uma Oya reservoir. Also the siltation of the Uma Oya reservoir is expected to be very high. At present this is due to traditional method of man made cultivation and improper land use pattern. Therefore an efficient silt ejection system in the dam has been proposed with a sound system of land use pattern for the catchment.

Downstream of the dams will no longer receive the natural inflows but will depend on the low level releases and spillage during the periods of flooding. However the effect of this will be significant only for the few kilometres downstream of the Uma Oya dam as there are several downstream tributaries such as Haloya, Madulu Oya with substantial yields. But dams will have an impact on the ecosystem downstream.

5.5 Impact to Environment from Uma Oya Power Tunnel and Power House

As all works are underground, there will be very little physical impact and environmental disturbances. There are so many suitable spoil tip areas to dispose excavated material from tunnel and power house construction. It is seen from past experience that construction of such tunnels could effect the water table and water quality in close proximity to and to some distance to the tunnel. This would result in the drying up and changes in water qaulity of wells, streams etc. along the surrounding areas of the tunnel. The Uma Oya power tunnel has been located just outside the southern border of the famous Rawana Ella forest reserve in Sri Lanka to reduce the impact on the Faune and Flora of the reserve.

6 CONCLUSION

The proposed Uma Oya multipurpose project is technically feasible and economicaly viable. No serious environmental problems are forseen. Therefore this project will help to meet future power and erergy demands in Sri Lanka and also will provide water to develop the most undeveloped South East Dry Zone of Sri Lanka.

REFERENCES

Central Engineering Consultancy Bureau 1991. Uma Oya Multipurpose Project. Pre-feasibility Study.

Jayasekara, H.B.1991. Environmental Changes due to Dam Construction. Seminar on Dam Safty & Reservoir Operation, Sri Lanka National Committee on Large Dams.

Hydropower'92, Broch & Lysne (eds) © 1992 Balkema, Rotterdam. ISBN 90 5410 054 0

Hydropower potentials in Greenland – Up to 70°N

Th.Thomsen
Nukissiorfiit, Greenland Energy Supply, Copenhagen, Denmark

Abstract: The hydropower potentials are part of the energy planning in Greenland. In the course of the years (since the seventies) areas have been localized, partly with a view to building hydropower plants for supplying energy to the towns, and partly "large" potentials designed to optimum the amount of energy produced.

The degree of specification of the localized plants varies a lot. The calculation of potentials of urban installations is based on measured values, whereas the potentials for the "large" installations are partly based on direct measurements, partly on interpolations.

The three areas for extension of the hydropower facilities in Greenland that are of the greatest current interest are all the small type of plants for supplying the towns.

Energy development

The awareness of energy has been strengthened since the sixties, and today everyone fully realizes that energy resources are an absolute must, if society is to function and expand.

Since the days of the energy crisis, the energy consumption in the industrialized world has diminished expressed in percent, but measured in absolute figures there is still a positive increase. This is not due to a correspondingly lower growth rate, but rather to the fact that the development today does not require as much energy as earlier. This development is preeminently based on new processes within the computer and processor technology which do not require as much energy as earlier types of development (machinery and engines).

The energy problem has changed character during the past decades, from being a resource problem to being a pollution problem, tabel 1.

Outlined in tabel 1 is the attitude basis of energy-political initiatives during the past 30 years. Prior to 1973 a barrel of oil cost 2 $ or less, and with a tripling of the prices, the attention started focusing on the resources. When prices were once more tripled in the late seventies, with a barrel oil costing upwards of 35-40 $, an economic crisis and recession also started in the western world, in addition to the resource crisis.

With this situation as the starting point, intensive hydropower investigations were initiated in Greenland during this period.

The price of oil has since then fallen to approx. 16 $ a barrel by the end of the eighties, and slightly more after that. In the same way there has been a change of attitude from the economic to the environmental aspect of the consumption of energy, so that

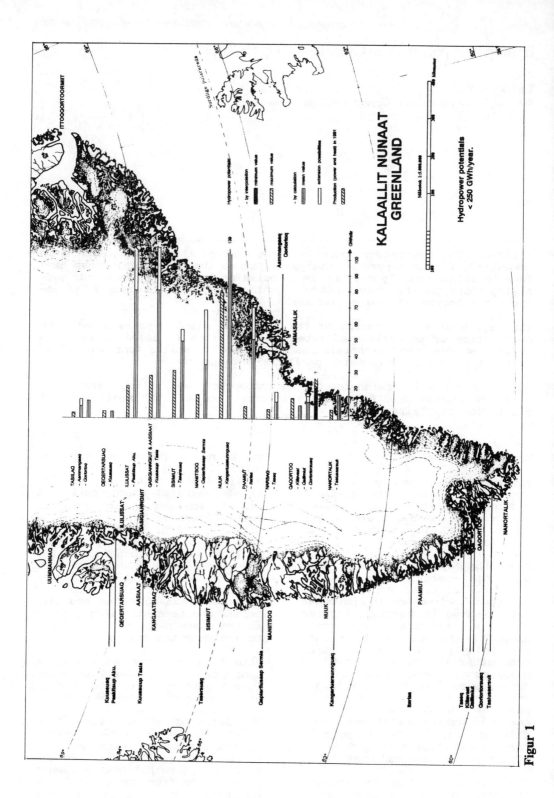

Figur 1

Tabel 1 The change character in the energy problem.

Year	Attitude
1960	H: No problem A: Free consumption.
1973	H: Energy crisis A: Nuclear power, coal. Energy restrictions.
1979	H: Energy for expensive A: Natural gas, coal. Social recessions.
1985	H: Energy pollution A: Wind, sun, natural gas. Purification introduced.
1990	H: CO_2, the real problem A: Brundtland plan of action CO_2 taxes.

H: Holding
A: Action

the energy problem today is primarily a CO_2 problem.

The known and estimated reserve (R) of oil in relation to the production (P) is today approx. 42 years. This R/P ratio has since the eighties been slightty increased, but at the same time there is growing recognition that the fossil fuels will not last indefinitely. So in the century to come we shall all see how the use of fossil fuels will peter out compared to other sources of energy. As reserves are gradually being exhausted, there will be an even greater geographical dependence on them in certain parts of the world.

The total energy consumption in the world has been and is still undergoing a change. So even if the oil consumption only accounts for a reduced part of the world's total energy turnover, there has nevertheless been an increase of

0.6% per annum, tabel 2. Hydropower, gas and above all nuclear power are the sources of energy that have taken over the energy production, subsidizing oil, and to some extent coal.

Tabel 2 World comsumption of energy sources and yearly change.

	1973 %	1990 %	±% p.a.
Oil	47	39	+ 0.6
Gas	18	22	+ 2.9
Coal	28	27	+ 1.6
Hydro	6	7	+ 3.0
Nuclear	1	6	+ 14.0

(Source: BP statistical review of world energy, june 1991).

It is not quite clear which alternative sources of energy will take over when the supply of fossil energy starts decreasing. No one believes in wind and solar energy, based on the expected development of the relevant technologies. Energy supply from these sources will only constitute a limited part of the total energy requirements. Alternative sources of energy will presumably be within the field of fission (nuclear power and breeder reactors), and whether fusion will ever become a reality is and remains the big question for many years to come.

The remaining known hydropower resources in the world will definitely be extended. Once it becomes possible to transport energy without loss, a number of the world's largest, remotely situated hydropower resources will become economically attractive.

Hydropower localization

Possible sites for the establishment of hydropower stations have been localized since the mid seventies. During this period there have been various motivations for localization of hydropower plant, so that we today distinguish between two types of power-plants.

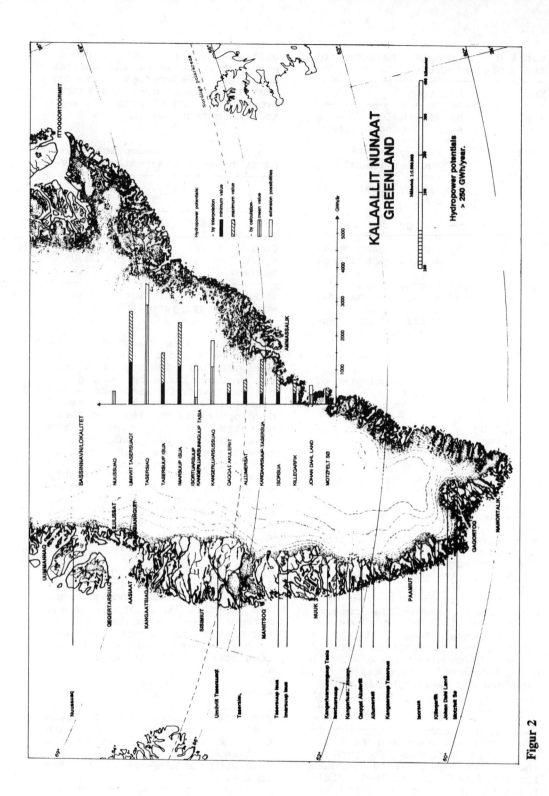

Figur 2

Small hydro-power plants for the supply of energy to towns are localized for the purpose of placing a hydropower plant as close as possible to a built-up area, with a potential that will be sufficient to supply the town with electricity and heating for the next 20 to 30 years.

The investment in hydropower plants must be profitable for the national economy, i.e. the total investments in hydropower facilities must be seen in relation to an extension and renewal of the traditional oil-based electricity-producing plant. The depreciation period for comparable investments is 20 years, based on an updating of the oil price and an internal interest rate of 7%.

The coming hydropower extension in Greenland will therefore depend on a renewal of the existing electricity-producing plant.

Localization and extension of small hydropower stations for supplying towns with energy are therefore a function of:

- the distance to the town in relation to the size of the potential (water volume and m. head)

- plant investment in relation to traditional diesel plant (storage facilities, distance to outflow and transmission line)

- extension potentials

- natural conditions

The localized potentials differ in their degree of specification, both with regard to the hydrological assumptions and the construction-technical extension potentials. The degree of specification will be extended successively, and when the economic conditions warrant an extension of the hydropower facilities, the project will be developed into a masterplan.

Localities for energy supply of the Greenlandic towns (<250 GWh/year) are shown in figur 1.)

The other type of plant are larger installations localized in order to establish a hydropower station with the greatest possible potential. The investment per produced unit must therefore be reduced to a minimum, so that energy-intensive production can take advantage of the "free" energy. Another way of exploiting it is by converting the produced energy into a hydrant or another energy-carrying substance capable of transporting the energy to another energy-consuming place. Energy losses are still considerable, but a feasiblity project is now underway which is to investigate the expected effectivity of an energy-storage system today and in the future, when hydrogen is used as energy carrier, and which type of energy carrier is most suitable for Greenland conditions, when energy has to be supplied to the towns in Greenland.

Localization of large hydropower installations for energy-intensive use is therefore a function of:

- optimization of the potential (great head, large water volumes, and transfer potentials)

- simple station design (storage facilities, tunnel lengths)

- logistics (navigation conditions during the year, access conditions)

- extension potentials

- natural conditions

Large localized hydropower potentials (>250 GWh/year) differ even more in degree of specification.

Today known localities and power potentials are shown in figur 2. The total know hydropower potential are 12 to 20 TWh.

Some figures from the localities in figur 2 are given i tabel 3.

Tabel 3: Basininformation for hydro potentials >250 GWh/year

Basin	Head (m)	Yearly run-off (mio m³)	Potential (GWh/year)		Tunnellength (km)
Nuussuaq[A]	250	599	365		18
Umiiviit Tasersuaqt	300	1500–3700	1100–2700		13
Tasesriaq[B]	600	1940–∞	2860–∞		25
Imarsuup Isua	650	1100–2400	1700–3700		15
Tasersuup Isua	65	2200–4000	350–640		8
Imarsuup + Tasersuup + Søndre Isortup Isua	65	3700–9600	600–1500		
Isortuarsuup Tasia	200	805		325	
+ Kangerluarsussuaq	255	1140		716	
2 anlæg i alt[C]			1111		25
Kangerluarsussuaq[D]	280	1560	1070		
	220	1280		690	
	90	500		110	
2 anlæg i alt			1870		20
Qaqqat Akuleriit	450	300–550	330–600		5
Allumersat	720	200–400	350–700		6
Kangaarsuup Tasersua	150	1600–3500	590–1300		5
Isorsua	700	200–500	340–850		30
Killeqarfik	510	250–500	310–625		6
Johan Dahl Land[E]	650	172–450	270–555°		15
Motzfeldt sø	120	800–1650	230–485		9

The yearly runn-off is estimeted maximum and minimum values given by Greenland Geological Surveys (pers. communication) except localities A to E.

A Run-off measured in the periode 1980–84

B Run-off measured in the periode 1975–p.t.
   ∞ depending on the watervolumen in the icedammed Lake 860

C Run-off measured in the periode 1981–p.t.

D Run-off measured in the periode 1975–87

E Run-off measured in the periode 1976–p.t.
   ° inclusive extension from drainage basin Thor sø, Odin sø and Hullet.

Total hydro-power potential know to-day: 12-20 TWh

Hydropower extension

The three areas for an extension of the hydropower facilities in Greenland that are of the grea-test current interest are all the small type of power plant for supplying the towns. Based on economic considerations it has been decided to build a hydropower plant at Nuuk where the Kangerluarsunnguaq reservoir will be exploited.

Calculations and evaluations are now being carried out for a hydropower station at Sisimiut where the Tasersuaq reservoir will be exploited. If the calculation and quotation come out positive for the local community, the project will be submitted for political evaluation in the autumn of 1992. The building of a power station in South Eastern Greenland at Qorlortorsuaq that is to supply energy to the towns Narsaq and Qaqortoq is likewise being investigated. The project is still at a stage where the technical, and especially the hydrological, assumptions have not yet been fully clarified.

Nuuk, Kangerluarsunnguaq

The hydropower station is situated some 50 km south/south east of Nuuk and has been established with a main reservoir at Kangerluarsunnguup Tasia that flows into the Kangerluarsunnguaq Fiord (Buksefjorden) see figur 3. The reservoir lake Kangerluarsunnguup Tasia will be dammed to 11 metres above the natural threshold, and with a submerged intake the reservoir's volume is estimated at a little more than 1,500 million m³. This seen in relation to a produced water volume at max. effect (28 MW) of 318 million m³ will provide regulating possibilities making the hydrological operating conditions of less crucial importance.

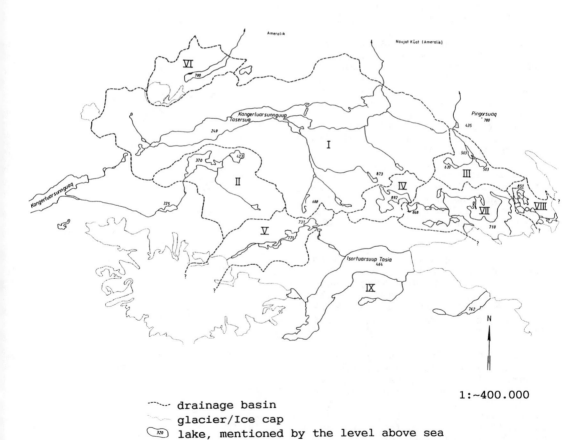

1:~400.000

⌒〜 drainage basin
〜〜 glacier/Ice cap
⌒³²⁰⌒ lake, mentioned by the level above sea

Fig. 3. Drainage basin, Kangerluarsunnguaq

The reservoir is a precipitation reservoir without glaciers of any significance. The effective mean precipitation in the area is 370 mm per year. The time series analysis of the precipitation from Nuuk has not demonstrated any statistically significant changes during the past 100 years, but with an absolute volume of "only" 360 mm/year, a relatively small change of the amount of precipitation might have a correspondingly greater relative effect.

The station will be established with runoff besides basin I from catchment areas II, IV and V, so that the annual flow to the power station will be approx. 300 million m³ and the energy production 199 GWh.

There will be numerous possibilities of extending the power station with catchment areas III, VI, VII, VIII and IX, figur 3.

Catchment area VII is an ice-dammed lake, the water volume of which has been estimated conservatively. Catchment area IX is the lake Isortuarsuup Tasia which is dominated by melt water from the inland ice (ablation). Catchment areas IV, V, VII and VIII under natural conditions flow to catchment area IX.

If all catchment areas are transferred to Kangerluarsunnguaq Tasia, the annual water volume will amount to 1,140 million m³, with a potential production of 715 GWh/year.

If a power station is established between Isortuarsuup Tasia at elevation 464 m and Kangerluarsunnguaq Tasia at elevation - 255 m, the energy potential will amount to 1,100 GWh/year (1.1 TWh).

During the operating stage a measuring system will be established that will make it possible partly to monitor the relevant hydrological conditions for the operation, and partly to set up a hydrological time series for the area, if hydropower had not already been established.

Both in relation to the small amount of precipitation in absolute figures and the hydrological operation, the data can also be used for assessment of long-term changes of the hydrological differences - not least changes arising in connection with a future change of climate caused by the increased greenhouse effect.

References

Run-off data are from Greenland Home Rule's hydrological database and Greenland Geological surveys.

Hydropower'92, Broch & Lysne (eds) © 1992 Balkema, Rotterdam. ISBN 90 5410 054 0

The Andhi Khola project, Nepal – Hydropower in development

G.A.Weller
Sir William Halcrow and Partners, UK (Formerly: BPC Hydroconsult, Nepal)

T.Skeie
Himal Hydro and General Construction Ltd, Nepal

D.Spare
United Mission to Nepal, Nepal

ABSTRACT: The Andhi Khola 5.1MW hydropower project in west Nepal is a milestone in a strategy for sustainable development of hydropower in Nepal and a Nepali hydropower industry. It has been constructed wholly by Nepali companies, making maximum use of local resources and is one of the cheapest sources of hydroelectricity in Nepal. Nationally it has led to the establishment and/or significant growth of companies able to design, construct, operate and maintain small and medium sized hydropower plants. In the project area it has been responsible for a sustainable improvement in the standard of life for the population through the implementation of programmes such as rural electrification, irrigation and water supply. It is considered that this is a good model for hydropower development which could be replicated elsewhere.

1. INTRODUCTION

Nepal is located to the north of India and to the south of China, stretching along the Himalayan mountains. Within its compact area of 147,000km^2 the land rises from the Ganges plain at elevation below 100m to the high Himalayas, including Everest, above elevation 8500m. This dramatic terrain is a major handicap to the economic development of the nation with many communities isolated from the centre. Yet it also provides a key to the successful development of the nation being the source of the country's one major renewable natural resource, water.

There are four major tributaries of the Ganges river system, one of which is shared with India. The annual runoff from the watersheds of these is approximately 200 x 10^{9}m^3, equivalent to about 14% of the total flows discharged from all the rivers of India and about 17% of those of Bangladesh. Warnock (1989).

The theoretical hydro-electric potential of Nepal is 83,000MW of which approximately 45,000MW is considered to be economically exploitable at the present time. This is far in excess of domestic need and could become a valuable export. It is most important, however, that the development of hydropower is carried out with great care to bolster the fragile economy of Nepal. If developed appropriately the benefits to the country could include:

- Major export earnings
- A basis for industrial development
- A major industry in itself
- Ecological improvements by replacement of fuelwood as main source of energy (currently over 80% of total energy consumption in Nepal).

Nepal is one of the poorest countries in the world, so foreign aid is needed to carry out hydropower development. However this needs to be carefully managed because conditions may be attached which require the use of expertise and equipment from the donor country. Projects are completed as quickly as possible to maximise profits and often there is little or no transfer of expertise at a national level or consideration given at a local level to the project area.

Recognising these problems the United Mission to Nepal (UMN) became involved in hydropower development in Nepal in the 1960s. UMN is a non-government organisation of professionals

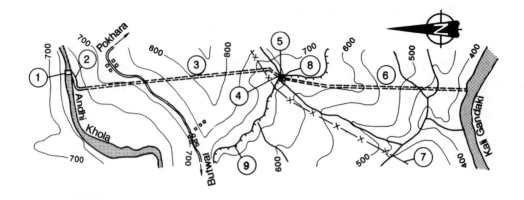

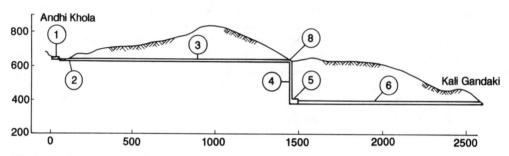

Fig.1 Site layout and profile of the Andhi Khola hydropower project, where:
1 = dam; 2 = desilting facility; 3 = headrace tunnel; 4 = shaft; 5 = powerhouse; 6 = tailrace tunnel;
7 = transmission line; 8 = irrigation scheme offtake; 9 = irrigation canal

from over 40 Christian missions. Working in partnership with His Majesty's Government of Nepal (HMGN) it is developing small and medium sized hydropower, and a national industry to carry this out.

The HMGN/UMN approach has been to establish private Nepali companies to: design, construct and maintain hydropower plants; provide on and off the job training; carry out research and development for appropriate technologies; and, eventually hand these over to Nepali management. Supported by expertise and funding from Norway, this approach is proving successful with 1MW and 5.1MW schemes completed, a 12MW scheme under construction and a 45MW scheme at feasibility stage.

2. THE ANDHI KHOLA HYDROELECTRIC PROJECT

The 5.1MW Andhi Khola Hydroelectric Project is located about 80km south of Pokhara in western Nepal. It was completed in mid 1991 and has been a key element in the HMGN/UMN strategy for the development of hydropower in Nepal. At all stages of its construction consideration has been given to the benefits of the project at national and local level.

At a national level, it has been demonstrated that schemes of this size can be constructed at low cost by Nepali companies. The project has also been a catalyst for the establishment and/or significant growth of the following companies in the field of hydropower:

- Butwal Power Company Pvt Ltd (BPC), Owner and Manager, 1966
- BPC Hydroconsult (BPCH), Consulting

Engineers, a division of Butwal Power Company, 1986

- Himal Hydro and General Construction Pvt Ltd (Himal Hydro), Civil Contractors, 1978
- Nepal Hydro and Electric Pvt Ltd (NHE), Electro-mechanical plant manufacturers, steel fabricator, 1986.

The impact on the project area has been considered and programmes implemented to increase local capabilities, thereby being of lasting benefit. Accordingly, alongside the hydropower development, rural electrification, irrigation, drinking water, sanitation, non-formal education and resource conservation schemes have been carried out.

The project has been financed by Norwegian Government aid (NORAD), which shares HMGN/UMN's aim to develop a Nepali hydropower industry. Grants of approximately US$5 million have been given towards the power and irrigation schemes.

2.1 Power scheme

During construction of the 5.1MW power scheme, emphasis was placed on training personnel and building up the companies involved. Local resources were used as far as possible and the project implemented in a way that could be replicated elsewhere in Nepal.

Figure 1 shows the site layout. The scheme utilises a difference in water level of about 250m between the Kali Gandaki river and its tributary, the Andhi Khola river, at a point where the two rivers come within a distance of about 2km of each other. The project is a run-of-river scheme with daily storage provided during the dry season by flashboards on top of a 6m-high dam. The dam is a 60m-long concrete gravity type with an ogee crest and flip bucket spillway. It is located at a point on the river where there is a narrowing and visible rock.

Sediment transport is a common major problem. A sluiceway with a 2 x 1.5m radial gate maintains free flow to the side intake, which incorporates an undersluice and gravel trap. An open canal leads to a desilting basin, where there

is a fish ladder to allow migrating fish to pass the dam. A 1.3km-long headrace tunnel leads a maximum $4m^3/s$ of water from the desilting basin through the 300m-high ridge separating the Andhi Khola and Kali Gandaki valleys. Water is then conveyed to an underground powerhouse through a 250m-long vertical, 1.05m maximum diameter welded steel penstock, to three 1.7MW pelton turbines, and then along a 1km-long tailrace tunnel to the Kali Gandaki river.

The vertical steel penstock is located in a 4m diameter shaft which is also the main access to the underground powerhouse.

The turbines are double jet peltons, each being designed for a flow of $0.9m^3/s$. The turbines are directly connected to three generators, each with a capacity of 2.2MVA. Power is carried up through the access shaft via three-phase 5.3kV busbars to the ground surface where 5.3kV/33kV transformers are located for distribution to the grid. The main control facility is also located here.

Almost all the electro-mechanical equipment has been obtained from Norway at virtually no cost, apart from packing and shipping. The three

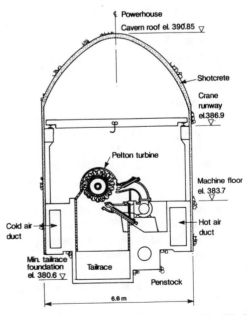

Fig.2 Cross section of the Andhi Khola Powerhouse

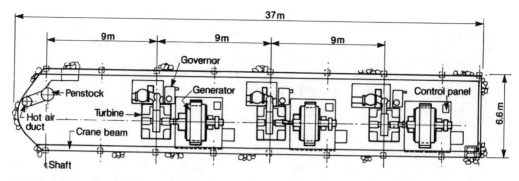

Fig.3 Plan of the Andhi Khola powerhouse

1.7MW pelton turbine sets were installed in the Mesna power station, Lillehammer between 1919 and 1926 and removed when the station became redundant. The crane at the shaft head had been ordered for a shipbuilding yard, but was never put to use. It has been modified in Nepal to meet project requirements.

Because of the highly unstable slopes of the Kali Gandaki valley, it was decided to locate a large part of the works underground. The underground works were not without problems, however, because the rock is highly fractured phyllite which is relatively unstable after excavation. Therefore the width of the powerhouse was minimized: it is 6.6m wide and 37m long.

At every stage of design and construction, care was taken to use the most appropriate technology. Because of the low cost of labour, labour-intensive methods were generally found to be cost effective. At its peak, the workforce was approximately 450. There was a lack of skilled labour. Good tunnellers were not available and experience had to be gained at the project itself. As a result there is now a pool of skilled Nepali tunnellers.

Stone rubble masonry was used extensively in the intake works and for lining the tunnels. The gravity dam includes 30 to 35 percent rock aggregate of size 200-300mm diameter, with a 300mm surface layer of high grade concrete, thus reducing the use of relatively expensive cement.

In the powerhouse, shotcrete and rockbolts have had to be used because of the poor rock quality. These have been valuable new techniques for Himal Hydro to learn.

Two ropeways have been constructed to supply the tailrace portal with materials, one from the shaft site for cement and equipment and another crossing the Kali Gandaki for aggregate. Ropeways are often a cost-effective alternative to construction roads in the steep, unstable hills of Nepal, providing year-round access with minimal environmental impact.

2.2 Rural electrification

One of the main purposes of the Andhi Khola project is provision of electricity to a target population of about 150,000 in the surrounding area. Two 33kV high tension lines have been constructed, one of which goes south to connect to the National Grid for the sale of surplus electricity. For the local supply 1kV distribution lines are used as an intermediate voltage to minimise overall cost.

To date about 450 households in two villages are connected to the local grid as a pilot project. Non-formal education is being undertaken to teach the new consumers how to use electricity. Active participation by the local people is encouraged in the construction of distribution lines, operation of the system and collection of charges.

Electricity is generally used for lighting, but an increasing number of consumers is investing in low wattage cookers, developed by UMN. This latter development is very important if a reduction in the use of fuelwood as a primary energy source is to be achieved. Domestic tariffs are simple and subsidised to place electricity within the reach of poor farmers. Most households subscribe to a certain load at a fixed rate per month regardless of use, eg 250 watts at 70 rupees per month. No meter is needed but a cut-out switch prevents the subscribed watts being exceeded.

2.3 Irrigation

The project area is food deficient. Therefore one of the highest priorities of residents in the area is irrigation. Accordingly it has been agreed to allow excess water from the power scheme to be used for irrigation, transforming nearly three hundred hectares of dry land.

The offtake for the irrigation scheme is located between the headrace tunnel and vertical steel penstock. The 7km long main canals of the scheme traverse some very steep slopes and considerable slope stability and soil conservation works have been required. Innovative use has been made of inverted syphons incorporating suspended pipe crossings, of up to 110m span, to cross deep gullies. Secondary canals flow down the steep hillsides and research is being carried out to find the most appropriate construction technologies and methods for these. New designs have been produced for simple proportional division structures which can be fabricated locally and operated by the farmers.

A central aim of the scheme is to distribute the benefits of the irrigation system more widely than to only those farmers who have substantial amounts of land. A users association has been formed which all 1,300 households in the area have been invited to join, whether they have land or not, earning water shares by the contribution of labour. Five days work, or 165 rupees, earns one of the 25,000 shares available. Distribution of water is according to each household's number of shares, and each family can purchase only a limited number. Land redistribution is a necessary part of this. A certain amount of land has been defined as necessary for self-sufficiency. Those with more are required to sell 10% of the excess at a fixed price to the users association, for re-sale to those who have less. Not surprisingly landowners were initially opposed to the scheme,

but have now changed their views as they see the benefits to themselves with greatly improved yields from their land and advantages for the whole community. The scheme is managed by the farmers and it is hoped that it will be copied elsewhere, allowing whole communities to benefit from irrigation schemes.

2.4 Drinking water and sanitation

Good quality drinking water is vital to the health of a community but rarely available in rural Nepal. In response to the requests of local residents the project is developing drinking water systems and communities are being encouraged to plan, build and maintain their own drinking water and sanitation systems. The community is responsible for supply of labour, local materials and maintenance. The project provides pumps, pipes and maintenance training. Before construction starts, each household is required to build a latrine.

2.5 Non-formal adult education and resource conservation

Non-formal education is considered essential to make rural electrification effective in raising standards of living and general welfare in the area. Young local people are trained to be trainers: visiting homes and villages, talking with people and teaching through drama.

Increasing people's capacity for choice is one goal of literacy. Since illiteracy is so prevalent in rural Nepal adult literacy education programmes have been developed. The emphasis has been mostly on women and the classes utilize materials that focus on subjects relevant to their activities and development needs. The method has been to train "facilitators" who are already resident in the participating villages to conduct night classes with women.

A primary emphasis in classes has been on resource conservation, the effects of indiscriminate fuelwood harvesting and positive alternatives such as the creation of well-managed farmer or community forests. Methods of soil conservation in this mountainous region of generally unmanaged gullies are also taught.

In addition to classes, the resource conservation programme has assisted some local farmers in the establishment of private nurseries for the propagation of trees, grasses and shrubs. The programme has also helped local farmers try new ideas to increase their productivity.

3. NATIONAL INDUSTRIAL/COMPANY DEVELOPMENT

Before the Andhi Khola hydroelectric project there were no Nepali companies with the expertise or experience to construct a scheme of this size and complexity. This was realised at the outset of the project and a key objective was therefore the development of such Nepali companies.

The Andhi Khola project is one step in the development of a competitive Nepali hydropower industry. It is only after gaining experience that a company can become competitive. Major funding support by donor agencies, such as NORAD for this project, is essential to provide work and take risks supporting new inexperienced companies. It has enabled the vicious circle of no experience, no jobs, no experience to be broken by the integration of construction projects with industrial development. Such development does not occur as the result of one project, but only through continued support over many years. NORAD have supported UMN/HMGN in this since 1966 and are continuing to do so through a US$19 million grant for the construction of a 12MW hydropower scheme to be completed in 1994.

The success of this strategy is becoming apparent. All the companies involved have been able to grow significantly, developing the necessary technical and management skills. Confidence in the companies has been established as they are now recognised as the leading companies in their field in Nepal and have assignments from national and international government and non-government organisations.

The catalyst effect Andhi Khola has had on the parties involved can be seen from figure 4 below which contrasts the growth of Himal Hydro with that of the average for the construction sector in

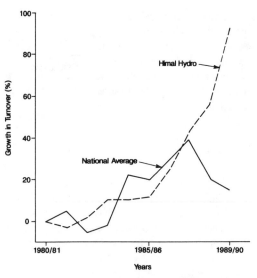

Fig.4 Growth in turnover: Himal Hydro compared to national average for construction industry. HMGN (1991)

Nepal. Andhi Khola has been the main contributor to the sharp increase in turnover and performance of NHE, BPC Hydroconsult and Himal Hydro.

3.1 Himal Hydro and General Construction

Himal Hydro is among the 3 largest civil contractors in Nepal and the leading one within hydropower construction, tunnelling and transmission line building.

Valuable skills acquired during the Andhi Khola project included high quality concreting, underground excavation and stabilisation, and construction of long span transmission lines.

The company has recently completed the construction of a 2MW hydel scheme in west Nepal for HMGN and is now constructing the Jhimruk 12MW Hydro-electric project securing work up to mid 1994 for most of its workforce.

3.2 Nepal Hydro and Electric

Nepal Hydro and Electric was established in 1986 for the purpose of manufacturing, installing, refurbishing and maintaining machinery and equipment for electric power generation, transmission and distribution.

At present it is the only company in Nepal capable of producing hydroturbines larger than 100kW. Through long term cooperation with Kvaerner Eureka of Norway it is developing the technology and skills for the manufacture of turbines of up to 10,000kW capacity.

The Andhi Khola project has been the secure market essential for the growth of NHE, allowing skills to be acquired such as:

- Refurbishment, installation and maintenance of hydroturbines, generators and associated electrical equipment.
- Fabrication and installation of steel penstocks, requiring high quality welding.
- Manufacture of lattice towers and tubular poles for transmission lines, in modular components which can be easily transported over the mountainous terrain.
- Hot-dip galvanising. A galvanising plant has been constructed for the corrosion protection of the transmission line towers and poles and other steel parts.

It would have been cheaper and quicker to have purchased most of the transmission line towers and poles from outside Nepal, but the short term loss will be more than compensated for in the long term in that Nepal now has a competitive company in this field.

3.3 Butwal Power Company

BPC was the original company established by HMGN/UMN in the field of hydropower founded in 1966 to carry out the construction of a 1MW hydropower scheme in West Nepal. Its primary role has been to manage hydropower projects. This has now been enlarged to include consultancy services with the establishment of BPC Hydroconsult in 1986 and power station operation with the commissioning of Andhi Khola in 1991. The company is unique in Nepal and it is not possible to compare it with other Nepali organisations. However, its development can be judged by the continued support of HMGN and foreign donors to its projects. Also BPC Hydroconsult is now recognised to be the leading Nepali consultant in hydropower, currently undertaking studies and designs for clients, national and international, for schemes of up to 50MW.

4. CONCLUSIONS

The approach adopted for the Andhi Khola hydroelectric project appears to be successful nationally and locally. Construction of the power scheme has been completed by Nepali organisations and rural electrification, irrigation and other programmes are continuing. As a result of the project there are now viable Nepali companies of good quality in hydropower and there has been a sustainable improvement in the standard of life for the project area population. It is considered that this is a good model for hydropower development which could be replicated elsewhere.

REFERENCES

HMGN, Central Bureau of Statistics, 1991. *Statistical year book of Nepal 1991.*
HMGN, Ministry of Finance, 1991. *Economic survey fiscal year 1990-1991.*
Warnock, J.G. 1989. The hydro resources of Nepal. *International Water Power and Dam Construction: Vol 41, No 3: 26-30.*

Hydropower'92, Broch & Lysne (eds) © 1992 Balkema, Rotterdam. ISBN 90 5410 054 0

Hydropower and the Lesotho Highlands Water Project

Nigel J.Widgery & Peter J.Mason
Sir Alexander Gibb & Partners Ltd, UK

Ben Rafoneke
Lesotho Highlands Development Authority, Lesotho

ABSTRACT: Various hydropower options considered during the planning of the Lesotho Highlands Water Project are described with details of the optimisation studies leading to the recommended 72 MW layout for Phase I. Hydraulic design and transient analyses are described with particular reference to the implications of the hydropower plant being supplied directly from a 45km long tunnel. The hydropower potential, included pumped storage, of later phases is indicated.

1 INTRODUCTION

The Lesotho Highlands Water Project (LHWP) is a multi-purpose project to develop in successive phases the water resources of the highland region by a series of dams, tunnels, pumping stations and hydro-electric works. Specifically the objectives are:

(a) To redirect a portion of the water, which presently leaves Lesotho in the Senqu (Orange) river, towards population centres in the Republic of South Africa (RSA) north of Lesotho, by transferring water from the Senqu catchment to the Vaal catchment;

(b) To generate hydro-electricity in Lesotho, utilising the water transfers and the difference in elevation between the Senqu and the Vaal basins;

(c) To provide water supply, irrigation and regional development in Lesotho.

The overall layout of the LHWP is illustrated on Figure 1. Construction of Phase IA, comprising Katse dam, and the first transfer and delivery tunnels, commenced in 1991. With a shorter implementation period, construction of the 'Muela Hydropower Project and tailpond dam will follow in time for overall commissioning of Phase IA in 1996.

This paper describes the re-optimisation studies and tender design for the hydropower works carried out in the period 1987 to 1989, and concludes with an indication of longer term hydropower potential.

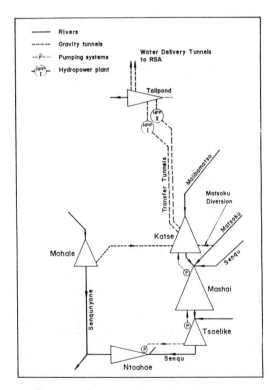

Fig.1 Overall layout of LHWP

2 COMPARISON OF HYDROPOWER OPTIONS

The hydropower design contract required the re-appraisal of two earlier proposals for hydropower generation which used the Phase I flows and head available between Katse reservoir and an appropriate tailpond. These were as follows:

(a) A low plant factor scheme with an installed capacity of 110 MW. This scheme incorporated a headpond near the village of Sentelina in the upper Nqoe valley, and an underground power station discharging to a tailpond adjacent to the village of Tlhaka in the Hololo valley. This scheme was investigated in detail in the Feasibility Study (1983 - 1986).

(b) A high plant factor scheme with an installed capacity of 50 MW, incorporating a surface power station discharging to a tailpond adjacent to the village of 'Muela in the Nqoe valley. This scheme did not include a headpond.

An assessment of the two schemes indicated that further variants on them were also feasible. Various elements within each scheme, such as the design and location of the headpond and tailpond, the choice between underground or surface power station and the installed capacity could be separately varied.

The comparison of hydropower options leading to the recommended layout was undertaken in a series of steps:

- Preliminary screening, particularly of alternative headpond layouts;
- Selection of tailpond and power station layout;
- Optimisation of transfer tunnel diameter, turbine and generator capacity, waterway sizing and tailpond full supply level.

2.1 Headpond arrangement

The layout with a low headpond at FSL+1960 in the Sentelina valley and an installed capacity of 110 MW would result in total costs of about M440 million at 1987 price levels discounted to the date of commissioning. (M = Maloti on par with South African Rand: M2.0 = US$1.0.) Comparable costs for an alternative layout with no headpond and an installed capacity of 70 MW would be about M300 million. The difference in hydropower benefits would be small, being about M390 million for the alternative layout and only M400 million for the low headpond arrangement,

primarily due to significantly lower energy generation (358 GWh/annum compared with 530 GWh/annum).

An alternative with a high headpond further up the Sentelina valley at FSL+2025 and 110 MW installed capacity was considered which showed a significant advantage over the low headpond layout. However, compared with the no headpond layout, total costs would be about M120 million more, whereas the increase in hydropower benefits would be only M70 million. A similar comparison of options with and without a high headpond as part of the Phase II development showed a cost-benefit ratio of 2.4 for the high headpond layout.

2.2 Tailpond location

After eliminating alternative sites higher up the Nqoe Valley, primarily on grounds of siltation risk, alternative tailpond locations at Tlhaka and at 'Muela were compared on the basis of options that were technically comparable and that produced similar hydropower benefits. For the overall project including Phase II the difference in net benefits for the base case comparison was relatively small. However, the corresponding difference in Phase I capital costs was significant, with the costs of the Tlhaka option (M264 million at 1987 price levels) being more than 9 per cent greater than 'Muela (M241 million).

In making the selection of tailpond the following factors increased the advantage at 'Muela in addition to the economic and financial comparisons:

- There was considered to be greater potential for reducing costs during detailed design at 'Muela than at Tlhaka. In the event, it was possible to reduce the significant length of steel lining assumed in the costs of the transfer tunnel to 'Muela. Alternative methods of construction also led to lower costs for the section of the delivery tunnel from 'Muela, including the Hololo river crossing.
- Geological conditions appeared better at 'Muela resulting in fewer technical problems.
- Environmentally 'Muela would have less impact than Tlhaka. The reservoir area would be about half the size and the natural river flow is much lower.

In other respects, such as access and sedimentation, the two sites are very similar.

2.3 Power station

It was concluded that the power station should be located underground, in the spur running northward from the end of the sharply defined ridge between the Nqoe and Khukhune river valleys, near the village of 'Muela. Topographical considerations precluded a surface installation at this site. Detailed studies of the earlier proposals for a surface power station on the opposite bank indicated that the site would be subject to unacceptable siltation risk in the medium to long term.

The optimum installed capacity was shown to be between 60 and 70 MW for the base case economic parameters. Within this range the net hydropower benefits were relatively constant. Several relevant unquantifiable factors were identified which overall would tend to increase the optimum installed capacity. It was concluded during a Donors' Conference in March 1988 that the tender design of the transfer tunnel, powerhouse, waterways and turbines should be based on a nominal installed capacity of 70 MW, i.e. towards the upper end of the optimum range identified. However, the possibility of installing higher capacity generators on these turbines was to be considered.

2.4 Optimisation of selected layout

Having established the basic layout of the 'Muela Hydropower Project, detailed optimisation studies were carried out as part of the tender design to finalise the leading dimensions of a number of key elements. The extreme length and associated headlosses and transient response of the transfer tunnel at 'Muela resulted in a greater degree of inter-relationship than normal between the various elements to be optimised. Based on the nominal capacity of 70 MW, a progressive optimisation procedure was adopted covering the ranges indicated in Table 1.

In each case a narrower range was identified as optimum by comparing incremental capital costs with hydropower benefits. The optimum value was selected taking account of alternative assumptions regarding discount rate and effective electricity tariff.

3 RECOMMENDED HYDROPOWER SCHEME

The recommended hydropower scheme is illustrated on Figure 2. The downstream end of the transfer tunnel runs in a relatively narrow rock ridge and hydrofracture testing was used to determine the local rock confining pressure and hence tunnel lining requirements.

It was shown that, in general, a concrete lining incorporating a plastic waterproofing membrane would be adequate in the ridge. The surge shaft for the hydropower complex was located at a sensible limit to that condition, after which steel lining would be needed.

The location of the tailpond dam was fixed by the natural narrowing of an already narrow gorge on the nearby Nqoe river and as the result of a cost comparison with alternative locations. The distance from the reservoir thus formed to the surge shaft was approximately 2 km. An optimisation process followed in which both surface and underground power stations were examined in various locations and with various locations of tailrace outfall. It was shown that the optimum solution was an underground power station located approximately 430m horizontally from the surge shaft. A downstream surge chamber was also required to regulate conditions in the 1.5 km long tailrace tunnel.

The upstream surge shaft was designed to accommodate the transient behaviour in the very long transfer tunnel and hydropower plant. It comprised a 13m internal diameter concrete and plastic membrane lined shaft with an expansion chamber above level +2067. The shaft was connected to the main penstock via a 3m diameter and 36m long steel lined spur shaft.

Table 1. Range of elements to be optimised

Transfer tunnel diameter (lined)	4.15	-	4.55m
Corresponding values for:			
Turbine design head/rated head	210	-	243m
Max turbine discharge	39.5	-	34.1m³/s
Surge shaft diameter	19.9	-	15.6m
Penstock guard valve diameter	2.1	-	2.9m
Penstock steel lining diameter	2.5	-	3.0m
Horseshoe tailrace tunnel "diameter"	3.5	-	4.5m
Station Rated Output	70	-	80 MW
Corresponding values for:			
Rated net head	236	-	266m
Max turbine discharge	35.2	-	37.4m³/s
Upstream expansion chamber diameter	29.5	-	31.4m

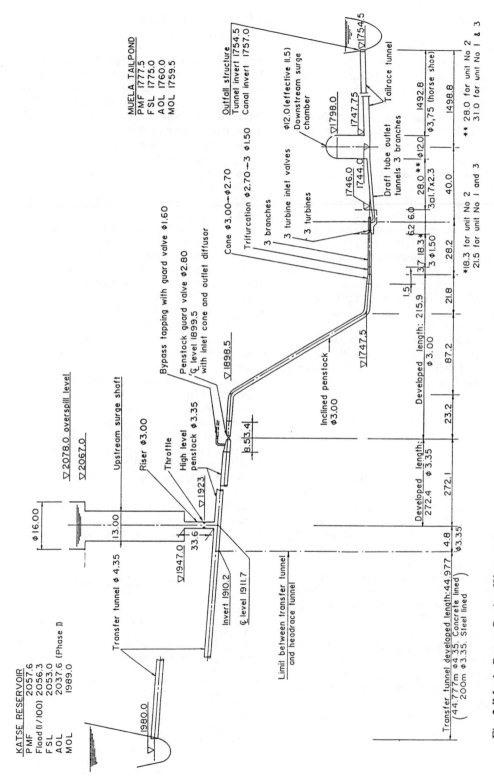

Fig. 2 'Muela Power Station Waterways

KATSE RESERVOIR
PMF 2057.6
Flood (1/100) 2056.3
FSL 2053.0
AOL 2037.6 (Phase I)
MOL 1989.0

MUELA TAILPOND
PMF 1777.5
FSL 1775.0
AOL 1760.0
MOL 1759.5

Outfall structure
Tunnel invert 1754.5
Canal invert 1757.0

∇2078.0 overspill level
∇2067.0

Upstream surge shaft

Riser φ3.00

Throttle

High level penstock φ3.35

Bypass tapping with guard valve φ1.60

Penstock guard valve φ2.80
℄ level 1899.5
with inlet cone and outlet diffusor

∇1898.5

Cone φ3.00→φ2.70

Trifurcation φ2.70→3 φ1.50

3 branches

3 turbine inlet valves

3 turbines

φ12.0 (effective 11.5)
Downstream surge chamber

∇1798.0

∇1754.5

1747.75

Tailrace tunnel

Draft tube outlet
tunnels 3 branches

1746.0

1744.0

∇1947.0

∇1923

Invert 1910.2
℄ level 1911.7

Transfer tunnel φ4.35

∇1980.0

φ16.00

13.00

33.6

B.53.4

∇1747.5

Inclined penstock φ3.00

Limit between transfer tunnel and headrace tunnel

Developed length: 272.4 φ3.35

Developed length: 215.9 φ3.00

Transfer tunnel developed length: 44.977
(44.777m φ4.35. Concrete lined)
(200m φ3.35. Steel lined)

4.8 φ3.35

272.1

23.2

87.2

1.5

28.2

21.8

37 18.3

3 φ1.50

6.2 6.0

3 φ1.7x2.3

28.0 **

40.0

φ12.0

1492.8
φ3.75 (horse shoe)

1498.8

*18.3 for unit No 2
21.5 for unit No 1 and 3

** 28.0 for unit No 2
31.0 for unit No 1 & 3

738

It was envisaged that surge shaft construction would probably be by raise-boring. The shaft was set off line from the main transfer tunnel so that the two construction operations did not conflict. Access to the base of the surge shaft was therefore via the upper level penstock and penstock guard valve chamber at approximately level +1900. The upper penstock was entirely steel lined and extended 300m downstream of the surge shaft. All the works at this level and above were in competent, but jointed, basalt containing perched water tables.

At 300m from the surge shaft the penstock was inclined at 60 degrees to the horizontal down to approximately level +1748. This was also the broad level of the main power station cavern, downstream surge chamber and tailrace. All these works were sited in the lower Clarens sandstone. This is a fine grained, massive and relatively hard and unjointed sandstone. It has, however, a high clay content and particular care was used to assess long term durability.

The main cavern excavation was 58m long by 27m high and 14.5m wide. Sizing was based on machine and other ancillary equipment requirements. Both primary and secondary access tunnels were provided together with a vertical shaft link to an operations and switchyard complex sited on the hillside above. The power station contained 3 Francis turbine generating sets, each with a nominal capacity of 24 MW under a rated head of 236m. Provision was made in the design of both the cavern and access to permit future extension for any Phase II development.

The downstream surge chamber was located 40m downstream of the cavern centreline. It was designed as a 12m diameter and 54m high concrete lined structure containing three individual draft tube gates and associated operating equipment designed to allow dewatering of the powerhouse waterways. The tailrace tunnel was also lined in view of doubts over the long term durability of the Clarens sandstone.

The location of the tailrace outfall was set so as to minimize any affects from siltation in the tailpond.

Various types of tailpond dam were designed and costed. In view of the local geology and material availability these were principally concrete faced rockfill and concrete gravity, arch and arch-gravity dams. The narrow gorge and significant river design flood resulted in the concrete dam solutions with incorporated overspills proving the most economic. Various factors such as the flood passage requirements and limits on the permissible foundation hydraulic gradients resulted in a curved concrete gravity dam being selected for tender design purposes. The tailpond also housed a bellmouth intake to the 37 km long delivery tunnel running north to the Ash river.

The operating level of the tailpond was determined largely by the long term hydraulic conditions on the delivery tunnel, with allowances for daily flow regulation.

4 HYDRAULIC DESIGN AND TRANSIENT ANALYSES

4.1 Friction factors

With the 45 km long transfer tunnel supplying water directly to the power station, particular care was necessary to ensure a satisfactory hydraulic design over all conceivable operating conditions. The tunnel is being constructed by tunnel boring machines (TBM) and it is expected that a significant proportion will be left unlined. Three alternative assumptions on friction factors were assumed as follows:

Minimum - representative of initial years of operation and assuming a largely unlined tunnel with minimal intake and local losses;

Average - typical values to be expected during Phase IA - i.e. the first 12 years of operation;

Maximum - long term conditions including allowances for full intake and local losses.

Friction factors were calculated using both Colebrook-White and Manning formulae. The factors shown in Table 2 were adopted.

Table 2. Friction factors adopted

Lining Type	Condition	K (mm)	Equivalent 'n'
Concrete)	Minimum	0.06	0.0110
cast in situ)	Average	0.30	0.0123
normal aging)	Maximum	1.50	0.0145
TBM -)	Minimum	1.5	0.015
Gunited or)	Average	5.0	0.016
unlined)	Maximum	7.5	0.017
	Invert debris	15.0	

At the nominal turbine discharge of 35m³/s, these friction factors give the following total headlosses:

Minimum	32.6m
Average	41.7m
Maximum	56.6m

4.2 Turbine operating range

The turbine operating range was determined from 10 different operating points as set out in Table 4. With the possible level variation of both Katse and 'Muela reservoirs, the steady state net head can vary from 173m to 295m (i.e. 73% to 125% of rated head.)

4.3 Hydraulic surge criteria

The hydraulic surge criteria adopted are summarised in Table 5.

4.4 Sizing of upstream surge shaft

The criteria for sizing the upstream surge shaft proved to be of particular interest. The minimum drawdown level was set at +1947 corresponding to a minimum total transient power output of about 42 MW. Due to the very long period of oscillation of the upstream surge shaft water level the minimum transient power would extend over some 10 minutes. To ensure maximum flexibility of operation it was considered desirable to provide for full load acceptance in 50 seconds. However, the application of this criteria at low Katse reservoir levels resulted in very large surge shafts and, close to minimum operating level, became quite impractical. This effect is illustrated on Figure 3, based on preliminary analyses without a throttle. The criterion adopted was that full load acceptance in 50 seconds should be available for at least 95% of the time.

With the tender design layout incorporating a 13m diameter surge shaft with throttle the load acceptance performance is expected to be as indicated in Table 3.

Table 3. Load acceptance performance

Katse Level	Frequency of Exceedance	Load acceptance time (sec)
1989*	>99%	175⁺
1989*	>99%	570
2008.5	95%	50

* Minimum operating level
⁺ Simple load acceptance

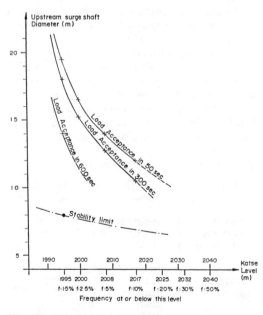

Fig.3 Relationship between surge shaft diameter, Katse reservoir level and load acceptance time.

5 FUTURE HYDROPOWER POTENTIAL

During subsequent phases of the Lesotho Highlands Water Project it is expected that the average water delivery will increase from 29.6m³/s in Phase I up to a maximum of 70m³/s, with further major storage reservoirs. During Phase II a second transfer tunnel, hydropower complex and delivery tunnel will be constructed parallel to the Phase I installations. An additional installed capacity of 100-120 MW is envisaged, utilising the tailpond constructed in Phase I.

However, a much greater hydropower potential in the form of pumped storage will be available between the major reservoirs in later phases as indicated in Table 6. Although the available heads are relatively low, the large reservoir surface areas and the major pumping installations and connecting tunnels required for water transfer suggest that pumped storage should be seriously considered, particularly in Phases III and IV where the pumping tunnel length is short.

Table 4. Turbine operating range

Operation No.	Type of operation	Katse Reservoir Level (masl)	'Muela Tailpond Level (masl)	Gross Head (m)	Head Losses (m)	Net Head (m)	Total Discharge (m³/sec)	Shaft power of each Turbine (kW)	Output of each Generator KW	Power Station Total output KW
1	Minimum Net Head for maximum friction factor, 3 units in operation	MOL: 1989	FSL: 1775	214	41	173	29.75	14 750	14 308	42 495
2	Minimum Net Head with average friction factor, 3 units in operation	MOL: 1989	FSL: 1775	214	32.5	181.5	30.80	16 425	15 930	47 320
3	Rated Net Head Average friction factor 3 units in operation At maximum wicket gates opening	Range from FSL: 2053 to 2040 MOL	1773 1760	280	44	236	35.90	25 000	24 250	72 000
4	Max Net Head for min friction factor 1 unit in operation at rated output	FSL: 2053	MOL: 1760	293	3	290	9.84	25.000	24 250	24 000
5	Overload operation Average friction factor 3 units in operation At maximum wicket gates opening	Range from FSL: 2053 to 2050	1762.5 MOL: 1760	290	47	243	37.20	26 560	25 760	76 500
6	Max. Net Head. Average friction factor 3 units in operation at overload	FSL 2053	MOL: 1760	293	45	248	36.30	26 560	25 760	76 500
7	Exceptional Maximum Net Head for Average friction factor 3 units in operation at overload	PMF: 2057.6	MOL: 1760	298	42	256	35.00	26 560	25 760	76 500
8	Exceptional Maximum Net Head for Minimum friction factor 3 units in operation at overload	PMF: 2057.6	MOL: 1760	298	30	268	33.30	26 560	25 760	76 500
9	Exceptional Maximum Net Head for Minimum friction factor 1 unit only in operation at overload	PMF: 2057.6	MOL: 1760	298	3	295	10.30	26 560	25 760	25 500
10	Transient Maximum Net Head for average friction factor and 1 unit in operation during upstream surge shaft upsurge	Transient level in u/s surge shaft 2078	MOL: 1760			317	9.72	26 560	25 760	25 500

Table 5. Hydraulic surge criteria

Location	Surge Level Considered	Manoeuvre and Timing	Katse Reservoir Level (masl)	'Muela Tailpond Level (masl)	Tunnel roughness	Hydraulic Surge Criteria
Upstream Surge Shaft	Maximum	Full-load rejection in 5 sec	PMF 2057.6	PMF 1777.5	Minimum	Maximum upsurge level < 2078 Freeboard: 0.5m
	Level	Full-load acceptance in 50 sec followed by Full-load rejection in 5 seconds initiated under the most adverse conditions	PMF 2057.6	PMF 1777.5	Minimum	Maximum pumped upsurge level < 2078 Freeboard: 0.5m
	Minimum	Full-load acceptance in a time to be defined to fulfil the criteria	MOL 1989	MOL 1760	Maximum	Minimum downsurge level > 1947 Freeboard: 0.5m
	Level	Full-load rejection in 5 sec followed by full acceptance, initiated under the most adverse conditions (initiation time t_0 and load acceptance time t_i to be defined)	MOL 1989	MOL 1760	Maximum	Minimum pumped downsurge level > 1947 Freeboard: 0.5m
Downstream Surge Chamber	Maximum	Full-load acceptance in 50 sec	PMF 2057.6	PMF 1777.5	Maximum	Maximum upsurge level < 1798 Freeboard: 0.5m
	Level	Full-load rejection in 5 sec followed by Full-load acceptance initiated under the most adverse conditions	PMF 2057.6	PMF 1777.5	Maximum	Maximum pumped upsurge level < 1798 Freeboard: 1m
	Minimum	Full-load rejection in 5 sec	2040	MOL 1760	Minimum	Minimum downsurge Level > 1747.7 Freeboard: 1m
	Level	Full-load acceptance in 50 sec followed by load rejection of 2 units over 3 in 5 seconds initiated under the most adverse conditions	2040	MOL 1760	Minimum	Minimum pumped downsurge level > 1747.7 Freeboard: 1m

Table 6. Pumped storage potential

	Phase II	Phase III	Phase IV
Planned Year in Service	2008	2017	2020
Upper Reservoir	Katse	Mashai	Tsoelike
Full Supply Level (FSL)	2053	1887	1757
Average Operating Level (AOL)	2041	1867	1742
Minimum Operating Level (MOL)	2003	1835	1725
Lower Reservoir	Mashai	Tsoelike	Ntoahae
FSL	1887	1757	1634
AOL	1867	1742	1620
MOL	1835	1725	1610
Length of Pumping Tunnel (km)	19	4.4	1.3
Planned Pumping Capacity (m^3/s)	50	25	8
maximum power(MW)	135	52	14
average power (MW	90	23	6
Gross Head Absolute maximum	218	162	147
Normal maximum	209	157	142
Average	174	125	122
Normal minimum	154	110	115
Minimum Active Storage Volume (MCM)	61	57	44
with ± 1m level variation*	Katse	Tsoelike	Ntoahae
Potential Pumped Storage Capacity (MW)	920	260	460
± 1m level variation*			

* Minimum active storage volume and potential pumped storage capacity
would increase in proportion to level variation assumed available for
power. Relatively small increases in dam height may be economically
justified for this purpose.

6 CONCLUSIONS

Detailed review of earlier proposals for
hydropower development as part of the Lesotho
Highlands Water Project led to an alternative
formulation with substantially lower capital costs,
higher energy output, similar hydropower benefits
and a significantly improved economic rate of
return.

Extensive studies demonstrated that satisfactory
operation over a very wide variation of head can
be achieved. The surge shaft and downstream
surge chamber can be sized in a cost effective
manner to provide a high degree of flexibility of
operation, although it is appropriate to accept
constraints on load acceptance times at extreme
low levels of Katse reservoir.

The hydropower potential of later phases of the
LHWP is substantial with the possibility of
pumped storage installations between the large
reservoirs scheduled for commissioning in 15 to
25 years' time.

Hydropower'92, Broch & Lysne (eds) © 1992 Balkema, Rotterdam. ISBN 90 5410 054 0

Greenland's first hydroelectric power scheme

K.T.Wæringsaasen & O.Stokkebø
Norconsult International A.S., Oslo, Norway

ABSTRACT: Based on a policy of shifting from total dependency on oil imports for electricity production, the Home Rule Government of Greenland decided to investigate the possibility of utilising the country's untouched hydropower resources.
As a result of these analyses, it was decided to build Greenland's first hydropower plant in Buksefjorden, some 60 km southeast of Nuuk. Due for commissioning in October 1993, the plant is designed and constructed by the Nuuk Kraft ANS consortium of which Norconsult International A.S. is a member. The environment in the project area is extremely vulnerable, the logistics difficult and the climatic conditions rough. This has been reflected in the chosen design and construction methods used in the implementation of the Buksefjord Hydropower project. Socio-economic problems have also been highlighted in the development. This has led to the requirement that local services, subcontractors and manpower should be involved in the project whenever possible.

1. INTRODUCTION

Greenland's electricity production is for the time being totally dependent on thermal power generation using imported oil.

In order to reduce this dependency, the Home Rule Government investigated the possibility of utilising Greenland's untouched hydropower resources. As a result of these investigations, it was found feasible to construct the Buksefjorden hydropower project to supply the capital city of Nuuk. In 1990 Nuuk Kraft ANS was awarded a turn-key contract for design and construction of the project and an operation contract for the first 15 years operation. Nuuk Karft ANS is a consortium of companies established for the Buksefjorden Hydropower Project (Veidekke, Atcon, Betonmast, Elektrisk Bureau, Kværner Eureka, Norconsult International and Selmer Anlegg) and represents uparalleled experience in planning and construction of hydropower developments and transmission lines and delivery and erection of power generation machinery and equipment.

As a member of Nuuk Kraft ANS, Norconsult is responsible for planning, design, quality assurance, construction supervision and co-ordination of all civil and electro/mechanical components, and for the environmental assessment of the project.

2. SITE SELECTION

As one of four feasible sites investigated for hydropower projects, Buksefjorden was selected by the Greenland Home Rule Government represented by the Greenland Energy Supply Authority. The distance between population centres is so vast in Greenland, and Buksefjorden's vicinity to the capital Nuuk (Godthåb) with population of 13000 was the main reason for this site selection.

The power station is situated at the head of the fjord, Buksefjorden, about 60 km southeast of Nuuk. The mountain lake of Kangerdluarssungup Taserssua, about 10 km inland, forms a large reservoir volume for hydropower production.

3. PROJECT DESCRIPTION

The powerplant utilises the head between the huge mountain lake Kangerdluarssungup

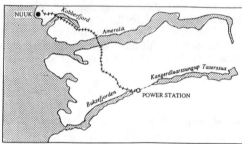

Fig. 1, Site Allocation

Table 1

PROJECT KEY DATA

Client	Greenland Home Rule Government, represented by Nukissiorfiit Greenland Energy Supply
Construction period	1990-1993
Capacity	2 x 15 MW (provisions made for a third 15 MW unit)
Turbines	Vertical Francis
Gross head	261 m
Powerhouse	Underground
Total catchment are	813 km^2
Annual runoff	311 mill. m^3
Reservoir area	75 km^2
Net reservoir capacity	1950 mill. m^3
Annual production	180 GWh
Total length of tunnels	15 km
Length of 132 kV transmission line	56.5 km
Two fjord crossings, of which one is the world longest span	(5376 m)

Taserssuua (Kang) and the sea at Buksefjorden.

The huge natural reservoir of Kang lake has a surface area of 75 km^2. Consequently only a 12 m high concrete dam and a lake piercing at 16 m below normal lake surface level were enough to create a usable reservoir of almost 2000 million m^3. This storage will be used as a carry over reservoir for equalising the variation of precipitation and electricity demand from year to year.

In addition to the Kang catchment, water from three neighbouring catchment areas will be transferred by means of short rock tunnels and a canal into the reservoir.

Water is led from Kang through a 8600 m long and 28 m^2 headrace tunnel, a 1900 m long unlined pressure tunnel and a 25 m long concrete embedded steel penstock to the underground power station.

The intake in Kang reservoir will be established by an underwater piercing, a well known Norwegian speciality. A short distance downstream of the lake tap is the intake gate shaft, where two gates will be installed. An upstream sliding gate for maintenance purposes, and a roller gate for emergency closing of the tunnel. The hoist machinery is in a rock cavern above HWL with access from outside through a short tunnel.

The trash rack is placed downstream of the gates in order to facilitate cleaning

This will be done very seldom due to the clean water and lack of vegetation in the area.

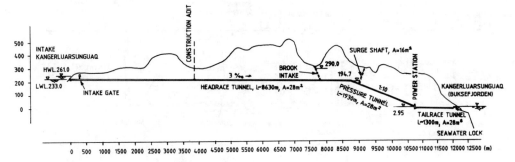

Fig. 2, Longitudinal Section of the Watersystem

The longitudinal section, Fig.2 , shows the water system sloping downstream all the way to the power station. This is done to facilitate drainage and avoid air pockets at high breaks.

A small catchment area is captured by a stream intake.

The 16 m² surge shaft is placed near the upstream end of the pressure tunnel, with its top 2 m above the highest surge level. This accommodates sureges and ensures the hydraulic stability of the system.

A stone and sand trap is placed upstream of the penstock. Only the lower half of the inlet cone is covered by a trash rack due to the very clean water in the reservoir with no trees and very little human activities and littering.

The construction adit will be closed by a concrete plug with a hinged bulkhead. In this area much attention is paid to the possibility of hydraulic splitting and water leakage in the rock.

During planning well known theories have been used for safe placement of the unlined high pressure tunnel. In addition hydraulic splitting tests have been carried out to confirm the theoretical calculations. To check water-tightness, water leakage tests are performed ahead of excavation. Although no leakage above 0.5 Lugeon has been measured, a systematic grouting fan will be established

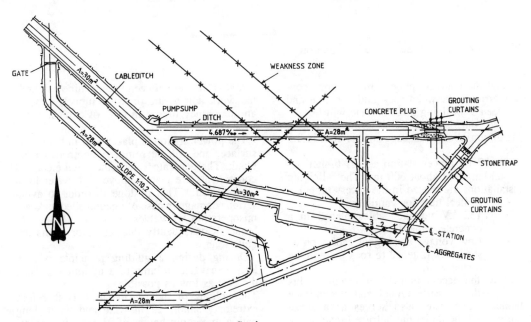

Fig. 3, Plan Tunnelarrangement at Power Station

745

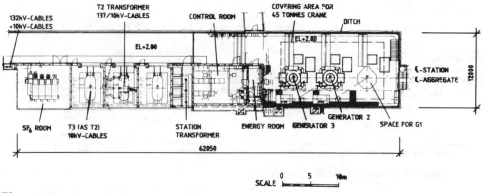

Fig. 4, Plan View of the Power Station
EL + 2.80

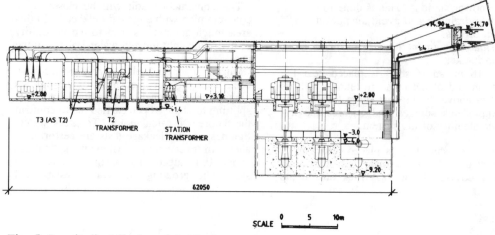

Fig. 5, Longitudinal Section of the Station

around the concrete plugs. In addition contact grouting between rock and concrete and between steel lining and concrete will be carried out using special grouting hoses. Maximum allowed leakage in this area is 0.1 l/sec.

The 32 m² power station access tunnel has a total length of about 600 m. The following installations are placed in the access tunnel:
- A steel tube for ventilation
- 132 kV voltage cables placed in a concrete cable duct
- Low voltage and signal cables on a steel rack fastened to rock bolts in the tunnel wall.

From the access tunnel construction adits lead to the pressure tunnel and the tailrace tunnel. The latter also serves as a surge tunnel. At the outlet the tailrace tunnel has a salt water barrier.

A short tunnel between the machine hall and the work adit to the pressure tunnel, acts as an emergency tunnel in case of explosion or fire.

The power station is placed 250 m below the surface, and will be remotely controlled from Nuuk. The generators and other vital electrical components are placed above high water level in the fjord. This is done in order to avoid serious damage in case of uncontrolled leakage, misoperation of valves or a pipe failure or burst. Consequently the generator shafts are quite long.

During design a guiding principle was to keep excavated volume to a minimun to keep the cost as low as possible.

The volume of the powerstation is therefore extremely small, 12 m wide and 63 m long with a maximum height of 21 m. All necessary items in the station are therefore compressed together as much as possible.

Three transformers, but only two generating sets are installed in the first building phase. The third transformer will therefore serve as a reserve.

As there is only one passage to outside, there is an emergency room near the control room close to the emergency tunnel. The room will be equipped with oxygen masks and other smoke protection equipment.

For erection and maintenance, a rock bolt suspended crane beam is installed in the machine hall. The crane has a lifting capacity of 45 tonnes.

A service building is placed at the entrance to the access tunnel. This contains garage, workshops, an emergency backup generator set, ventilation system for the power station and facilities for the service personnel. Several electrical items and towers for the transmission line are placed on the roof.

The energy will be transferred from the portal building to the transformer station in Nuuk through a 56.5 km long 132 kV transmission line.

The transmission line has three conductors suspended on 206 steel towers. Due to the great climatic variation along the line, two different types of towers will be used, see figs. 6 and 7.

Free-standing towers where heavy load conditions occur and guyed towers other places:

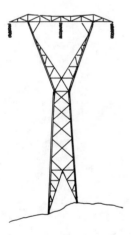

Fig. 6, Free-standing Tower

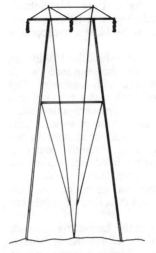

Fig. 7, Guyed Tower

The line crosses two fjords, one of which (the Ameralik fjord crossing) holds the world record with a free span of 5376 m. The towers on the south side of this fjord span will be at an elevation of 1013 m and on the north side at an elevation of 444 m. Minimum height above the sea will be 65 m. The design of this span has been one of the most challenging tasks for the planners. Due to the environmental aspects the last part of the power transmission into Nuuk will be as an underground cable.

4. LOGISTICS

The only access to the hydropower plant is by boat, helicopter or seaplane. This makes transport of materials and workers extremely costly and caused problems in periods with bad weather conditions and severe frost (down to 40° C has been measured).

A small quay and a permanent road to the service building have been established in Buksefjorden. From the service building to the construction adit and intake area the construction road will be maintained after commissioning.

5. ENVIRONMENTAL IMPACT ASSESSMENT

During planning, design and construction of Buksefjorden Hydropower project, the environmental aspects have been highlighted. Investigations started already in 1982 to evaluate the consequences of building a hydropower plant here.

Buksefjorden is a thresholdfjord with an underwater ridge near the entrance which prevents normal mixture with seawater. The increased amount of fresh water released into the fjord from the hydropower outlet could reduce the vertical circulation of water. This might affect the biological conditions in the bottom zone of the fjord. Another outcome will probably be that the fjord will become ice covered at an earlier stage than normal.

The reservoir, Kang and its outlet, does not contain trout stock, and the effect of regulation will be minor.

50 % of the water in the Eqaluit river, south of Buksefjorden, will be led into Kang through a tunnel. In this river there is quite an amount of trout and the effect of the flow reduction is presently being investigated and some remedial action will be taken.

In the project area there are about 2000 reindeer. The transmission line will cross their areas and a minor part of their grazing land will be submerged by the reservoir. The coastal area west of the transmission line is however not so much used that this will have any influence if the reindeer are not willing to cross under the transmission line. In addition, experience from Norway shows that transmission lines do not normally cause problems for reindeer.

In the reindeer calving area east of Kang, flying and driving has been prohibited between mid-May and the end of June.

It is impossible to construct a hydropower plant without influencing the local environment to some extent. However, by using the underground waterway systems and installations as described here, and by using a landscape architect experienced in landscape forming using tunnel spoils, the impact has been kept to a minimum.

6. CONCLUSION

The experience so far in construction of the Buksefjord Hydropower project has been very good. The rock quality has been excellent and the announced permafrost has only been observed in the first 60 m of the access tunnel and partly in the dam foundation.

Despite the difficulties attached to building a hydropower plant in this environment with such extreme climatic conditions, all the construction elements have been tackled. The more than 100 years of experience in Norway with hydropower construction have made this possible.

Late papers

Hydropower'92, Broch & Lysne (eds) © 1992 Balkema, Rotterdam. ISBN 90 5410 054 0

A study on the mechanism of the interaction of the lining and the rock of the pressure tunnel

Xie Moweng
Wuhan University of Hydraulic and Electric Engineering, People's Republic of China

ABSTRACT: Based on the response of the surrounding rock and the reinforced concrete lining of the pressure tunnel in the full strength model test, this article discusses the interaction and the factor of affecting the interaction of the lining and the rock under the effect of the internal water pressure, and investigates the combined action of the lining and the rock. The conclusion from the study provides us the convincing argument to consider the nonlinear characters of the interaction of the lining and the rock.

1. INTRODUCTION

It is very important for designing the pressure tunnel safely and economically to decide the capability of the surrounding rock of bearing the internal water pressure. Thus, it's necessary to study the interaction of the lining and the rock. The complicated surrounding rock had to be simulated, so the geomechanical model test is always used to study the pressure tunnel. But the scale of stress of the geomechanical model is ordinarily smaller than 1, it's very difficult to choose the model material to simulate the prototype on the whole stress-strain path, and the model can not reflect the feature of the prototype directly and truely. The simulants which have the same or near mechanical parameters of the prototype are used in the full strength model test, so the model can tell the feature of the prototype under all conditions, and the results from the full strength model test is direct and reliable. Now, the gypsum model and the geomechanical model have been replaced gradually by the full strength model in the study of the hydraulic structure, but the full strength model is rarely adopted in the study of the underground structure and the rock because of the difficulty of simulating the rock. Herbert E. Lindberg (1982) had used the full strength model to study the effect of the rock joint and the weak plane, and he found that the results from the 1:28 model test were same as

the results from the prototype test in this article, the 1:19 full strength model has been used to study the interaction of the surrounding rock and the lining and the satisfied results have been gotten from the test.

2. BASIC INFORMATION OF THE MODEL TEST

2.1 Outline of the pressure tunnel

The tunnel we concern is 9.5 meter in internal diameter and 0.9 meter in thickness of the lining. After poured the 400# reinforced concrete lining, the backfill and consolidation grouting would be taking, and the pressure of the consolidation grouting has been designed as 1.0MPa. The thickness of burden is 77 meter. The geostress is only caused of the dead load and the geostress coefficient of confinement pressure K is 0.37. The normal water head of the pressure is 65 meter and that of the water hammer is 35 meter. The surrounding rock is consisted of the limestone, relaxed zone and fault F_{26}. The main mechanical parameters of the rock and the lining are listed in Table I.

2.2 Similar conditions

The full strength model must have these similar conditions as follow.
1. Geometric similarity: Choose the suitable ratio of the geometric similarity C according to the testing conditions

Table I Parameter of prototype and model

Material	Parameter	Elastical modulus x10 MPa	Tensile strength MPa	Notation
Lining (400# concrete)	prototype	33.00	3.10	Same percentage of reinforcement 1.85%
	Model	30.00	2.97	
Surrounding rock (limestone)	Prototype	15.00	1.70	
	Model	14.40	1.68	
Relaxed rock (grouted rock)	Prototype	12.00	1.35	
	Model	11.40	1.37	
Fault	Prototype	1.50	0.35	Shear parameter is P: c=0.1MPa, =21.8° M: c=0.11MPa, =24.2°
	Model	1.60	0.30	

and the testingequiment,

(1) $C_L = \dfrac{Lp \text{ (linear size of prototype)}}{Lm \text{ (linear size of model)}}$

$= 19$

2. Similarity of the physical and mechanical parameter: The ratio of similarity of the elastical modulus is

(2) $C_E = \dfrac{Ep \text{ (E of prototype)}}{Em \text{ (E of model)}} = 1$

The ratio of the similarity of the stress is

(3) $C_\sigma = 1$

In the light of the stress-changing and cracking characteristics of the lining and the rock, we choose the elastical modulus and the tensile strength to concent prioritily the similar condition. So the ratio of the similarity of the strength is

(4) $C_{Rt} = \dfrac{(Rt)p(Rt \text{ of prototype})}{(Rt)m(Rt \text{ of model})} = 1$

3. Conditions of the similarity of the load: For the pressure tunnel, the lining and the rock bear the geostress and the internal water pressure. The ratio of the similarty of geostress is

(5) $C_{Pv,Ph} = \dfrac{(Pv,ph)p(prototype)}{(Pv,Ph)m(model)} = 1$

The ratio of the similarty of the internal water pressure is

(6) $C_{po} = \dfrac{(Po)p(Po \text{ of prototype})}{(Po)m(Po \text{ of model})} = 1$

The aforementioned similar conditions show that the testing loads are about the actural loads when the main mechanical parameters of the model are nearly same to that of the prototype.

4. Similarity of the boundary condition: Generally speaking, the distance of the boundary of the tunnel to the boundary of the geostress load is about (1.5-2.5)D (D is diameter of tunnel). In our study, the range of the model is about 3.5D, and the finite elements analysis had showed that this range of the model can content to the demand of the precision.

2.3 Model material

In order to content to the main similar conditions, after many times tests, we choose the ceramsite concrete to simulate the limestone and the relaxed grouted rock and the air-entraining concrete to simulate the fault F_{2b}. The main parameters of the model materials are listed in table I for comparing. From the Table I, we can see that the main parameters of the model are very near to those of the prototype.

2.4 Design of the model

In the light of the ratio of the geometric similarity, the (length x width x height) of the model is (2.0 x 2.0 x 0.276) meter. The internal diameter of the lining is 0.5 meter and the thickness of the lining is about 0.058 meter. The depth of the relaxed grouted zone is about 0.08 meter and the width of the fault is abiut 0.02meter. The plane sketch of the model is shown in Fig.1. The test is taken under the plane state of stress.

2.5 Loading and measuring technique

The jacks which are supported by the rigid tank are used to exert the geostress load (Pv=2.0MPa, P =0.74MPa). The whole testing time is about 50 days, and the geostress load must be holded in these days, so we

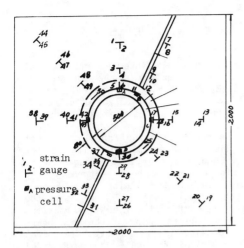

Fig.1 The plane sketch of the model and the distribution of the gauges

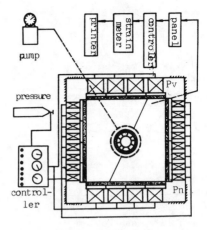

Fig.2 The sketch of the testing equipment

use the WY-300 presure controlling equipment to control the pressure of the jacks.

A special circumferential loading circle is made to exert the internal water pressure.

The sketch of the testing equipment is shown in Fig.2.

A special measuring system of bar which is burned in the model and can eliminate the rigid body's displacement is used to tell the dis placement of the interface of the rock and the lining, and the miniature pressure cells are placed on the interface to measure the pressure there. The strain of the rock is measured by the special strain blocks, and the strain of the lining is determined by the strain wire gauges.

2.6 Testing steps

According to the actual construction procedure and work condition, the test is done by the progress as follow:

1. Pour the surrounding rock and the lining according to the prescription, and burn various gauges.

2. After the model gets its strength, exert the geostress load (Pv=2.0MPa, Ph=0.74MPa) and measure the displacement and the strain.

3. After the displacement of the rock become steady, put the prepoured lining in the hole and grout the interface of of the rock and the lining.

4. After the grouted cement mortar gets it designing strength, exert the internal water pressure until the model damaging.

3. MAIN RESULTS FROM TEST

Because the lining doesn't bear the rock pressure, the interaction of the lining and the rock only concern the resisting force of the rock.

In the light of the result of the measuring pressure P_i of the interface of the lining and the rock, the curve of P_i/P_o (it is actually the part of the rock of bearing the internal water pressure, where P_o means the internal water pressure) and P_o can be gotten as Fig.3. It can be seen from Fig.3 that: With the internal water pressure goes up, P_i/P_o gogoes up too; But after the internal water pressure 0.75 MPa, P_i/P_o goes down with the internal water pressure goes up: When the internal water pressure is in 1.0-1.3MPa, P_i/P_o is almostly unchanging, and P_i/P_o goes up suddenly at the internal water pressure 1.3MPa. At other hand, P_i/P_o of different parts are different, but they have same tendency.

We can get the brief account of the curve of Fig.3 as follow: When the pressure of the internal water is relatively low, the gap between the lining and the rock is gradually closed by the effect of the internal water pressure, and the resistant force of the rock (to Lining) is gradually growing too, this means the P_i/P_o goes up; When P_o is 0.75 MPa, the the gap between the lining and the rock is completely closed; When P_o 0.75MPa the effect of the internal water pressure on the rock becomes move obvious, it results the radial pressive stress and the tangential tensile stress, so the state of the stress changes to the ten-

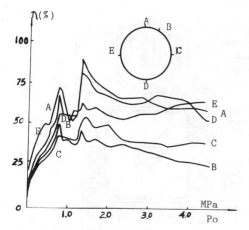

Fig.3 Curve of Pi/Po and Po

$$(7) \quad \Lambda = \frac{\dfrac{2a^2}{E_1(b^2 - a^2)}}{\dfrac{m_2+1}{m_1} + \dfrac{(m_1-1)b^2+(m_1+1)a^2}{m_1E_1(b^2 - a^2)}}$$

where: a. b.— the internal and external
radius of the lining,
E_1, E_2 & m_1, m_2 — the deformation
& the reciprocal of the poison
ratio of the lining and the rock

This is clear from the formula (7) that
is only relates to the size of the
lining and the elastical parameters of
the lining and the rock. Actually, the
interaction of the lining and the rock
is very complicated. Only we know the
factor of affecting the interaction,
we can decide the part of the rock bearing the internal water pressure. From
the result of the full strength model
test and other investigation, the main
factors of affecting the interaction
of the lining and the rock mechanical
characteristics of the lining and the
rock, the state of the secondary stress
of the rock, whether the lining cracked
or not and the internal water pressure.

4.1 Mechanical characteristics of the
lining and the rock

The deformation modulus is the important factor of affecting the interaction.
If the pressure tunnel we concerned in
the full strength model test has the
condition which the formula (7) required,
set a=4.75m, b=5.65m, m =6 and m =5,
we get the curve of Λ (Fig.4) by the
different E_1 & E_2.
The tensile strength effects not only
the cracking of the lining and the rock,
but also their interaction. When the
tensile strength of the rock is very
low, the cracking range of the rock is
expanding with E goes up. The higher
the tensile strength of the rock, the
the biger the resistant force of the
rock, and this effect is more obvious
with the E goes up.

4.2 Contact condition of lining and rock

The contact conditions of the rock of
the pressure tunnel are classified the
partical contact(with fissure), the complete contact and the pre-stress contact
by the different grouting conditions.
For the concrete or reinforced concrete
lining with the backfill and consolidated
grout, the contact condition is generally
the partial contact, and this gap is
caused by the construction the difference
in temperature.

sile—pressive stress state with the
internal water pressure goes up, and
this change results the decline of the
resistant force of the rock and the value
of Pi/Po; Certainly, this effect is con-
cered with the secondary stress of the
rock. If the geostress is very large,
and the internal water pressure is relatively small, this effect is relatively
small; When Po is 1.0MPa—1.3MPa, Pi/Po
almost changes slightly, it is caused
by the development of cracking in the
lining, the development of the plasticity
of the rock and the change of stress
state of the rock (to the unfavourable
tensile--pressive state). After the crack
ran through the lining, the internal
water pressure would be beared by the
bar of the lining and the rock, so pi/po
goes up suddenly at Po=1.3MPa. Then,
because of the effect of the bar, the
unfavourable change of the stress state
and the development of cracking in the
rock, Pi/Po is goes down with the interal
water pressure goes up. On the other
hand, the uneven distribution of Pi is
caused by the different position and
the uneven stress of the rock.

4. MAIN FACTORS OF AFFECTING INTERACTION
OF LINING AND ROCK

The interaction of the lining of pressure
tunnel and the rock is always expressed
by the   (the part ofthe rock of bearing
the internal water pressure). If the
lining and the rock act completely combined, we get:

754

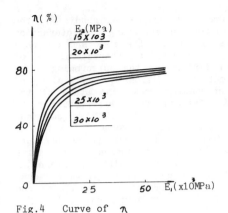

Fig.4   Curve of λ

The poor grout between the lining and the rock is one of the major reasons which deform the pressure tunnel. Generally, there is the uneven gap between the lining and the rock, so we can't assume lindly their complete combined action.

4.3 Secondary stress of rock

Acturally, the mechanical characteristics of the rock is largely concerned with its stress state. The value of stress of the rock controls the initial level of the stress and the cracking of the rock, and the stress distribution controls the distribution of the resistant force.

4.4 Whether the lining cracked or not

After the lining were cracked, the internal water pressure is beared by the bar and the rock, at the same time, the cracking of the lining changes the characteristics of the seepage. The out seep of the internal water may be very dangerous. See Table II, when the internal water pressure $Po=0-0.8MPa$, the contact stress (σ_z)MPa of the lining and the rock has the change of pressive--tensile stress, and the tangential tensile stress (σ_θ)MPa of the internal layer of the lining is gradually becoming high.

4.5 The internal water pressure

With the different internal water pressure, the interaction of the lining and the rock may indicate the periodic change. The internal water pressure decides the working condition of the lining, the stress state of the rock, the characteristics of the seepage and the degree of the combined action.

5. THE COMBINED ACTION OF THE LINING AND THE ROCK

Under the effect of the internal water pressure, the degrees of the combined action of hte lining and the rock different with the different working conditions. Set   to indicate the degree of the combined action, λ will show the periodic change with growth of the internal water pressure(Po)

On the light of Fig.3, the combined action of the lining and the rock can be divided to four stages:

1. The stage of condensing of the interface of the lining and the rock: When the internal water pressure is 0.75MPa, the lining and the rock combined act completely, and the internal water pressure now is called the critial internal water pressure Pocr of the complete combined action of the lining and the rock. The internal water pressure of this stage is 0-0.75MPa (0-Pocr).

2.The stage of the complete combined action: In this stage, Pi/Po goes down with the internal water pressure goes up, it is because of the change of the stress state of the rock (the effect of the internal water pressure is to cause the tangential tensile stress and the radial pressive stress in the rock, so the stress state changes to the unfavourable tensile--pressive stress state). On the other hand, the development of the plasticity of the rock is also to cause Pi/Po(λ) to go down. The internal water pressure of this stage is 0.75-1.0MPa(Pocr-Pz).

3.The stage of cracking of the lining: In the prograss of cracking of the lining,

Table II   Effect of seepage

p_o	0	0.1	0.2	0.3	0.4	0.5	0.6	0.7	0.8
σ_z	-0.52	-0.42	-0.32	-0.22	-0.13	-0.03	+0.07	0.17	0.26
σ_θ	1.51	1.54	1.57	1.59	1.62	1.65	1.67	1.70	1.72

changes slightly , it is caused by the development of cracking the development of the plasticity of the rock and the change of the stress state of the rock to the unfavourable tensile-pressive state.After the crack ran through the lining, the internal water pressure would be beared by the bar of the lining and the rock, so goes up suddenly at the internal water pressure 1.3MPa. The internal water pressure of this stage is 1.0-1.3MPa(P_I-P_{II}).

4. The stage of after-cracked of the lining: After the lining cracked, the bar of the lining is to bear the internal water pressure gradually, so λ goes down with the internal water pressure goes up. When the internal water pressure is 4.5MPa, the bar of the lining is still in the state of elasticity. The internal water pressure of this stage is 1.3-4.5MPa.

From the above-mentioned analysis, we can see that the combined action of the lining and the rock is in the different stage with the change of the internal water pressure. So, we must understand the stage of the combined action of the lining and the rock in the practice analysis of the pressure tunnel.

Just as we can analysis the feature of the stress and the fracture of the rock by using the relationship of the stress and the strain of the rock, we can also analysis the pressure tunnel by using the curve of the interaction of the surrounding rock and the lining, and we can use a certain nonlinear element—interaction element to simulate the interaction in the analysis, and the feature of this element is expressed by the curve of λ-Po, such as the curve of Fig.3.

6. CONCLUSION

1. The analysis based on the full strength model test of the lining and the surrounding rock is feasible, the results from the full strength model test is more direct and more reliable than that from the gypsum and geomechanical model test, and the key to the question is to decide the simulating material of the surrounding rock. In this test, using the ceramsite concrete and the air-entraining concrete to simulate the surrounding rock is successful. In particular, the analysis of the weak plane is very difficult for designer, and the analysis based on the full strength model test privides the direct and accurate basis for designer.

2. Under the concerned condition, the main factors of effecting the interaction of the lining and the rock are the physical and mechanical characteristics of of the lining and the surrounsing rock, the contact condition of the lining and the surrounding rock, the state of the secondary stress of the surrounding rock, whether the lining cracked or not and the internal water pressure.

3. With the internal water pressure change, the combined action of the lining and the rock is in the different stage.

4. With the curve of λ-Po, we can analysis the interaction of the lining and the rock of the pressure tunnel.

REFERENCES

G.M.Sabiulise, 1989. Structure model and testing technique. Railway Publishing House of China. U.S.A. (Chinese)

Prabht Kumar. 1988. Effects of rock mass anisotropy on lining medium interaction in power tunnel. International Symposium. Jan. India.

E.Fumagalli 1973. Statical and geomechanical models. Springer verlag, Wien.

Herber E. Linderg. 1982. Tunnel response in modeled jointed rock. Issues in rock Mech.

Moweng Xie. 1990. A study on the mechanism of the interaction of the lining and the rock of the pressure tunnel. Thesis of Master Degree of Wuhan Univ. of Hydraulic & Electric Engr. China.

R.Koopans & R.W.Hughes, 1986. The effect of stress on the determination of deformation modulus, 27th U.S.Symp. on Rock Mech.

Hydropower'92, Broch & Lysne (eds) © 1992 Balkema, Rotterdam. ISBN 90 5410 054 0

Modern design method of submerged tunnel piercing in China

Zhao Zong-Di
Tianjin Investigation & Design Institute of MWR, People's Republic of China

Gao-Xiang
Department of Hydraulic Engineering, Tsinghua University, People's Republic of China

ABSTRACT: Varying design methods of lake tap in China are breifly described in this paper. A detailed description of row blasting with large diameter boreholes applied in lake tap is also given. Meanwhile, the seismic action and variation law of earthquake strength after blasting are discussed as well.

1 INTRODUCTION

Lake tap, a submerged tunnel piercing, is the process of connecting a tunnel system to a reservoir or a recipient. During the last 20 years, it has played moderately important role in Chinese hydropower construction. Although this method of lake-tap was adopted quite late in China, China has obtained a lot of ripe experiences in mastering this kind of method through practising it in 9 domestic hydraulic schemes, and especially has more experiences and developments in this area of applying row blasting with 100 mm diameter boreholes to lake tap. The author has plentiful experiences in design and construction of underwater tunnel piercing. Combined with some examples of lake tap in China, this paper will introduce the design method of adopting row blasting with large dia-meter(100 mm ϕ) boreholes in construction, also the seismic action after blasting.

2 BREIF INTRODUCTION OF LAKE TAPS IN CHINA

Since the early 1970's, 9 successful submerged tunnel piercings have been performed in China. At present, at three existing reservoirs, lake taps have been authorized to be planned, while lake taps at Fenghe Reservoir and Xianghongdian Reservoir will be giant ones in the world in the early 1990's. A selection of lake taps performed and planned in China is presented in Table 1.

Table 1. shows that, in the early periods, chamber blasting + debris collection was quite often used in lake taps in China. Through practising, Chinese engineers have developed some new approaches for lake tap, the method of row blasting with large boreholes + debris discharging was adopted for submerged piercing at Xiangshan Reservoir in

1976, the in-situ test was carried out in 1978. The success of lake tapping at Xiangshan provided valuable reference for Miyun scheme where this method was adopted for the flood discharge tunnel piercing. This new method has been mastered and frequently applied in Chinese hydraulic schemes.

3 DESIGN OF ROW BLASTING ADOPTED IN PLUG BLAST

3.1 Features of row blasting design method

Since 1979, five lake taps with row blasting have been successfully performed in China, such as Miyun Reservoir scheme etc. These practices show that, this kind of method has many advantages, such as safe construction, convenient for opera-ting machine, dispersion of explosive, weak seismic shock after blasting, even dimension of debris, etc.. Meanwhile, the thickness of rock plug can be known precisely by trial boreholes which penetrate the plug. According to statistics of several domestic lake taps, the cost per cubic meter plug for lake taps with row blasting is 50~60 percent of that with chamber blasting. However, this method of row blasting still has some shortcomings, such as quite large amount of work and highly demanded constructing techniques for drilling, charging and laying blast circuit.

3.2 Determination of plug dimension

1. The diameter of the opening

The bottom face of plug is determined by the tunnel cross-section $w(=\pi D^2/4)$ and shrink ratio of cross-section, the flow velocity should not be larger than the scouring velocity for rock or concrete near the intake.

2. Thickness of the plug

The prevailing approach adopted in determining

Table 1. Selection of submerged piercings in China

Location	Function	Rock type	Depth (m)	Diameter of plug (m)	Thickness of plug (m)	Blasting pattern	settlement of debris	Year
No. 211	water supply	granite-gneiss	18	6.0	7.5	chamber blasting	collection	1971
7.1Reservoir	generating & irrigation	argillaceous shale	24	3.5	4.2	same above	discharging	1972
No. 310	diversion & generating	diorite	23	8×9	8.0	same above	collection	1975
Xiangshan	flood-discharge	coarse-granite	30	3.5	4.52	row blasting	discharging	1979
No. 250	same above	metamorphic brecia	19.8	11.0	15.0	chamber blasting	discharging +collection	1979
Miyun	same above	granite-gneiss	34	5.5	4.54	row blasting	discharging	1979
Meipu	same above	limestone	10.3	2.6	3.6	same above	discharging	1979
Hengjing	same above	mixture rock	26	6.0	9.0	chamber blasting	discharging +collection	1984
Shuichaozi	sediment-flushing	basalt	30	4.5	3.4	row blasting	discharging	1988
Miyun	water supply	granite-gneiss	30.4	5.0	6.5	same above	same above	1993
Fenghe	flood-discharge	amphibole-schist	22.4	8.0	8.0	same above	same above	1993
Xianghong-dian	diversion & generating	brecia	30	9.0	8.0	row + chamber	collection	1993

Table 2 Plug thickness in partial Chinese hydraulic schemes

Location	Diameter D(M)	Thickness H plan (m)	Thickness H real (m)	H/D plan	H/D real	tension stress (Mpa)	compression stress (Mpa)	shear stress (Mpa)	safety factor against slipping
Xiangshan	3.5	5.0	4.52	1.43	1.29				
Miyun(1)	5.5	5.0	4.54	0.91	0.825	0.21	1.30	0.65	2.20
Shuichaozi	4.5	4.6	3.40	1.02	0.76	0.84	1.12	0.50	4.60
Miyun(2)	5.0	6.5		1.30		0.00	Kc=44.5	Ks=5.34	4.47
Fenghe	8.0	8.0		1.00		0.25	1.14	0.57	5.88

the thickness of plug is, referring to experiences of other projects, then checking by theoretical calculation. In China, for the early piercings, the ratio of the plug thickness, H, to its span (or diameter D), varies between 1.0 and 1.5. Through practising lake taps with row blasting in late 1970's, we propose that the ratio of H to D be 0.85~1.0, i.e. the plug thickness can be determined in terms of the following empirical formula:

$$H=(1.0 \sim 0.85) D$$

3.3 Arrangement of boreholes

In chinese lake taps, 100 mm diameter boreholes are used for plug blast, the 40 mm diameter boreholes are used as peripheral presplit holes. Therefore, the arrangement rule of boreholes for the plug blast is different from that for common tunnel excavation. We suggest that, 4 cutting-holes be located around the center, large bore-holes be located on circles spaced at an spacing of 50~100 cm along the radius. In general, for this case, D<5 m, 3 circles of boreholes are needed, the number of boreholes on each circle is 4,6,12 or 4,8,12 respectively, the total number is 22~24; when D>5m, 4~6 circles of boreholes are enough, the distribution of boreholes is 4,8,12, 16 or 4,8,12,16,22,30 respectively, then the total number is 40~92.

For borehole arrangement, no matter how the plug diameter varies, the row blasting factor η should be controlled in this range between 0.25 and 0.34 m/m³ (that is, for per cubic meter plug, 100 mm diameter boreholes with length of 25~34 cm should be drilled)

1. Cutting holes, i.e. the boreholes used to break through the center of plug. In general, they are located on a small circle around the plug center, and arranged according to following rules:
radius = 0.25~0.30 m
borehole spacing = 0.39~0.47 m

2. Key boreholes, i.e. the boreholes with the diameter of 100 mm located in plug, the borehole spacing should meet following demands, first, boreholes on a circle should be located at same spacing; secondly, boreholes on different circles should be arranged as hexagon. At the same time, the determined spacing of boreholes should not affect the blasting function of boreholes on the adjacent circles, and the rock plug should be splitted by the joint action of all the boreholes. Of course, the boreholes spacings should be adjusted to promote the blasting efficiency, in terms of the face of plug face.

3. Peripheral presplit holes. In order to control effectively the outline of the opening and debate the shock to surrounding rock, a circle of presplit boreholes are arranged on the plug periphery. Results of blasting show that, presplit holes have played obviously positive role in guaranteeing explosive forming of the opening and reducing the shock action to sur-rounding rock. By adopting the presplit holes, the velocity and acceleration amplitude of seismic wave will be reduced by 6.7% and 31% respectively.

4. Central uncharged borehole, an uncharged borehole is arranged in the center of plug, with a diameter of 100 mm.

In order to fix the location and direction of boreholes, except the cutting holes parrallel to axis of the plug, the other large boreholes and peripheral presplit holes are radially drilled, and intersected with the plug axis at one point. Details about borehole arrangement in several schemes are listed in Table 3 and showen in Fig.1.

3.4 Calculation of explosive amount

Up to now, no precise formula is used to estimate the explosive amount for the final blast. On the basis of studying the basic kinetic law of plug blast and taking part in the design and constru-tion work of several domestic lake-taps, the author has adopted a valid calculation method just according to the plug volume. This method is based

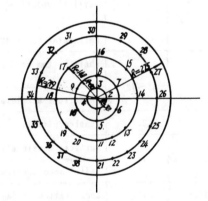

Fig.1   Arrangement of boreholes in rock plug

Table 3   Examples of borehole arrangement

Loca-tion	plug diameter (m)	central uncharged hole		cutting-holes		key--holes		Peripheral presplit hole	
		diameter (mm)	number of holes	diameter (mm)	number	diameter (mm)	number	diameter (mm)	number of holes
Miyun	5.5	100	1	100	4	100	3	40	60
Shui-chaozi	4.5	100	1	100	4	90~100	3	60	36
Fenghe	8.0	100	1	100	4	100	4	60	50

on this fact, under a condition of explosive and rock type, rock volume to be broken is directly proportional to the explosive amount. The total charge can be determined by this equation, Q=KV. In order to control the dimension of rock debris, the total charge will be distributed to every borehole, and the explosive loaded in boreholes can be determined by the following equation.

$$q = \frac{\pi \; d^2}{4} \; L \; \triangle$$

At the same time, this amount of explosive must meet another equation, that is

$$q = K \, W \, a \, H$$

where:
q, explosive amount loaded in a borehole (kg)
d, diameter of explosive cartridge (m)
L, loadding length (m)
$\triangle$, charge concentration (kg/m³)

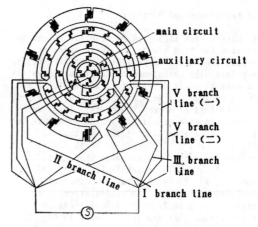

main circuit
auxiliary circuit
V branch line （一）
V branch line （二）
Ⅲ. branch line
Ⅱ branch line
Ⅰ branch line
S

Fig.2　Layout of construction circuit

Table 4　statisitcal table of blasting seismic parameter value and its damagement in some Chinese projects

position	distance	parameter	unit	maximun	damage
fundation	20	velocity	cm/s	20.9	no damage
rock	10	velocity	cm/s	74.3	no damage
tunnel	24.6	vertical velocity	g	5.24	no damage to crack with nice filling
body		horizontal velocity	g	10.12	no damage
gate-	65～72	shaft bottom acceleration	g	1.78	few pices of falling rock
shaft		shaft roof acceleration	g	2.47	no damage
tunnel roof	8.1	tension straiin	μ ε	+146	no damage
ground	71	vertical acceleration	g	5.54	no damage
surface		radial acceleration	cm/s	16.4	no damage
concrete gra-vity dam	280	acceleration at dam bottom parallel to river	g	0.159	no damage
		acceleration at dam top parallel to river	g	1.33	no damage
	290	displace of culvert parallel to river	mm	0.157	no damage
		displace of dam top parallel to river	mm	1.25	no damage
roller-compucted	138	perpendicular	cm/s	0.50	no damage
earth dam		acceleration parallel to river	g	0.09	no damage
		vertical displacement	mm	0.098	no damage

K, unit explosive consumption (kg/m²)
W, minimum (m)
a, borehole spacing (m)
H, length of borehole (m)

From the two equations previously mentioned,
another equation can be derived, That is

$$\frac{\pi\,d^2}{4}\,L\,\triangle\,=\,K\,W\,a\,H$$

This equation can assure the amount of explosive.

3.5 Design of blasting circuit

Electric blasting circuit is often applied to
lake-taps. Then, the compound circuit is often
adopted. Two sets of parallel circuits are adopted
for guaranteeing successful blast. For each
circuit, two detonators in loaded holes are
connected in parallel, then, these parallel
detonators with same interval are connected in
series to form a branch line, then all these
branch lines are connected in parallel after
resistance equillibrium. The demand of resistance
equillibrium must be met for this compound
circuit in order to assure the safe and accurate
blast, i, e. The total resistance of detonators,
and conducting wires of each branch line should
be same. Fig.2 shows a typical layout of circuit.

4 SEISMIC ACTION AFTER PLUG BLAST

The reason for studying and analysing the seismic
action caused by plug blast is to evaluate the
shock influence upon other hydro-structures and
surrounding rock as the lake-tap being performed,
and to provide reference data for other underwater
piercing design.

4.1 Destruction range of plug blast.

Extensive in-situ observations on the surface and
underground influential field and the blasting
earthquake strength at the foundation of hydro-
structure have been carried out in the hydraulic
projects where lake-tap is adopted. that has
practical value for evaluting the seismic action
of row blasting. Some related data and seismic
influence observed in partial Chinese schemes are
presented in Table 4. Table 4 proves that the
destruction range is very narrow as adopting the
method of row blasting.

4.2 Variation of earthquake with the depth

The conclusion that surface shock effect is
stronger than the shock effect upon the tunnel
consists with the observed results carried out in
domestic schemes. In order to study the variation
of blasting earthquake with the depth, the attenuation factors of blasting earthquake varied
with depth have been obtained by calculation for
varying blasting patterns, in terms of the observed data, see Table 5. Table 5 shows, the total
attenuation of surface and underground blasting
earthquake of row blasting is 20% to 50% larger
than that of chamber blasting.

Table 5   Comparision of attenuation factorsb in plug blast

project	blast pattern	kenetic parameter			average value of shock ρ	value of shock	attenuation value
Miyun	row blasting	velocity cm/s	radia	surface	0.017	0.322	78.57
				underground	0.017	0.069	78.57
			vertical	surface	0.017	0.156	67.27
				underground	0.017	0.054	67.27
		acceler- ation	radia	surface	0.017	0.086	80.23
				underground	0.017	0.017	80.23
			vertical	surface	0.017	0.056	83.93
				underground	0.017	0.009	83.93
310 pro- ject	chamber blasting	velocity cm/s	radia	surface	0.0735	2.0871	51.11~60.92
				underground	0.0735	0.9581	51.11~60.92
		acceler- ation	radia	surface	0.1625	4.3198	57.59~60.92
				underground	0.1625	1.6712	57.59~60.92
			vertical	surface	0.085	0.5465	16.4~46.23
				underground	0.085	0.3312	16.4~46.23

761

Table 6 Experienced formula of earthquake in different type of dam

project	type of dam and blasting environment	physical index		experienced formula coefficient K	index α	suited range ($\rho = Q^{1/3}/R$)
Fengman	concrete gravity dam, metamorphic conglomerate, the amount of charges is 4076.5kg, among which Q=197kg.	V (cm/s)	parallel to river	907	2.17	0.000~0.06
			vertical to river	341	2.02	0.000~0.06
		a(g)	parallel to river	178	2.34	0.006~0.06
			vertical to river	282	2.46	0.006~0.06
		A(mm)	parallel to river	741	2.55	0.000~0.06
Miyun	sloping core compacted embankment dam. mixed rock of granite and gneiss, the amount of charges is 738.2kg, among which, Q=310.5kg.	V (cm/s)	vertical to river	135.9	1.81	0.016~0.20
			radia	40.2	1.42	0.016~0.25
		a(g)	vertical to river	35.5	1.86	0.016~0.25
			radia	17.7	1.69	0.016~0.25
		A(mm)	vertical to river	4.78	1.99	0.016~0.30
			radia	0.49	1.45	0.016~0.30
xiang-shan	masonry gravity-arch dam. granite rock. The amount of charges is 256kg, among which Q=106kg.	V (cm/s)	vertical to river	164.9	1.50	0.0159~0.0875
			radia	256.3	1.61	0.0159~0.0875
		a(g)	radia	300	2.48	0.015~0.05
				918.8	2.59	0.003~0.07
			vertical to river	554.4	3.27	0.003~0.07

4.3 Attenuation law of blasting earthquake near dams

The blasting earthquake upon the dam nearby is quite concerned in design and construction of lake tap. After analyzing the observed data from domestic schemes, the author suggests empirical formula of blasting earthquake strength for varying dam types. See Table 6.

5 CONCLUSION

The method of row blasting with large diameter boreholes adopoted in lake-tap, just like other technology and methods, also has a lot of questions needed further studying by successively accumulating observeed data, so that it can be improved in theory and practice.

Hydropower'92, Broch & Lysne (eds) © 1992 Balkema, Rotterdam. ISBN 90 5410 054 0

Diversion tunnel from Luanhe River into Tianjin City

Hu Ji-Min
Water Resource Design Institute of Tianjin, People's Republic of China

Liang Hai-Bo
Hydraulic Engineering Department, Tsinghua University, People's Republic of China

ABSTRACT: A new type of tunnel lining and some successful experiences on design, excavation of tunnel in the diversion project from Luanhe River to Tianjin are introduced in this paper. Tunnel lining, excavation of inlet and inclined construction shafts are involved in this paper.

1 INTRODUCTION

Tianjin is the third largest city in China, with the population of 10 million. The diversion project from Luanhe River to Tianjin which was expected to relieve the shortage of water supply in Tian-jin, is located along the dividing crest zone between the Daheiting Reservoir on Luanhe River and Jieguanting village near Nihe River. This project includes diversion project, diversion channel, open-cut tunnel, cavity-cut tunnel, flood control sluice and energy dissipator at outlet, etc., with a total length of 12.39 km. Among it the main parts are 1724 m long open-cut tunnels and 9666 m long cavity-cut tunnels, with a cross-section of city-gate shape (i.e. arch+straight wall), 5.7 m wide and 6.25m high, and with a design discharge of 60m³/s, maximum discharge of 75m³/s and annual diversion volume of 1.0×10⁹ cubic meters. The tunnels go through low hilly lands, more than 20 valleys are distributed along the tunnel line, the ground sur-

face elevation varies between 120 m and 300 m, while the elevation of vally floor varies between 120 m and 150 m. Correspondingly, most tunnels are located underground at depth of 40~60 m, few of them reaches 100m, while only about 10 m at low concave areas. The topographic profile along the tunnel line is shown in Fig.1.

Except quartzites at some sections, most parts of tunnels are located in slightly weathered rock formations. About 176 scattered large or small faults are presented along the tunnels, see Table 1, most of them are of high-angle compresso-shearing. In general, more than 3 sets of joints occur simultaneously. The unfavourable geological conditions, such as faults and joints, bring about severe questions to tunnel excavation, therefore, a series of measures have been adopted to overcome these difficulties in the course of design and construction, then to assure the diversion project completed on schedule.

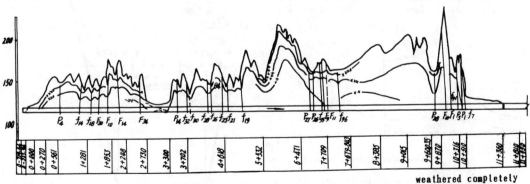

Fig.1 longitudinal profile of topography

weathered completely
weathered intensely
weathered slightly

Table 1 Statics of faults

section	station number	length (m)	number of faults		
			total	width>1m	influential
eastern section	0+268~2+730	2462	14	9	4
middle section	3+302~6+650 6+650~9+210	3348 2560	149	6	1
western section	9+210~10+506	1296	13	31	6
sum		9666	176	52	16

2 EXCAVATION OF TUNNEL'S INTAKE

Geological conditions at location of intake were extremely bad. So some necessary measures were adopted correspondingly.

2.1 Excavation of tunnel's intake

The intake-site was a low gentle slope, with exposed bedrock. 8~20 thick overburden rock were intensely or completely weathered. Considerable jionting and fissuring were observed, mainly there were four sets of jionts with spacing of 20~30m. The geological profile is showed in Fig.2.

Between station number 0+268 and 0+284, lots of underground water existed, which appeared as state of drip above the tunnel, was linear flow near faults. Some flow of 0.4 L/s penetrated partial rockwall. Severe rock weathering results in that rock solidity factor f was less than one during the 18m long section between 0+268 and 0+286, and it was only 2 between 0+286 and 0+330. Therefore, the intake had been suggested by geologists to be located at 0+330, but, that would need enormous open cut volume of 42,000 cubic meters, and more work days. Finally, the intake was decided to be located at 0 + 268, some open-cut was abandoned instead of cavity-cut with adopting smooth blasting and shotcrete & bolting support. Good results were obtained

by adopting these measures. Some details are introduced as following:

1.Consolidating slope above tunnel before tunnelling. The situations that the overburden was 15.87m in thickness, less than double tunnel diameters and intensely or completely weathered, were quite unfavorable to the stability of slope. Therefore, one 3~5 cm thick concrete was firstly shot on the slope, then the permenant bolts with diam. of 22mm and 3.0m in length were pressed into the rock slope. The spacing of bolts varied as 1.0×1.0m, 0.8×0.8m and 0.6×0.6m respectively from upper parts to down, see Fig.3. As these bolts were welded with steel meshes comprising 22mm diameter steel bars with the spacing of 0.4×0.4m, finally, another layer of concrete of 5cm thick was shot on the steel meshes. At the same time of consolidating the slope, one 14m long open-cut tunnel before the inlet was deposited by concrete to support the slope above the intake.

2.Consolidating the intake face before tunnelling. Due to loose rock, anxiousness for collapse after blasting still existed as the tunneling was being conducted under the consolidated slope, therefore, cage type support was adopted for

Fig.2 portal geology

—×××—weathered completely

—×××—weathered intensely

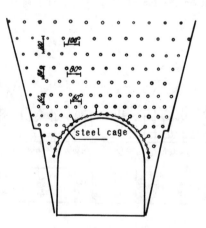

Fig.3 sketch of slope's consolidating

consolidating in advance, see Fig.4.

Boreholes with the depth of 3.0m and spacing of 20~30cm were drilled along the periphery of the arch roof and parallelto the axis of tunnel. A 22mm diameter bolt was installed whenever one borehole was drilled. The exposed parts of bolts were welded by 10mm diameter steel bars, then welded with the steel meshes on the slope to form one cage.After the cage was formed, the tunneling was conducted with upper stage excavating method, and 1m per round. After two rounds, another cage was installed. This process was repeated during the first tunneling section of 9m length.

3.Excavating the intake section

In the course of excavating the intake section, these rules of short round, light breakout, timely shotcrete and rockbolt support were persisted. The arrangement of boreholes was same as for conventional smooth blasting, the drilled length was 1.0m long, spacing of borehole varied between 0.20m and 0.25m. Partial smooth blasting holes around the arch roof were unloaded to reduce blasting shock. During the tunneling in intake section, following excavating and support procedures were adopted, see Fig.4.

1.After one 5m long upper stage was excavated, the tunnelling work for upper stage and down stage was carried out simultaneously, and the tunnelling length per round was 1.0m and 1.5m respectively, but the lap of 3~5m between the upper and down stage should be assured.

2.After each blasting, a layer of shotcrete and steel fibres were used to support the surrounding rock, followed by drilling and installing radia bolts and advanced bolts which were permanent. The advanced rock bolts were 22mm in diameter, 3m in length, 0.6m in spacing. The overlap of two row of rockbolts was about 1m. These advanced rockbolts took a role of consolidating the surrounding rocks in advance.

3.One inner layer of steel meshes were hung on the rockwall, melted with the radial and advanced bolts.

4.The second layer concrete was shot to make a tatal 10~15cm thick shotcrete.

5.When the rockbolts were installed, the down stage was drilled after the upper stage had been drilled. Then the next round began.

Due to the measures mentioned above, the excavation of the 33m long intake section was completed in 30 days, ahead of the expected time according to open-cut method. This successful tunnelling provided valuable experiences for excavating and supporting by shotcrete and bolts in the areas where intensely weathered rocks, weak surrounding rocks, faults and crushed zones were presented.

3 COMPOUND LINING

What is called compound lining is that, shotcrete and bolting first, then lining, see Fig.5. In the outer layer, the method of shotcrete and bolting was to support, while in the inner layer, the method of depositing concrete was to smooth the surface. The measuring value of the roughness, n, is from 0.0120 to 0.0123.

3.1 Stability analysis of thin lining and corresponding technical measures.

To assure the stability of the thin side-wall, following analysis and measures were carried out in design and construction.

1.Analysis of stability of rude tunnels, comprises two steps. First, the stress field of surrounding rock was calculated by FEM (Finite Element Method). The calculating results showed that no tension stress existed. The value of compression stress varied between -12 and -58 kg/cm^2, less than the allowable compression stress of surrounding rock; Second, observation in the inclined construction shaft showed the nice stability of rude shaft where the cross-section was same as that of the diversion tunnel. Therefore, the calculation was proved, and we could conclude that the surrounding rock of rude tunnel was stable in the whole.

2.The main measures adopted to assure the stability of tunnel were presented as following: first, after the rude tunnel was formed, timely

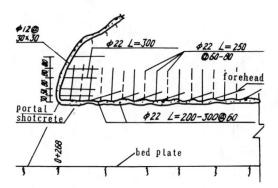

Fig.4 cage type support

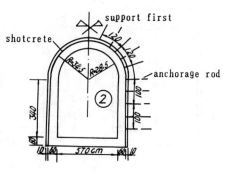

Fig.5 compound lining

shotcrete was carried out to consolidate the rockwall and restrict its deformation. Second, steel meshes of 1×1.2m were installed on the thin side wall, welded with the outer head of rock bolts.

3.2 Advantages in technology and economy

1. Fully utilize the confidence of surrounding rock to reduce the thicknees of linling. As compared with conventional design, the thickness of sidewall lining was reduced by a half, only 25~30cm, while the thickness of roof lining was reduced by three quarters, see Fig.6.

2. The timely shotcrete on the roof of tunnel was the effective measure to safeguard during construction, and was one part of the permanent lining as well.

3. The combination of thin lining and shotcrete not only smoothed the tunnel surface, but also reduced the excavated cross-secrion.

4. The compound lining regarded the surrounding the surrounding rock, shotcrete and thin lining as an organic whole. The surrounding rock had the ability of self-support in the second stress field. It was the basis of the stability of tunnel. Shotcrete firstly consolidated the surrounding rock in time and restrict the deformation of rock. Secondly combined the surrounding rock and thin lining as a whole. While, the thin lining firstly provided protection for surrounding rock, secondly smoothed the tunnel surface.

3.3 Compound lining in tunnel passing through big fault zone

The longitudinal profile of the unusual big fault zone is shown in Fig.7. This fault zone was 213 m long along the tunnel line.

Due to this fault zone, the 470m long tunnel with horseshoe shaped cross-section was originally planned. This type of cross-section was quite suitable to external force, but unfavourable for construction. Therefore, the arch-straight type of cross-section was finnaly adopted. This change based on the following reasons:

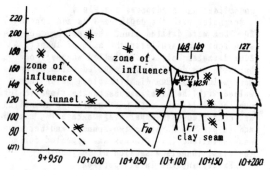

Fig.7 unusual big fault zone geology profile

1. Extra geological exploration during construction showed that the supply spring of ground water was limited. If the drainage measures were adopted in time, the stability of rude tunnel would be improved.

2. The successful lining in other sections with unfavourable geological conditions indicated the possibility of adopting partial-face excavation and compound lining in the fault zone.

During construction, not only the effective construction methods (such as small blasting, short round, early support, embedding surveying instruments to monitor the status of surrounding rock and timely adjusting blasting parameters) were carried out, but also some new materials and advanced tachniques (such as quick-strength aluminate cement, TS1 quick-setting agent, new type drainage materials, quick-cement rockbolts and quick-setting & strength shotcrete) were adopted. adopted.

The procedure of the compound lining support in partial big fault zone was described as the following:

1. Shoting the first layer of concrete of 3~5 cm in thickness.

2. Installing the radial rockbolts and the first layer of steel meshes.

3. Shoting the second layer of concrete to make

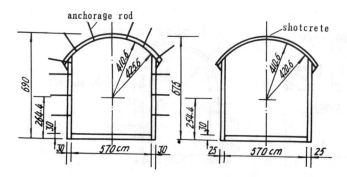

Fig.6   weak adjioning rock   solid adjioning rock

766

the total shotcrete 8~12cm thick.

    4.Installing the second layer of steel meshes.

    5.Shoting the third layer of concrete to reach 10~15 cm in thickness.

    This procedure was referred to as 'one bolt, two mesh & three shoting', and was only adopted at F1 fault zone, while at F10 fault zone, only one layer of steel meshes was installed, so the procedure was called as 'one bolt, one mesh & three shoting', see Fig.8.and Fig.9.

    Due to measures mentioned above, the tunnel successfully passed through the big fault zone of 213m in length.

    Similarly, at the other fault zones corresponding measures were adopted as well, so that 47 gradually varied sections planned originally were cancelled, and a common cross-section was adopted for the tunnels with a total length of 11.39 km.

4 EXCAVATION OF INCLINED SHAFTS

Due to the topographic features, 17 inclined construction shafts with a cross-section of 3.9×3m were located along the tunnel line, so that 36 working faces were formed, an average excavation length of 268.6m was attributed to each working face.

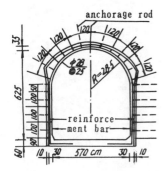

Fig.8 adjoining rock in the fault zone

4.1 Construction of entrance of shaft

Because entrances of most inclined shafts were located in completely or intensely weathered rock formations, collapse would occur due to any maltreating. Therefore 3 kinds of methods were adopted in order to guarantee the stability and safety of shaft entrance.

    1.Dig small lead tunnel to the inclined shaft and enlargen the shaft after tunnelling 8~12 m, then conduct lining from the lead face to the entrance.Outside the entrance there was a 2~4m long open tunnel. If the location of the entrance was smooth terrain, the lining for the open tunnel was up to ground surface. The open tunnel played the role of supporting the slope dug in open. Above the entrance, masonry was adopted as protective revetment to further stabilize slope.

    2.As the solid factor of rock at the entrance of inclined shaft was 2 or 3, steel or wooden rib was used to suport the shaft as soon as it had been excavated by full-face method. Because self-stable period of surrounding rock limited, short round and quick-movement of debris were needed to assure the completion of supporting work before the rock falled.

    3.As the un-stable rock existed at the shaft entrance, the shaft was excavated under the protection of the open tunnel with a length of 2~4 m.

    Construction results proved these measures effective.

4.2 Tunnelling of inclined shaft

Because of the complicated supporting work, for the sake of the safety, it was necessary to make the explosive formation sucessful at one time and assure the fully excavating of the bottom per round. Drilling direction of boreholes must be parallel to the slope of inclined shaft. According to experiences, the borehole should be deepen by 0.2~0.3 m, its dip angle should 2~3 degrees bigger than that of the inclined shaft. To prevent collapse, unit charge and arrangement

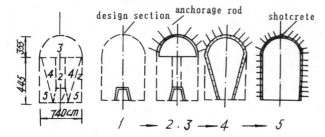

Fig.9 constructing process

of cutting holes were precisely controlled. All
these measures guaranteed the successful tunnelling of inclined shaft.

5 CONCLUSION

Based on embedded reinforcement bar stress gauge
at typical section, the measured value of stress
varied between 20 and 50 MPa.

The value of stress varied with the temperature, stress falled as temperature rised, so
did that conversely. This phenomenon proved
compression stress of adjoining rock small-fry,
the calibra-tion thin type lining and compound
lining were rational.

It took only 14 months to complete this
project. A good hydraulic condition and an
increasement by 25% for the diversion capacity
have been obtained. Hence, the tunnel lining
is of high speed and top quality.

These experiences of design and construction
are spread and absorbed by Yantian diversion
tunnel from Zhujiang River to Shenzhen Special
Zone and Hongkong, this tunnel is 7 km in length,
5.6 m in width and 7 m in height.

Hydropower'92, Broch & Lysne (eds) © 1992 Balkema, Rotterdam. ISBN 90 5410 054 0

Author index